Learning and Applying SolidWorks 2013-2014 Step-by-Step

L. Scott Hansen, Ph.D.

Department Chair of Engineering Technology and Construction Management and Associate Professor of Engineering Technology

College of Science and Engineering

Southern Utah University
Cedar City, Utah

Industrial Press

32 Haviland Street

South Norwalk, CT 06854

ISBN 9780831134839
Seventh Edition, August 2013

Sponsoring Editor: John Carleo
Cover Design: Janet Romano

Industrial Press Inc.
32 Haviland Street,
South Norwalk, CT 06854

10 9 8 7 6 5 4 3 2 1

Note to the Reader

This book provides clear and concise applied instruction in order to help you develop a mastery of *SolidWorks*. Almost every instruction includes a graphic illustration to aid in clarifying that instruction. Software commands appear in **bold** or in "quotation marks" for anyone who prefers not to read every word of the text. Most illustrations also include small pointer arrows and text to further clarify instructions.

This book was written for classroom instruction or self-study, including for individuals with no solid modeling experience at all. You will begin at a very basic level, but by the time you finish you will be completing complex functions.

For any organization requiring additional help, I am available for onsite training. Please contact me at hansens@suu.edu

Scott Hansen
Cedar City, Utah

Table of Contents

Chapter 1 Getting Started

Objectives:

- Create a simple sketch using the Sketch commands
- Dimension a sketch using the Smart Dimension command
- Extrude a sketch using the Extruded Boss/Base command
- Create a hole using the Extrude Cut command
- Create a fillet using the Fillet command
- Create a counter bore using the Hole Wizard

Chapter 1 includes instruction on how to design the part shown below.

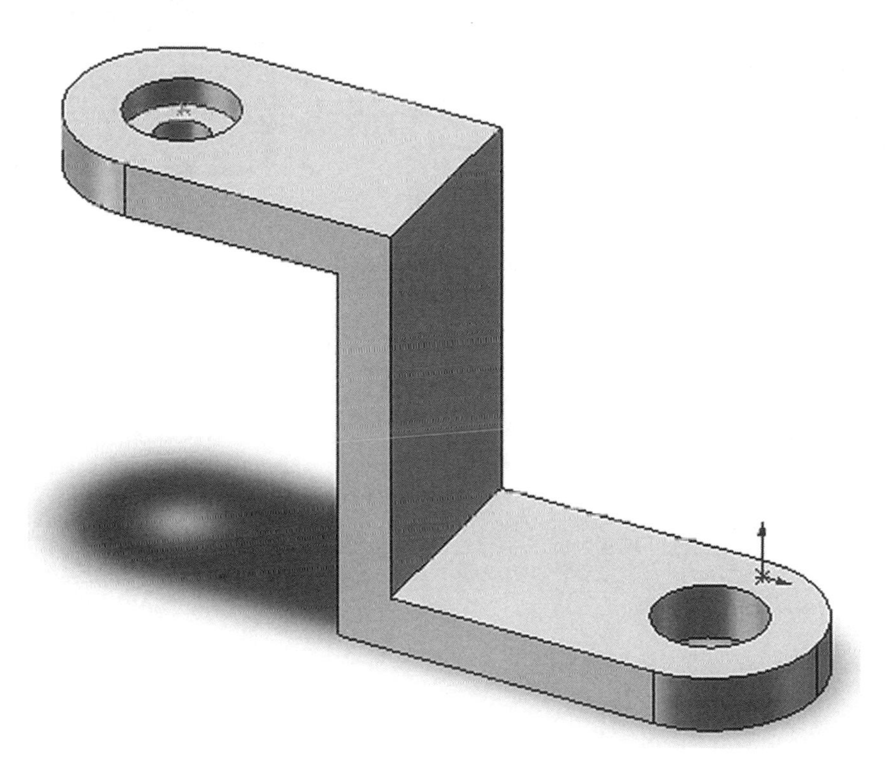

1. Start SolidWorks by moving the cursor to the **start** button in the lower left corner of the screen. Click the left mouse button once.

2. A pop up menu of the programs that are installed on the computer will appear. Scroll through the list of programs until you find "SolidWorks".

3. Move the cursor over the text "SolidWorks" and left click once as shown in Figure 1.

Figure 1

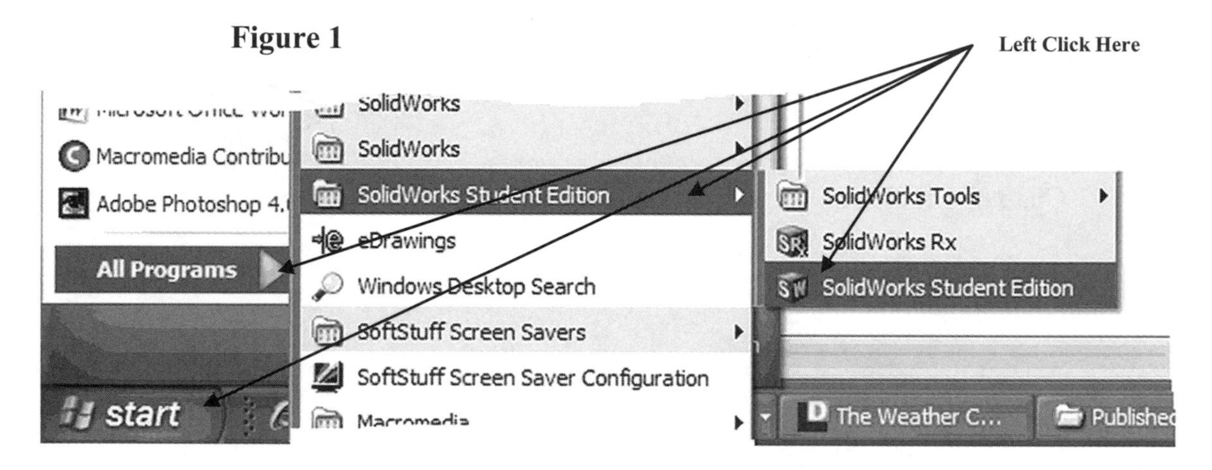

4. SolidWorks will open (load up and begin running).

5. Move the cursor to the upper left corner of the screen and left click on the "New Document" icon as shown in Figure 2.

Figure 2

6. The New SolidWorks Document dialog box will appear. Left click on **Part** as shown in Figure 3.

Figure 3

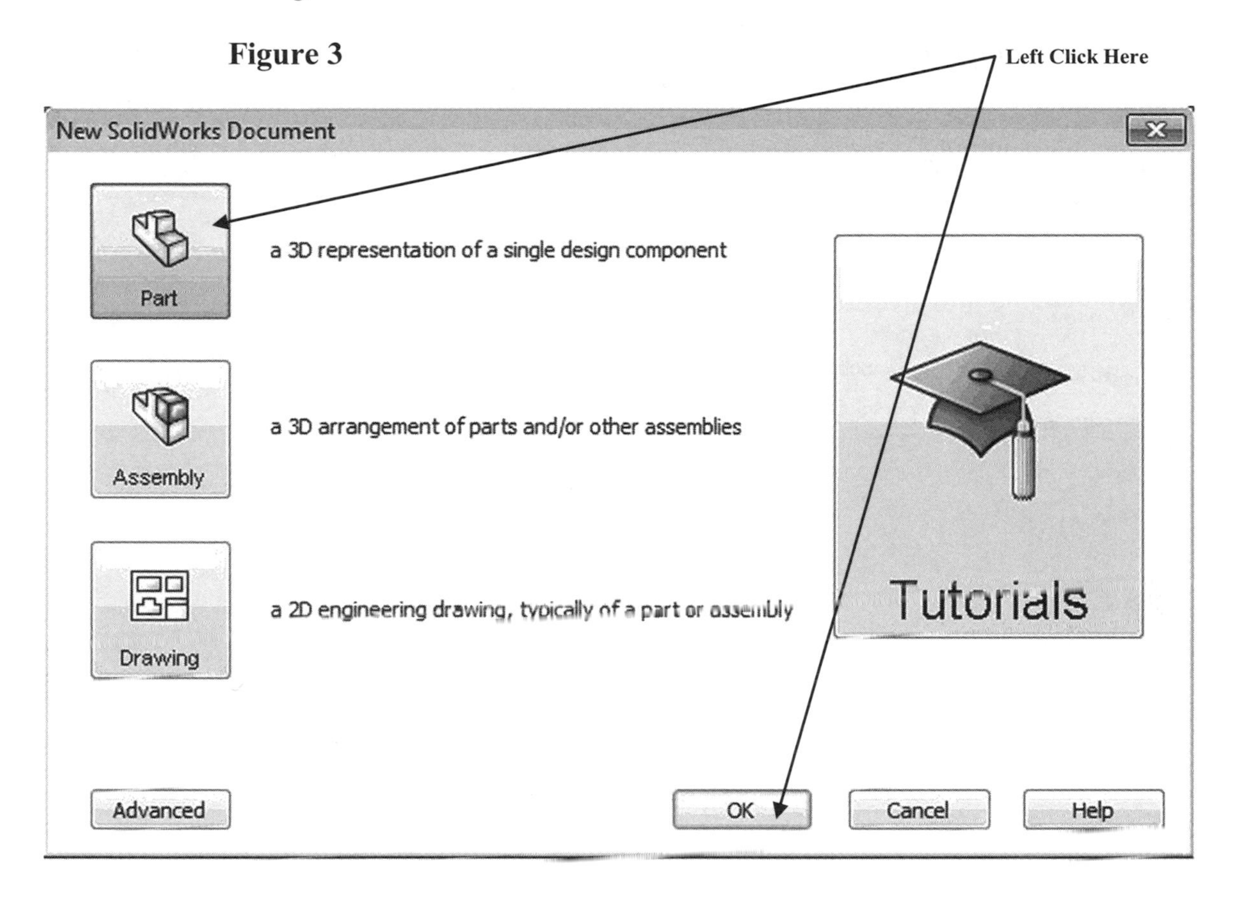

7. Left click on **OK**.

8. The screen should look similar to Figure 4. Left click on the **Sketch** icon. If the Sketch tools are not visible, left click on the **Sketch** tab as shown.

Figure 4

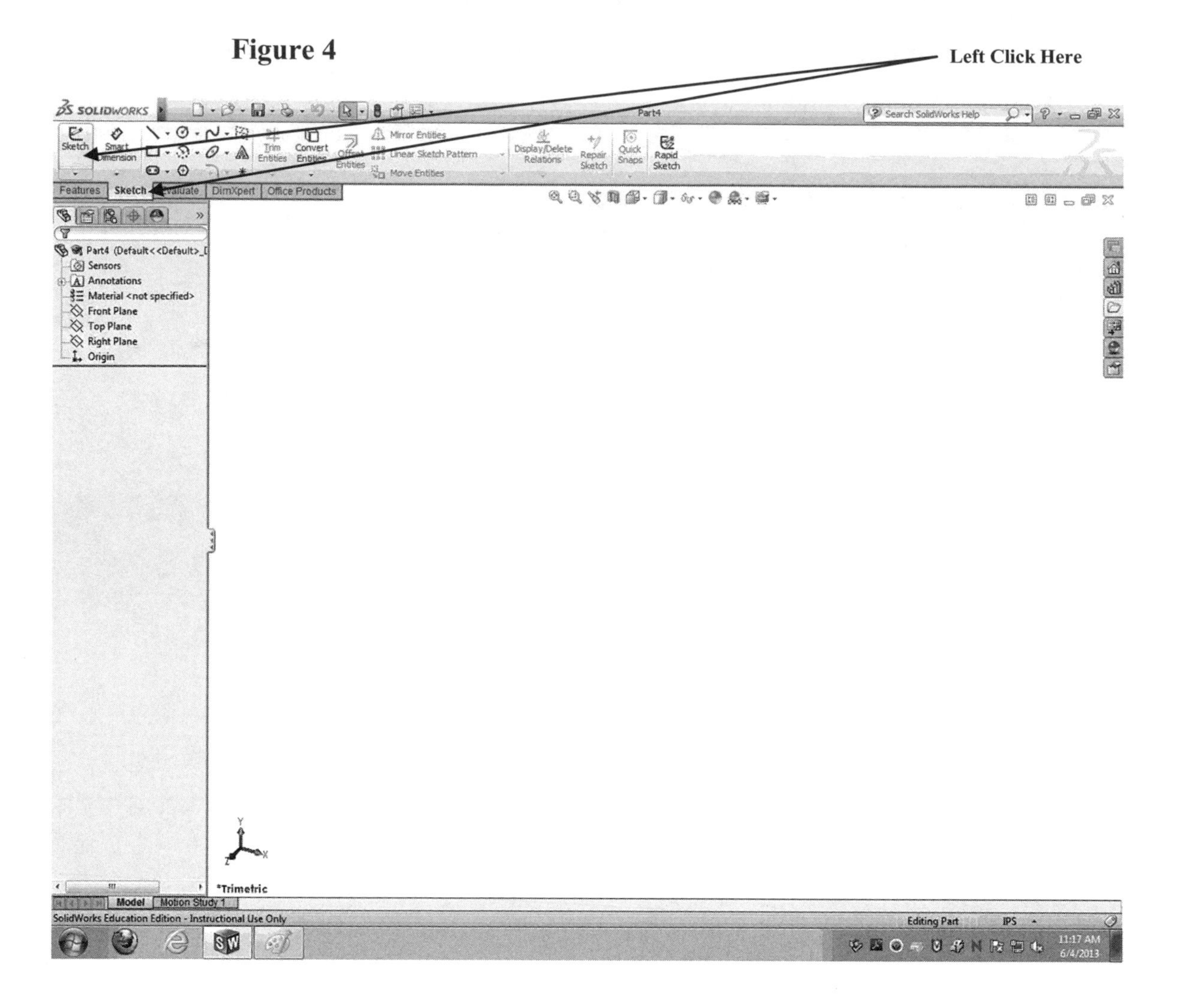

9. SolidWorks will display three different work planes. Move the cursor over the “Front Plane” until it turns red. Left click once as shown in Figure 5. SolidWorks will provide a perpendicular view of the front plane.

Figure 5

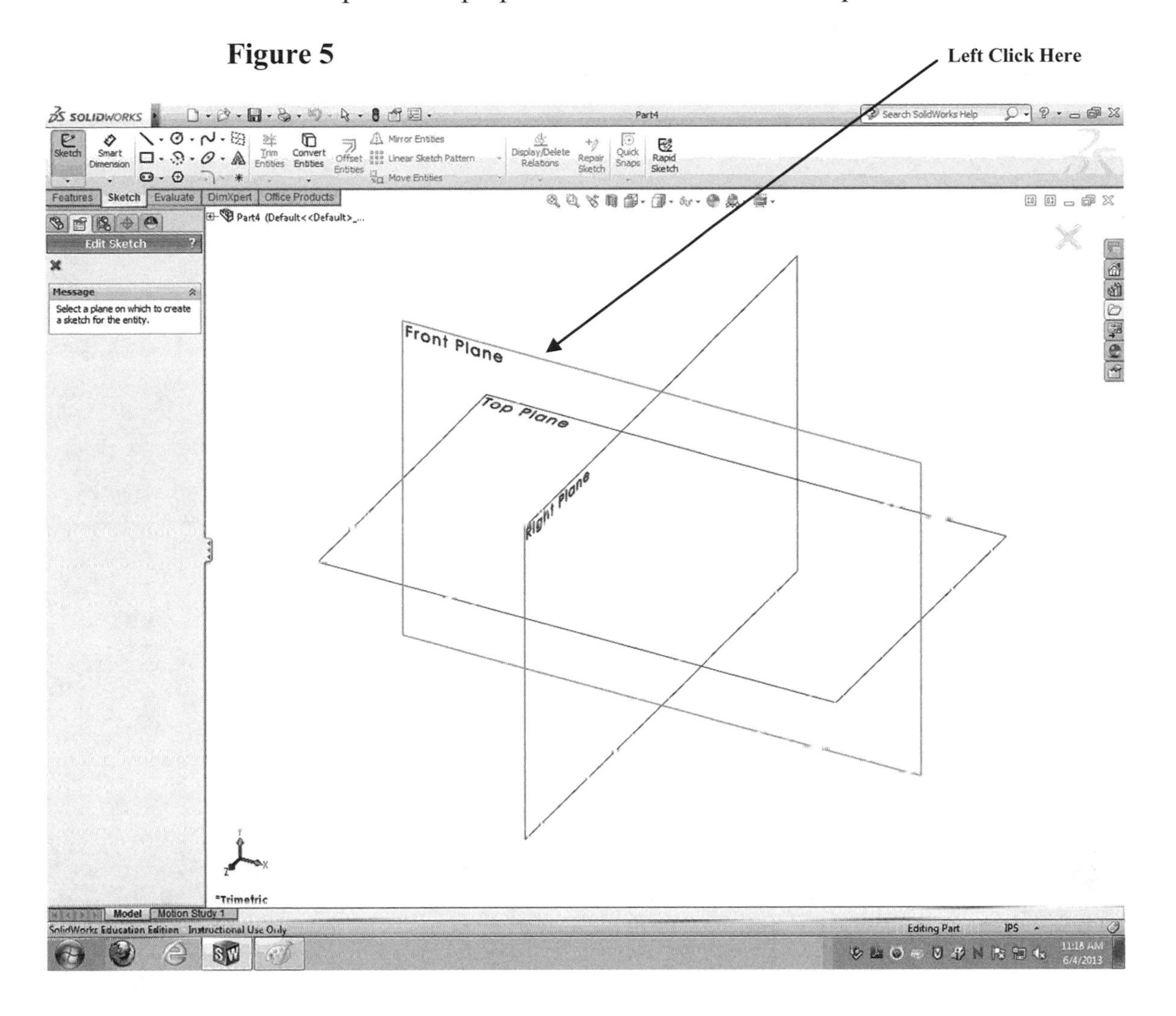

10. Your screen should look similar to Figure 6.

Figure 6

11. If the grid is displayed, skip to instruction number 12. If there is no grid visible, continue following the next steps.

NOTE: Move the cursor to the middle of the screen and right click once. A pop up menu will appear. Scroll to the bottom and left click on Display Grid. SolidWorks will turn the grid on.

Figure A

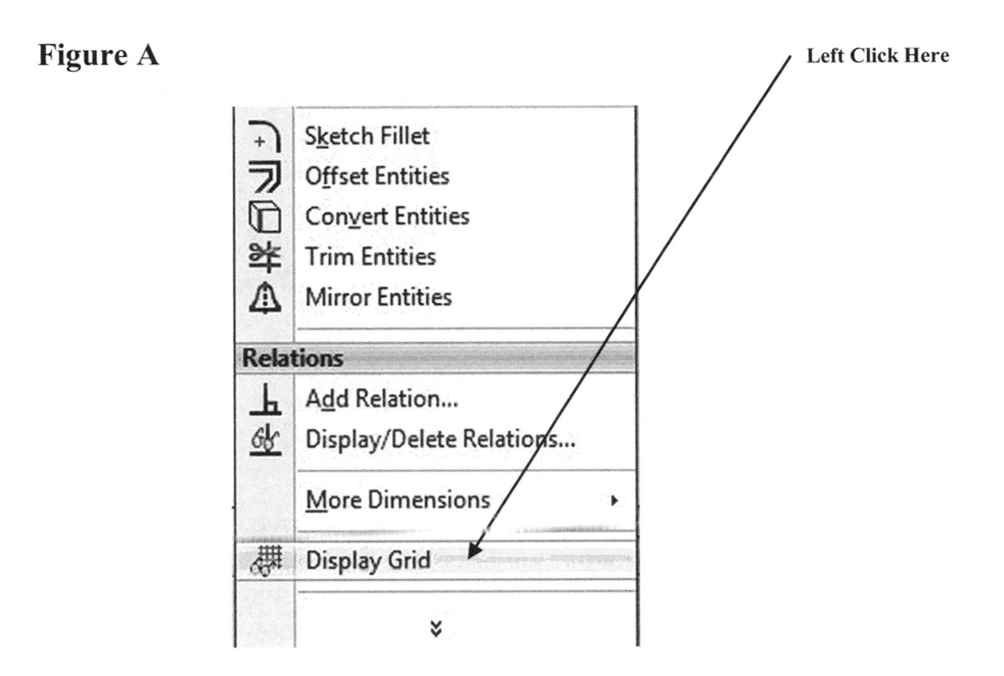

Another method of turning on the Grid is to left click on the arrow to the right of the SolidWorks text in the far upper left corner of the screen. A drop down menu will appear. Left click on **Options** as shown in Figure B.

Figure B

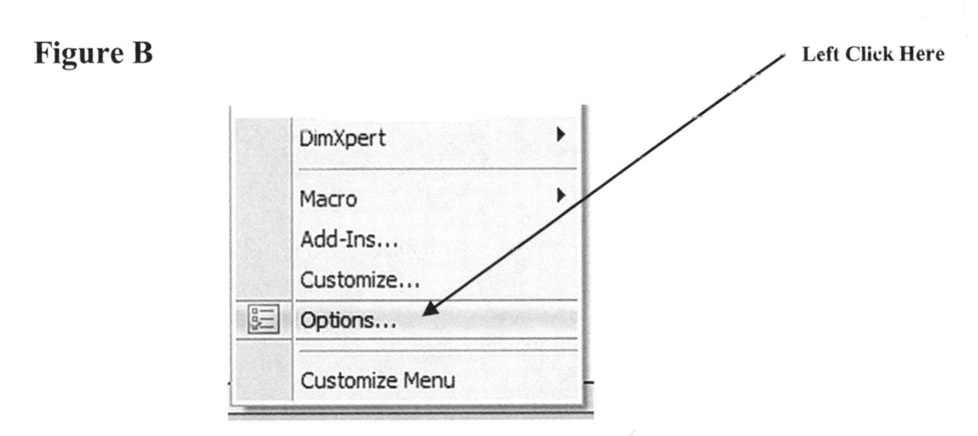

The Document Properties – Drafting Standard dialog box will appear. Left click on the **Document Properties** tab at the upper left portion of the dialog box as shown in Figure C.

Figure C

Left Click Here

Document Properties - Drafting Standard

System Options | Document Properties

Drafting Standard
Annotations
Dimensions
Virtual Sharps
Tables
Detailing
Grid/Snap
Units
Colors
Material Properties
Image Quality
Plane Display

Overall drafting standard
ANSI
Rename | Copy | Delete
Load From External File...
Save to External File...

Move the cursor down to **Grid/Snap** and left click once. The Grid/Snap properties will appear. Place a check in the box (with the left mouse button) next to Display Grid as shown in Figure D. To change drawings units, (not shown) left click on **Units** and place a dot to the left of (IPS), inch pound, second.

Figure D

Left Click Here

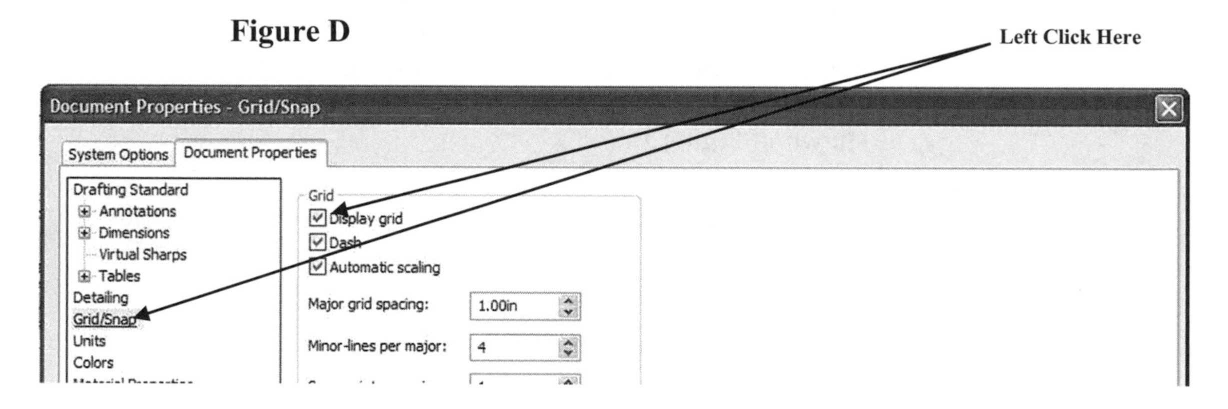

Left click on **OK** as shown in Figure E.

Figure E

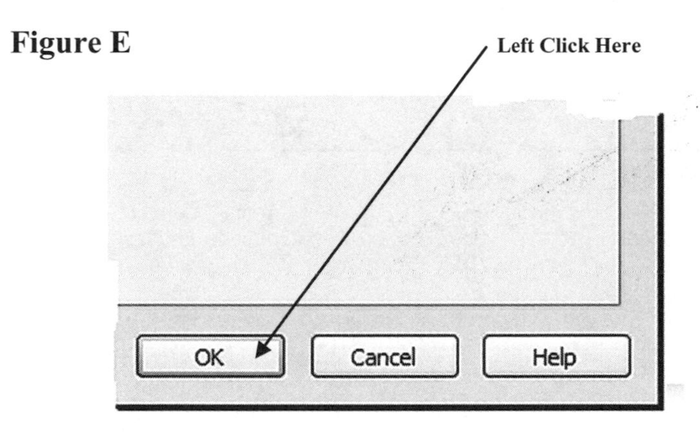

SolidWorks should now display a visible grid as shown in Figure F.

Figure F

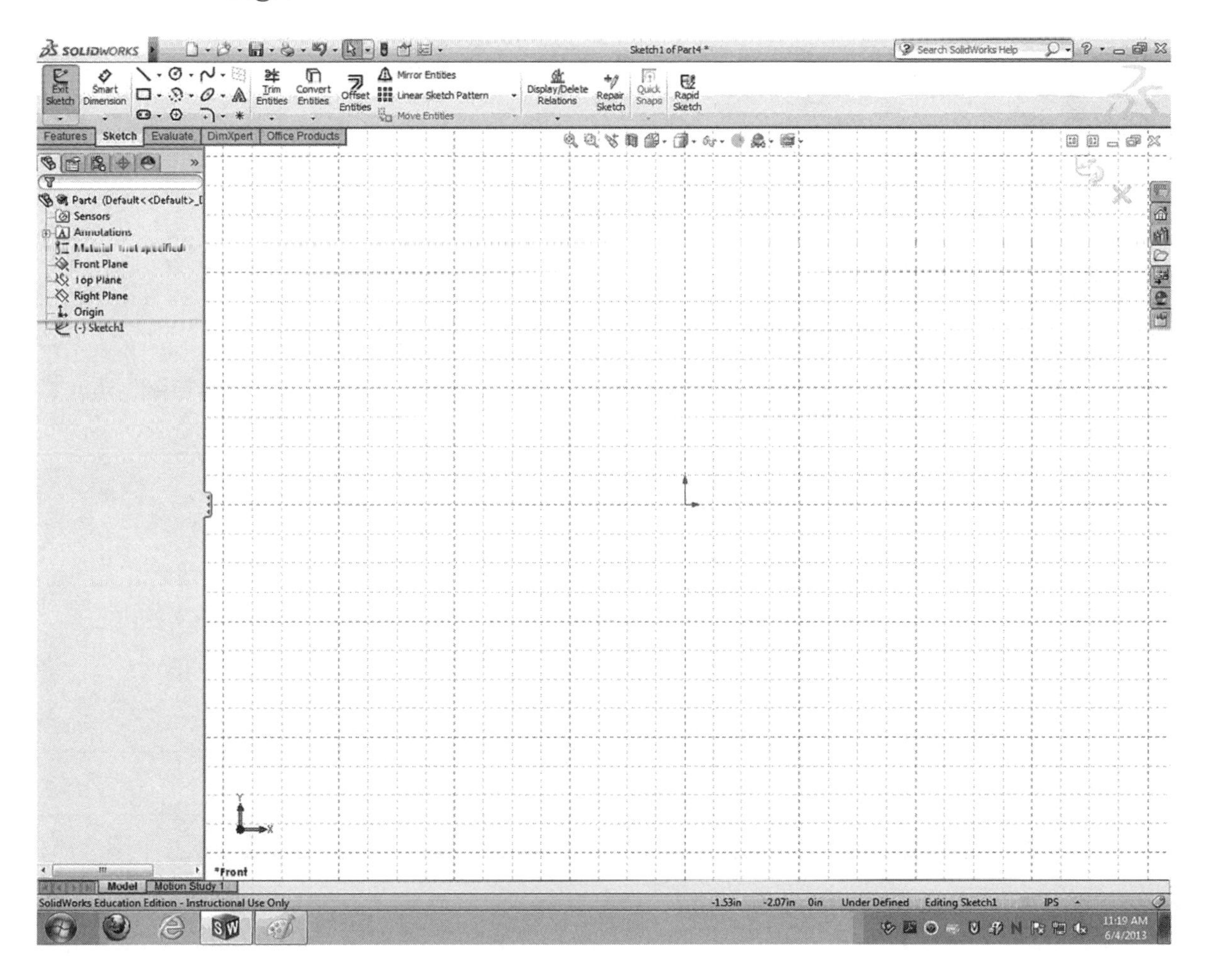

12. SolidWorks is now ready for use. Left click on **Line** as shown in Figure 7.

Figure 7

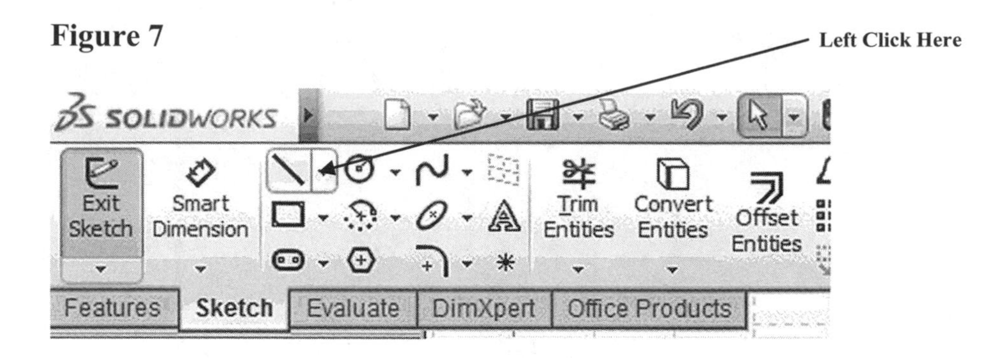

13. Move the cursor in the lower left portion of the screen and left click once. This will be the beginning end point of a line as shown in Figure 8.

Figure 8

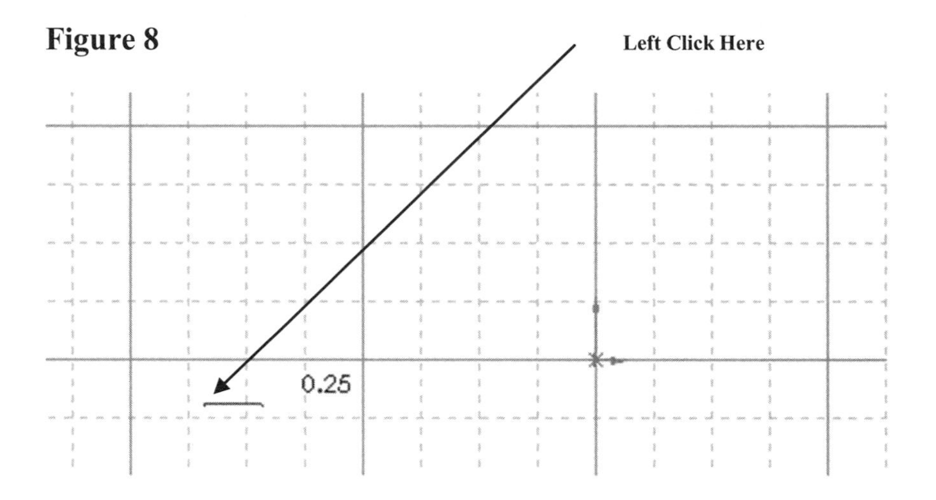

14. Move the cursor towards the lower right portion of the screen and left click once as shown in Figure 9.

Figure 9

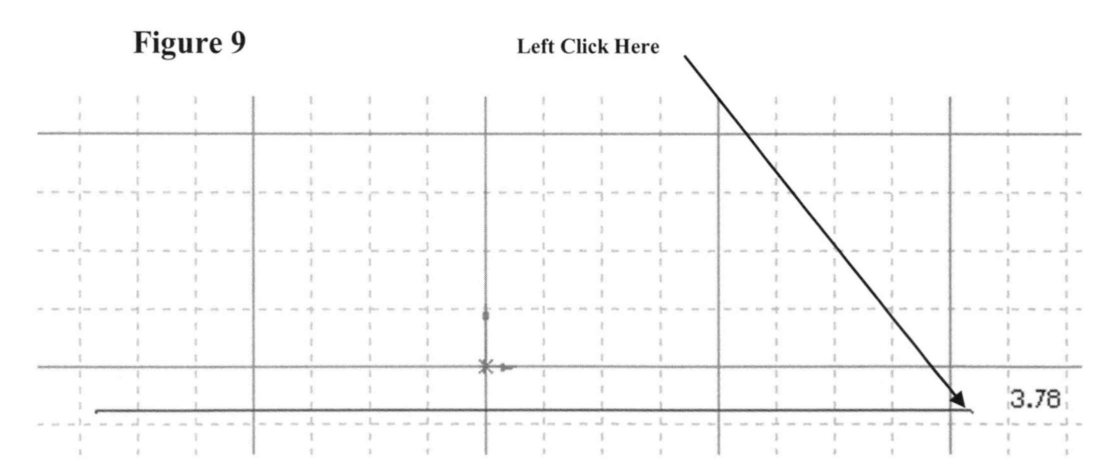

15. While the line is still attached to the cursor, move the cursor towards the top of the screen and left click once. Notice the length of the line is attached to the cursor at the right as shown in Figure 10.

Figure 10

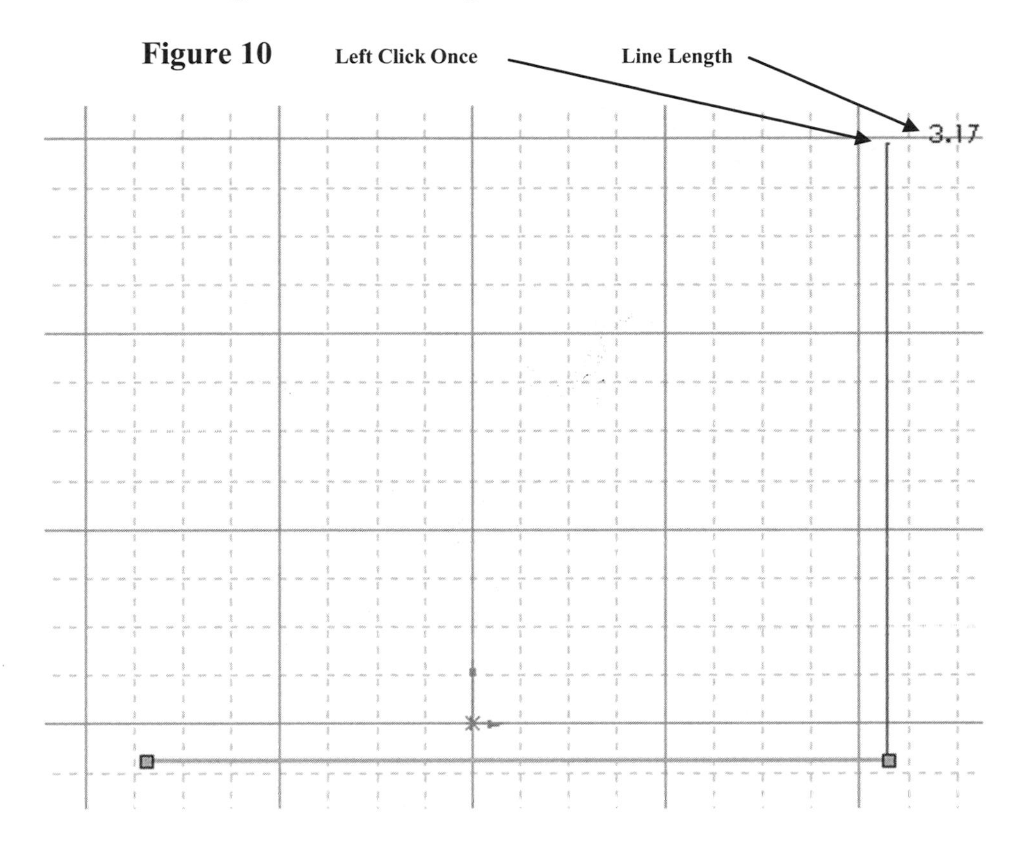

16. With the line still attached to the cursor, move the cursor towards the left side of the screen. Notice the line of small dots connecting the first and last point together. Left click once when the small dots appear as shown in Figure 11.

Figure 11

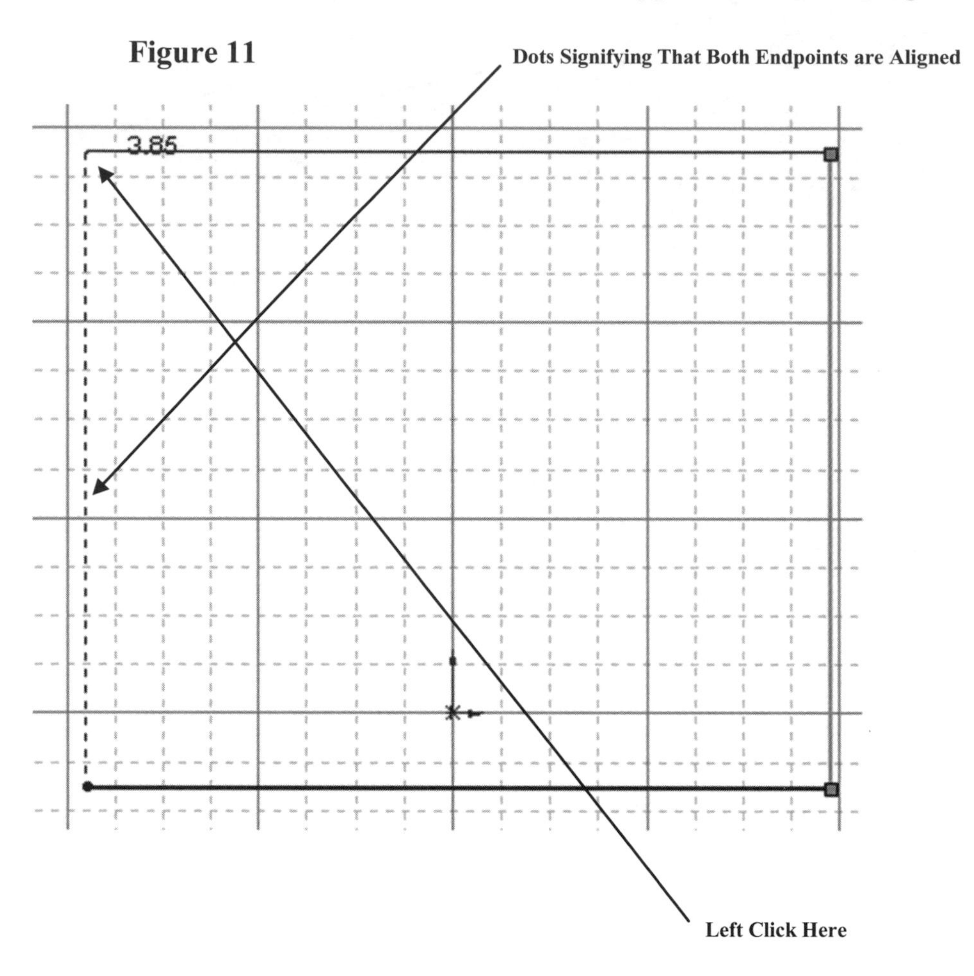

17. Your screen should look similar to Figure 12.

Figure 12

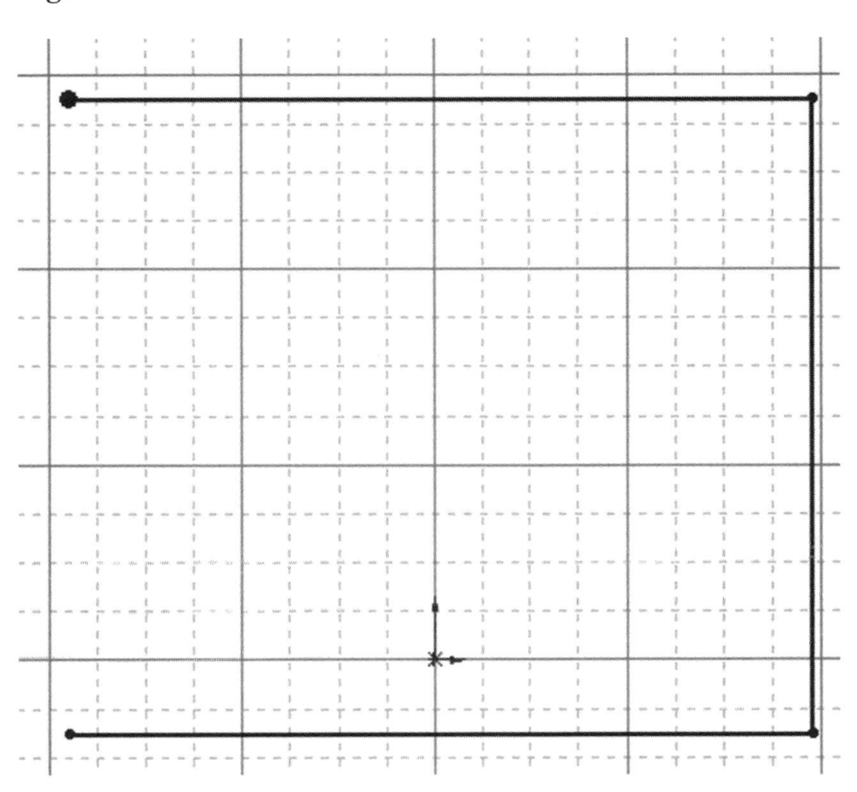

18. Move the cursor down towards the original starting point. Ensure that a red dot appears at the intersection of the two lines. This indicates that SolidWorks has "snapped" to the intersection of the lines. After the red dot appears, left click once. This will form a 90 degree box as shown in Figure 13.

Figure 13

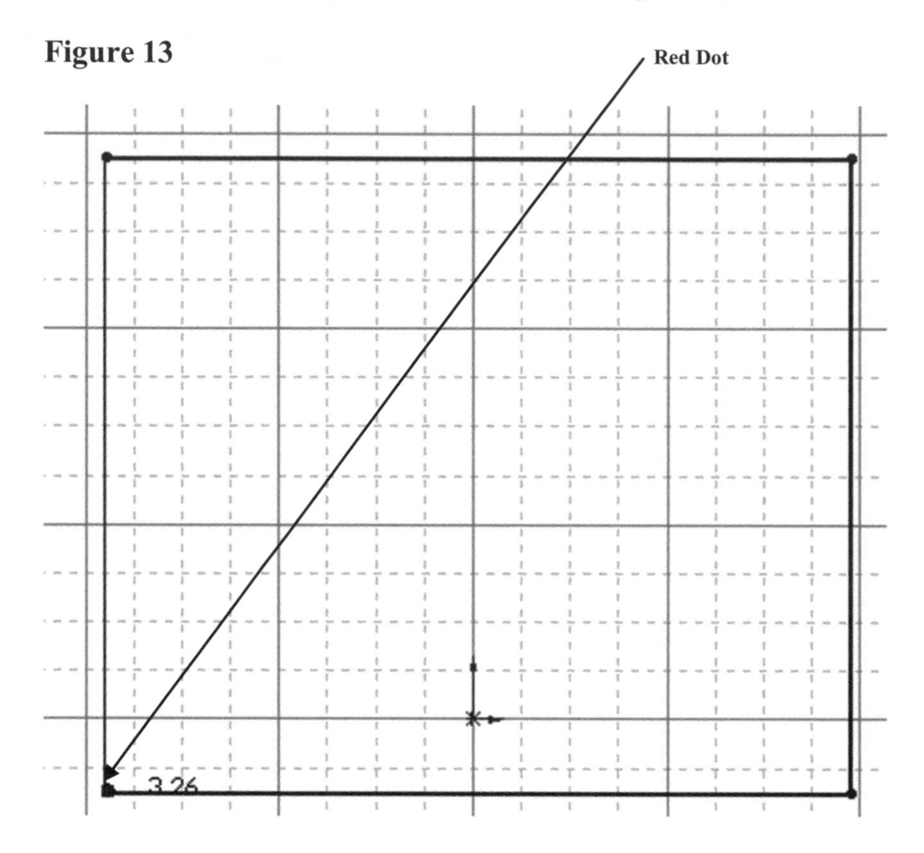

19. Your screen should look similar to Figure 14. After you have closed the box, right click around the drawing. A pop up menu will appear. Left click on **Select** as shown in Figure 14. This verifies that no commands are active.

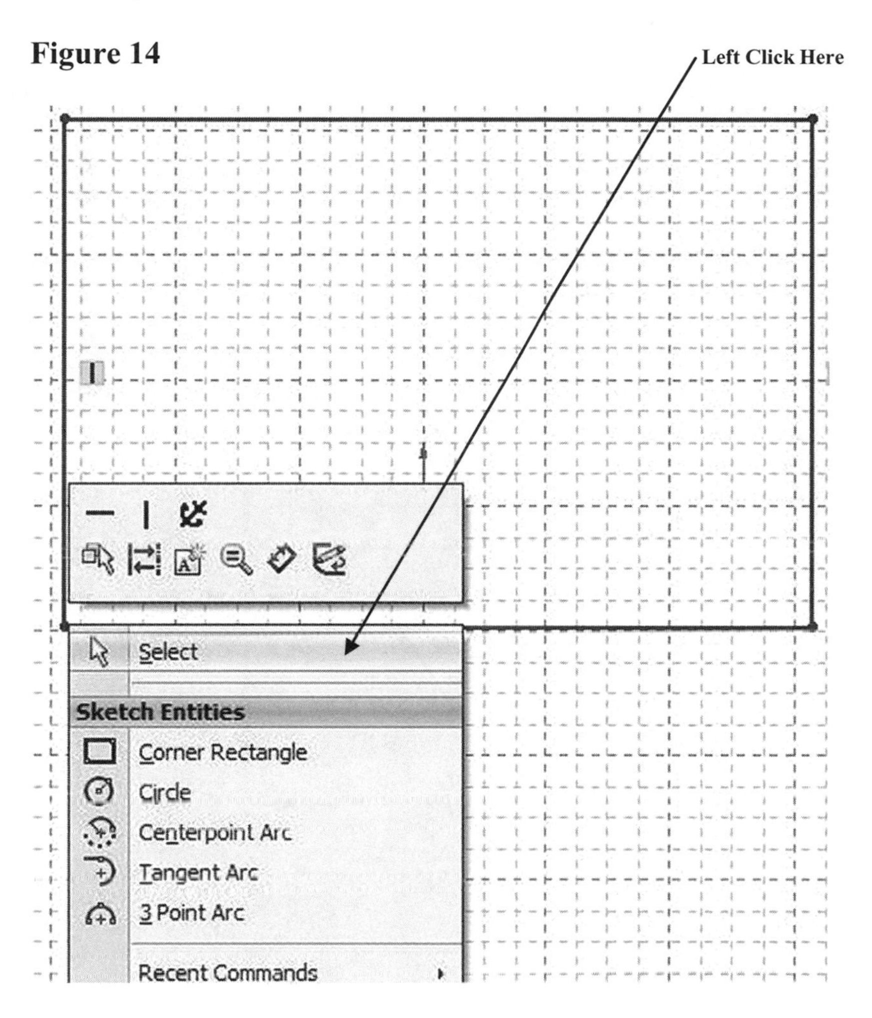

Figure 14

20. Move the cursor to the upper left portion of the screen and left click on the arrow to the right of the SolidWorks text (not shown). Left click on **Tools**. A drop down menu will appear. Left click on **Options** located at the bottom of the menu as shown in Figure 15.

Figure 15

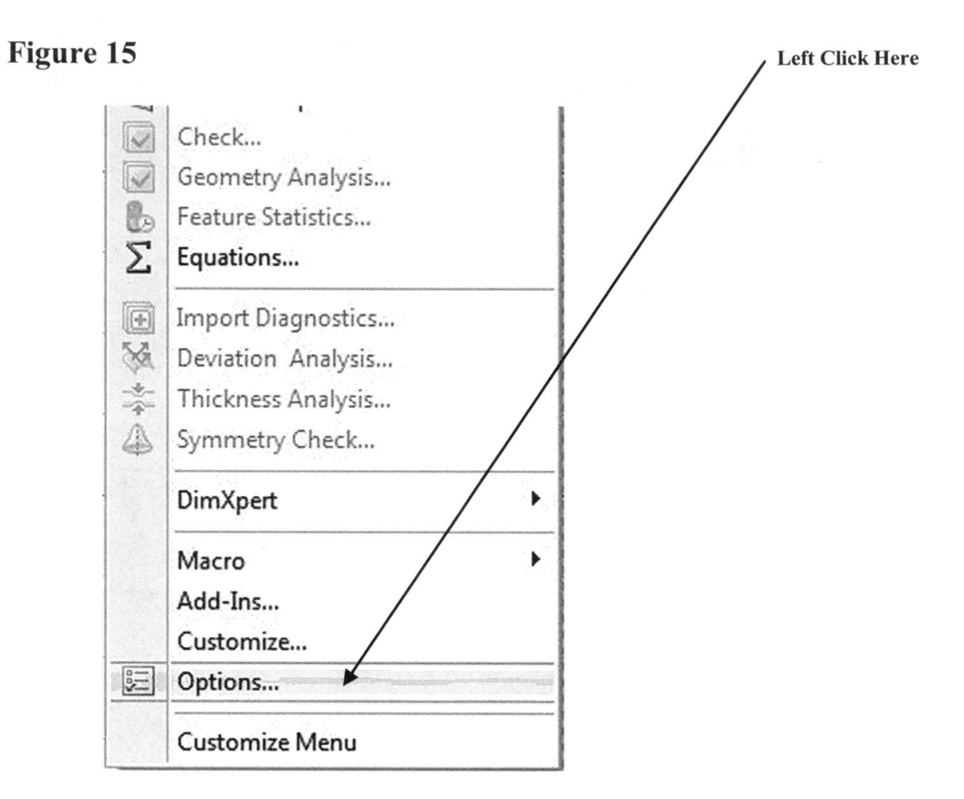

21. The Document Properties – Drafting Standard dialog box will appear. Left click on the **Document Properties** tab. Left click on the drop down arrow to the right of "ANSI". A drop down menu will appear. Left click on **ANSI** as shown in Figure 16 (ANSI may already be selected).

Figure 16

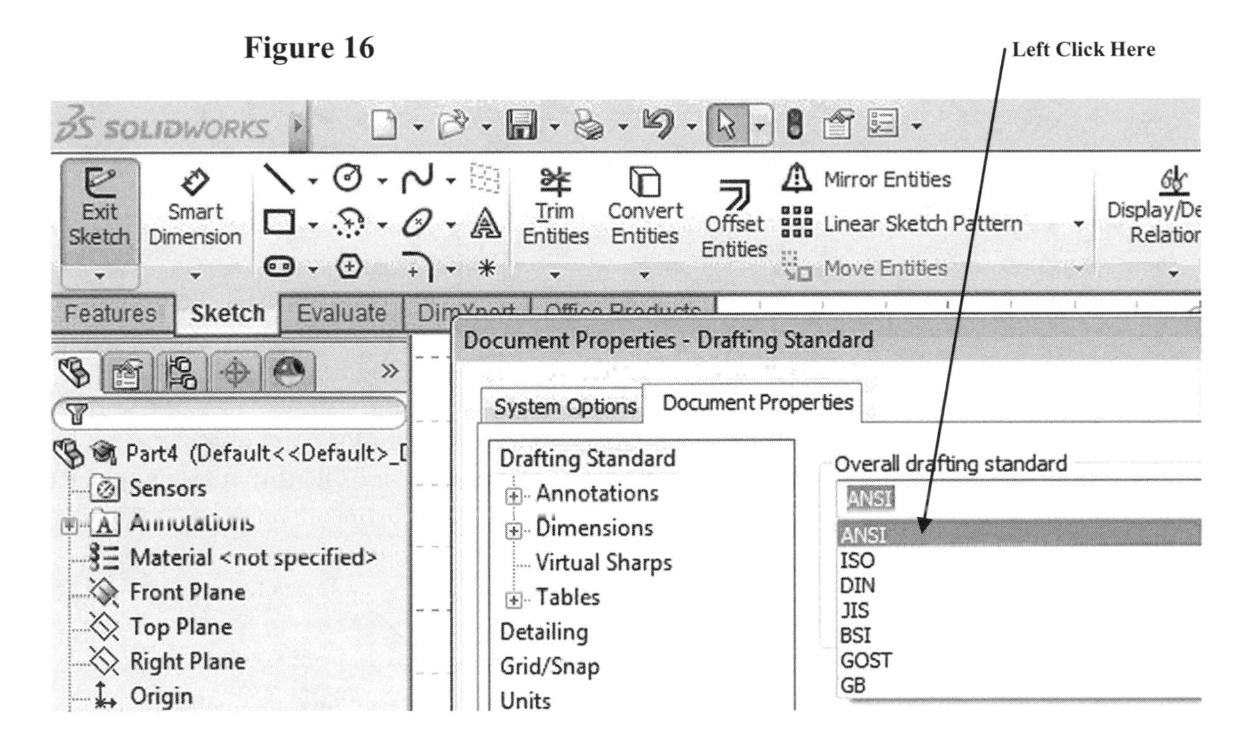

22. Left click on **OK** at the bottom of the screen

23. Move the cursor to the upper left portion of the screen and left click on **Smart Dimension** as shown in Figure 17.

Figure 17

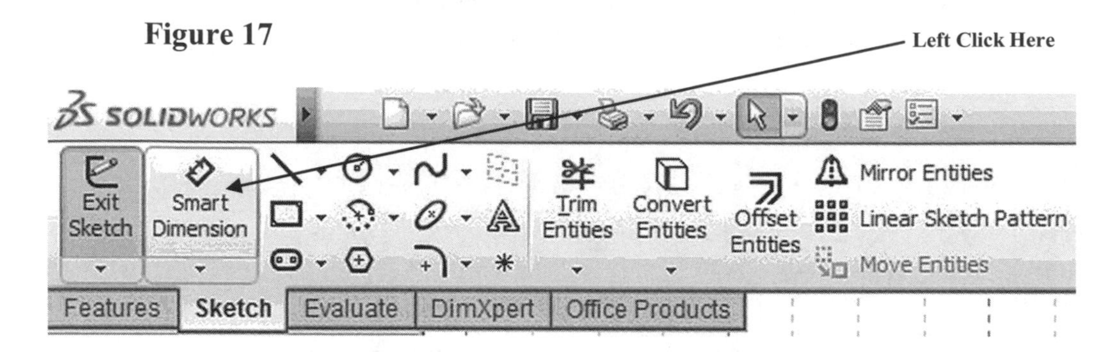

24. Move the cursor over the bottom horizontal line until it turns red as shown in Figure 18. Select the line by left clicking anywhere on the line **or** on each of the end points. To use the end points of the line, move the cursor over one of the end points. A small red dot will appear. Left click once and move the cursor to the other end point. After the red dot appears, left click once. The dimension will be attached to the cursor.

Figure 18

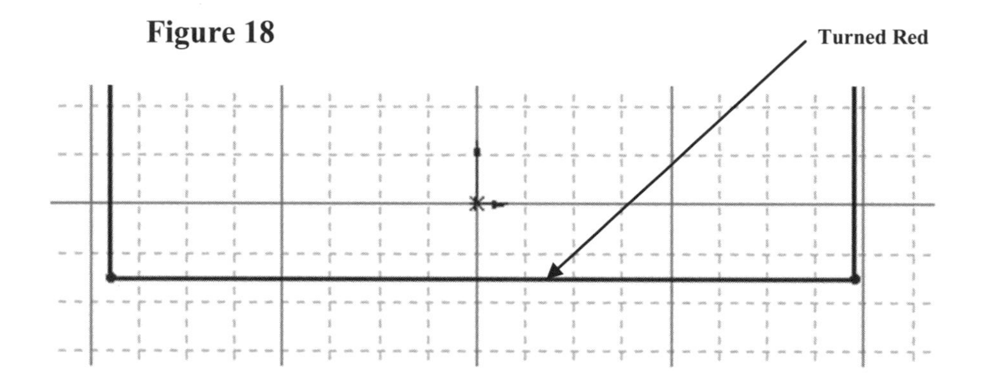

25. Move the cursor down to where the dimension will be placed and left click once as shown in Figure 19.

Figure 19

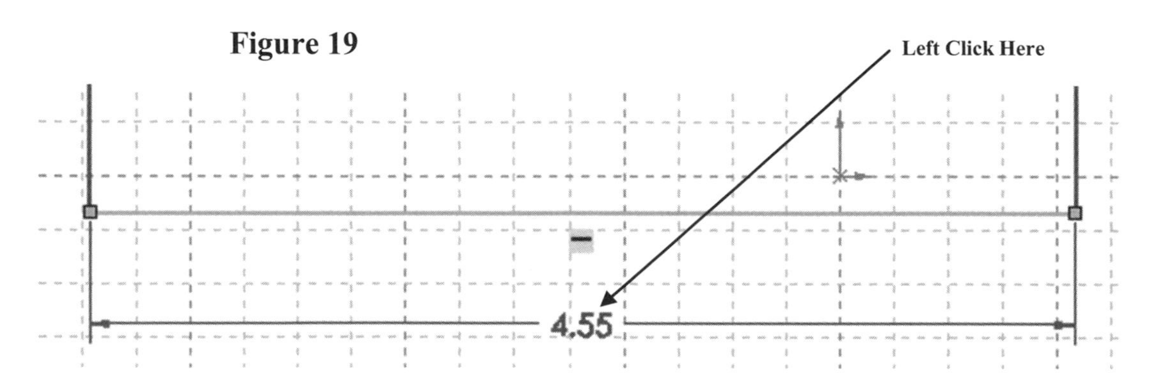

26. The Modify dialog box will appear as shown in Figure 20.

Figure 20

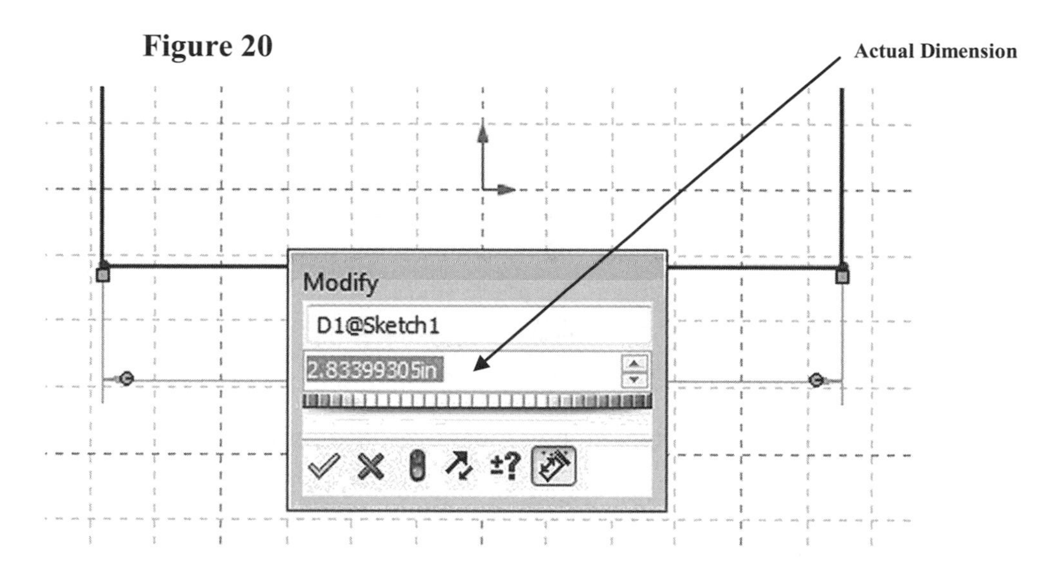

27. To edit the dimension, enter **2.00** in the Modify dialog box (while the current dimension is highlighted) and left click on the green checkmark as shown in Figure 21.

Figure 21

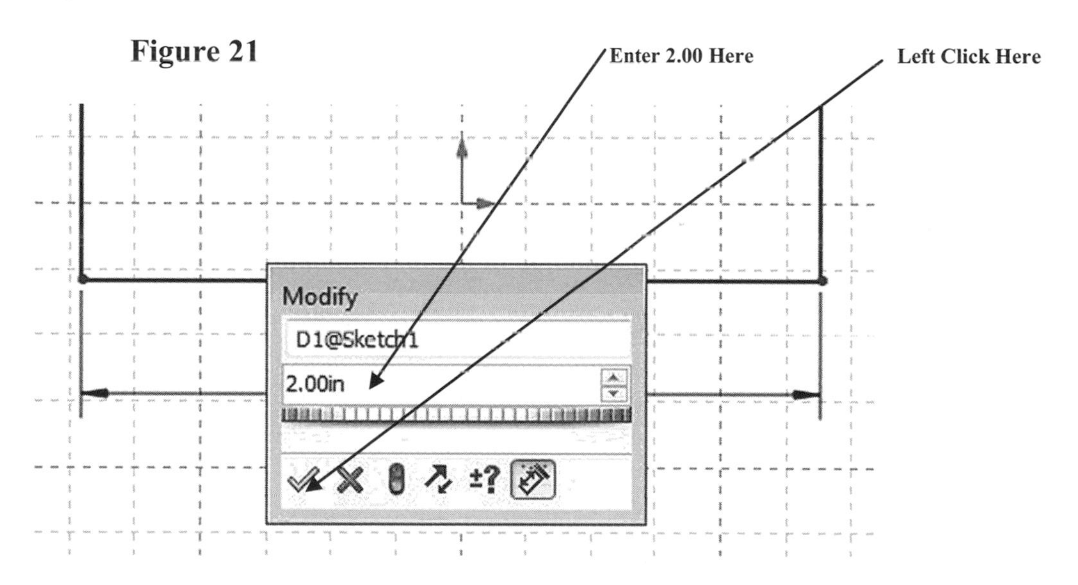

28. The dimension of the line will become 2.00 inches as shown in Figure 22.

Figure 22

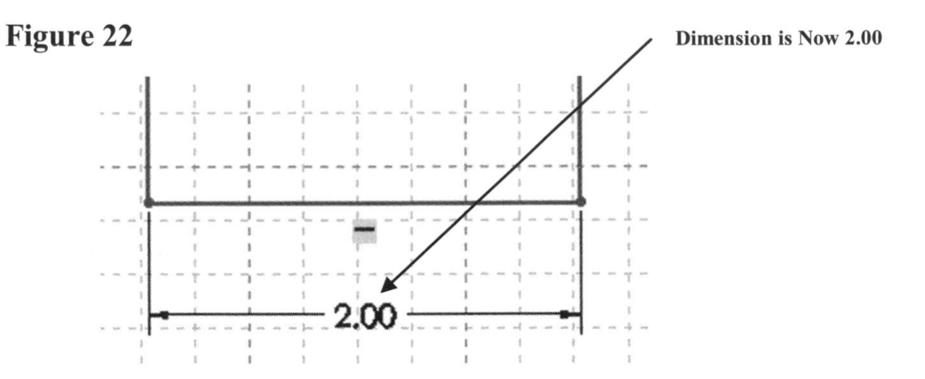

29. Move the cursor over the right side vertical line until it turns red as shown in Figure 23. Left click once.

Figure 23

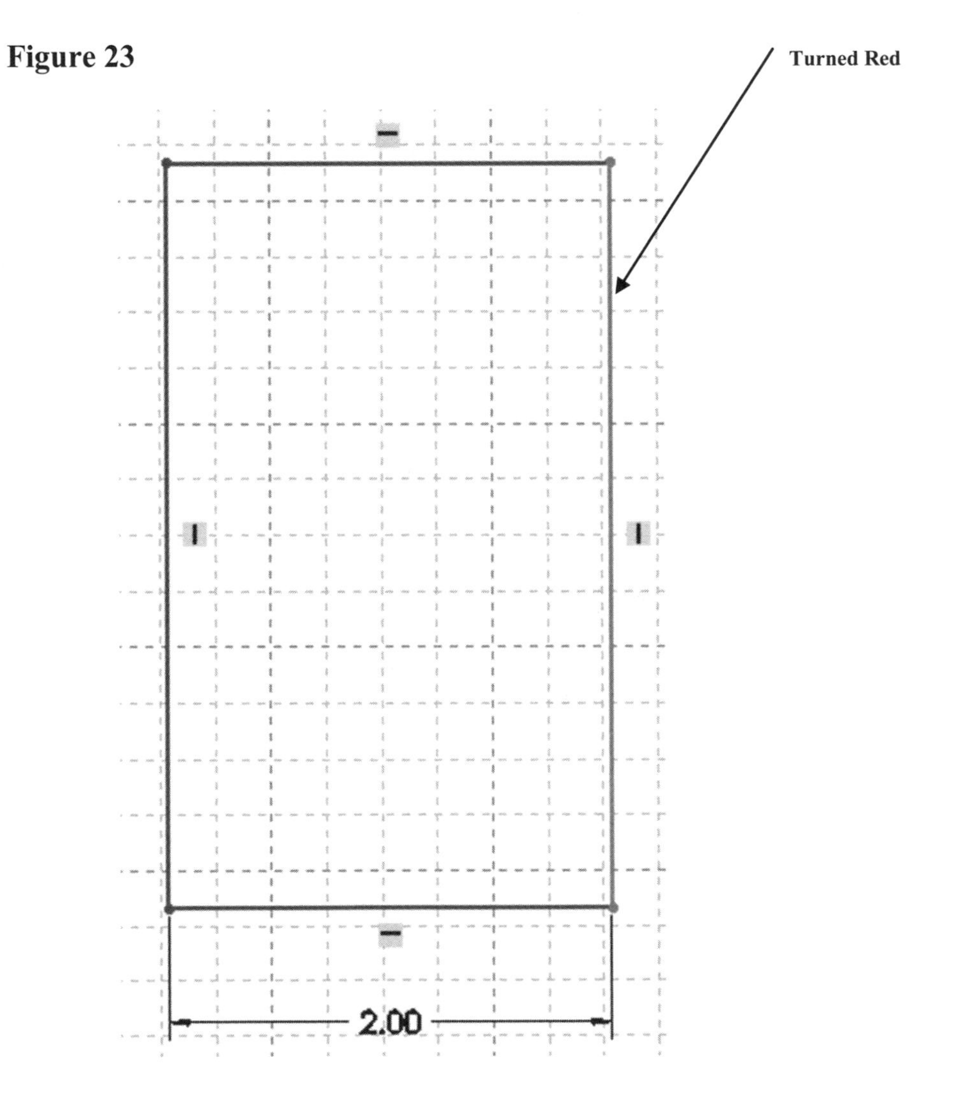

30. Move the cursor to where the dimension will be placed and left click once as shown in Figure 24.

Figure 24

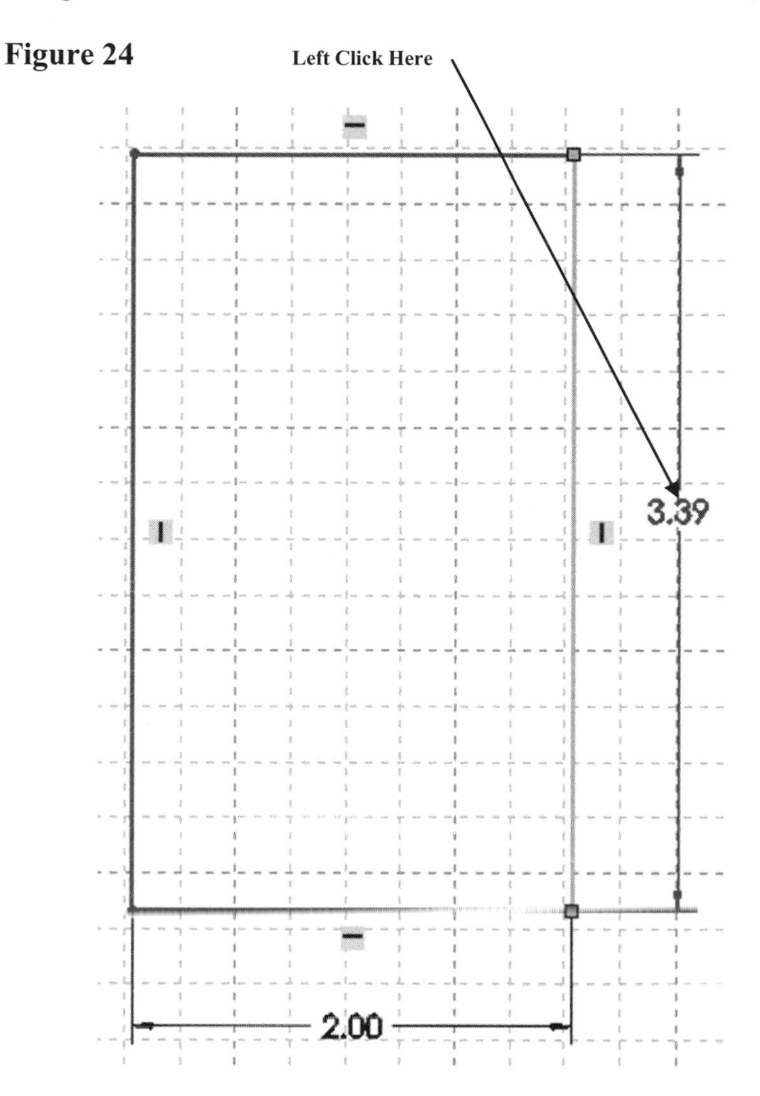

31. The Modify dialog box will appear as shown in Figure 25.

Figure 25

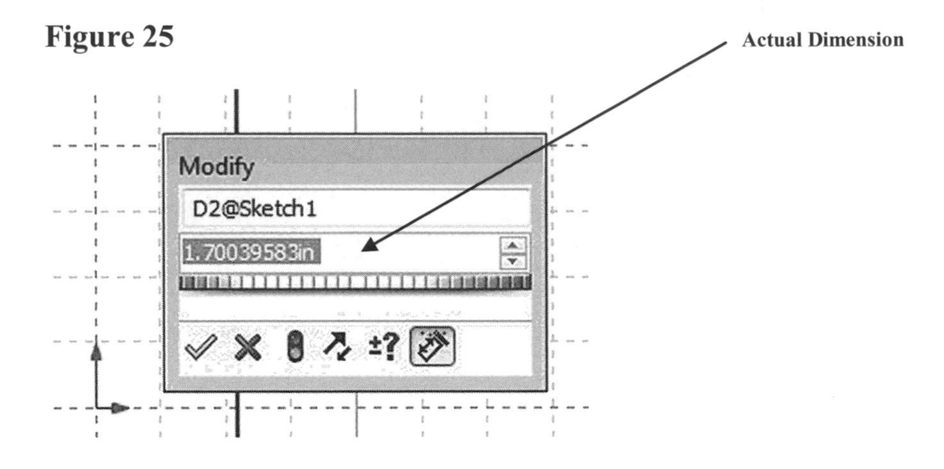

32. To edit the dimension, enter **.25** in the Modify dialog box (while the current dimension is highlighted) and left click on the green checkmark as shown in Figure 26.

Figure 26

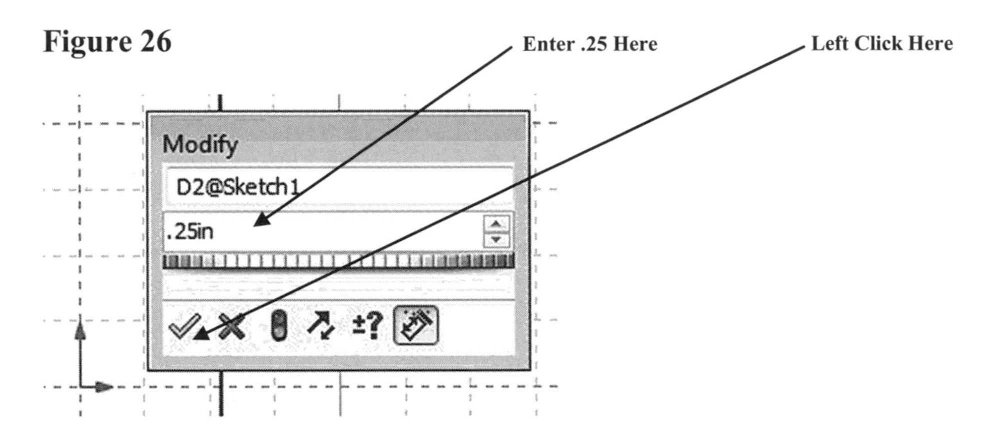

33. The screen should look similar to Figure 27.

Figure 27

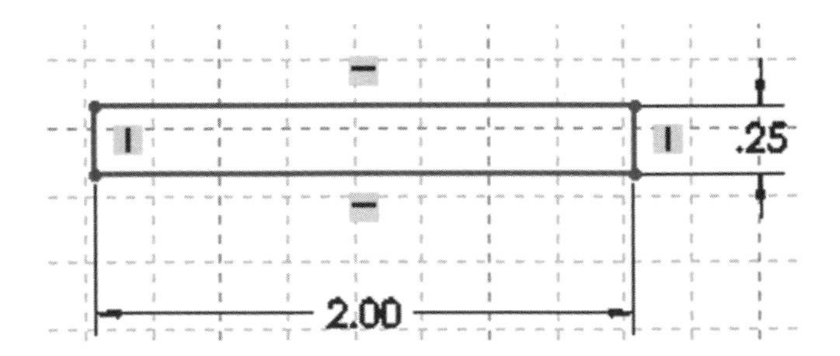

34. Right click anywhere around the drawing. A pop up menu will appear. Left click on **Select** as shown in Figure 28. This verifies that no commands are active.

Figure 28

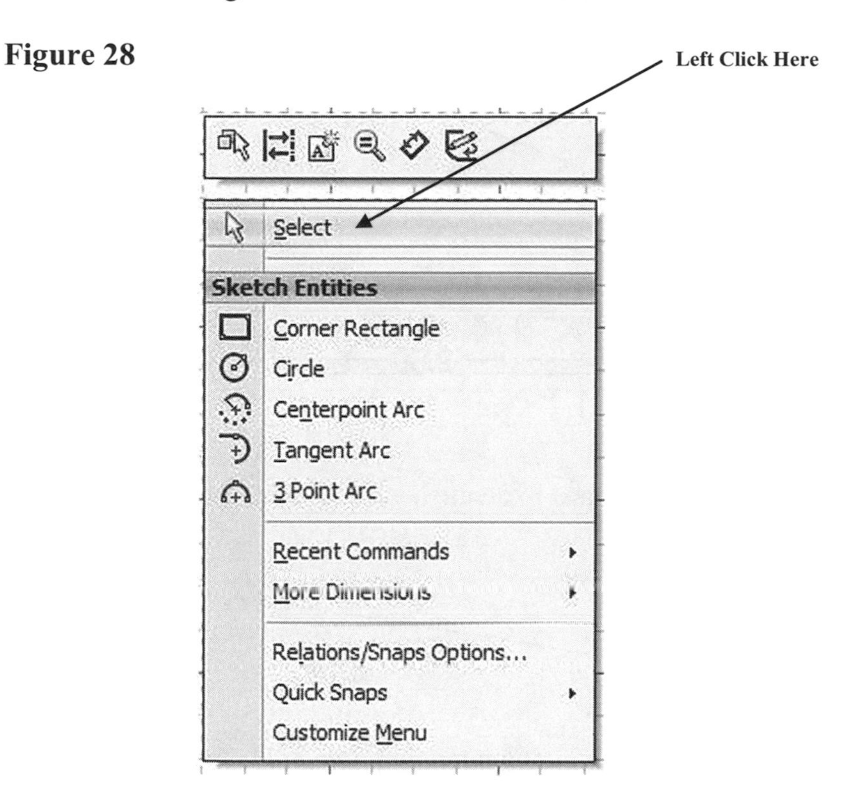

35. Move the cursor to the upper left corner of the screen and left click on **Line** as shown in Figure 29.

Figure 29

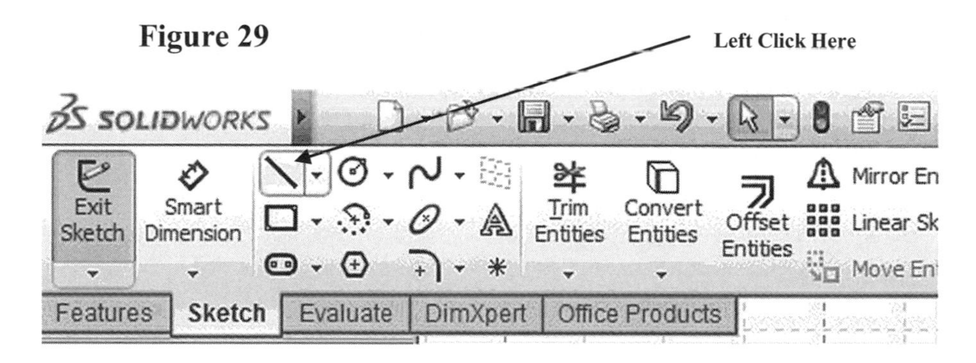

36. Move the cursor to the upper left corner of the box as shown in Figure 30 and left click once.

Figure 30

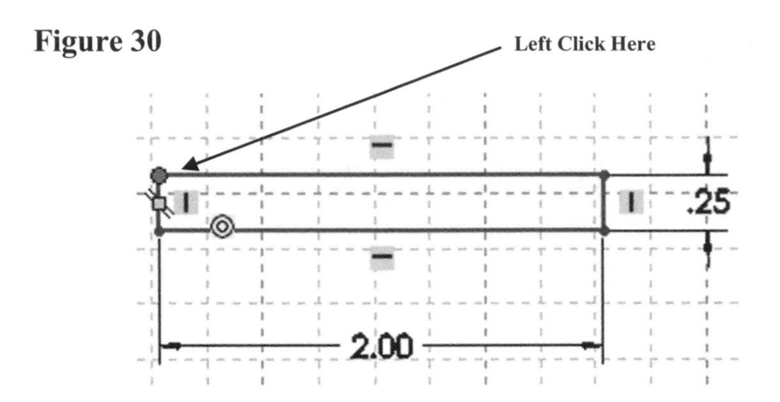

37. Move the cursor upward to create a vertical line and left click once as shown in Figure 31.

Figure 31

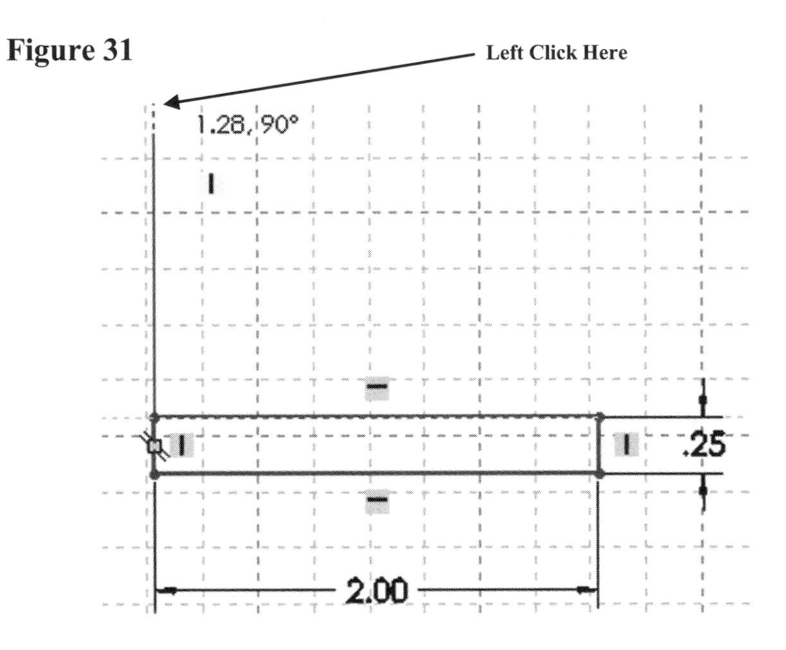

38. Right click the mouse. A pop up menu will appear. Left click on **Select** as shown in Figure 32. This verifies that no commands are active.

Figure 32

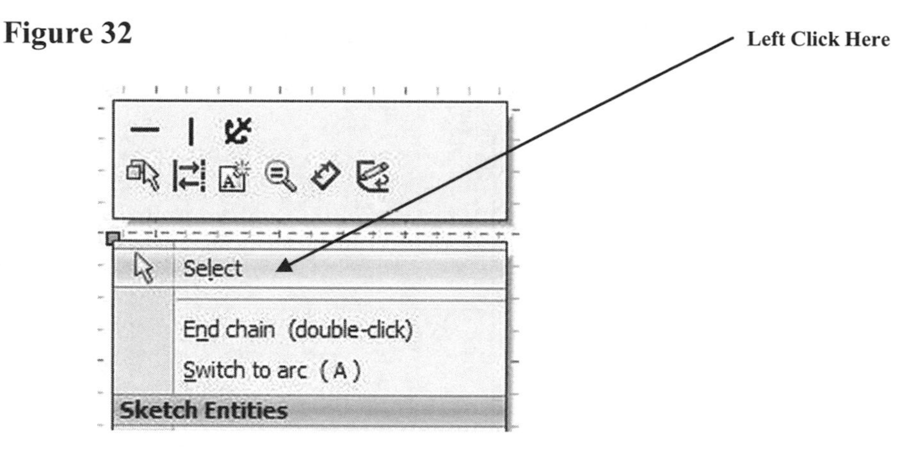

39. You may have to zoom out to leave enough room to construct the following lines. To zoom out use the scroll wheel on the mouse or move the cursor to the upper middle portion of the screen and left click on the "Zoom to Fit" icon as shown in Figure 33.

Figure 33

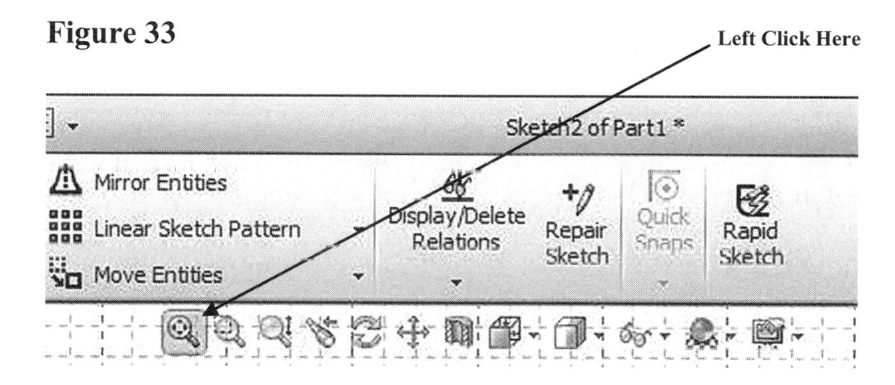

40. Holding down the mouse wheel will allow you to rotate the sketch/part.

41. Move the cursor to the upper left corner of the screen and left click on **Line** as shown in Figure 34.

Figure 34

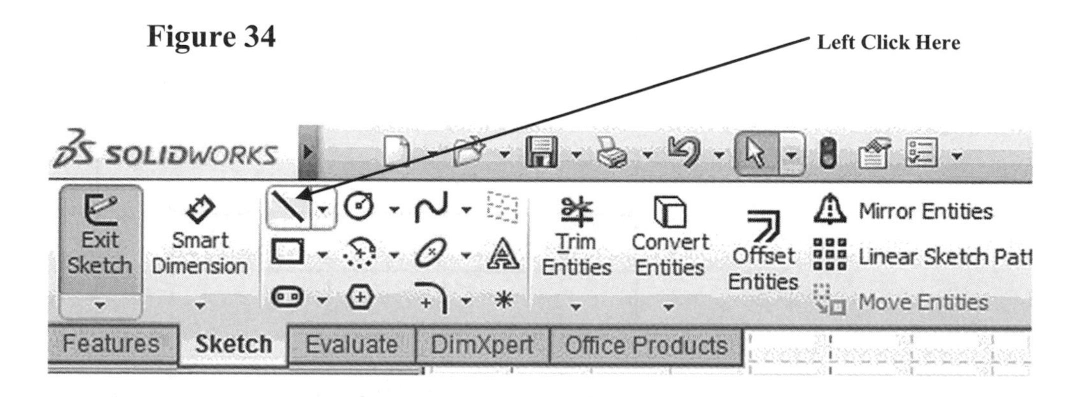

42. Left click on the upper endpoint of the line. Move the cursor to the left and left click once as shown in Figure 35.

Figure 35

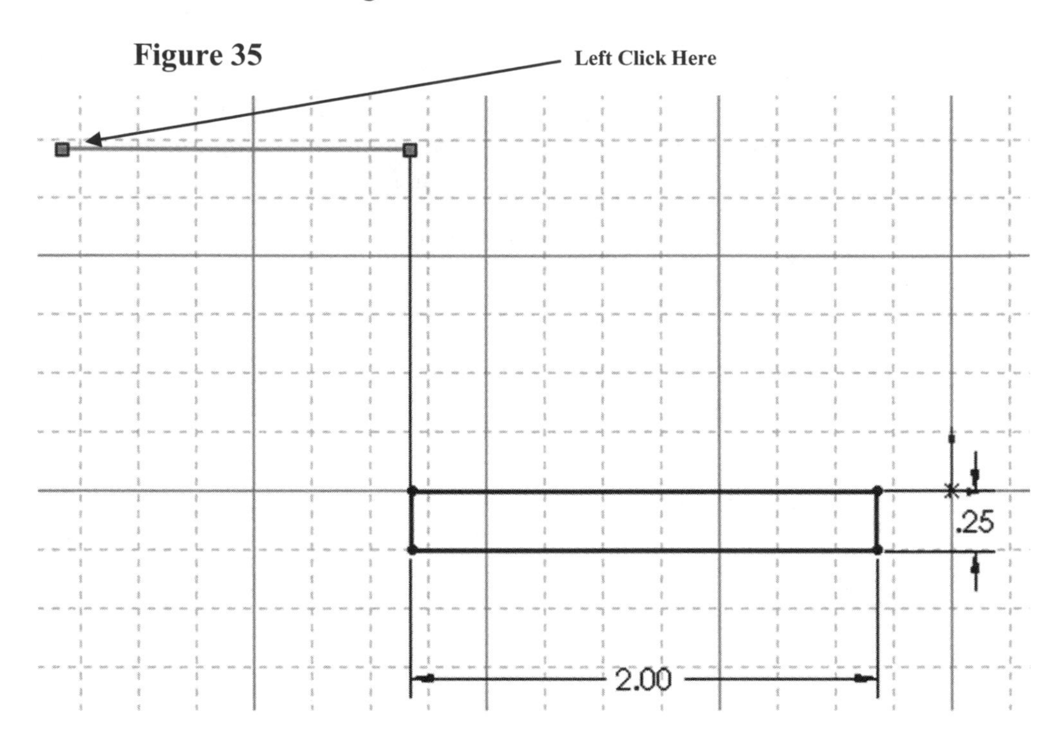

43. With the line still attached to the cursor, move the cursor up and left click once as shown in Figure 36.

Figure 36

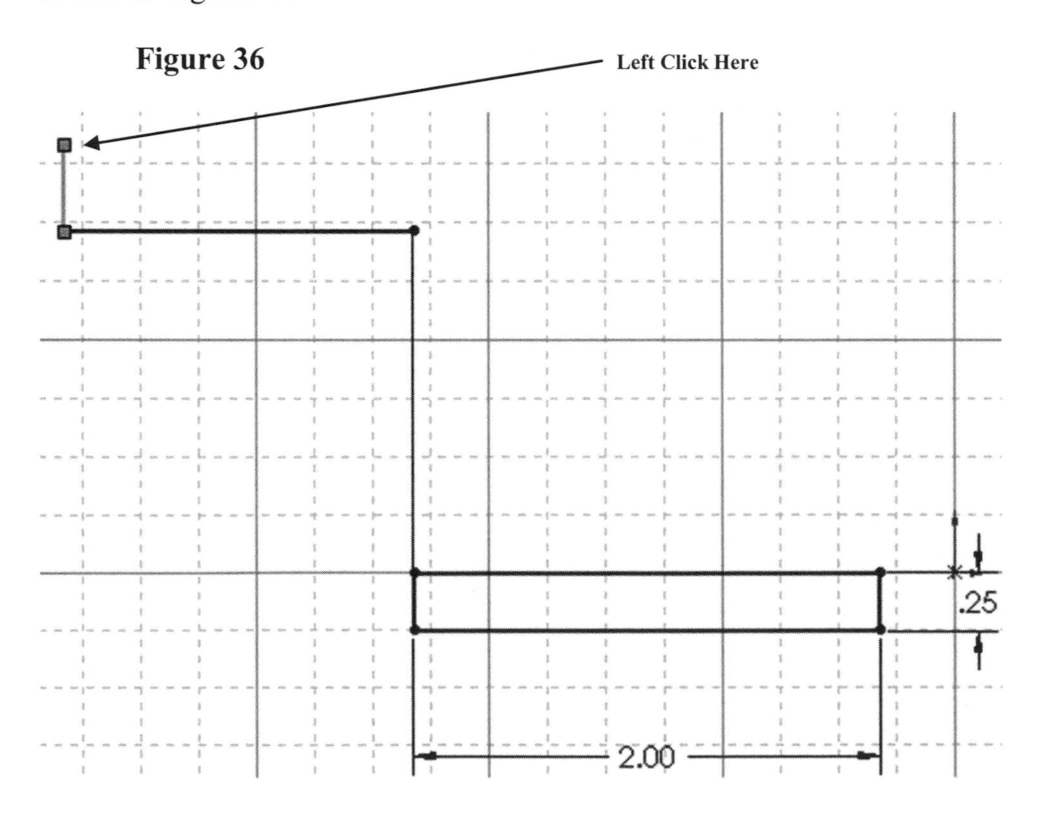

44. With the line still attached to the cursor, move the cursor to the right side of the screen and left click once as shown in Figure 37.

Figure 37

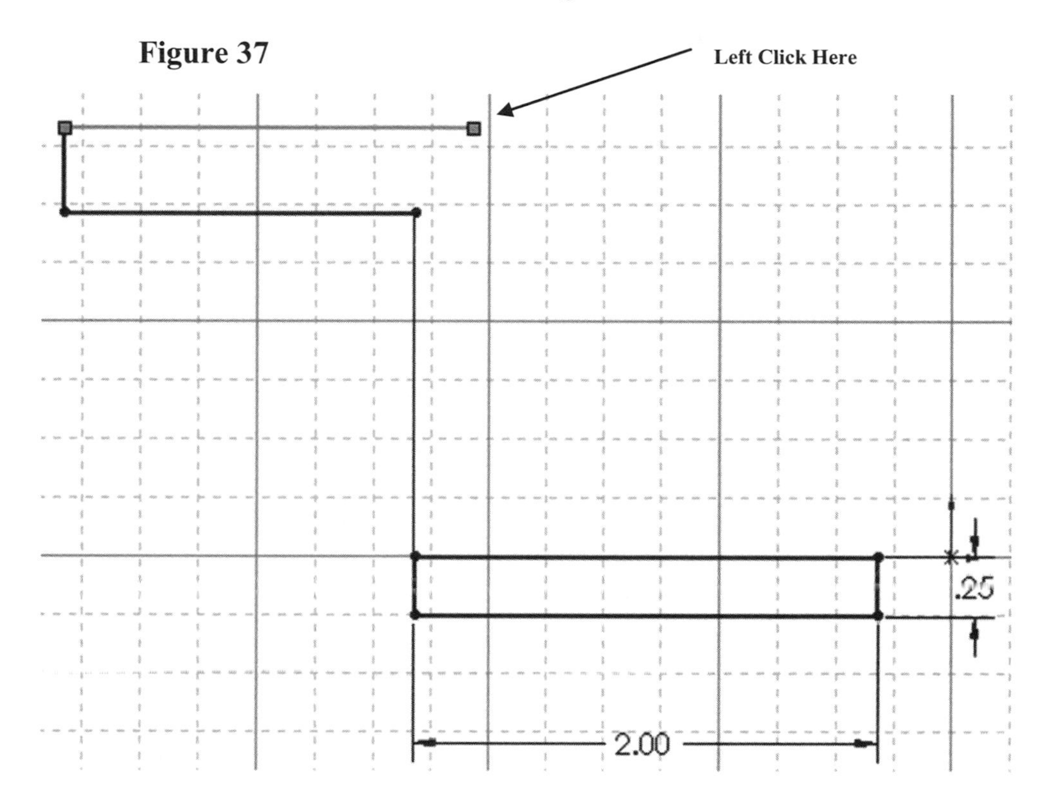

45. With the line still attached to the cursor, move the cursor down towards the bottom of the screen and left click once as shown in Figure 38. Notice the small green square appearing at the intersection of the two lines. After the left mouse button is clicked the green square will change to a green dot. This indicates that the lines are all connected.

Figure 38

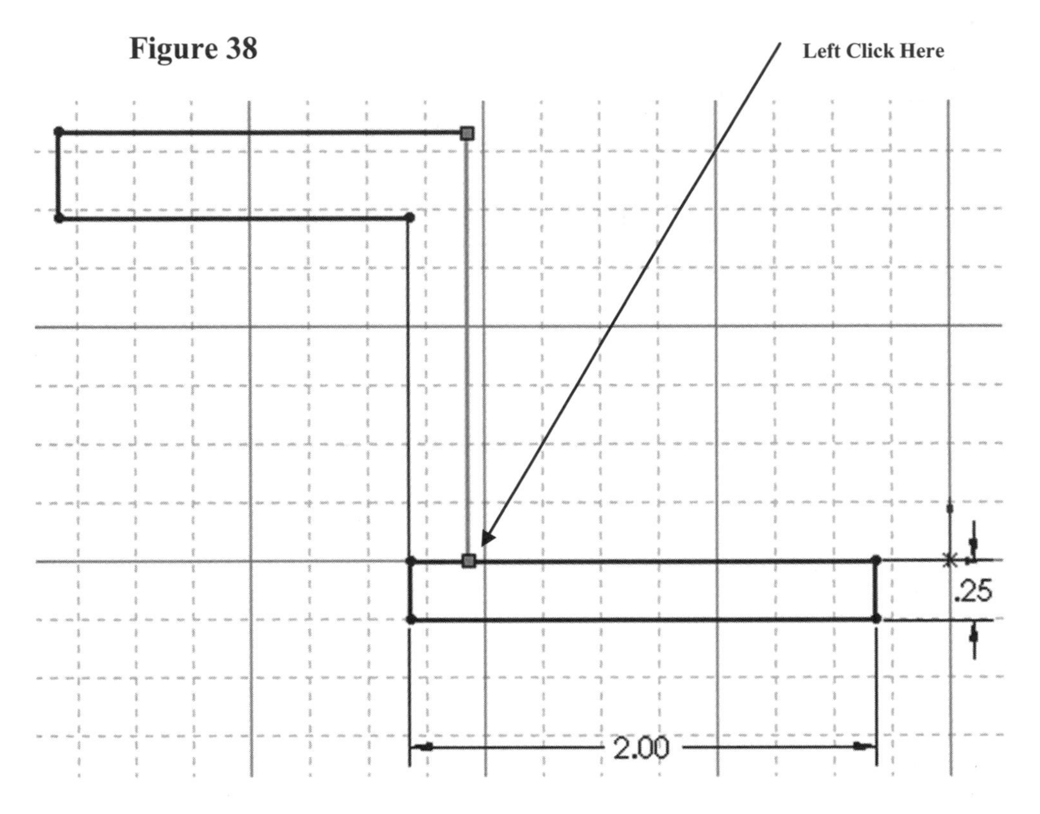

46. With the line still attached, right click the mouse. A pop up menu will appear. Left click on **Select** as shown in Figure 39. This verifies that no commands are active.

Figure 39

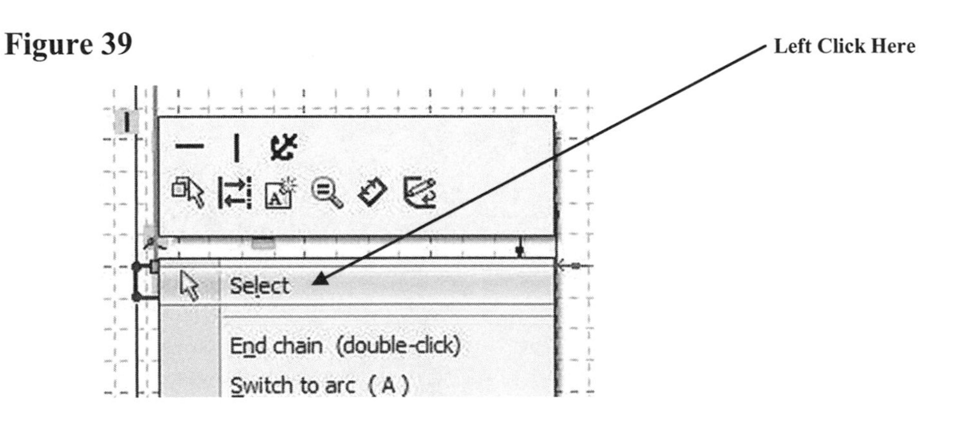

47. Move the cursor to the upper middle portion of the screen and left click on the the arrow below Trim Entities as shown in Figure 40.

Figure 40

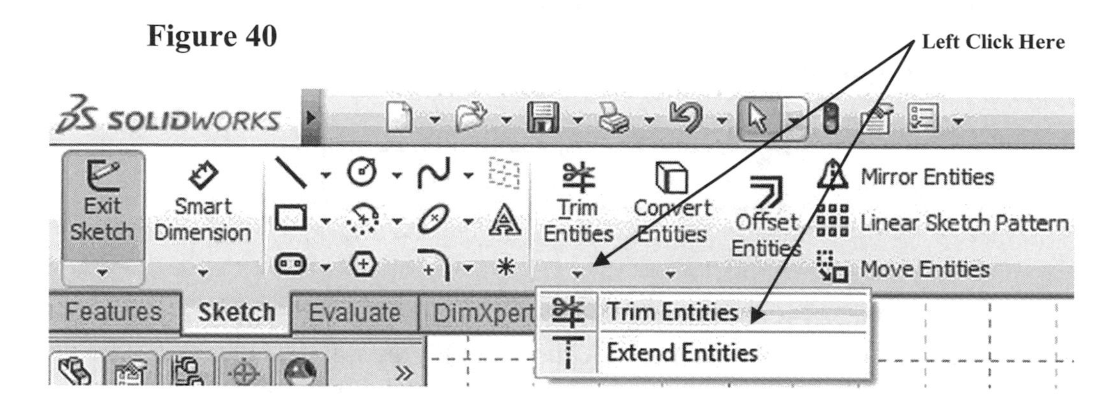

48. More icons will appear. Left click on **Trim Entities** as shown in Figure 40.

49. After selecting **Trim Entities** move the cursor to the middle left portion of the screen and left click on **Trim to closest**. Now move the cursor over the line that will be trimmed causing it to turn red and left click once as shown in Figure 41.

Figure 41

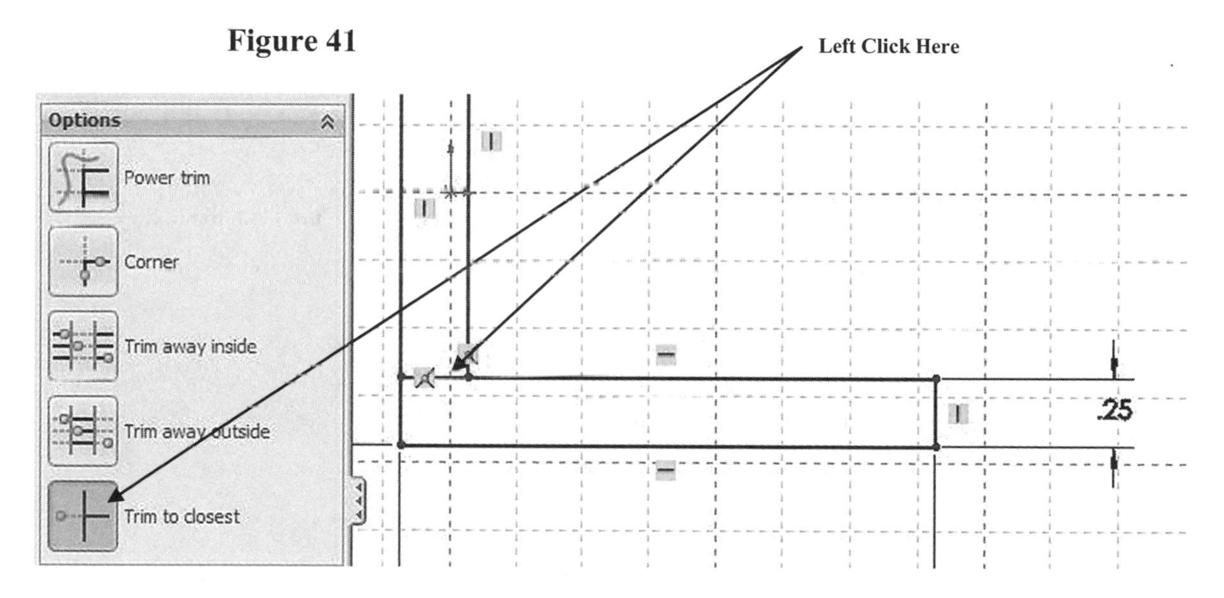

50. The line will disappear as shown in Figure 42. Left click on the green check mark under Trim or press the **Esc** key once. This verifies that no commands are active.

Figure 42

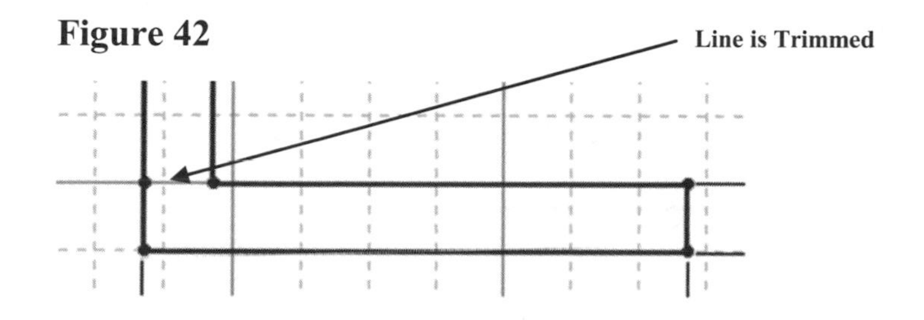

51. Move the cursor over the line in the lower left corner of the drawing as shown in Figure 43. The line will turn red. This particular line must be deleted so that the line above can be extended the full length.

Figure 43

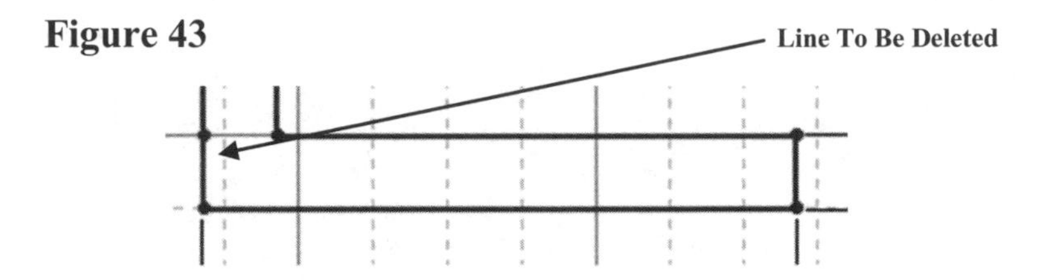

52. Left click once on the line that will be deleted. The line will turn red or green with small squares at each end as shown in Figure 44.

Figure 44

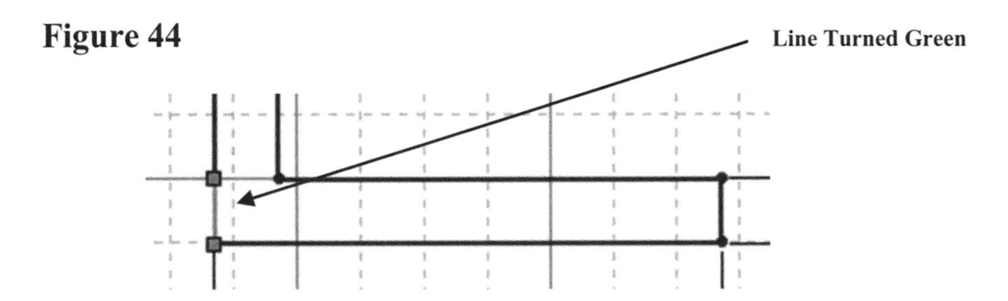

53. After the line is highlighted (turned red or green) right click once. A pop up menu will appear. Left click on **Delete** as shown in Figure 45.

Figure 45

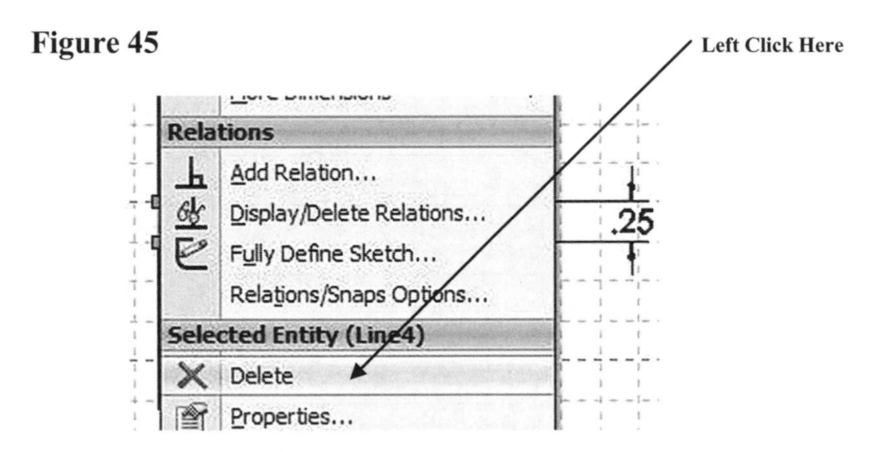

54. The line will be deleted as shown in Figure 46.

Figure 46

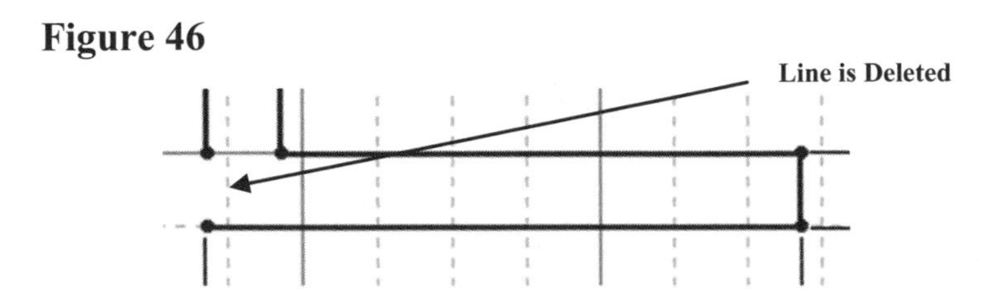

55. Move the cursor to the upper middle portion of the screen and left click on **Trim Entities** as shown in Figure 47.

Figure 47

56. There are several different methods of trimming or extending a line.

57. Move the cursor over the line to be extended. It will turn red as shown in Figure 48.

Figure 48

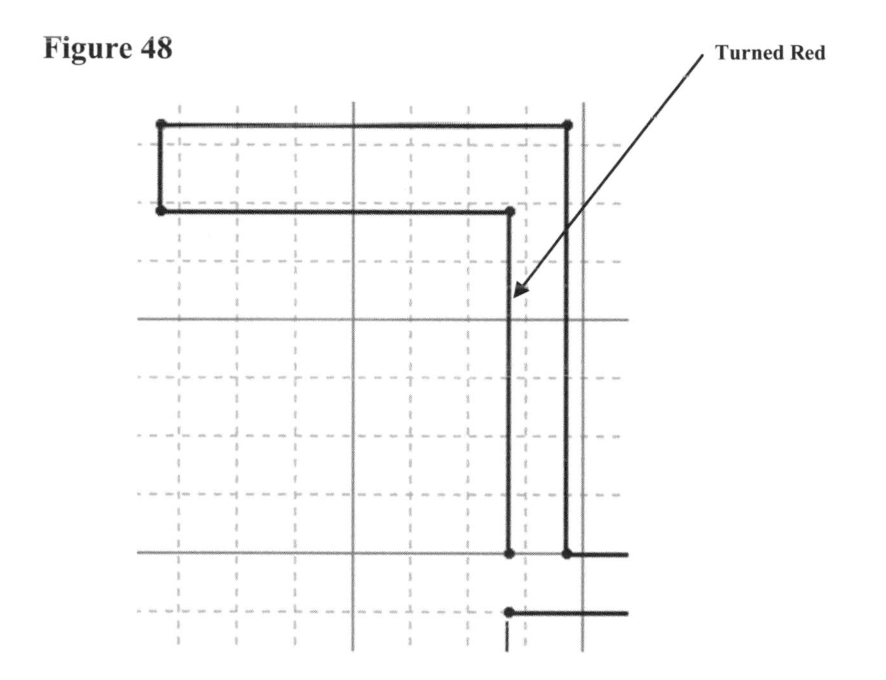

58. After the line has turned red, right click once. A pop up menu will appear. Left click on **Extend Entities** as shown in Figure 49.

Figure 49

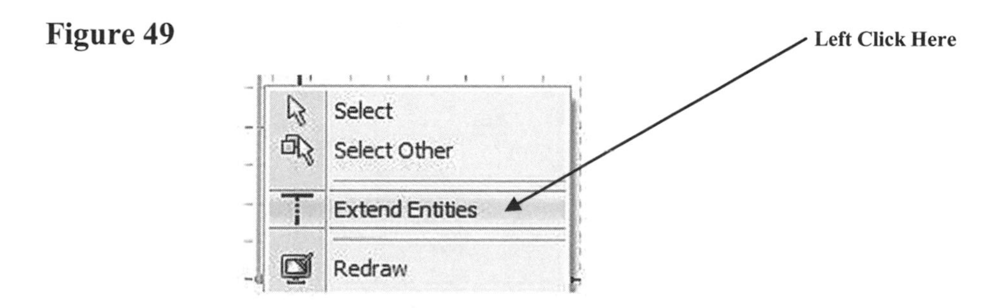

59. Move the cursor over the line to be extended. The line will extend downward forming one continuous line. Left click once as shown in Figure 50.

Figure 50

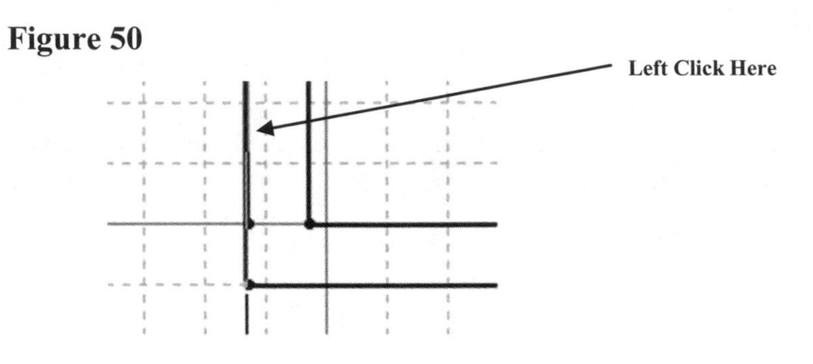

60. Your screen should look similar to Figure 51.

Figure 51

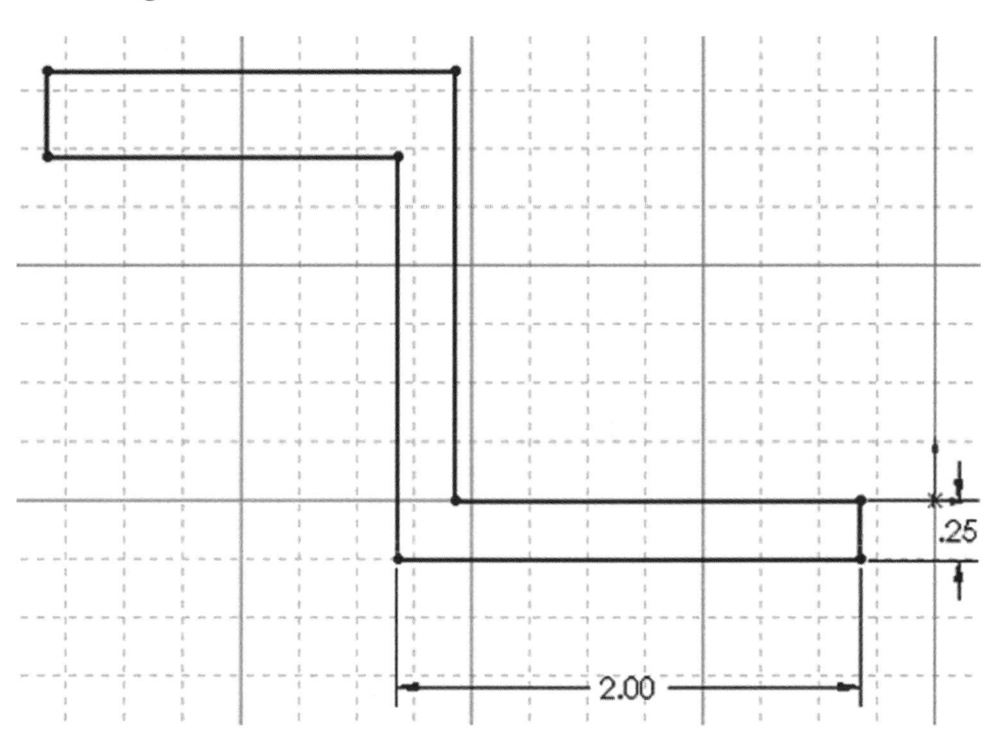

61. Move the cursor to the upper left portion of the screen and left click on **Smart Dimension** as shown in Figure 52.

Figure 52

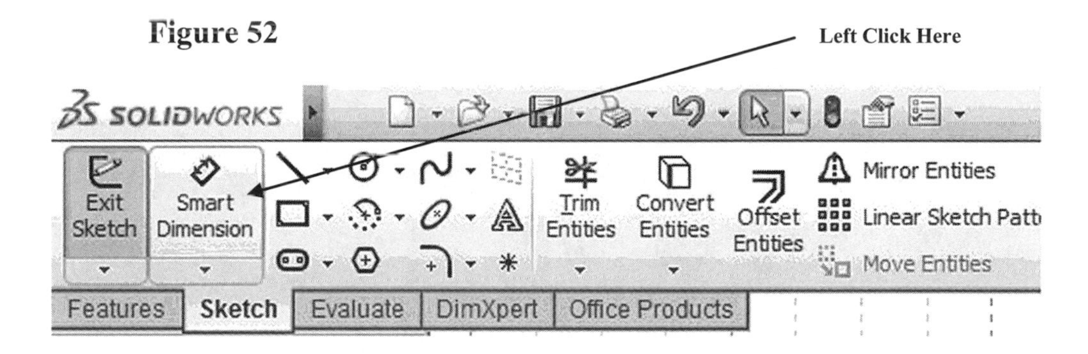

62. After selecting **Smart Dimension** move the cursor over the left side vertical line. The line will turn red as shown in Figure 53. Left click on the line.

Figure 53

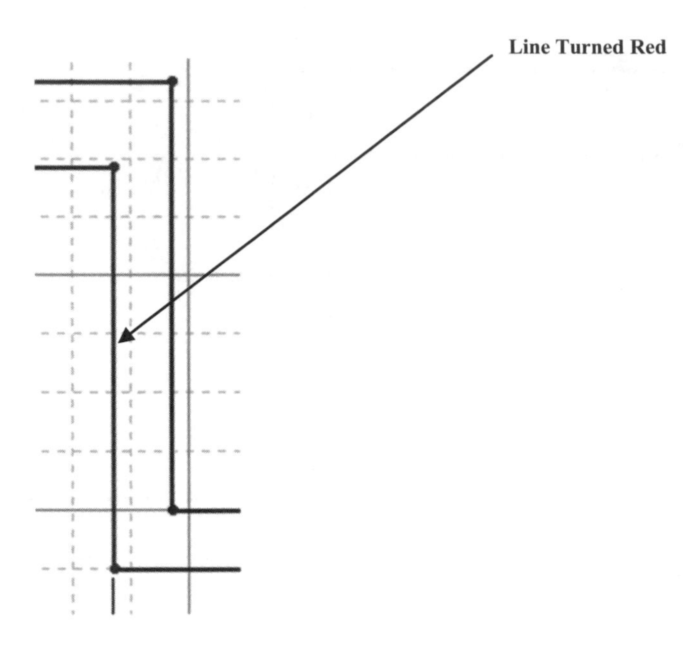

63. The dimension is attached to the cursor. Move the cursor to where the dimension will be placed and left click once. The Modify dialog box will appear as shown in Figure 54.

Figure 54

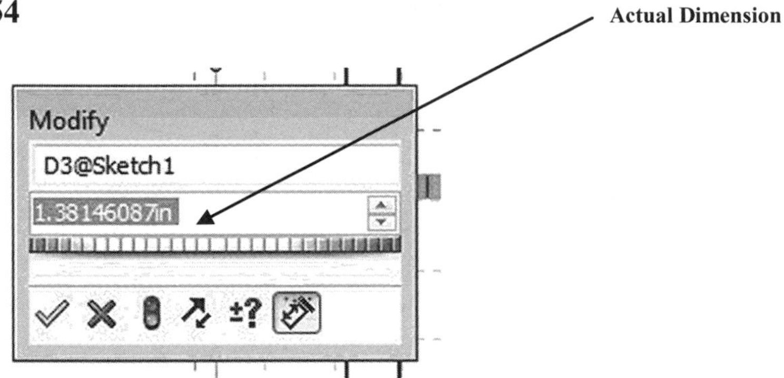

64. To edit the dimension, enter **1.75** in the Modify dialog box (while the current dimension is highlighted) and left click on the green checkmark as shown in Figure 55.

Figure 55

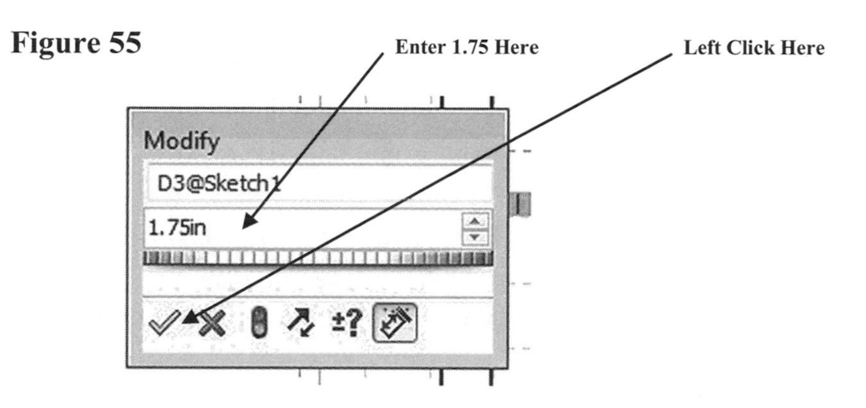

65. The dimension is now 1.75 as shown in Figure 56.

Figure 56

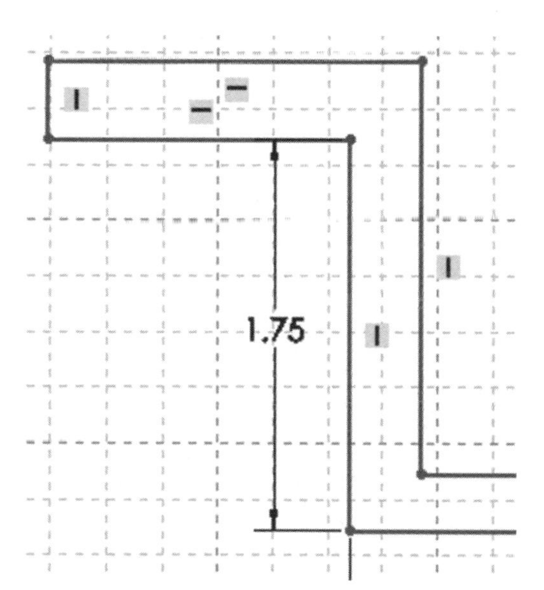

66. Move the cursor over the left side vertical line until it turns red as shown in Figure 57. Left click on the line.

Figure 57

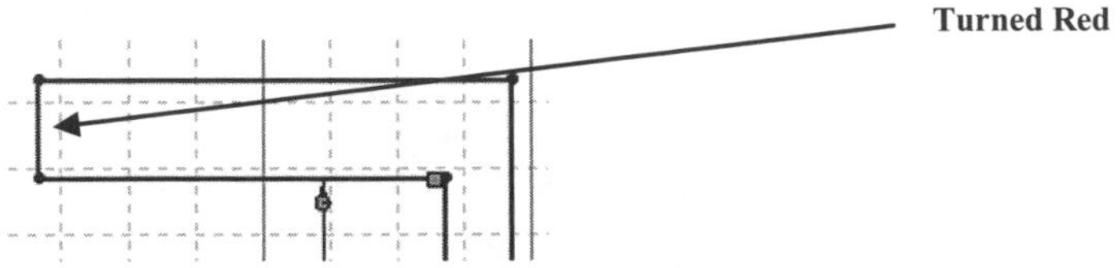

Turned Red

67. The dimension is attached to the cursor. Move the cursor back and forth to verify it is attached. Move the cursor to where the dimension will be placed and left click once. The Modify dialog box will appear as shown in Figure 58.

Figure 58

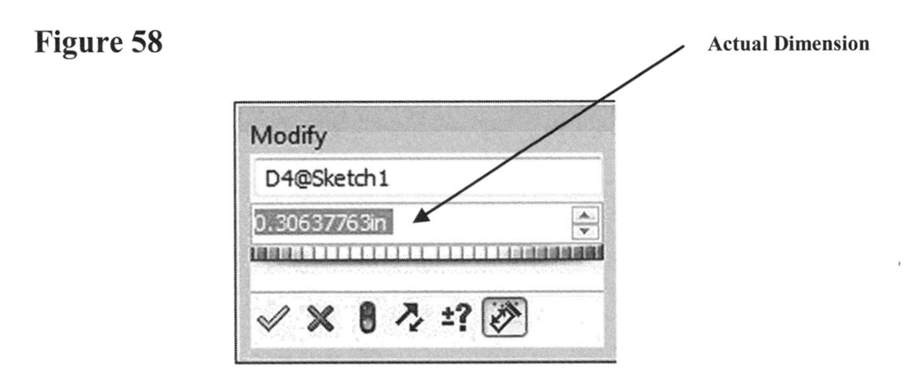

68. To edit the dimension, enter **.25** in the Modify dialog box (while the current dimension is highlighted) and left click on the green checkmark as shown in Figure 59.

Figure 59

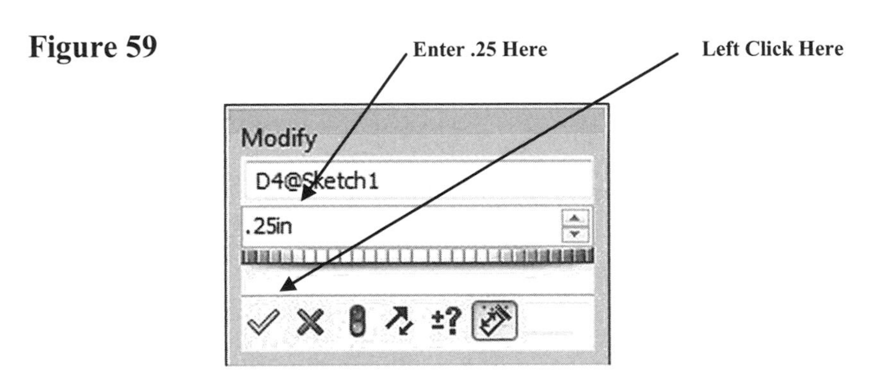

69. The dimension is now .25 as shown in Figure 60.

Figure 60

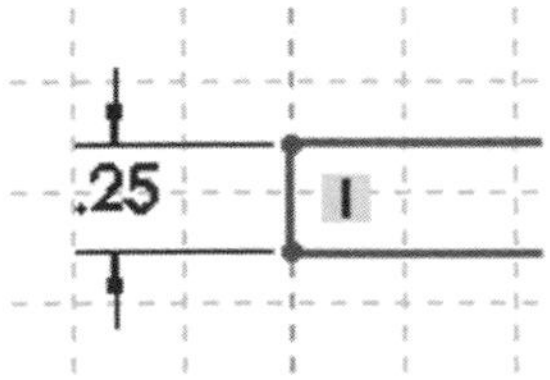

.25

70. Move the cursor over the top horizontal line until it turns red as shown in Figure 61. Left click on the line.

Figure 61

Turned Red

.25

71. The dimension is attached to the cursor. Move the cursor up and down to verify it is attached. Move the cursor to where the dimension will be placed and left click once. The Modify dialog box will appear as shown in Figure 62.

Figure 62

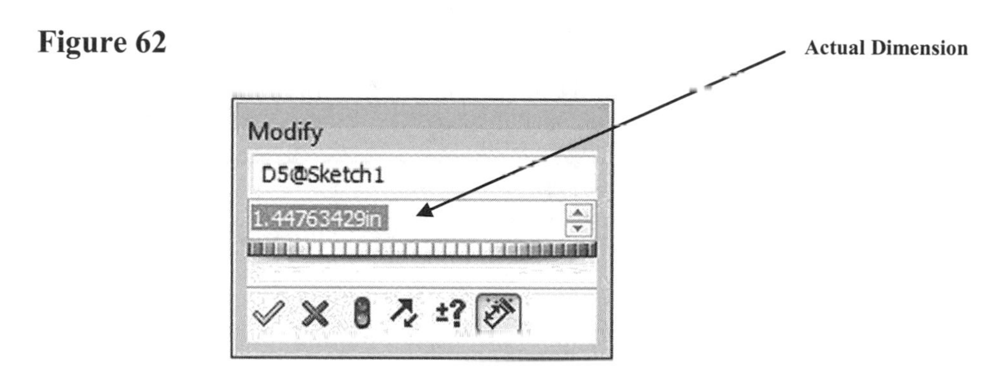

72. To edit the dimension, enter **1.75** in the Modify dialog box (while the current dimension is highlighted) and left click on the green checkmark as shown in Figure 63.

Figure 63

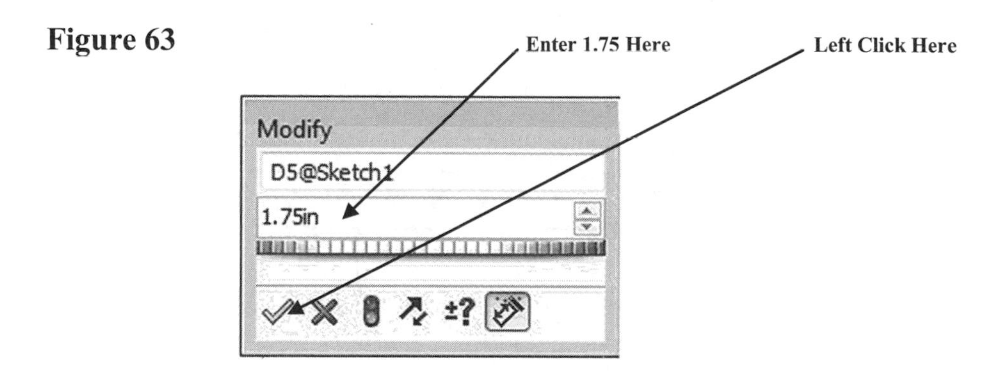

73. The dimension is now 1.75 as shown in Figure 64.

Figure 64

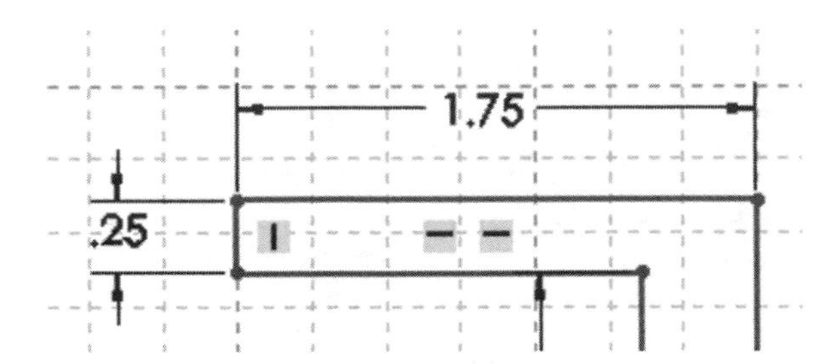

74. Move the cursor over to the left side vertical line until it turns red and left click as shown in Figure 65.

Figure 65

Left Click Here

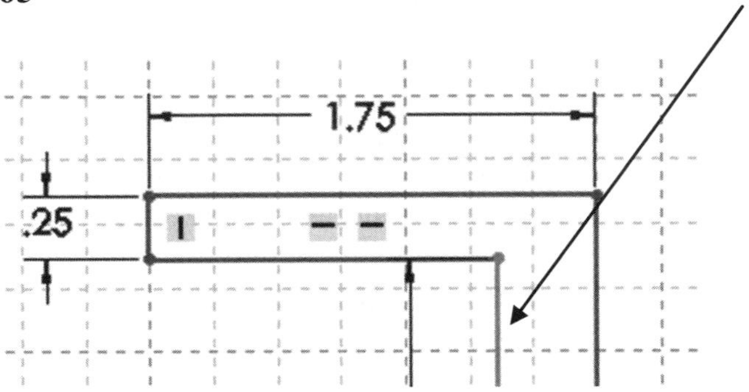

75. Move the cursor to the other vertical line until it turns red and left click as shown in Figure 66.

Figure 66

Left Click Here

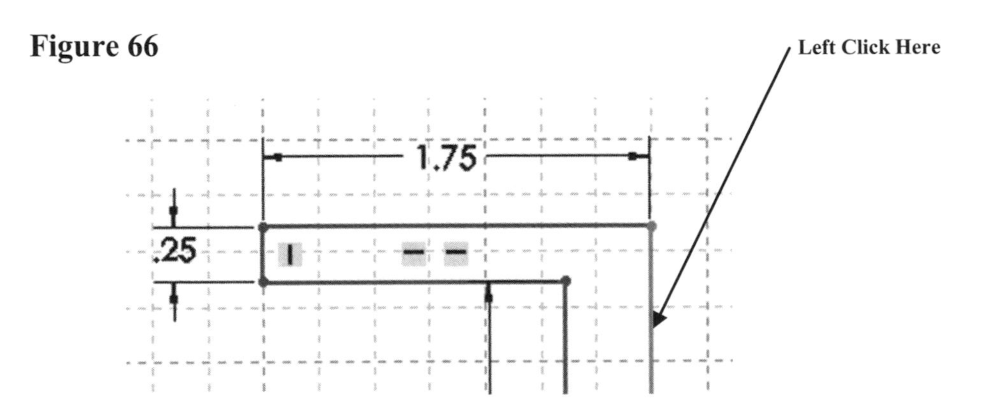

76. The dimension is attached to the cursor. Move the cursor back and forth to verify it is attached. Move the cursor to where the dimension will be placed and left click once. While the dimension is highlighted, left click the mouse once. The Modify dialog box will appear as shown in Figure 67.

Figure 67

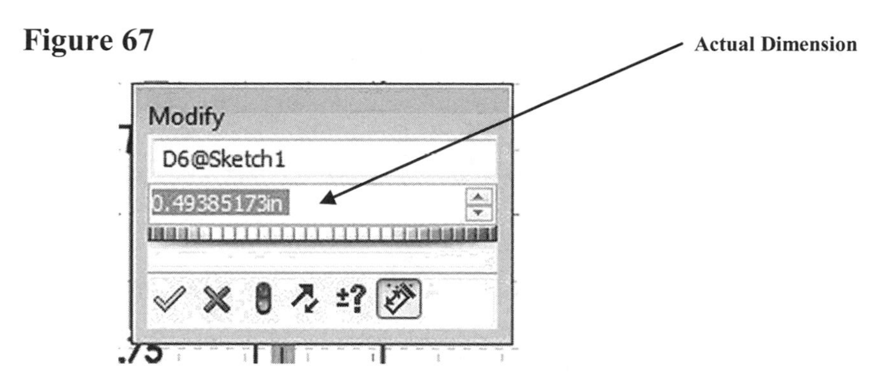

77. To edit the dimension, enter **.25** in the Modify dialog box (while the current dimension is highlighted) and left click on the green checkmark as shown in Figure 68.

Figure 68

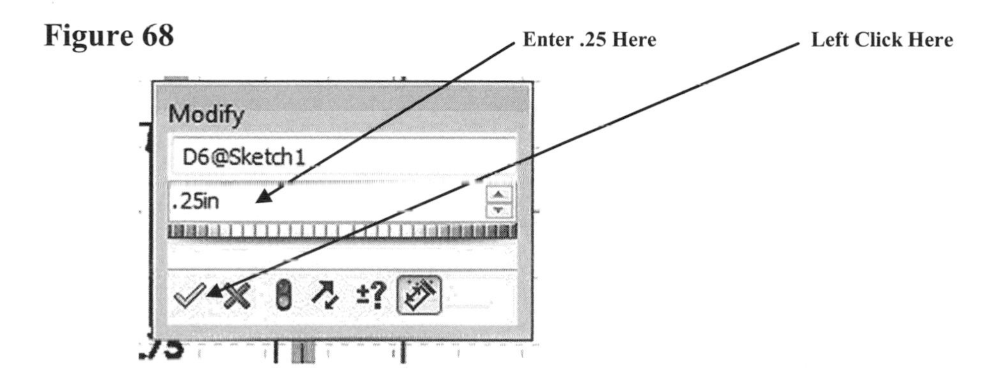

78. Your screen should look similar to Figure 69.

Figure 69

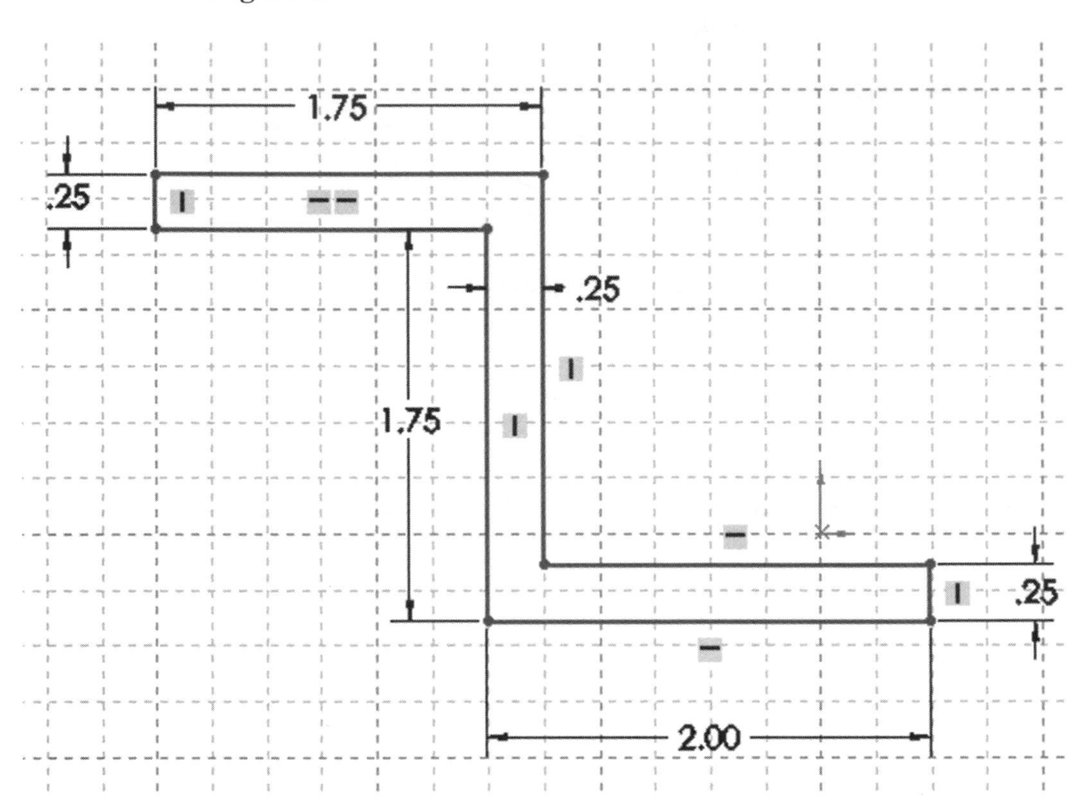

79. Right click near the drawing. A pop up menu will appear. Left click on **Select** as shown in Figure 70. This will ensure that no commands are still active.

Figure 70

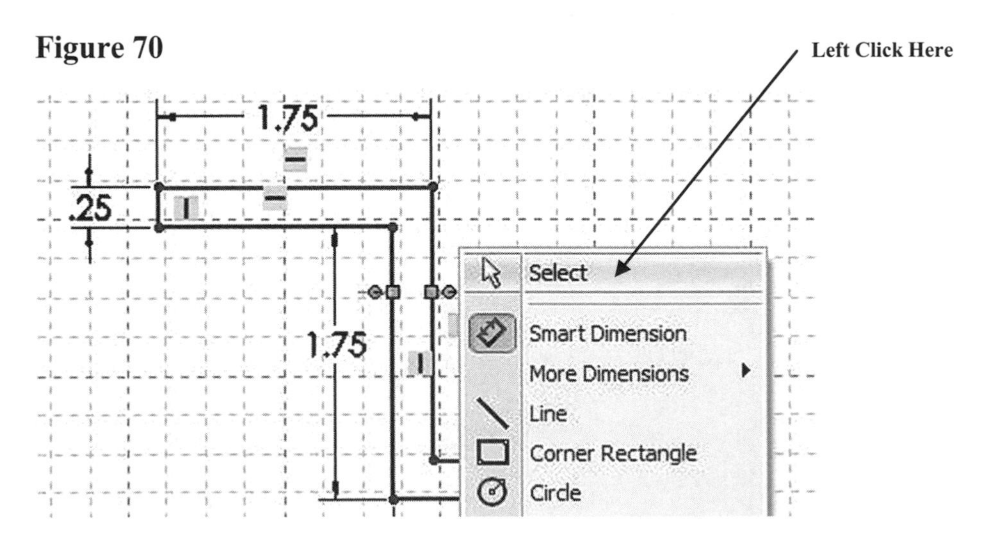

80. Move the cursor to the upper left portion of the screen and left click on **Exit Sketch** as shown in Figure 71.

Figure 71

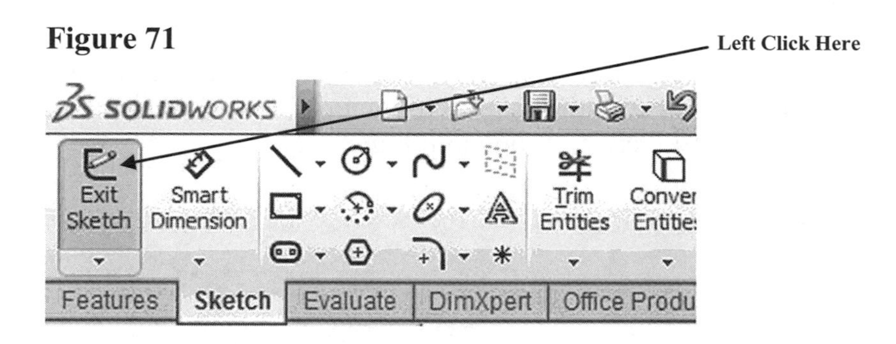

81. Your screen should look similar to Figure 72.

Figure 72

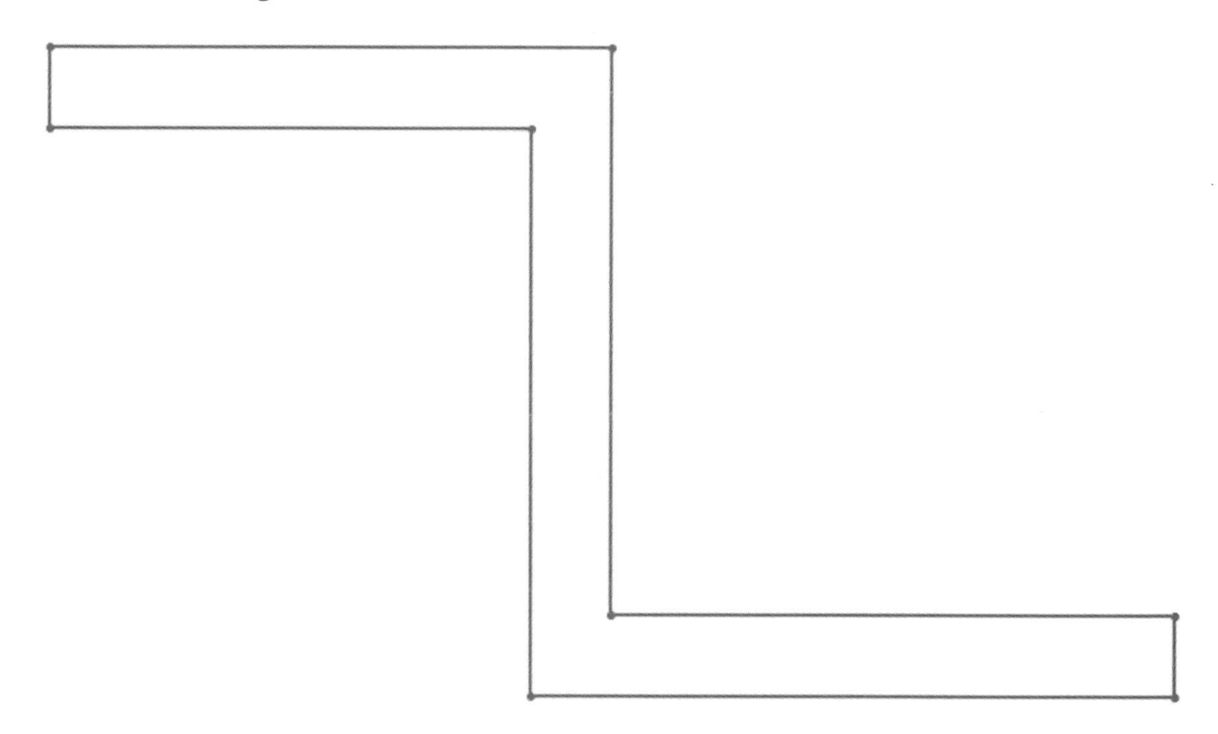

82. To utilize the Features commands, a sketch must be present and have no opens (non-connected lines). If there are any opens in the sketch an error message will appear.

83. Move the cursor to the upper middle portion of the screen and left click on the drop down arrow next to the "View Orientation" icon. Left click on the arrow next to the isometric cube as shown in Figure 73.

Figure 73

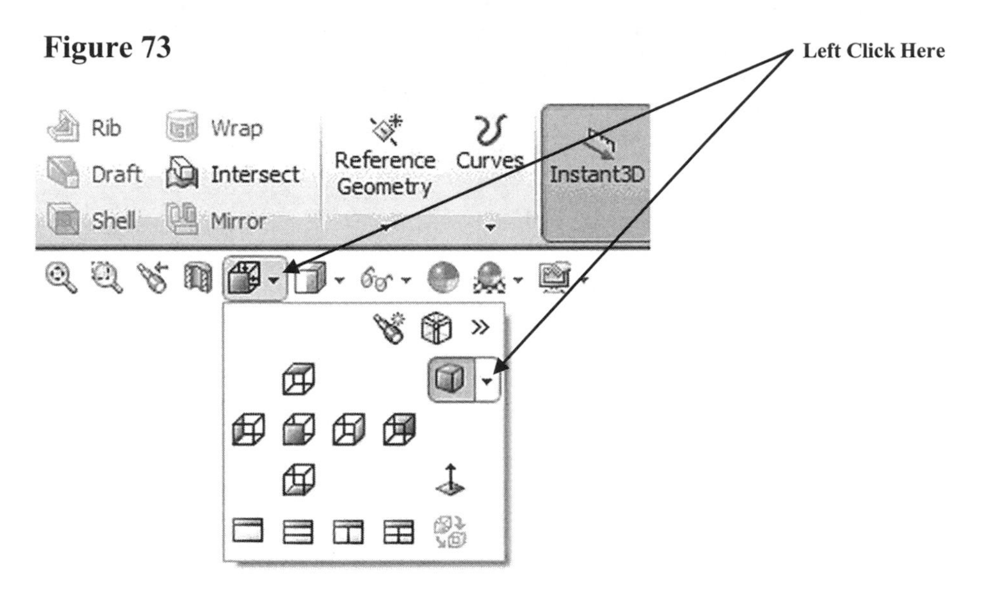

84. A drop down menu will appear. Left click on **Trimetric** as shown in Figure 74.

Figure 74

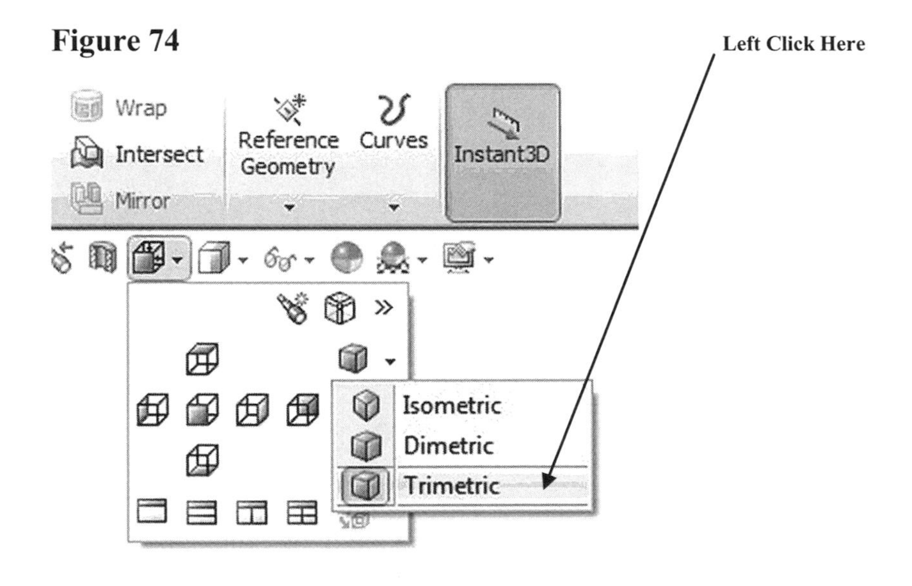

85. The view will become trimetric. If needed use the scroll wheel to zoom out as shown in Figure 75.

Figure 75

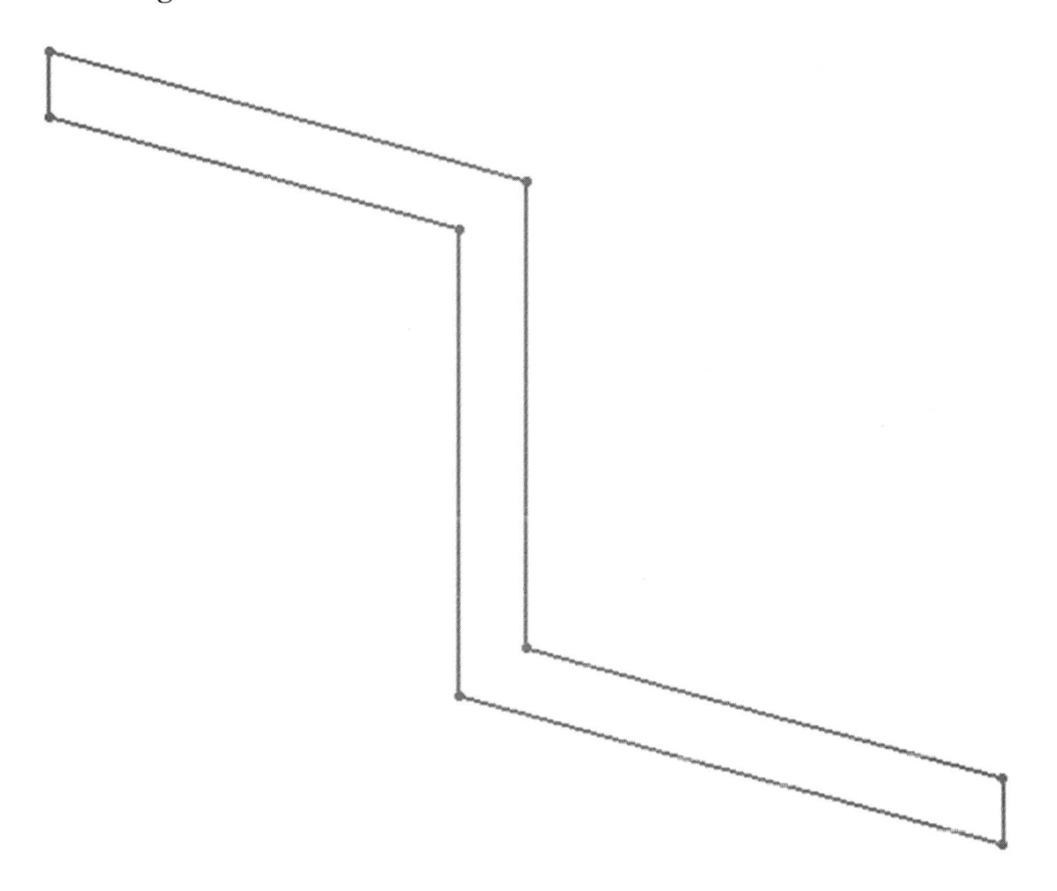

86. Move the cursor to the upper left portion of the screen and left click on **Extruded Boss/Base**. If the Extruded Boss/Base icon is not visible, left click on the **Features** tab as shown in Figure 76.

Figure 76

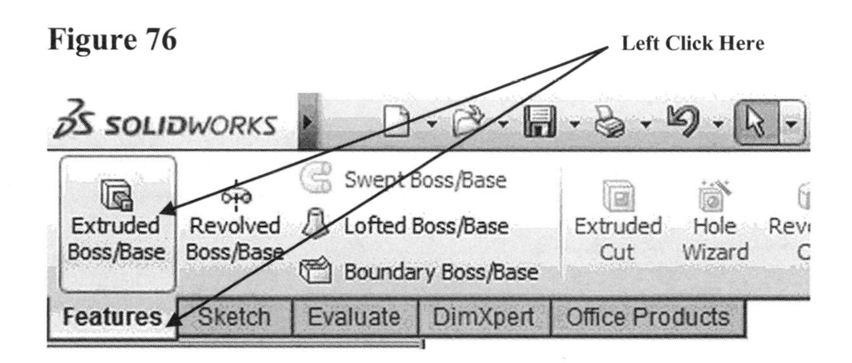

87. The Extrude dialog box will appear. SolidWorks also provides a preview of the extrusion. If SolidWorks gave you an error message, there are opens (non-connected lines) somewhere on the sketch. Check each intersection for opens by using the Extend Entities and Trim Entities commands. Your screen should look similar to Figure 77. If the Extrude dialog box did not appear, left click on any line making up the sketch as shown in Figure 77.

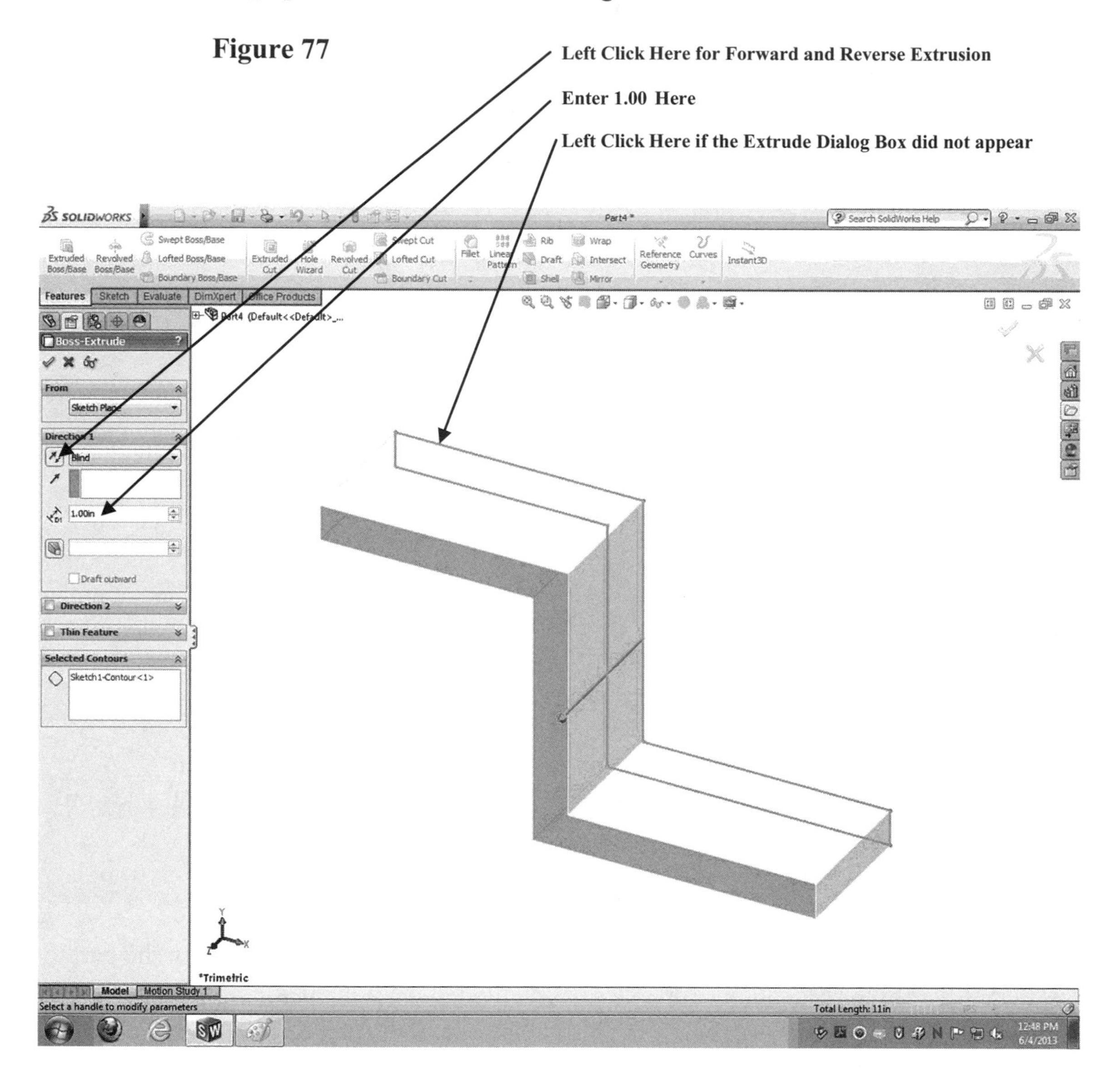

Figure 77

88. Left click on the icon under Direction, located next to the text "Blind". The extrusion will change directions. Ensure that the extrusion is facing forward. While the text located next to "D1" is still highlighted, enter **1.000.** To preview the extruded part, left click on the "Glasses" icon located at the top of the Extrude dialog box as shown in Figure 78. The part will become purple.

Figure 78

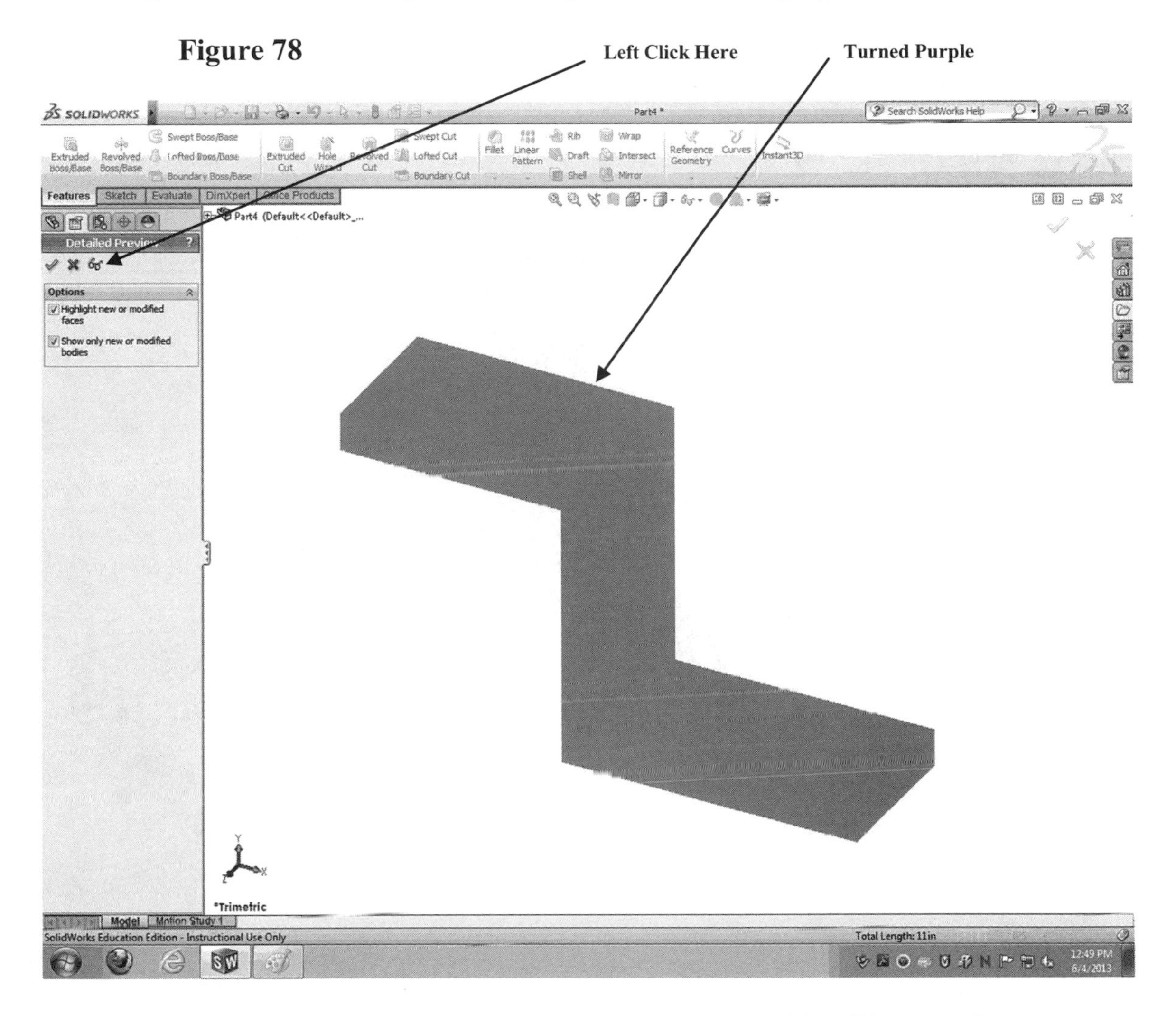

89. To end preview, left click on the "Glasses" icon again. This will return the part to the previous view.

90. Left click on the green checkmark as shown in Figure 79.

Figure 79

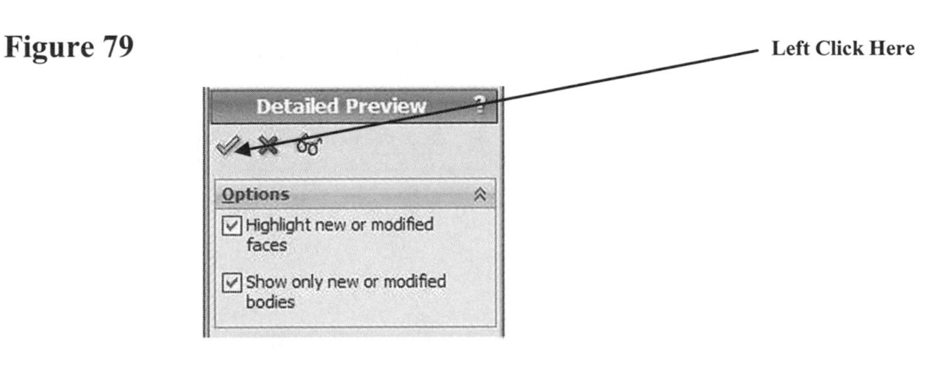

91. Left clicking on the green checkmark will create a solid from the sketch. Your screen should look similar to Figure 80. Left click anywhere around the drawing.

Figure 80

92. Move the cursor to the upper middle portion of the screen and left click on **Fillet** as shown in Figure 81.

Figure 81

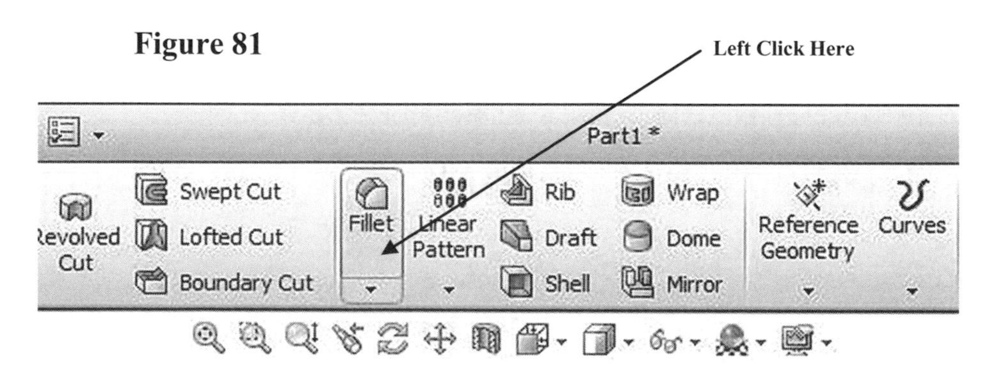

93. The menu at the left will change. Enter **.50** for the radius and select **Full Preview** as shown in Figure 82.

Figure 82

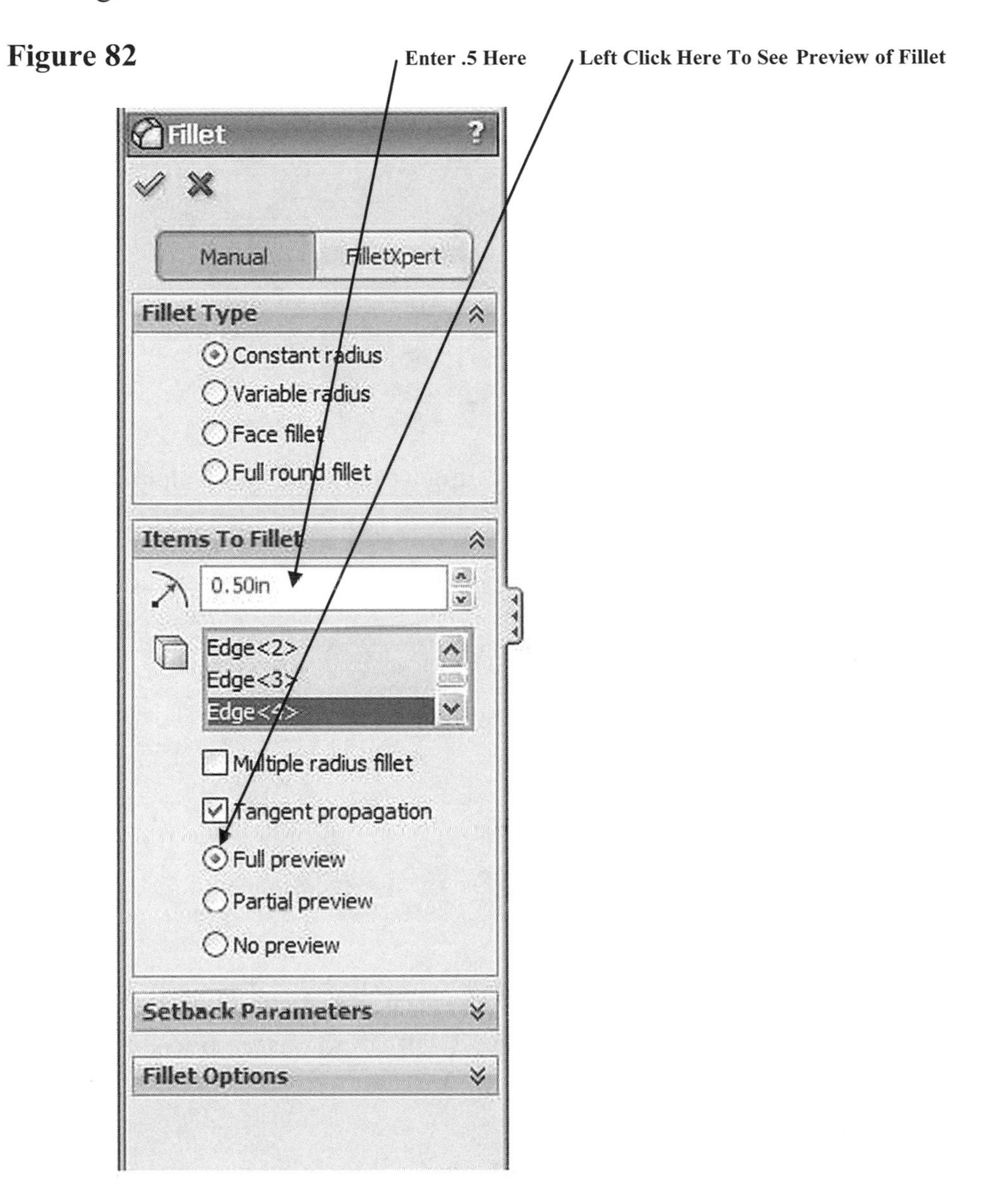

94. Move the cursor over the lower front edge causing it to turn red and left click once as shown in Figure 83.

Figure 83

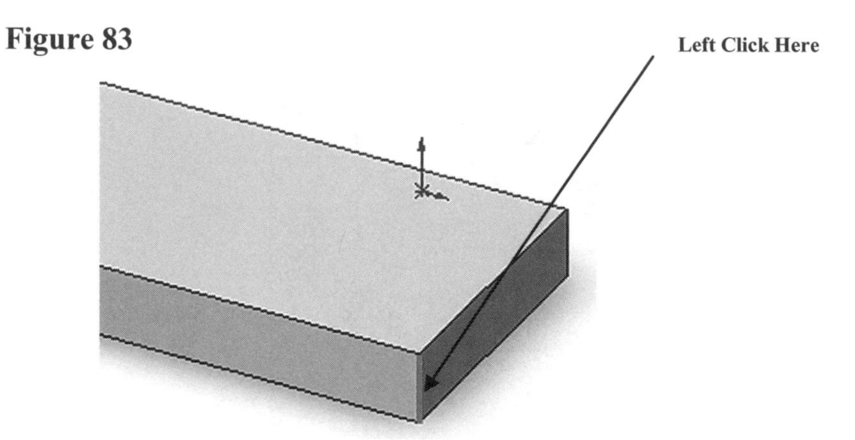

95. Notice the yellow lines illustrating a preview of the fillet as shown in Figure 84.

Figure 84

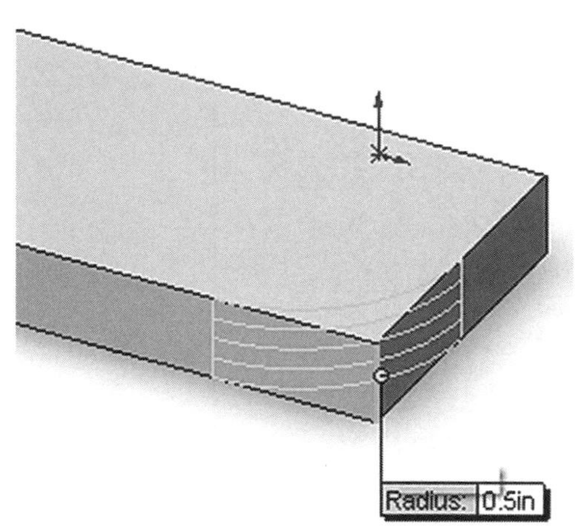

96. Move the cursor over the upper front edge causing it to turn red and left click once as shown in Figure 85.

Figure 85

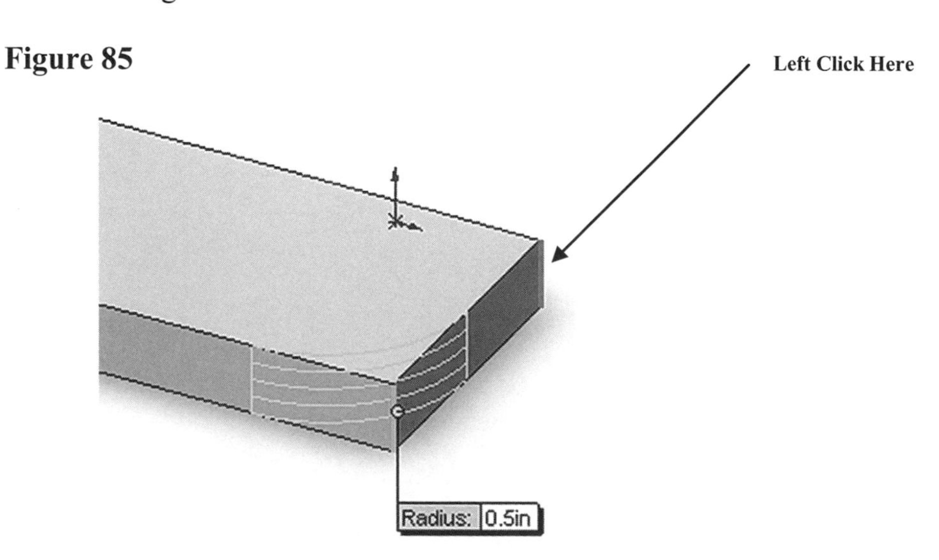

97. Notice the yellow lines illustrating a preview of the fillet as shown in Figure 86.

Figure 86

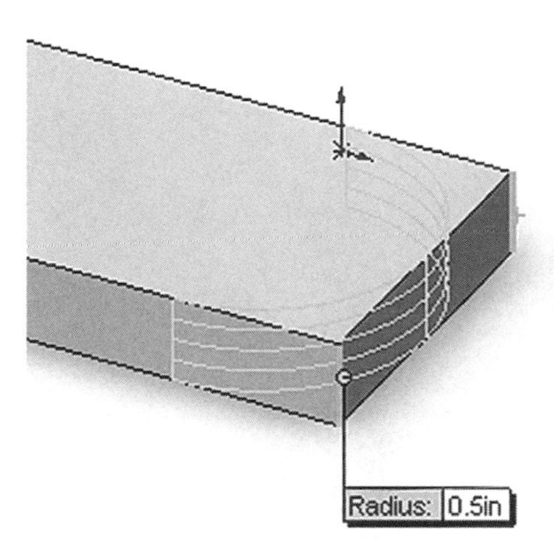

98. Left click on the two upper remaining edges. You may have to use the rotate command to select the upper edge that is not visible.

99. **Note: The following is for users of SolidWorks 2009 and earlier. If you are using version 2010 or newer then skip to step 100.** Move the cursor to the upper middle portion of the screen and Right click on the drop down arrow to the right of the View Orientation icon. A drop down menu will appear. Place a check in the box (with the left mouse button) next to the "Rotate" icon as shown in Figure 87. The Rotate icon will now appear with the Viewing tools located at the top of the screen in the drawing area. Once a check mark has been placed in the box, left click in the drawing area as shown in Figure 87. Pushing (and holding down) the scroll wheel while moving the mouse will also rotate the model.

Figure 87

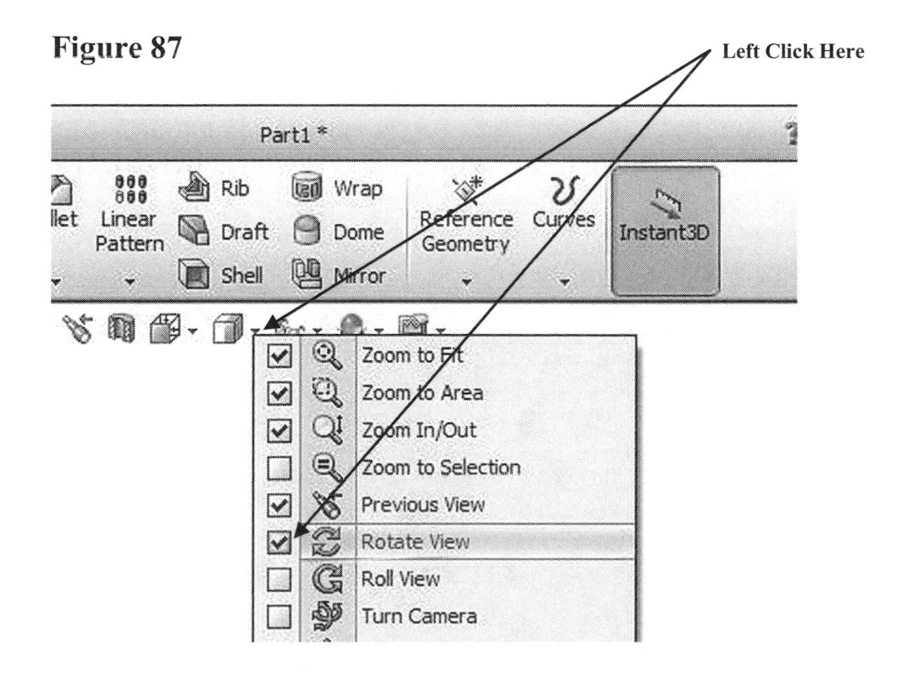

100. Left click on the left portion of the part (holding the left mouse button down). Drag the cursor to the right allowing access to the upper right corner of the part as shown in Figure 88.

Figure 88

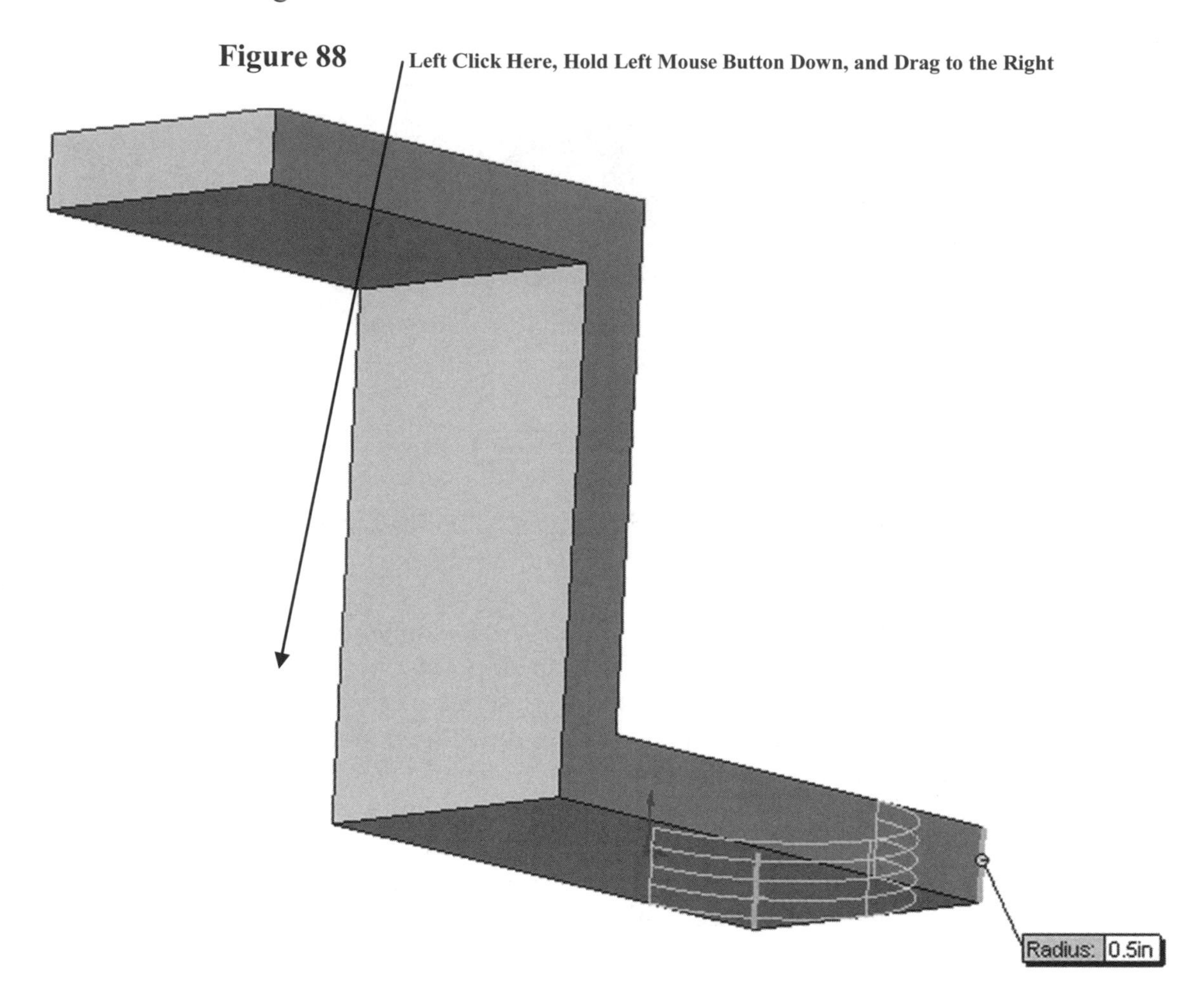

101. Move the cursor to the upper middle portion of the screen and left click on the "Rotate" icon again if the backside of the part is not visible as shown in Figure 89. Another method of activating the Rotate command is to push the mouse wheel down (holding the mouse wheel down) and move the mouse to the desired position.

Figure 89

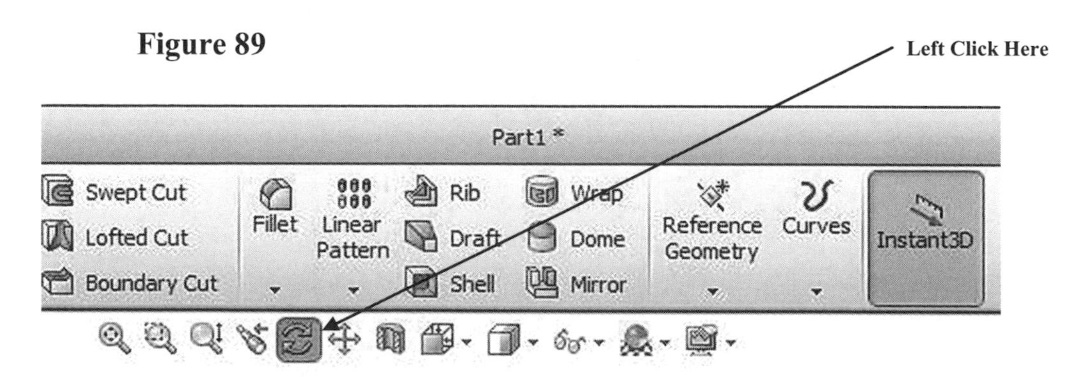

102. Move the cursor to the upper right corners of the part causing the edges to turn red. Left click once as shown in Figure 90.

Figure 90

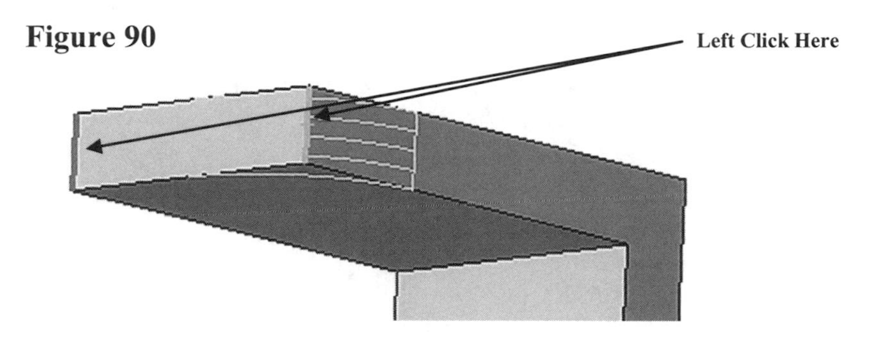

103. Your screen should look similar to Figure 91.

Figure 91

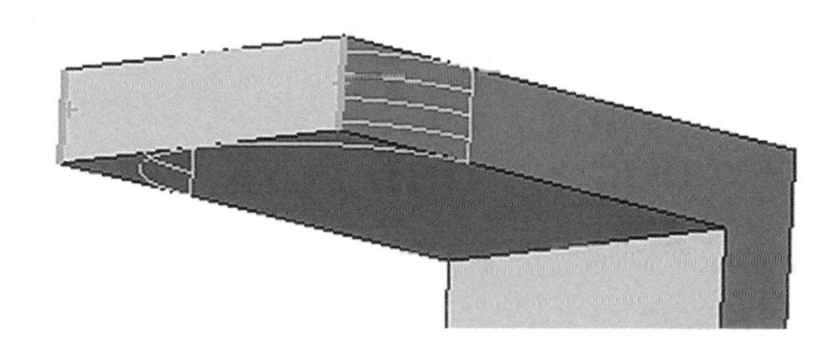

104. Move the cursor to the upper middle portion of the screen and left click on the drop down arrow next to the "View Orientation" icon. Left click on the arrow next to the isometric cube as shown in Figure 92.

Figure 92

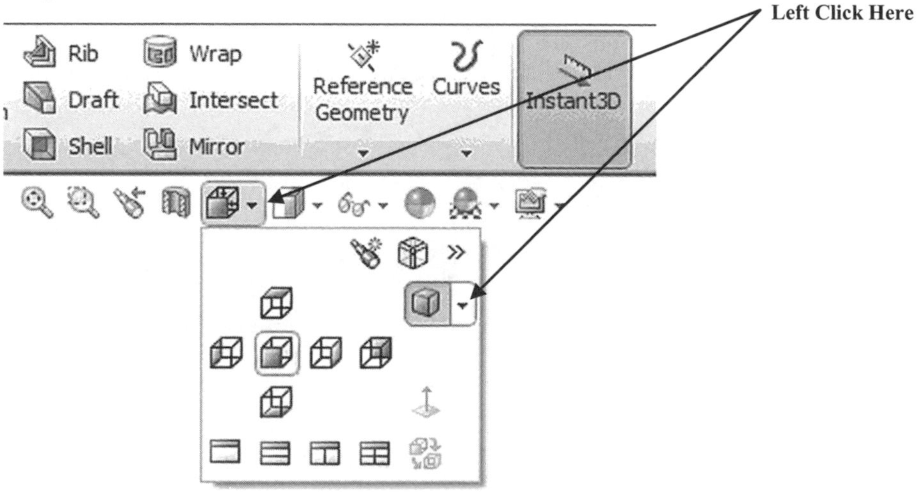

Left Click Here

105. A drop down menu will appear. Left click on **Trimetric** (if the view is not already in Trimetric) as shown in Figure 93.

Figure 93

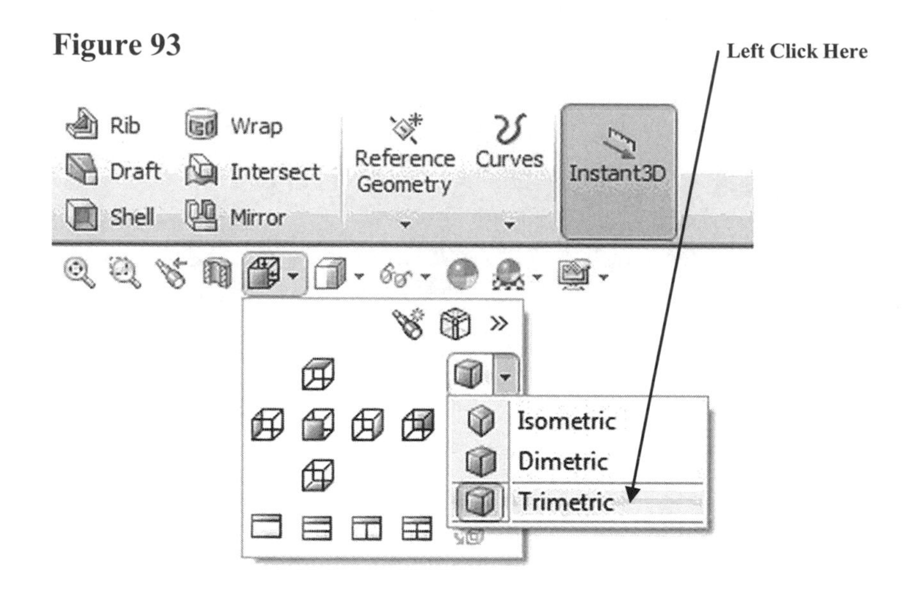

106. Your screen should look similar to Figure 94.

Figure 94

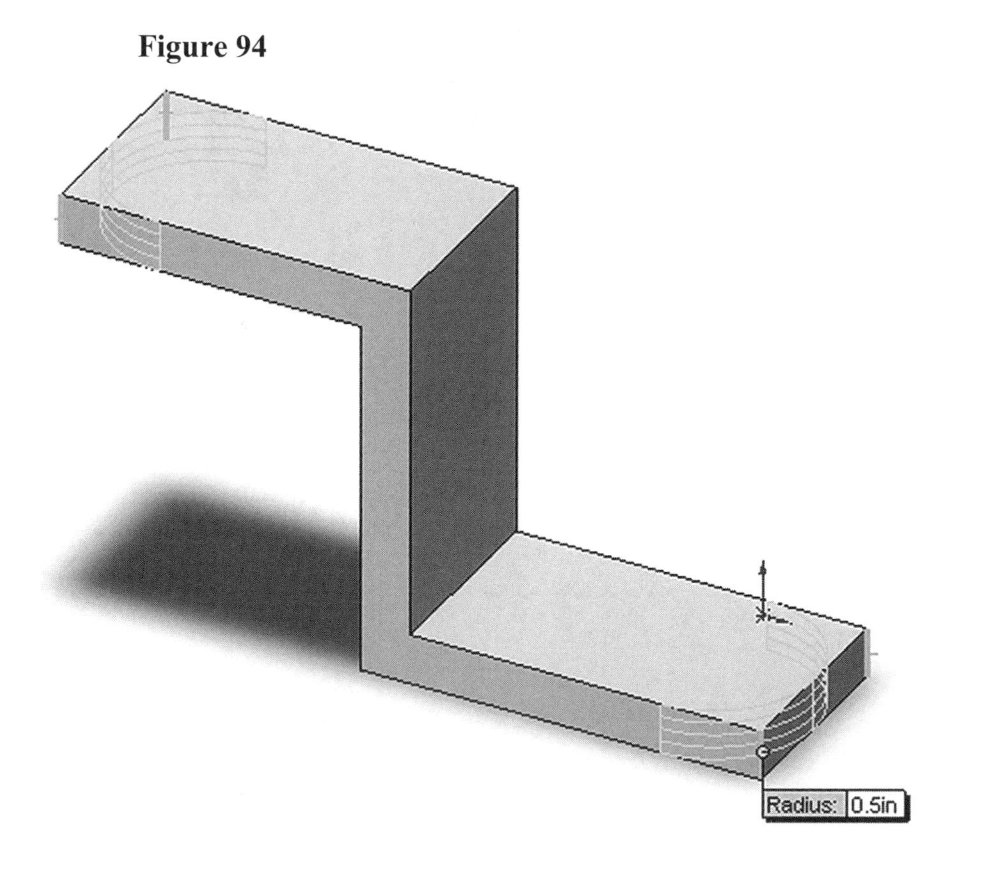

107. Left click on the green checkmark as shown in Figure 95.

Figure 95

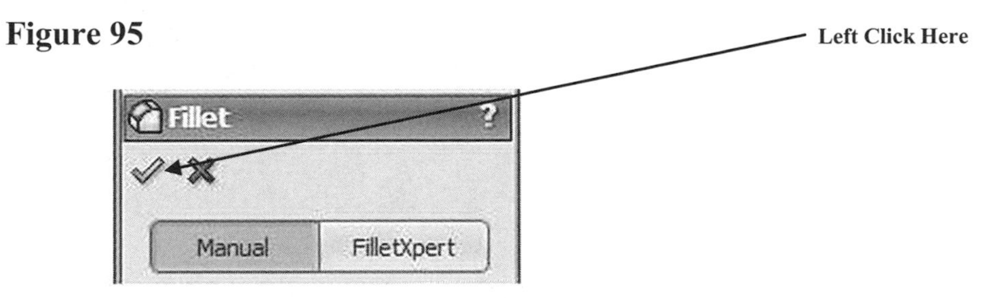

108. Your screen should look similar to Figure 96.

Figure 96

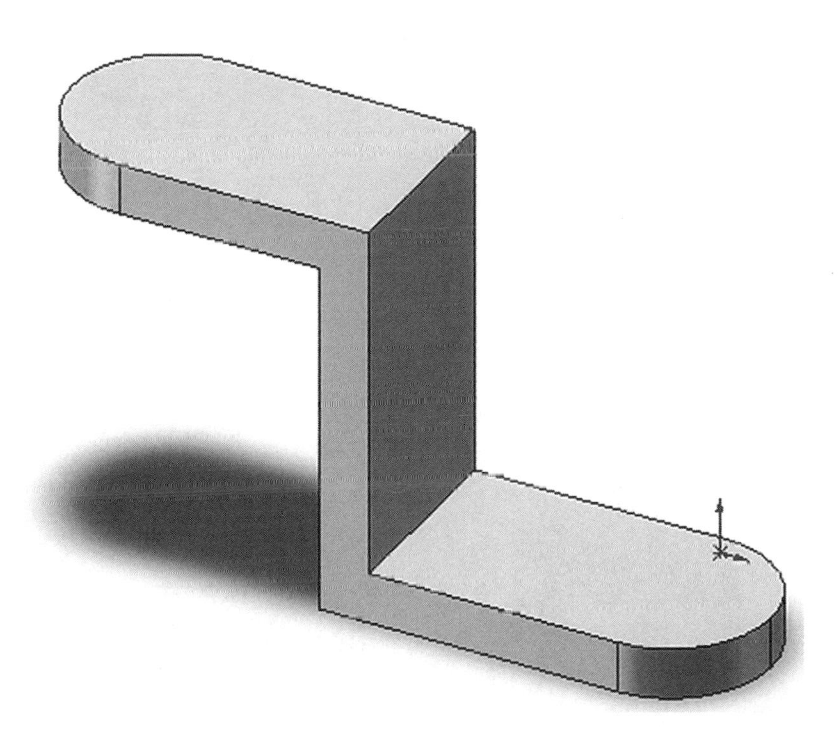

109. The next task will include cutting a hole in each of the ends. To do this, a sketch will need to be constructed on each surface. To begin a new sketch on any surface, move the cursor to the surface that will include the new sketch. Notice the edges of the surface are red. After the edges have turned red, right click on the surface as shown in Figure 97.

Figure 97

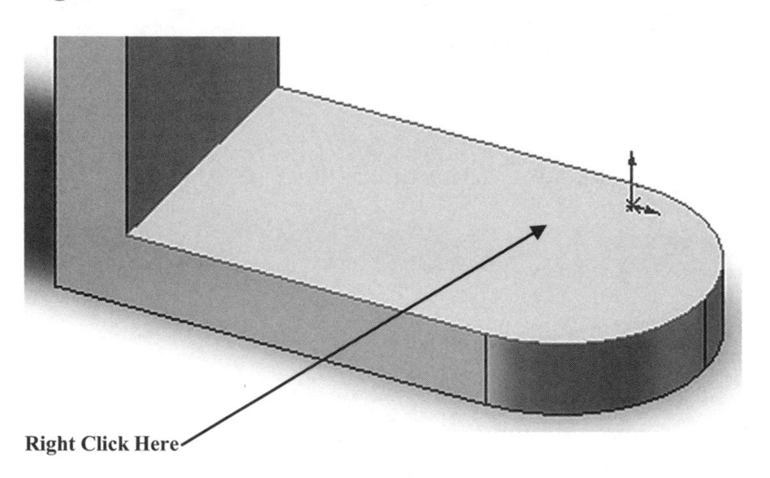

110. The surface will change color. A pop up menu will appear. Left click on **Insert Sketch** as shown in Figure 98.

Figure 98

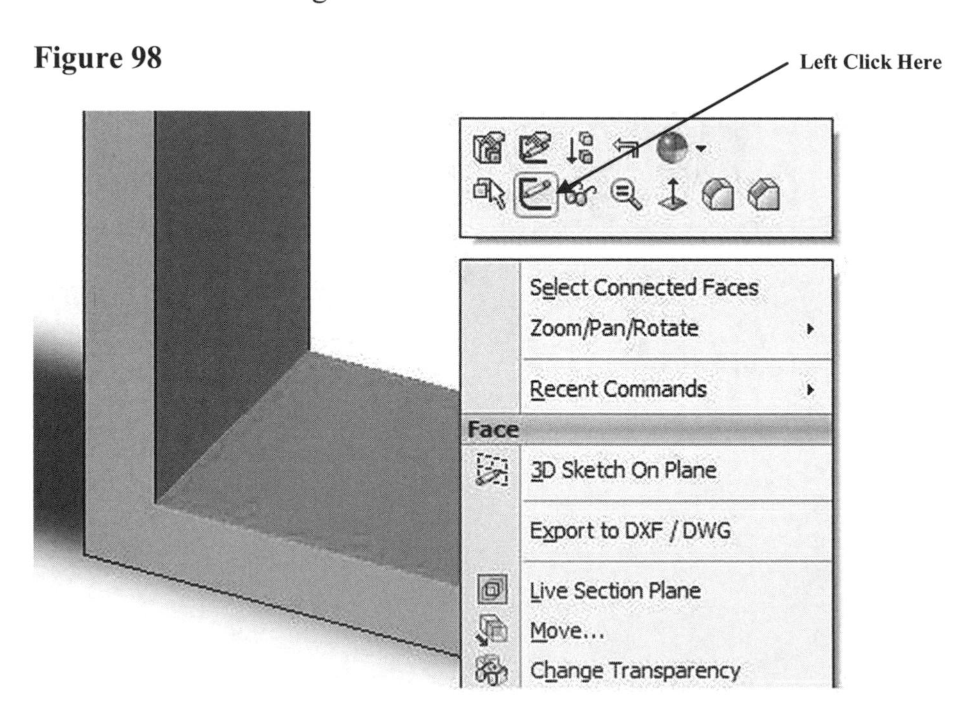

111. SolidWorks will create a "sketch" on that particular surface. Notice the toolbar at the top of the screen has changed back to the sketch commands.

112. Your screen should look similar to Figure 99.

Figure 99

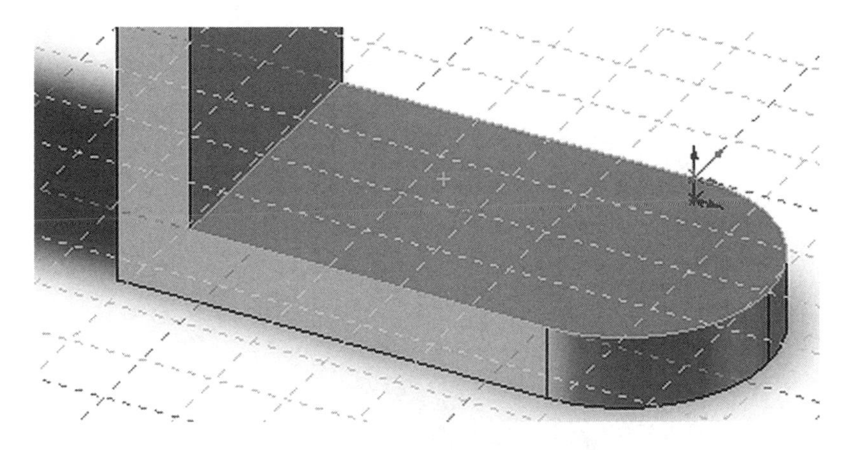

113. Move the cursor to the upper left portion of the screen and left click on **Circle** as shown in Figure 100.

Figure 100

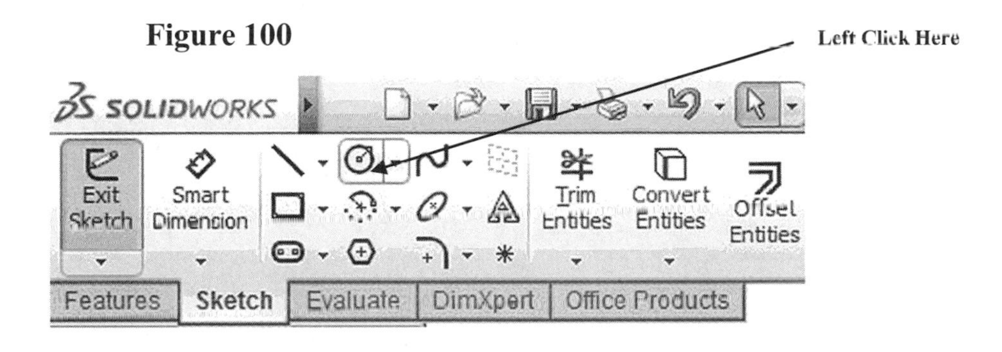

114. Move the cursor to the edge of the circle. The center point of the circle will appear as shown in Figure 101.

Figure 101

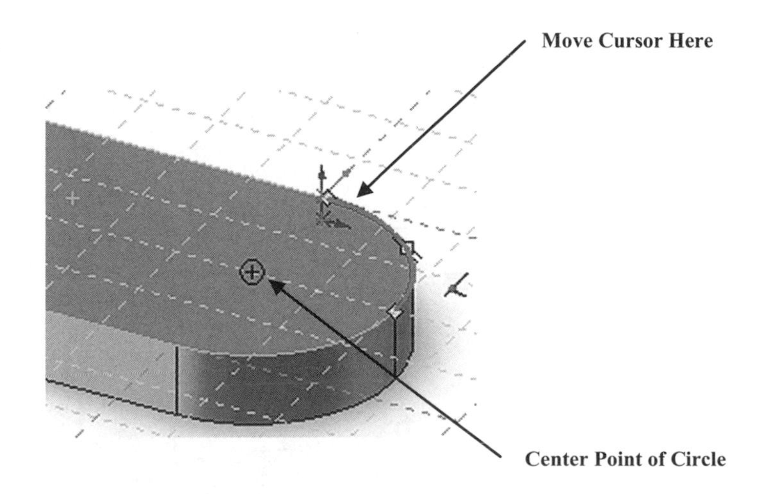

115. After the center marker appears, left click on it once. This will be the center of a circle, which will later become a thru hole. Move the cursor out to the side to make the hole larger. Move the cursor out far enough to create a hole size similar to Figure 102.

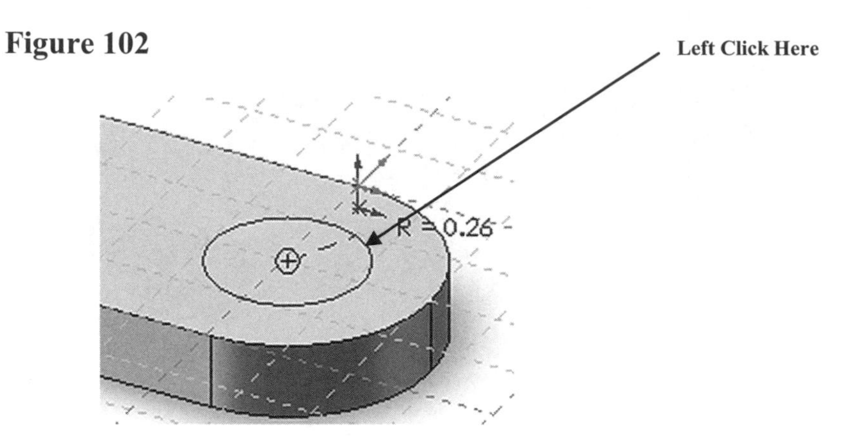

Figure 102

116. After the hole size looks similar to Figure 102, left click once. Left click on the green check mark at the middle left (not shown).

117. Move the cursor to the upper left portion of the screen and left click on **Smart Dimension** as shown in Figure 103.

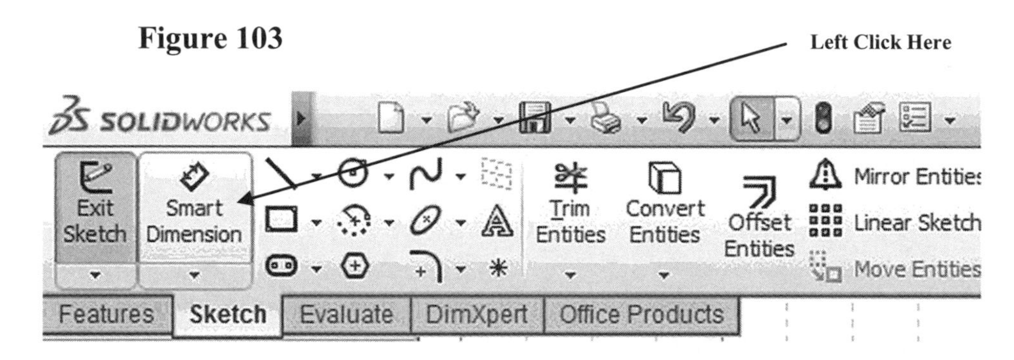

Figure 103

118. Left click on the edge (not the center) of the circle as shown in Figure 104. The circle will turn red.

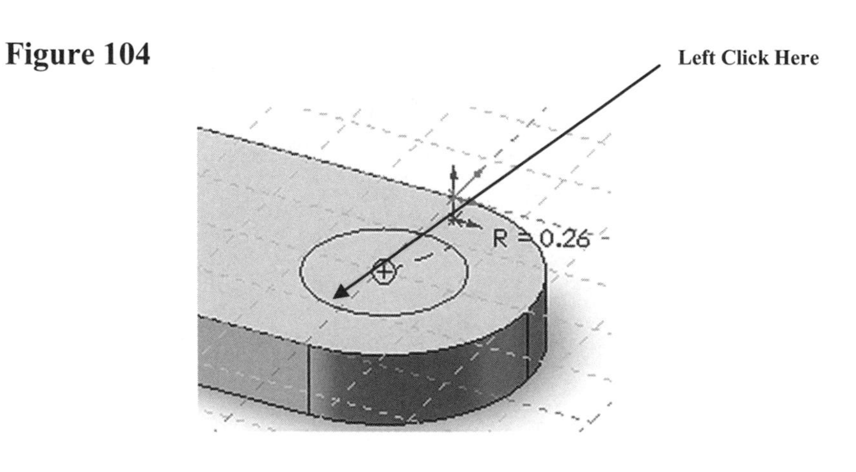

Figure 104

119. The dimension will appear attached to the cursor. Move the cursor up and down to verify it is attached. Move the cursor to where the dimension will be placed and left click once as shown in Figure 105.

Figure 105

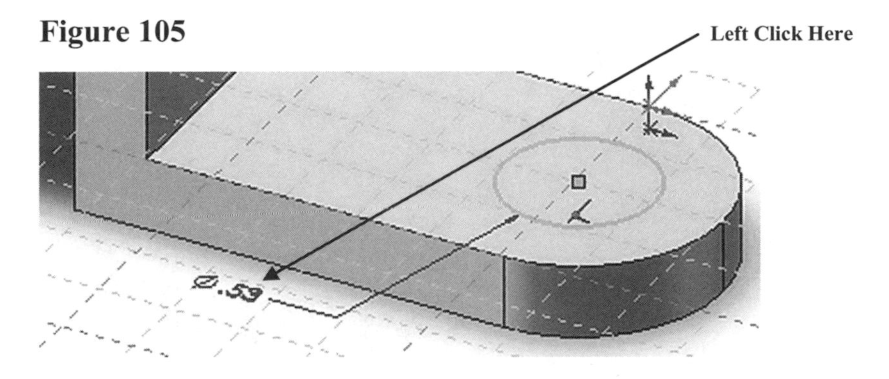

120. The Modify dialog box will appear as shown in Figure 106.

Figure 106

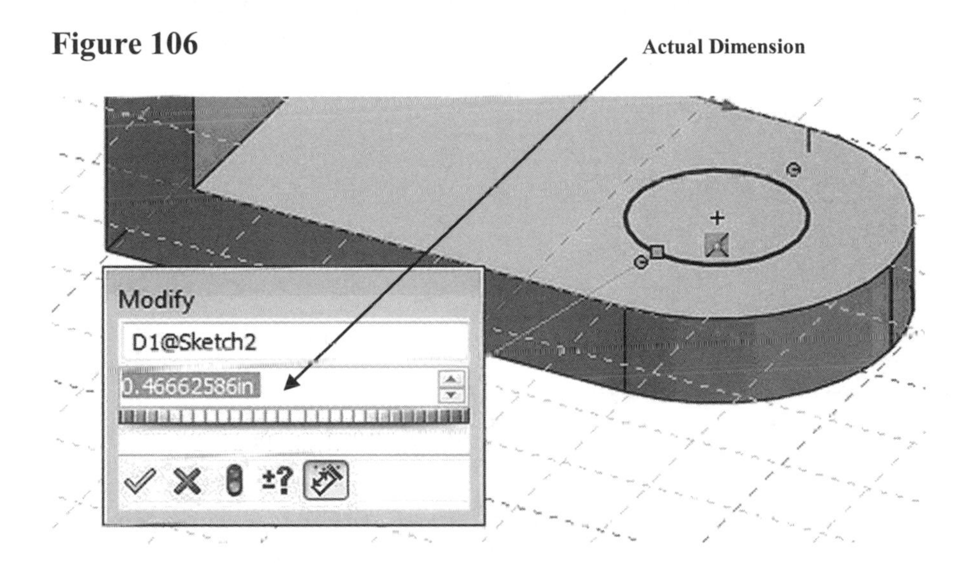

121. Enter **.500** in the Modify dialog box and left click on the green checkmark as shown in Figure 107.

Figure 107

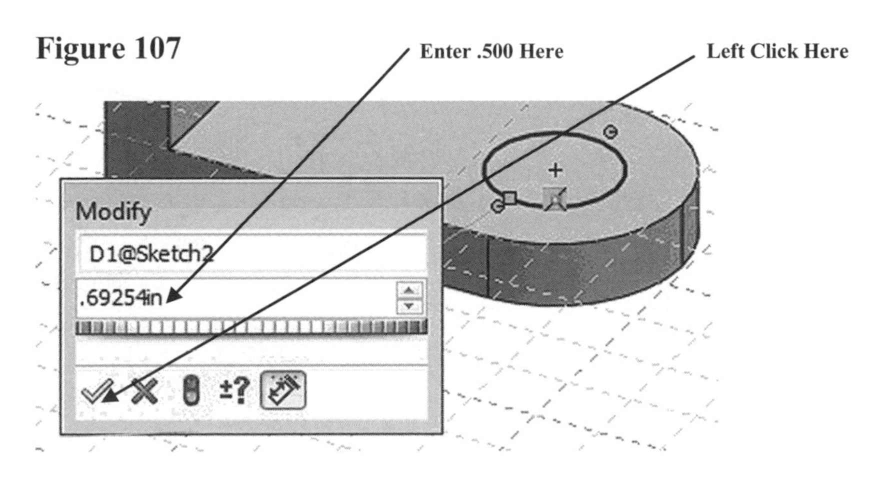

122. The diameter of the hole will become .500 inches as shown in Figure 108.

Figure 108

123. Right click near the drawing. A pop up menu will appear. Left click on **Select** as shown in Figure 109. This will ensure that no commands remain active.

Figure 109

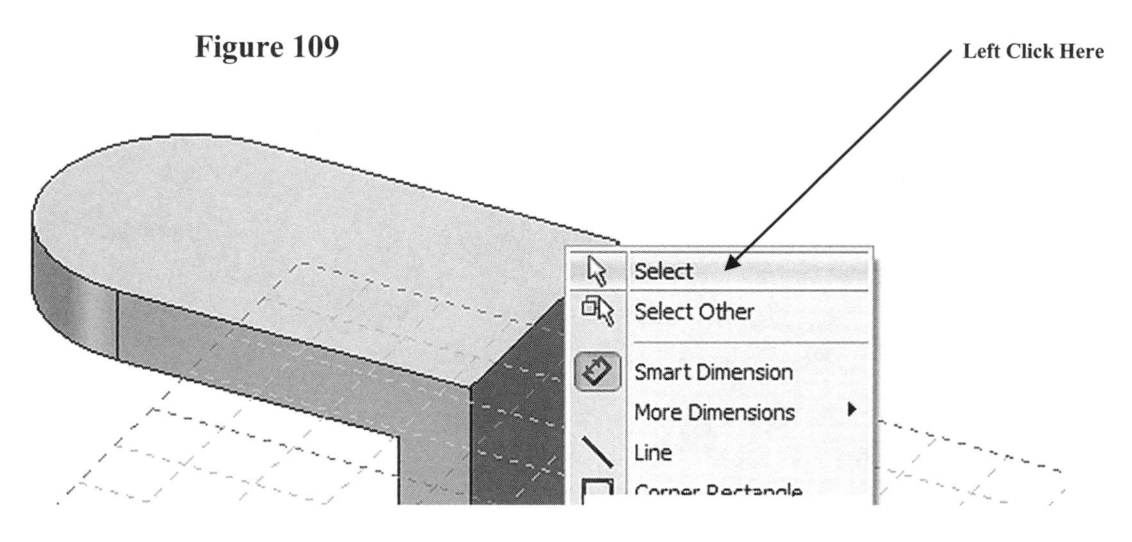

124. Move the cursor to the upper left portion of the screen and left click on **Exit Sketch** as shown in Figure 110.

Figure 110

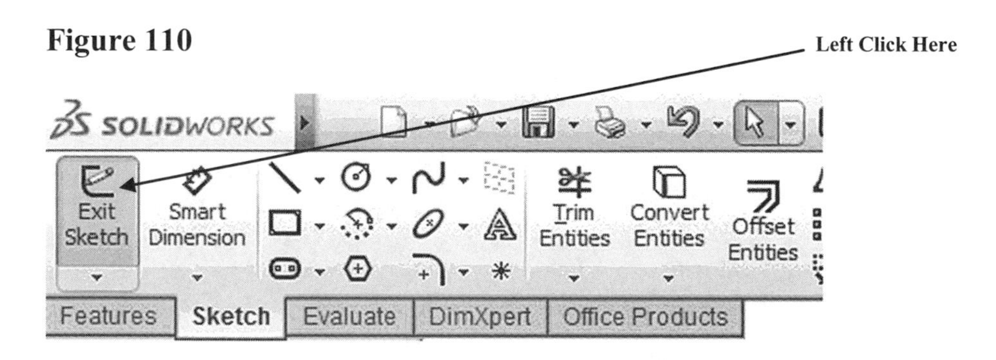

125. SolidWorks is now out of the Sketch commands and into the Features Commands. Notice that the commands at the top of the screen are now different. Also notice that the Features tab is now active as shown in Figure 111.

Figure 111

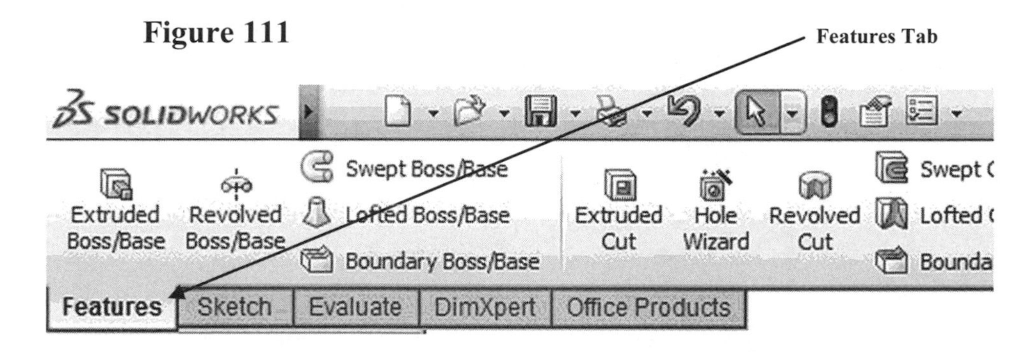

126. Move the cursor to the upper left portion of the screen and left click on **Extruded Cut** as shown in Figure 112. This time SolidWorks will extrude "space" or "air" rather than material as was done to create the bracket. If the command does not become active you will need to left click on the circle that was created in the sketch area.

Figure 112

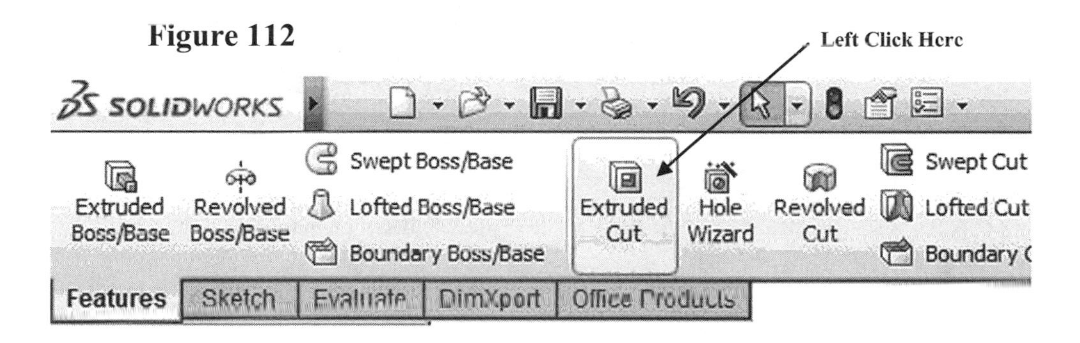

127. The menu at the left will change. Enter **.500** as shown in Figure 113.

Figure 113

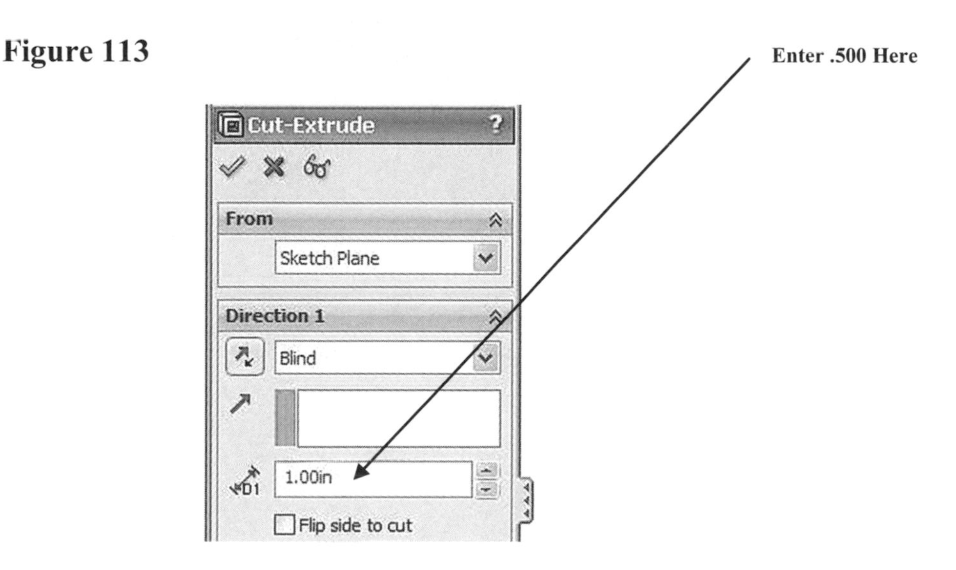

128. Move the cursor to the upper left portion of the screen and left click on the "Glasses" icon as shown in Figure 114.

Figure 114

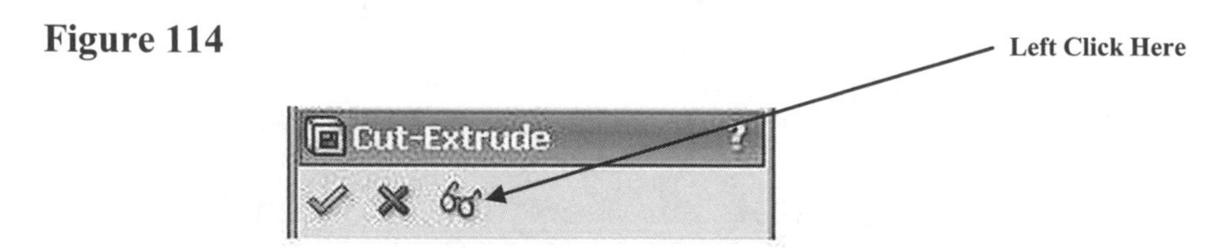

129. A preview of the part will be displayed as shown in Figure 115.

Figure 115

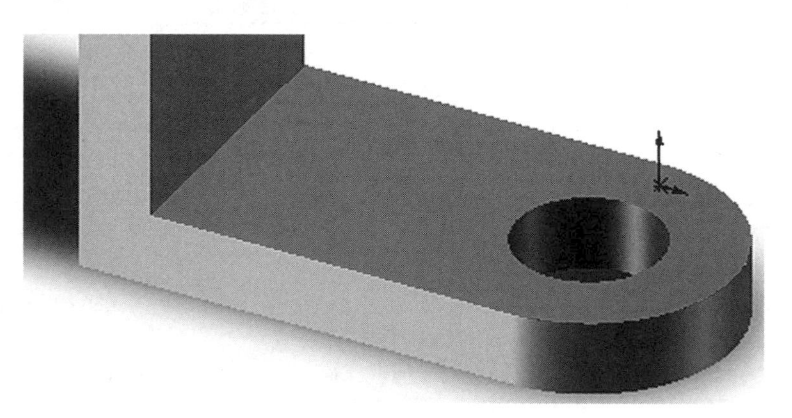

130. To return to the main menu, left click once on the "Glasses" icon.

131. Left click on the green checkmark as shown in Figure 116.

Figure 116

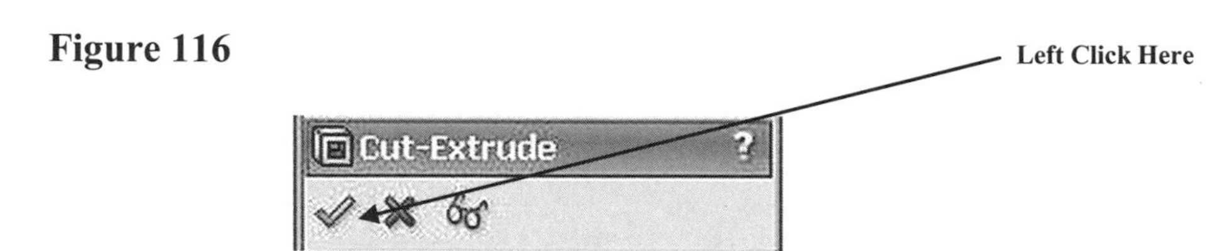

132. There should be a thru hole in the part similar to Figure 117. Left click anywhere around the drawing

Figure 117

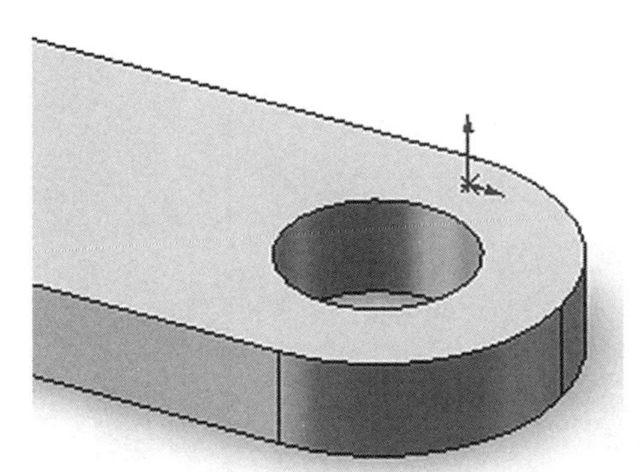

133. Another method of creating a hole is to use the Hole Wizard command.

134. To use the Hole Wizard command a "Point" must be constructed in the Sketch command. Move the cursor to the top portion of the part as shown in Figure 118. The outer edges of the part will turn red.

Figure 118

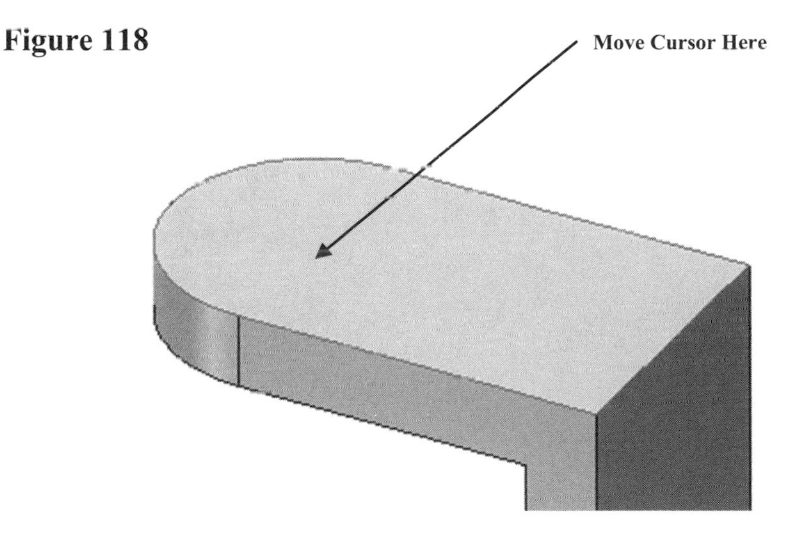

135. Right click on the surface. The surface will change color. A pop up menu will also appear. Left click on **Insert Sketch** as shown in Figure 119.

Figure 119

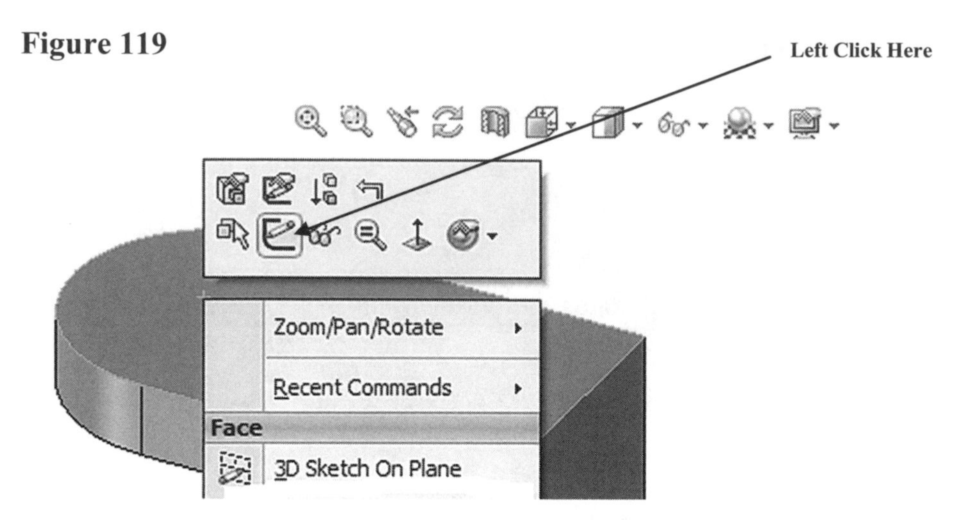

136. SolidWorks will return to the Sketch commands as shown in Figure 120.

Figure 120

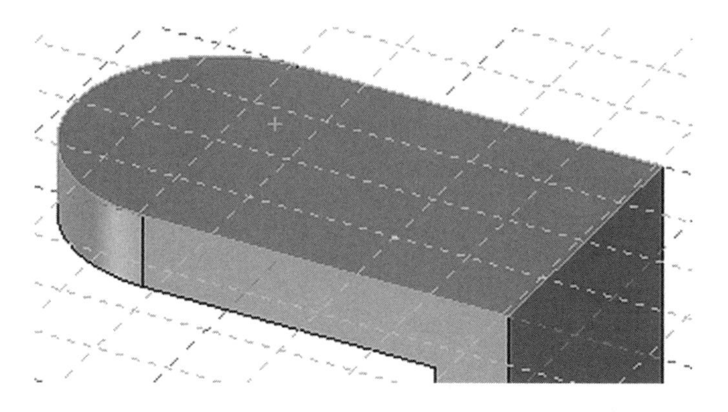

137. Move the cursor to the upper right portion of the screen and left click on **Point** as shown in Figure 121.

Figure 121

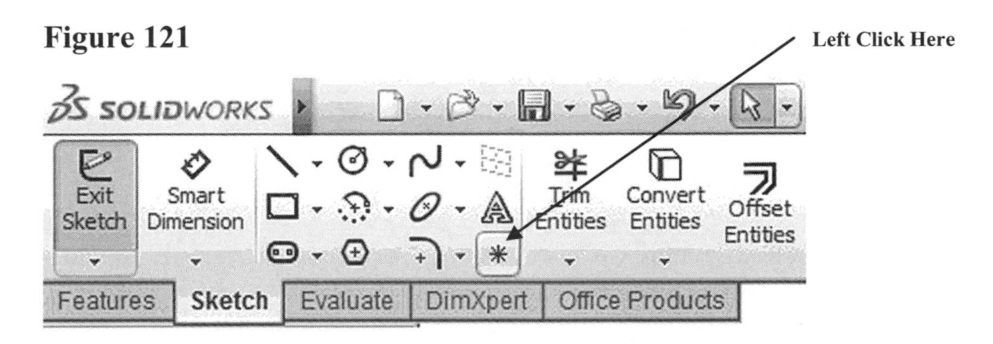

138. Move the cursor to edge of the part. A portion of the edge will turn red. The center point of the fillet radius will also appear. Left click once on the center point as shown in Figure 122.

Figure 122

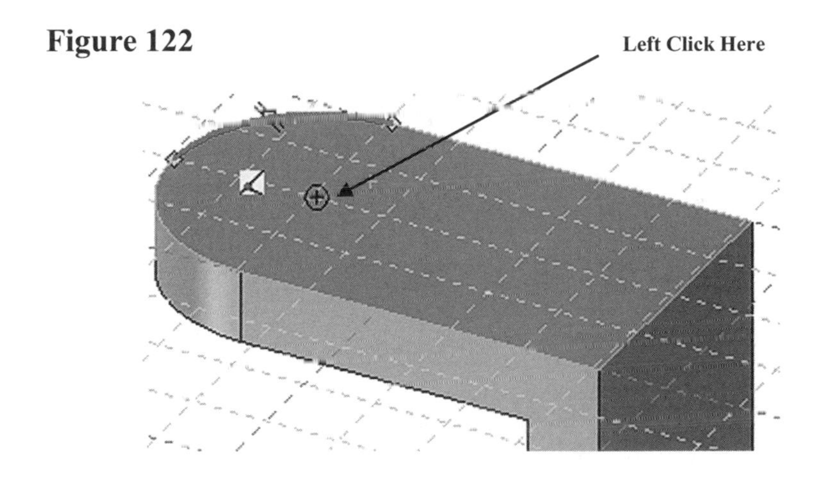

139. A small center marker will appear on the center of the fillet radius as shown in Figure 123.

Figure 123

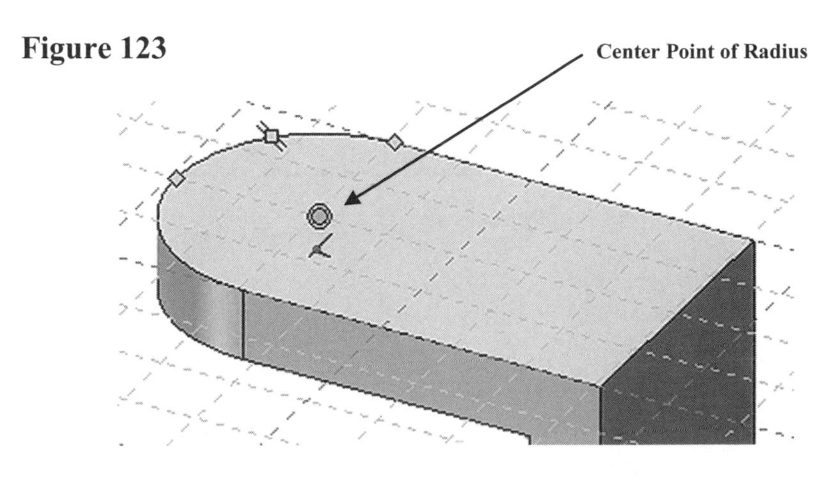

140. Move the cursor to the upper left portion of the screen and left click on the green checkmark as shown in Figure 124.

Figure 124

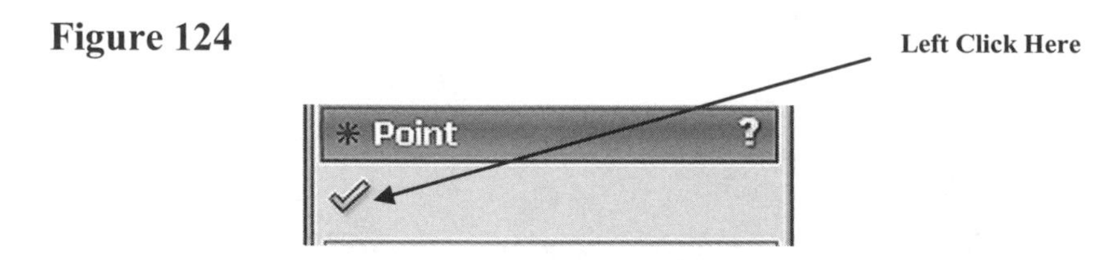

141. Right click anywhere on the drawing. A pop up menu will appear. Left click on **Select** as shown in Figure 125.

Figure 125

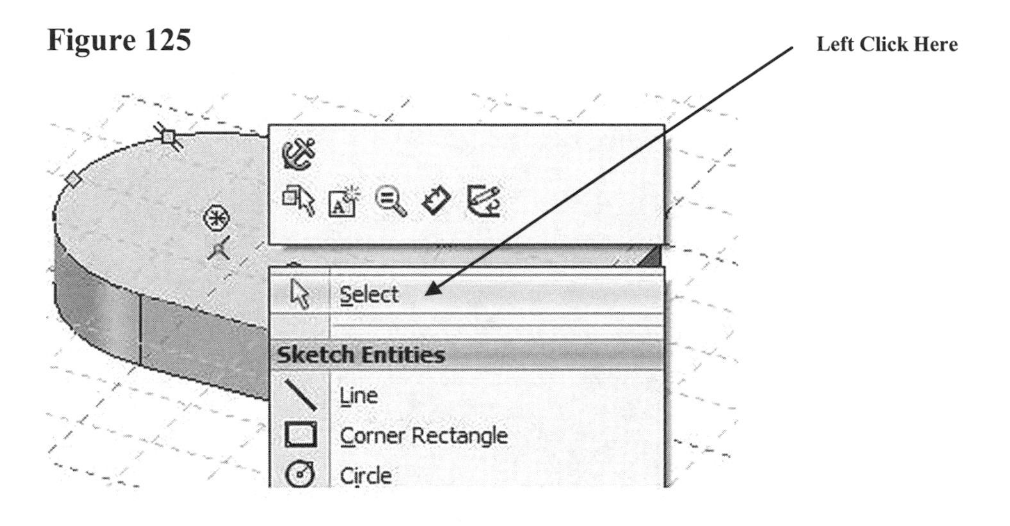

142. Right click anywhere on the drawing. A pop up menu will appear. Left click on **Exit Sketch** as shown in Figure 126.

Figure 126

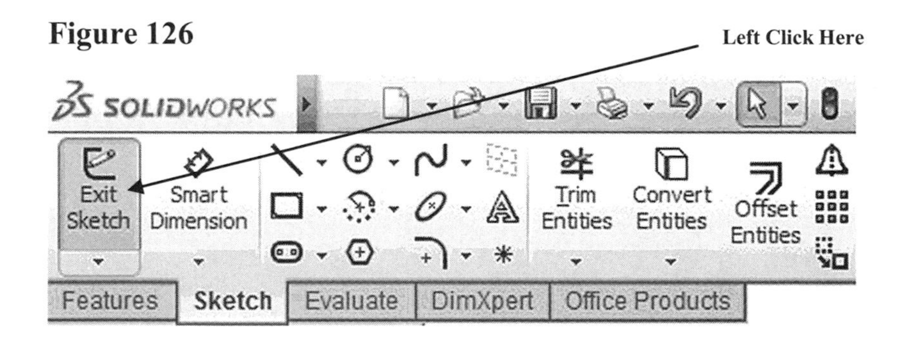

143. SolidWorks is now out of the Sketch commands and into the Features commands. Notice that the commands at the top of the screen are now different. Your screen should look similar to Figure 127.

Figure 127

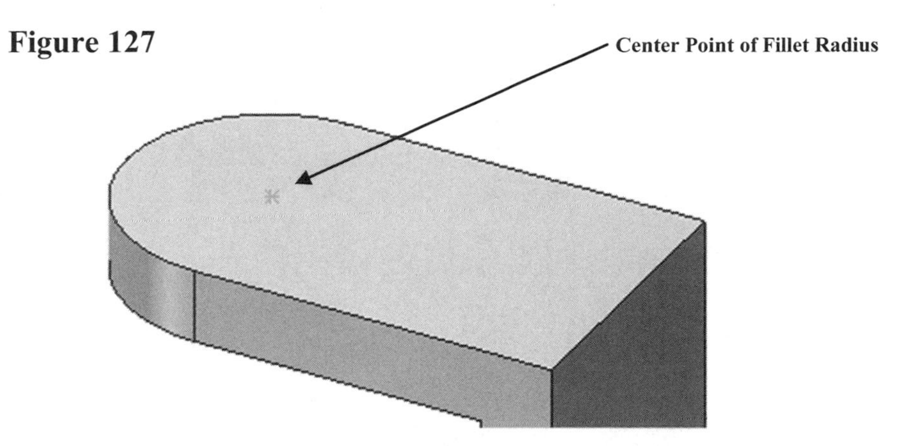

144. Move the cursor to the upper right portion of the screen and left click on **Hole Wizard** as shown in Figure 128.

Figure 128

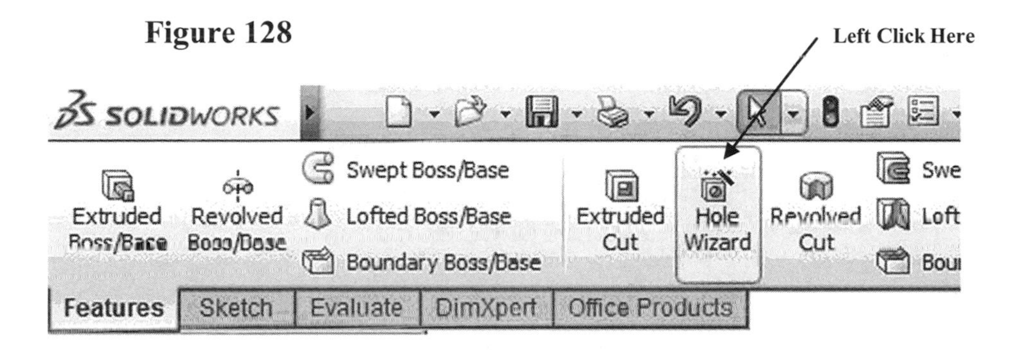

145. The Hole Specification dialog box will appear. Left click on the Counter Bore icon. Place a check in the "Show custom sizing" box. Below Custom Sizing enter **.250** for Hole Fit and Diameter, **.500** for C'Bore Diameter and **.125** for the C'Bore Depth as shown in Figure 129.

Figure 129

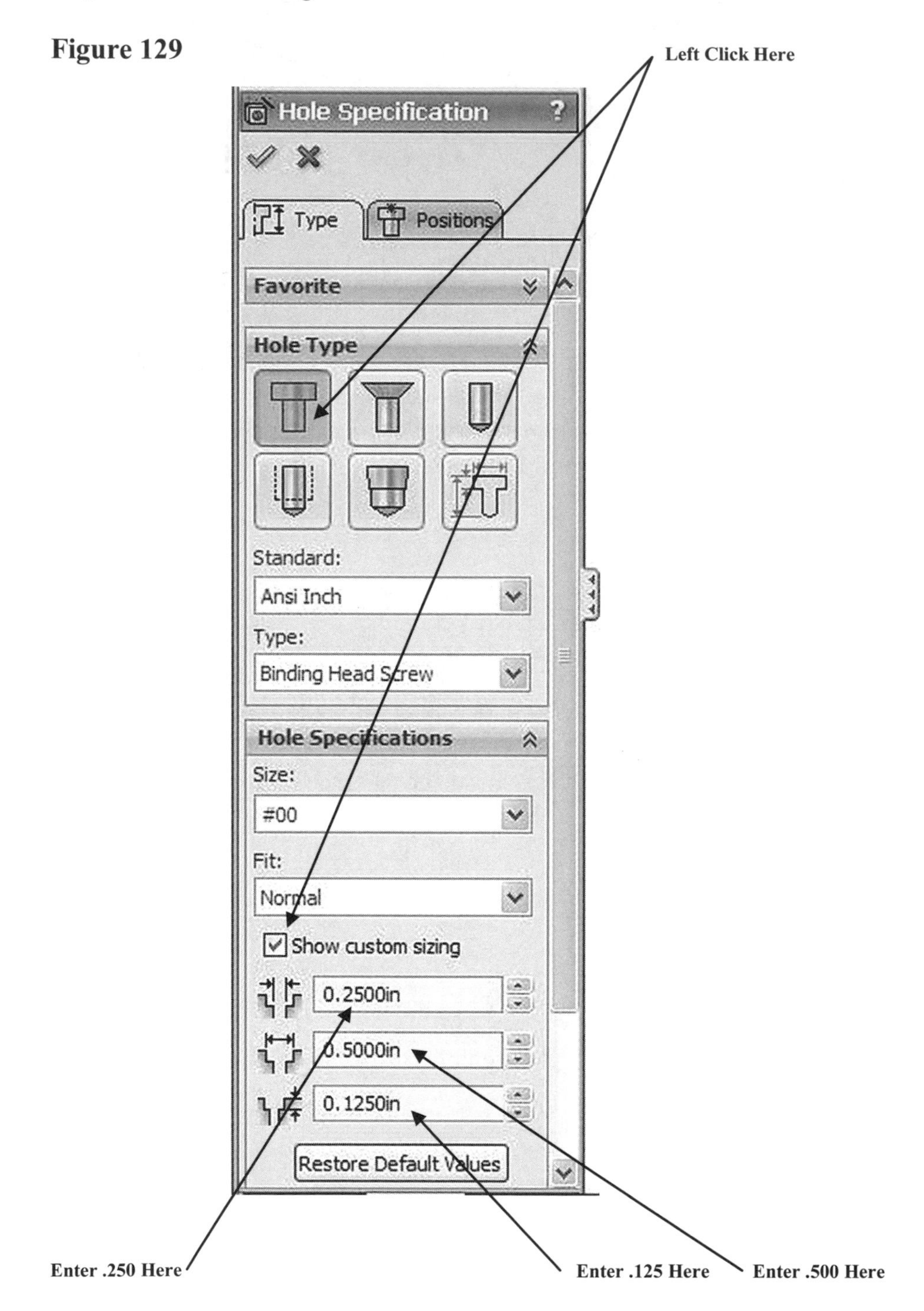

146. Move the cursor to the top of the screen and left click on the **Positions** tab. Left click on **3D Sketch** as shown in Figure 130.

Figure 130

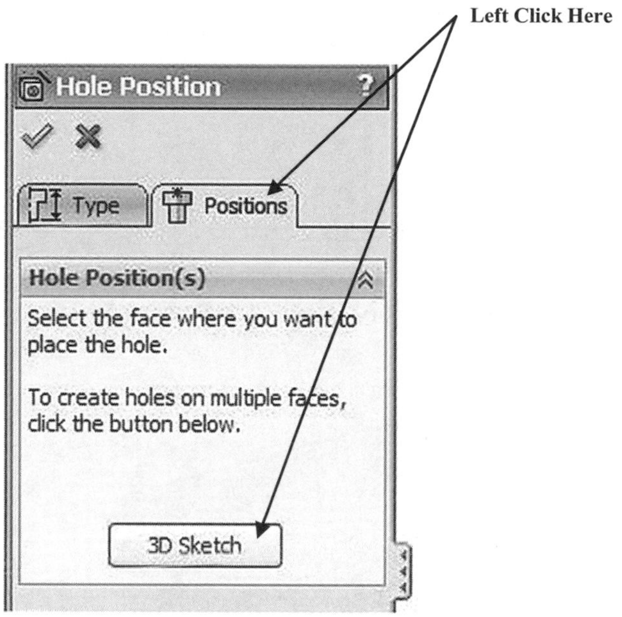

Left Click Here

147. Move the cursor to the center point of the fillet radius and left click once as shown in Figure 131.

Figure 131

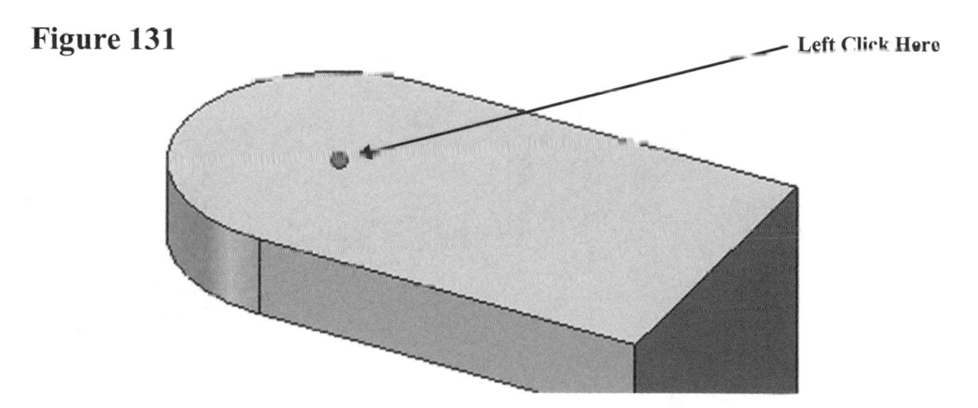

148. SolidWorks will provide a preview of the counterbore as shown in Figure 132.

Figure 132

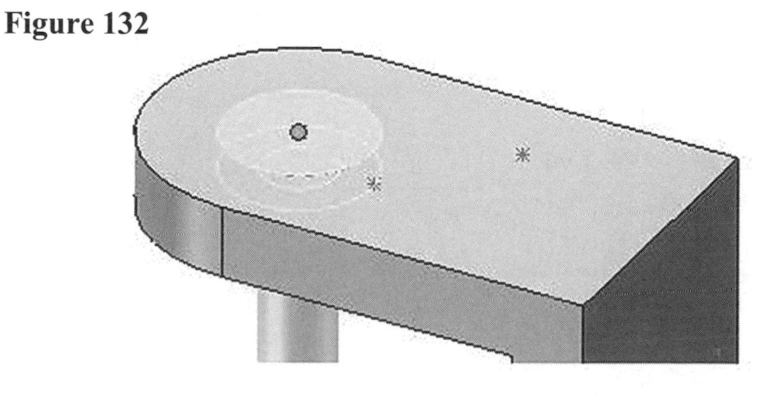

149. Left click on the green checkmark as shown in Figure 133. Left click anywhere around the drawing.

Figure 133

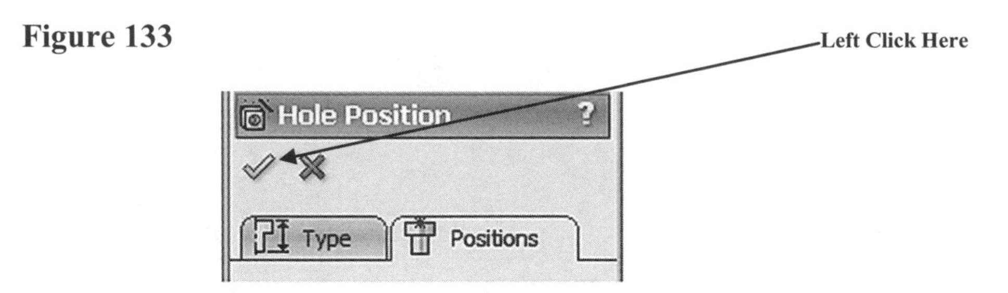

150. Your screen should look similar to Figure 134.

Figure 134

151. To ensure that the hole is correct push the mouse wheel down (holding it down) and move the cursor around. The model will be attached to the cursor. To gain a full view of the model, move the cursor to the top portion of the screen and left click on the "Zoom to Fit" icon as shown in Figure 135.

Figure 135

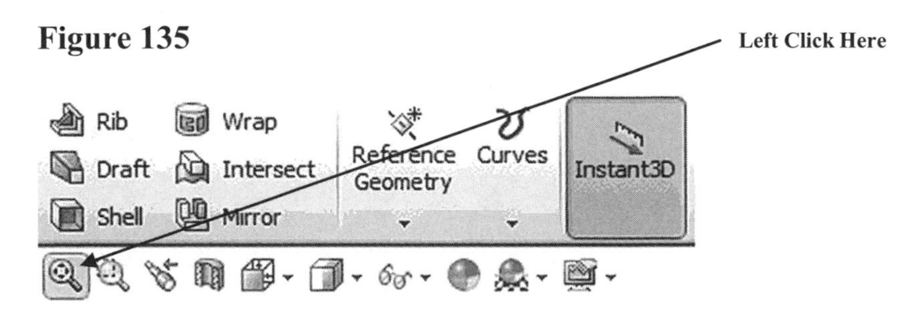

152. Left click anywhere around the part and hold the left mouse button down. Drag the cursor upward. The part will rotate upward as shown in Figure 136.

Figure 136

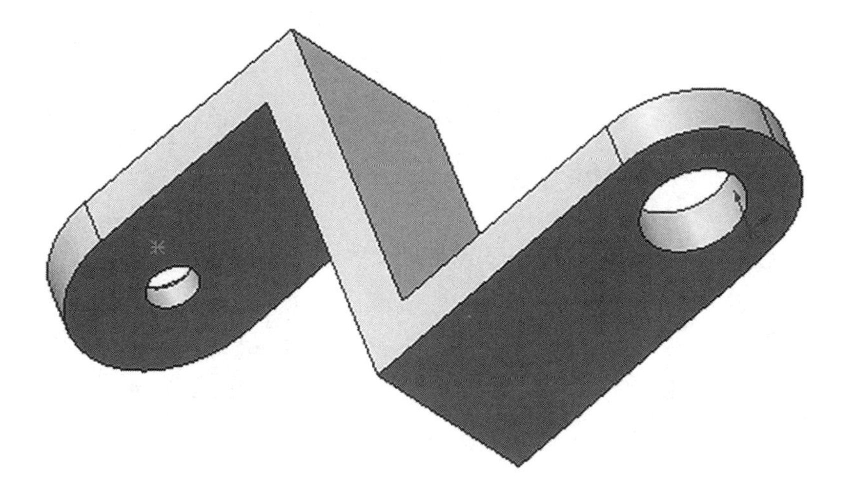

153. The best method of rotating the model around to view different sides of it is to push the mouse wheel down (holding the mouse wheel down) and move the cursor around to the desired position.

For SolidWorks 2009 and older users: Holding the left mouse button down keeps the part attached to the cursor. To view the part in Trimetric, right click anywhere on the screen and left click on **View Orientation** as shown in Figure 137. Scrolling the mouse wheel will also zoom the view in or out.

Figure 137

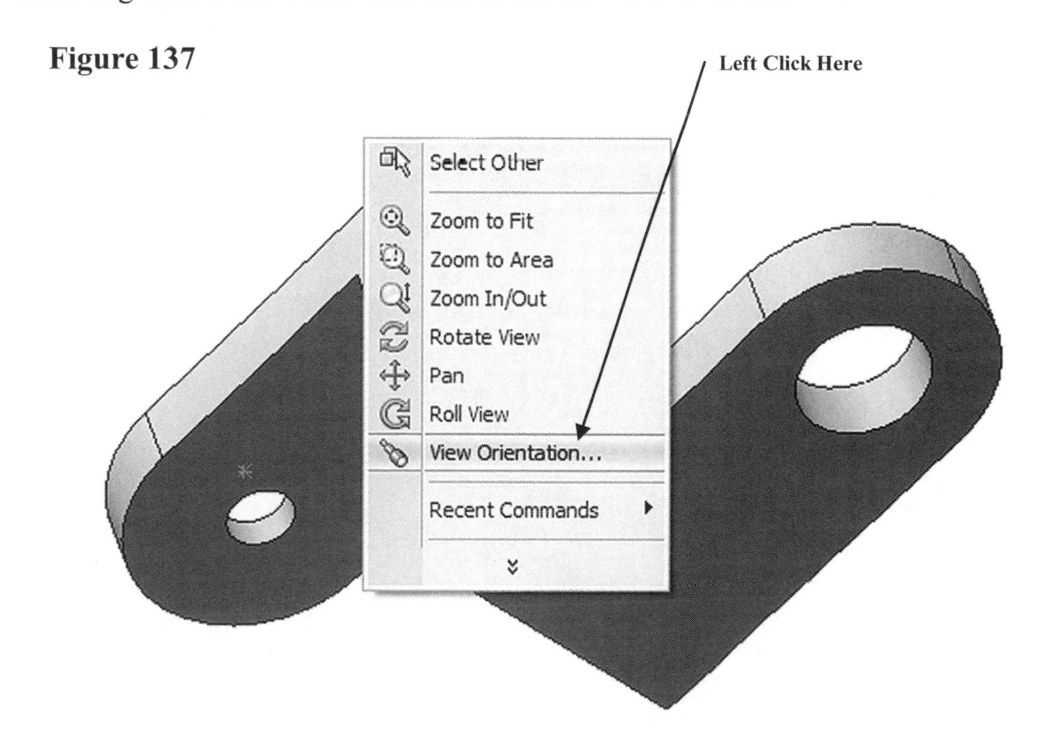

154. A drop down menu will appear. Double click on **Trimetric View** as shown in Figure 138.

Figure 138

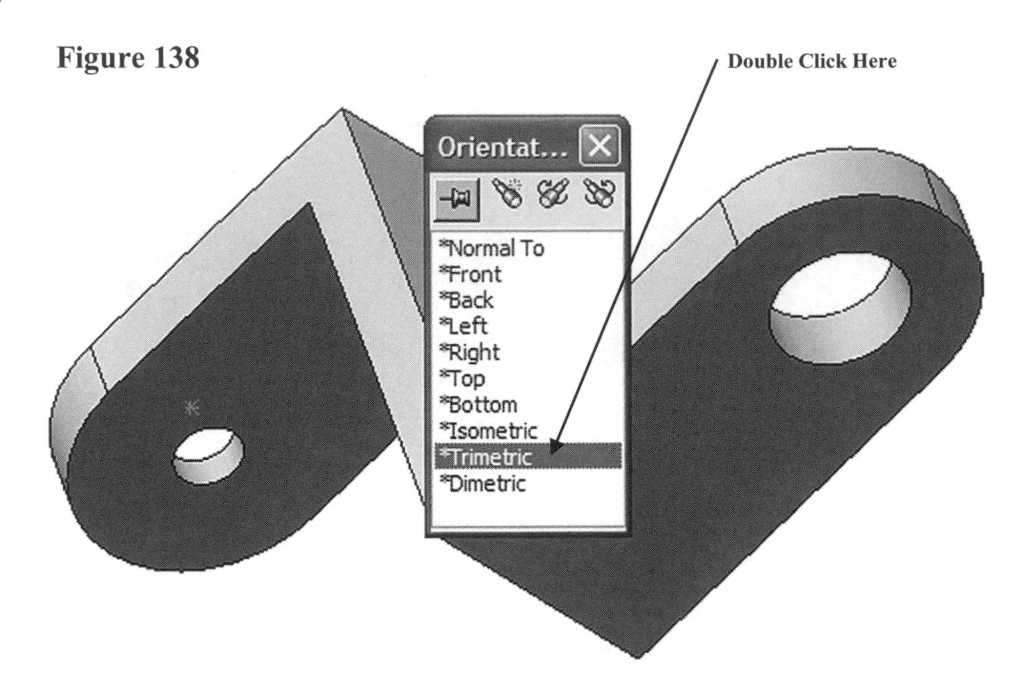

155. Other commands for viewing are located at the top of the screen as shown in Figure 139.

Figure 139

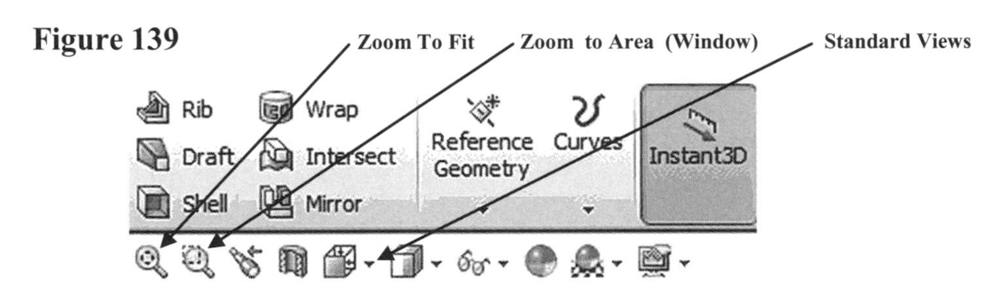

For SolidWorks 2009 and older users:

156. The Zoom to Area (Window) command works by using the cursor to draw a window around the area you want to zoom in on. After selecting the "Zoom Window" icon, hold the left mouse button down and drag a diagonal box around the desired area. Release the left mouse button when the proper amount of zoom is achieved.

157. The Zoom In/Out command works by scrolling the mouse wheel in either direction to achieve the desired view.

158. The Pan command works similar to the Zoom In/Out command. Start by selecting the "Pan" icon. If the Pan icon is not visible with the other Viewing tools it will need to be added by using the same steps outlined on page 49. Once the icon has been added to the Viewing tools, left click on the icon. Left click on the drawing and hold the left mouse button down while moving the cursor up and down or side to side. Release the mouse button after the desired view is achieved.

159. The Zoom to Fit command works by filling the screen so that the entire part is as large as possible while remaining visible in its entirety.

160. The same viewing options can be accessed by right clicking anywhere on the drawing. A pop will be displayed as shown in Figure 140.

Figure 140

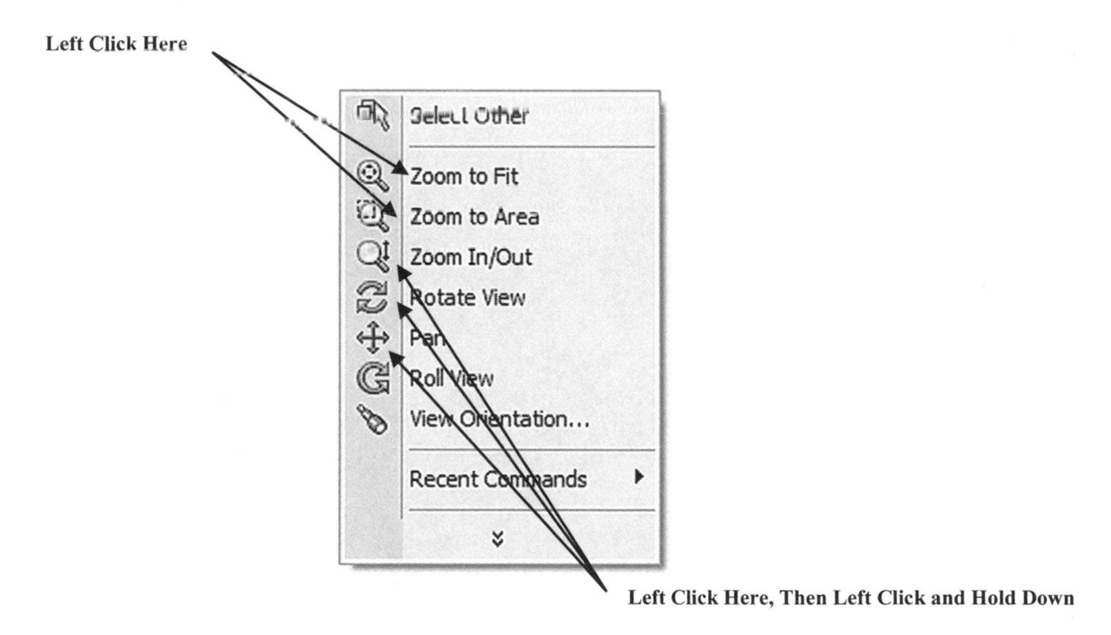

Drawing Activities

Problem 1

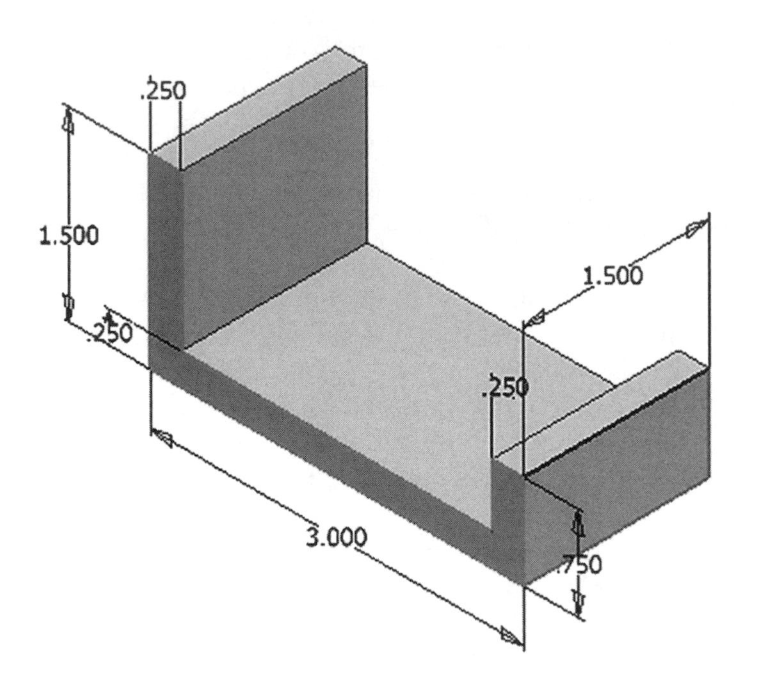

Problem 2

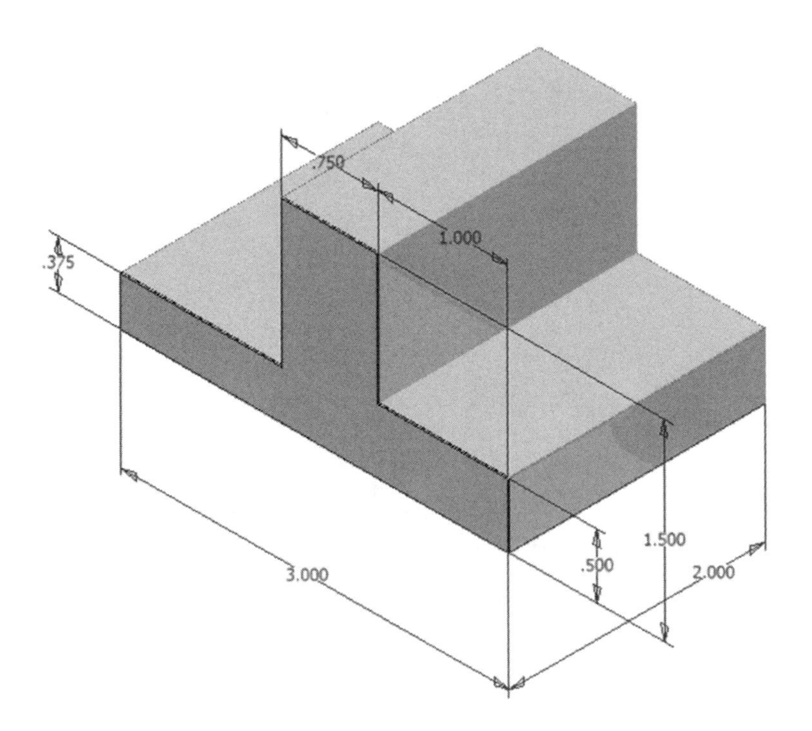

Problem 3

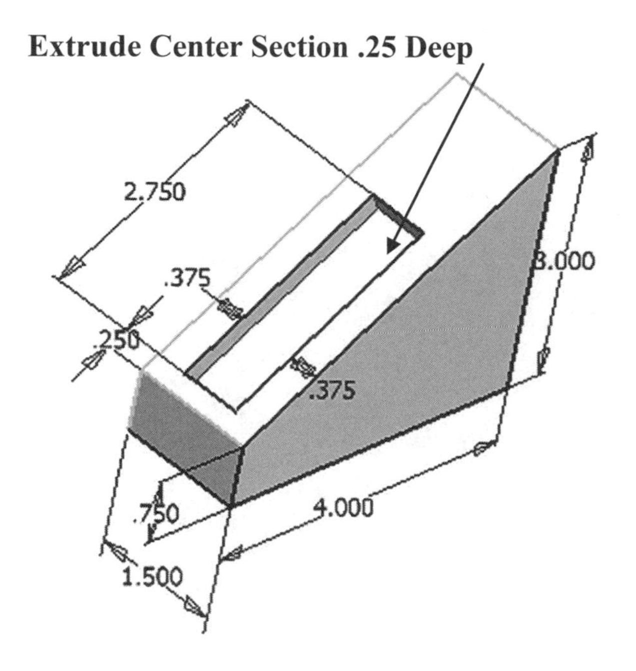

Problem 4

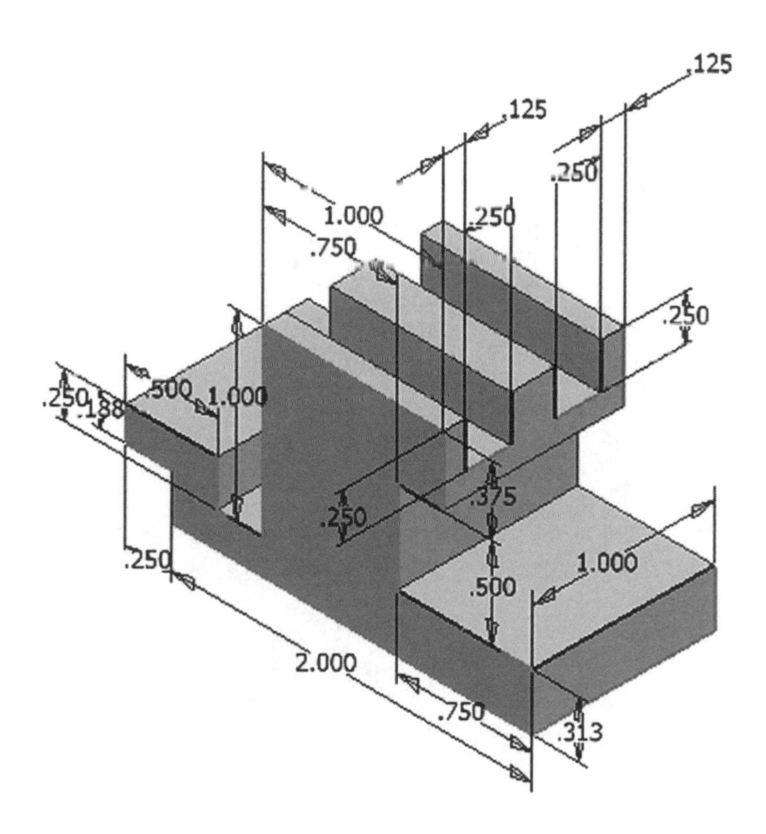

Problem 5

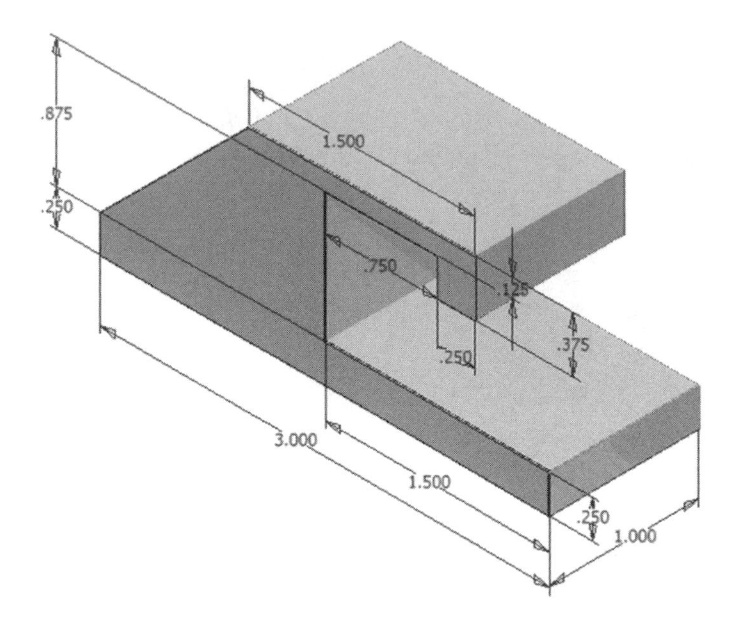

Problem 6

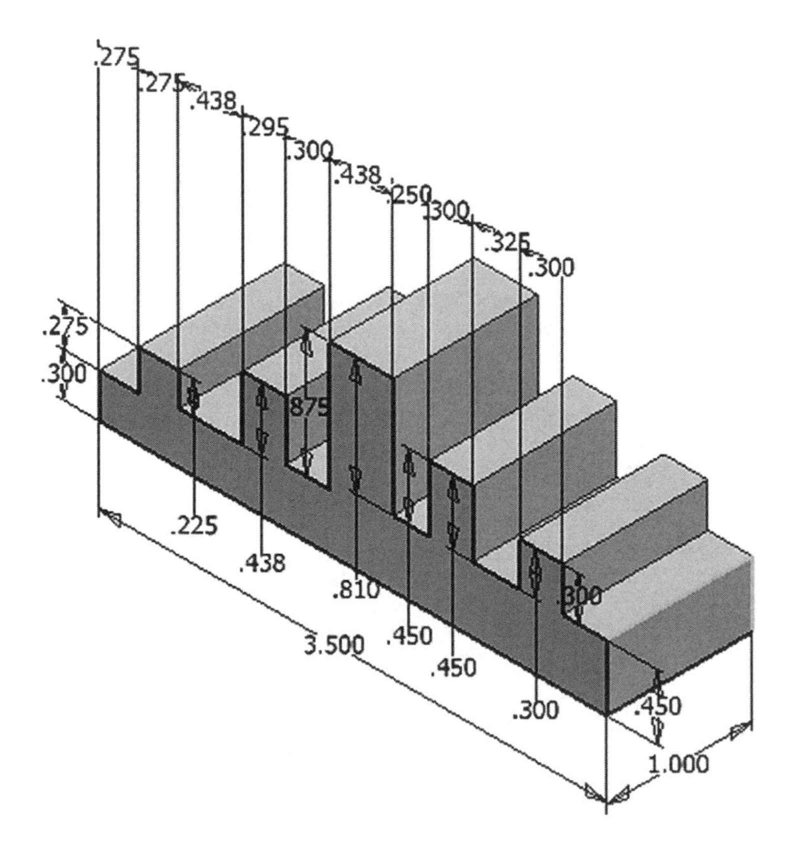

Problem 7

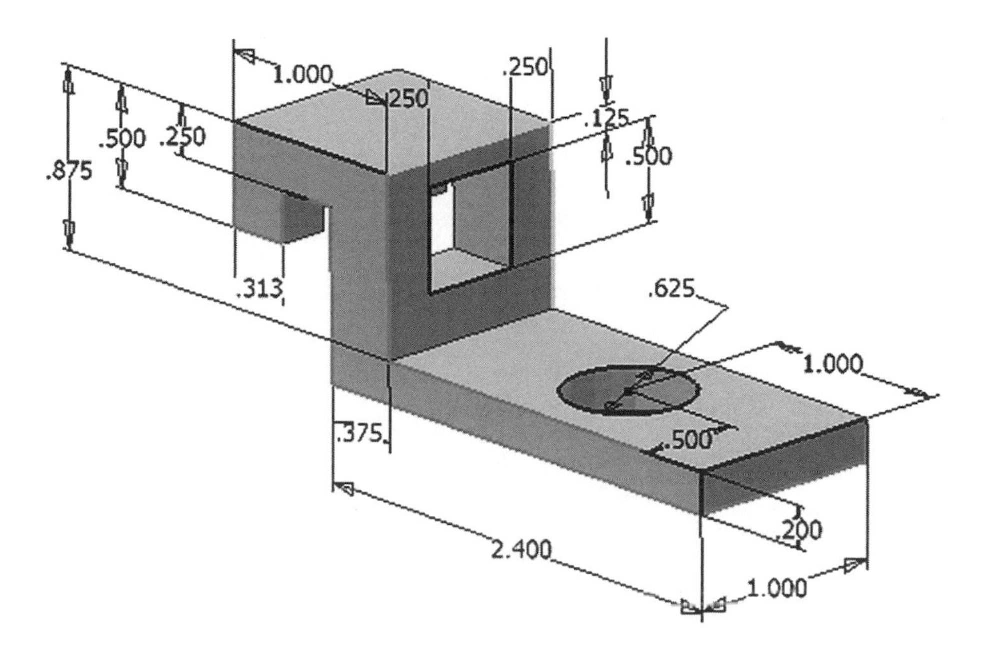

Problem 8

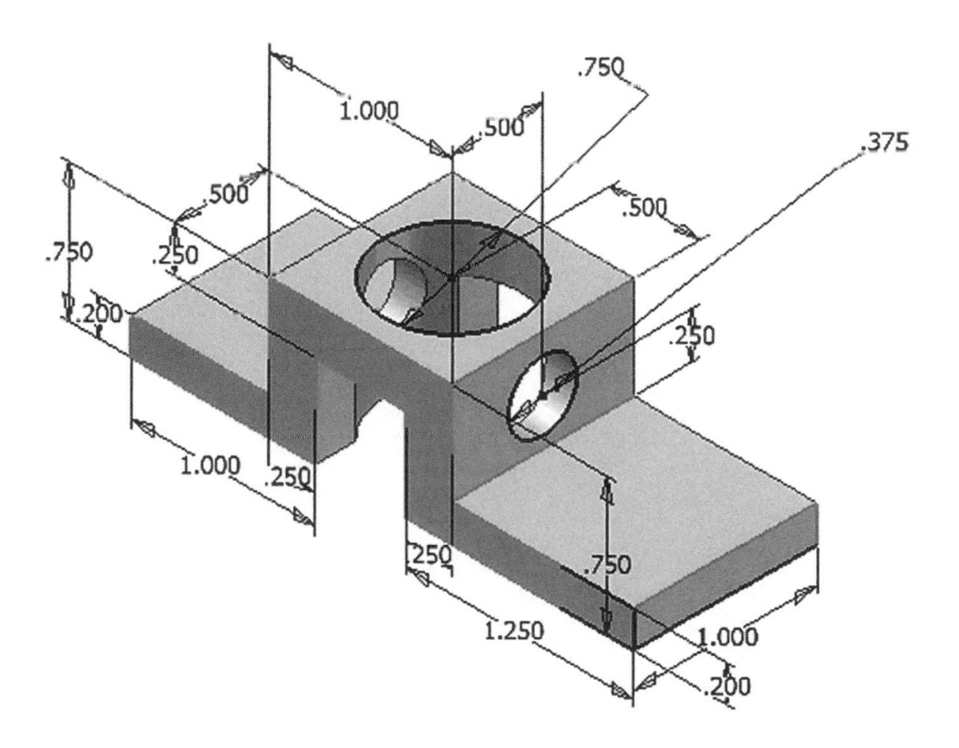

Chapter 2 Learning More Basics

Objectives:

- Create a simple sketch using the Sketch commands
- Dimension a sketch using the Smart Dimension command
- Revolve a sketch using the Revolved Boss/Base command
- Create a hole using the Extruded Cut command
- Create a series of holes using the Circular Pattern command

Chapter 2 includes instruction on how to design the part shown below.

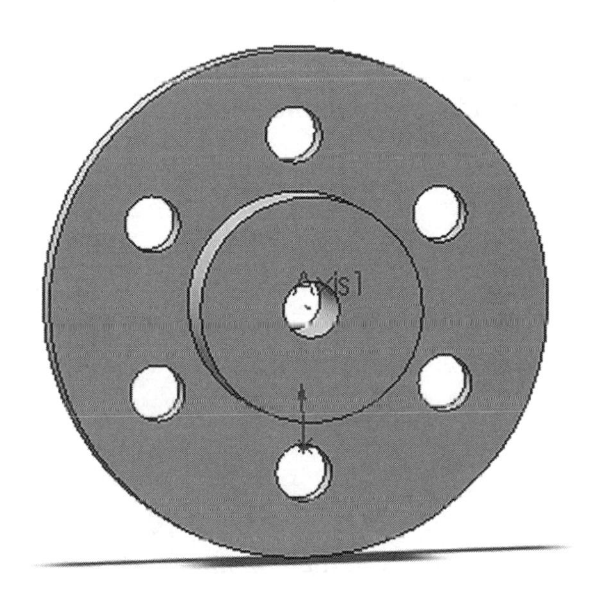

1. Start SolidWorks by referring to "Chapter 1 Getting Started".

2. After SolidWorks is running, begin a new sketch.

3. Move the cursor to the upper left corner of the screen and left click on **Line** as shown in Figure 1.

Figure 1

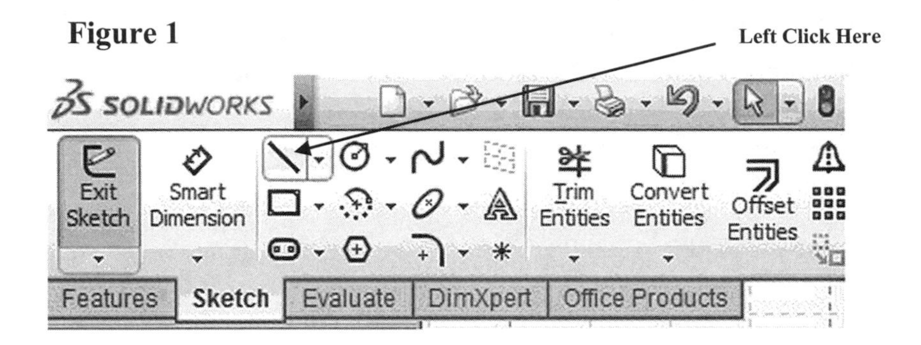

4. Move the cursor to the lower left portion of the screen and left click once. This will be the beginning end point of a line as shown in Figure 2.

Figure 2

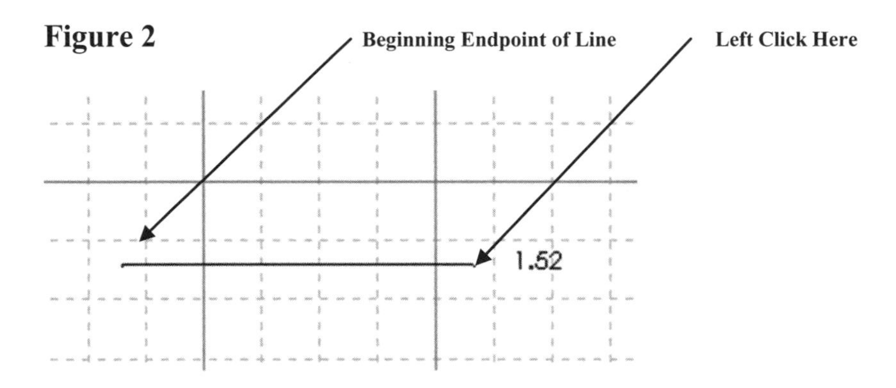

5. Move the cursor to the right and left click once as shown in Figure 2.

6. Move the cursor up and left click once as shown in Figure 3.

Figure 3

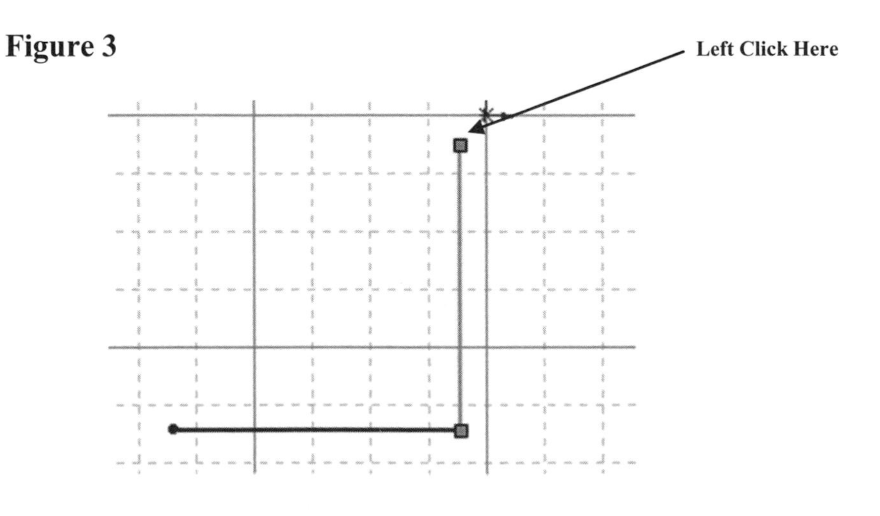

7. Move the cursor to the right and left click once as shown in Figure 4.

Figure 4

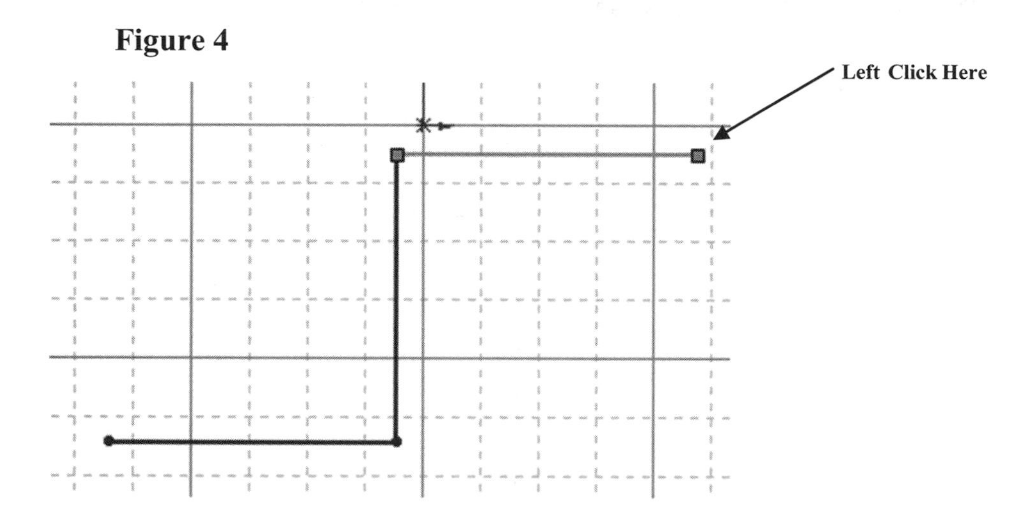

8. Move the cursor up and left click once as shown in Figure 5.

Figure 5

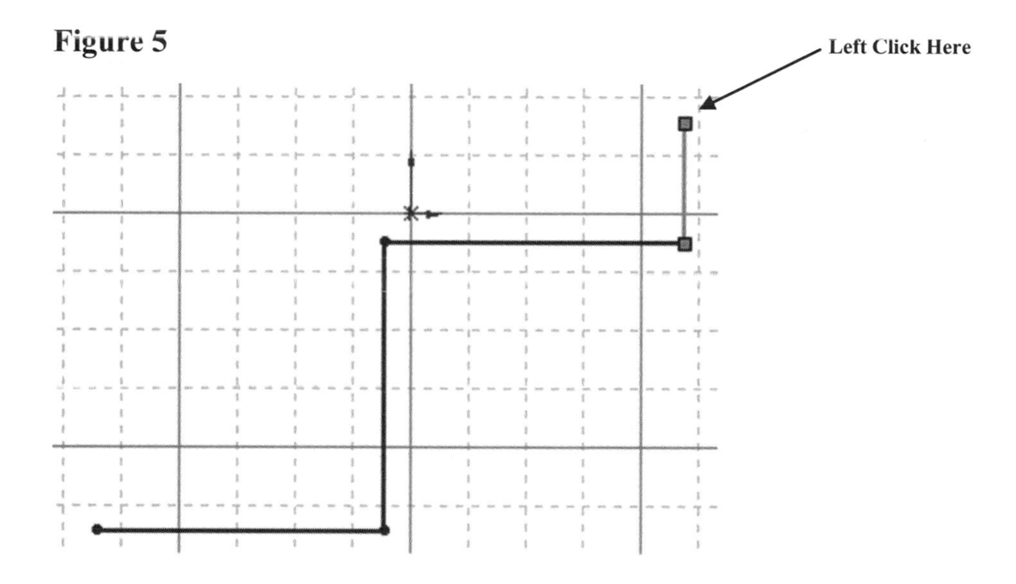

9. Move the cursor to the left. Ensure that the dots between the first end point and the last end point appear as shown in Figure 6. Left click once.

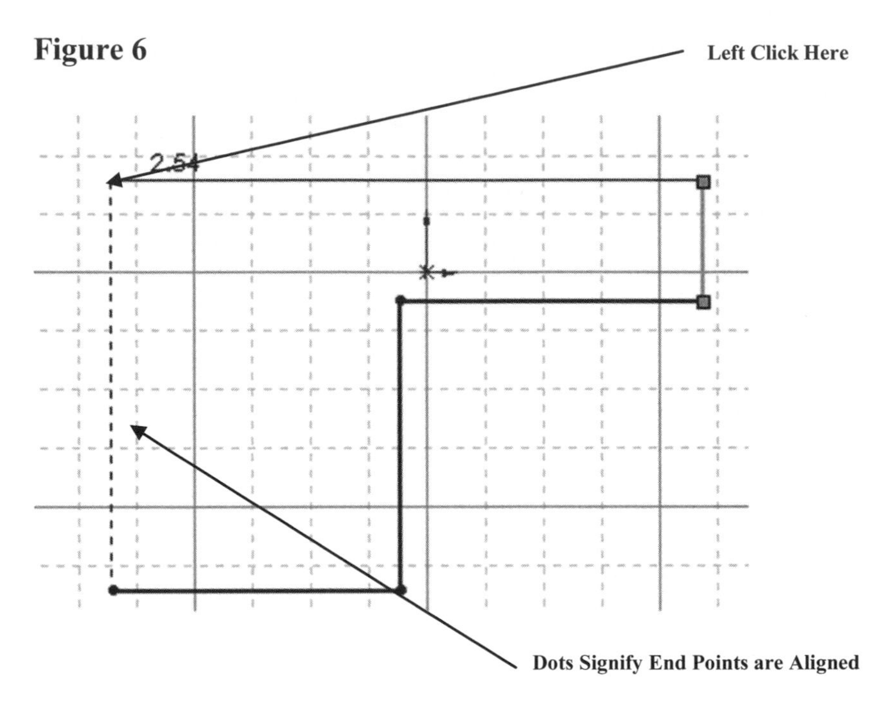

Figure 6

10. Move the cursor back to the original starting end point and left click once as shown in Figure 7. Press the **Esc** key once.

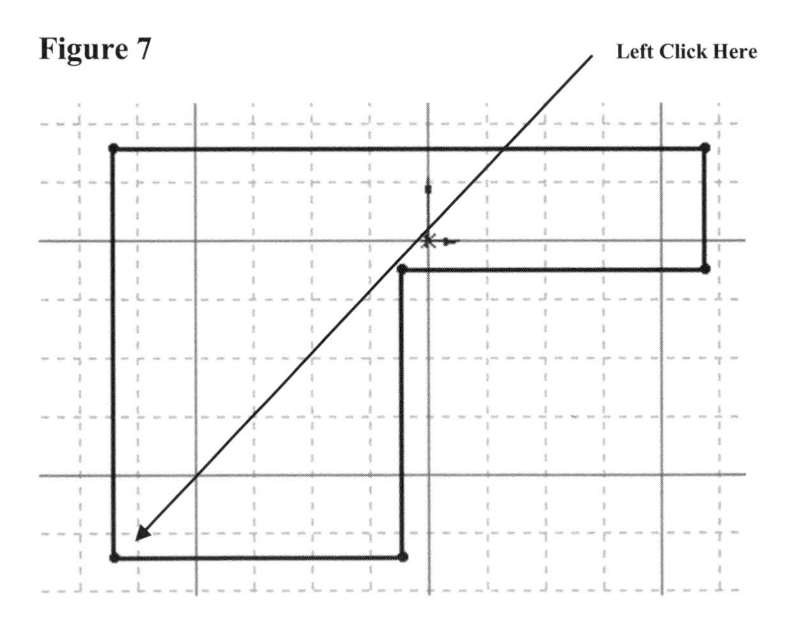

Figure 7

11. Move the cursor to the upper left portion of the screen and left click on **Smart Dimension** as shown in Figure 8.

Figure 8

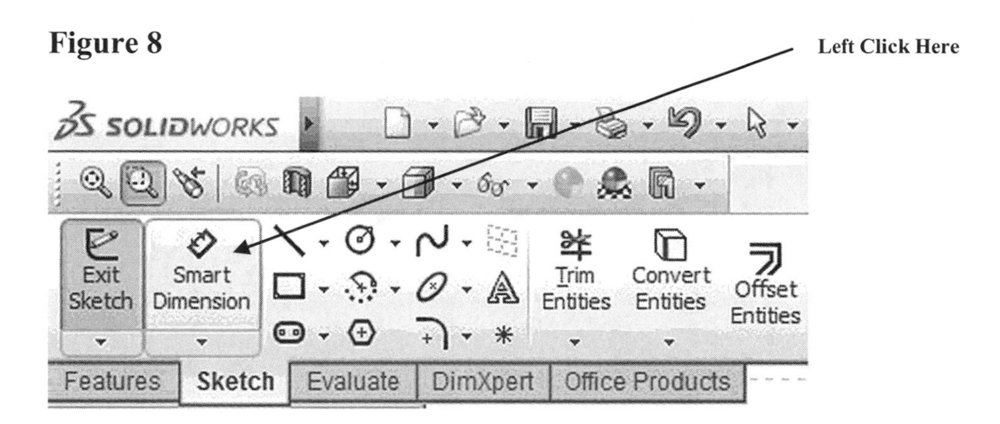

12. After selecting **Smart Dimension** move the cursor over the bottom horizontal line until it turns red as shown in Figure 9. Select the line by left clicking anywhere on the line **or** on each of the end points. To use the end points of the line, move the cursor over one of the end points. A small red dot will appear. Left click once and move the cursor to the other end point. After the red dot appears, left click once. The dimension will be attached to the cursor.

Figure 9

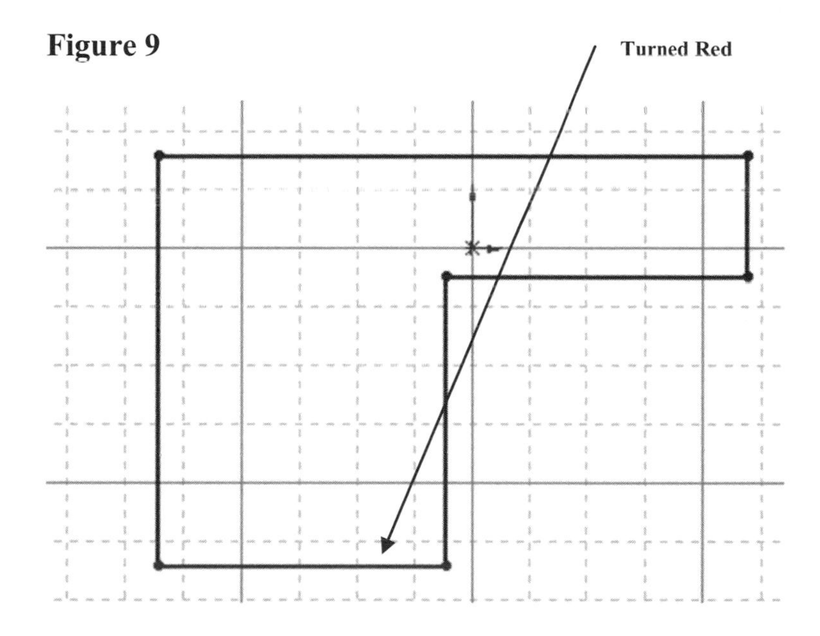

13. Move the cursor to where the dimension will be placed and left click once as shown in Figure 10.

Figure 10

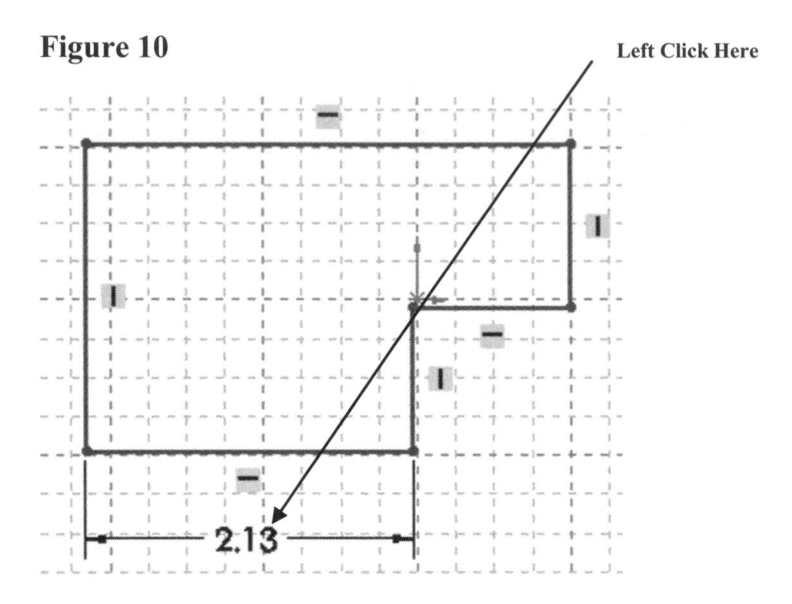

14. The Modify dialog box will appear as shown in Figure 11.

Figure 11

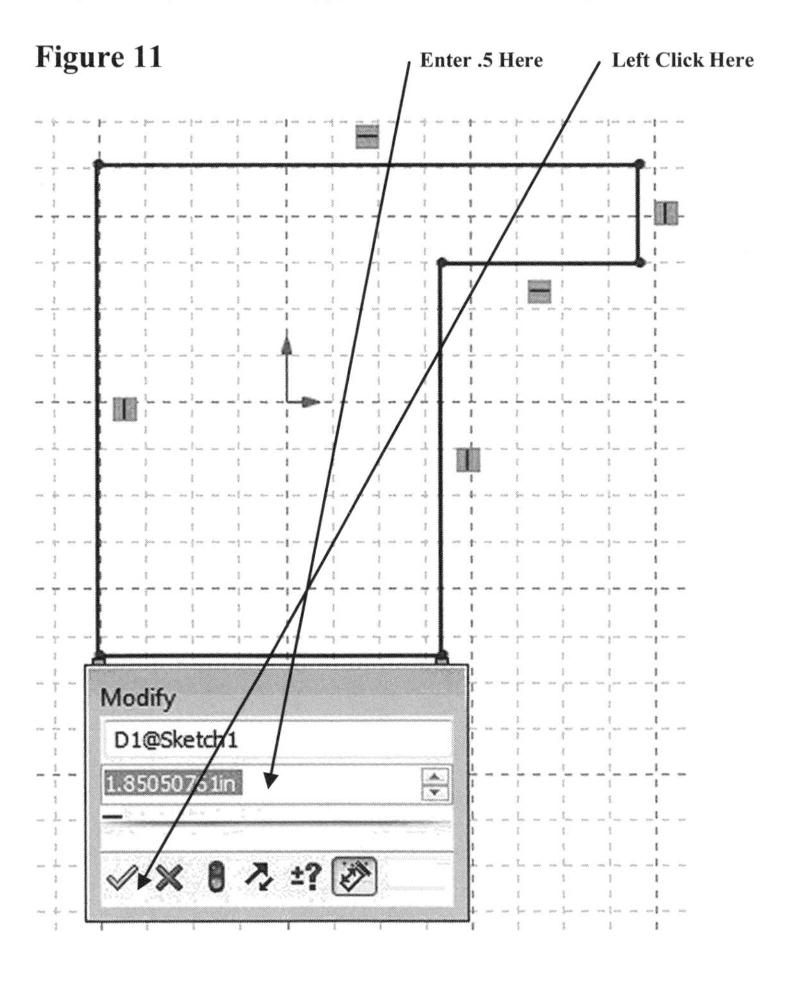

15. To edit the dimension, enter **.50** in the Modify dialog box (while the current dimension is highlighted) and either press **Enter** on the keyboard or left click on the green checkmark as shown in Figure 11.

16. The dimension of the line will become .50 inches as shown in Figure 12.

Figure 12

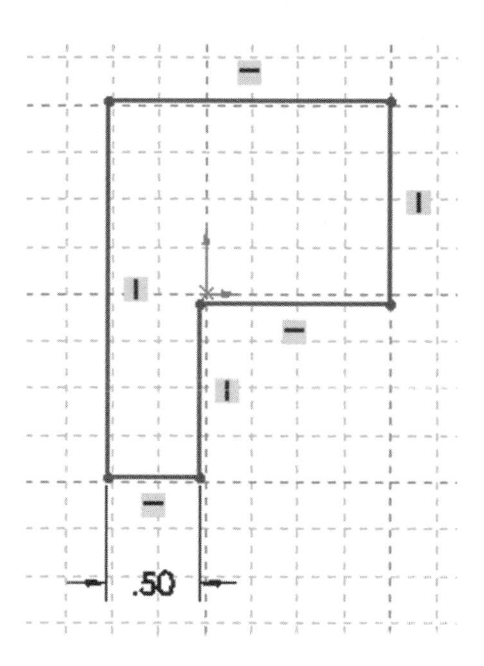

17. To view the entire drawing move the cursor to the upper middle portion of the screen and left click once on the "Zoom To Fit" icon as shown in Figure 13.

Figure 13

Left Click Here

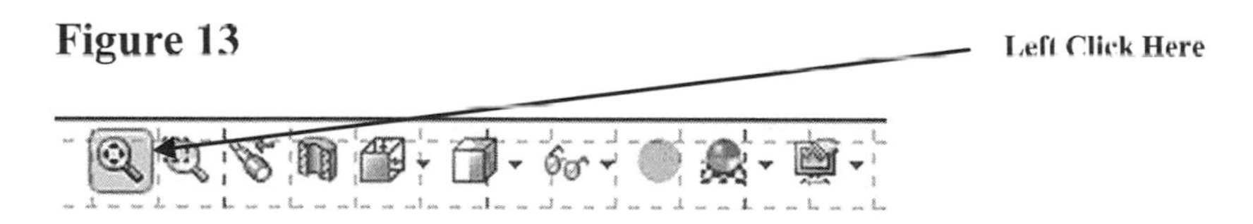

18. The drawing will "fill up" the entire screen. If there is a need to zoom in on one area of the drawing, left click on the "Zoom to Area" icon as shown in Figure 14. Simply drag a box or window around the area you want to zoom in on. SolidWorks will zoom in on that area.

Figure 14

Left Click Here

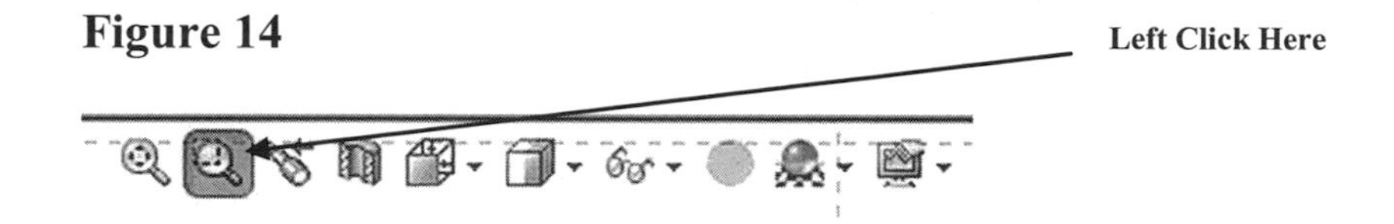

19. Move the cursor to the middle left portion of the screen and left click on **Smart Dimension** as shown in Figure 15.

Figure 15

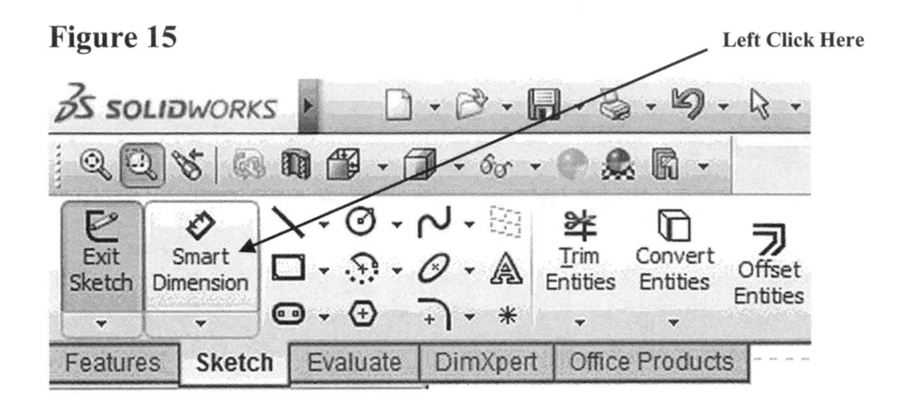

20. After selecting **Smart Dimension** move the cursor over the vertical line until it turns red as shown in Figure 16. Select the line by left clicking anywhere on the line **or** on each of the end points. To use the end points of the line, move the cursor over one of the end points. A small red dot will appear. Left click once and move the cursor to the other end point. After a red dot appears, left click once. The dimension will be attached to the cursor.

Figure 16

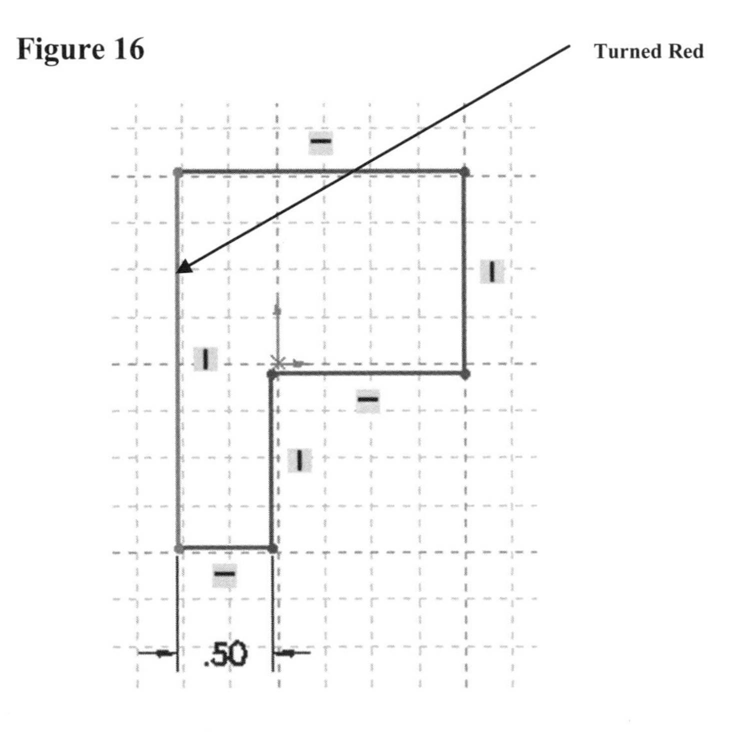

21. Move the cursor to where the dimension will be placed and left click once as shown in Figure 17.

Figure 17

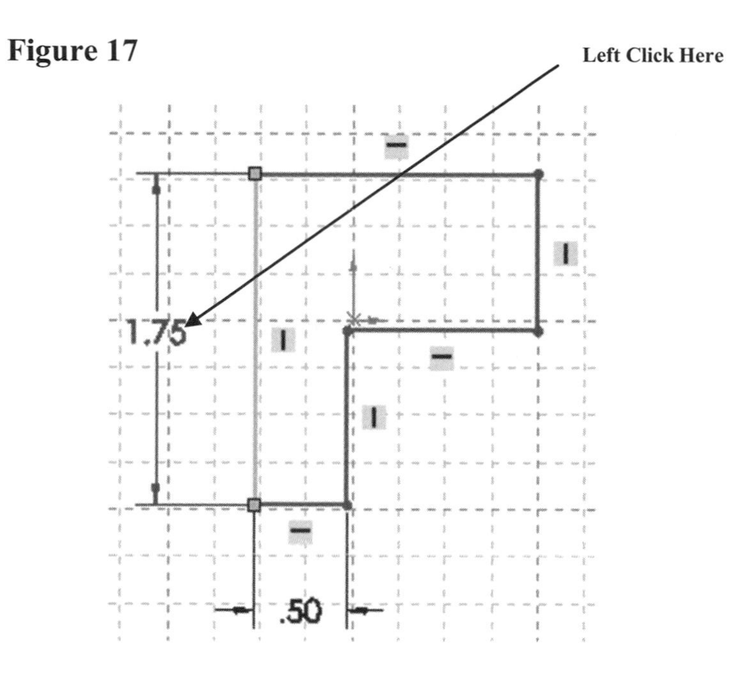

22. The Modify dialog box will appear as shown in Figure 18.

Figure 18

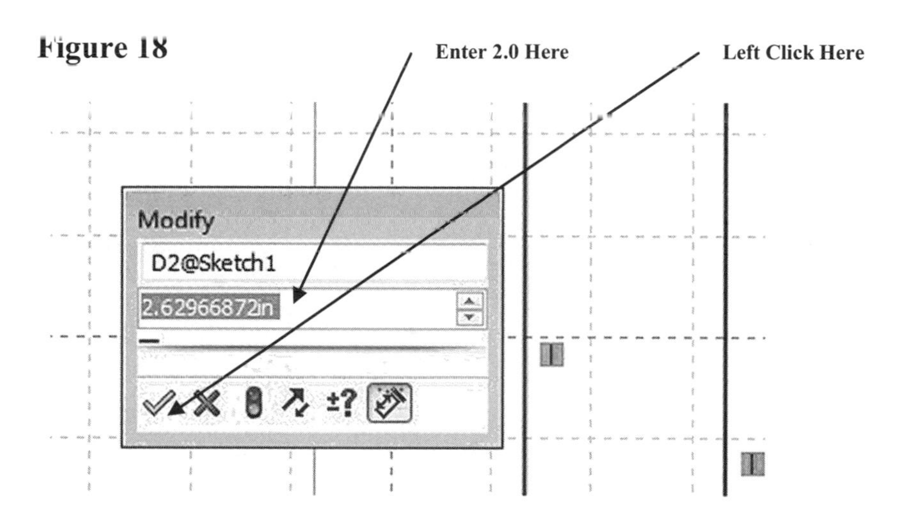

23. To edit the dimension, enter **2.0** in the Modify dialog box (while the current dimension is highlighted) and either left click on the green checkmark (as shown in Figure 18) or press **Enter** on the keyboard.

24. The dimension of the line will become 2.0 inches as shown in Figure 19. Use the Zoom icons to zoom out if necessary.

Figure 19

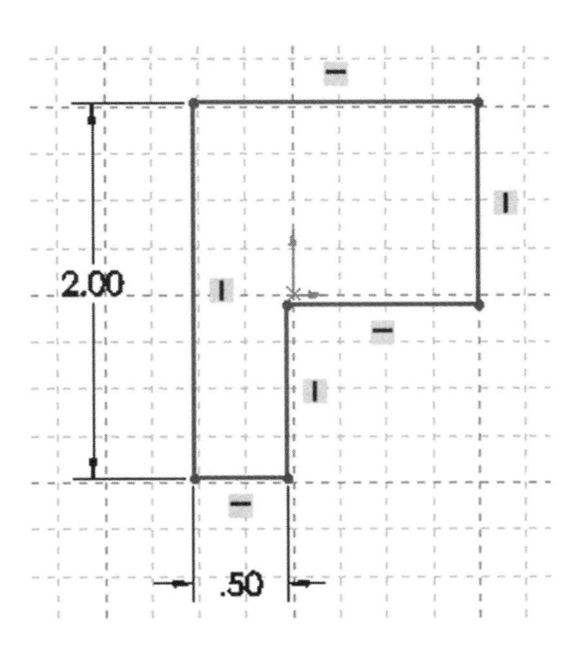

25. Select the next line by left clicking anywhere on the line **<u>or</u>** on each of the end points. The dimension will be attached to the cursor.

Figure 20

Line Turned Red

2.00

.50

26. Move the cursor to where the dimension will be placed and left click once as shown in Figure 21.

Figure 21

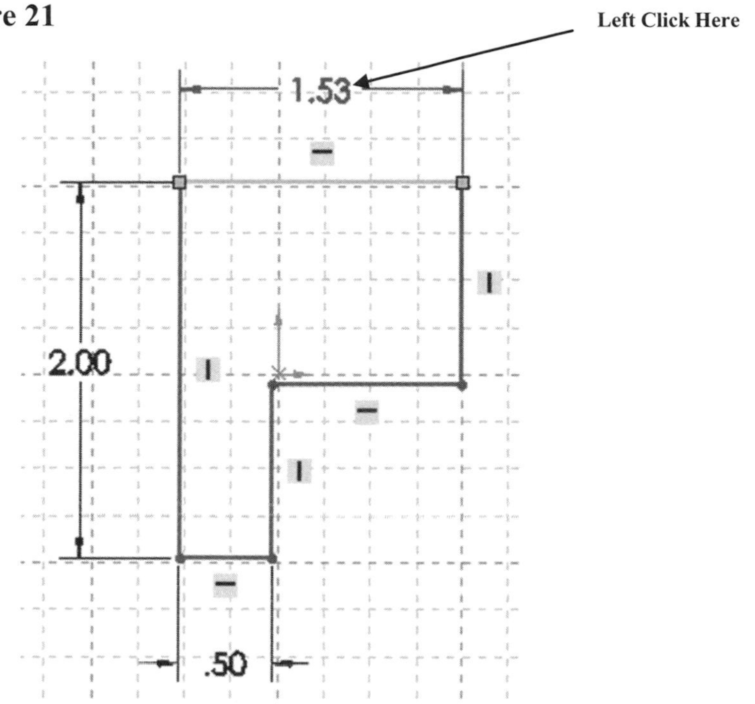

Left Click Here

1.53

2.00

.50

27. The Modify dialog box will appear as shown in Figure 22.

Figure 22

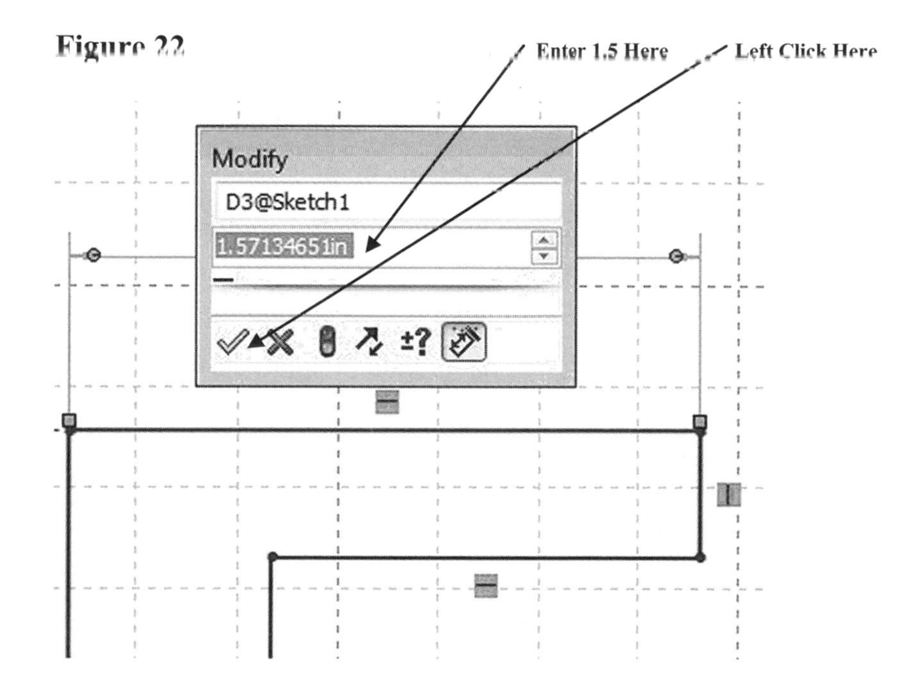

28. To edit the dimension, enter **1.5** in the Modify dialog box (while the current dimension is highlighted) and either left click on the green checkmark (as shown in Figure 22) or press **Enter** on the keyboard.

29. The dimension of the line will become **1.5** inches as shown in Figure 23. Use the Zoom icons to zoom out if necessary.

Figure 23

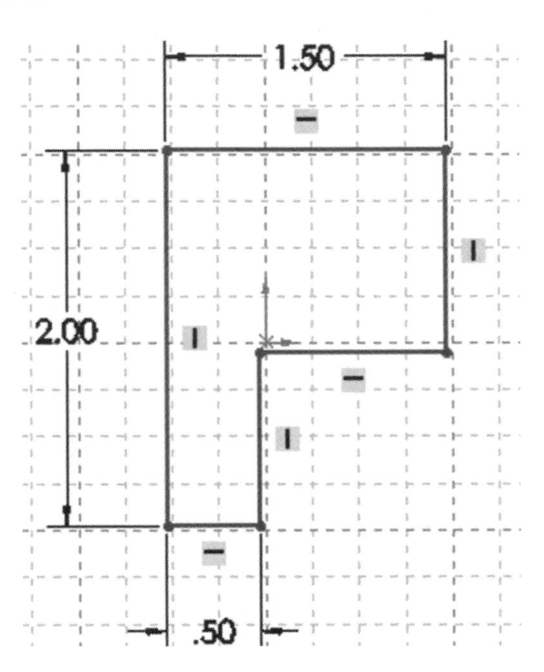

30. Select the next line by left clicking anywhere on the line **<u>or</u>** on each of the end points. The line will turn red as shown in Figure 24. The dimension will be attached to the cursor.

Figure 24

Turned Red

1.50

2.00

.50

31. Move the cursor to where the dimension will be placed and left click once as shown in Figure 25.

Figure 25

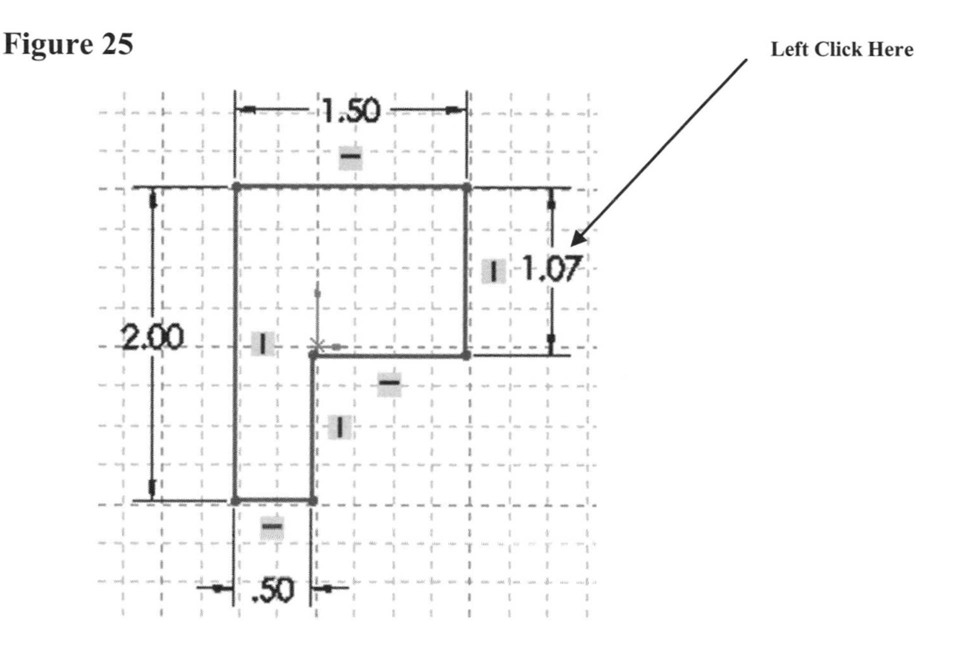

32. The Modify dialog box will appear as shown in Figure 26.

Figure 26

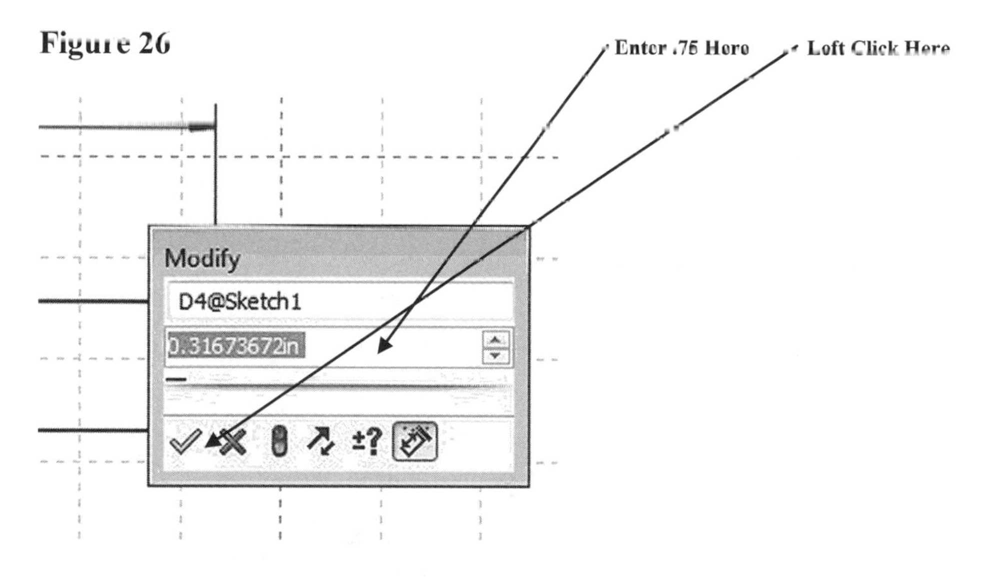

33. To edit the dimension, enter **.75** in the Modify dialog box (while the current dimension is highlighted) and either click on the green checkmark or press **Enter** on the keyboard.

34. The dimension of the line will become **.75** inches as shown in Figure 27. Use the Zoom icons to zoom out if necessary.

Figure 27

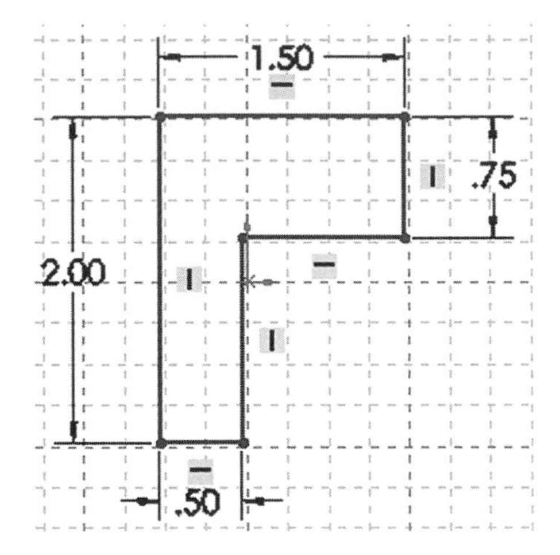

35. Move the cursor to the upper left corner of the screen and left click on **Line** as shown in Figure 28.

Figure 28

Left Click Here

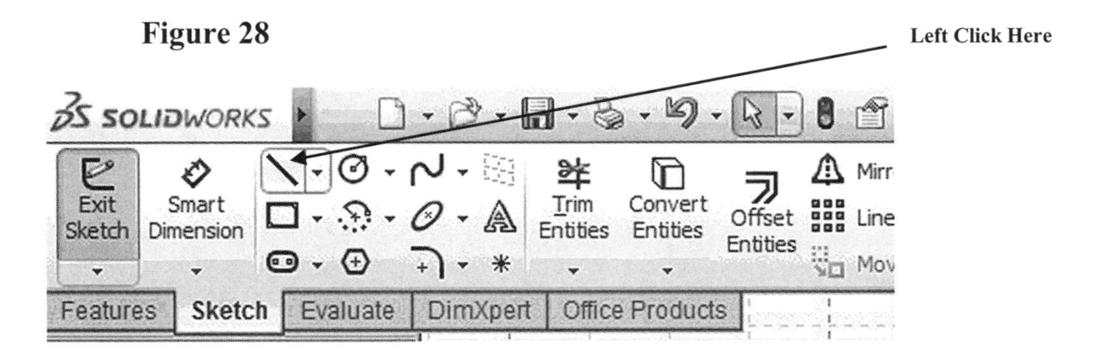

36. Draw a line parallel to the top horizontal line as shown in Figure 29.

Figure 29

Draw This Line

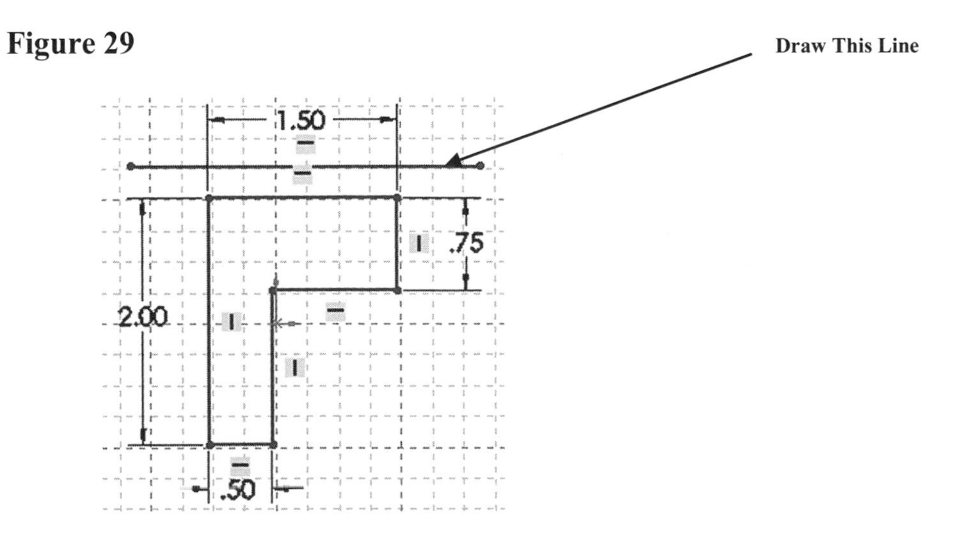

37. Move the cursor to where the dimension will be placed and left click once. The Modify dialog box will appear. To edit the dimension, enter **.25** in the Modify dialog box (while the current dimension is highlighted) and either click on the green checkmark or press **Enter** on the keyboard. Your screen should look similar to Figure 30.

Figure 30

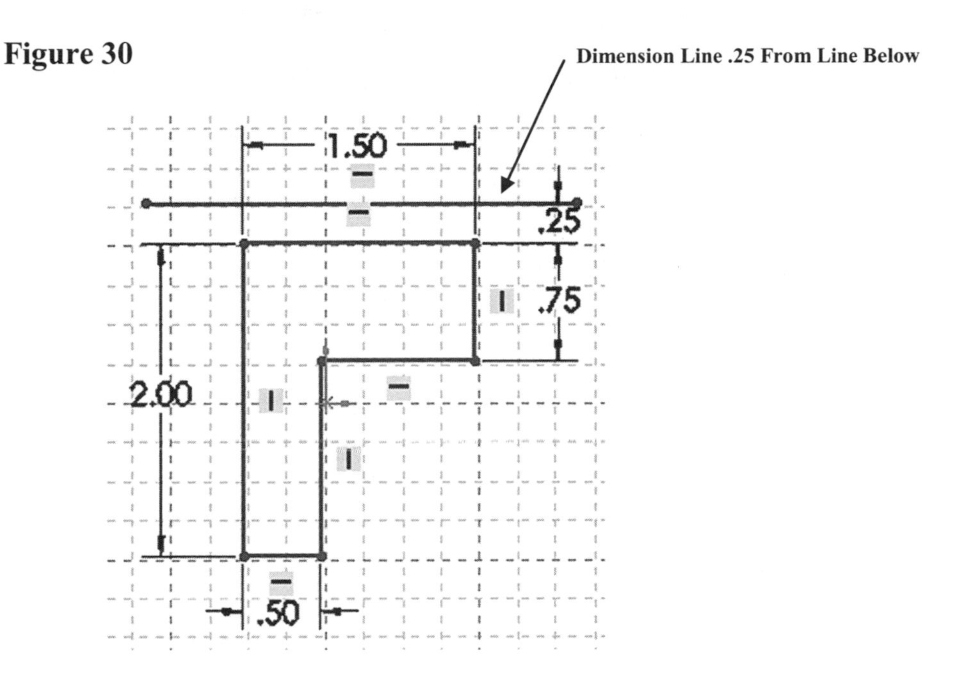

38. Right click anywhere on the drawing. A pop up menu will appear. Left click on **Select** as shown in Figure 31. This verifies that no commands are active.

Figure 31

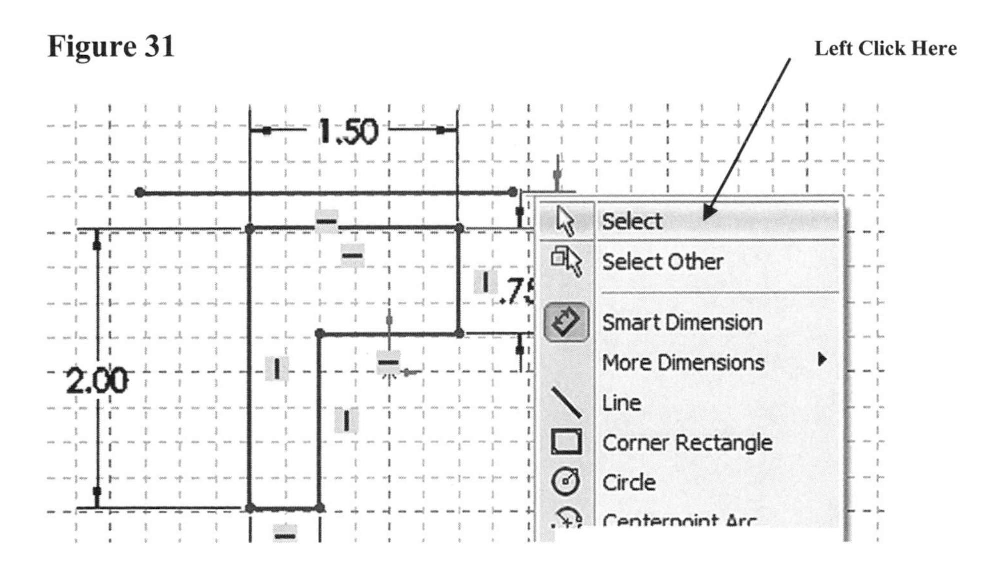

39. After the sketch is complete it is time to revolve the sketch into a solid.

40. Move the cursor to the upper left portion of the screen and left click on **Exit Sketch** as shown in Figure 32.

Figure 32

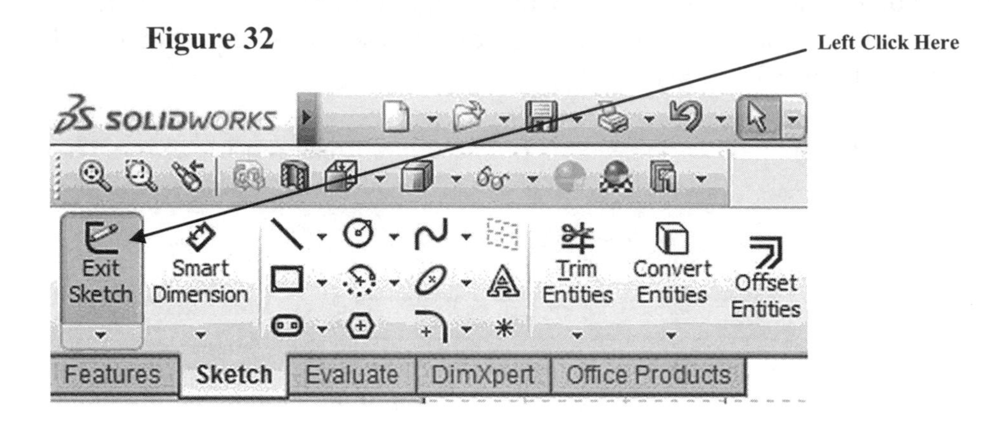

41. SolidWorks is now out of the Sketch commands and into the Features commands. Notice that the commands at the top of the screen are now different. To work in the Features commands a sketch must be present and have no opens (non-connected lines). If there are any opens in the sketch an error message will appear. Your screen should look similar to Figure 33.

Figure 33

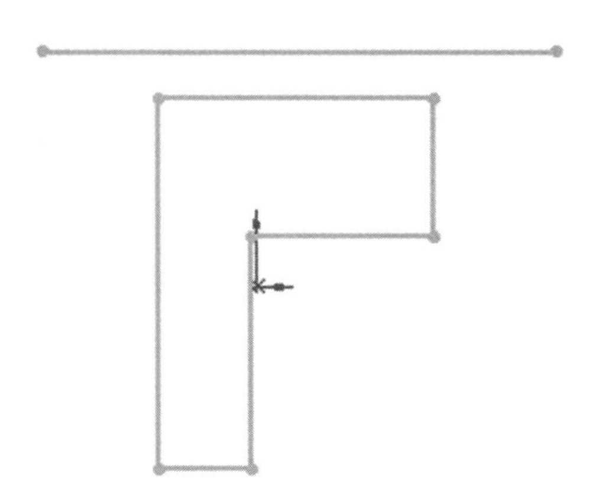

42. Move the cursor to the upper middle portion of the screen and left click on the drop down arrow next to the "View Orientation" icon. A drop down box will appear. Left click as shown in Figure 34.

Figure 34

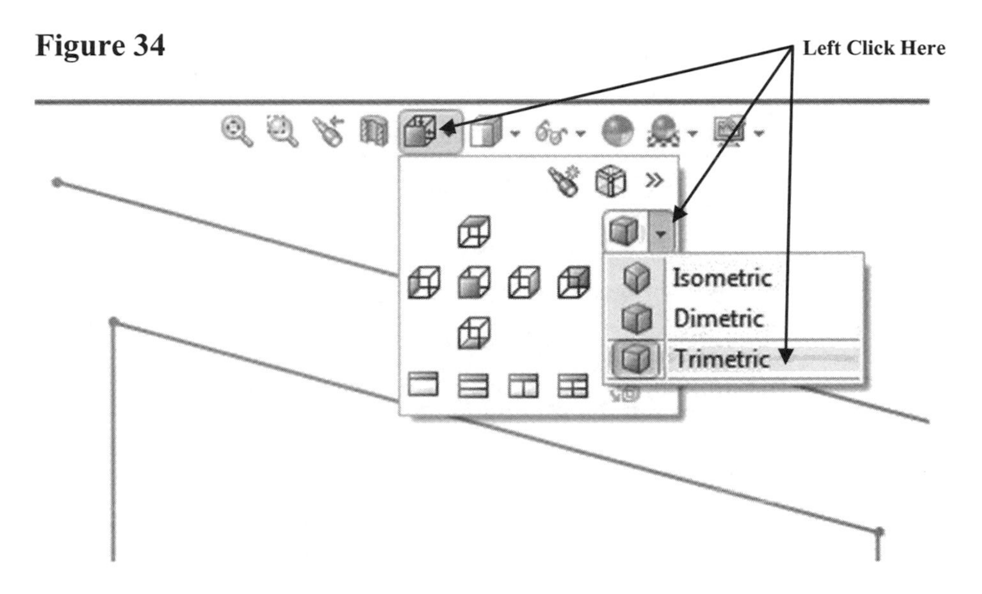

43. Left click on **Trimetric** as shown in Figure 34.

44. The view will become trimetric as shown in Figure 35.

Figure 35

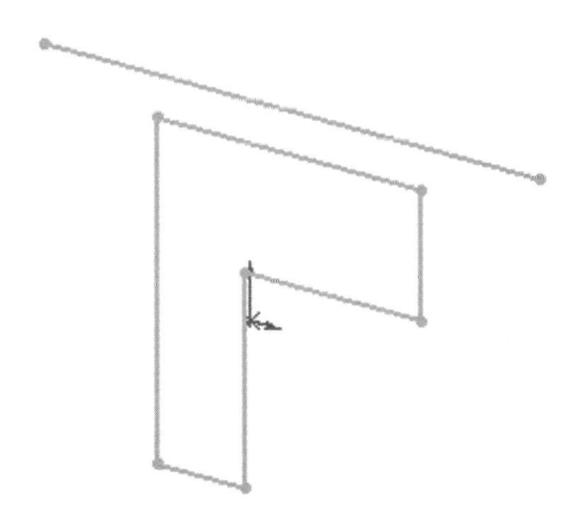

45. Move the cursor to the upper left portion of the screen and left click on **Revolved Boss/Base**. If the **Revolved Boss/Base** icon is not visible, left click on the **Features** tab as shown in Figure 36.

Figure 36

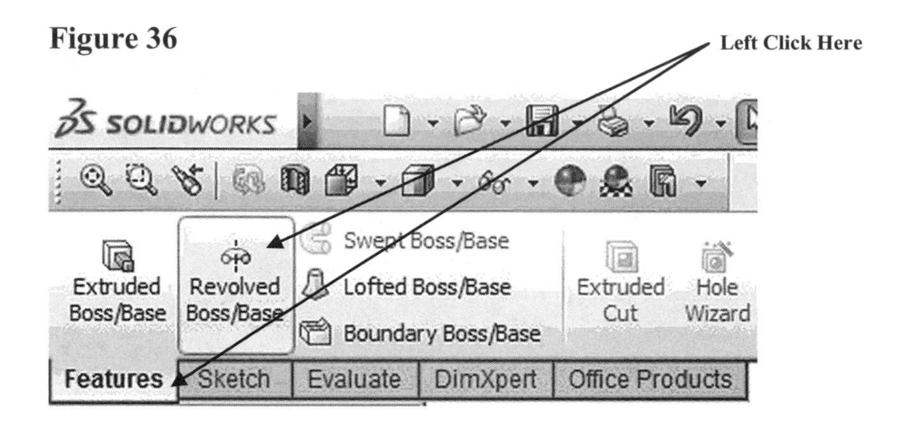

46. If the box to the right of the axis symbol contains text, move the cursor to the "axis" line that was drawn above the part and left click once. The line will turn red. If the box to the right of the axis symbol does not contain text, left click inside the box, then left click on the axis line as shown in Figure 37. It may take several attempts to complete the revolve.

Figure 37

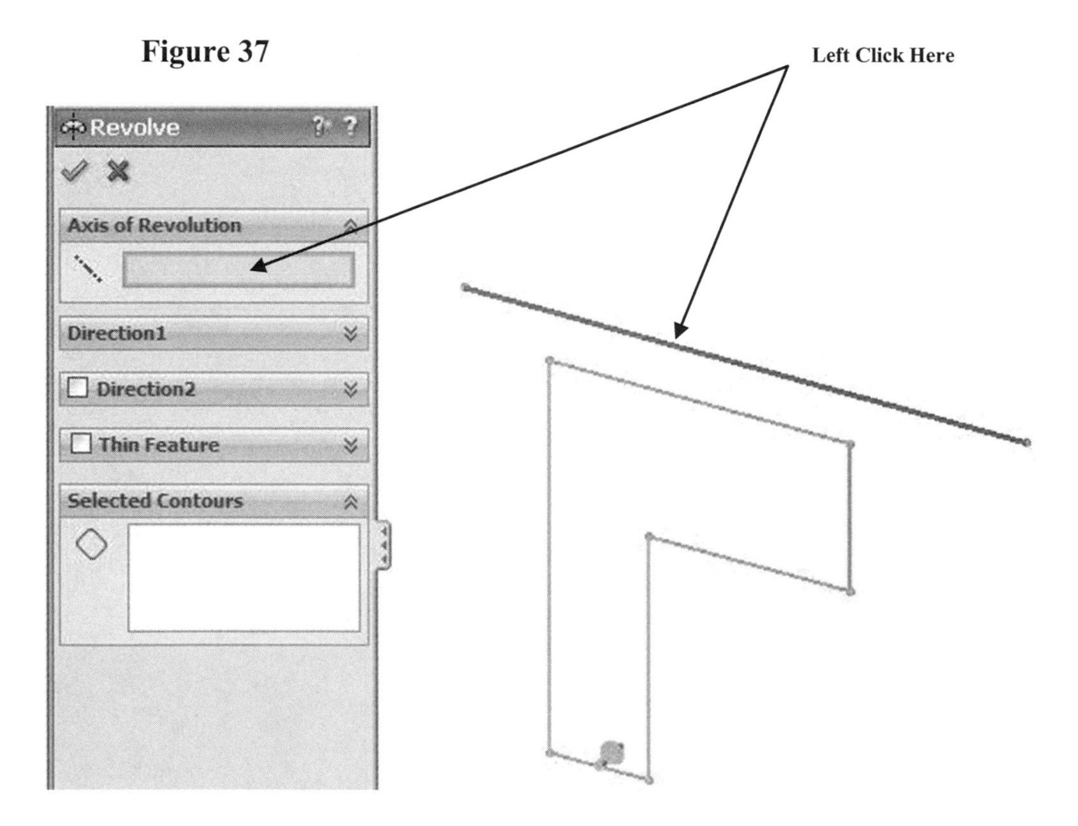

47. Move the cursor inside the profile causing it to turn red.

48. Move the cursor under "Selected Contours" and left click once. Left click inside the profile as shown in Figure 38.

Figure 38

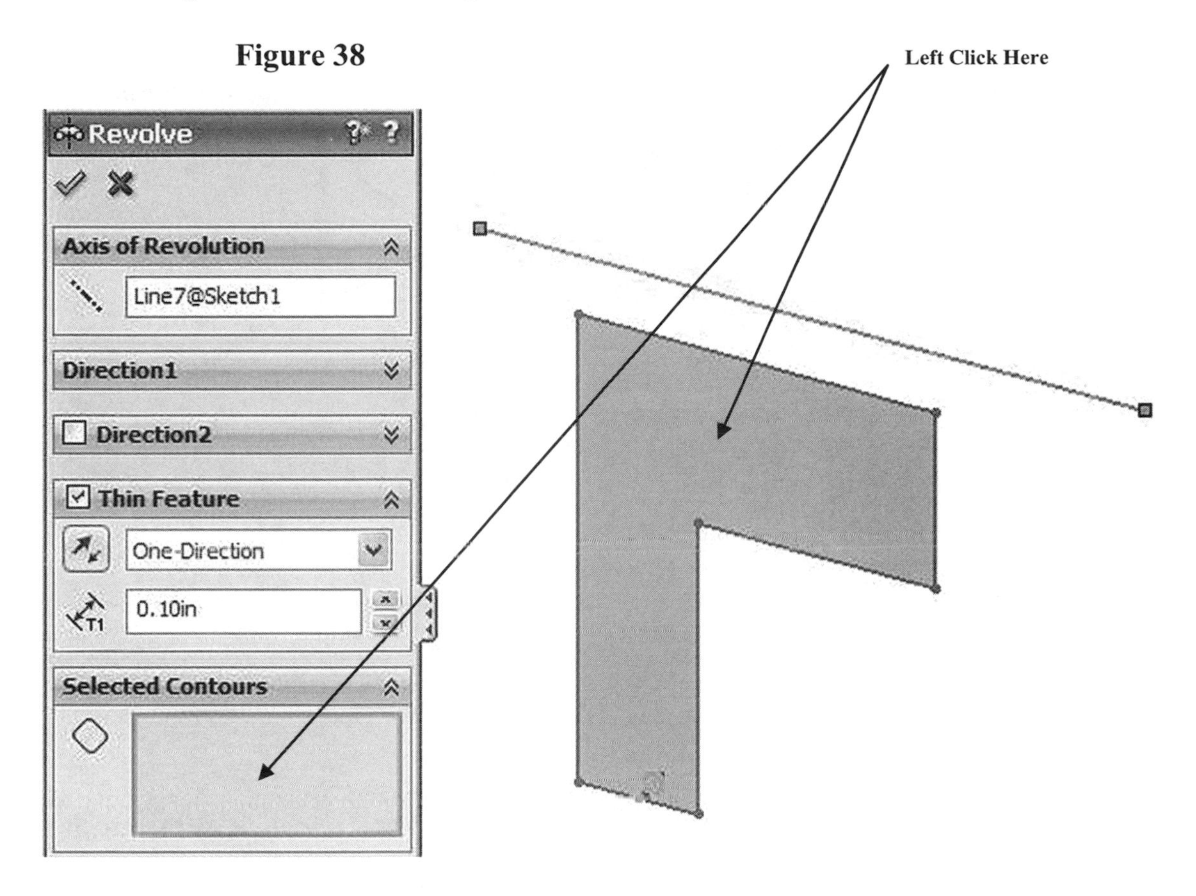

49. A preview of the revolve will appear as shown in Figure 39.

Figure 39

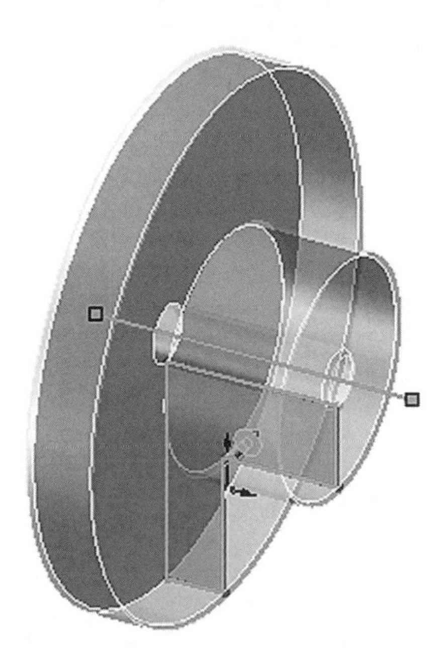

50. Move the cursor to the middle left portion of the screen and left click on the checkmark next to the text "Thin Feature". **Remove the checkmark** (if needed) from the box as shown in Figure 40.

Figure 40

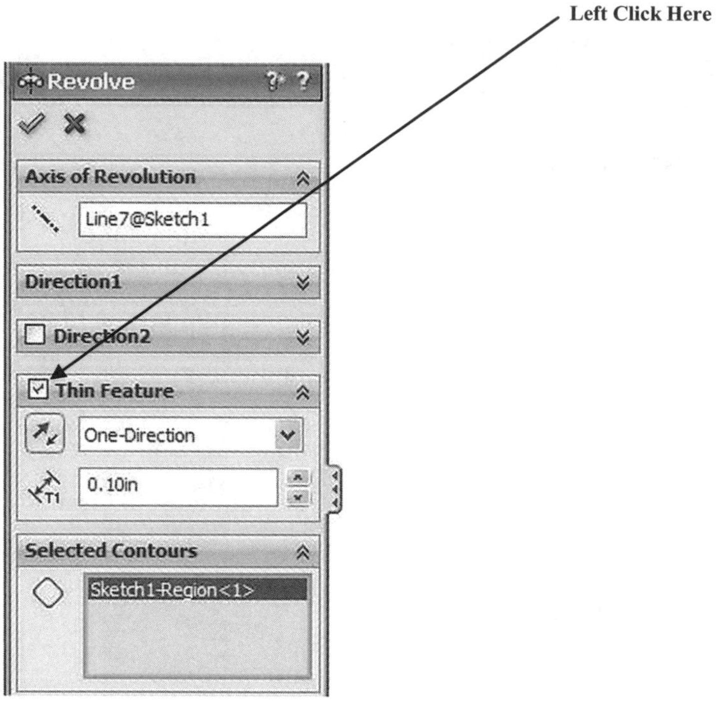

Left Click Here

51. Move the cursor to the upper left portion of the screen and left click on the green checkmark as shown in Figure 41.

Figure 41

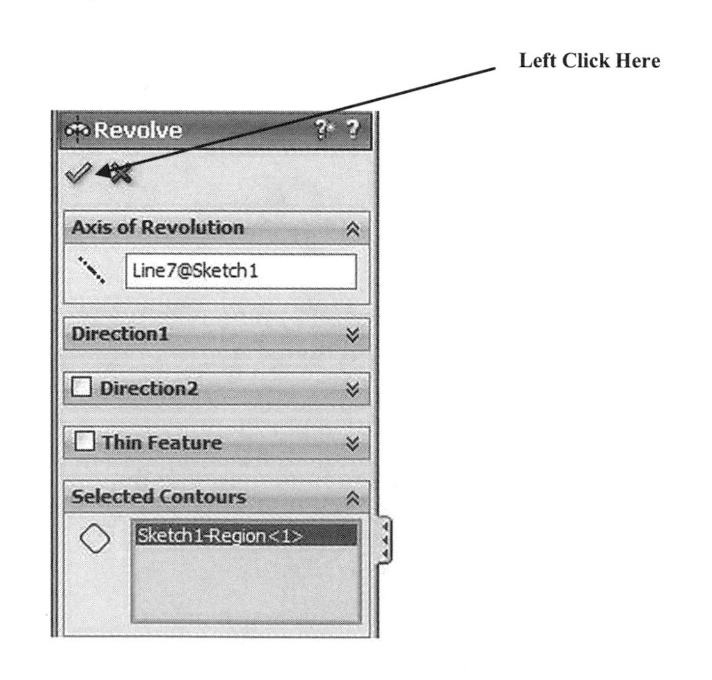

52. Your screen should look similar to Figure 42. You may have to use the zoom out command to view the entire part. Left click anywhere around the drawing.

Figure 42

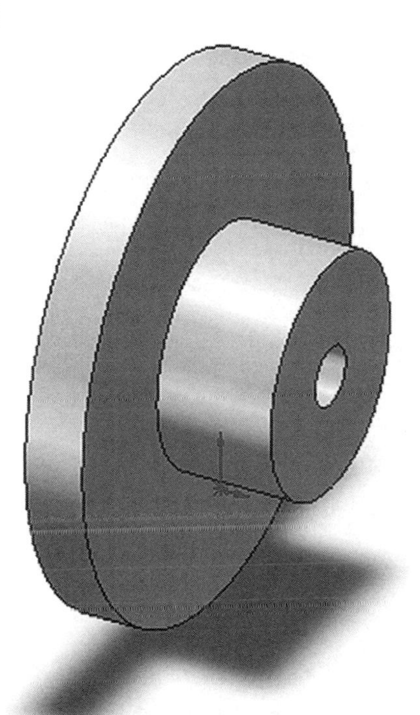

53. Move the cursor to the edge of the part causing the edges to turn red. After the edges become red, right click on the surface as shown in Figure 43.

Figure 43

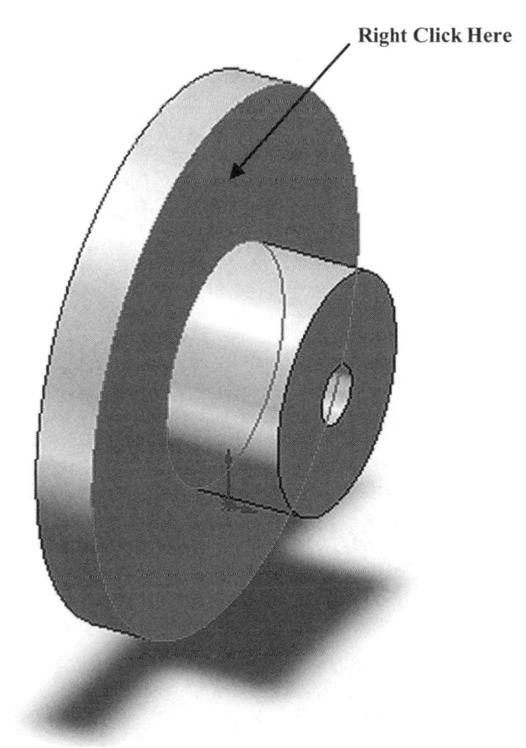

54. The surface will turn green and a pop up menu will appear. Left click on **Insert Sketch** as shown in Figure 44.

Figure 44

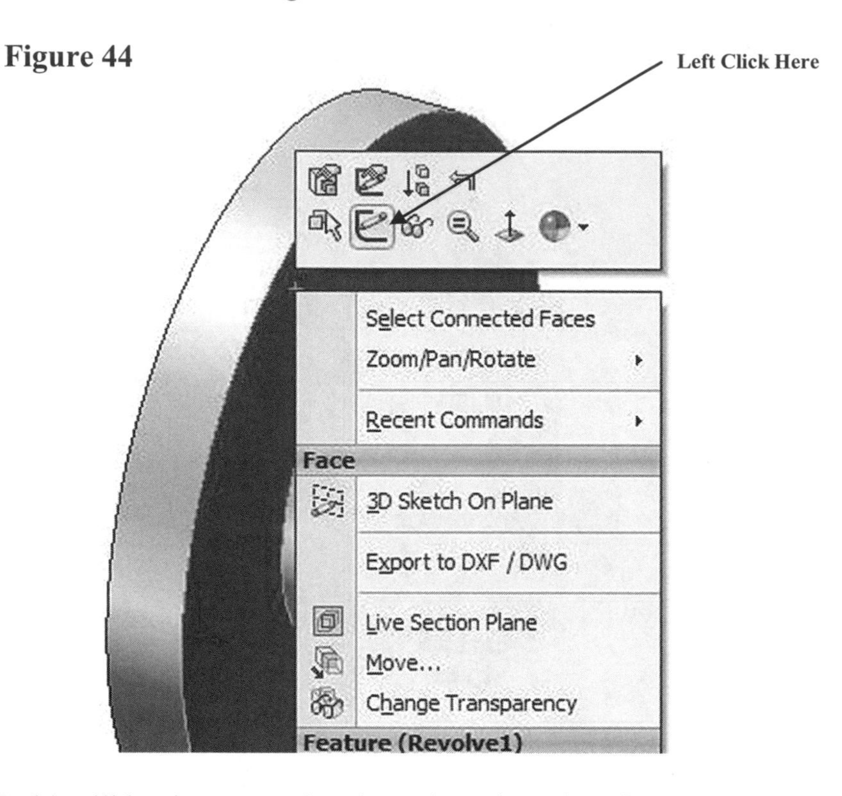

55. SolidWorks will begin a new sketch on the selected surface. Your screen should look similar to Figure 45.

Figure 45

56. Move the cursor to the upper right portion of the screen and left click on the drop down arrow next to the "View Orientation" icon. A drop down box will appear. Left click on **Right** as shown in Figure 46.

Figure 46

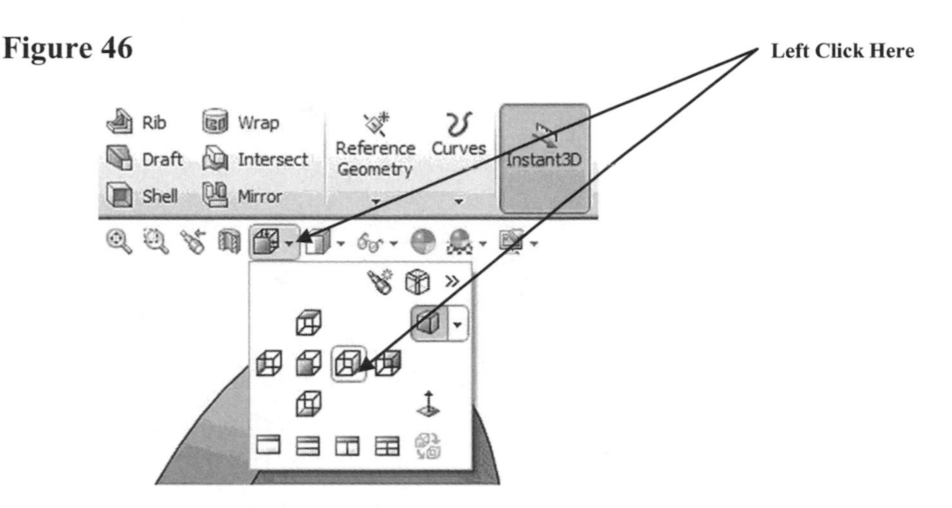

57. SolidWorks will rotate the part to provide a perpendicular view of the selected surface as shown in Figure 47.

Figure 47

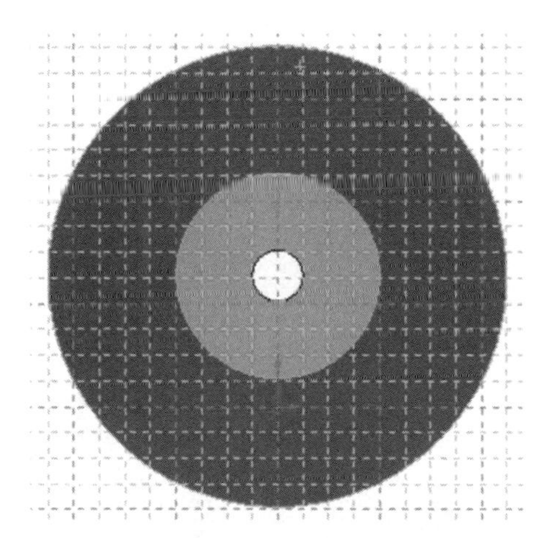

58. Move the cursor to the upper left corner of the screen and left click on **Line** as shown in Figure 48.

Figure 48

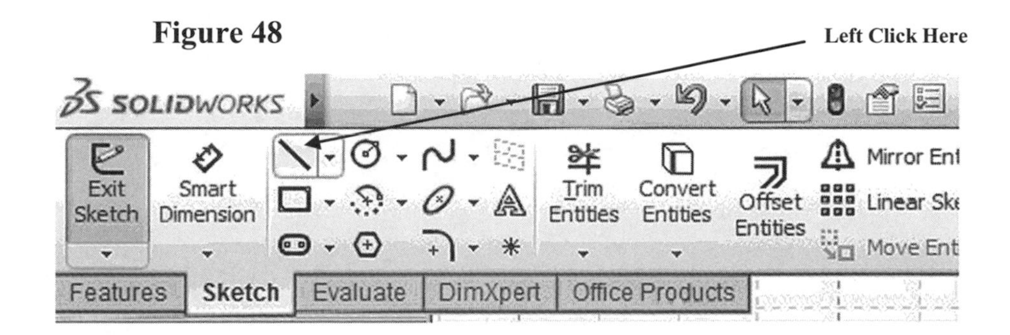

59. Move the cursor over the small circle and wait a few seconds. A center marker will appear. Left click on the center marker as shown in Figure 49.

Figure 49

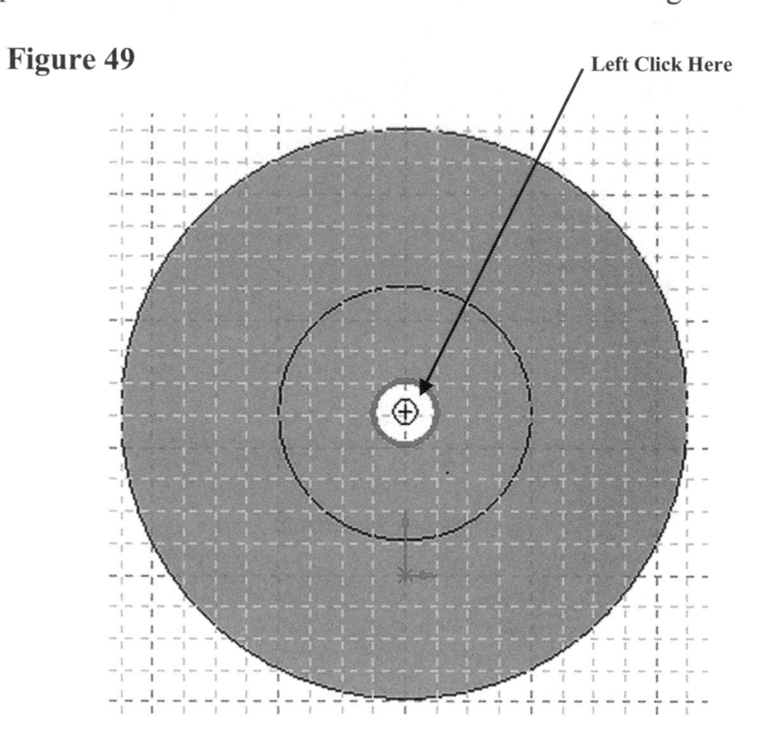

60. Move the cursor straight up and left click as shown in Figure 50.

Figure 50

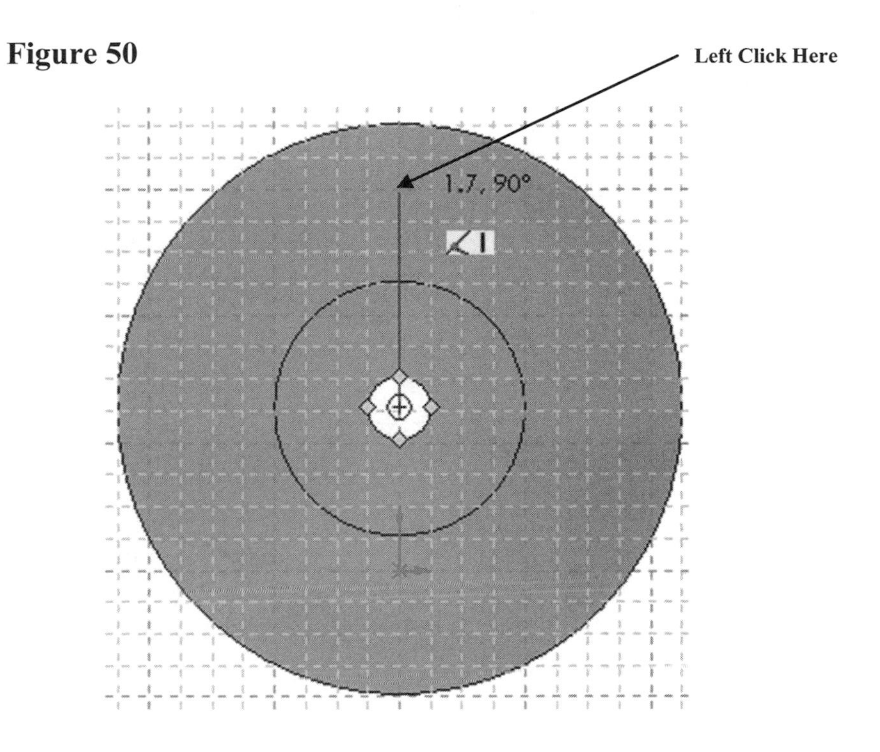

61. Now right click. A pop up menu will appear. Left click on **Select** as shown in Figure 51.

Figure 51

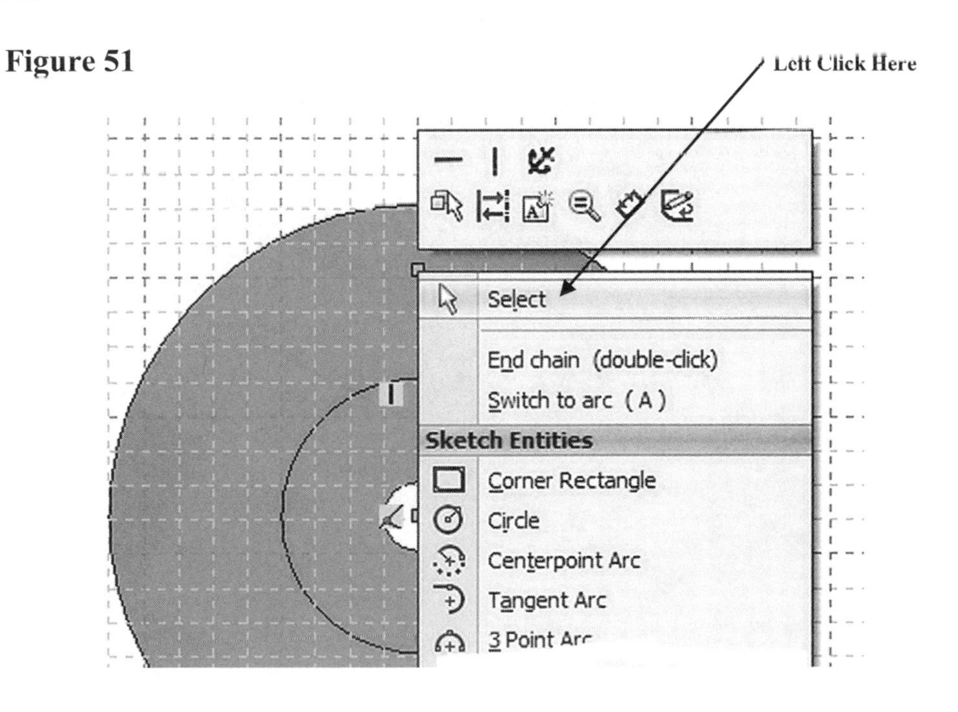

62. Move the cursor to the upper left portion of the screen and left click on **Smart Dimension** as shown in Figure 52.

Figure 52

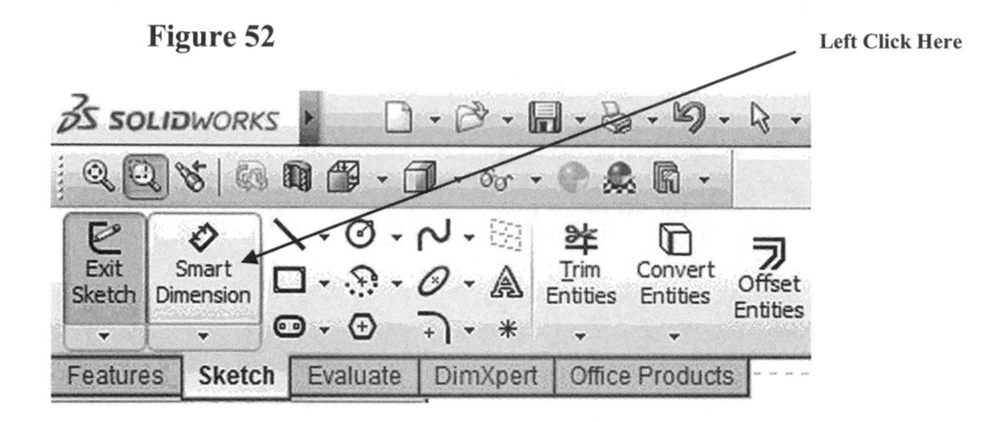

63. Move the cursor over the line that was just drawn. The line will turn red as shown in Figure 53. Select the line by left clicking anywhere on the line **or** on each of the end points. The dimension will be attached to the cursor.

Figure 53

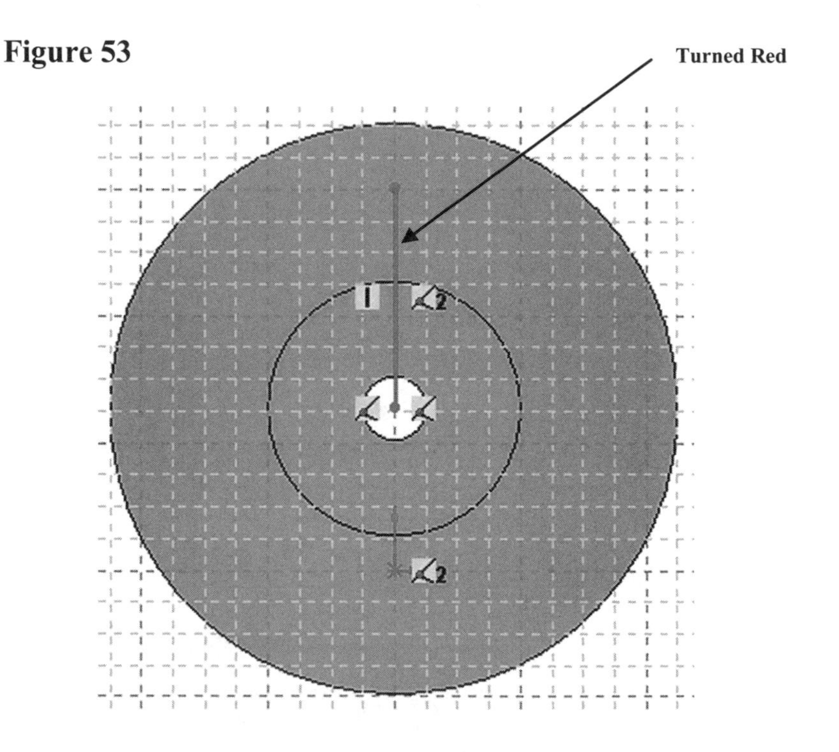

64. Move the cursor to where the dimension will be placed and left click once as shown in Figure 54.

Figure 54

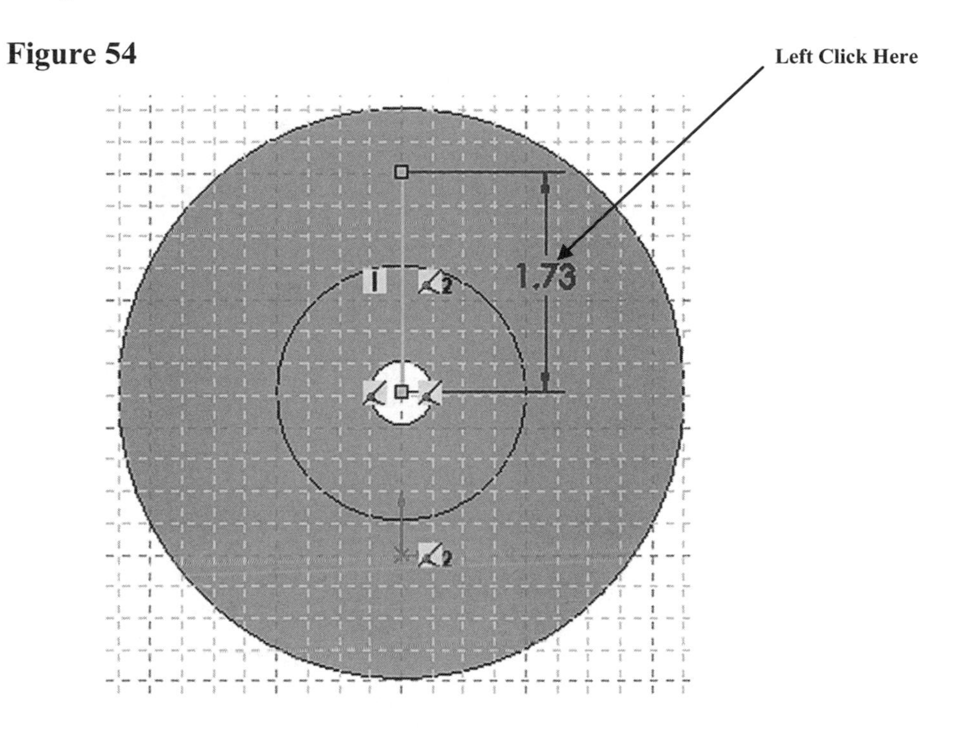

65. The Modify dialog box will appear as shown in Figure 55.

Figure 55

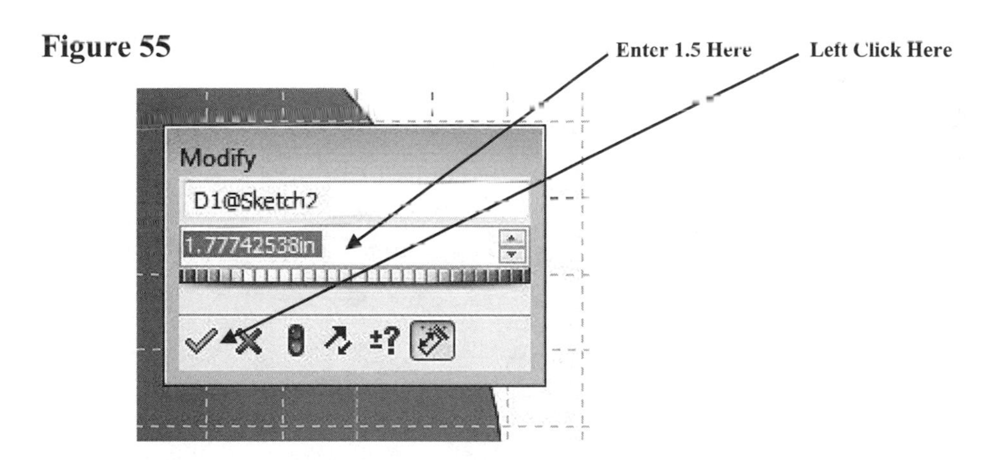

66. To edit the dimension, enter **1.5** in the Modify dialog box (while the current dimension is highlighted) as shown in Figure 55. Left click on the green checkmark or press **Enter** on the keyboard.

67. Right click anywhere on the screen. A pop up menu will appear. Left click on **Select** as shown in Figure 56.

Figure 56

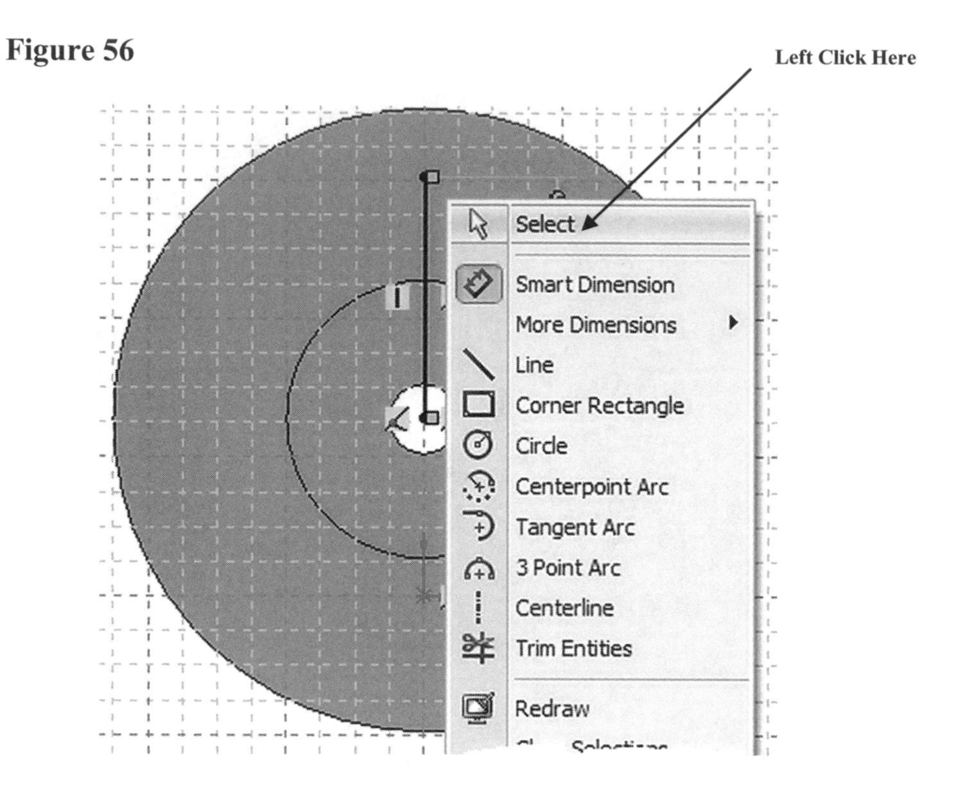

68. The dimension of the line will become 1.5 inches as shown in Figure 55. Use the Zoom icons to zoom out if necessary.

Figure 57

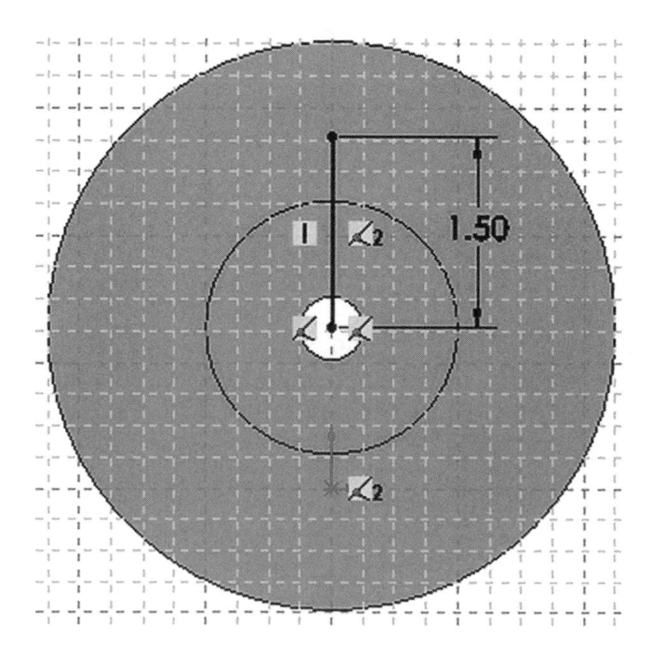

69. Move the cursor to the upper left portion of the screen and left click on **Circle** as shown in Figure 58.

Figure 58

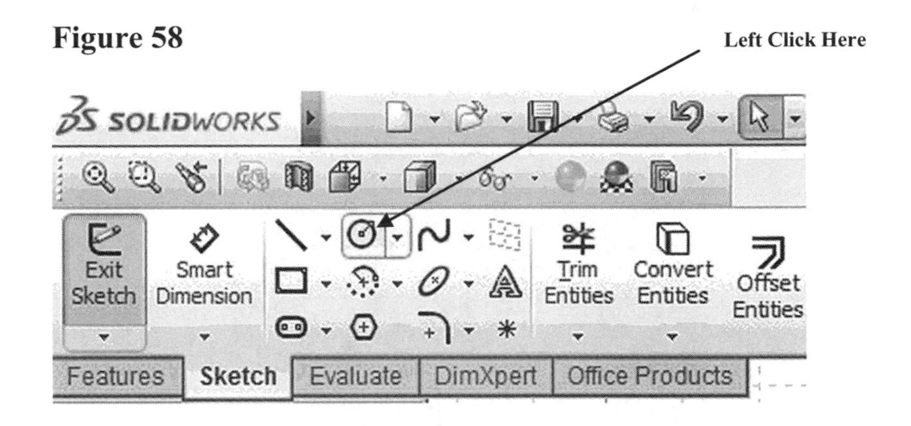

70. Left click on the endpoint of the line as shown in Figure 59.

Figure 59

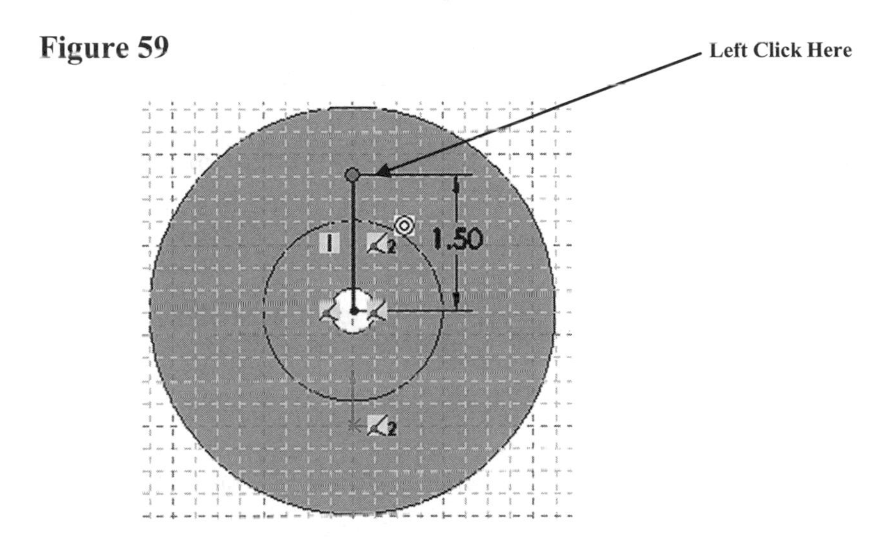

71. Move the cursor out to create a circle as shown in Figure 60.

Figure 60

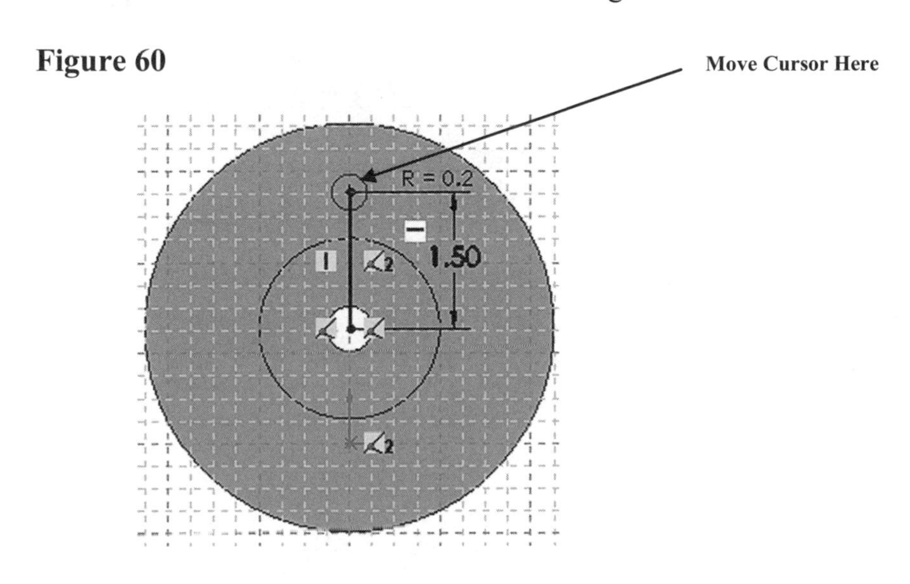

72. Left click as shown in Figure 61.

Figure 61

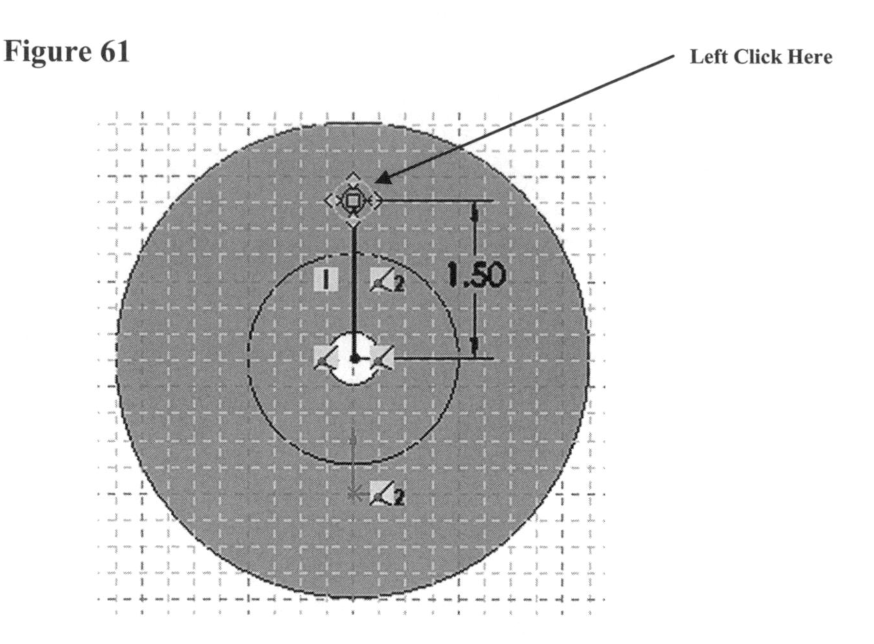

73. Move the cursor to the left portion of the screen and left click on the green checkmark as shown in Figure 62.

Figure 62

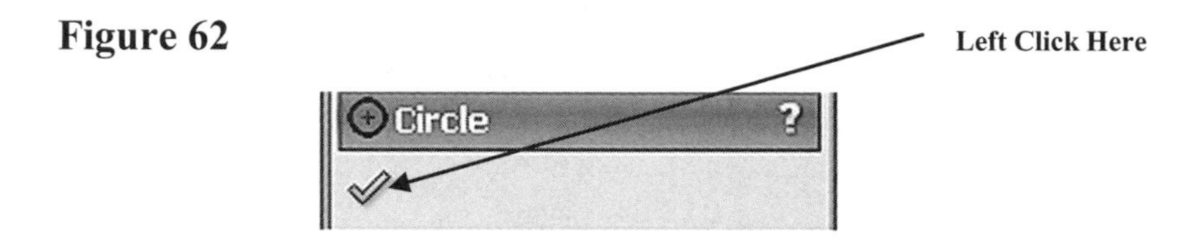

74. Move the cursor to the upper left portion of the screen and left click on **Smart Dimension** as shown in Figure 63.

Figure 63

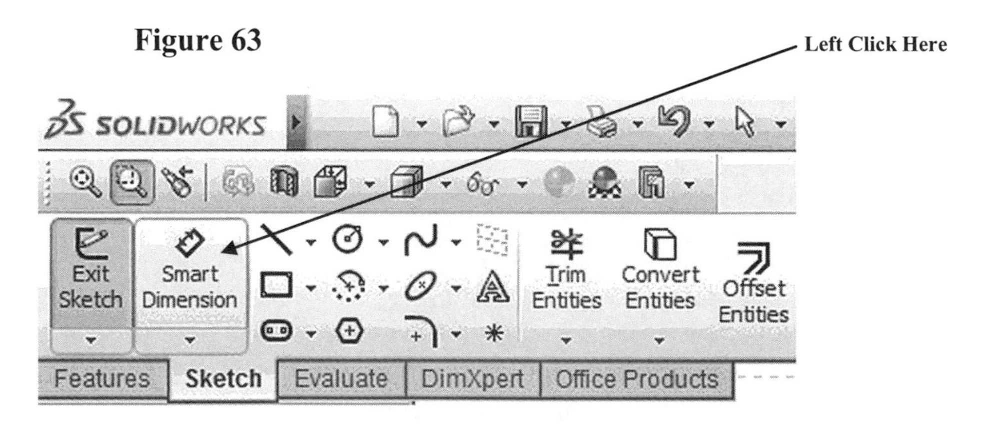

75. Move the cursor to the circle that was just drawn. The circle will turn red. Select the circle by left clicking anywhere on the circle (not the center) as shown in Figure 64. The dimension will be attached to the cursor.

Figure 64

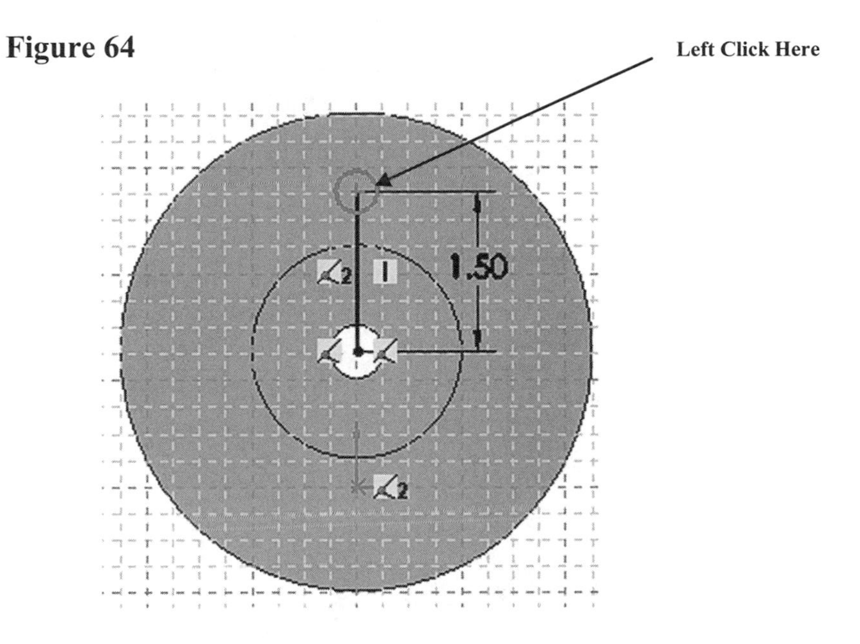

76. Move the cursor to where the dimension will be placed and left click once as shown in Figure 65.

Figure 65

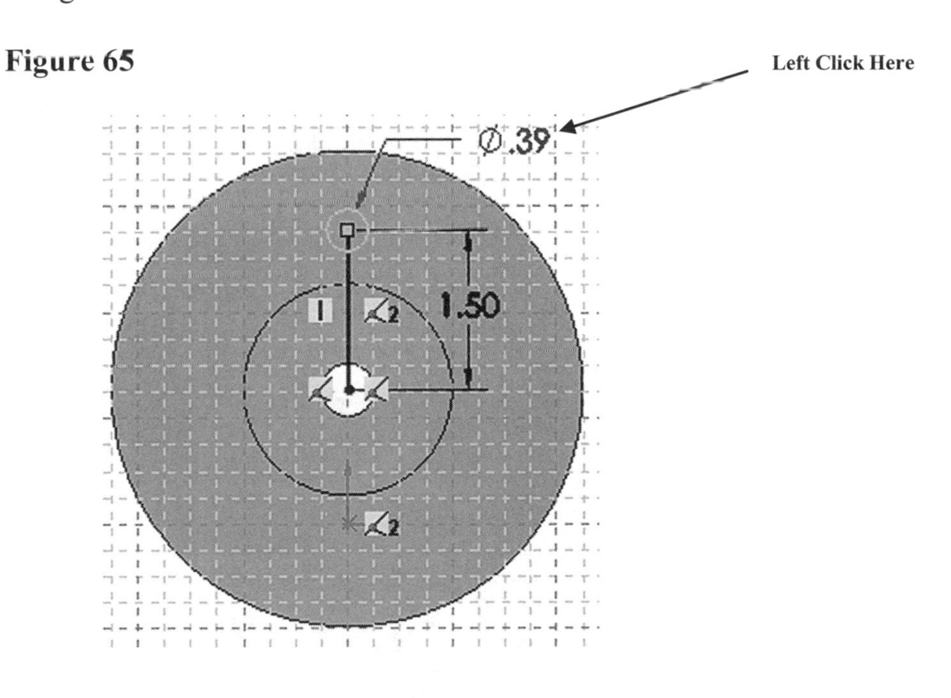

77. The Modify dialog box will appear as shown in Figure 66.

Figure 66

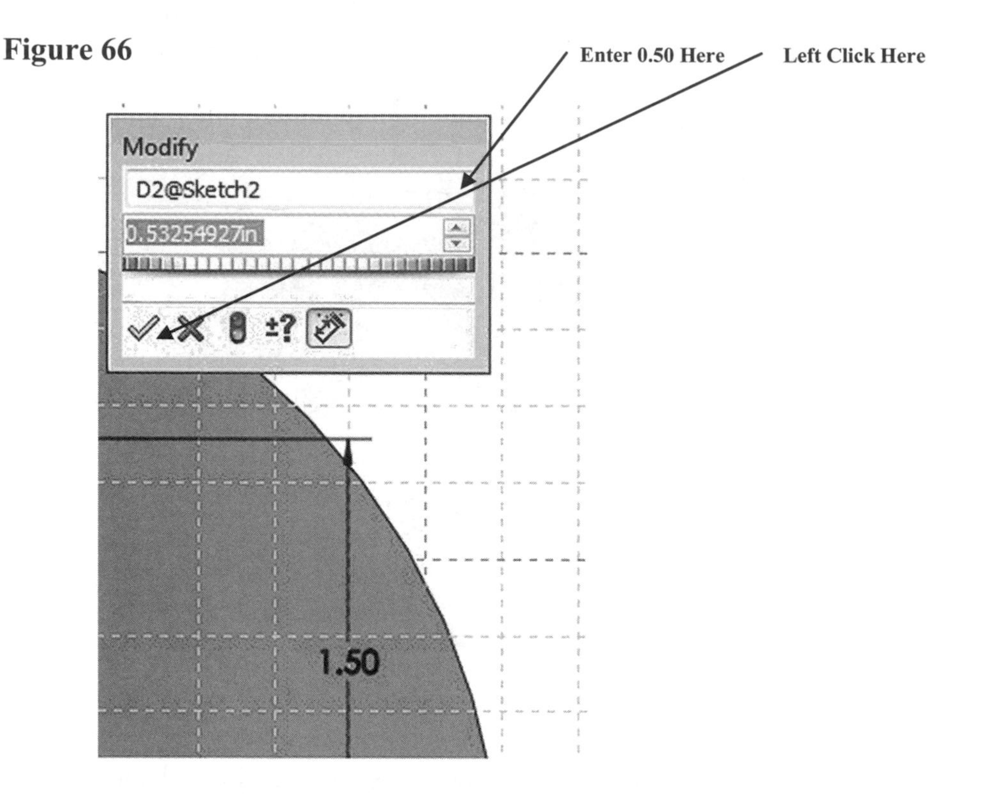

78. To edit the dimension, enter **0.50** in the Modify dialog box (while the current dimension is highlighted) and either press **Enter** on the keyboard or left click on the green checkmark as shown in Figure 66.

79. The dimension of the line will become **0.50** inches as shown in Figure 67. Use the Zoom icons to zoom out if necessary.

Figure 67

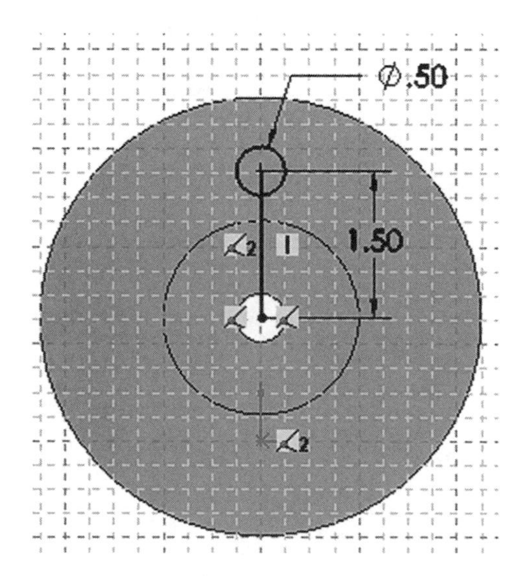

80. Right click anywhere on the screen. A pop up menu will appear. Left click on **Select** as shown in Figure 68.

Figure 68

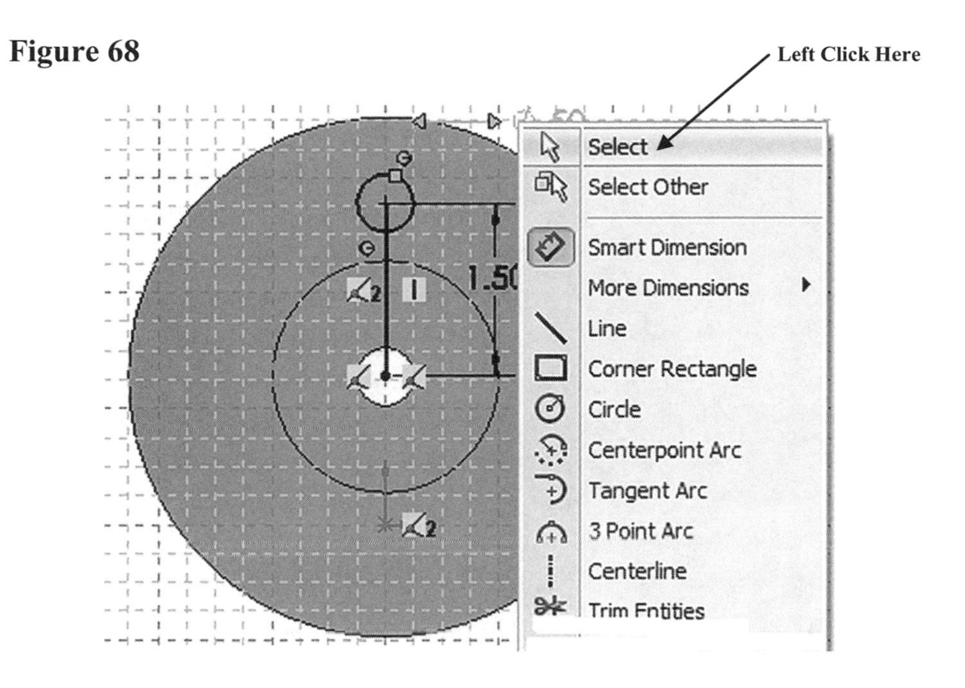

81. Move the cursor to the line that was used to locate the center of the circle. The line will turn red as shown in Figure 69.

Figure 69

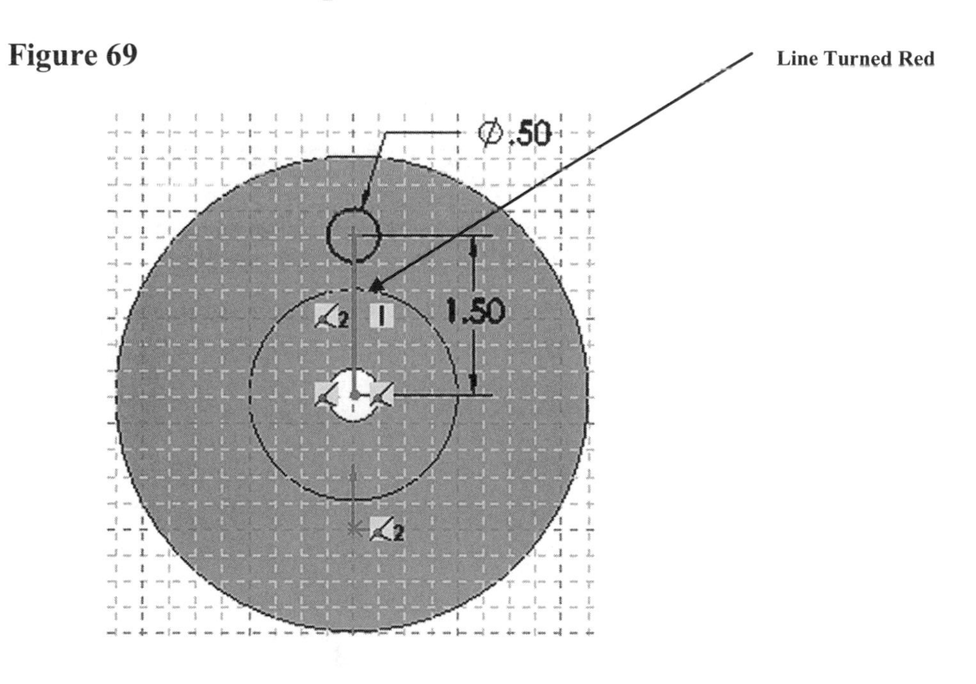

82. Right click on the line after it turns red. A pop up menu will appear as shown in Figure 70.

Figure 70

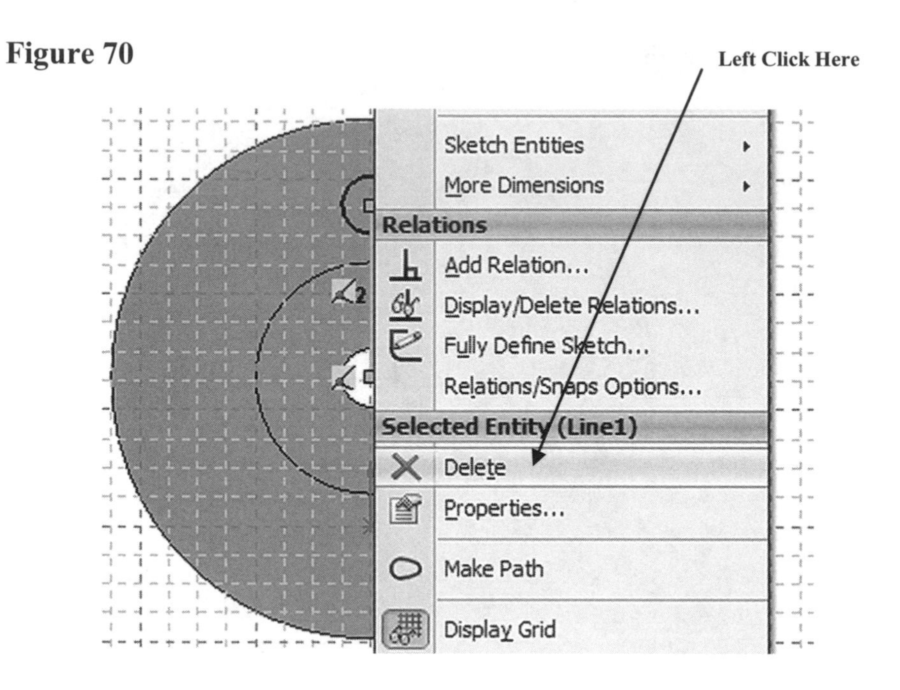

83. Left click on **Delete** as shown in Figure 70.

84. The Sketcher Confirm Delete dialog box will appear. Left click on **Yes** as shown in Figure 71.

Figure 71

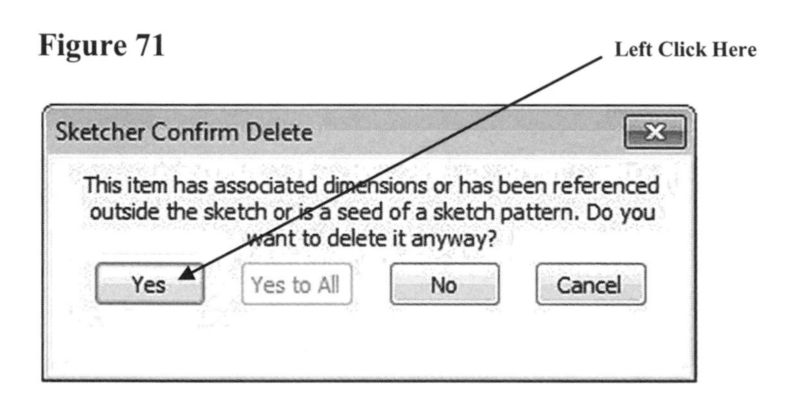

85. After you have verified that no commands are active, move the cursor to the upper left portion of the screen and left click on **Exit Sketch** as shown in Figure 72.

Figure 72

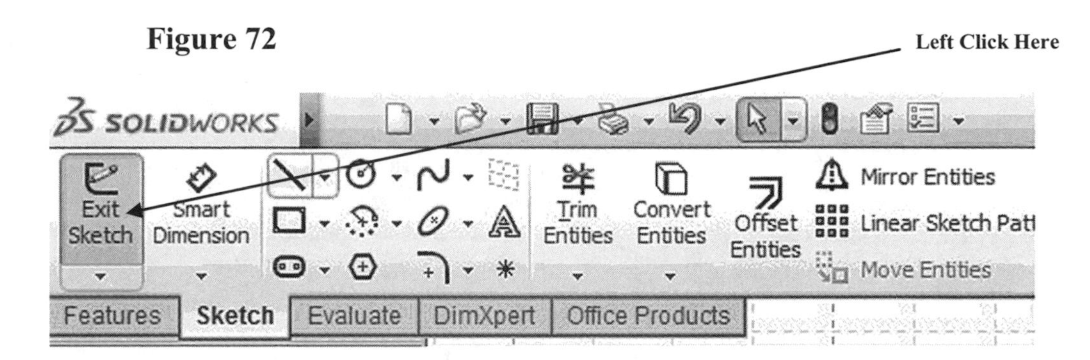

86. SolidWorks is now out of the Sketch commands and into the Features commands. Notice that the commands at the top of the screen are now different. To work in the Features commands a sketch must be present and have no opens (non-connected lines). If there are any opens in the sketch an error message will appear. Your screen should look similar to Figure 73.

Figure 73

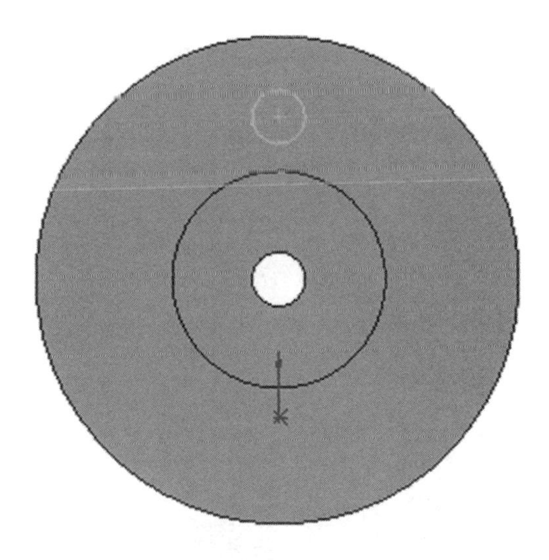

87. Move the cursor to the upper right portion of the screen and left click on the drop down arrow next to the "Views Orientation" icon. A drop down box will appear. Left click on **Trimetric** as shown in Figure 74.

Figure 74

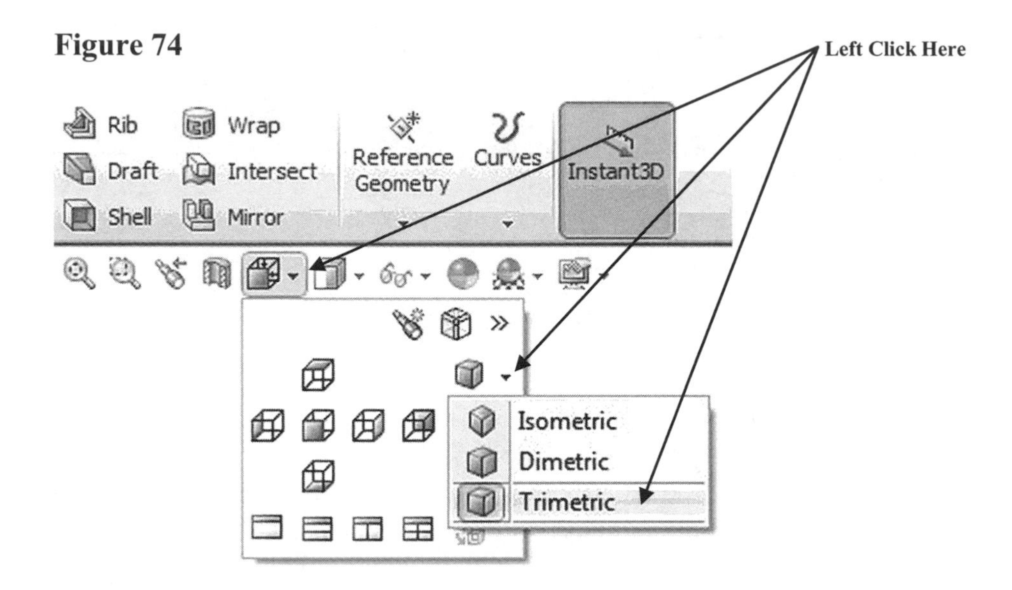

88. The view will become trimetric as shown in Figure 75.

Figure 75

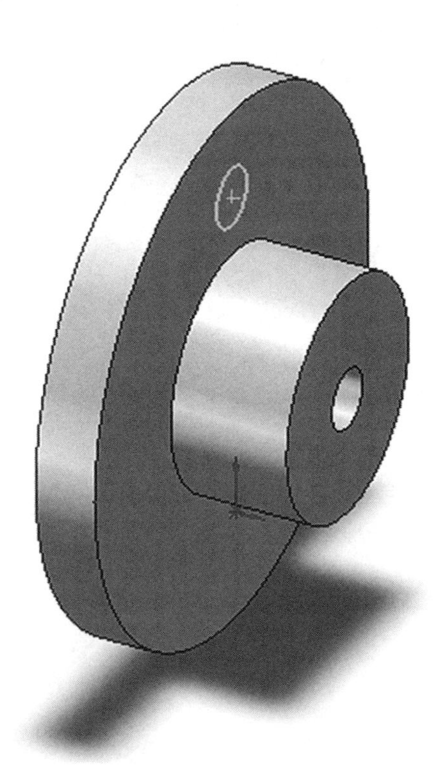

89. Move the cursor to the upper middle portion of the screen and left click on **Extruded Cut.** If the **Extruded Cut** icon is not visible, left click on the **Features** tab as shown in Figure 76.

Figure 76

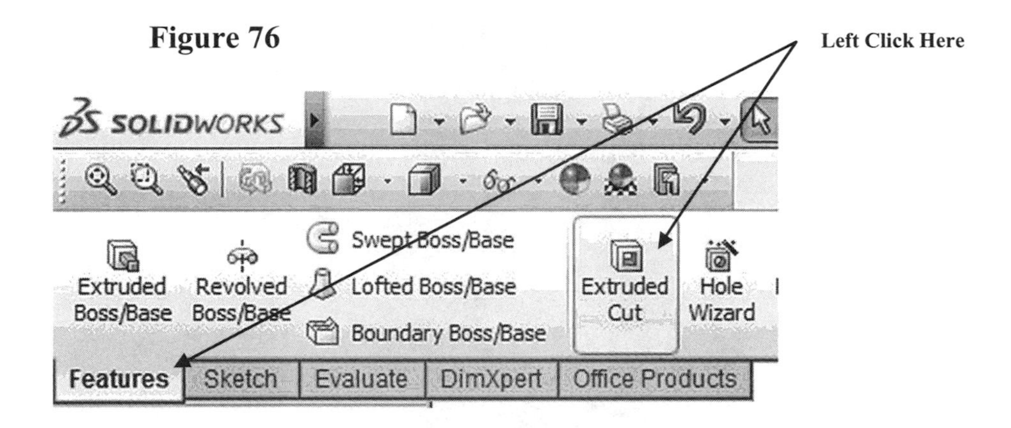

90. If SolidWorks did not provide a preview of the Extruded Cut, move the cursor to the edge of the circle. The edge will turn red as shown in Figure 77.

Figure 77

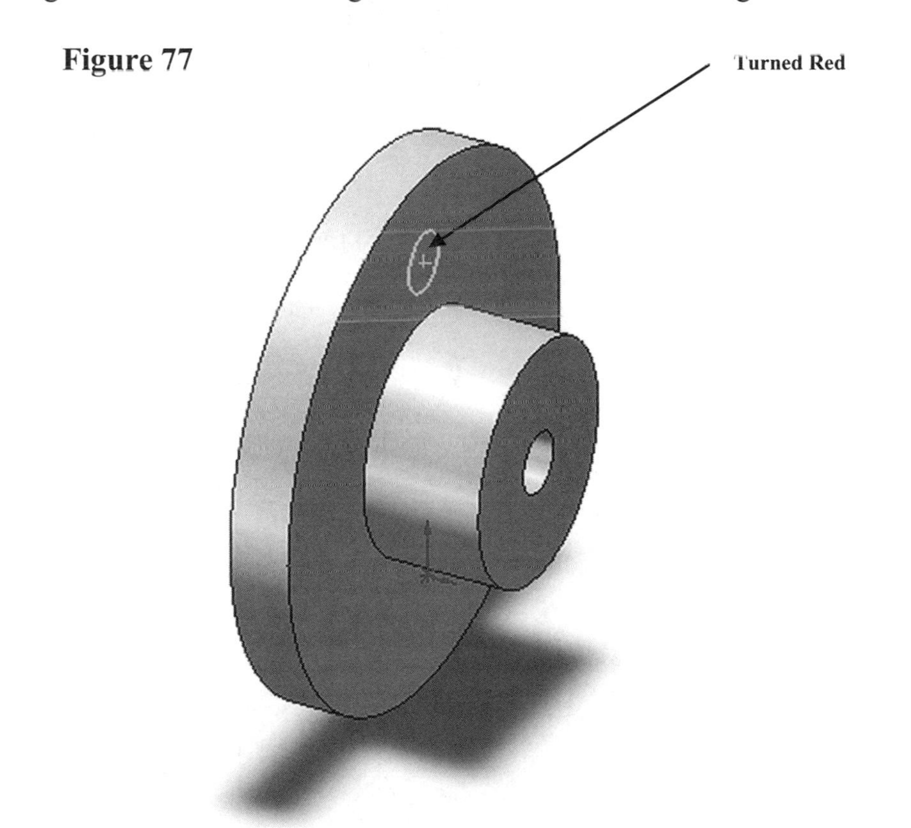

91. After the hole turns red, left click once.

92. A preview of the hole is displayed as shown in Figure 78.

Figure 78

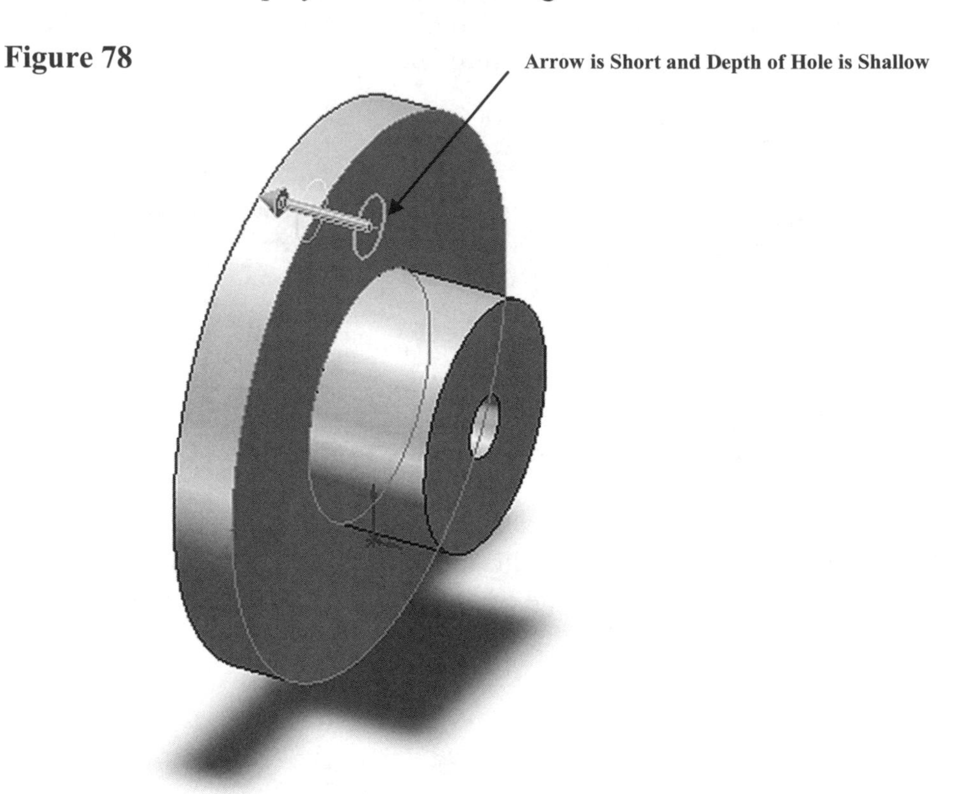

93. Move the cursor to the middle left portion of the screen. Highlight the text located next to D1 (as shown in Figure 79) and enter **.75** and press the enter key on the keyboard.

Figure 79

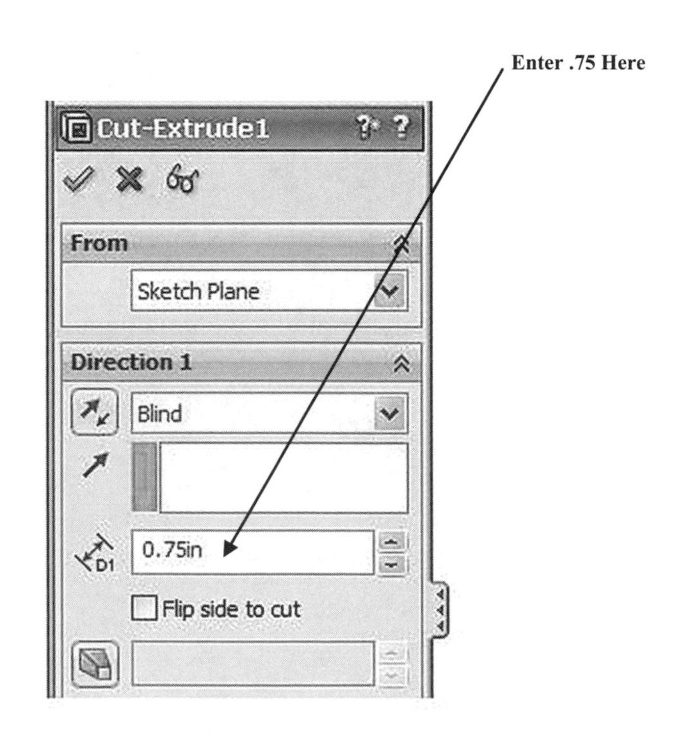

94. Notice that the depth of the hole has changed in the preview. The arrow illustrating the hole depth is now longer as shown in Figure 80.

Figure 80

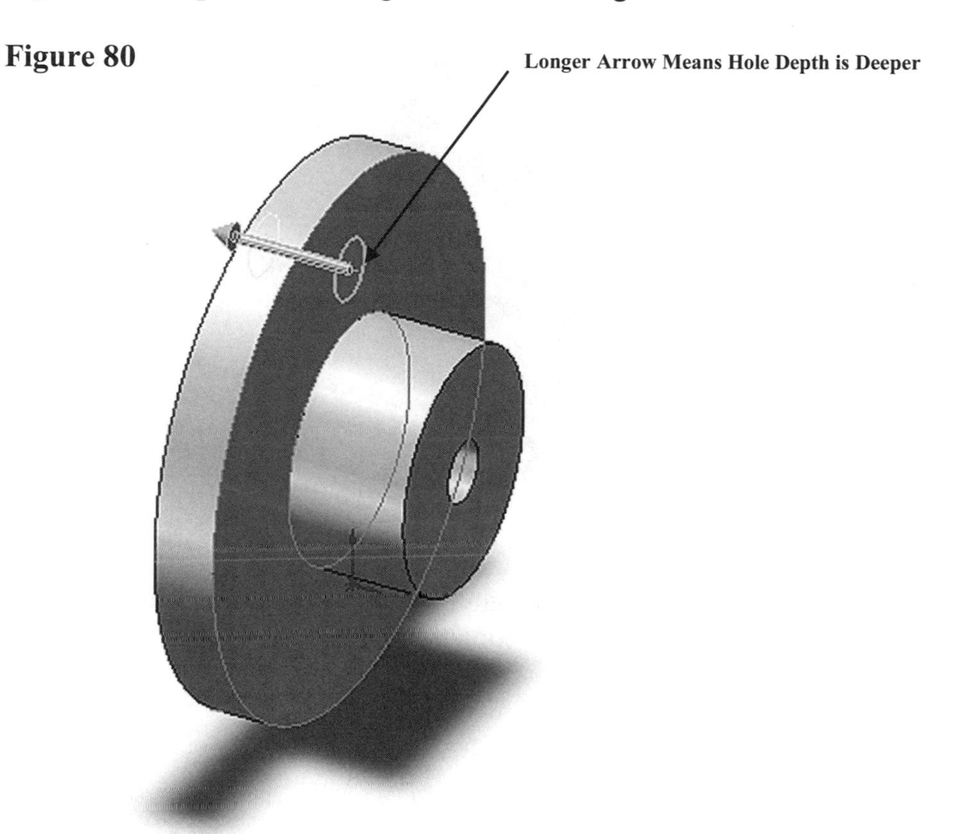

95. Left click on the green checkmark as shown in Figure 81.

Figure 81

96. Your screen should look similar to Figure 82.

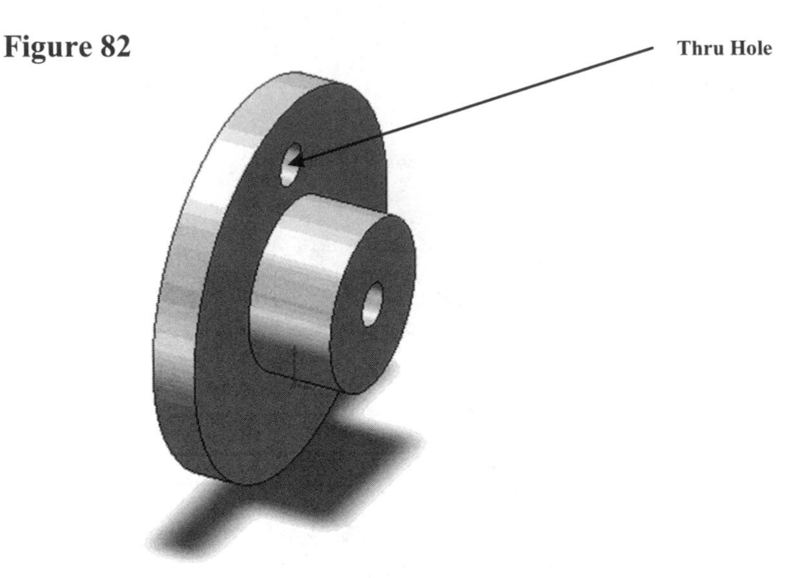

Figure 82

97. Move the cursor to the far right portion of the screen and left click on the arrow located below Reference Geometry as shown in Figure 83.

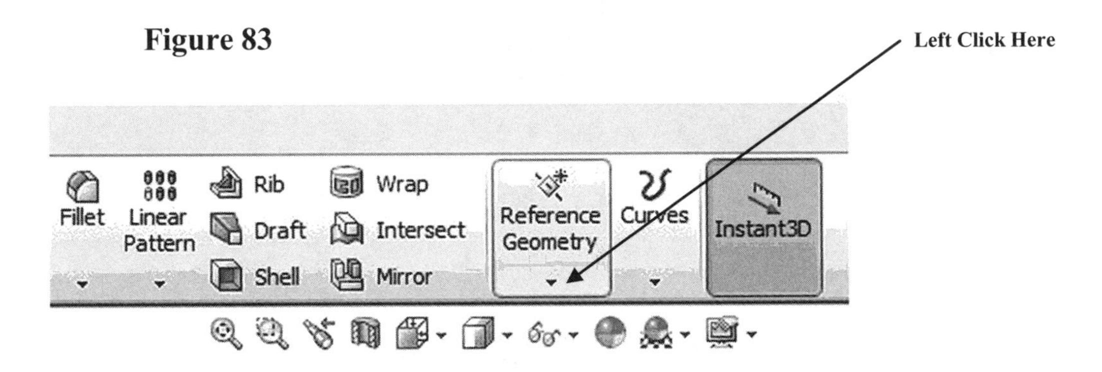

Figure 83

98. A drop down menu will appear. Left click on **Axis** as shown in Figure 84.

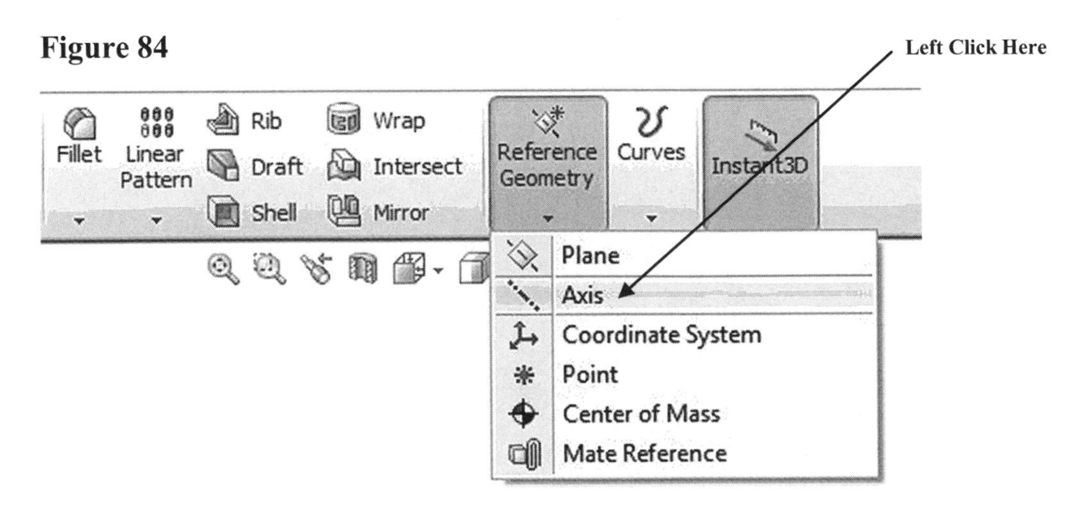

Figure 84

99. Move the cursor over the center of the center hole causing it to turn red as shown in Figure 85.

Figure 85

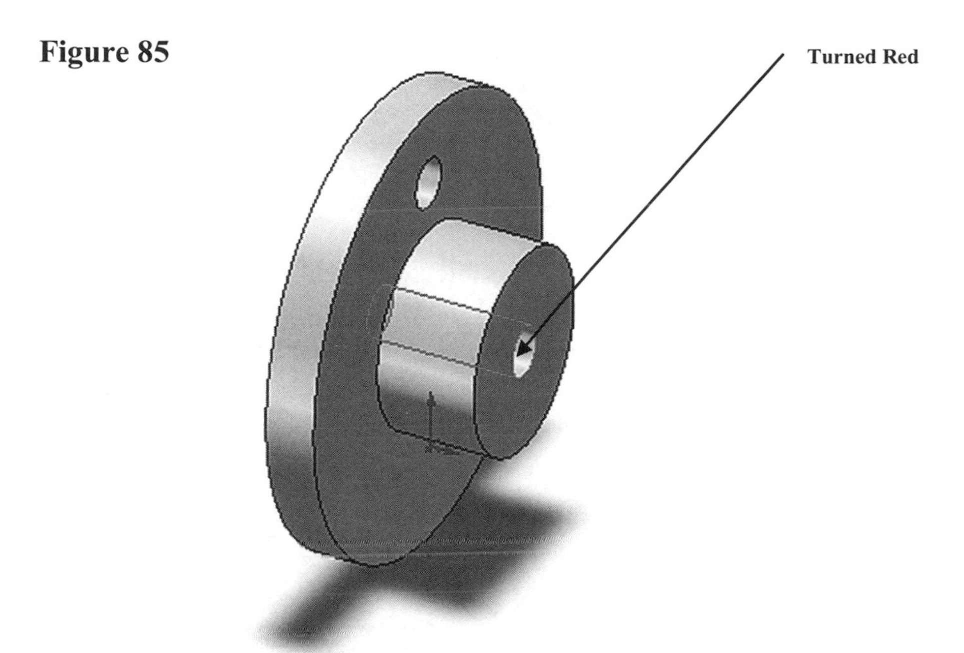

100. After the inside of the hole turns red, left click once. A center axis will be placed inside the hole as shown in Figure 86.

Figure 86

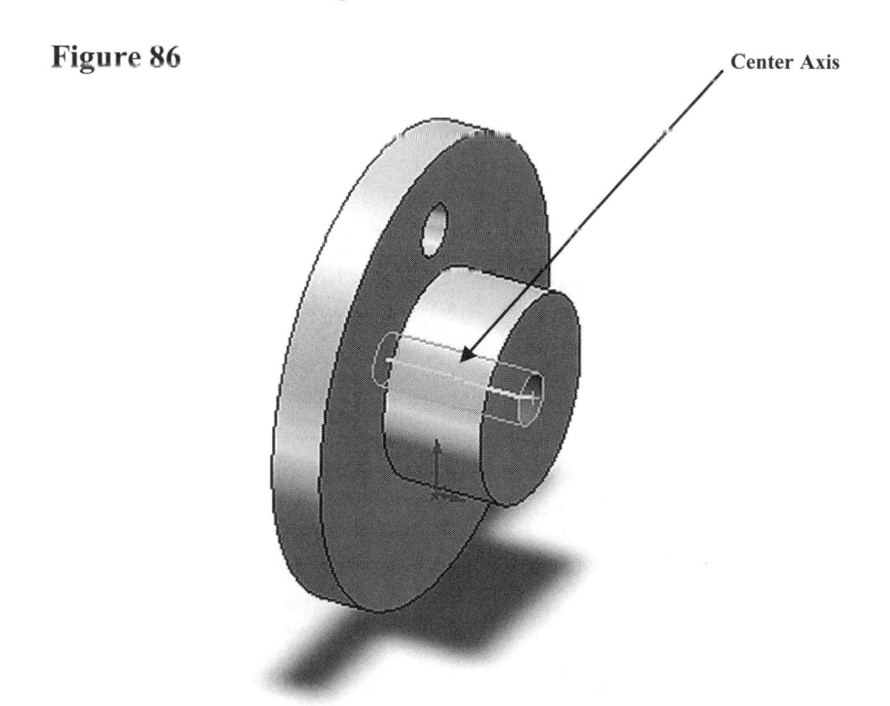

101. Move the cursor to the middle left portion of the screen and left click on the green checkmark as shown in Figure 87.

Figure 87

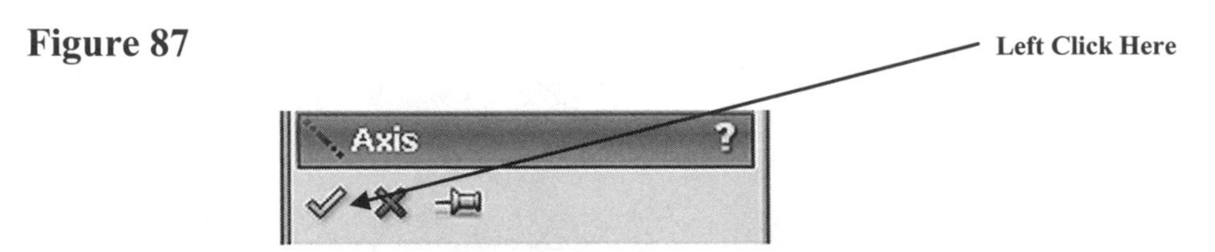

102. Move the cursor to the upper middle portion of the screen and left click on the drop down arrow located below Linear Pattern. A drop down menu will appear. Left click on **Circular Pattern** as shown in Figure 88.

Figure 88

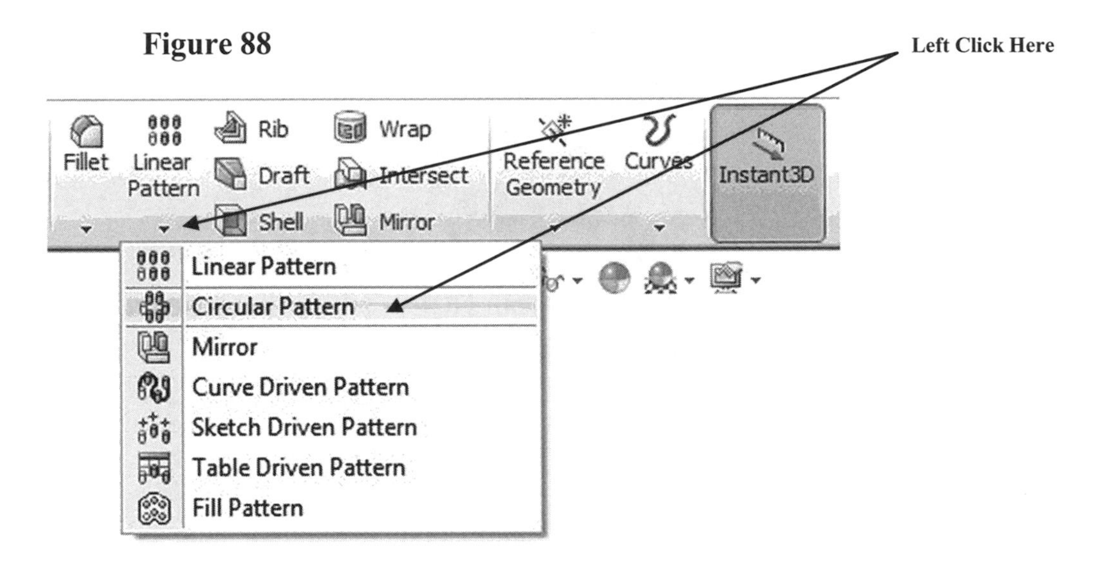

103. Move the cursor to the inside edge of the hole. The outside edge of the entire hole will turn red. Left click once. Left click on the axis as shown in Figure 89.

Figure 89

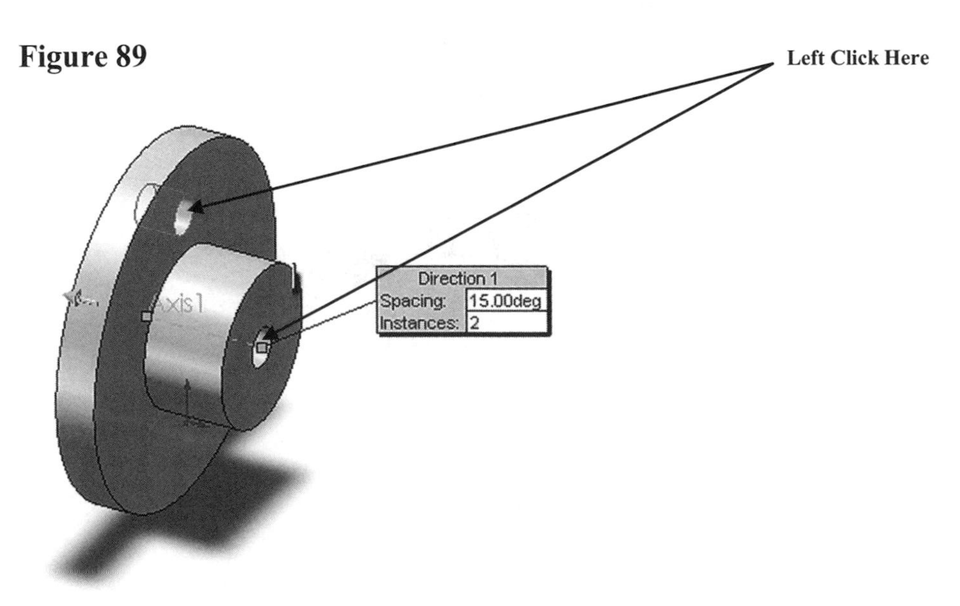

104. Move the cursor to the left portion of the screen. Place a checkmark in the box next to the text "Equal Spacing". The degree amount automatically changes to 360 degrees. Enter **6** for the number of holes directly below 360 and press **Enter** on the keyboard as shown in Figure 90.

Figure 90

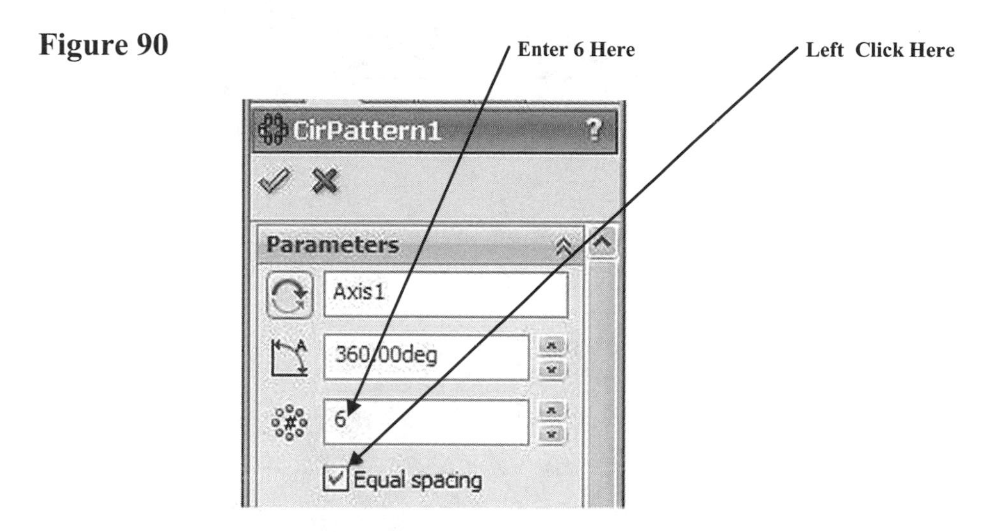

105. A preview of the circular pattern will be displayed as shown in Figure 91.

Figure 91

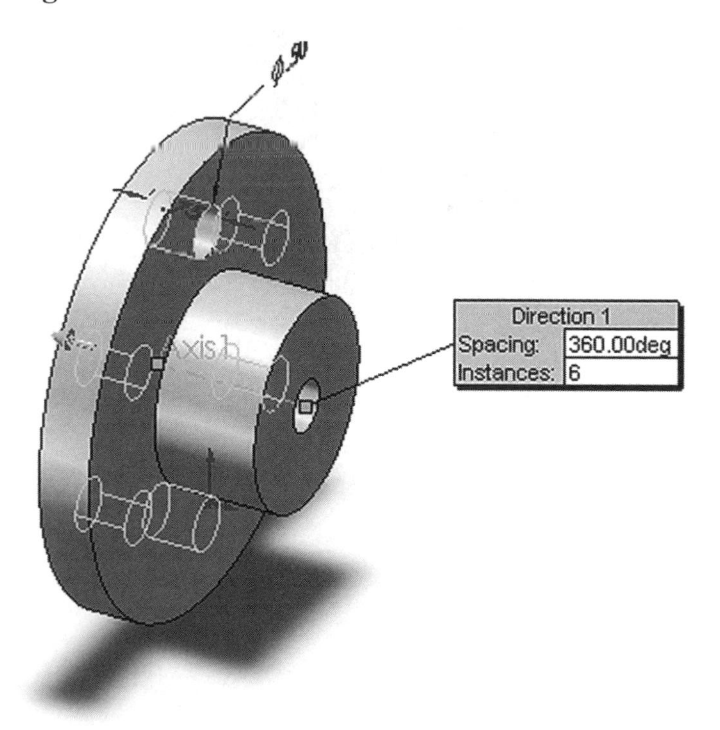

106. Move the cursor to the upper left portion of the screen and left click on the green checkmark as shown in Figure 92.

Figure 92

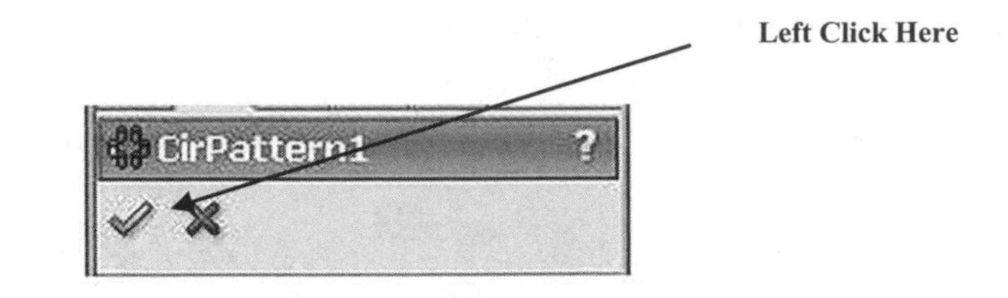

107. Your screen should look similar to Figure 93.

Figure 93

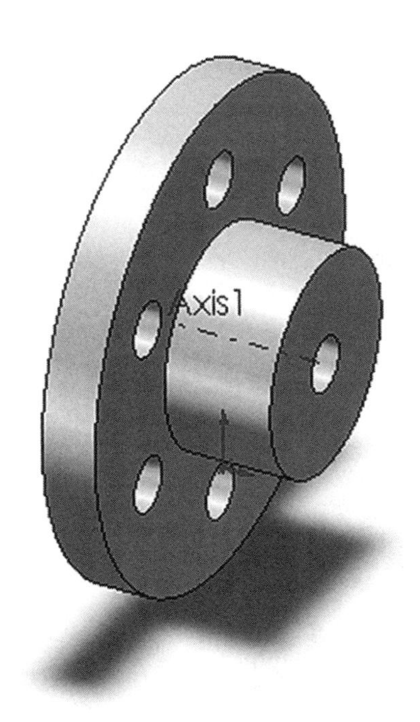

108. To ensure that the holes are correct, push the mouse wheel down (holding the mouse wheel down) and move the cursor around. The model will be attached to the cursor. For SolidWorks 2009 and earlier users, move the cursor to the upper middle portion of the screen and Right click on the drop down arrow to the right of the View Orientation icon. A drop down menu will appear. Place a check in the box (with the left mouse button) next to the "Rotate" icon as shown in Figure 94. The Rotate icon will now appear with the Viewing tools located at the top of the screen in the drawing area. The Rotate command can also be made active by pressing the mouse wheel down and moving the cursor to achieve the desired view.

Figure 94

109. The part will rotate as shown in Figure 95.

Figure 95

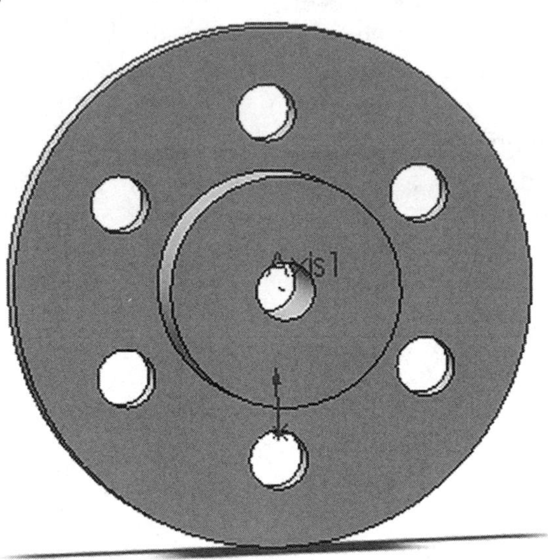

110. Right click around the part. A pop up menu will appear. Left click on **Select**. Other viewing options are shown in the pop up menu in Figure 96.

Figure 96

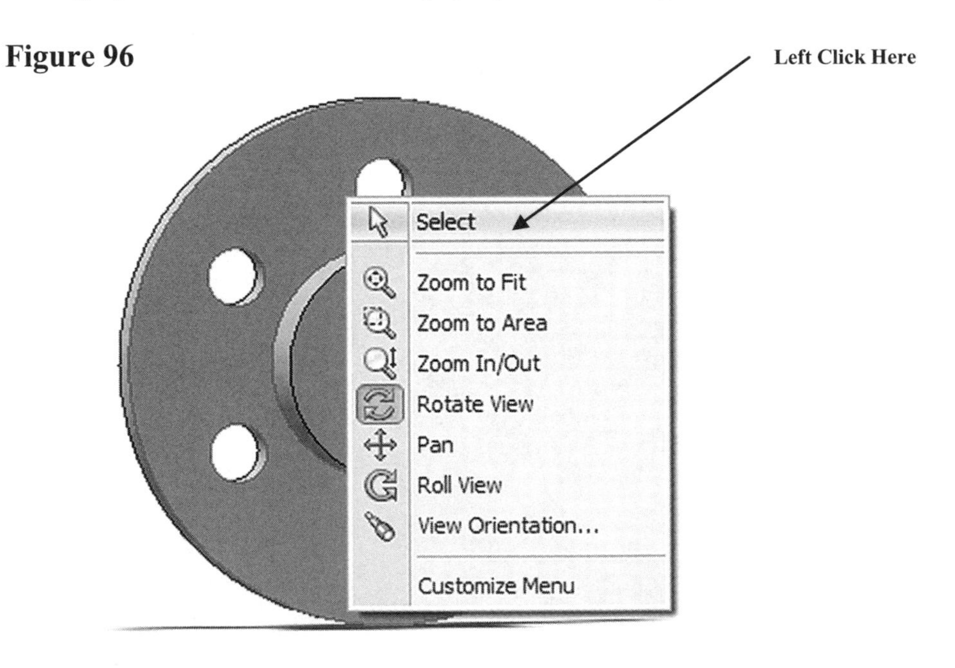

Drawing Activities

Problem 1

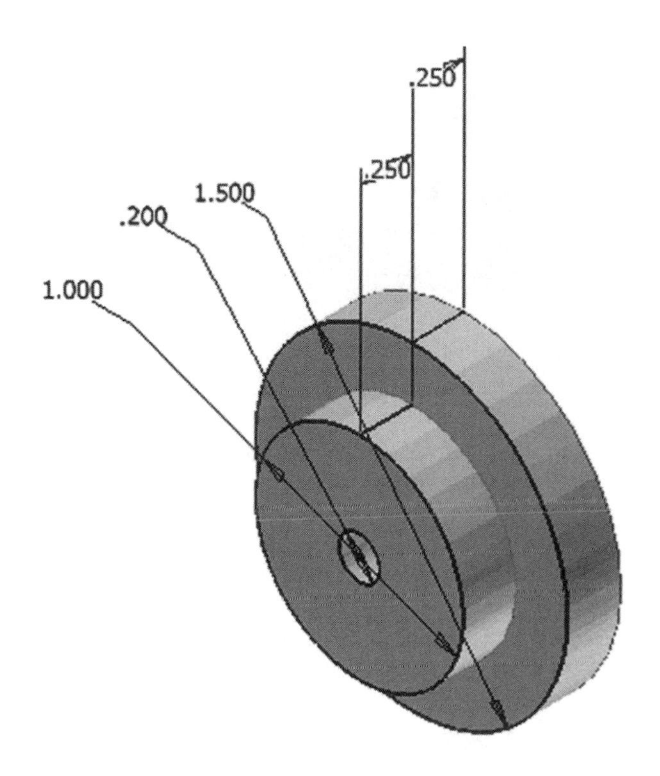

Problem 2

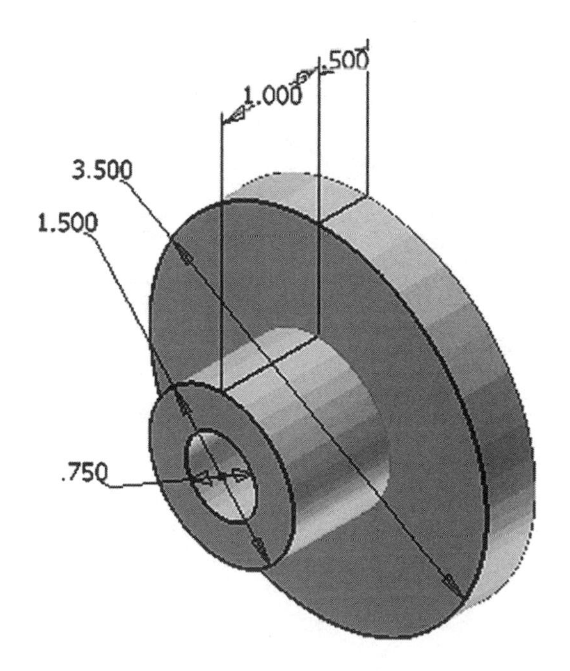

Problem 3

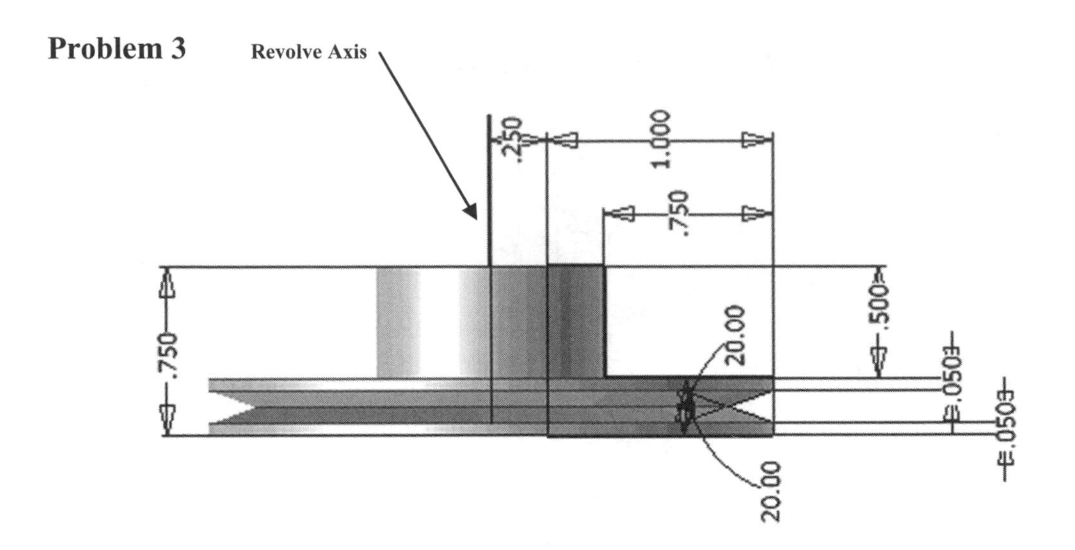

Problem 4

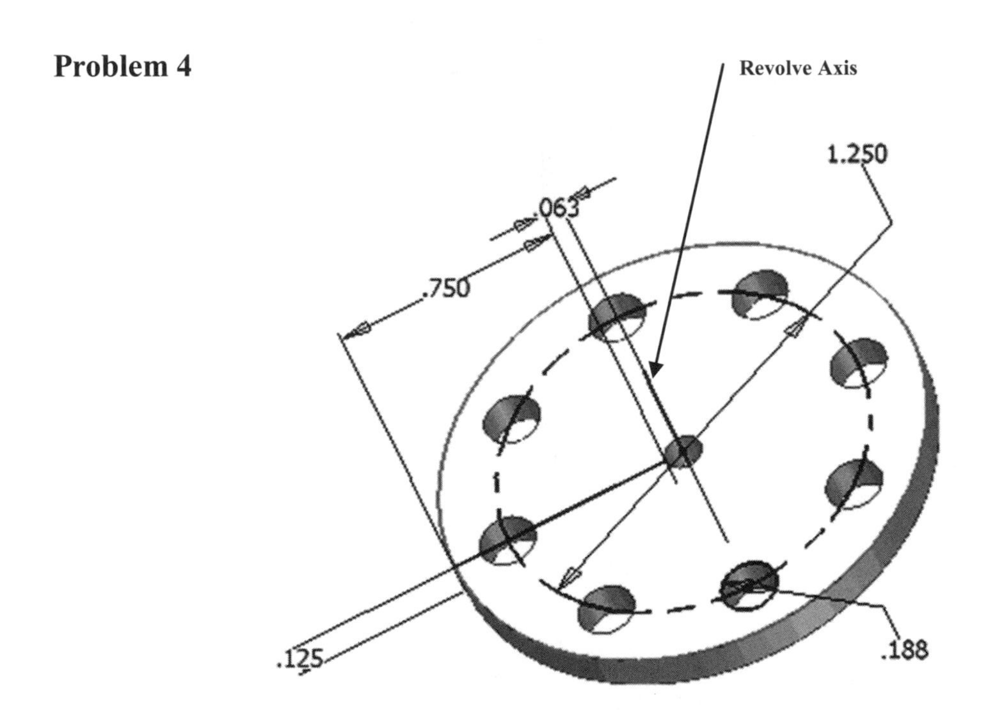

Problem 5

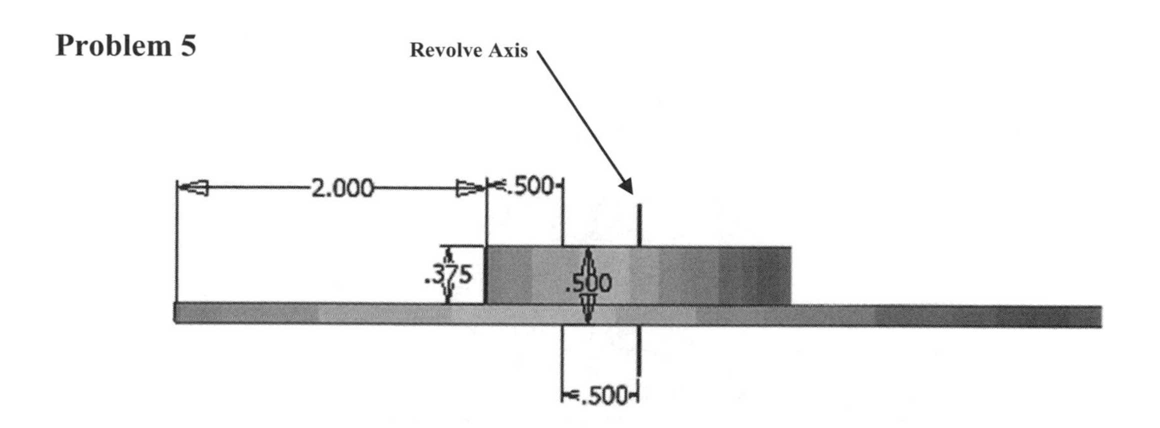

Problem 6

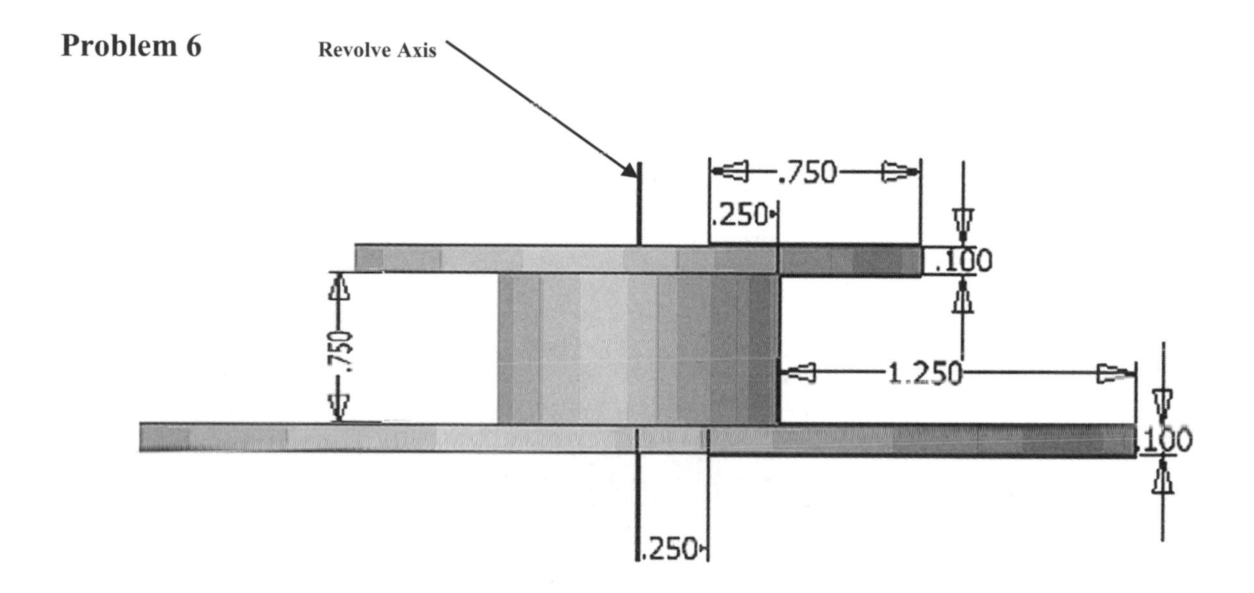

Problem 7

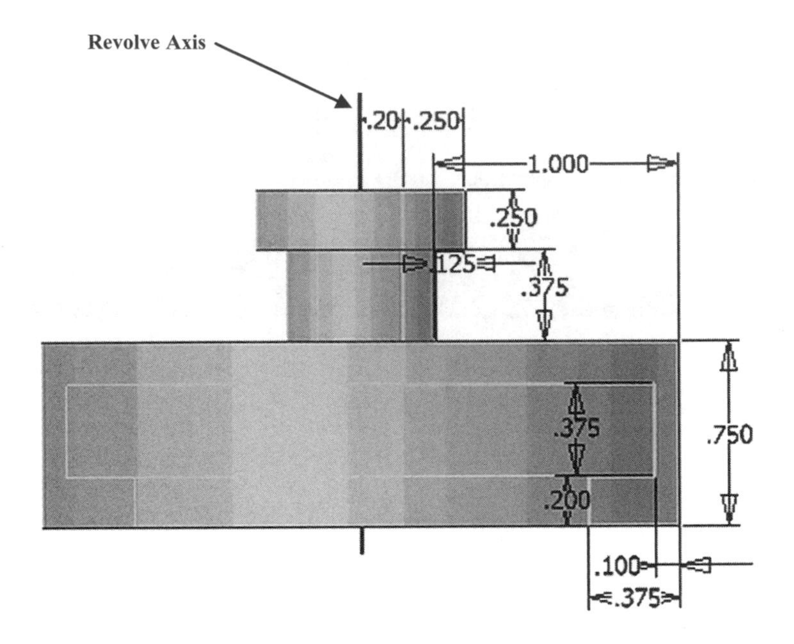
Revolve Axis
.20
.250
1.000
.250
.125
.375
.375
.750
.200
.100
.375

Problem 8

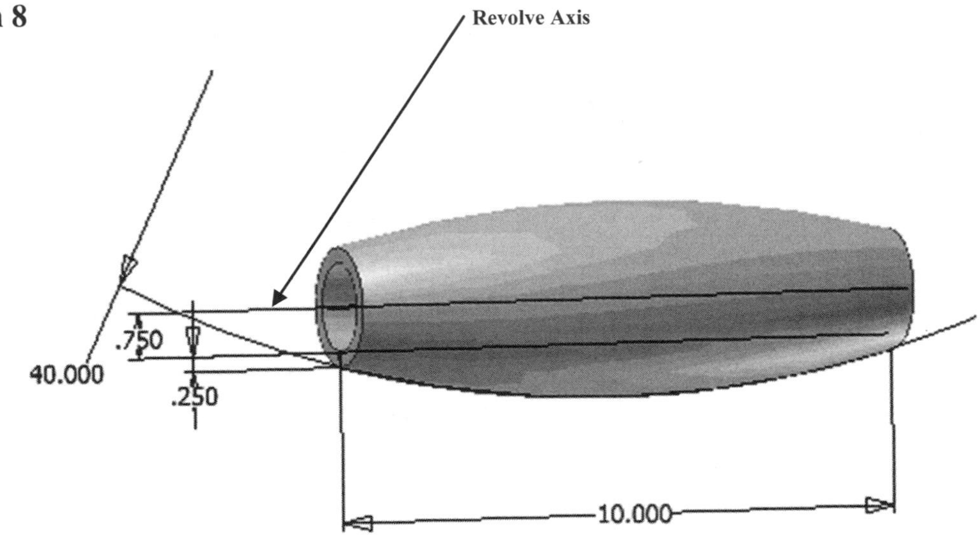
Revolve Axis
.750
40.000
.250
10.000

Chapter 3 Learning To Create a Detail Drawing

Objectives:

- Create a simple sketch using the Sketch commands
- Extrude a sketch into a solid using the Features commands
- Create an Orthographic view using the Drawing commands
- Create a Solid Model view using the Model View command

Chapter 3 includes instruction on how to design the parts shown below.

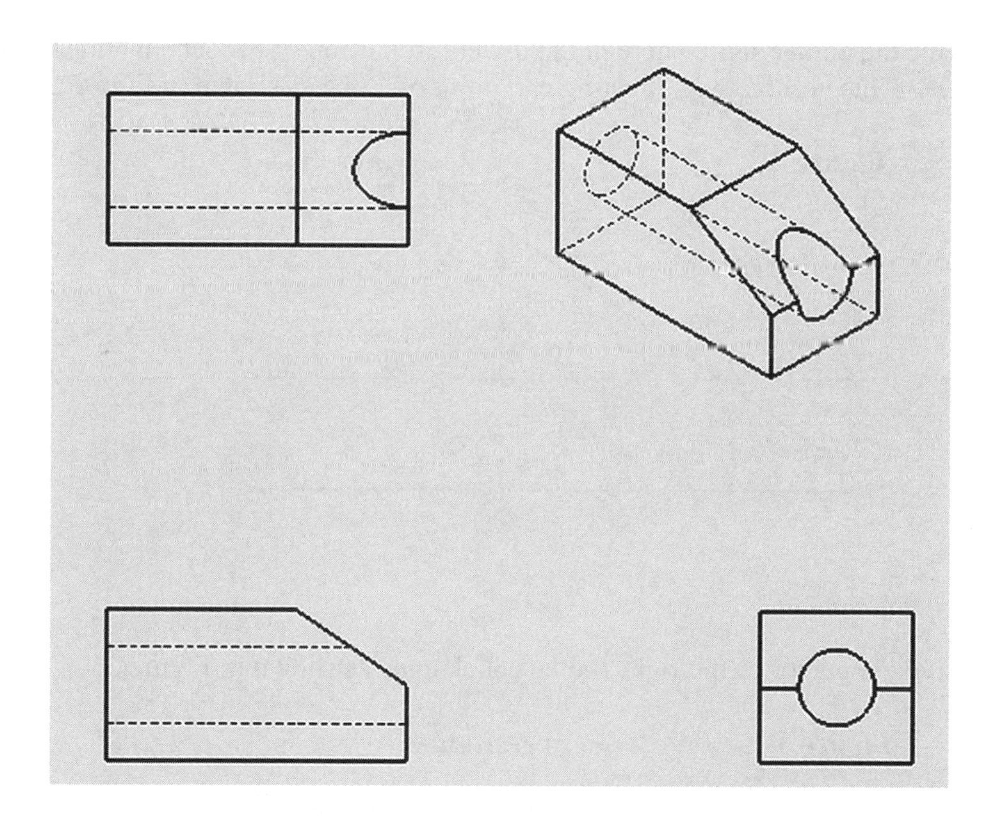

1. Start SolidWorks by referring to "Chapter 1 Getting Started".

2. After SolidWorks is running, begin a new sketch.

3. Move the cursor to the upper left corner of the screen and left click on **Line** as shown in Figure 1.

Figure 1

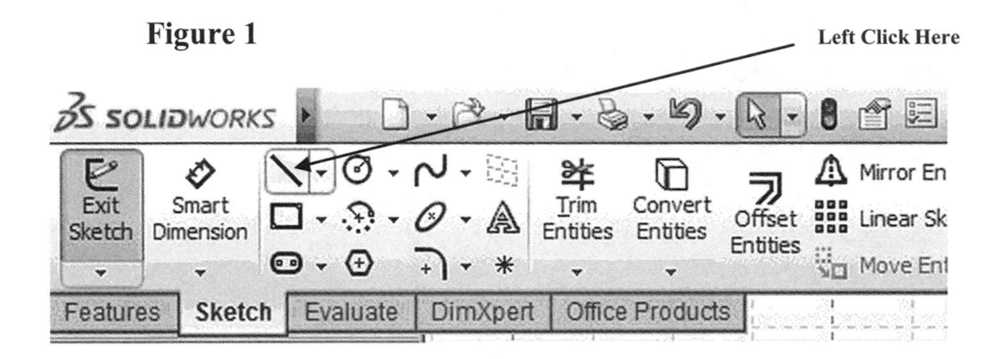

4. Move the cursor somewhere in the lower left portion of the screen and left click once. This will be the beginning end point of a line as shown in Figure 2.

Figure 2

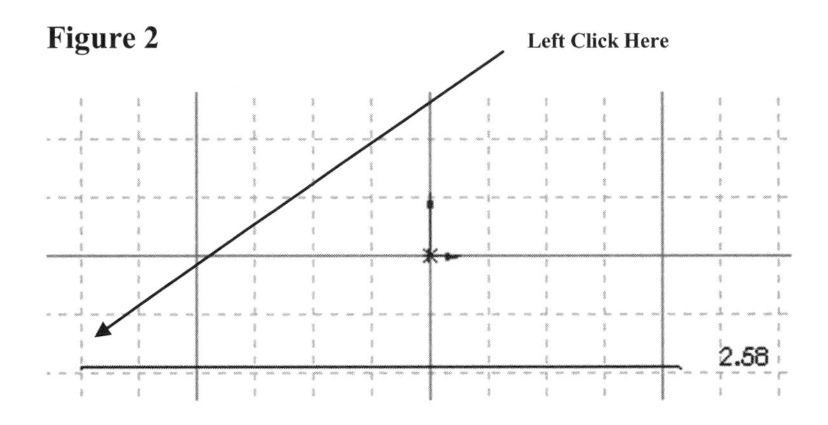

5. Move the cursor to the right and left click once as shown in Figure 3.

Figure 3

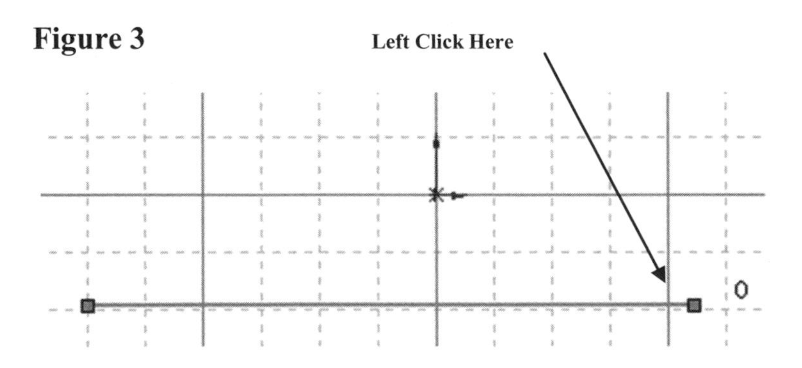

6. Move the cursor upward and left click once as shown in Figure 4.

Figure 4

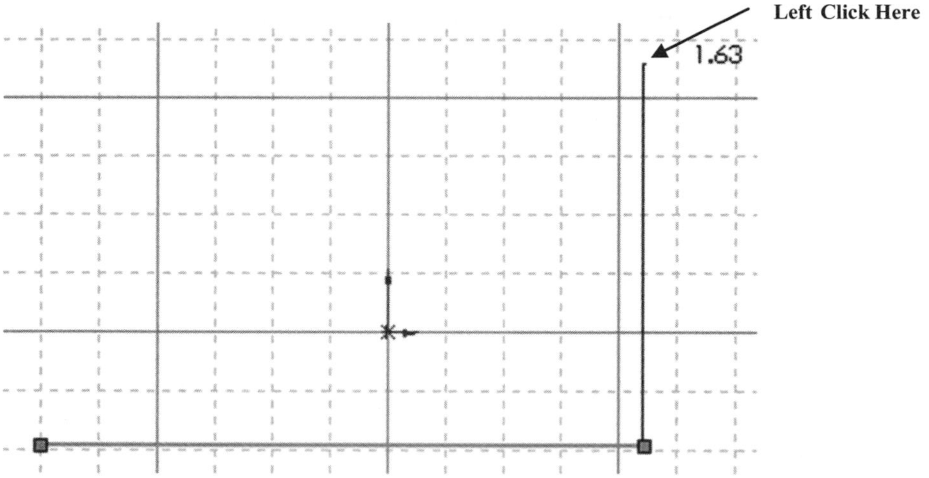

Left Click Here

1.63

7. Move the cursor to the left and wait for dots to appear. Left click once as shown in Figure 5.

Figure 5

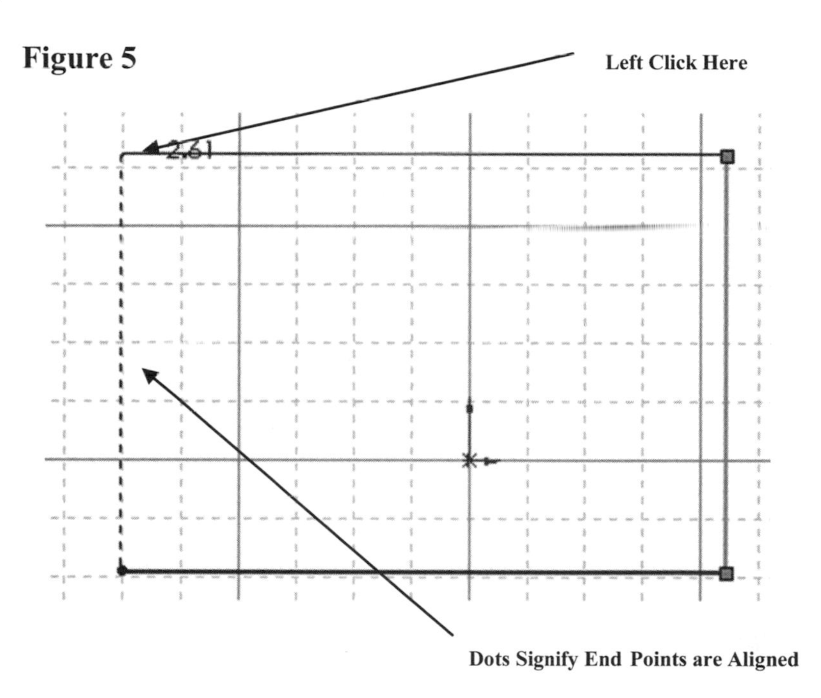

8. Move the cursor back to the original starting end point. A red dot will appear. Left click once. Your screen should look similar to Figure 6.

Figure 6

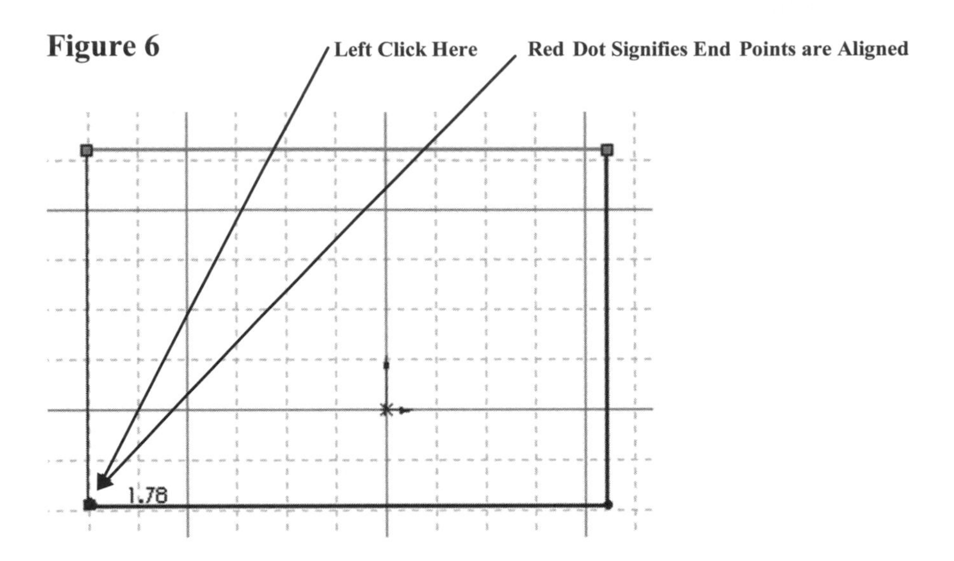

9. Right click anywhere on the screen. A pop up menu will appear. Left click on **Select** as shown in Figure 7.

Figure 7

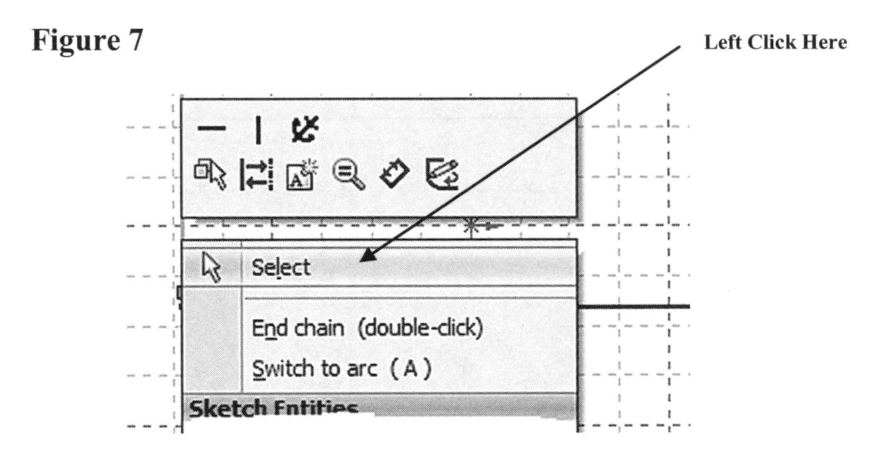

10. Move the cursor to the upper middle portion of the screen and left click on **Smart Dimension** as shown in Figure 8.

Figure 8

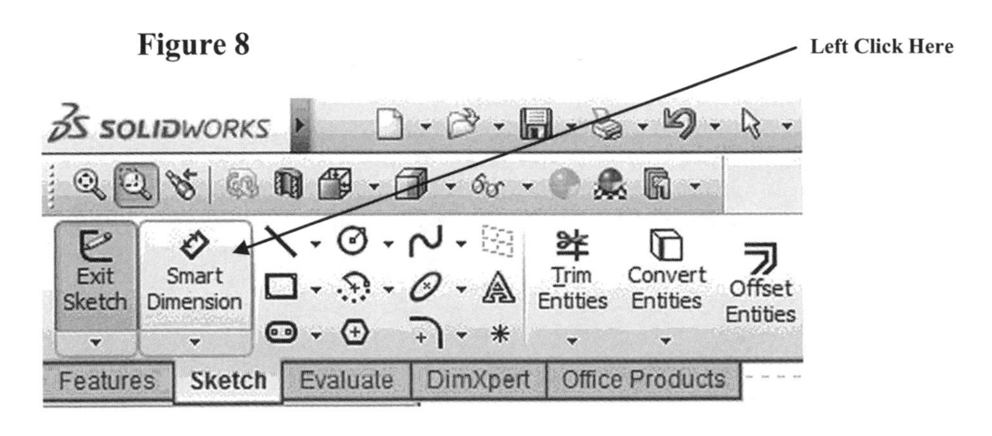

11. After selecting **Smart Dimension** move the cursor over the bottom horizontal line until it turns red as shown in Figure 9. Select the line by left clicking anywhere on the line **or** on each of the end points. The dimension will be attached to the cursor.

Figure 9

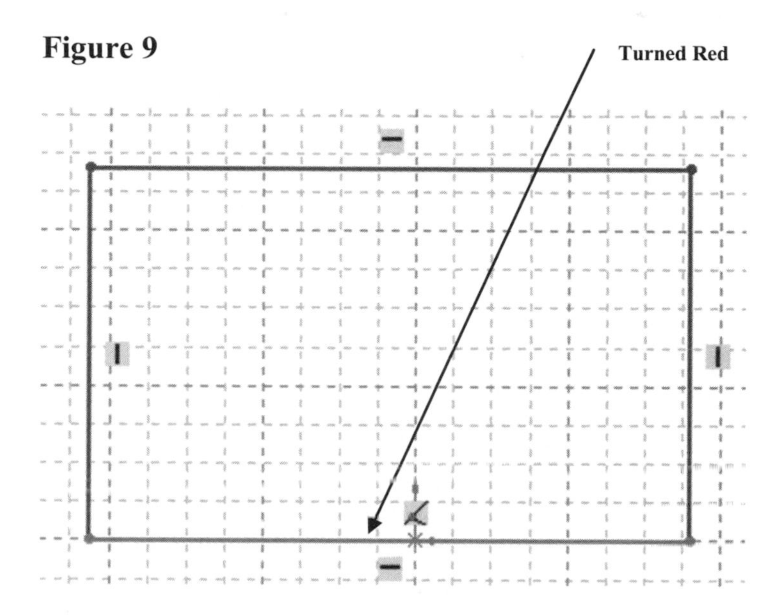

12. Move the cursor to where the dimension will be placed and left click once as shown in Figure 10.

Figure 10

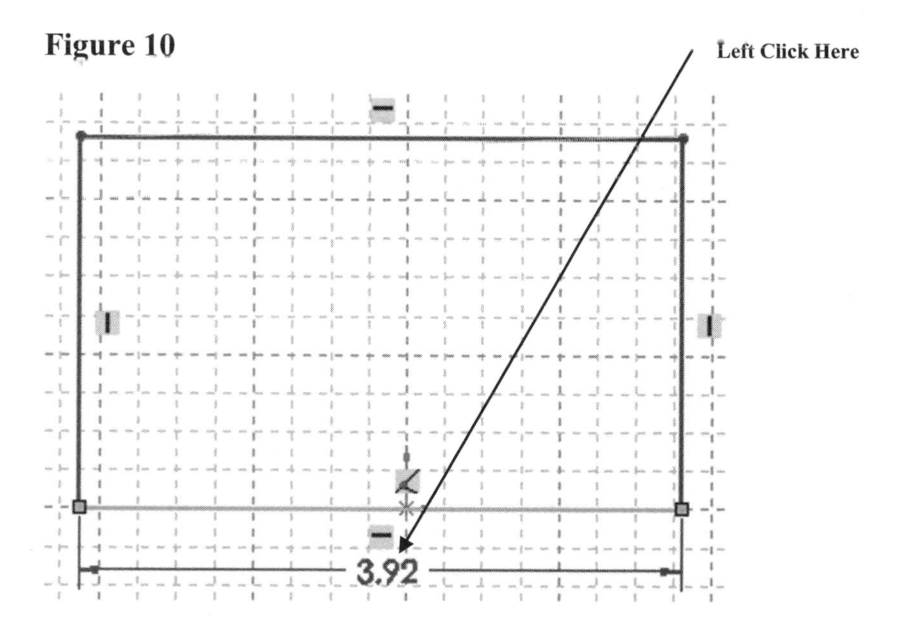

13. The Modify dialog box will appear as shown in Figure 11.

Figure 11

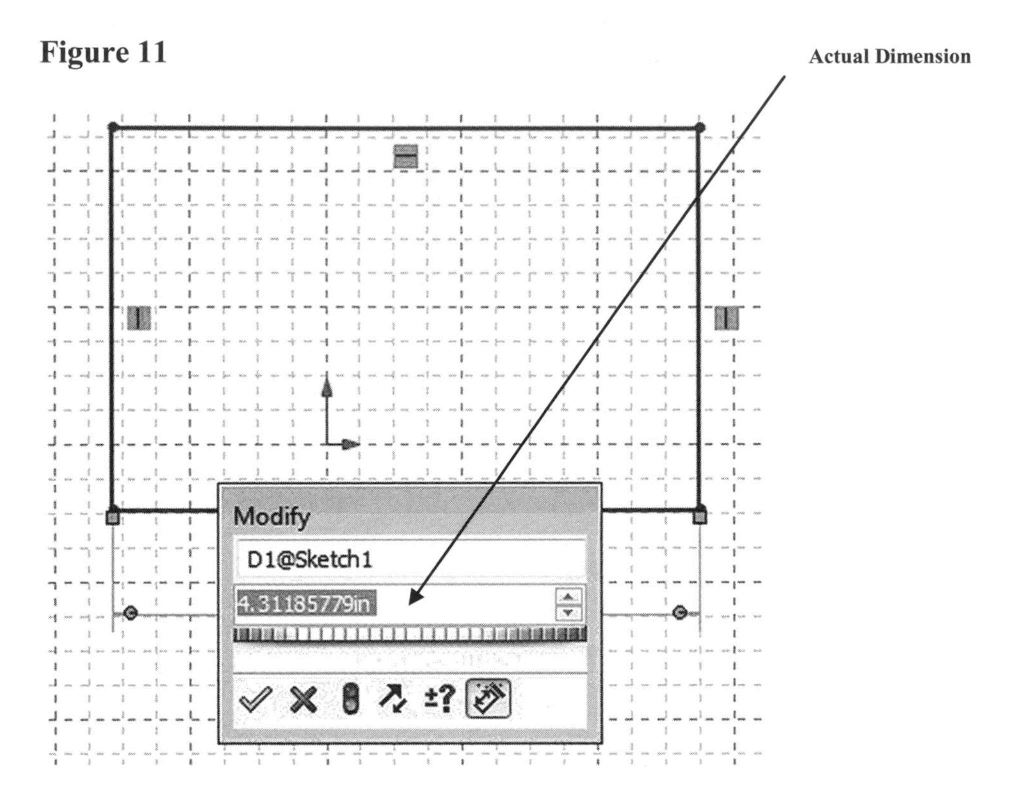

14. To edit the dimension, enter **2.00** in the Modify dialog box (while the current dimension is highlighted) and either press **Enter** on the keyboard or left click on the green checkmark as shown in Figure 12.

Figure 12

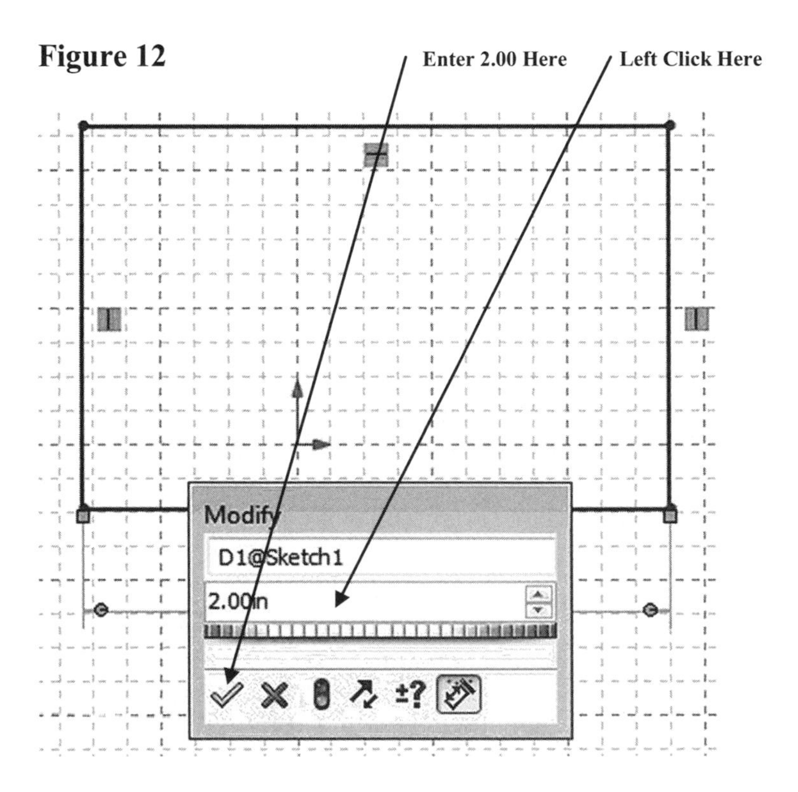

15. The dimension of the line will become 2.00 inches as shown in Figure 13.

Figure 13

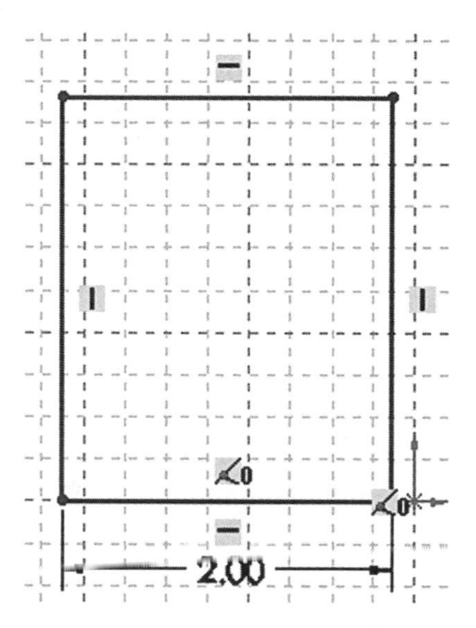

16. Right click anywhere on the screen. A pop up menu will appear. Left click on **Select** as shown in Figure 14.

Figure 14

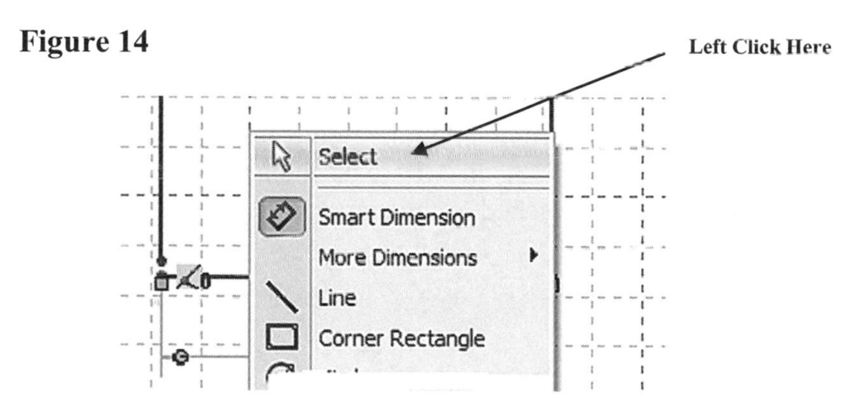

17. To view the entire drawing, move the cursor to the upper right portion of the screen and left click once on the "Zoom to Fit" icon as shown in Figure 15.

Figure 15

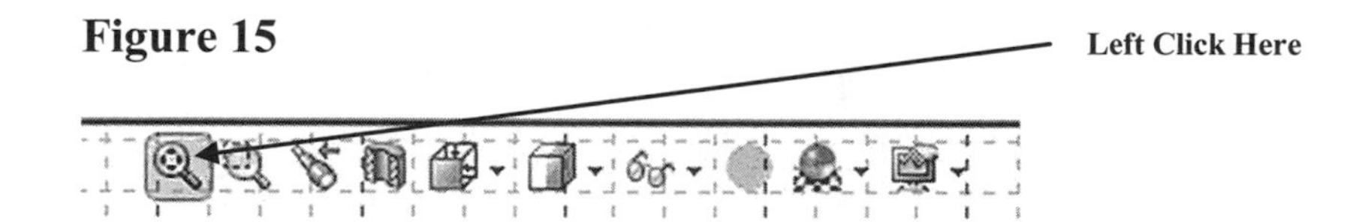

18. The drawing will "fill up" the entire screen. If an area of the drawing needs to be enlarged, left click on the "Zoom to Area" icon as shown in Figure 16. Hold the left mouse button down drag a window around the area to enlarge. Once the area is enlarged to the desired view, release the left mouse button.

Figure 16

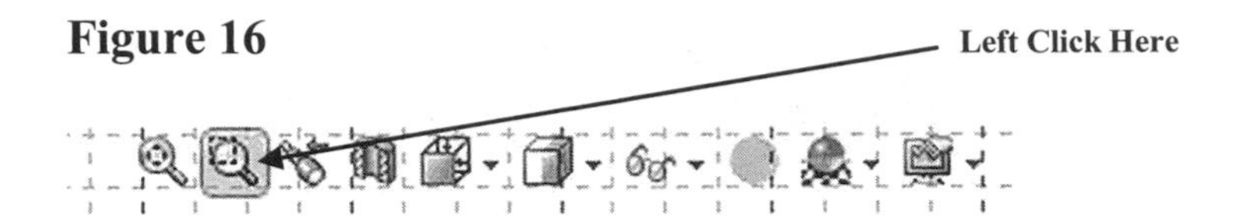

19. Move the cursor to the upper left portion of the screen and left click on **Smart Dimension** as shown in Figure 17.

Figure 17

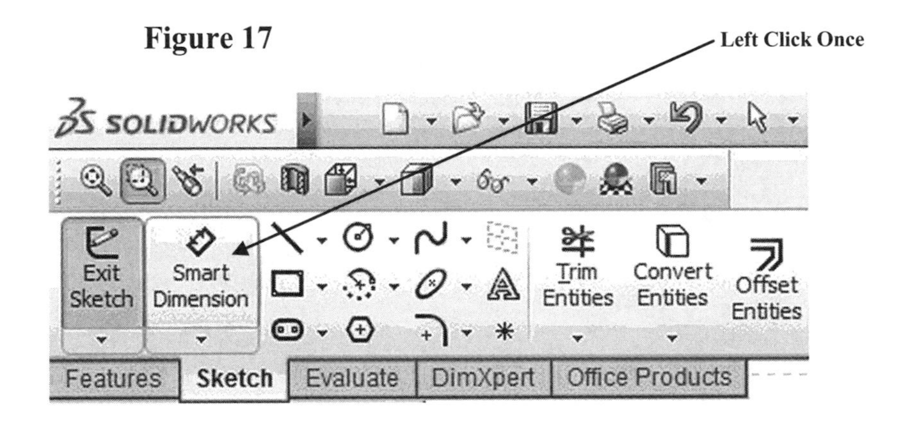

20. Move the cursor over the right side line until it turns red as shown in Figure 18. Select the line by left clicking anywhere on the line **or** on each of the end points. The dimension will be attached to the cursor.

Figure 18

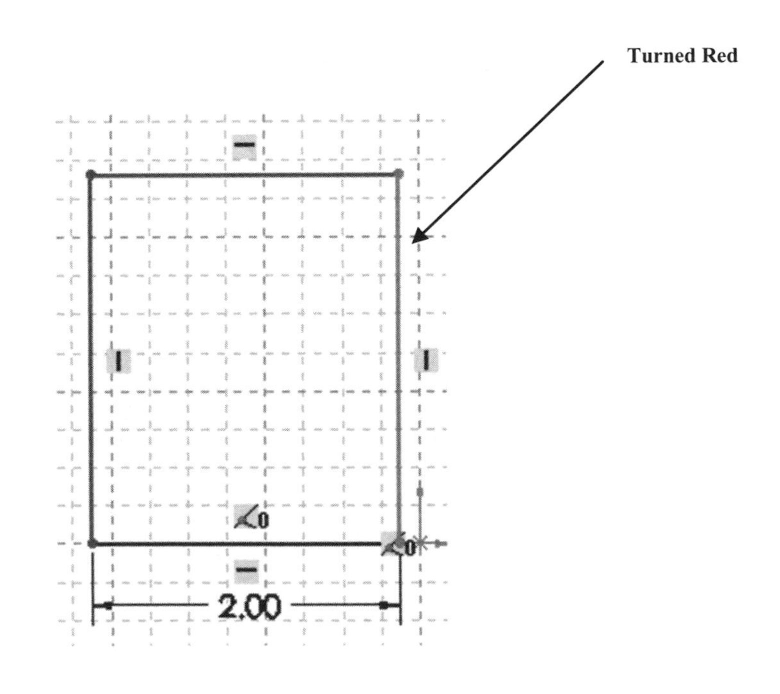

21. Move the cursor to where the dimension will be placed and left click once as shown in Figure 19.

Figure 19

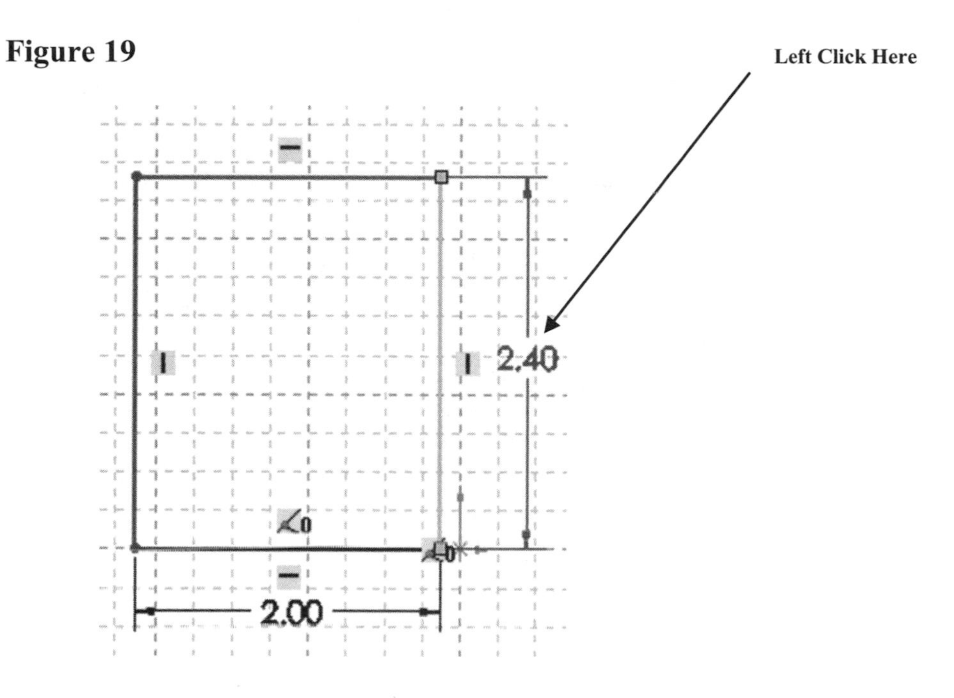

22. The Modify dialog box will appear as shown in Figure 20.

Figure 20

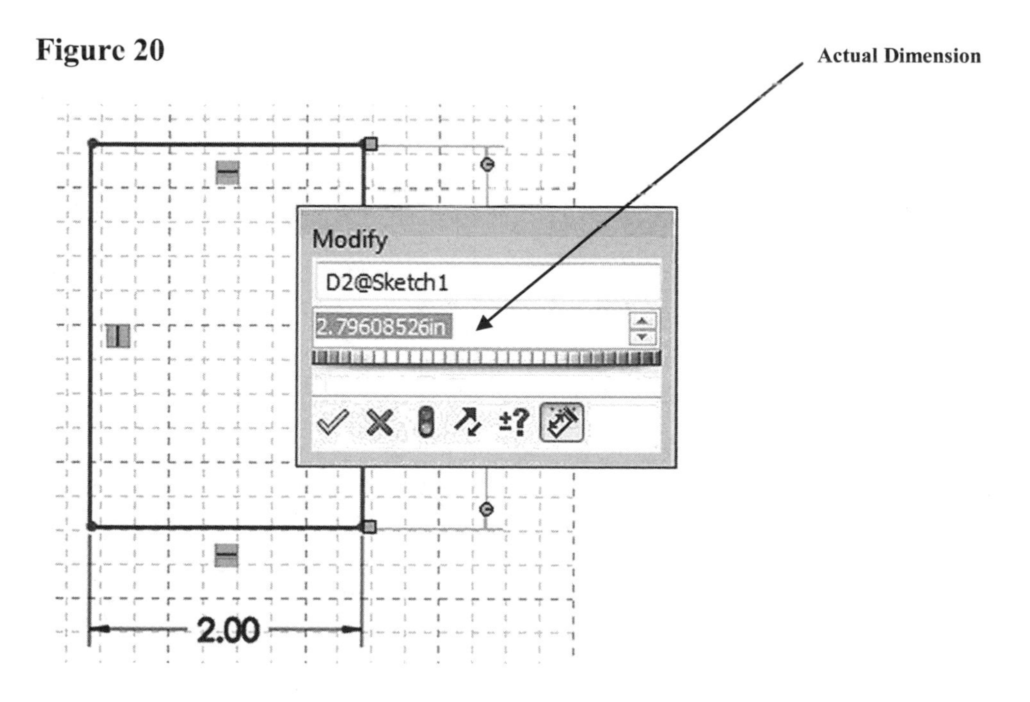

23. To edit the dimension, enter **1.00** in the Modify dialog box (while the current dimension is highlighted) and either press **Enter** on the keyboard or left click on the green checkmark as shown in Figure 21.

Figure 21

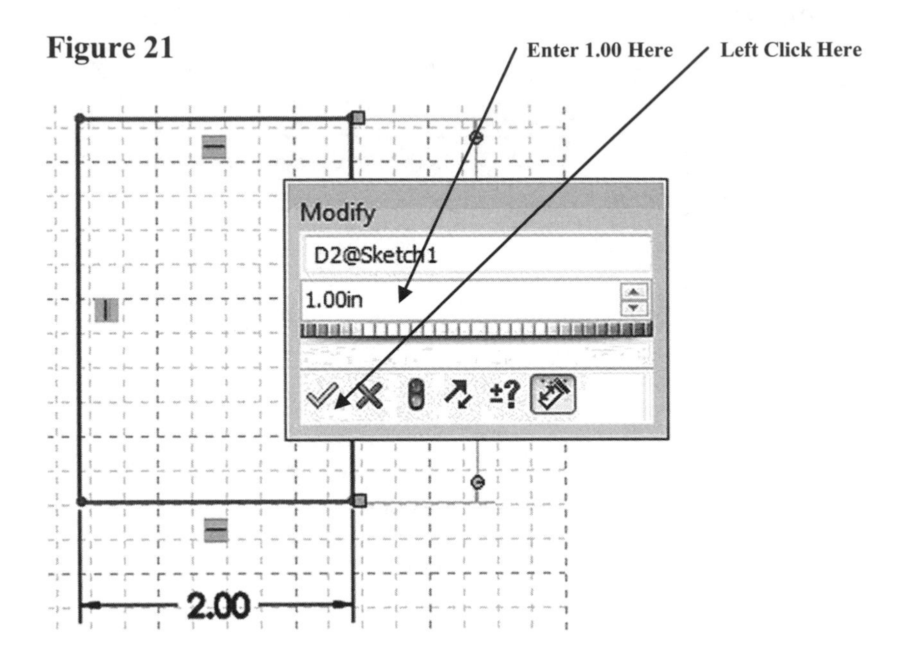

24. The dimension of the line will become 1.00 inches as shown in Figure 22. Use the Zoom icons to zoom out if necessary.

Figure 22

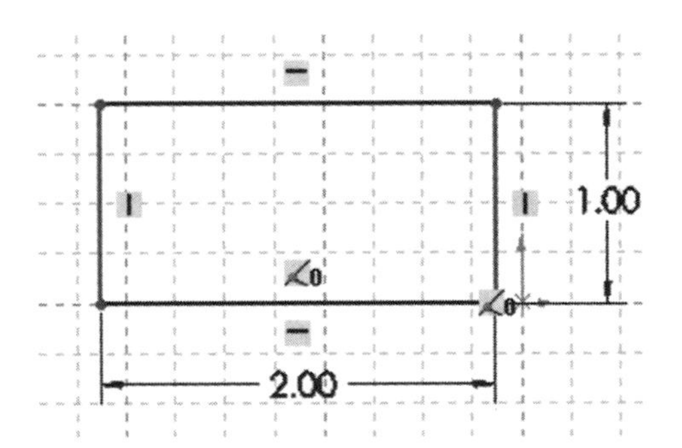

25. Right click anywhere on the screen. A pop up menu will appear. Left click on **Select** as shown in Figure 23.

Figure 23

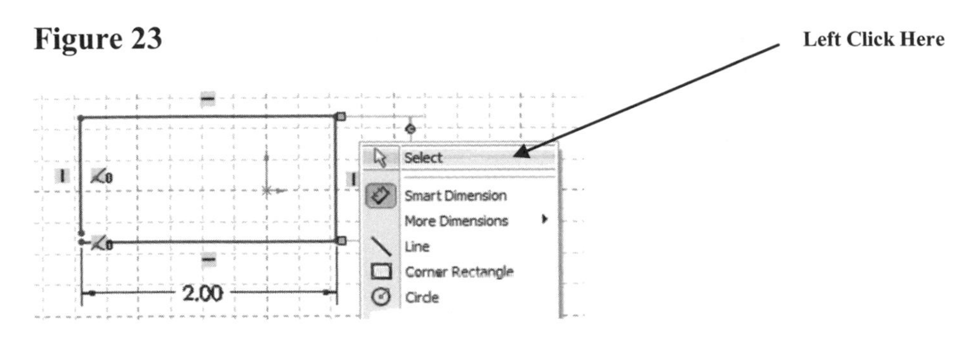

26. Move the cursor to the upper left portion of the screen and left click on **Smart Dimension** as shown in Figure 24.

Figure 24

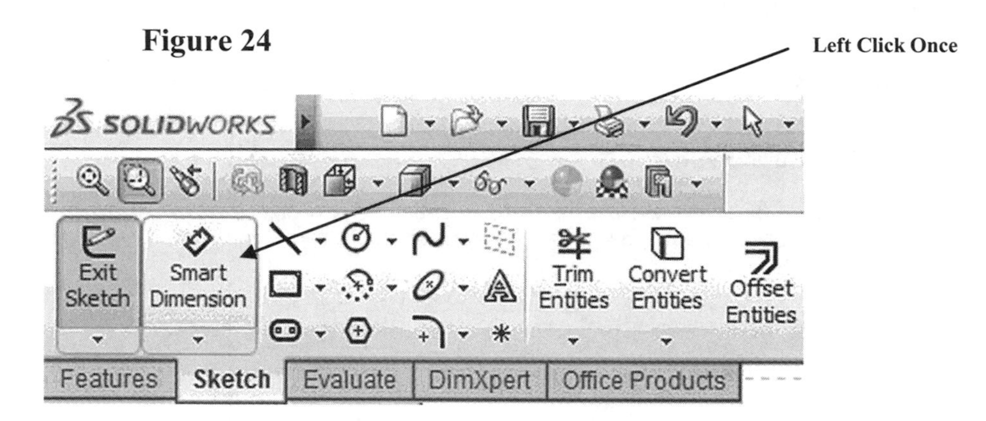

27. After selecting **Smart Dimension** move the cursor over the top horizontal line until it turns red as shown in Figure 25. Select the line by left clicking anywhere on the line **or** on each of the end points. This will cause the dimension to be attached to the cursor.

Figure 25

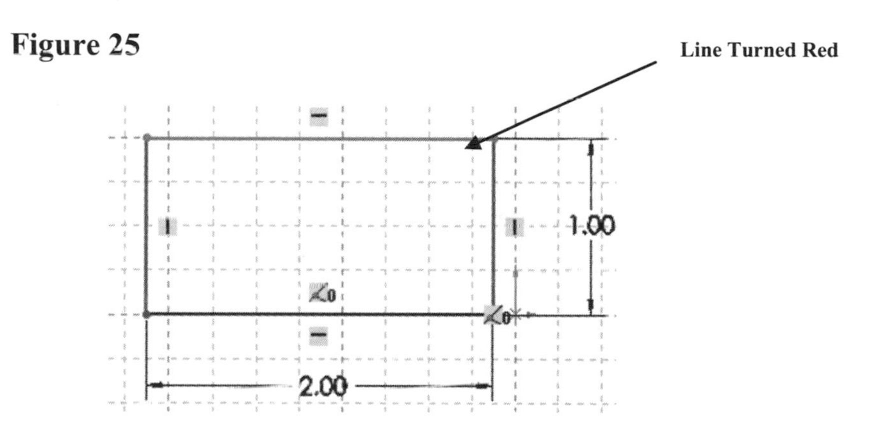

28. Move the cursor to where the dimension will be placed and left click once as shown in Figure 26.

Figure 26

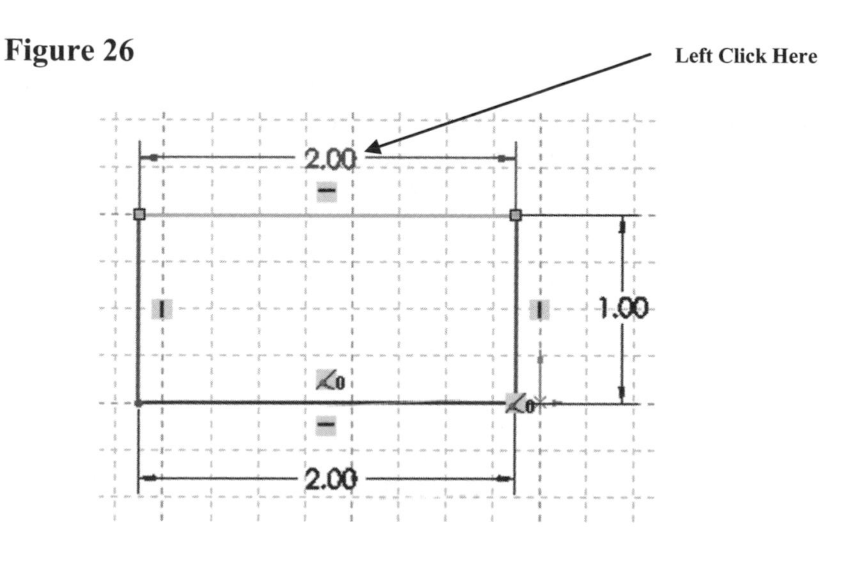

29. Notice that the dimension is exactly 2.000. Move the cursor to where the dimension will be placed and left click once. The Make Dimension Driven? dialog box will appear as shown in Figure 27.

Figure 27

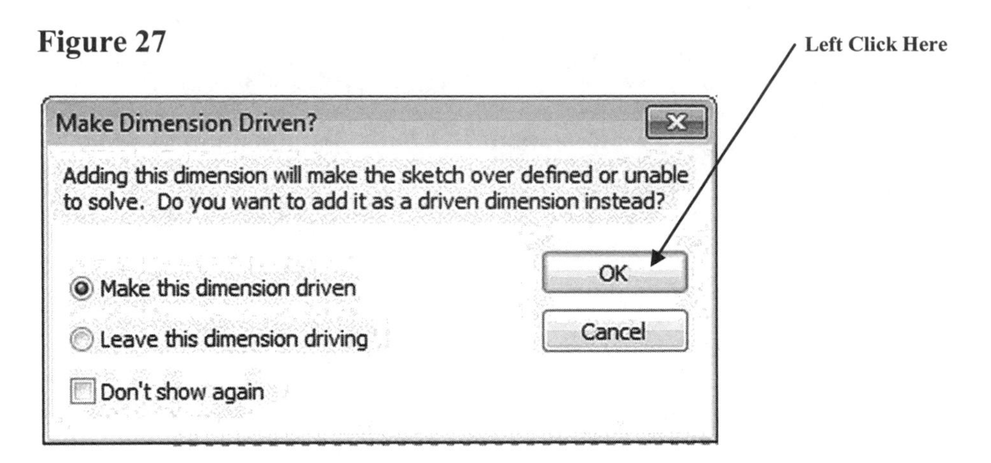

30. Right click anywhere on the screen. A pop up menu will appear. Left click on **Select** as shown in Figure 28.

Figure 28

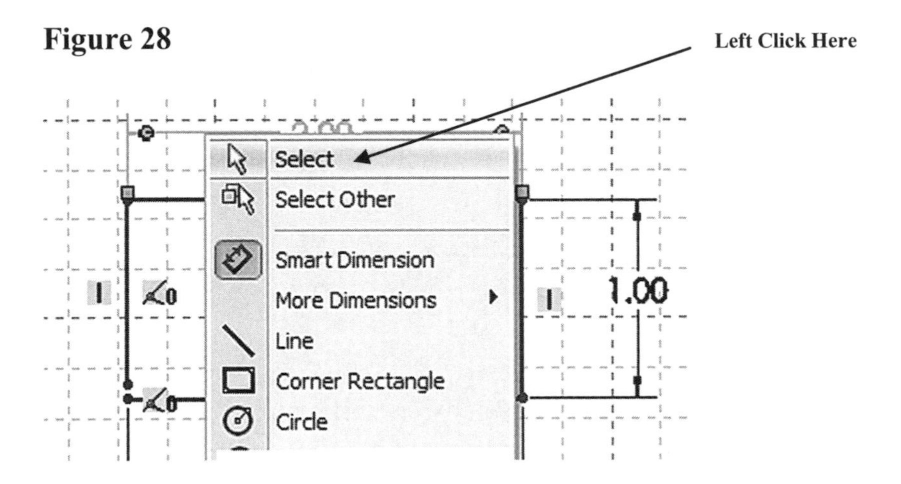

31. This dimension will over constrain the sketch because the sketch has been constrained with 90 degree angles when it was constructed. Left click on **OK**. The dimension will be driven meaning it cannot be used to edit or change the length of the line.

32. The driven dimension appears grayed out as shown in Figure 29.

Figure 29

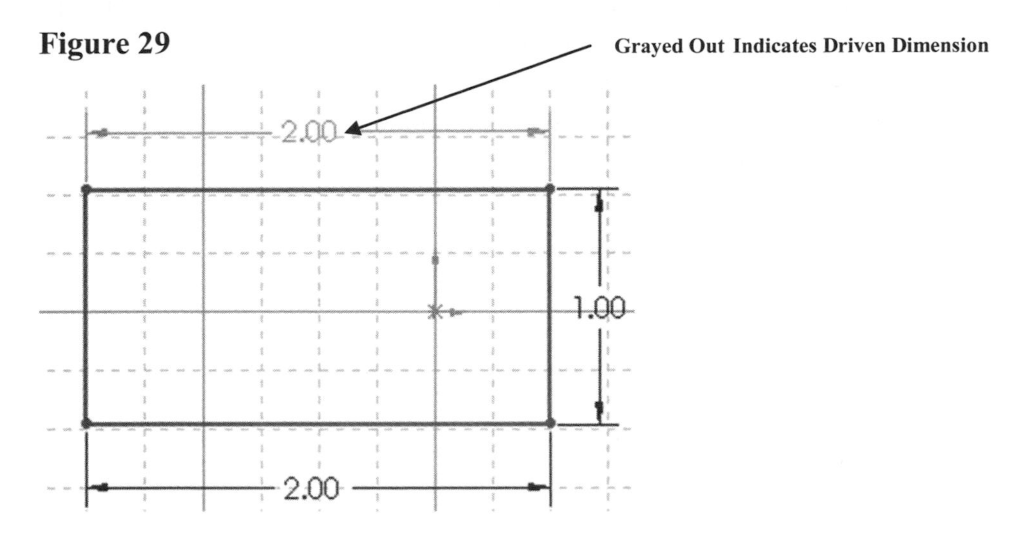

33. Dimensioning the far left line would also result in a driven dimension. Because of this, the dimensioning portion is complete.

34. After the sketch is complete it is time to extrude the sketch into a solid.

35. After you have verified that no commands are active, move the cursor to the upper left portion of the screen and left click on **Exit Sketch** as shown in Figure 30.

Figure 30

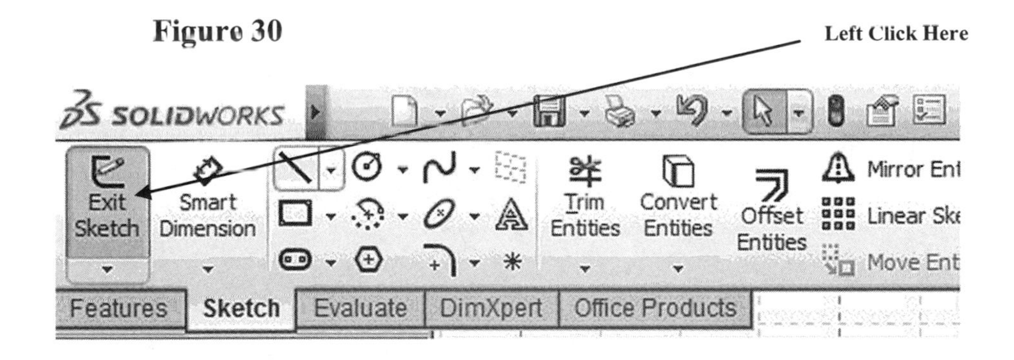

36. SolidWorks is now out of the Sketch commands and into the Features commands. Notice that the commands at the top of the screen are now different. To work in the Features commands a sketch must be present and have no opens (non-connected lines). If there are any opens in the sketch an error message will appear. Your screen should look similar to Figure 31.

Figure 31

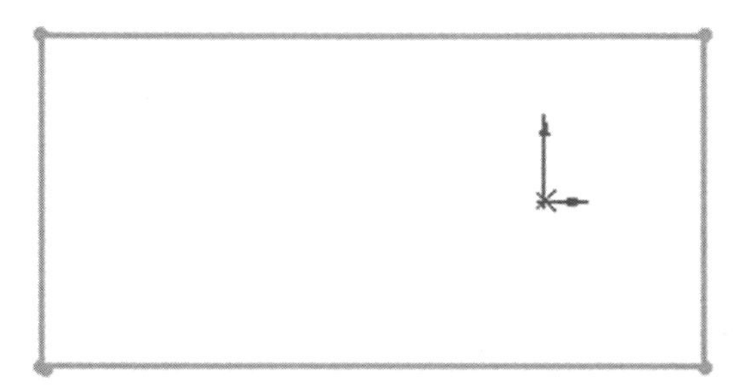

37. Move the cursor to the upper middle portion of the screen and left click on the drop down arrow next to the "Views Orientation" icon. A drop down menu will appear. Left click on **Trimetric** as shown in Figure 32.

Figure 32

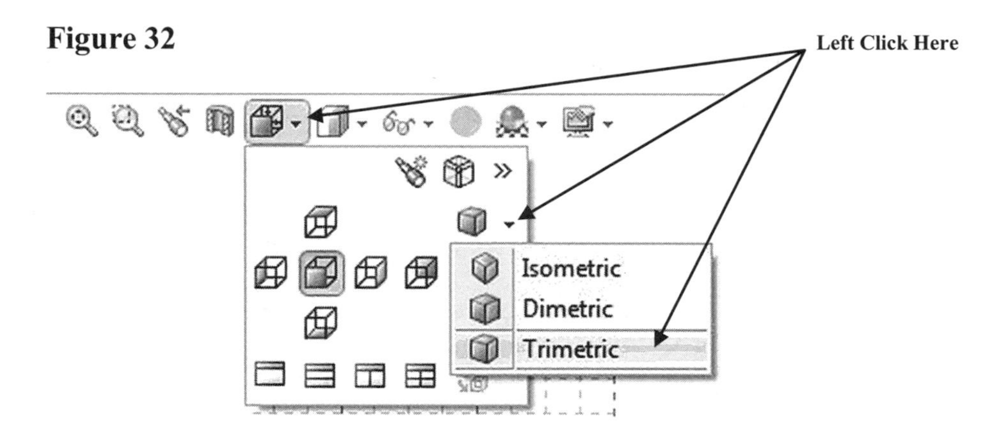

38. The view will become trimetric as shown in Figure 33.

Figure 33

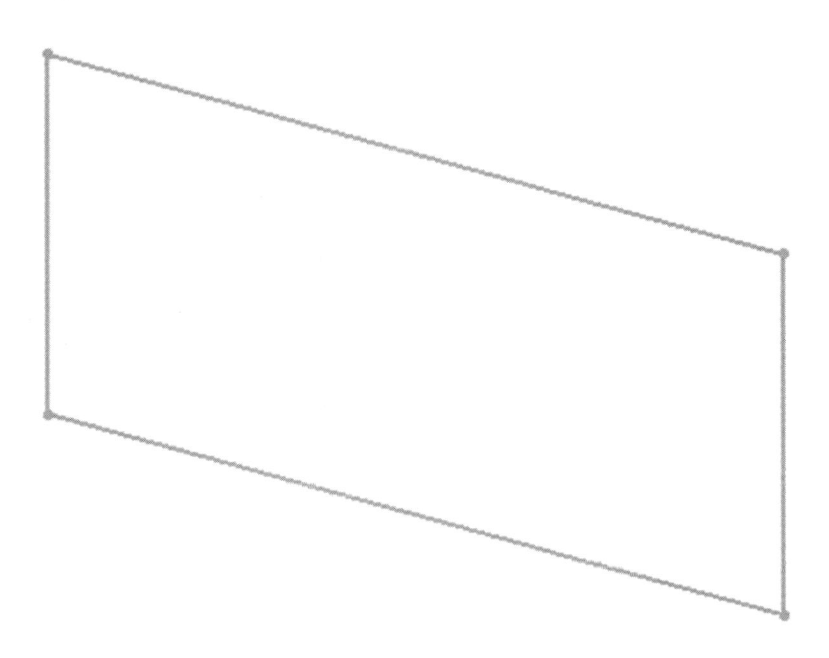

39. Move the cursor to the upper left portion of the screen and left click on **Extruded Boss/Base** as shown in Figure 34. If SolidWorks gave you an error message, there are opens (non-connected lines) somewhere on the sketch. Check each intersection for opens by using the **Extend** and **Trim** commands.

Figure 34

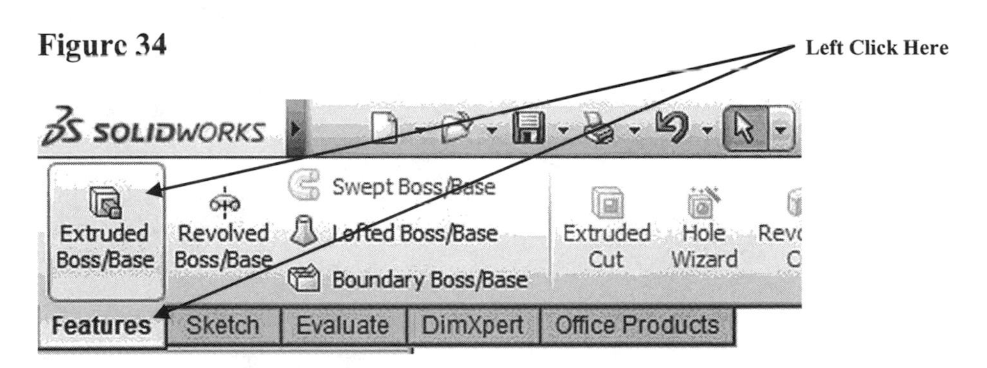

40. A preview of the extrusion will be displayed as shown in Figure 35.

Figure 35

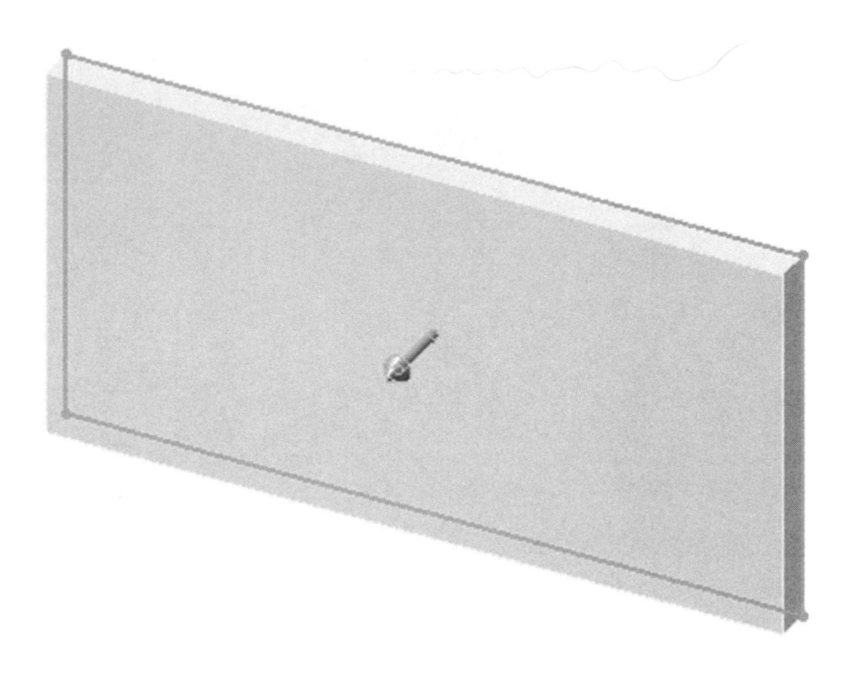

41. If a preview does not appear, move the cursor over any of the lines. The line will turn red as shown in Figure 36. Left click once.

Figure 36

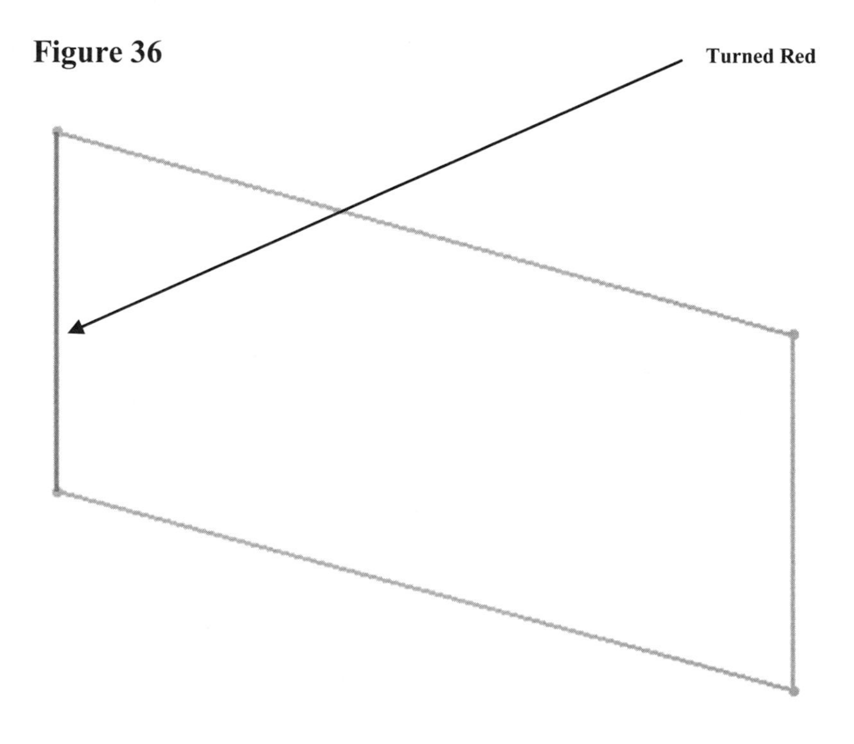

42. Enter **1.00** for D1 as shown in Figure 37.

Figure 37

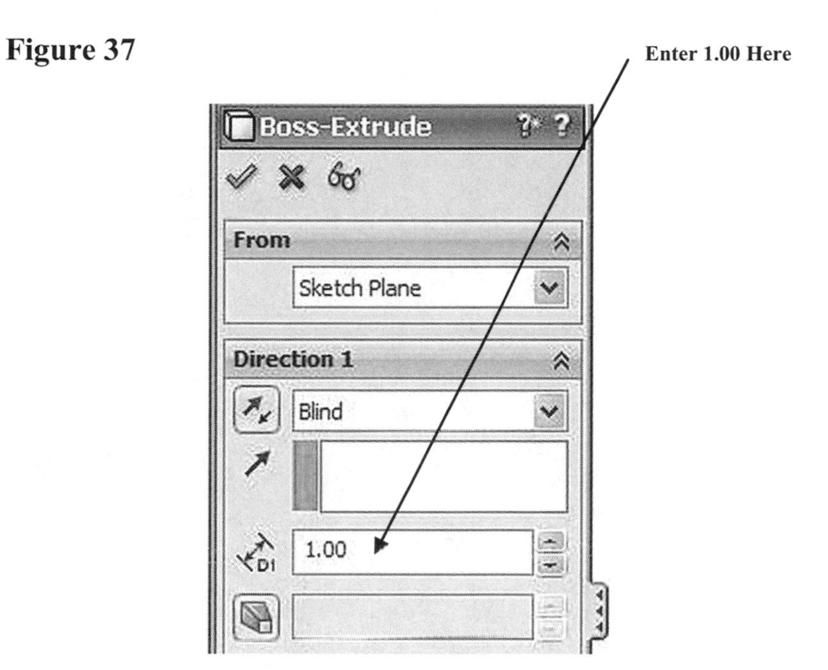

43. Move the cursor to the upper left portion of the screen and left click on the green checkmark as shown in Figure 38. Left click once anywhere around the drawing.

Figure 38

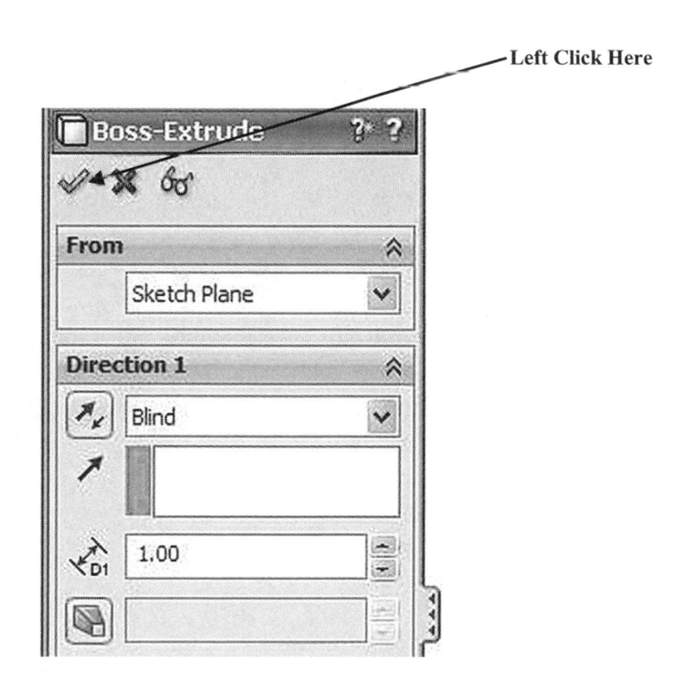

44. Your screen should look similar to Figure 39. You may have to use the Zoom Out command to view the entire part.

Figure 39

45. Move the cursor to the upper middle portion of the screen and left click on the drop down arrow located under Fillet. A drop down menu will appear. Left click on **Chamfer** as shown in Figure 40.

Figure 40

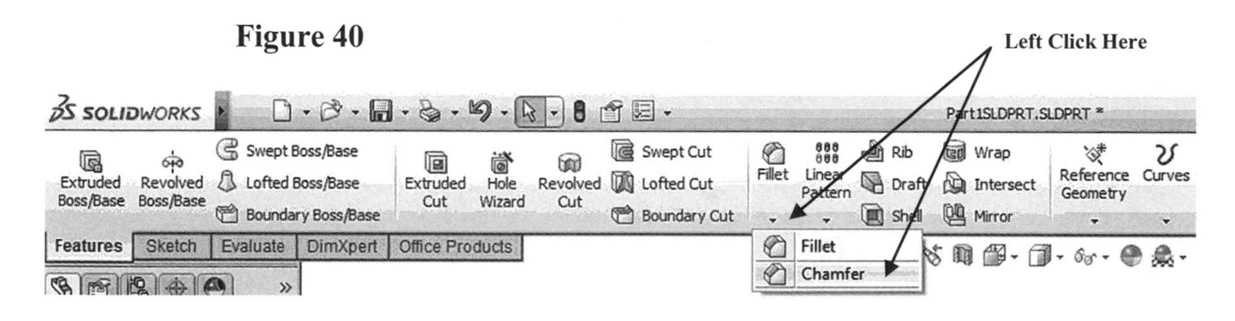

46. After selecting **Chamfer,** place a dot (left click) next to the text "Distance distance". Place another dot (left click) next to the text "Full preview" as shown in Figure 41.

Figure 41

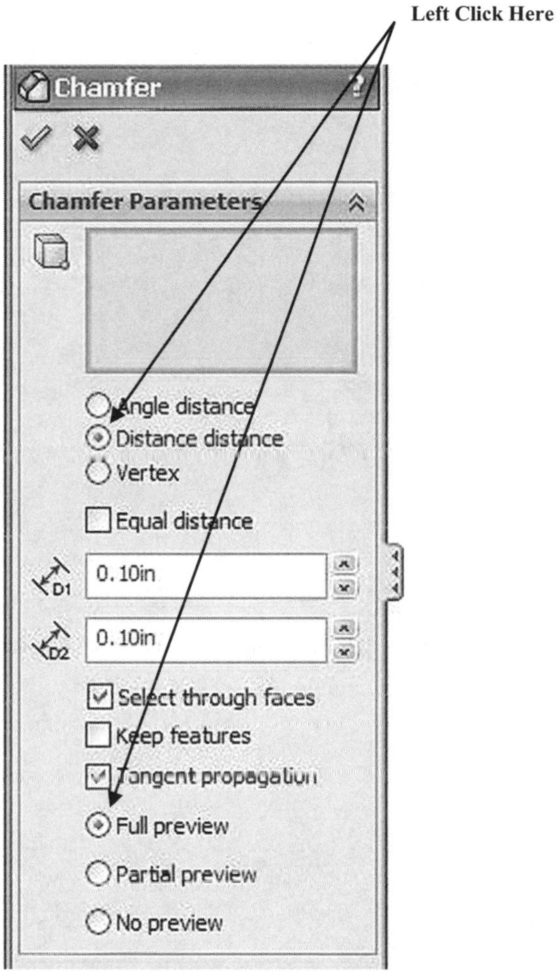

Left Click Here

47. Move the cursor to the front upper corner. A red line will appear as shown in Figure 42. Left click once.

Figure 42

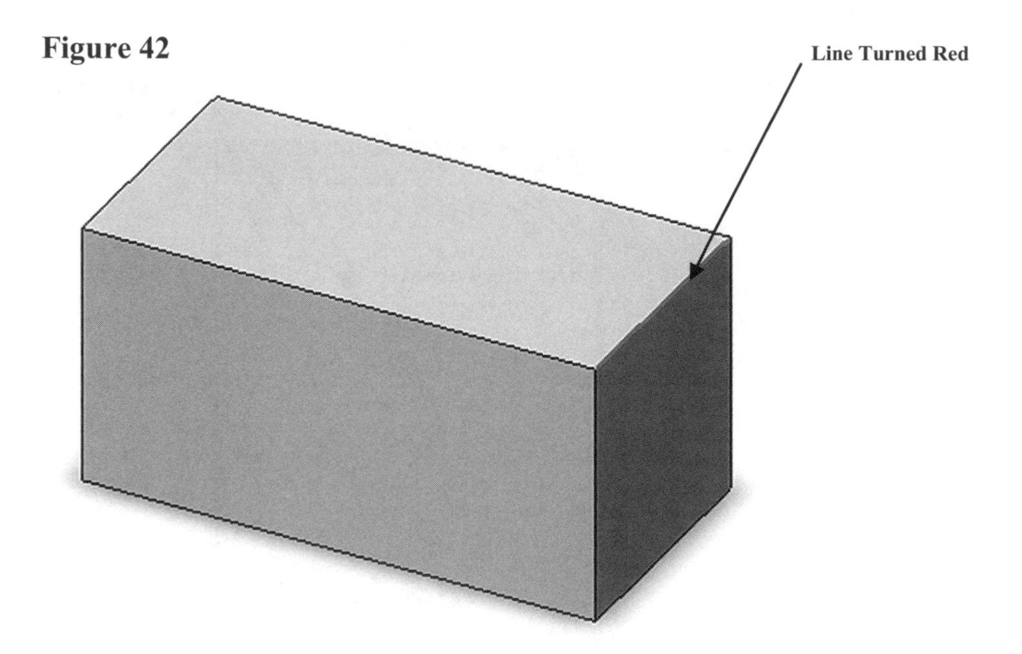

48. A preview of the anticipated chamfer will be displayed as shown in Figure 43.

Figure 43

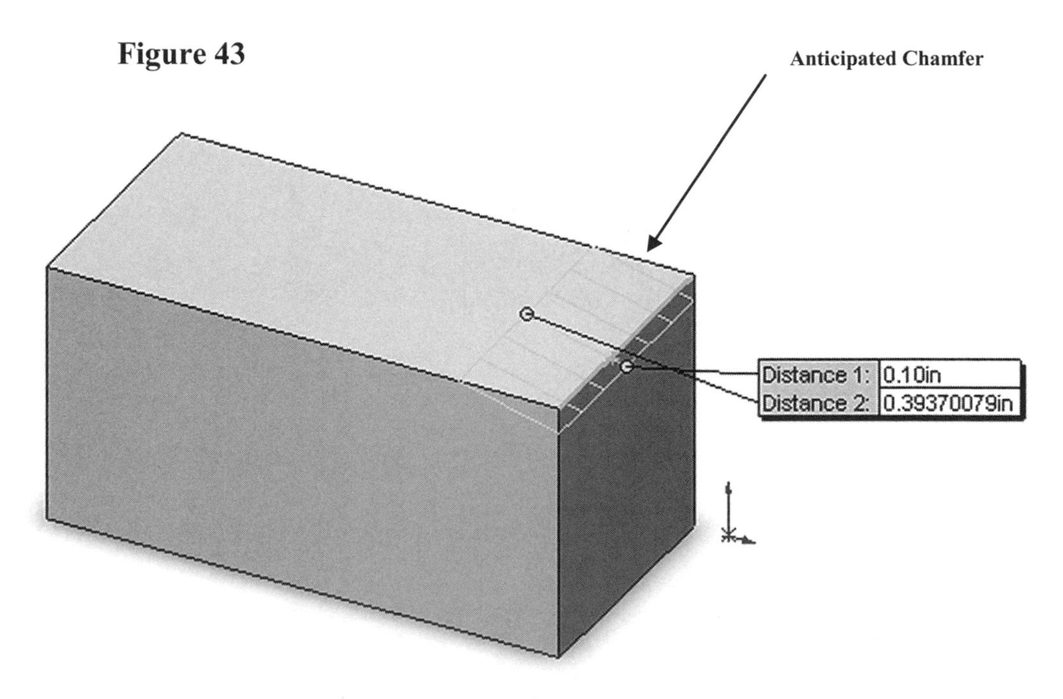

49. Move the cursor to D1 in the Chamfer dialog box and highlight the text. Enter **.50.** Move the cursor to D2 in the dialog box and highlight the text. Enter **.75** as shown in Figure 44.

Figure 44

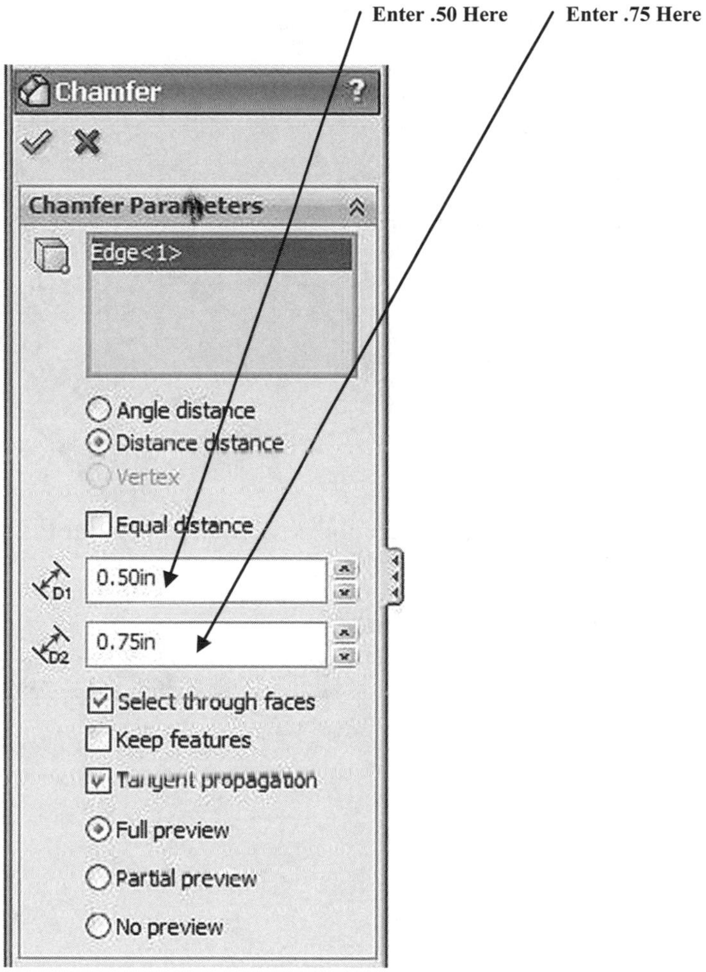

Enter .50 Here

Enter .75 Here

Chamfer

Chamfer Parameters

Edge<1>

Angle distance

Distance distance

Vertex

Equal distance

D1 0.50in

D2 0.75in

Select through faces

Keep features

Tangent propagation

Full preview

Partial preview

No preview

50. A preview of the chamfer will be displayed as shown in Figure 45.

Figure 45

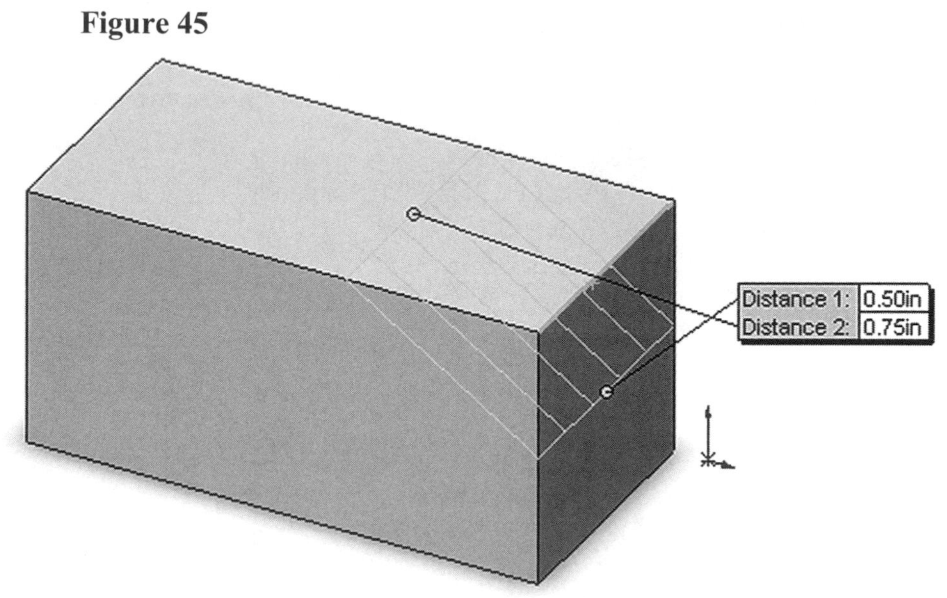

51. Left click on the green checkmark as shown in Figure 46.

Figure 46

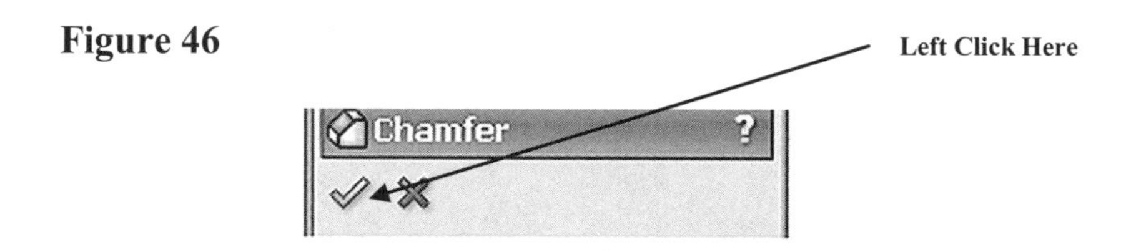

52. Your screen should look similar to Figure 47.

Figure 47

53. Use the **Rotate** command (by pressing down the mouse wheel and moving the mouse around) to rotate the part as shown in Figure 48.

Figure 48

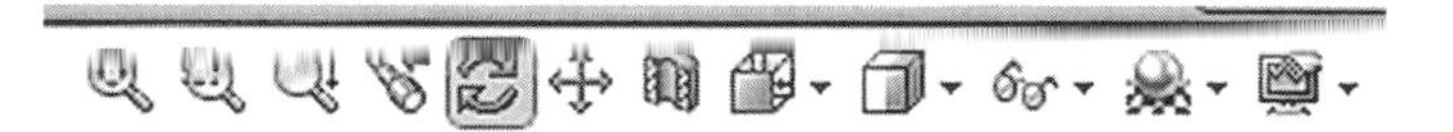

54. While holding the left mouse button down, drag the cursor to the right to gain access to the backside of the part as shown in Figure 49.

Figure 49

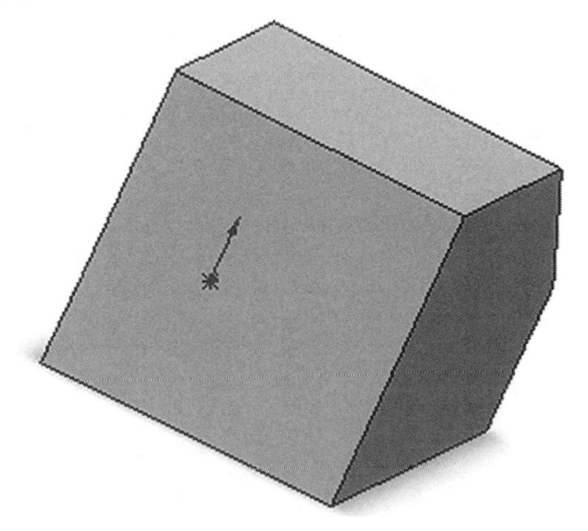

55. Right click anywhere on the screen. A pop menu will appear. Left click on **Select** as shown in Figure 50.

Figure 50

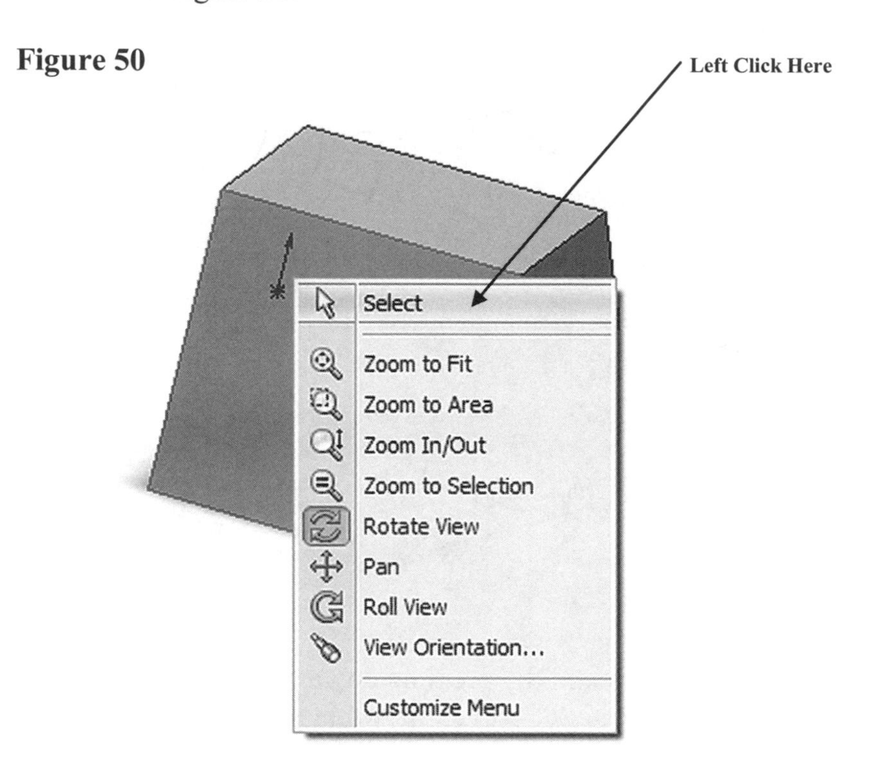

56. Move the cursor to the upper middle portion of the screen and left click on the drop down arrow next to the "Views Orientation" icon. A drop down box will appear. Left click on **Left** as shown in Figure 51.

Figure 51

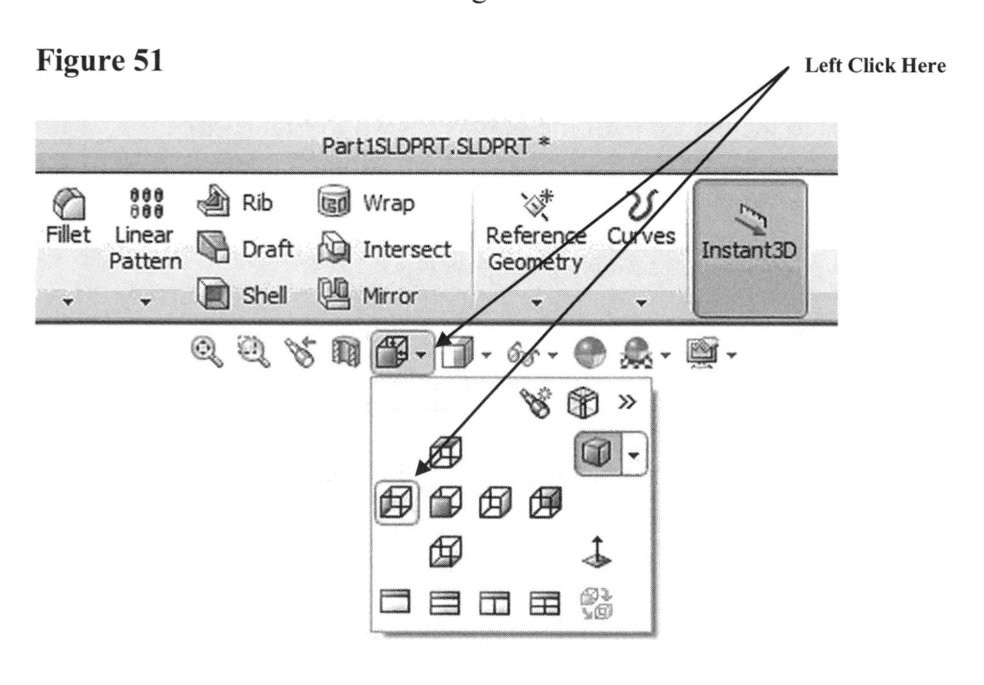

57. The part will rotate and provide a perpendicular view of the surface as shown in Figure 52.

Figure 52

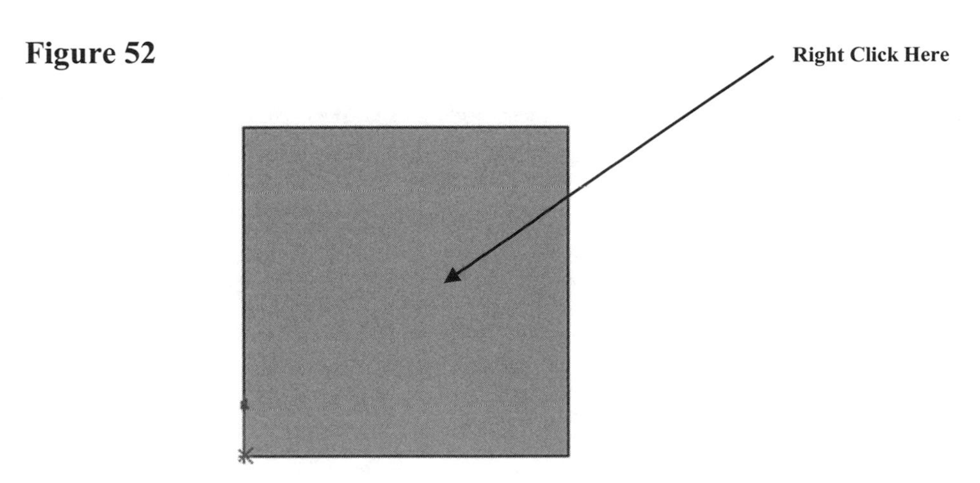

58. Right click anywhere on the surface. A pop up menu will appear. Left click on **Insert Sketch** as shown in Figure 53.

Figure 53

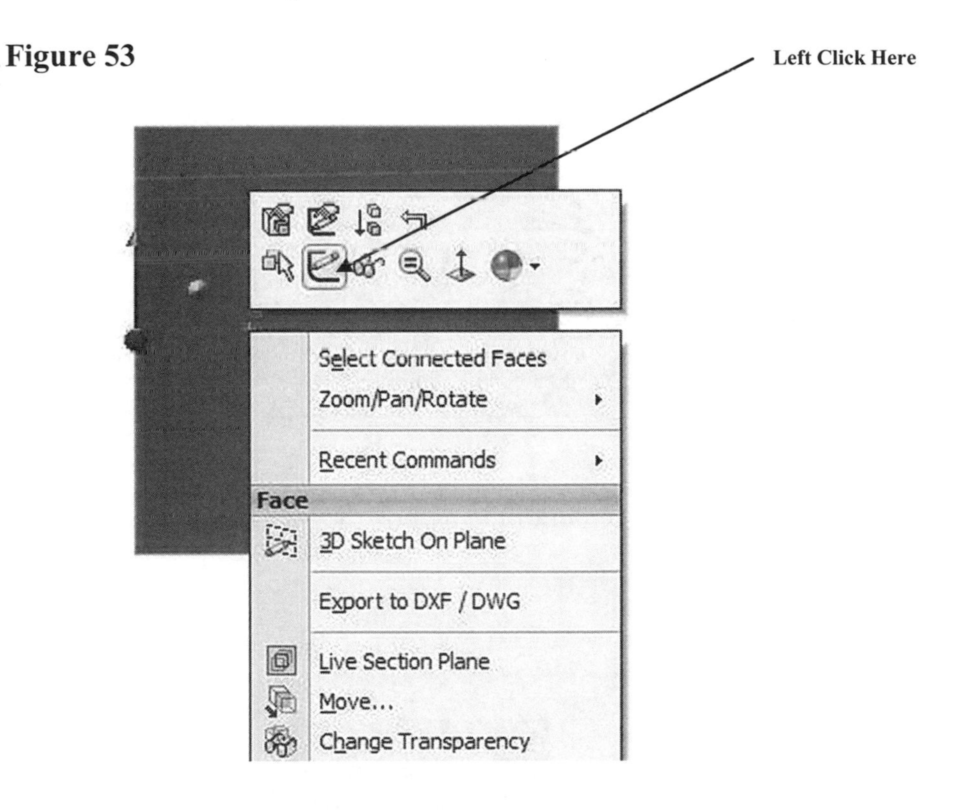

59. A new sketch will appear on the surface as shown in Figure 54.

Figure 54

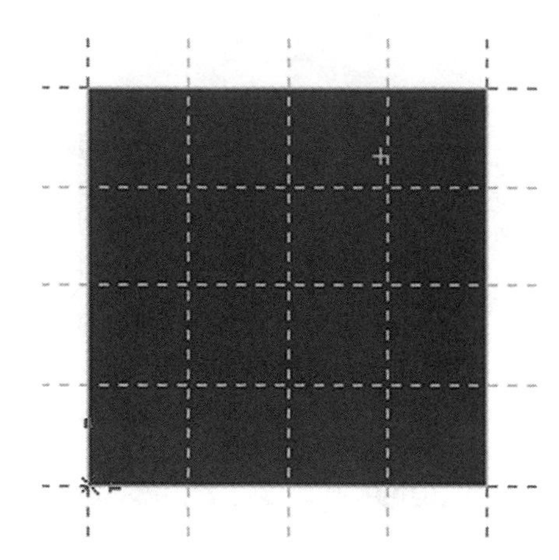

60. Move the cursor to the upper left portion of the screen and left click on **Circle** as shown in Figure 55.

Figure 55

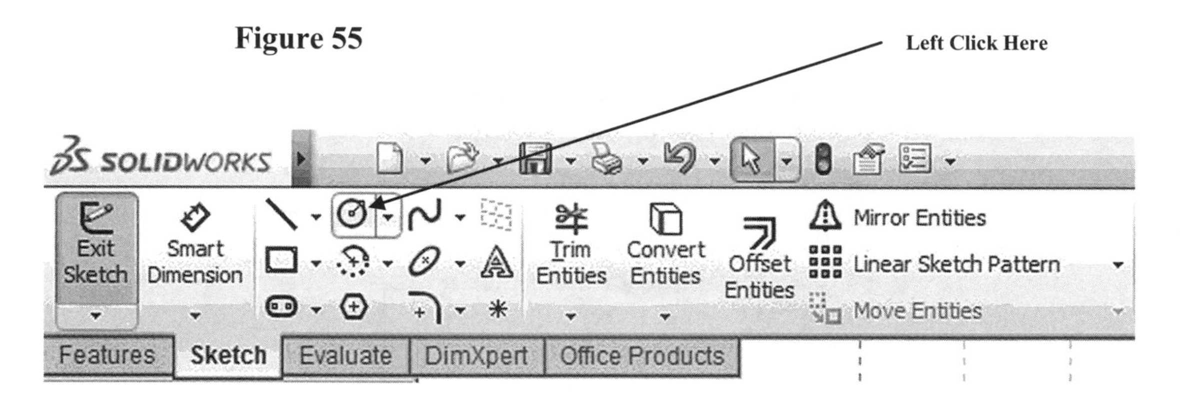

61. Left click near the center of the part on the backside surface as shown in Figure 56.

Figure 56

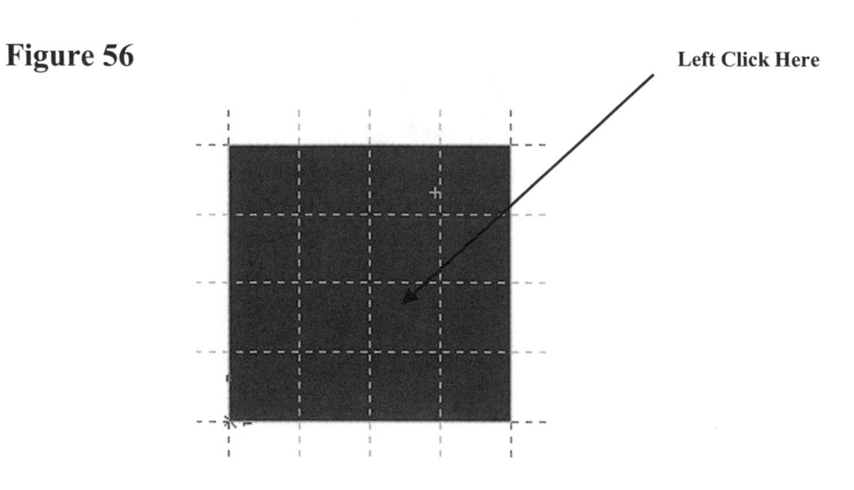

62. Move the cursor to the side forming a circle and left click as shown in Figure 57.

Figure 57

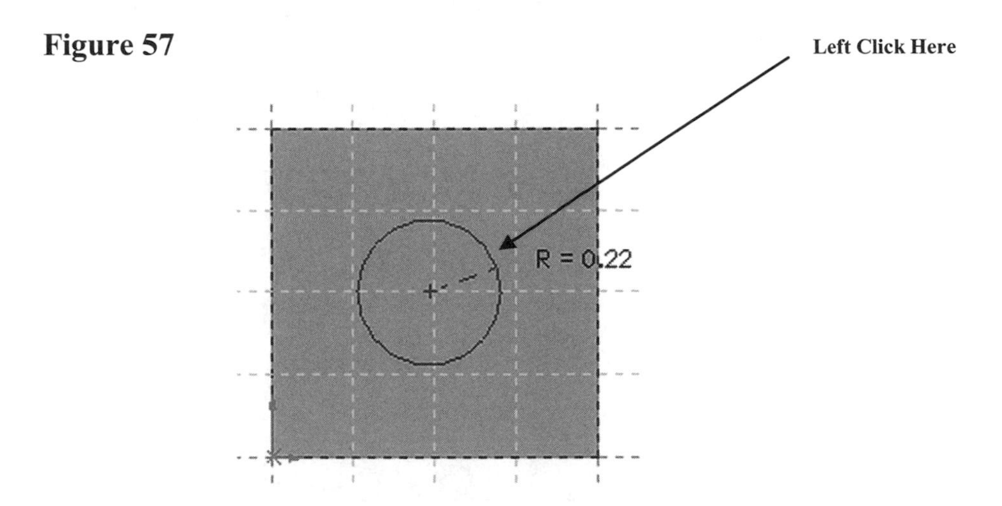

63. Right click anywhere on the screen. A pop up menu will appear. Left click on **Select** as shown in Figure 58.

Figure 58

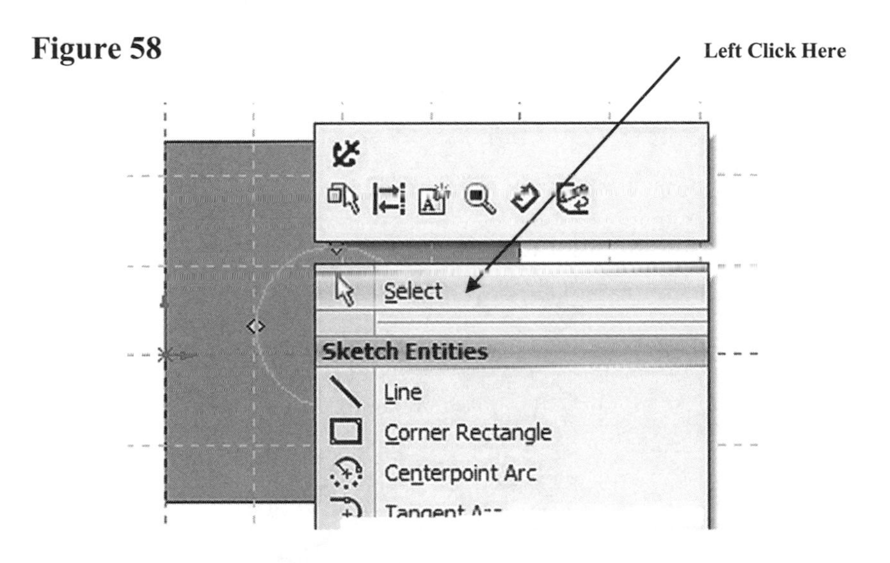

64. Move the cursor to the upper left portion of the screen and left click on **Smart Dimension** as shown in Figure 59.

Figure 59

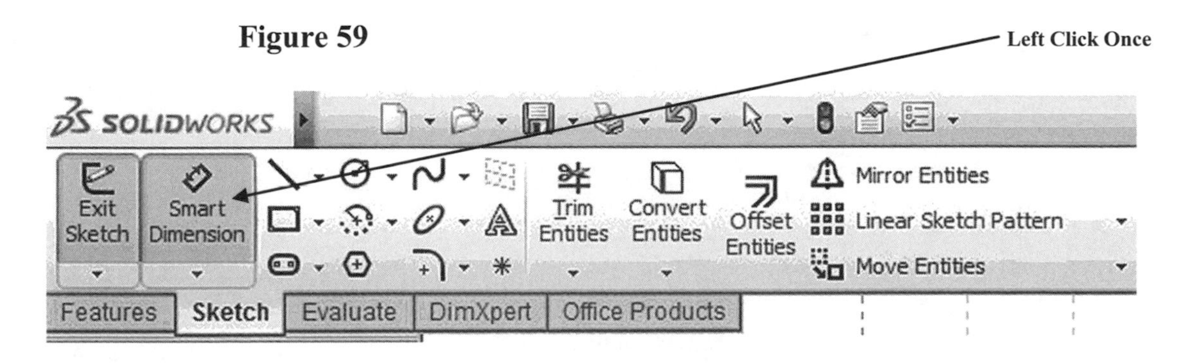

65. After selecting **Smart Dimension** move the cursor over the edge of the circle until it turns red as shown in Figure 60. Left click on the edge of the circle.

Figure 60

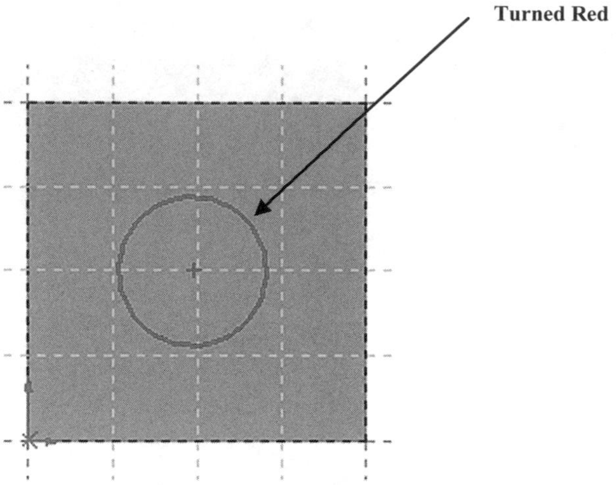

Turned Red

66. Move the cursor to where the dimension will be placed and left click once as shown in Figure 61.

Figure 61

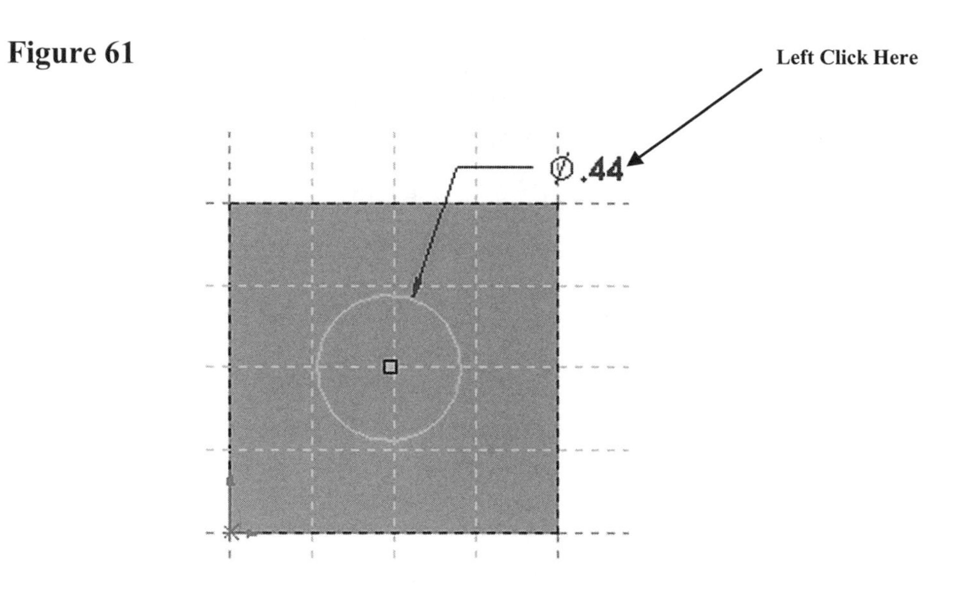

67. The Modify dialog box will appear as shown in Figure 62.

Figure 62

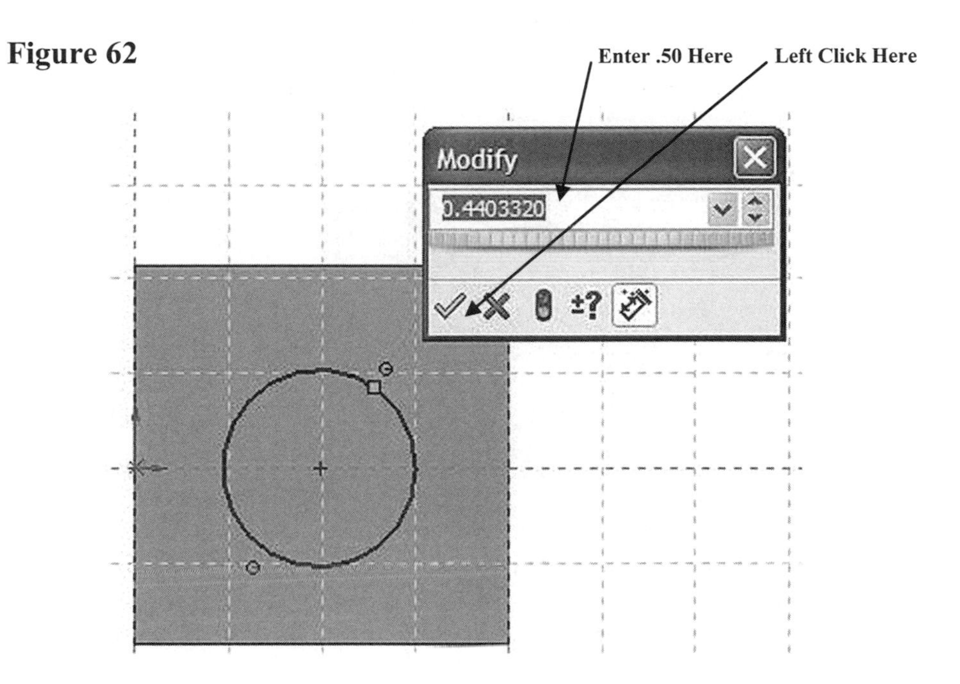

68. To edit the dimension, enter **.50** in the Modify dialog box (while the current dimension is highlighted) and either press **Enter** on the keyboard or left click on the green checkmark as shown in Figure 62.

69. Your screen should look similar to Figure 63.

Figure 63

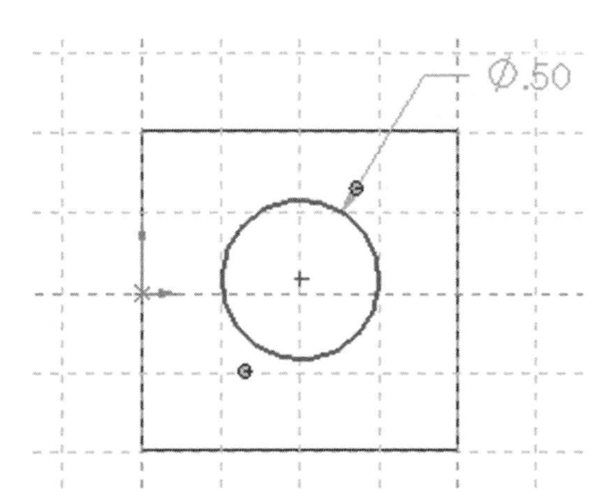

70. Move the cursor over the edge of the part until it turns red. Select the edge by left clicking anywhere on the edge as shown in Figure 64.

Figure 64

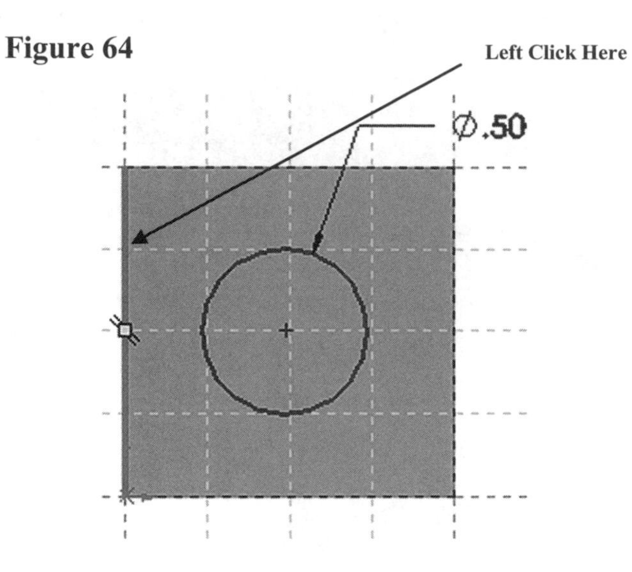

71. Move the cursor over the center of the circle until it turns red and left click as shown in Figure 65.

Figure 65

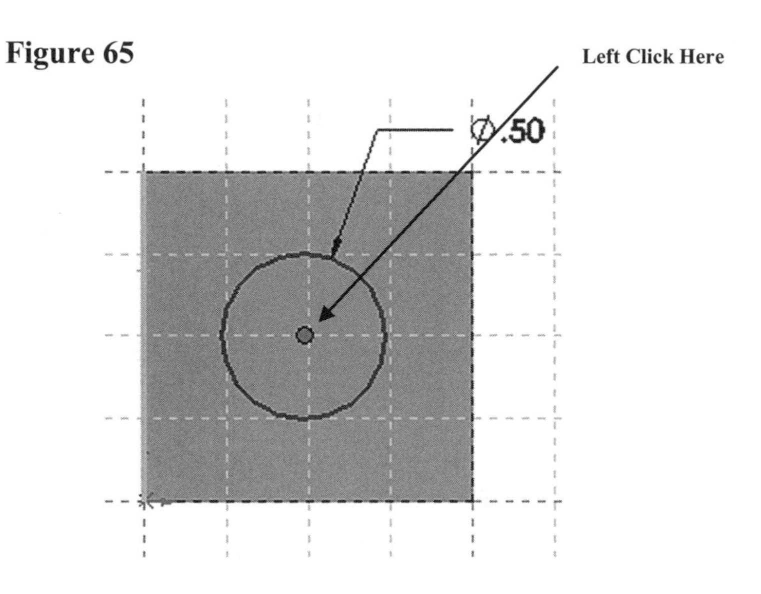

72. Move the cursor to where the dimension will be placed and left click once as shown in Figure 66.

Figure 66

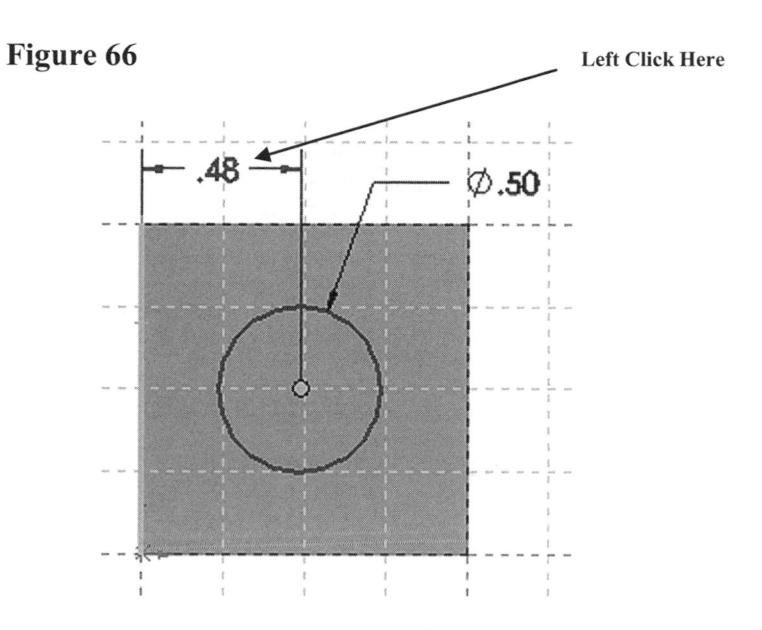

73. The Modify dialog box will appear as shown in Figure 67.

Figure 67

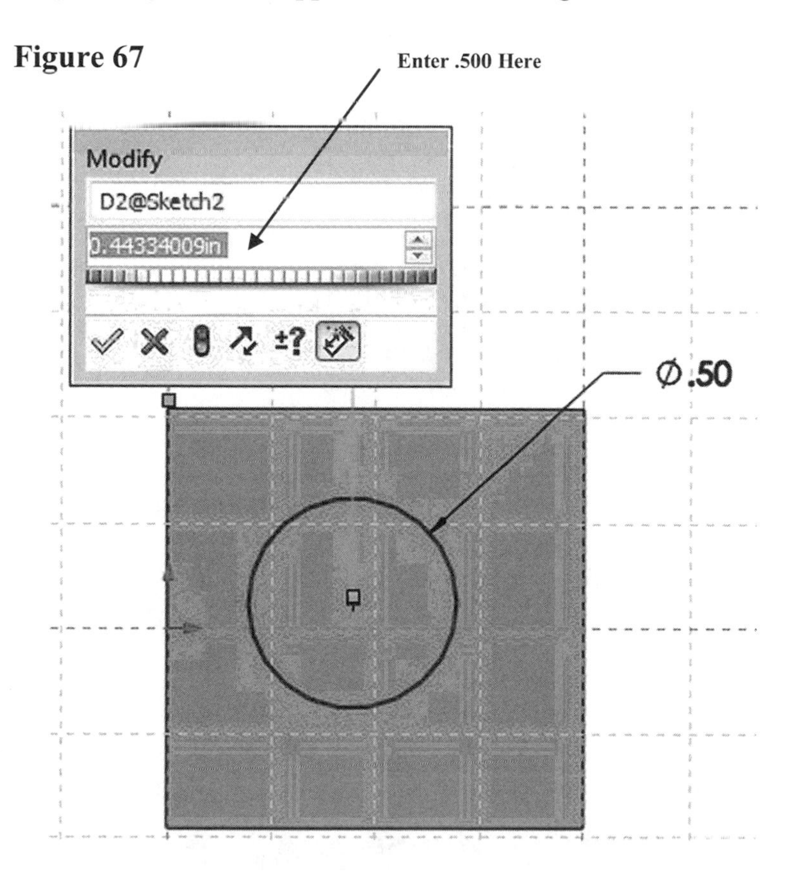

74. To edit the dimension, enter **.500** in the Modify dialog box (while the current dimension is highlighted) and either press **Enter** on the keyboard or left click on the green checkmark as shown in Figure 68.

Figure 68

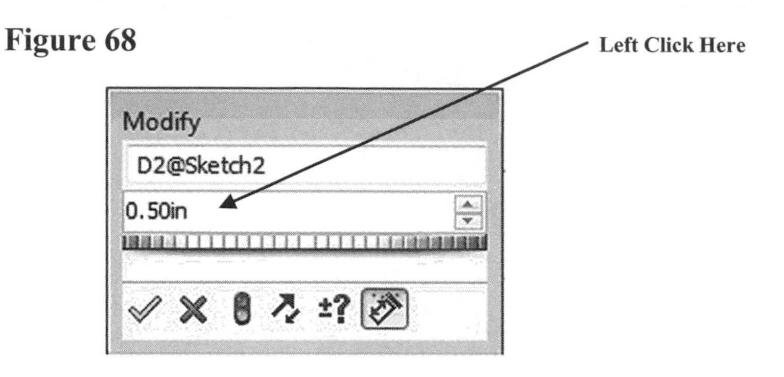

75. Your screen should look similar to Figure 69.

Figure 69

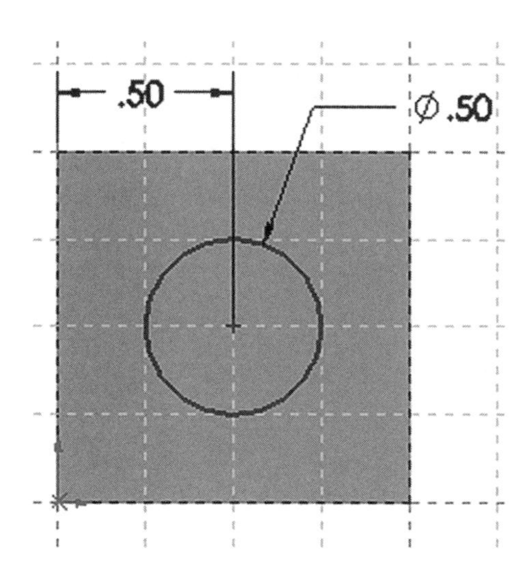

76. Move the cursor to the upper left portion of the screen and left click on **Smart Dimension** as shown in Figure 70.

Figure 70

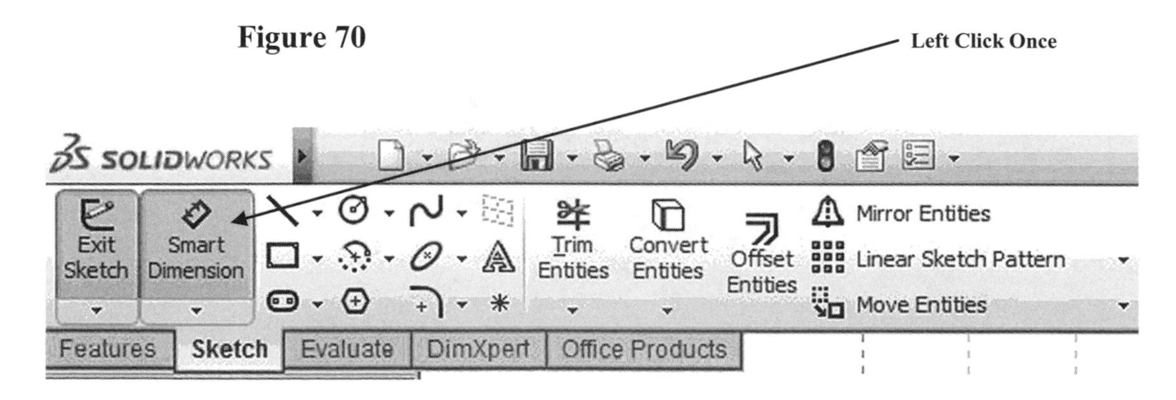

77. Move the cursor to the edge of the part until it turns red. Left click anywhere on the edge of the part as shown in Figure 71.

Figure 71

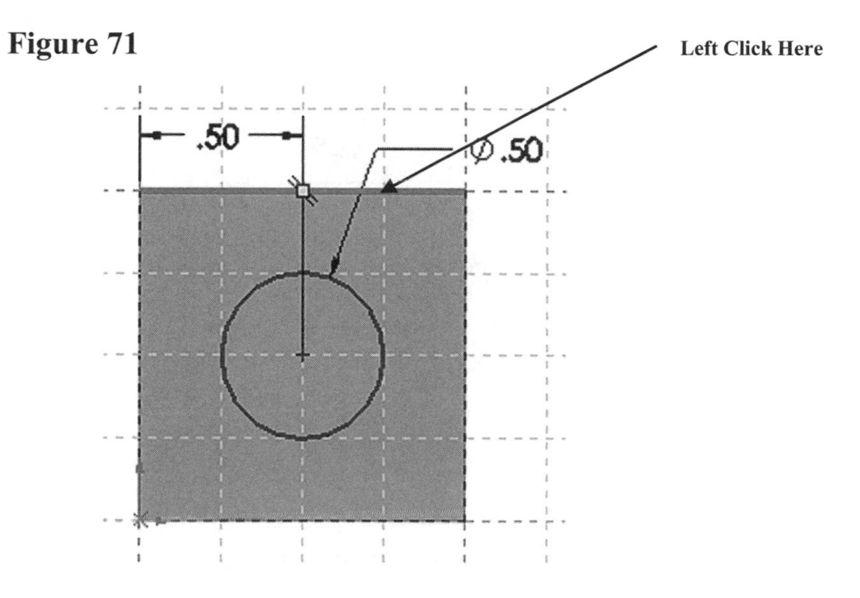

78. Move the cursor over the center of the circle until it turns red and left click as shown in Figure 72.

Figure 72

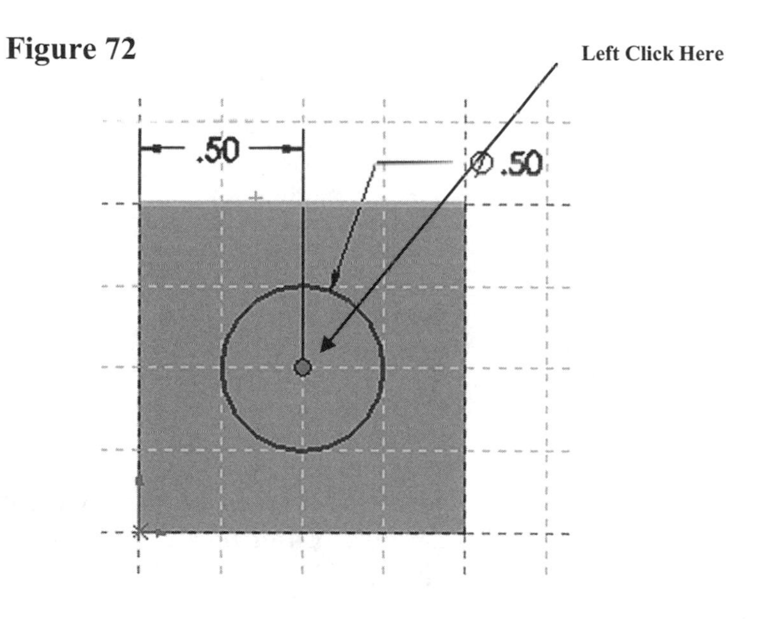

79. Move the cursor to where the dimension will be placed and left click once as shown in Figure 73.

Figure 73

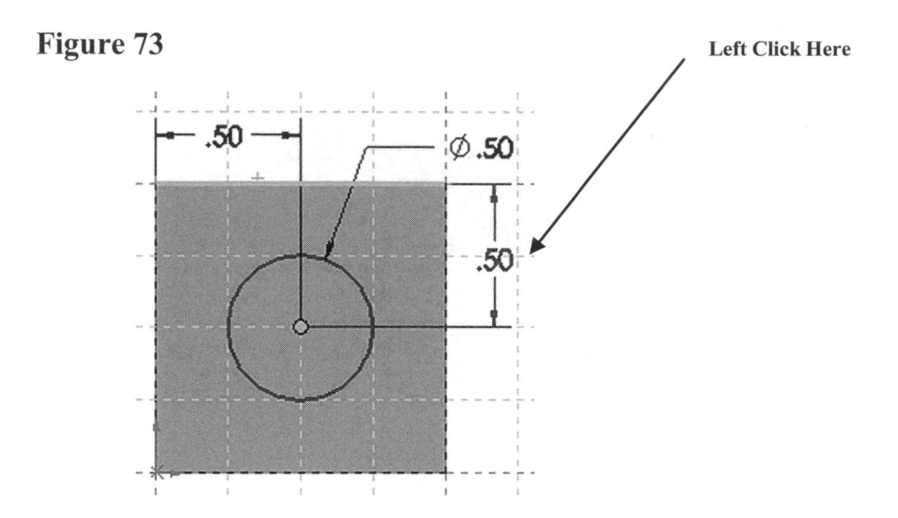

80. The Modify dialog box will appear as shown in Figure 74.

Figure 74

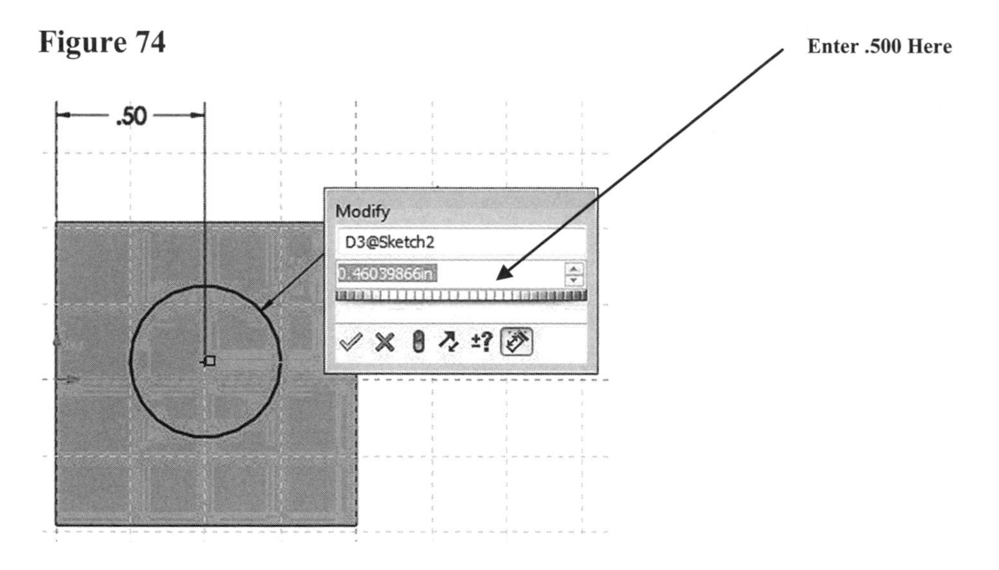

81. To edit the dimension, enter **.500** in the Modify dialog box (while the current dimension is highlighted) and either press **Enter** on the keyboard or left click on the green checkmark as shown in Figure 75.

Figure 75

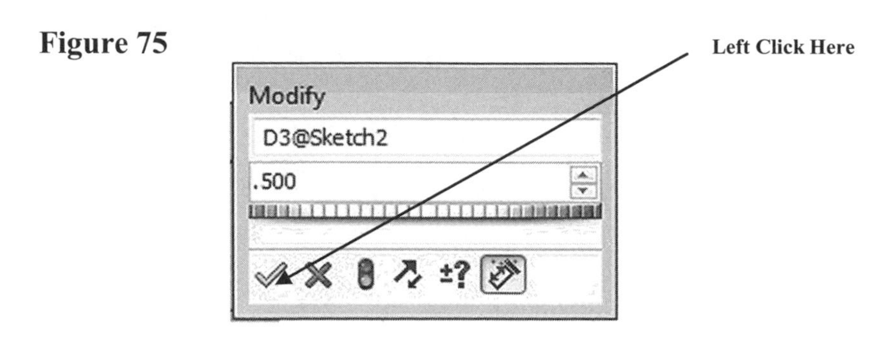

82. Your screen should look similar to Figure 76.

Figure 76

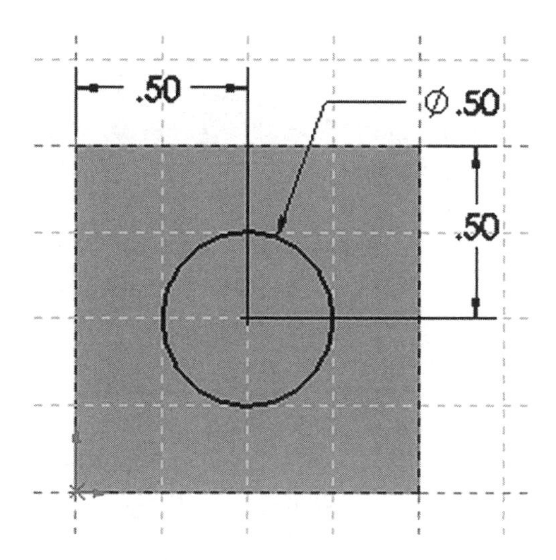

83. After the sketch is complete it is time to extrude a hole through the solid.

84. Right click anywhere on the screen. A pop up menu will appear. Left click on **Select** as shown in Figure 77. This will ensure that no commands are active.

Figure 77

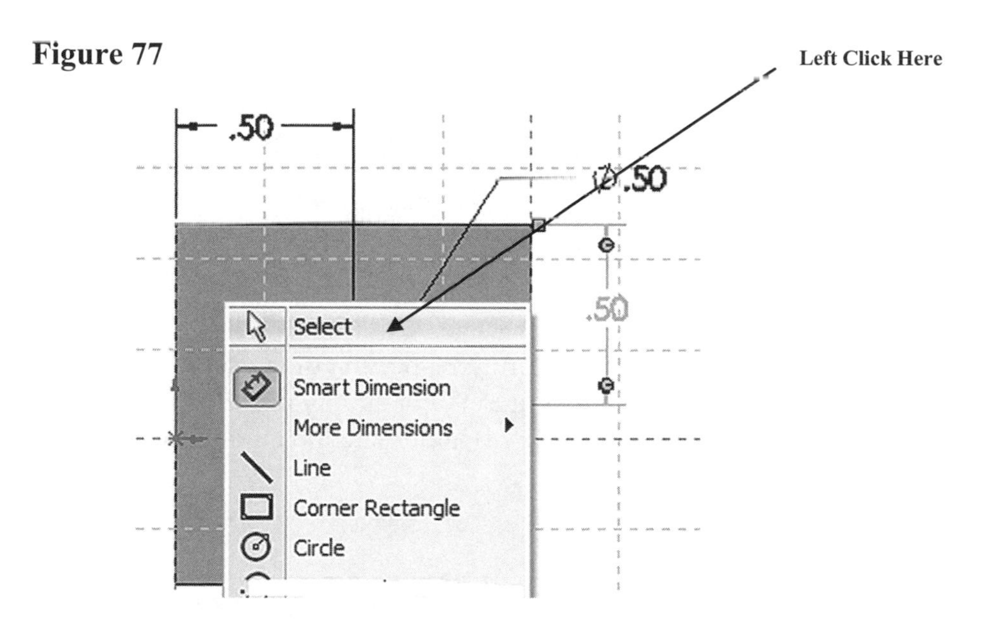

85. After you have verified that no commands are active, move the cursor to the upper left portion of the screen and left click on **Exit Sketch** as shown in Figure 78.

Figure 78

86. SolidWorks is now out of the Sketch commands and into the Features commands. Notice that the commands at the top of the screen are now different. To work in the Features commands a sketch must be present and have no opens (non-connected lines). If there are any opens in the sketch an error message will appear. Your screen should look similar to Figure 79.

Figure 79

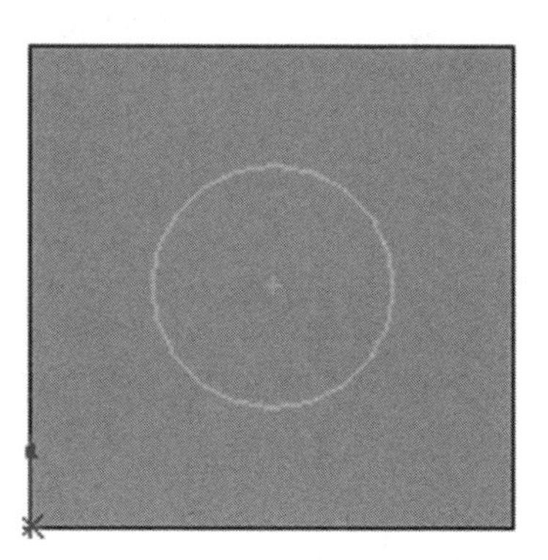

87. Move the cursor to the upper middle portion of the screen and left click on the "Rotate" icon as shown in Figure 80 or simply push the mouse wheel down (holding the mouse wheel down) and rotate the model off to the side.

Figure 80

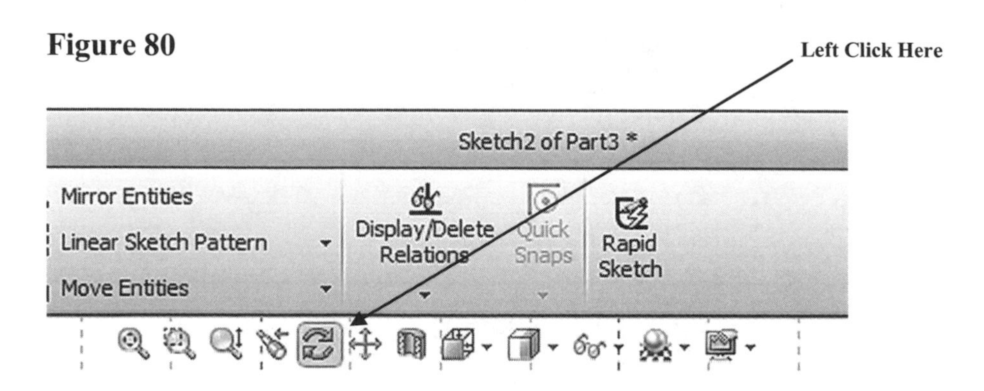

88. While holding the left mouse button down, drag the cursor to the right to gain an isometric view of the part as shown in Figure 81.

Figure 81

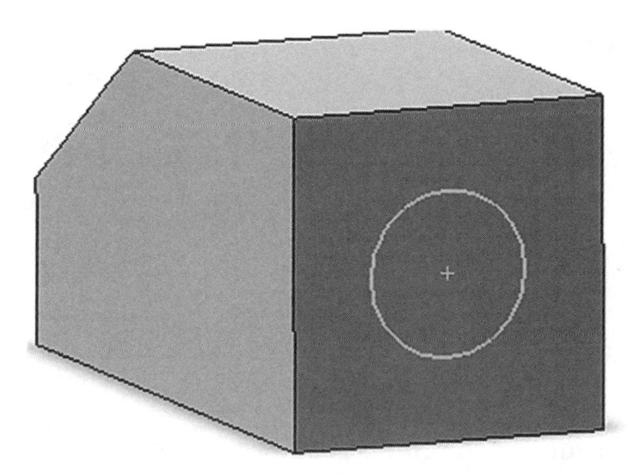

89. Right click anywhere on the screen. A pop up menu will appear. Left click on **Select** as shown in Figure 82.

Figure 82

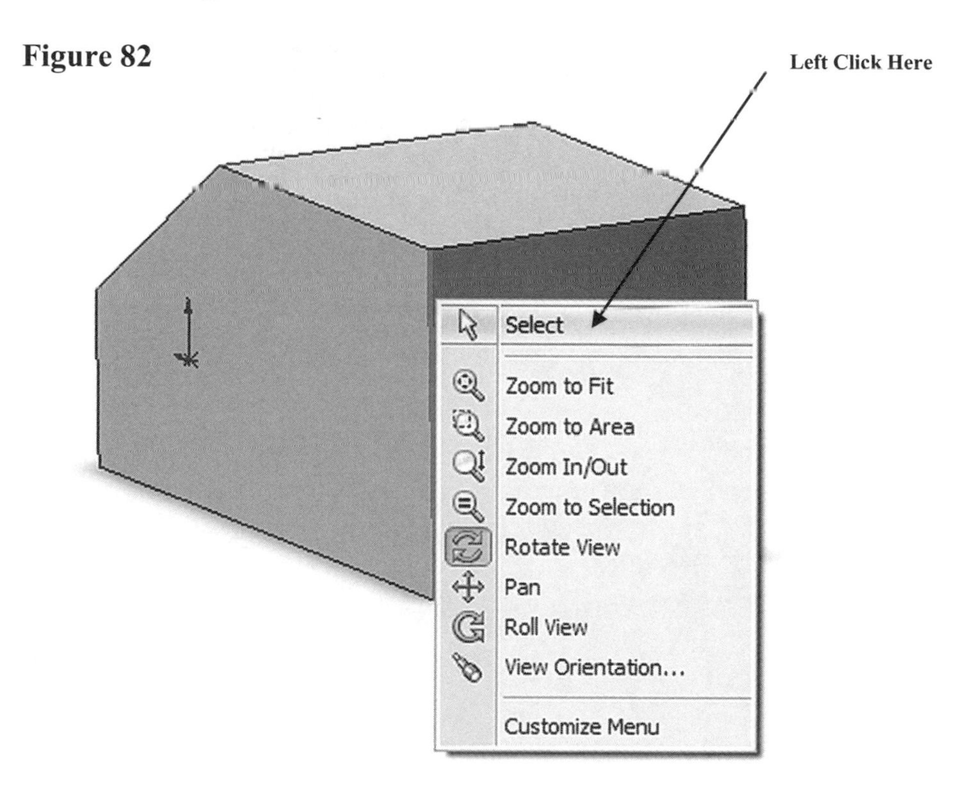

90. Move the cursor to the upper left portion of the screen and left click on **Extruded Cut** as shown in Figure 83. If you received an error message, there are opens (non-connected lines) somewhere on the sketch. Check each intersection for opens by using the **Extend** and **Trim** commands.

Figure 83

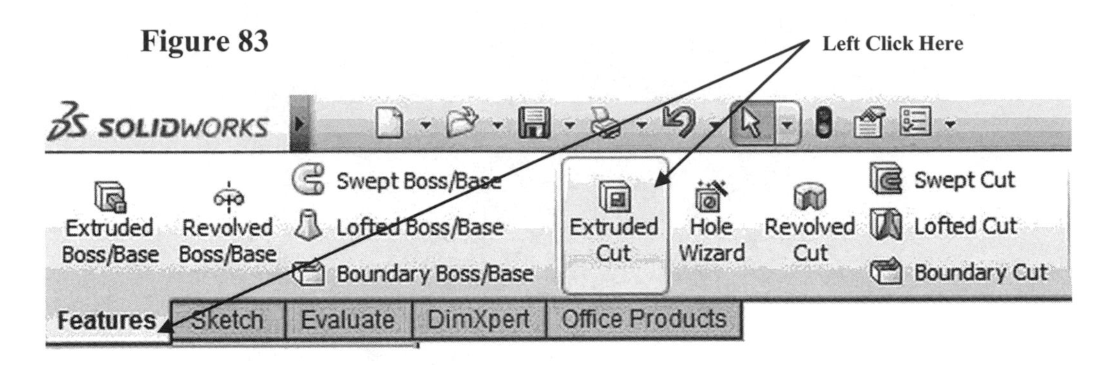

91. Left click on the edge of the circle as shown in Figure 84.

Figure 84

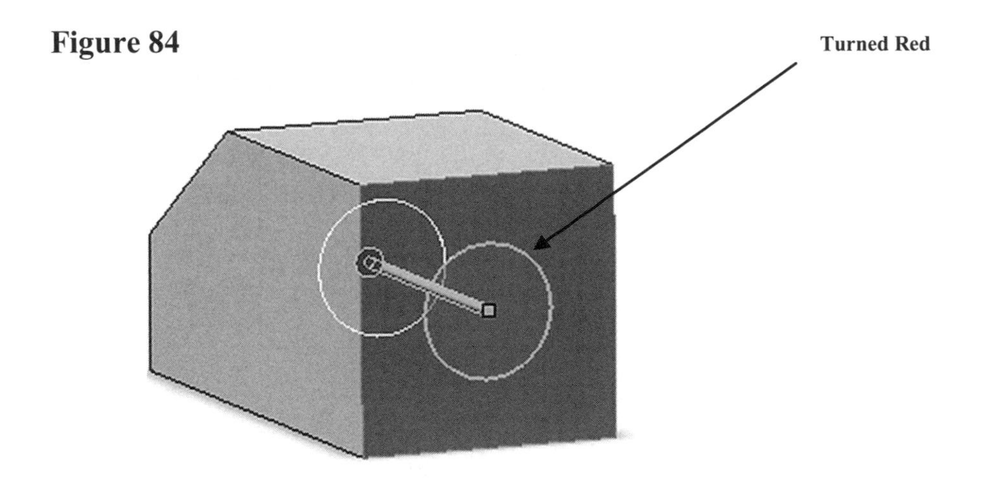

92. Enter **2.00** for D1 as shown in Figure 85.

Figure 85

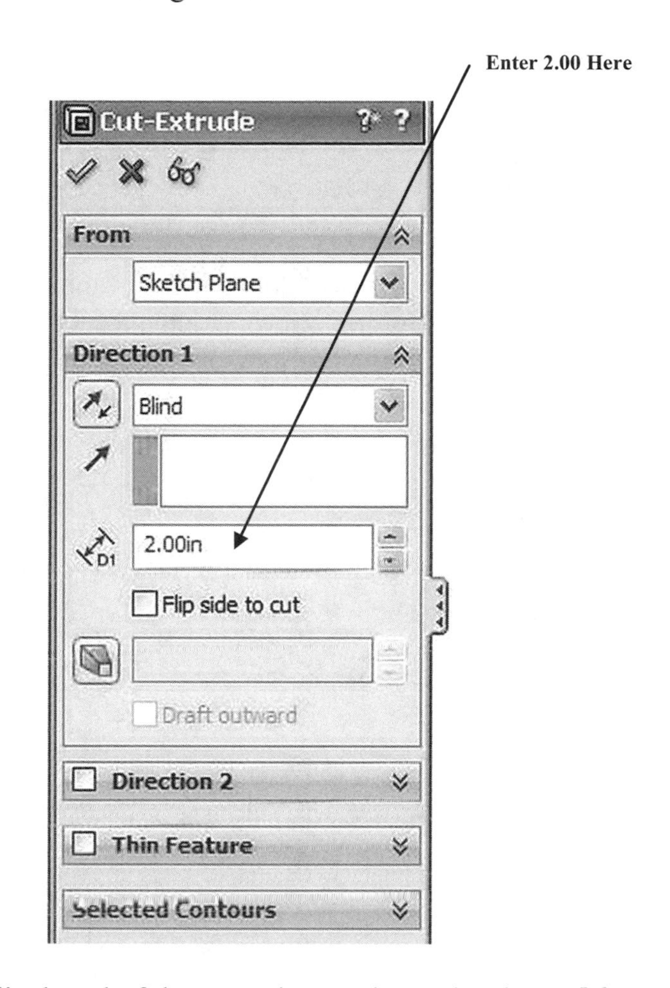

93. A preview will be displayed of the extrusion as shown in Figure 86.

Figure 86

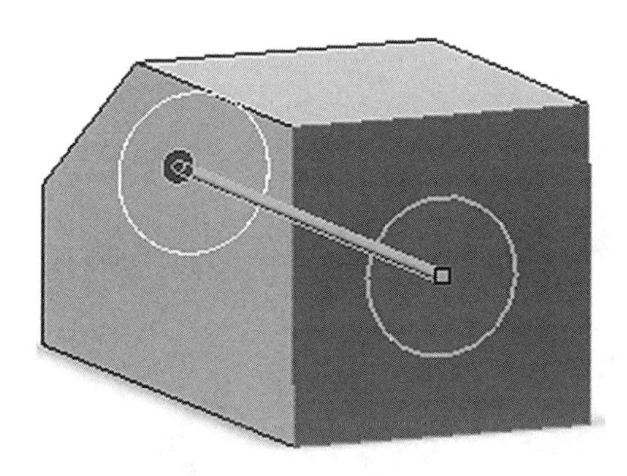

94. Left click on the green checkmark as shown in Figure 87.

Figure 87

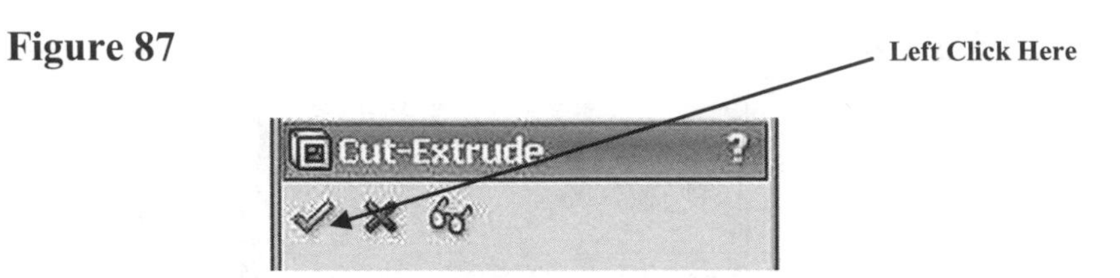

95. Move the cursor to the upper middle portion of the screen and left click on the drop down arrow next to the "Views Orientation" icon. A drop down menu will appear. Left click on **Trimetric** as shown in Figure 88.

Figure 88

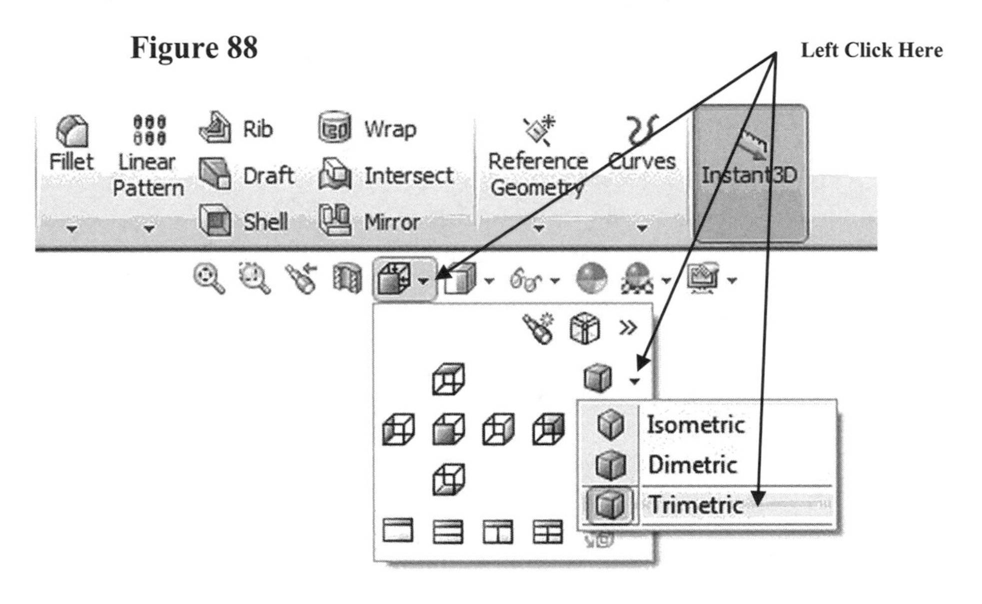

96. Your screen should look similar to Figure 89.

Figure 89

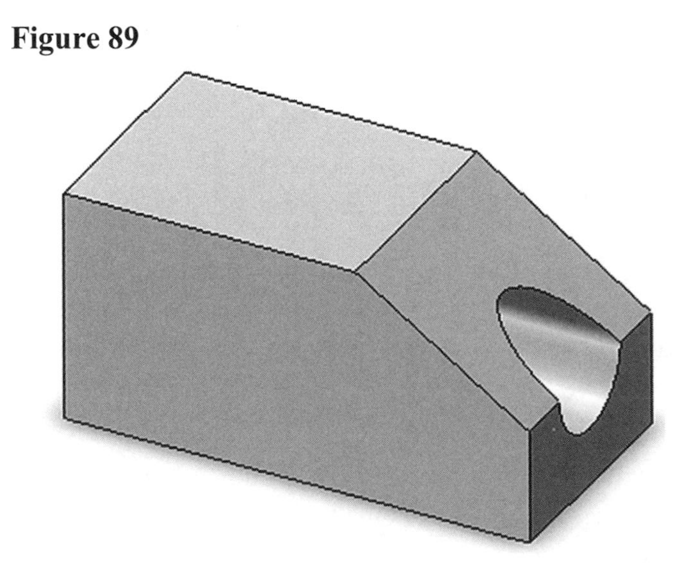

97. Save the part file for easy retrieval to be used in the following section.

98. After the part file has been saved, move the cursor to the upper left portion of the screen and left click on the drop down arrow to the right of the "New Drawing" icon. A drop down menu will appear. Left click on **Make Drawing from Part/Assembly** as shown in Figure 90.

Figure 90

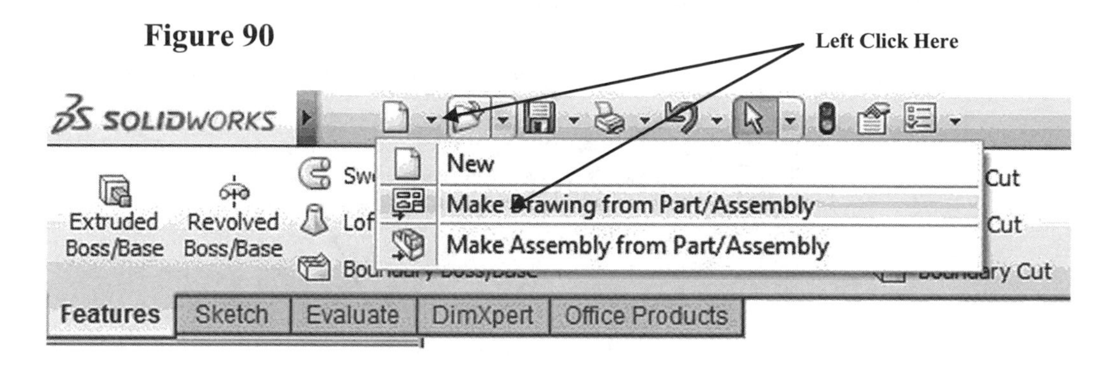

99. The Sheet Format/Size dialog box will open as shown in Figure 91.

Figure 91

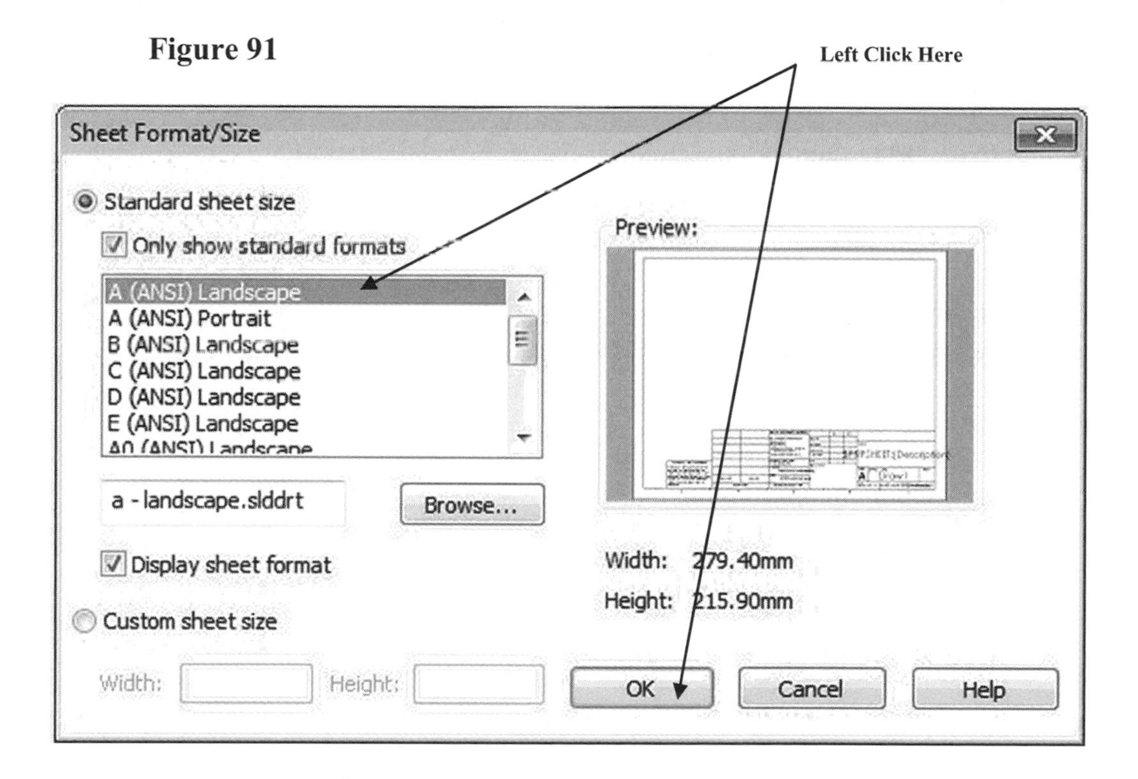

100. Left click on **A-Landscape** as shown in Figure 91.

101. Left click on **OK**.

102. Drag any view, that you want to be the Front view, off of the View Palette at the right (not shown) on to the sheet. After placing the Front view, move the cursor up and left click once. The Top view will be attached to the cursor and will appear. Continue placing the Side view which will also be attached to the cursor. After the views have been placed, move the cursor to the upper right portion of the screen and right click on **Sheet Format**. A pop up menu will appear. Left click on **Properties** as shown in Figure 92.

Figure 92

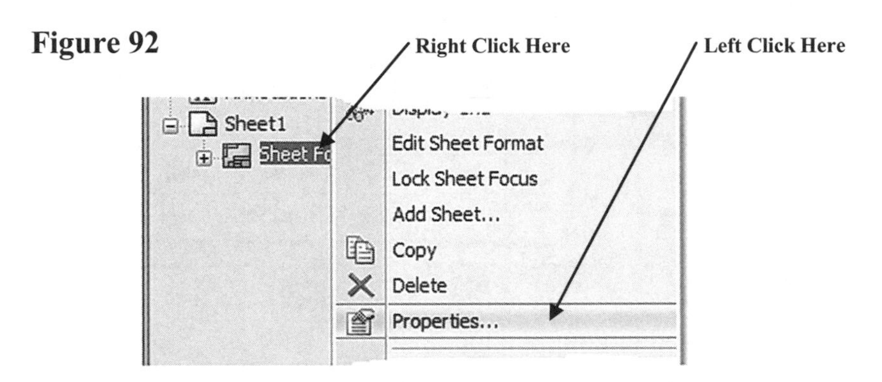

103. The Sheet Properties dialog box will appear. You may want to increase/decrease the scale to fit the work areas. Left click next to Third Angle Projection. Left click on **OK** as shown in Figure 93.

Figure 93

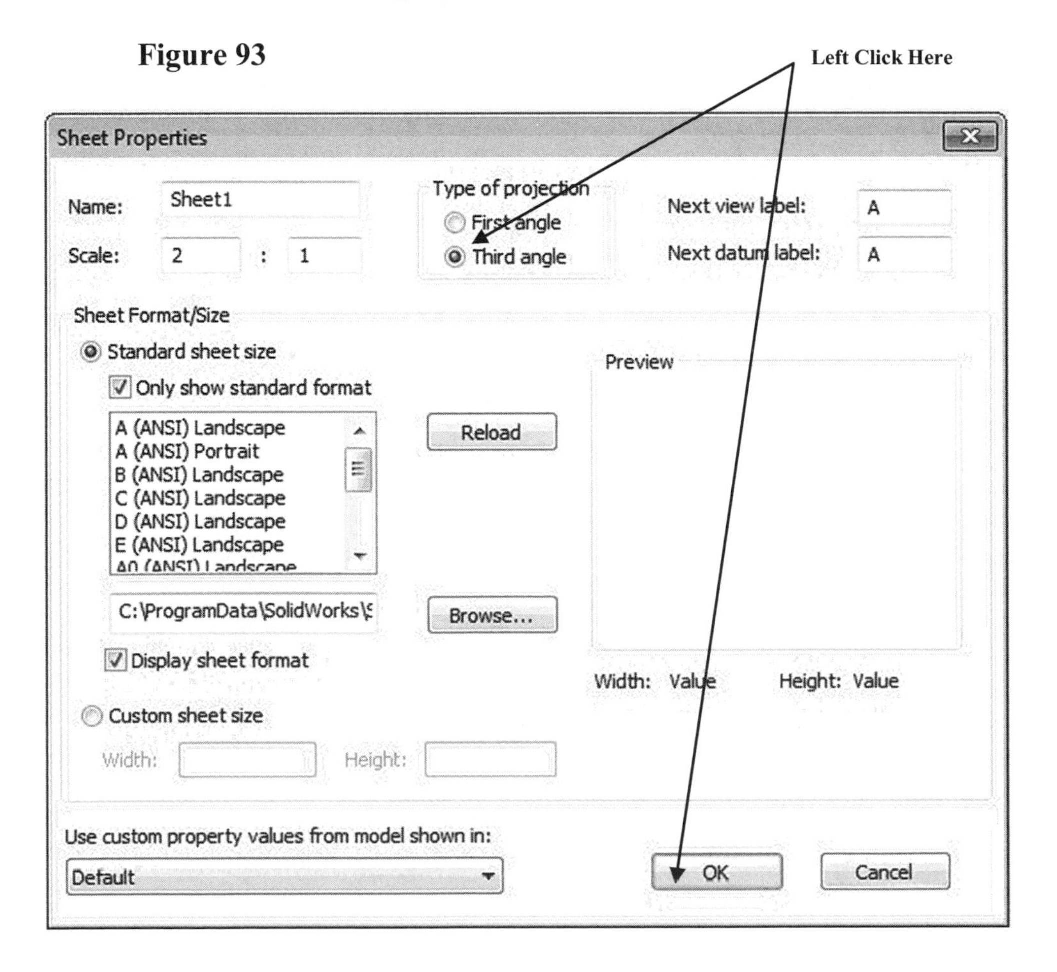

104. The following tools should be visible. If the tools are not visible, left click on the **View Layout** tab as shown in Figure 94.

Figure 94

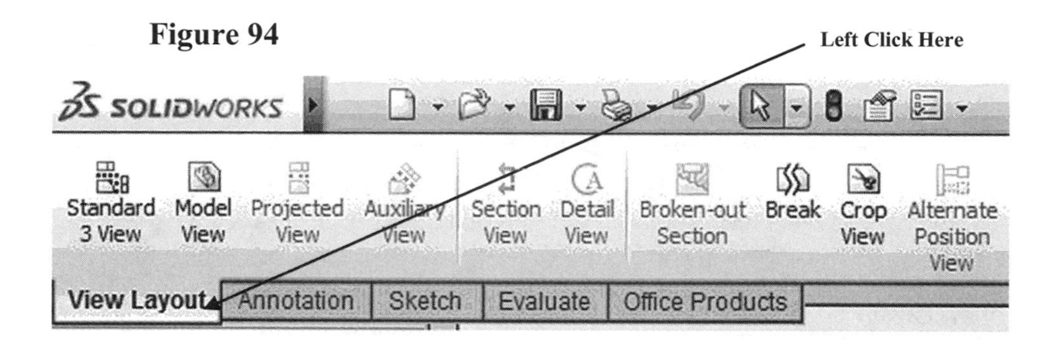

105. **NOTE: If you were able to create a 3 view drawing buy dragging views off the View Pallet then skip to step 107.** Move the cursor to the upper left portion of the screen and left click on **Standard 3 View** as shown in Figure 95.

Figure 95

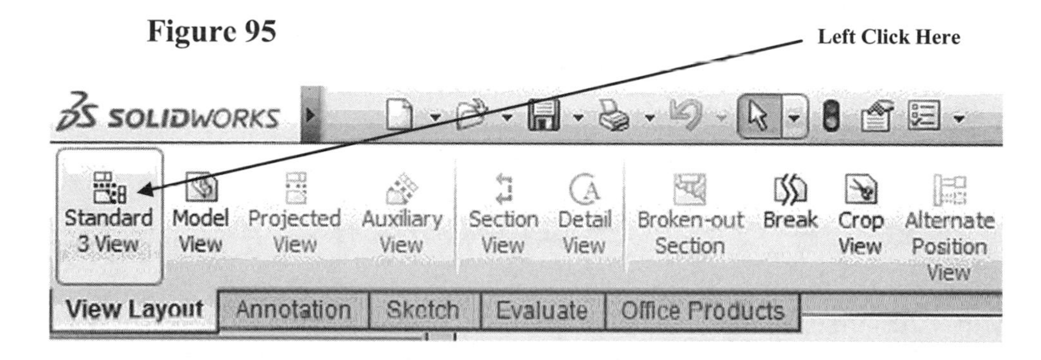

106. If the file name you have been working on appears as shown in Figure 96, move the cursor over the file name and double click. If the file name does not appear, use the Browse button to locate the file. SolidWorks will create a 3 view drawing.

Figure 96

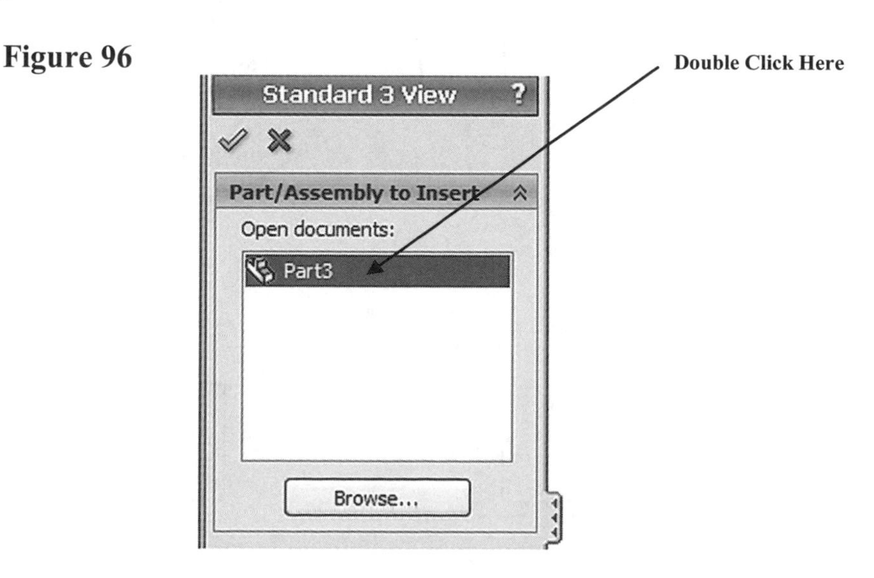

107. Move the cursor to the upper left portion of the screen and left click on **Projected View** as shown in Figure 97.

Figure 97

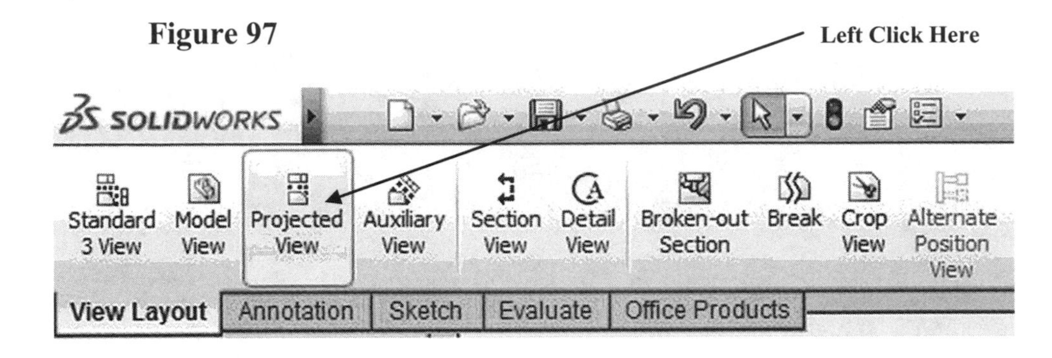

108. Move the cursor to the Front View causing a red dashed box to appear. Left click inside the box as shown in Figure 98.

Figure 98

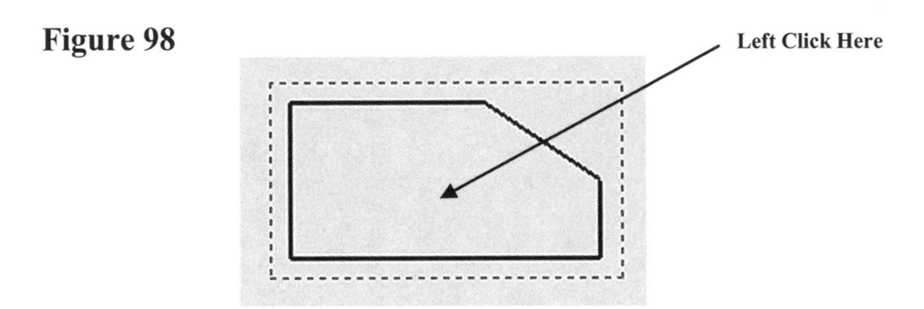

109. Move the cursor to the upper right portion of the screen. An isometric view of the part will be attached to the cursor. Left click as shown in Figure 99.

Figure 99

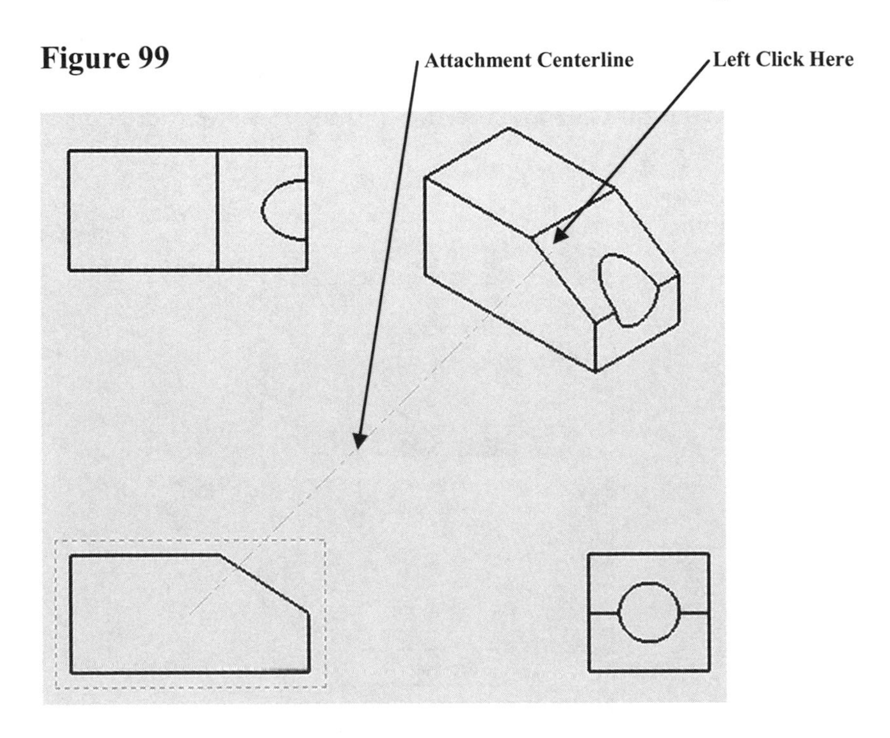

110. Move the cursor to the Front View causing a red dashed box to appear. Left click inside the box as shown in Figure 100. The Feature Manager will appear to the left.

Figure 100

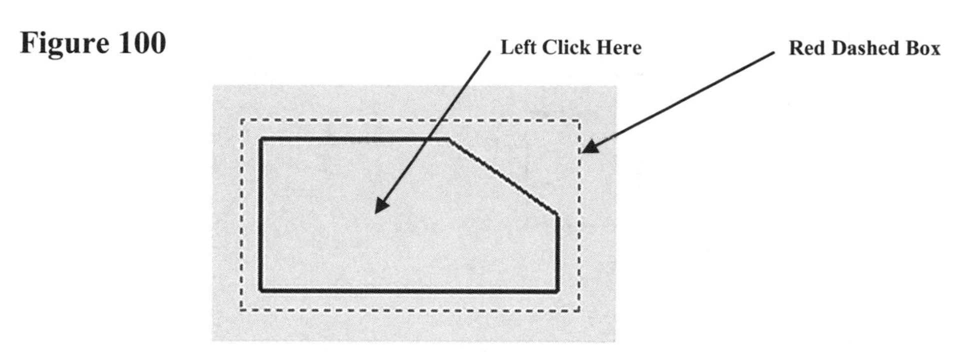

111. Move the cursor to the middle left portion of the screen under "Display Style" and left click on the "Hidden Lines Visible" icon as shown in Figure 101.

Figure 101

112. Move the cursor to the upper left portion of the screen and left click on the green checkmark as shown in Figure 102.

Figure 102

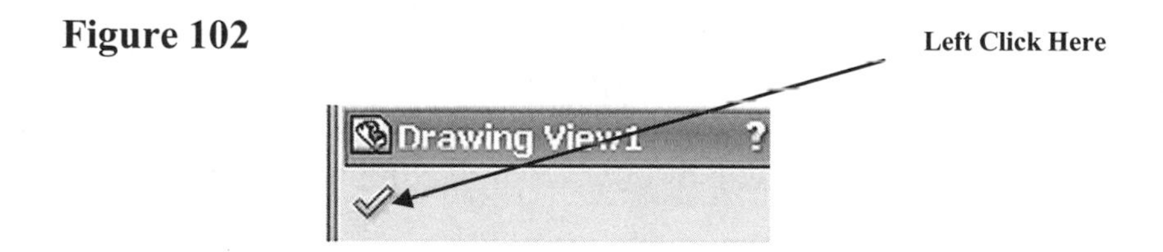

113. Your screen should look similar to Figure 103.

Figure 103

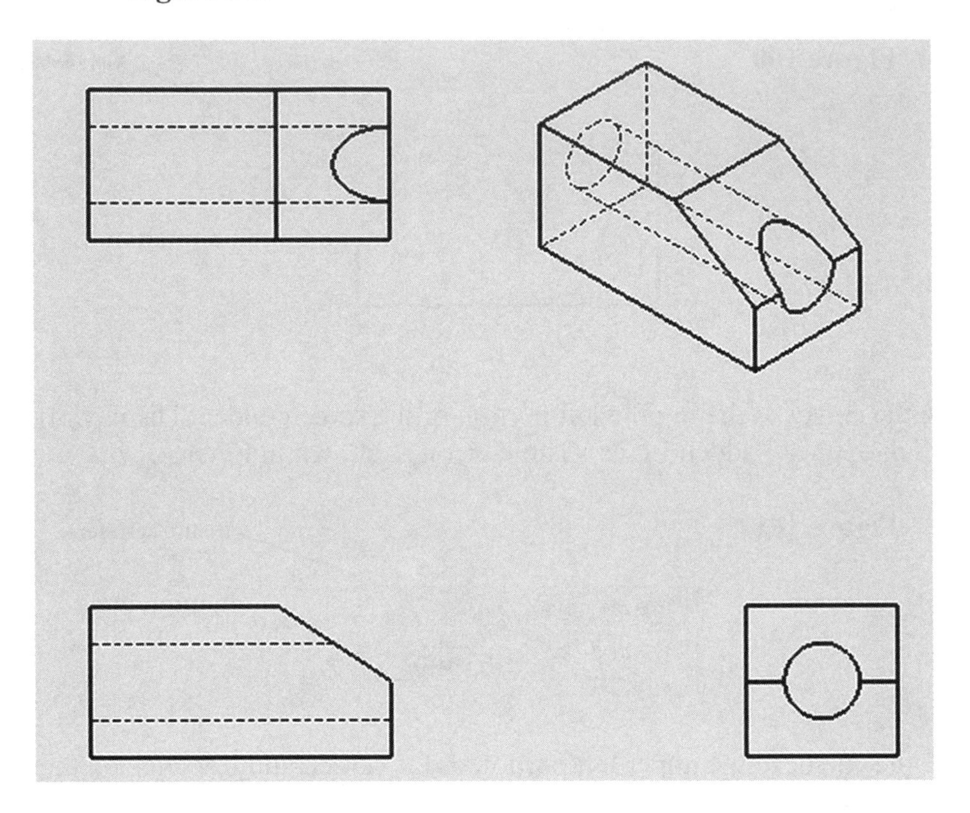

114. If for some reason the views need to be relocated, (moved closer together for example) move the cursor to any of the views (the top view in this case) causing red dots to appear and left click on the red dashed line (holding the left mouse button down). Drag the cursor down to move the view to the desired location as shown in Figure 104.

Figure 104

Left Click Here, Hold Down and Drag

115. Once the views have been moved closer together move the cursor to the upper left portion of the screen and left click on the green checkmark as shown in Figure 105.

Figure 105

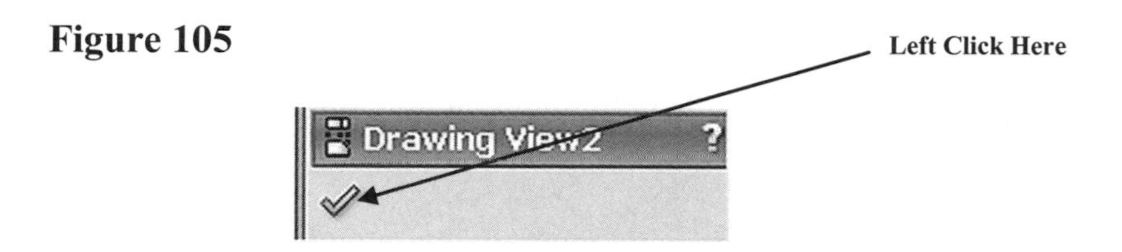

116. Your screen may look similar to Figure 106.

Figure 106

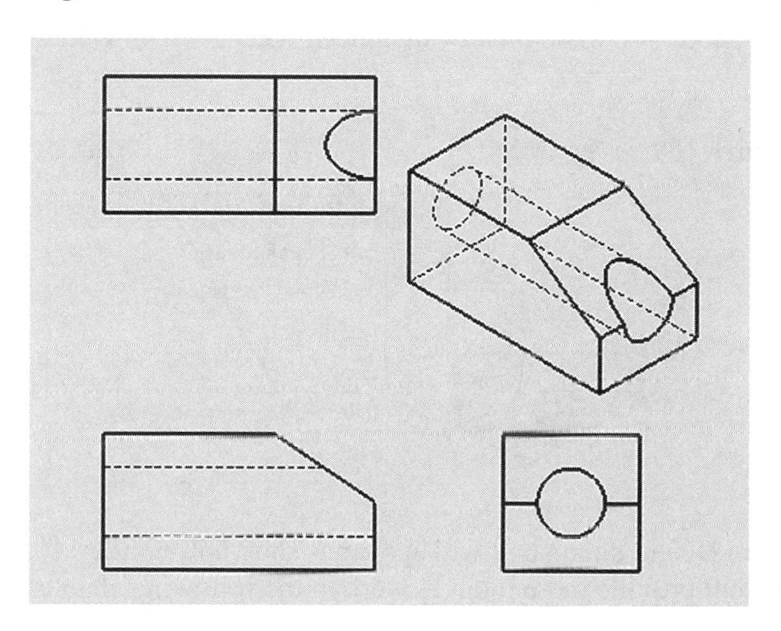

117. Move the cursor to the isometric view causing a red dashed box to appear around the part. Left click once. Move the cursor to the middle left portion of the screen and left click on the "Shaded with Edges" icon as shown in Figure 107.

Figure 107

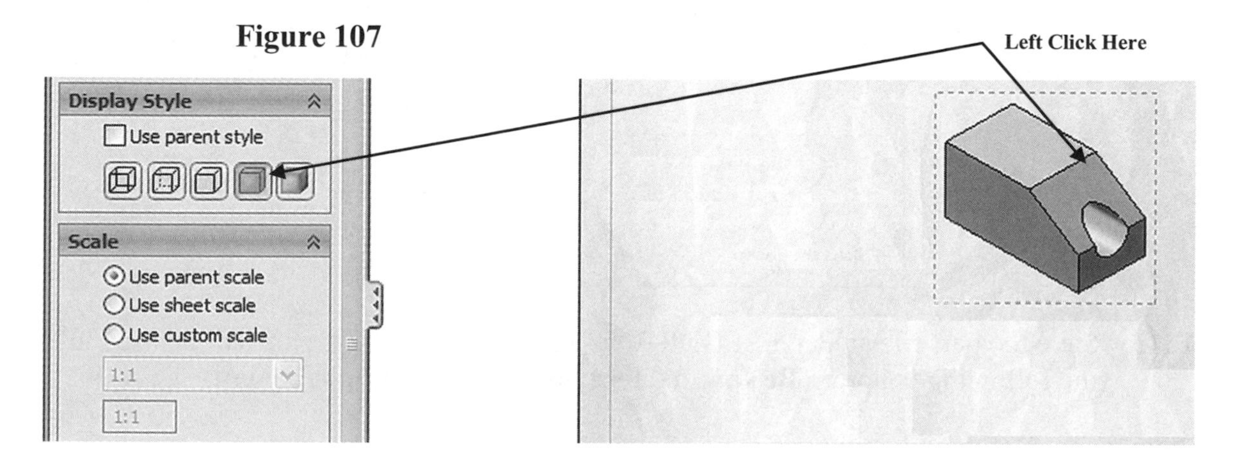

118. Once you have created the isometric view, it will need to be deleted in order to create more room on the drawing. Move the cursor around the isometric view causing a red dashed box to appear around the part as shown in Figure 108.

Figure 108

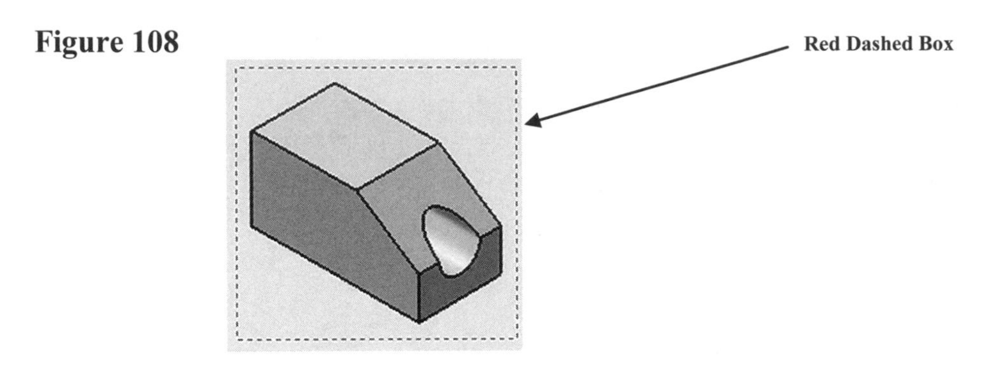

119. Right click once. A pop up menu will appear. Left click on **Delete** as shown in Figure 109.

Figure 109

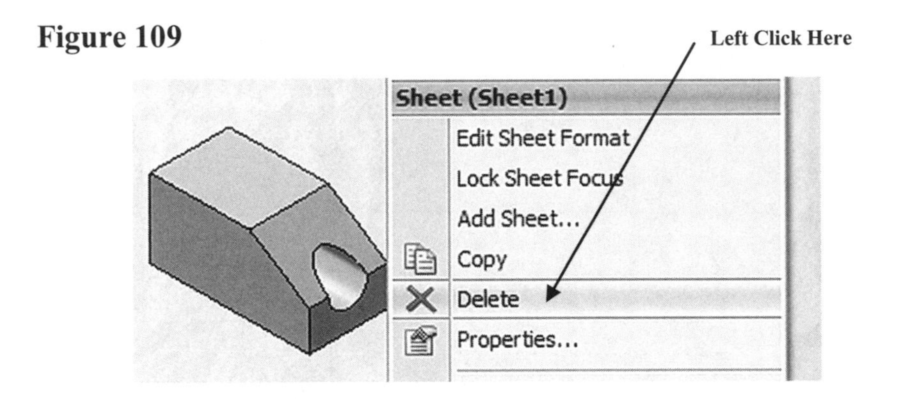

120 The Confirm Delete dialog box will appear as shown in Figure 110. Left click on **Yes**. This will provide more room to work in the following chapter.

Figure 110

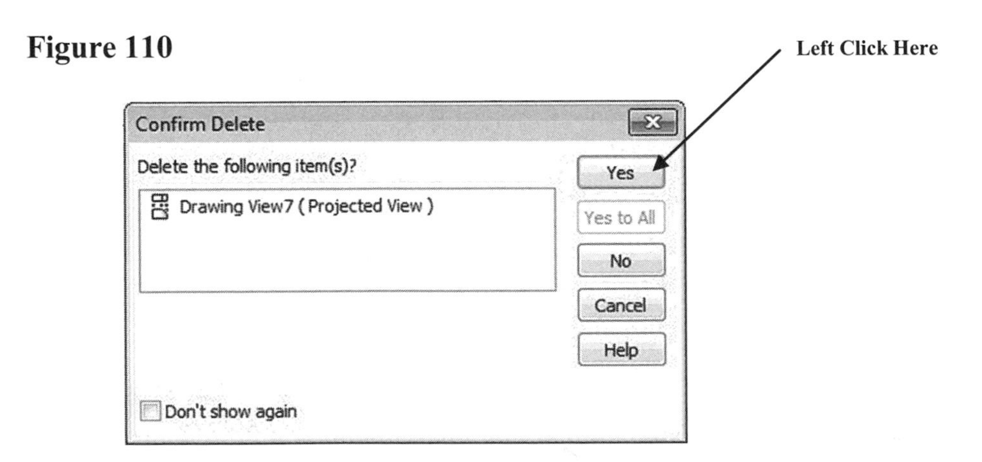

121. Save the part file as Part1.SLDDRW for easy retrieval. This part will be used in the following chapter. **Be sure to close SolidWorks completely out**.

Drawing Activities

Use these problems from Chapters 1 and 2 to create 3 view orthographic view detail drawings.

Problem 1

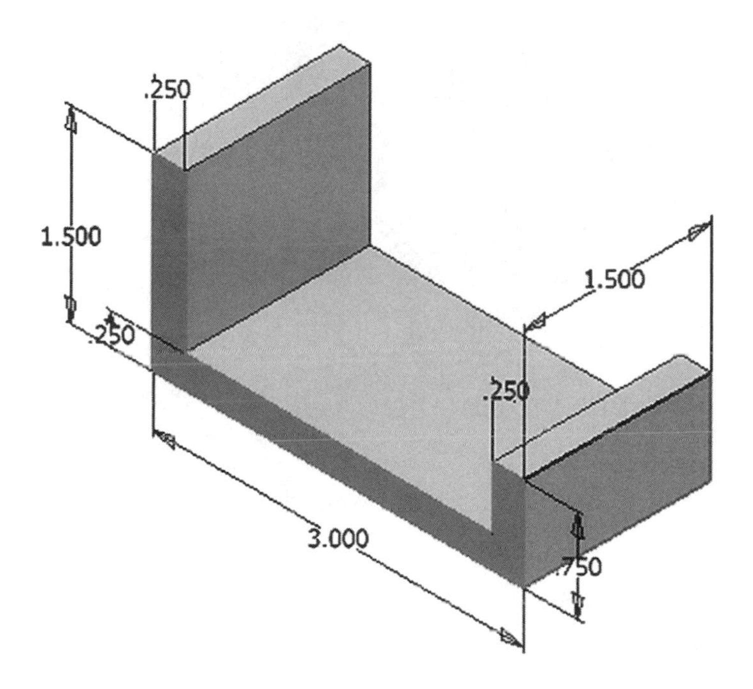

Problem 2

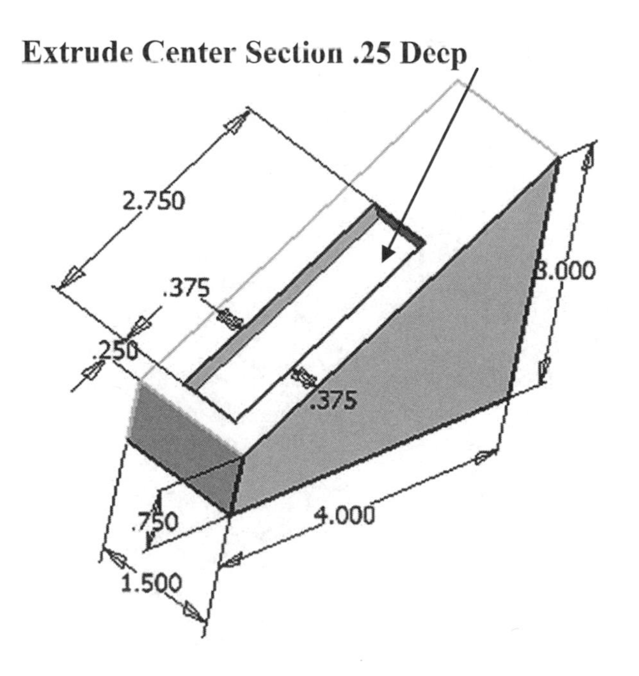

Problem 3

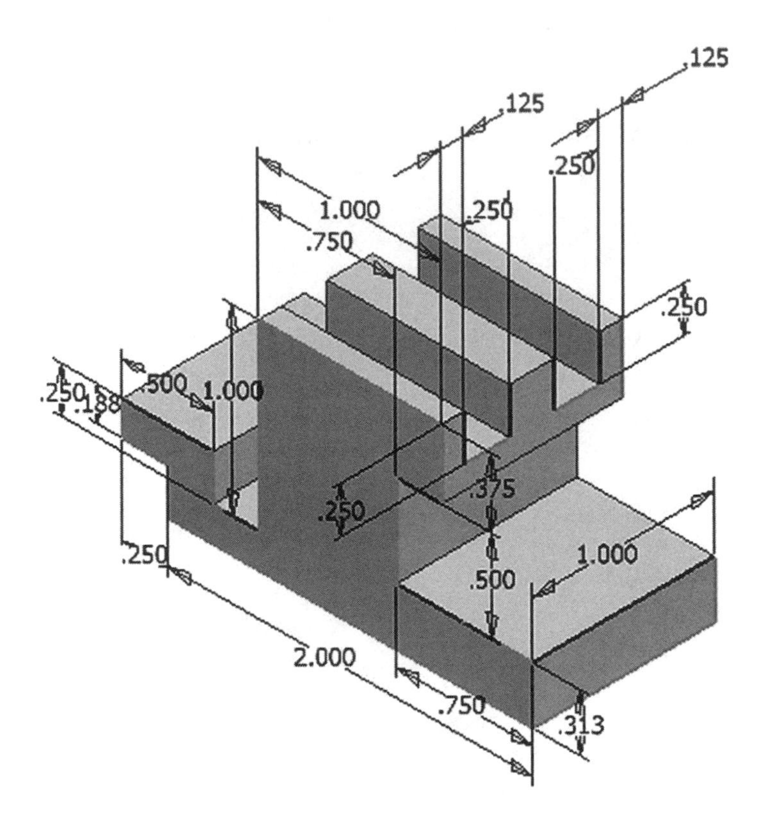

Problem 4

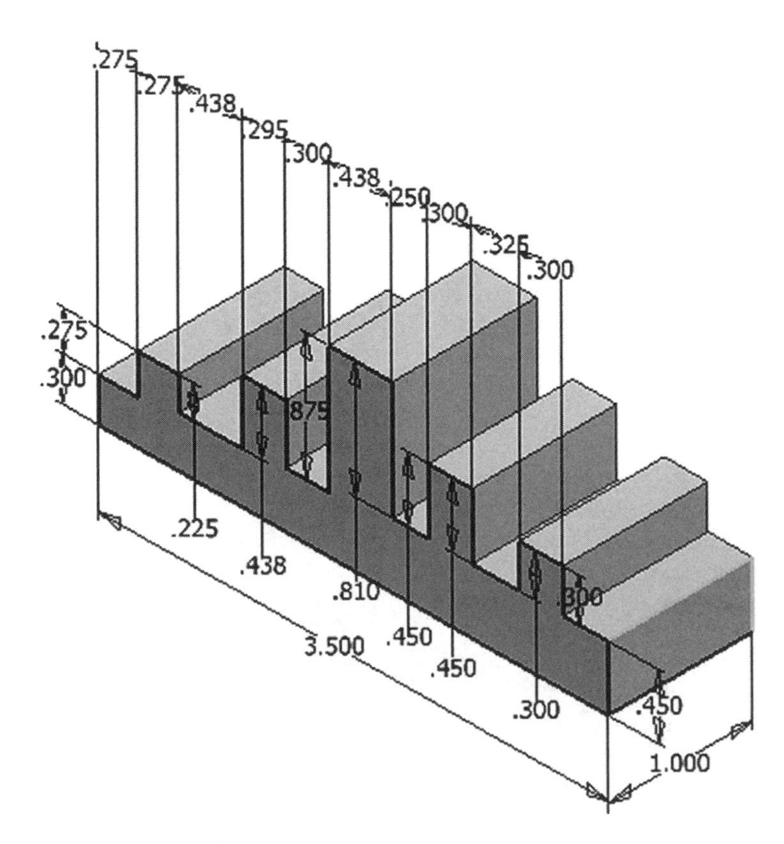

Problem 5

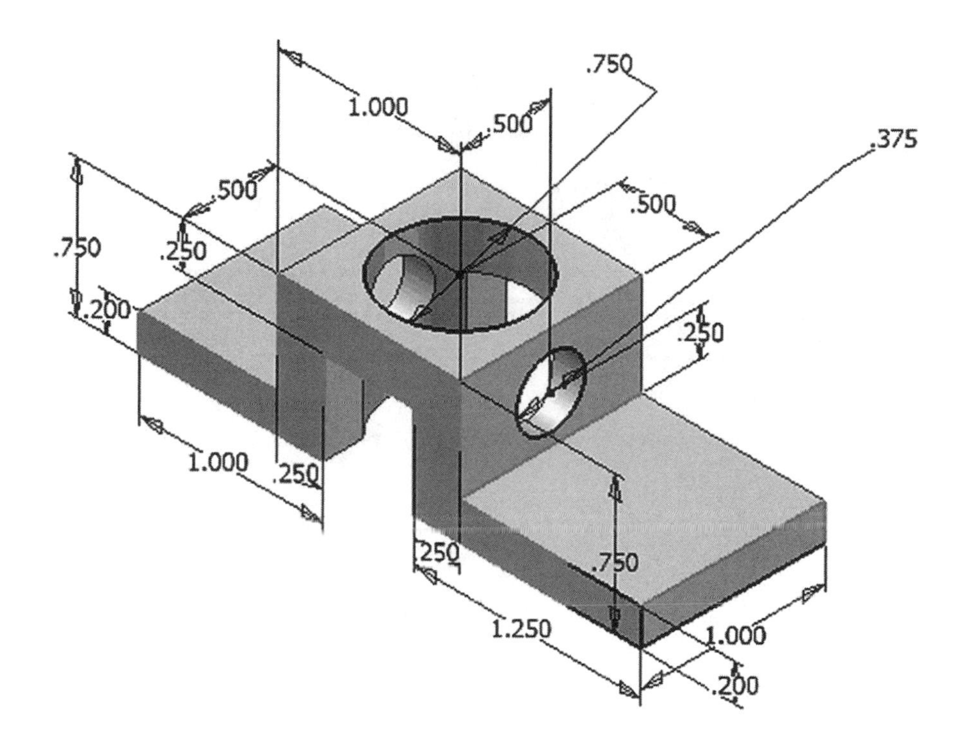

Problem 6

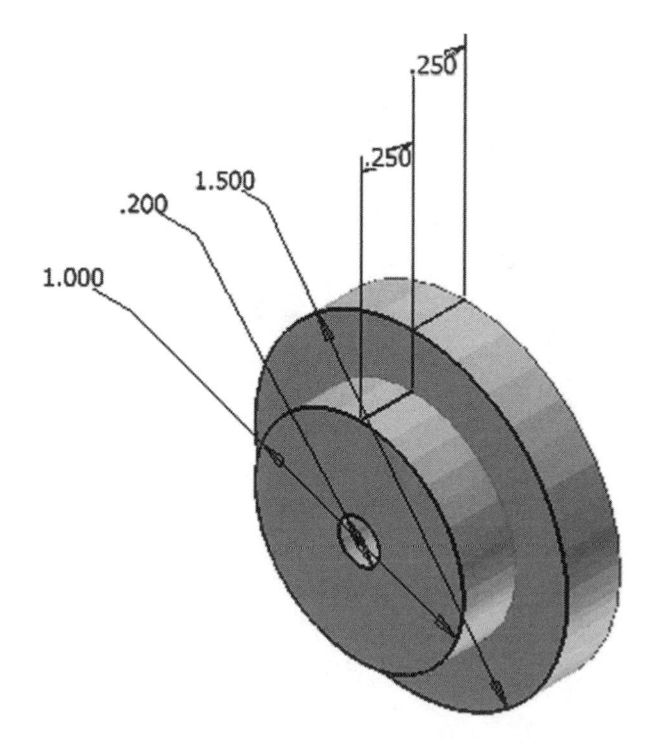

Problem 7

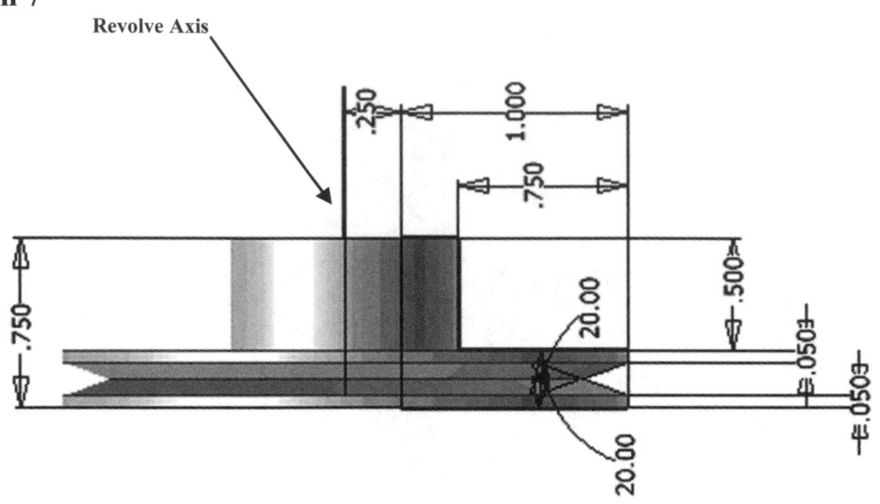

Problem 8

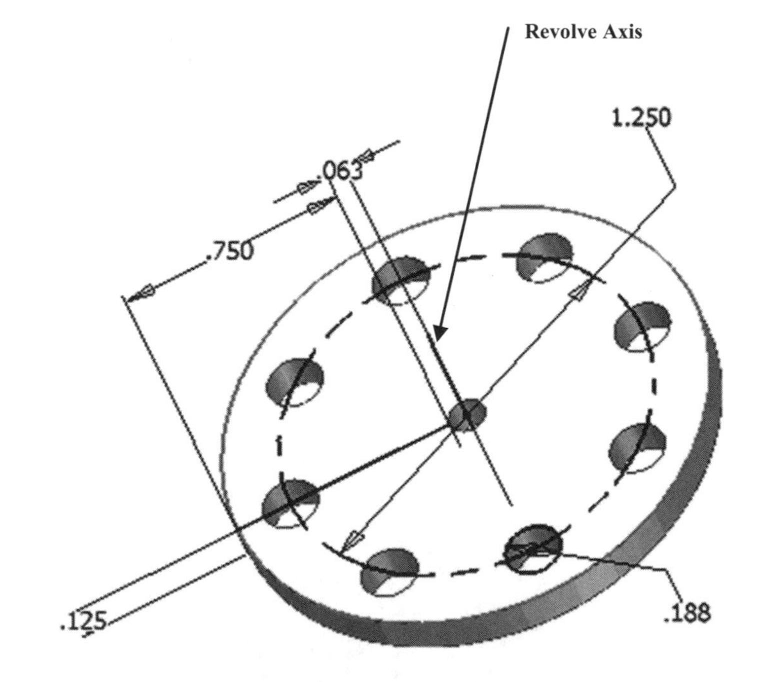

Chapter 4 Advanced Detail Drawing Procedures

Objectives:

- Create an Auxiliary View using the Auxiliary View command
- Dimension views using the Smart Dimension command
- Create a Section View using the Section View command
- Create Text using the Note command

Chapter 4 includes instruction on how to create the drawings shown below.

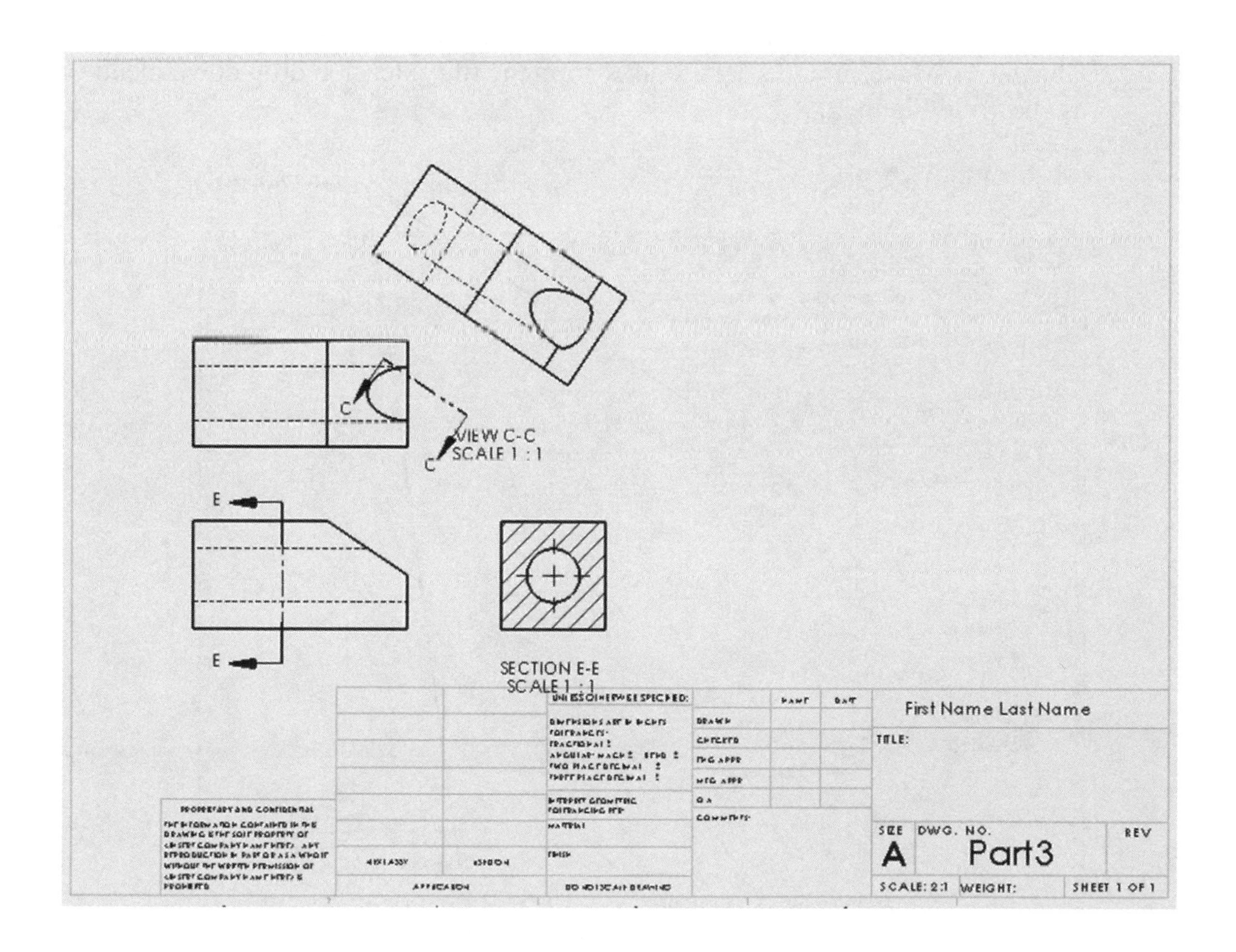

1. Start SolidWorks by referring to "Chapter 1 Getting Started". SolidWorks will have to be started up from "scratch" before you begin this chapter.

2. After SolidWorks is running, open the .SLDDRW file that was created in Chapter 3. Move the cursor to the upper left corner of the screen and left click on "Open" as shown in Figure 1.

Figure 1

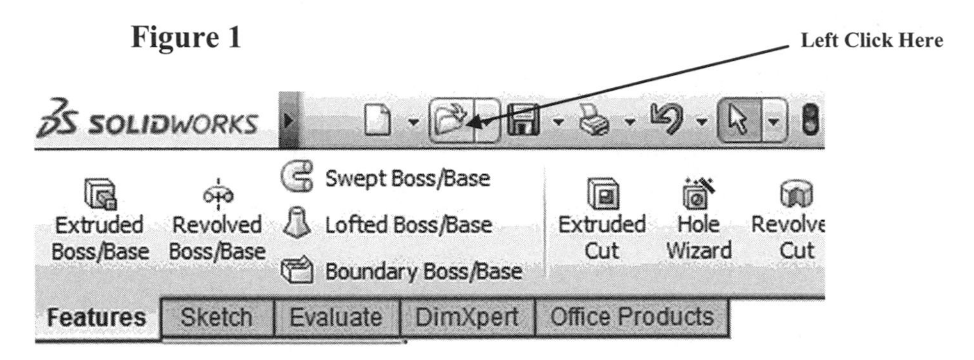

3. The Open dialog box will appear. Left click on the drawing that was created in Chapter 3. Make sure to select .slddrw from the file extension drop down menu as shown in Figure 2.

Figure 2

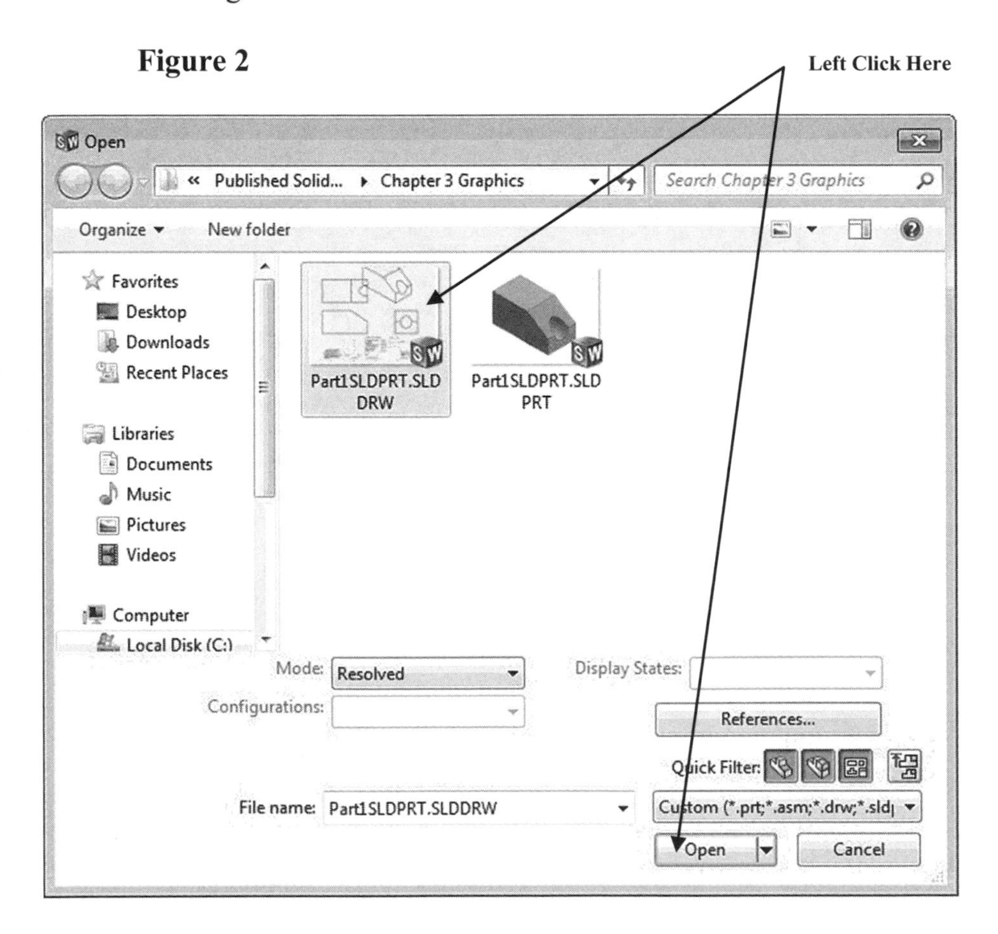

4. Left Click on **Open**.

5. After the .SLDDRW file is open, move the views closer to each other to provide additional room on the drawing. Start by moving the cursor over the top view. Dots will appear around the view. Left click (holding the left mouse button down) on the dots and drag the view down closer to the front view as shown in Figure 3.

Figure 3

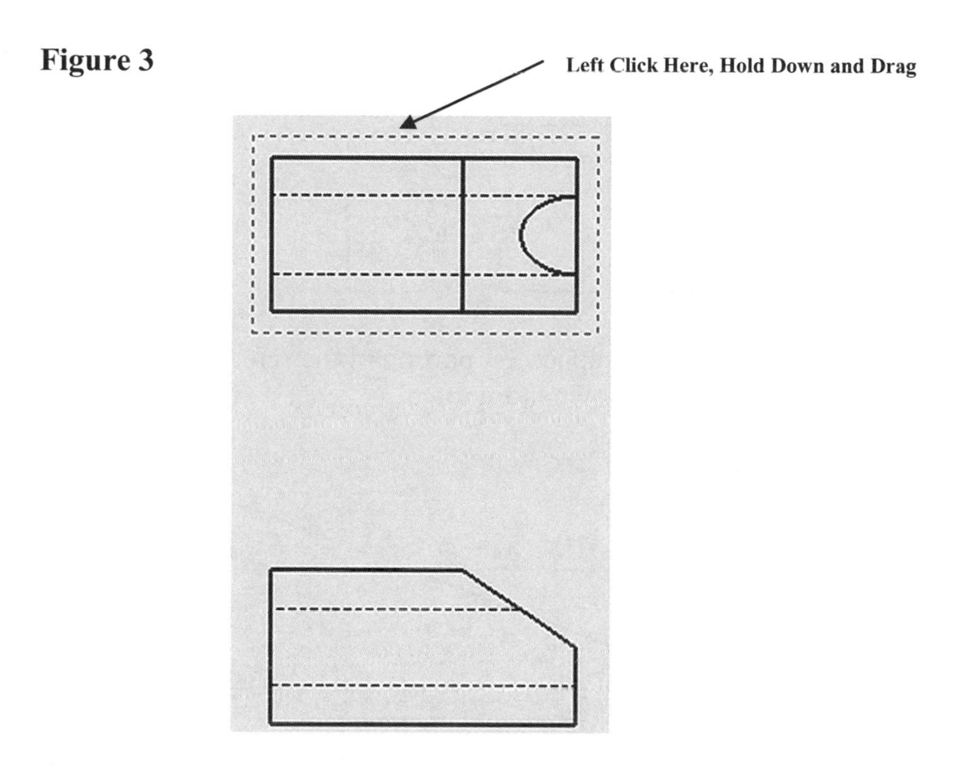

6. Move the side view closer to the front view. Start by moving the cursor over the side view. Dots will appear around the view. Left click (holding the left mouse button down) on the dots and drag the view closer to the front view as shown in Figure 4.

Figure 4

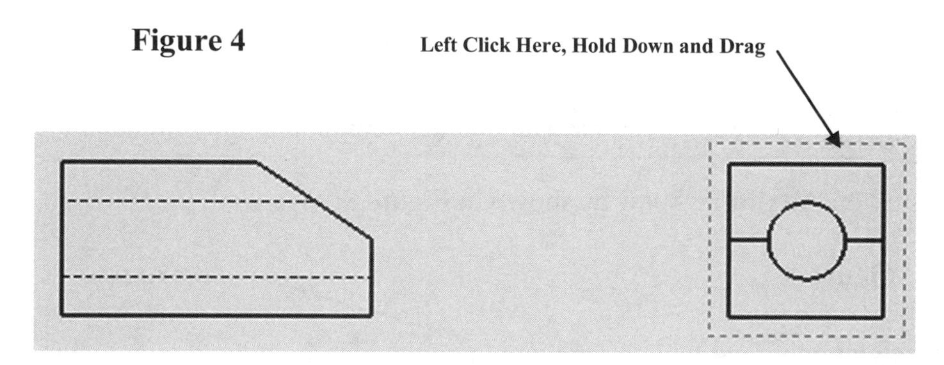

7. This will provide more room to work. Your screen should look similar to Figure 5. Press **ESC** on the keyboard.

Figure 5

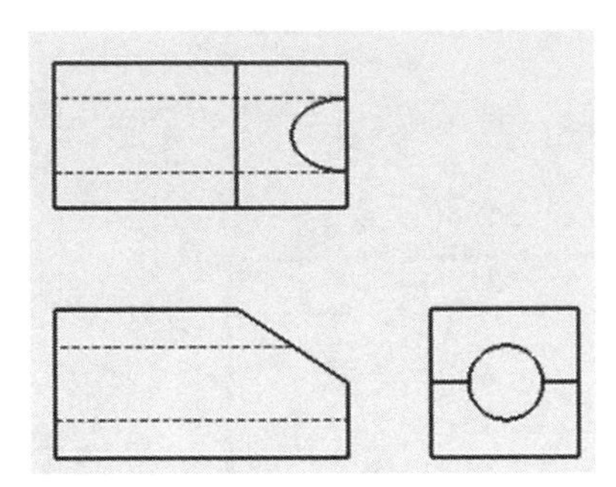

8. Move the cursor to the upper left portion of the screen and left click on the **View Layout** tab as shown in Figure 6.

Figure 6

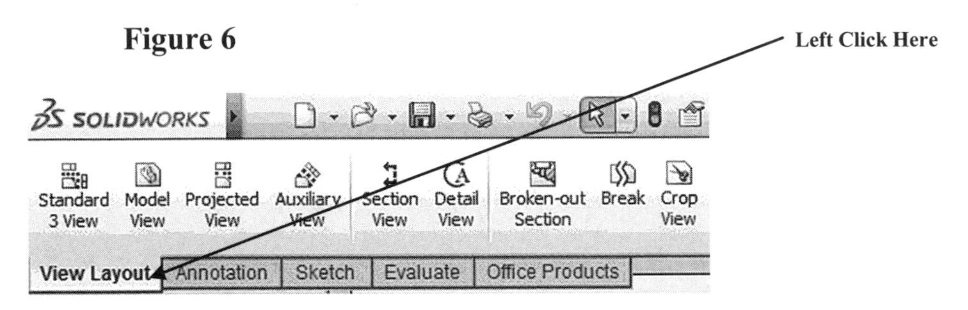

9. Notice that the commands at the top of the screen are now different as shown in Figure 7.

Figure 7

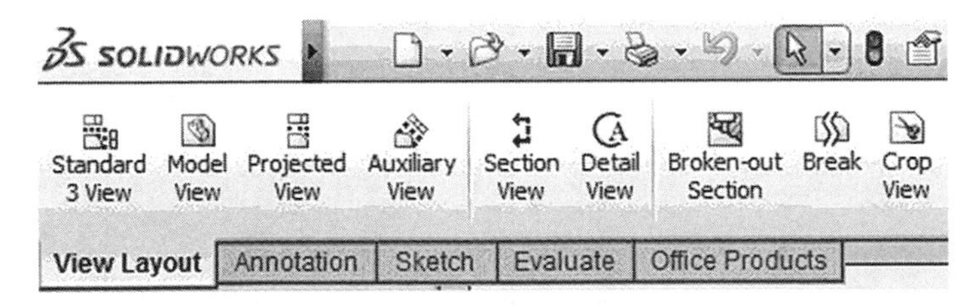

10. Left click on **Auxiliary View** as shown in Figure 8.

Figure 8

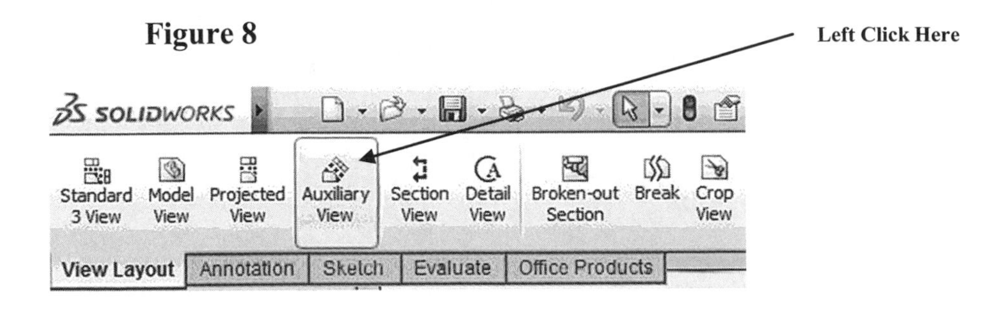

11. Move the cursor to the wedge line and left click as shown in Figure 9.

Figure 9

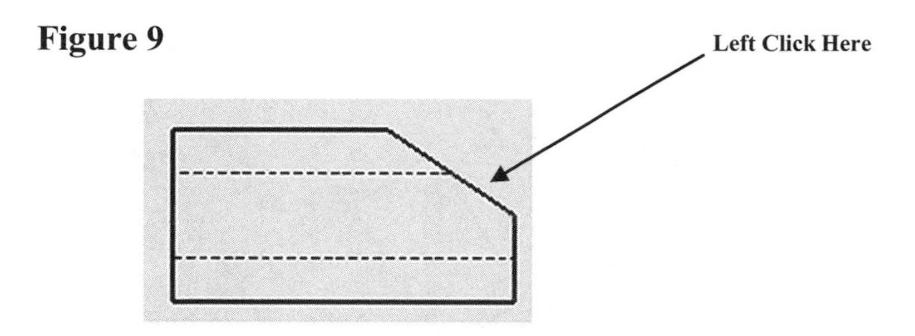

12. SolidWorks will create an auxiliary view from the selected surface. The view will be attached to the cursor as shown in Figure 10.

Figure 10

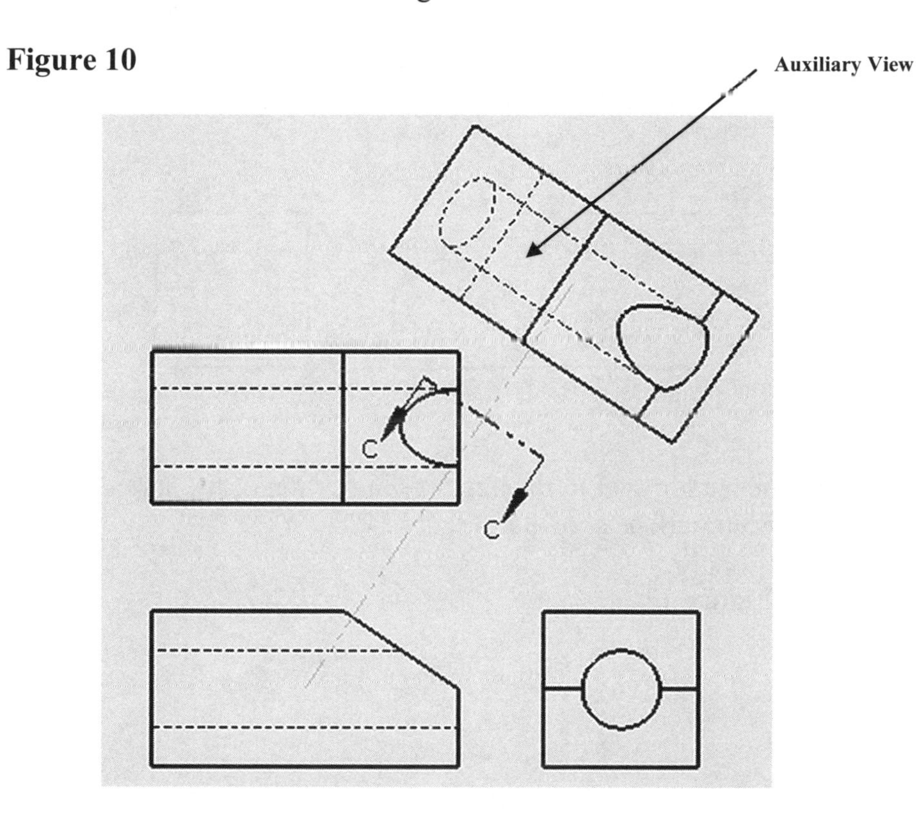

13. Drag the cursor towards the other views and left click as shown in Figure 11.

Figure 11

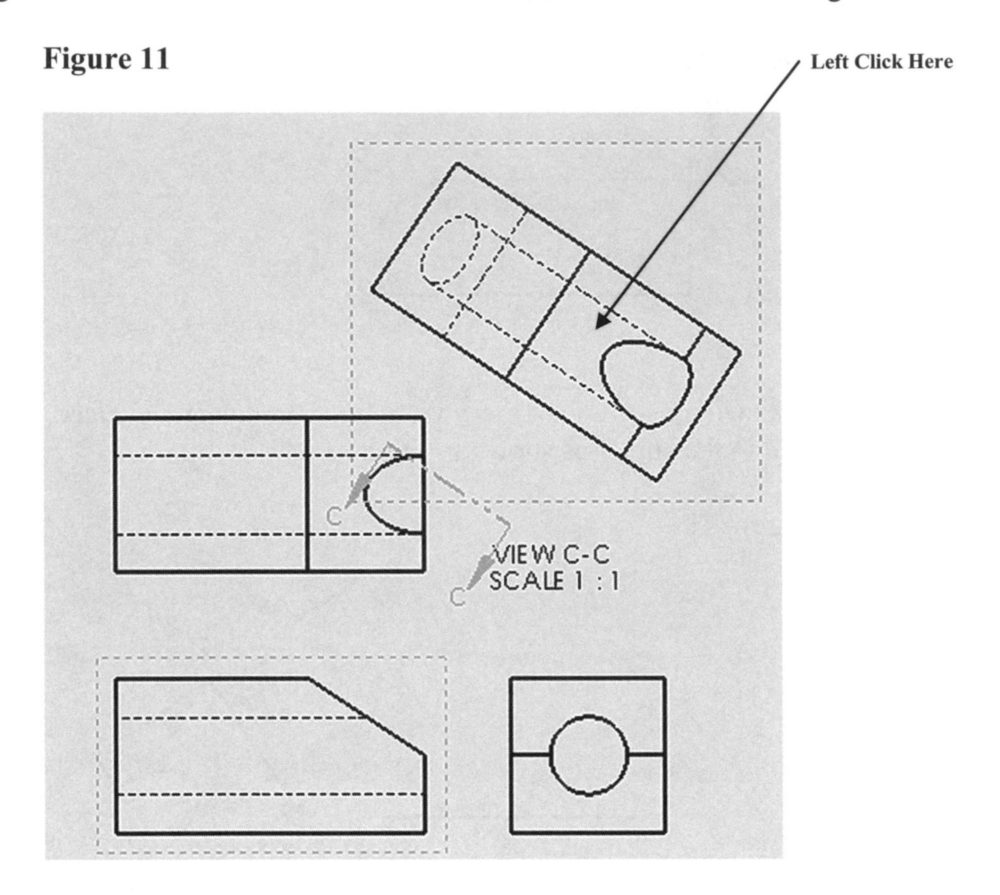

14. Move the cursor over to the upper right portion of the screen and left click on the green checkmark as shown in Figure 12.

Figure 12

15. Your screen should look similar to Figure 13.

Figure 13

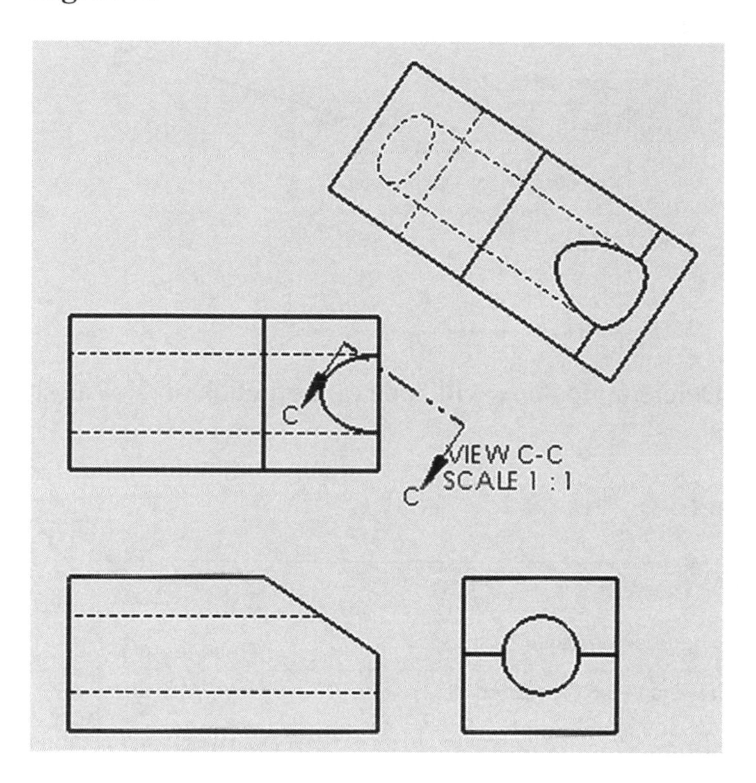

16. Move the cursor to the side view. Red dots will appear as shown in Figure 14.

Figure 14

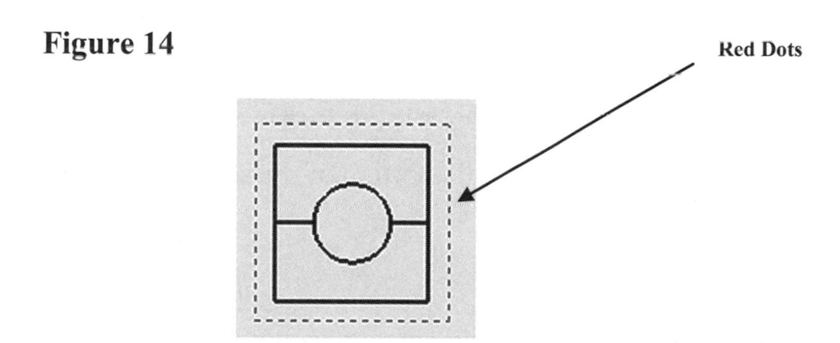

17. Right click on the view. A pop up menu will appear. Left click on **Delete** as shown in Figure 15.

Figure 15

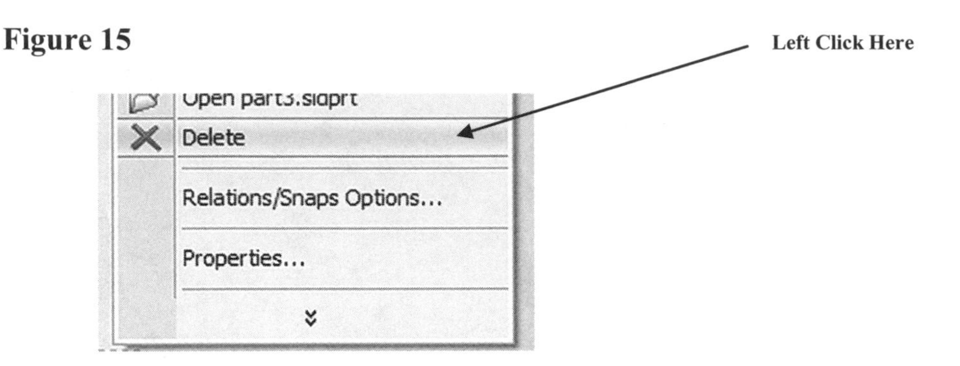

18. The Confirm Delete dialog box will appear. Left click on **Yes** as shown in Figure 16.

Figure 16

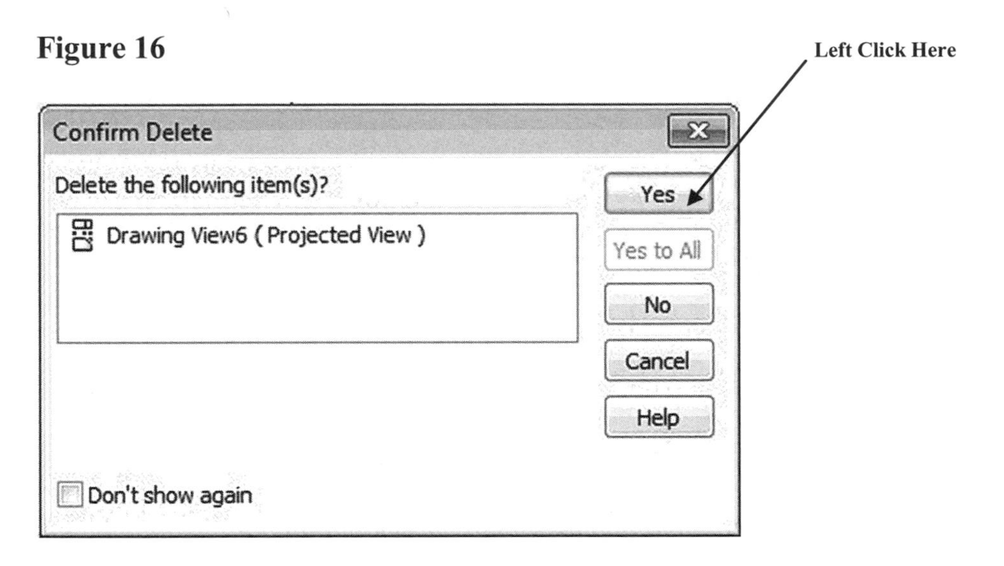

19. Move the cursor to the upper middle portion of the screen and left click on **Section View** as shown in Figure 17.

Figure 17

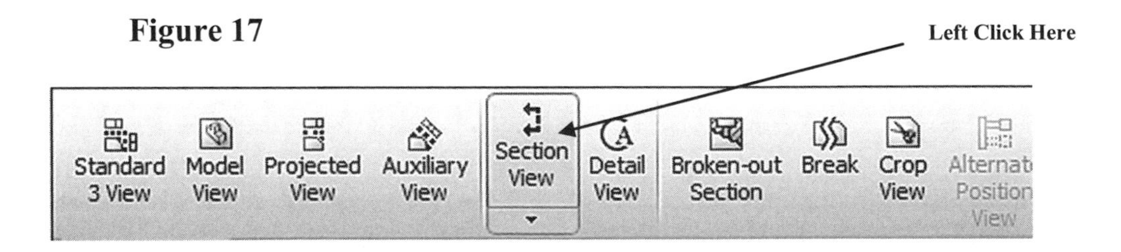

20. Move the cursor over the front view. Red dots will appear around the view. Left click above the top of the part as shown in Figure 18.

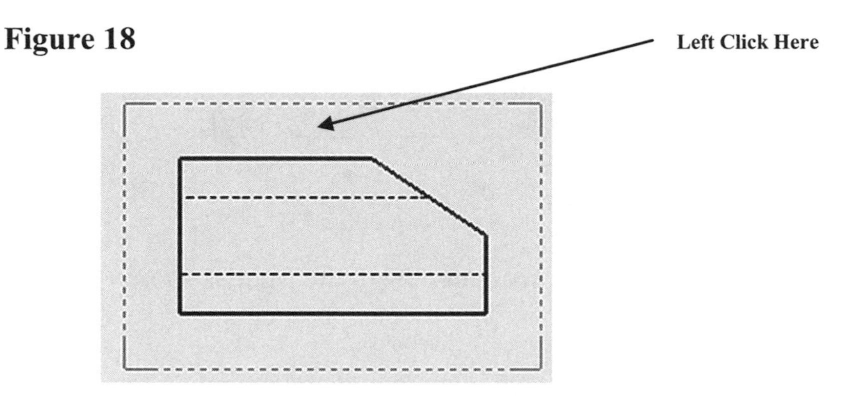

Figure 18

21. Drag the cursor down across the part and left click as shown in Figure 19.

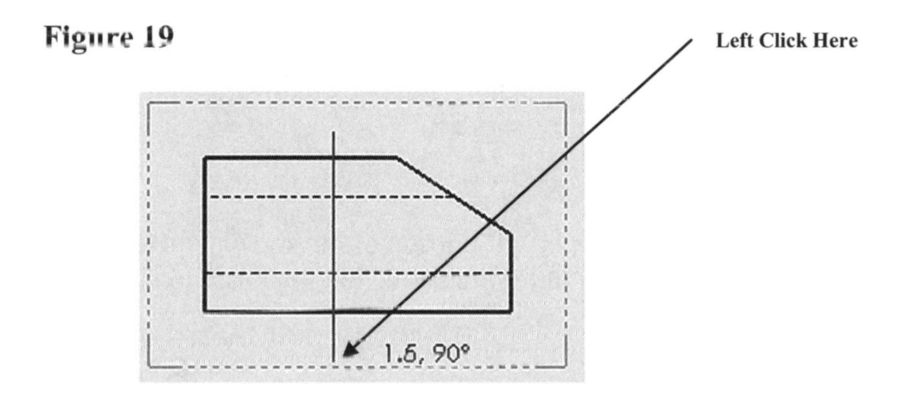

Figure 19

22. Drag the cursor to the right. The section view will be attached to the cursor. Place a check in the "Flip Direction" box if cutting plane arrows are pointing the wrong direction. Place the section view to the side of the front view and left click as shown in Figure 20.

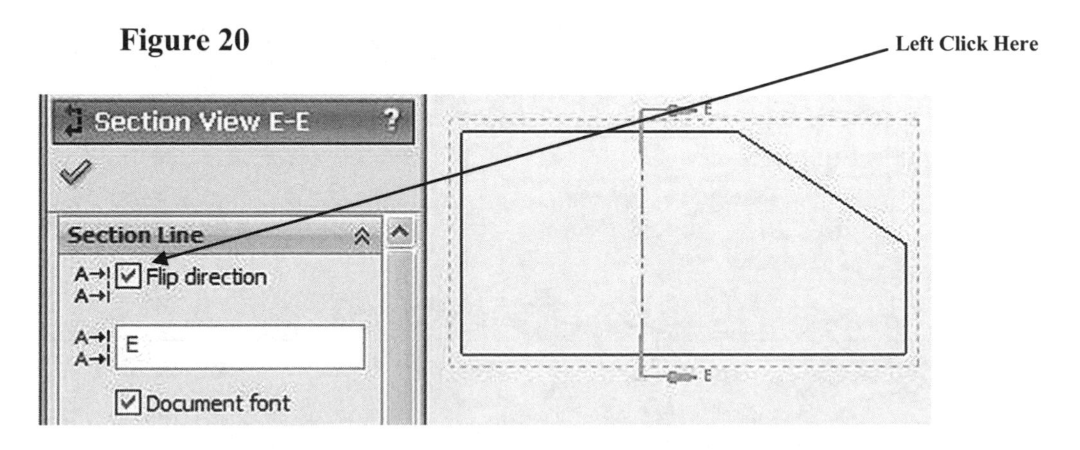

Figure 20

23. Move the cursor to the upper left portion of the screen and left click on the green checkmark as shown in Figure 21.

Figure 21

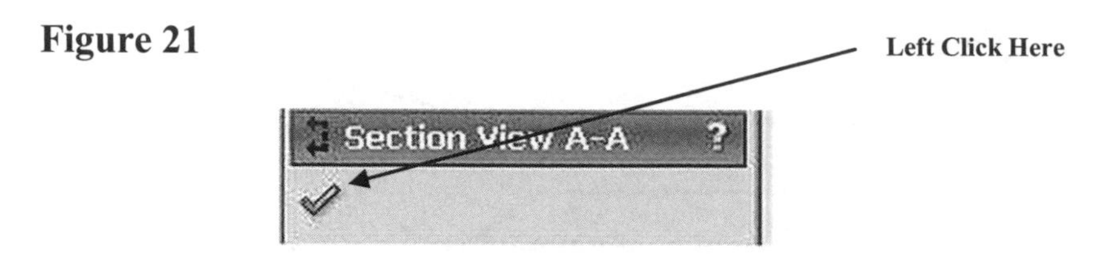

24. SolidWorks will create a section view to the right as shown in Figure 22.

Figure 22

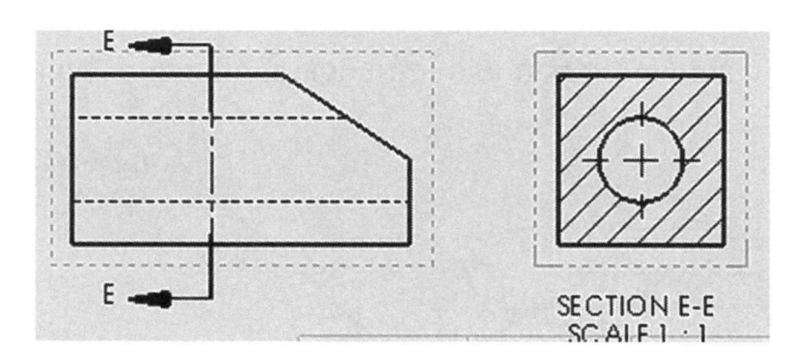

25. Move the cursor to the upper left portion of the screen and left click on the **Annotation** tab as shown in Figure 23.

Figure 23

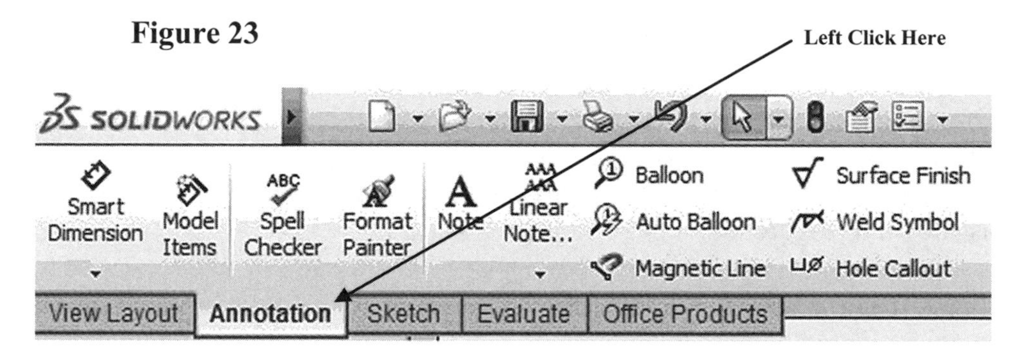

26. Notice that the commands at the top of the screen are now different as shown in Figure 24.

Figure 24

27. Annotations commands are typically where "drafting" activities are performed.

28. Move the cursor to the upper left portion of the screen and left click on **Smart Dimension** as shown in Figure 25.

Figure 25

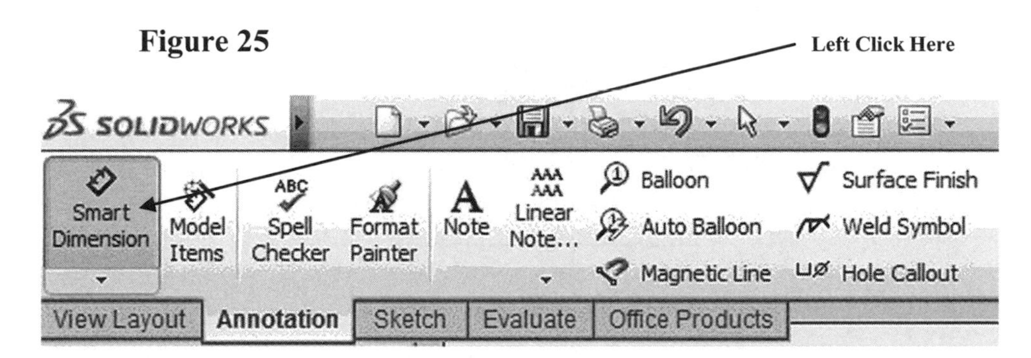

29. Move the cursor over the left side vertical line until it turns red as shown in Figure 26. Select the line by left clicking anywhere on the line **or** on each of the end points. To use the end points of the line, move the cursor over one of the end points. A small red dot will appear. Left click once and move the cursor to the other end point. After the red dot appears, left click once. The dimension will be attached to the cursor.

Figure 26

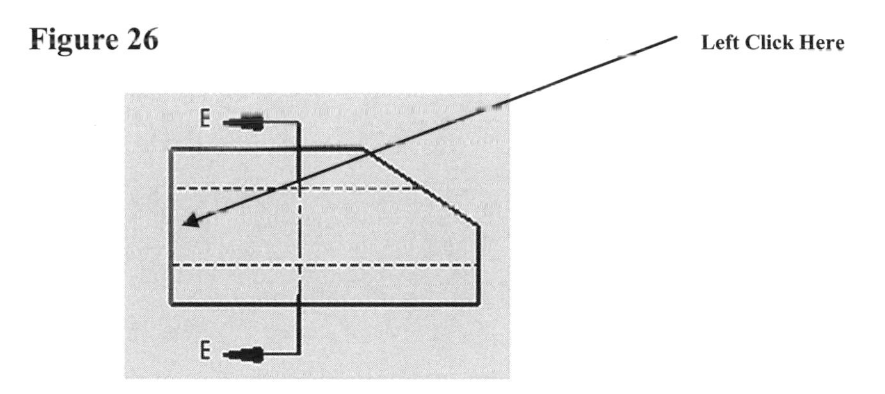

30. Move the cursor to where the dimension will be placed and left click once as shown in Figure 27.

Figure 27

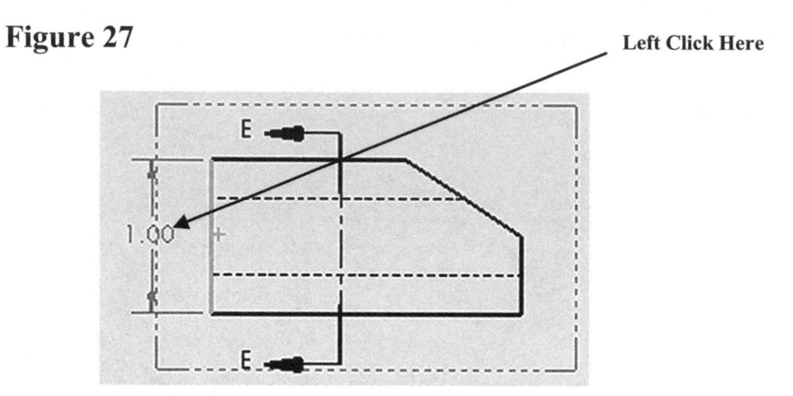

31. Move the cursor to the upper left portion of the screen and left click on the green checkmark as shown in Figure 28.

Figure 28

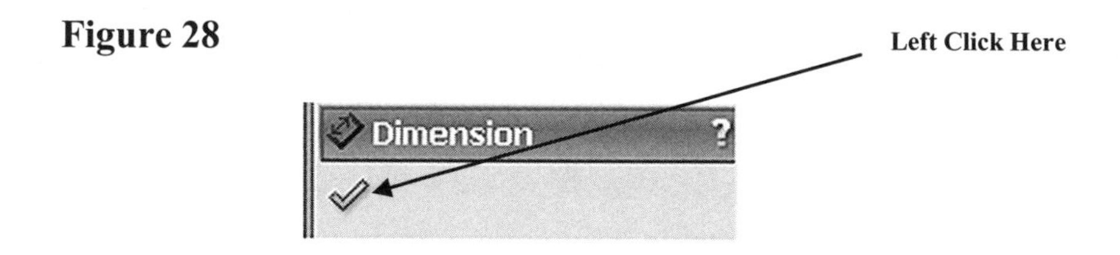

32. Your screen should look similar to Figure 29.

Figure 29

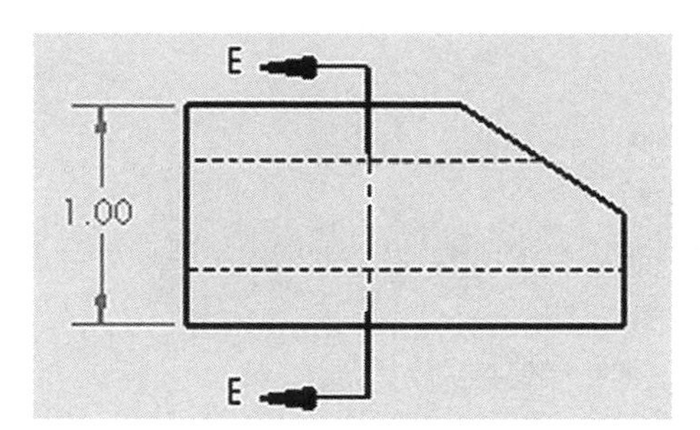

33. Finish dimensioning the part to your own satisfaction. When the part is satisfactorily dimensioned, save the file to a location where it can be easily retrieved.

34. **<u>If</u>** there is ever a need to hide any dimension, move the cursor over the dimension. The dimension will turn red as shown in Figure 30.

Figure 30

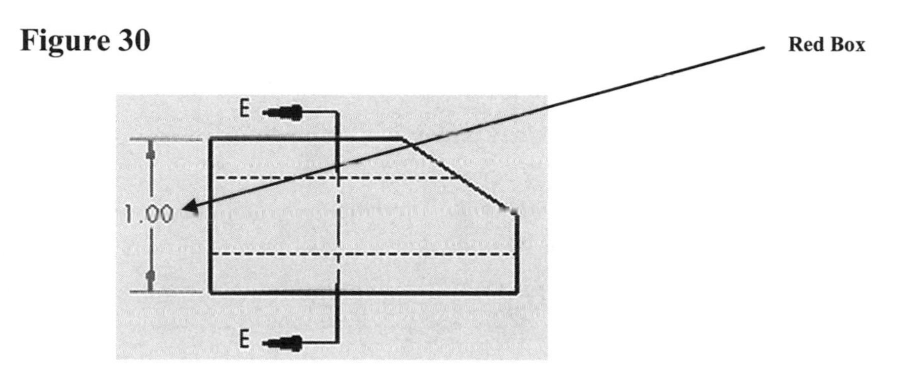

35. Right click on the dimension. A pop up menu will appear. Left click on **Hide** as shown in Figure 31. To retrieve the dimension, immediately left click on the **Undo** icon at the far upper middle portion of the screen.

Figure 31

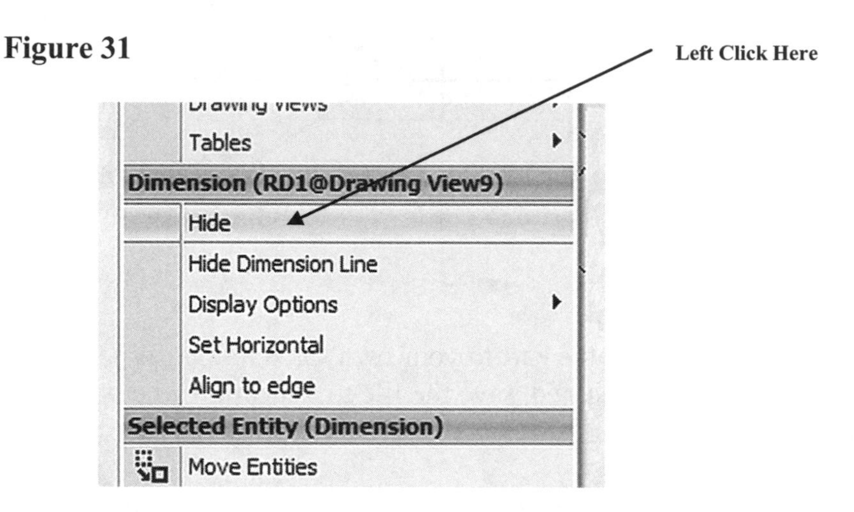

36. Left click on **Note** as shown in Figure 32.

Figure 32

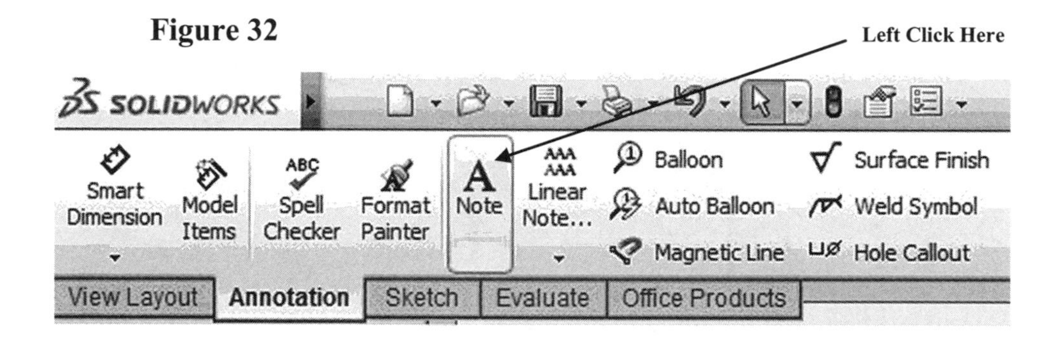

37. Move the cursor to the title block location. A small box will be attached to the cursor. Left click as shown in Figure 33.

Figure 33

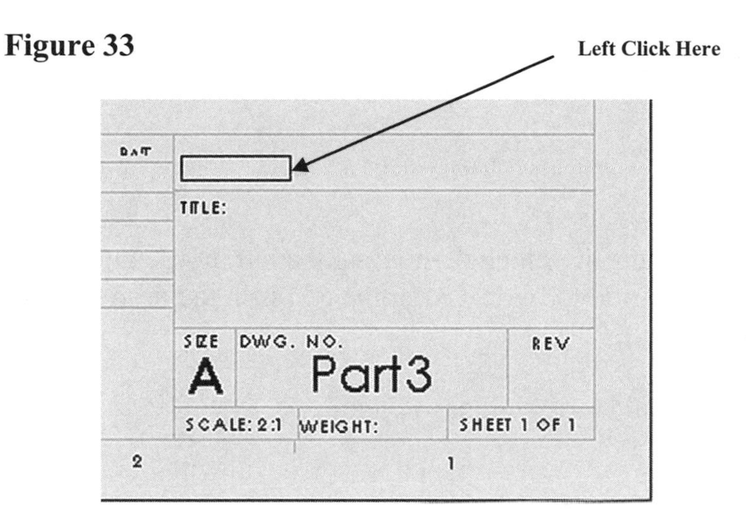

38. The cursor will begin blinking. Enter your first name and last name. After the text has been entered, move the cursor to the upper left portion of the screen and left click on the green checkmark as shown in Figure 34.

Figure 34

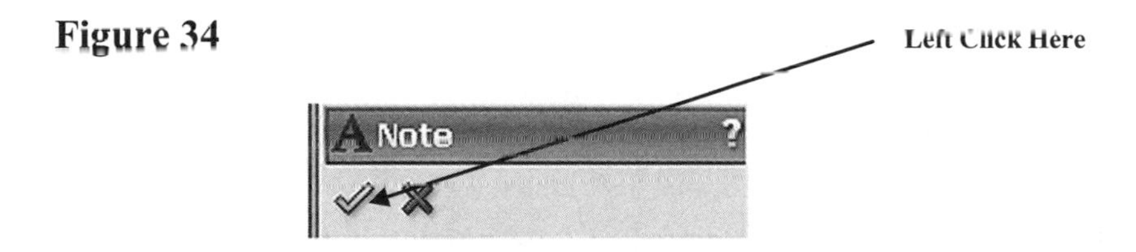

39. Text will appear in the title block as shown in Figure 35. Double clicking on the text will allow you to edit the text if needed.

Figure 35

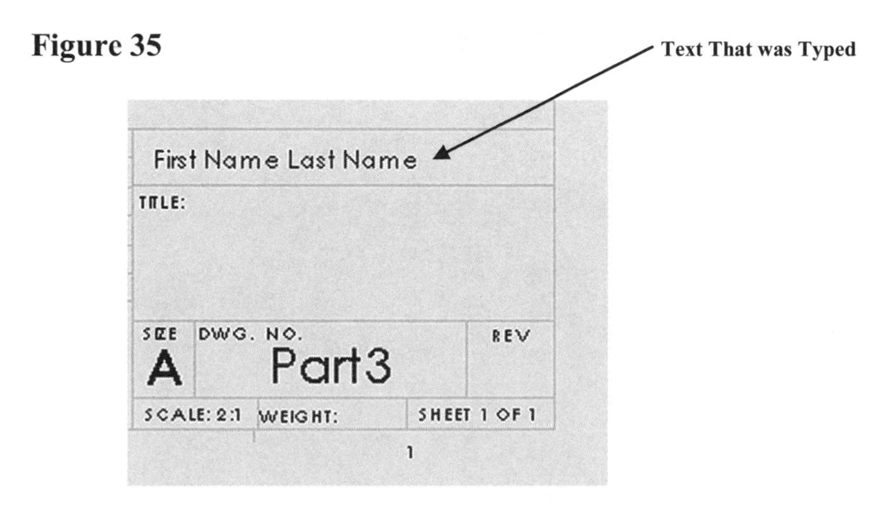

40. If the text needs to be moved, move the cursor over the text causing it to turn green as shown in Figure 36.

Figure 36

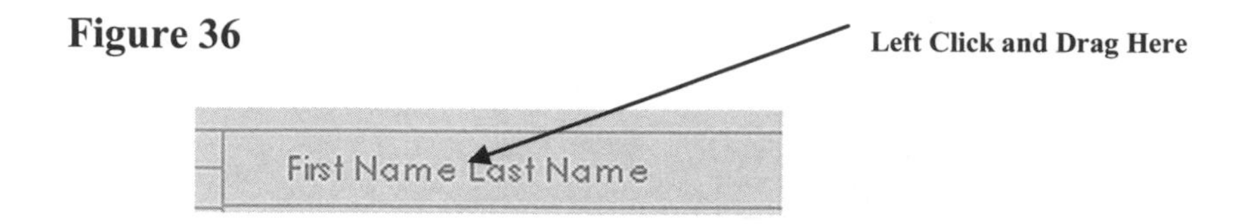

41. While the text is green, left click (holding the left mouse button down) and drag the text to the desired location. After the text is in the desired location, release the left mouse button.

42. Your screen should look similar to Figure 37.

Figure 37

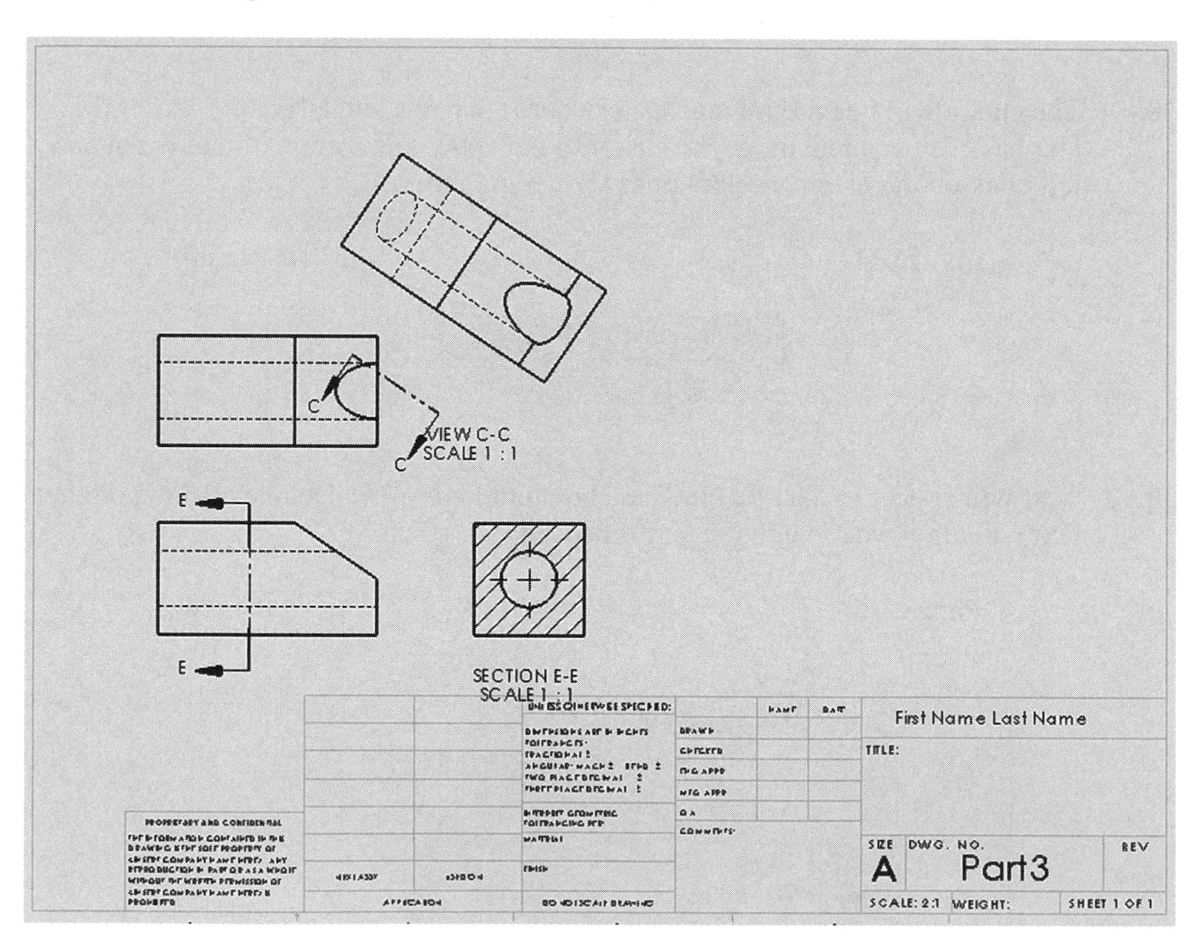

43. Save the current sheet where it can be easily retrieved.

Drawing Activities

Create Section View Drawings for the following:

Problem 1

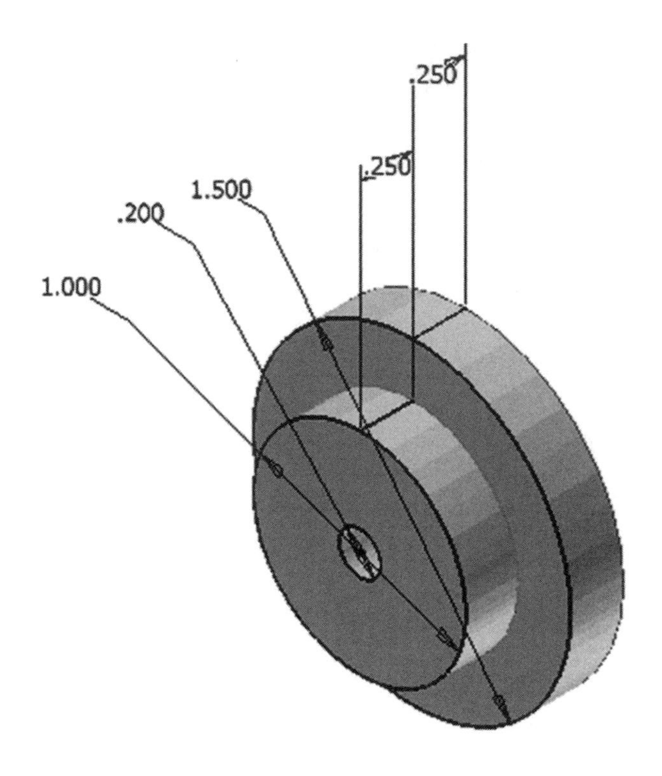

Problem 2

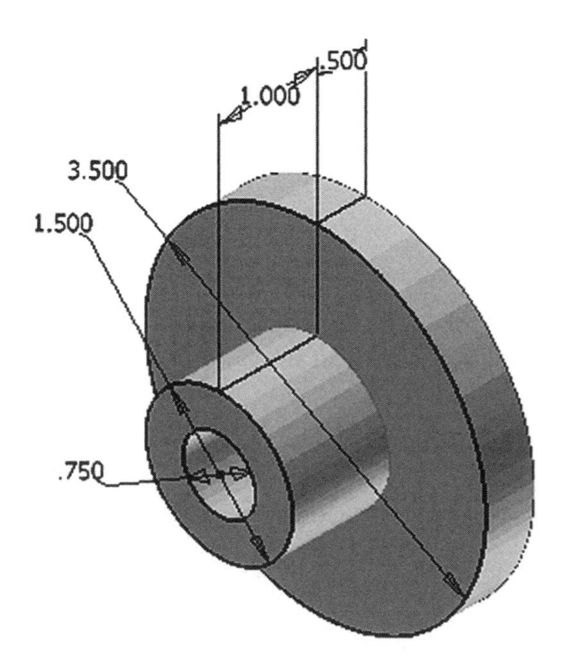

Create Auxiliary View Drawings for the following:

Problem 3

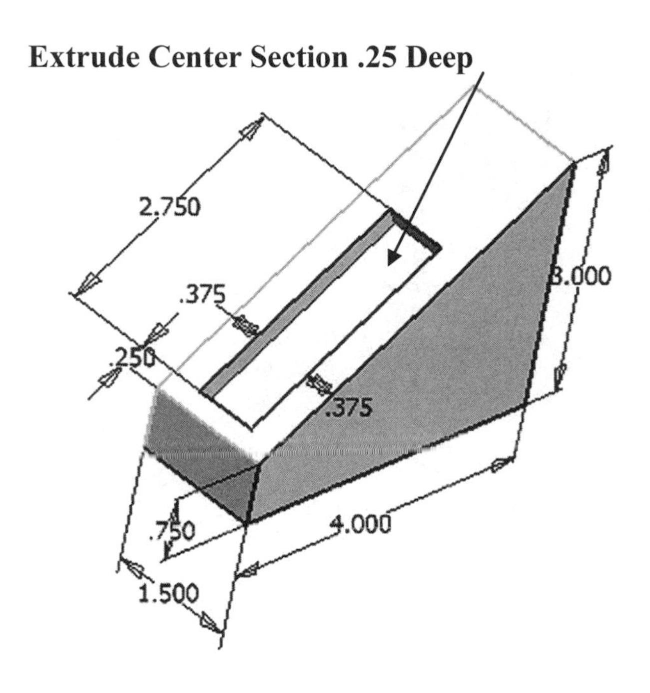

Problem 4

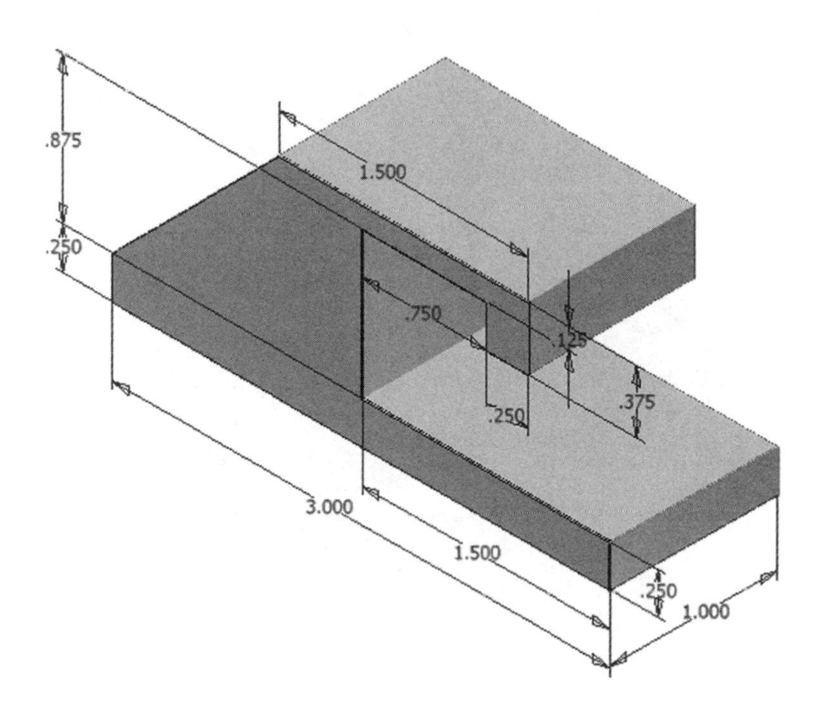

Create Section View Drawings for the following:

Problem 5

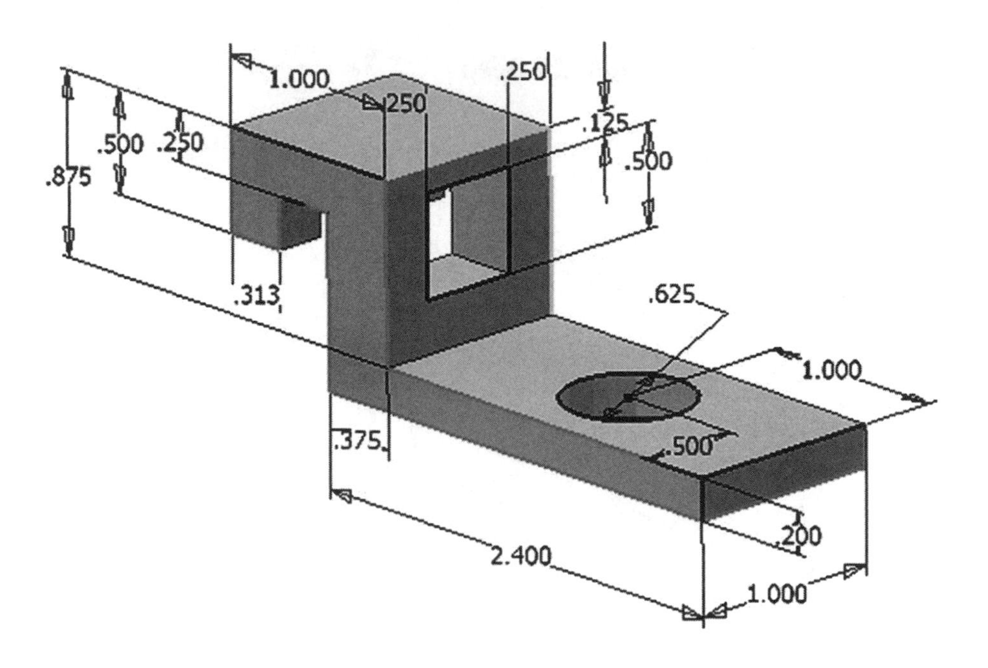

Problem 6

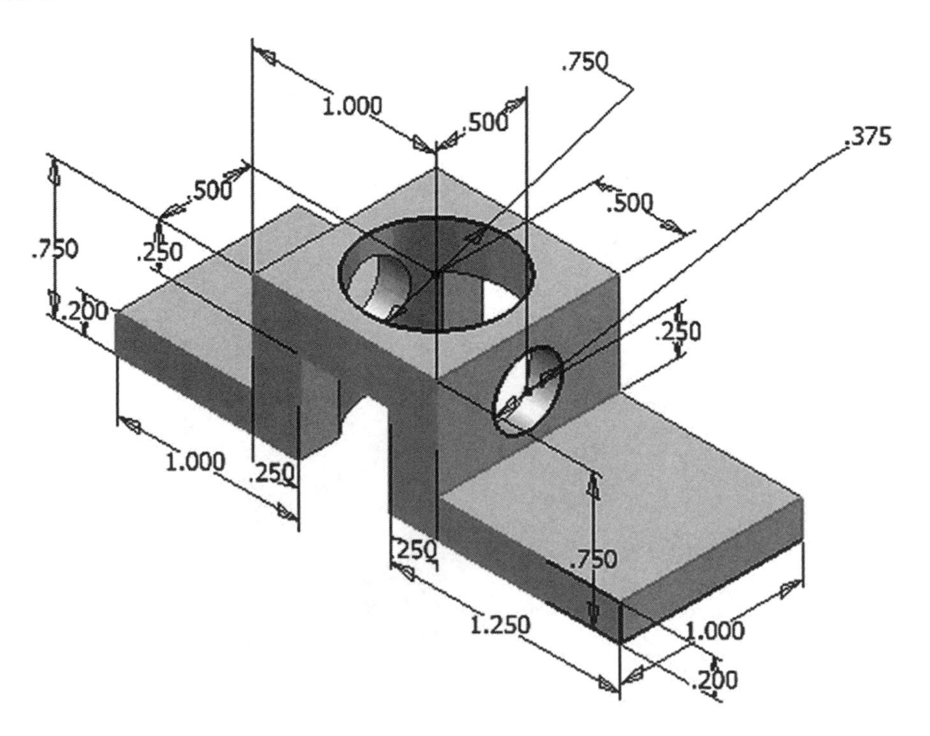

Create Section View Drawings for the following:

Problem 7

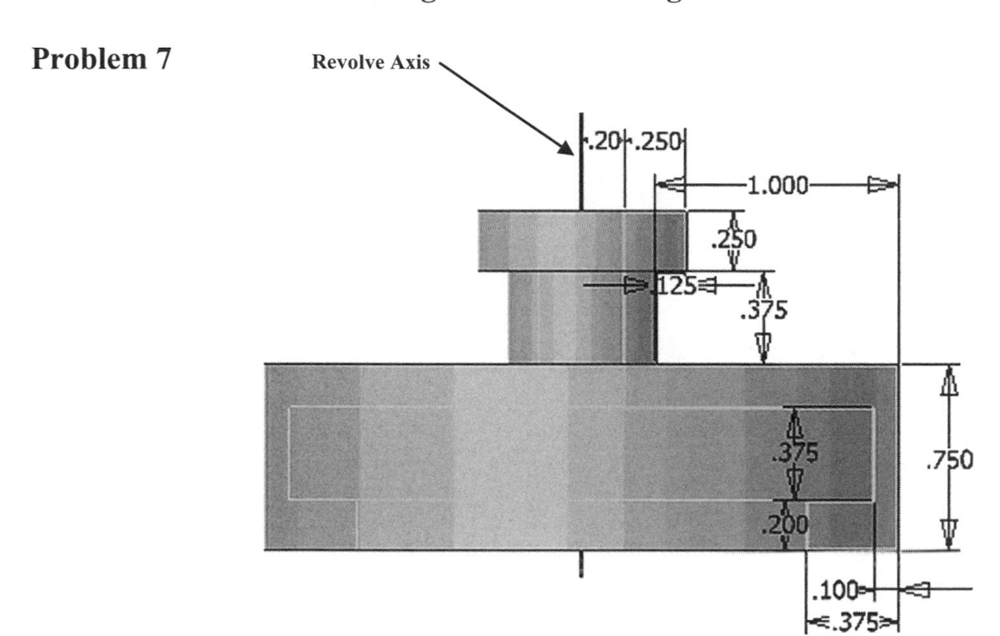

Problem 8

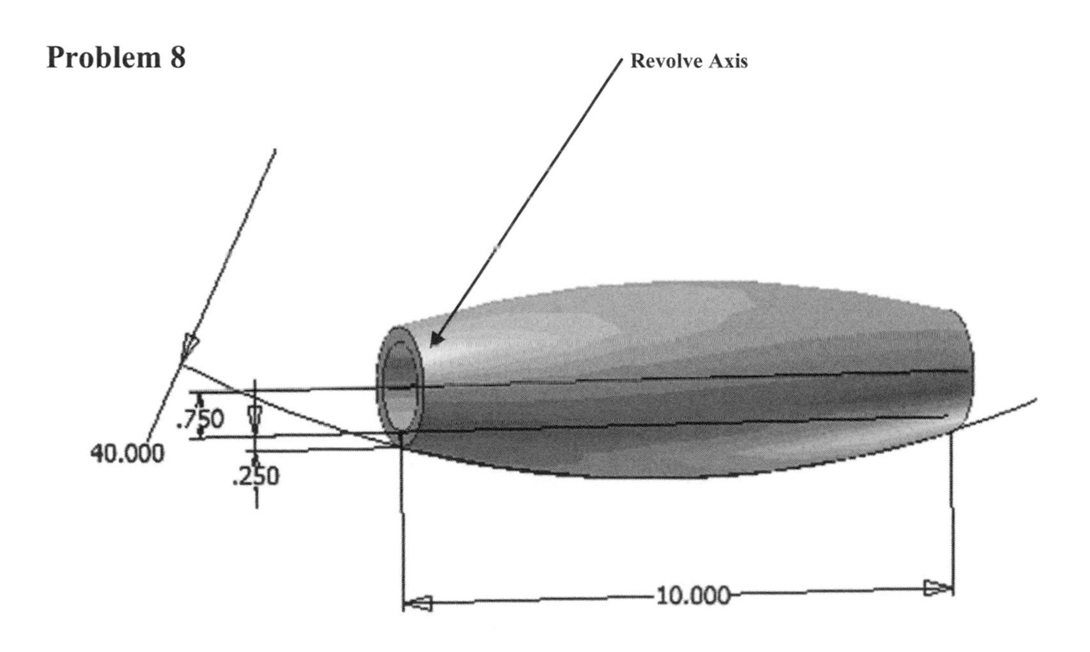

Chapter 5 Learning To Edit Existing Solid Models

Objectives:

- Design a simple part
- Learn to use the Circular Pattern command
- Learn to edit a part using the Circular Pattern command
- Edit the part using the Sketch command
- Edit the part using the Extruded Boss/Base command
- Edit the part using the Extruded Cut command
- Edit the part using the Chamfer command

Chapter 5 includes instruction on how to design and edit the part shown below.

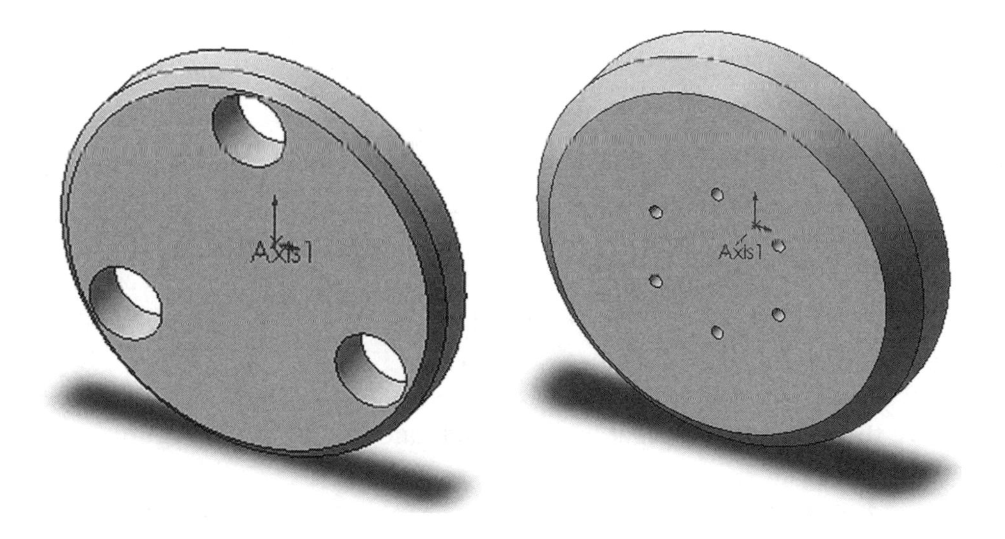

1. Start SolidWorks by referring to "Chapter 1 Getting Started".

2. After SolidWorks is running, begin a new sketch.

3. Move the cursor to the upper left corner of the screen and left click on **Circle** as shown in Figure 1.

Figure 1

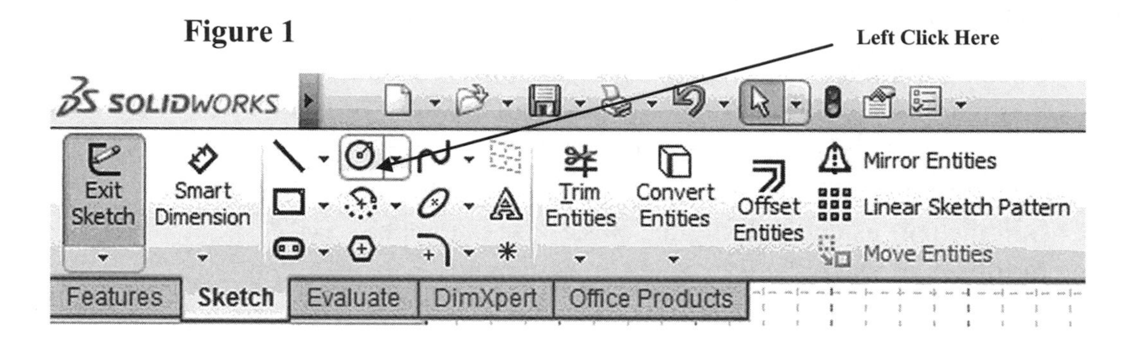

4. Move the cursor to the center of the screen and left click once. This will be the center of the circle as shown in Figure 2.

Figure 2

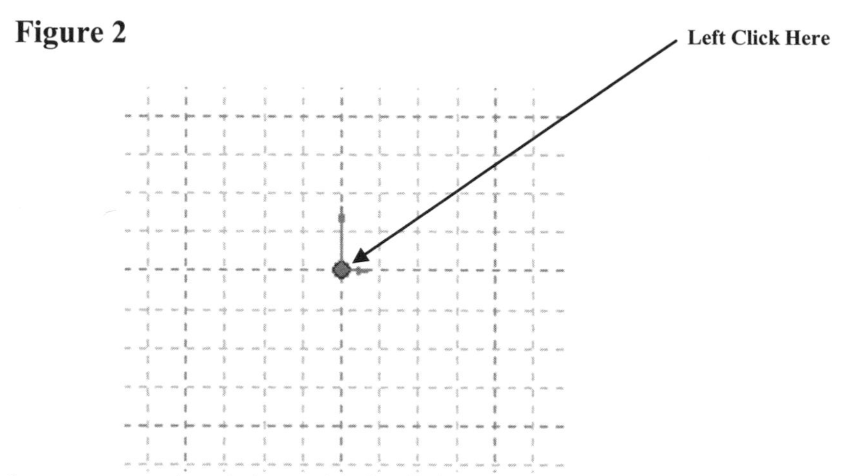

5. Move the cursor to the right and left click once as shown in Figure 3.

Figure 3

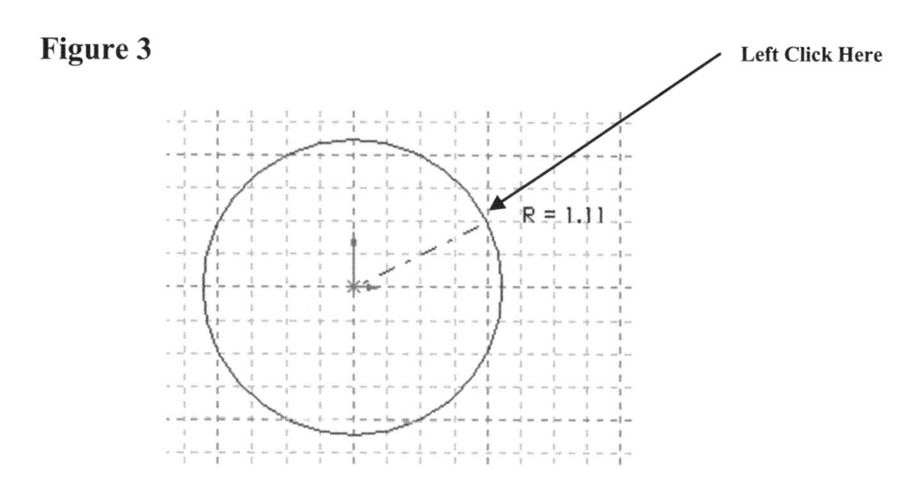

6. Right click anywhere on the drawing. A pop up menu will appear. Left click on **Select** as shown in Figure 4

Figure 4

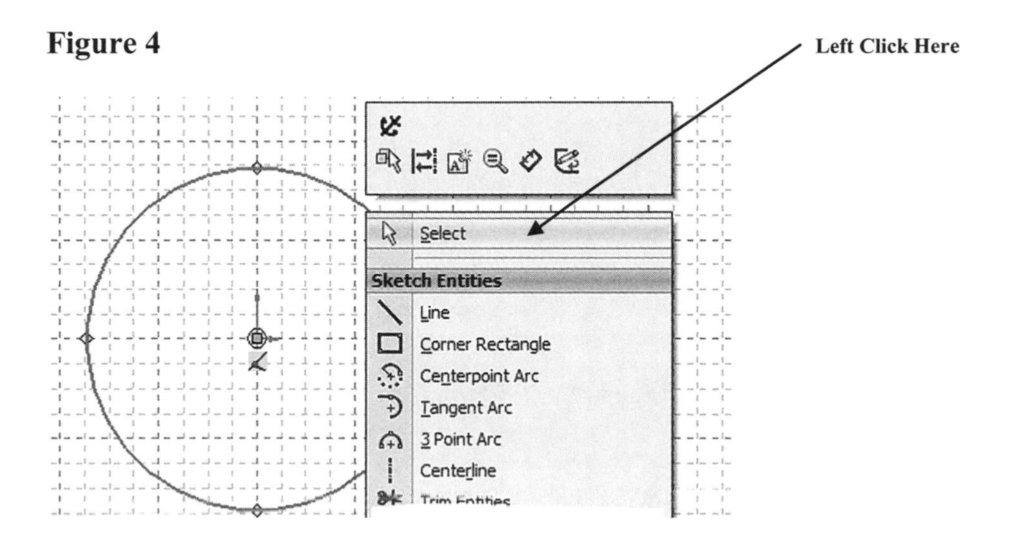

7. Move the cursor to the middle left portion of the screen and left click on **Smart Dimension** as shown in Figure 5.

Figure 5

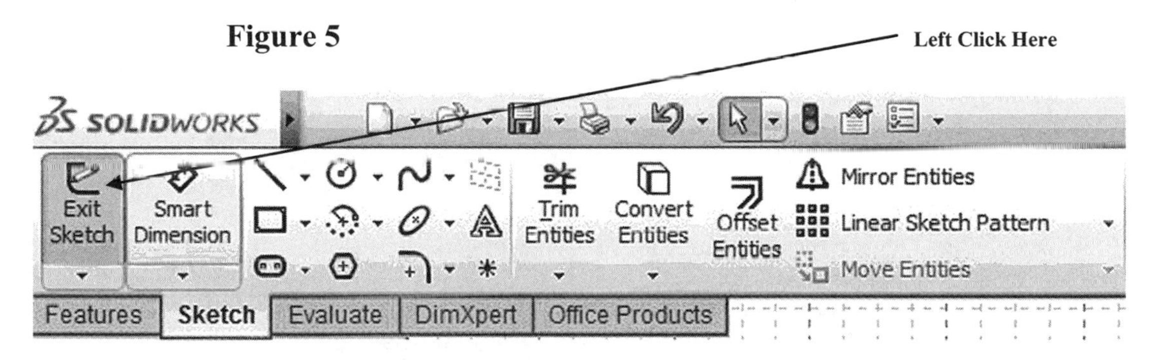

8. Move the cursor over the edge (not center) of the circle until it turns red. Left click once as shown in Figure 6. The dimension will be attached to the cursor.

Figure 6

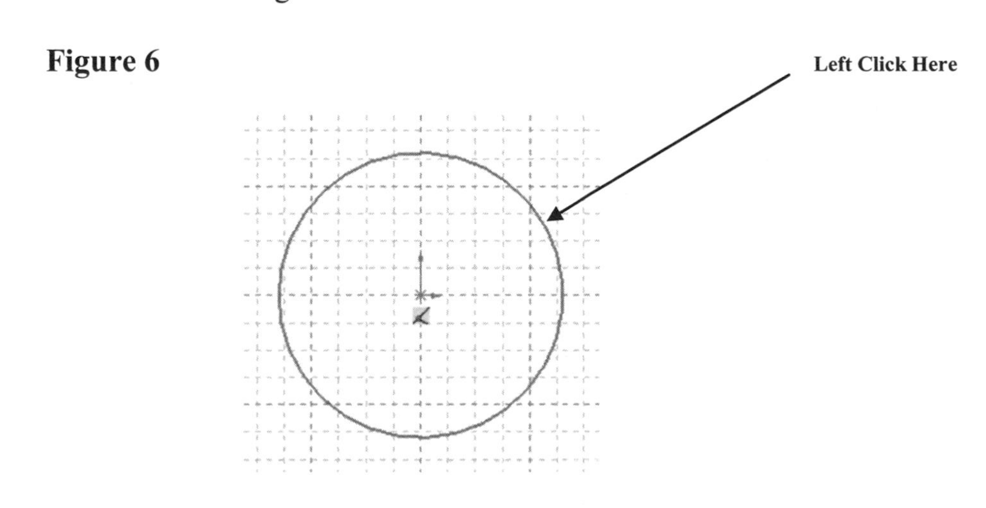

9. Move the cursor to where the dimension will be placed and left click once as shown in Figure 7.

Figure 7

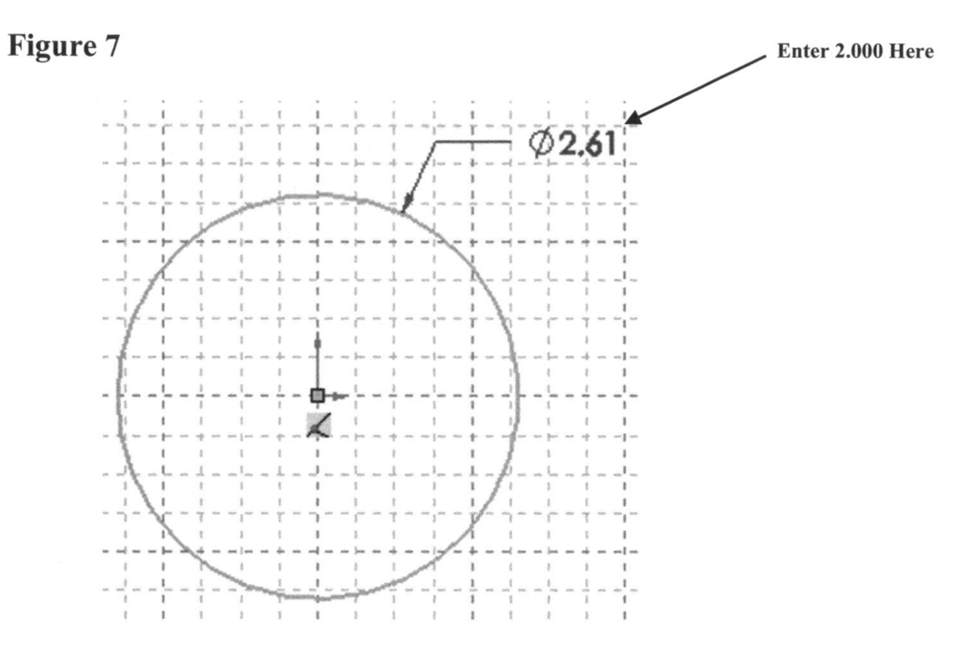

10. The Modify dialog box will appear as shown in Figure 8.

Figure 8

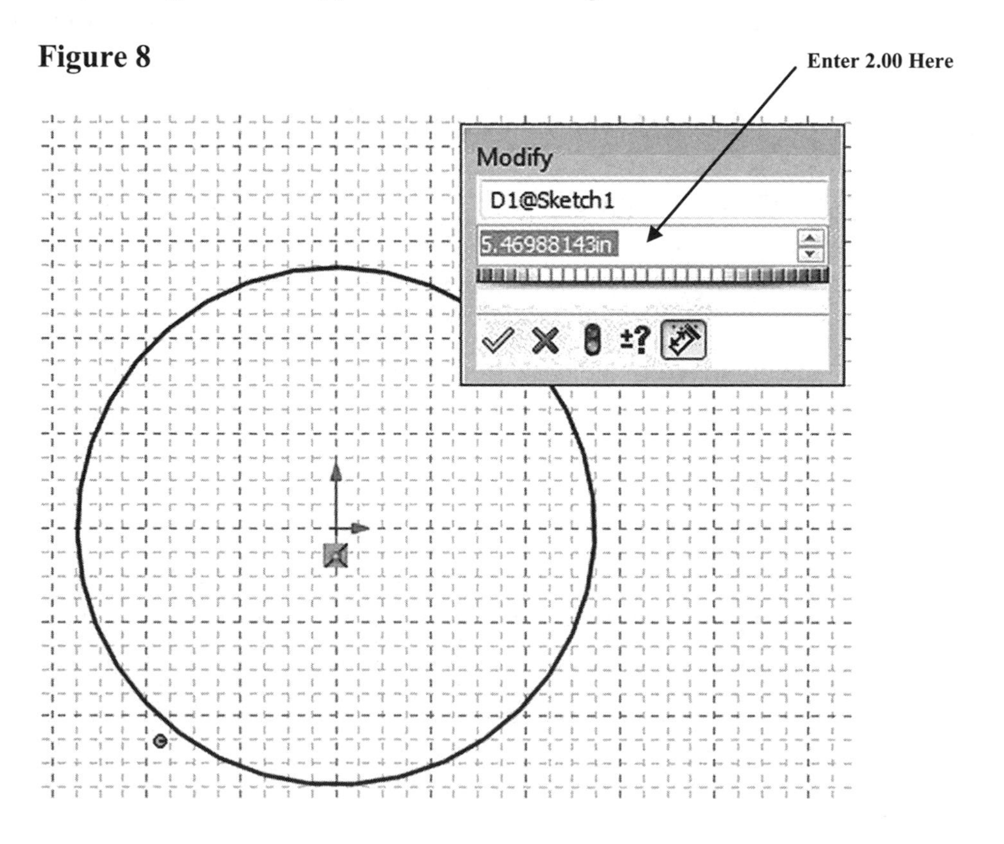

11. To edit the dimension, enter **2.00** in the Modify dialog box (while the current dimension is highlighted) and left click on the green checkmark as shown in Figure 9.

Figure 9

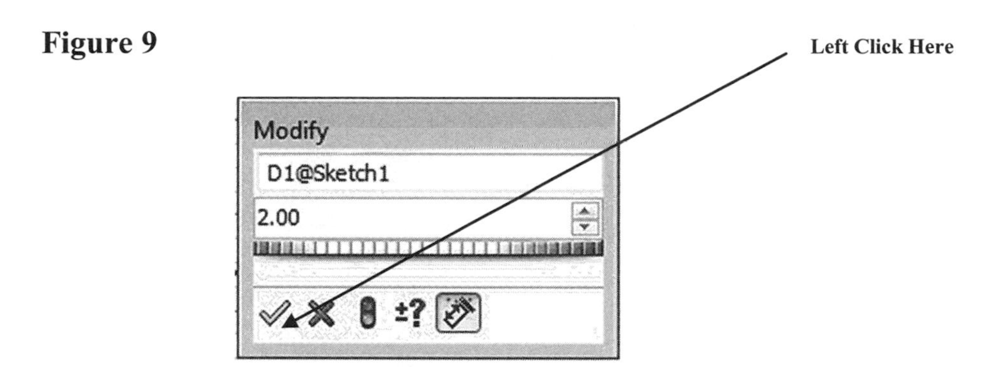

12. The dimension of the circle will become 2.00 inches as shown in Figure 10.

Figure 10

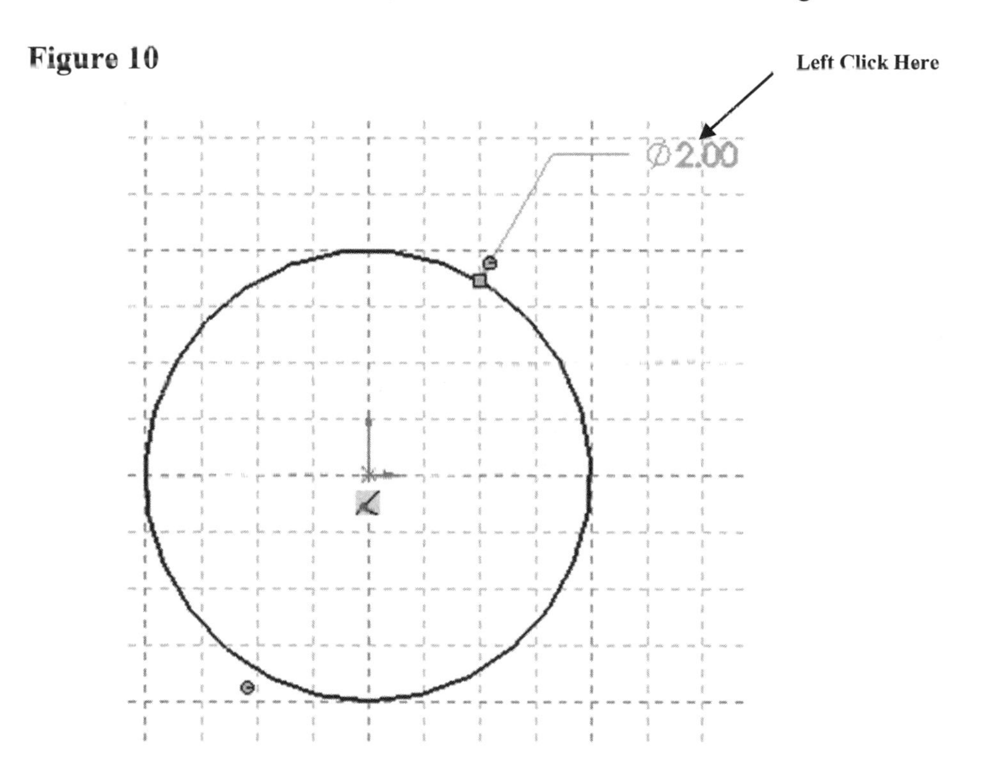

13. Right click anywhere around the drawing. A pop up menu will appear. Left click on **Select** as shown in Figure 11.

Figure 11

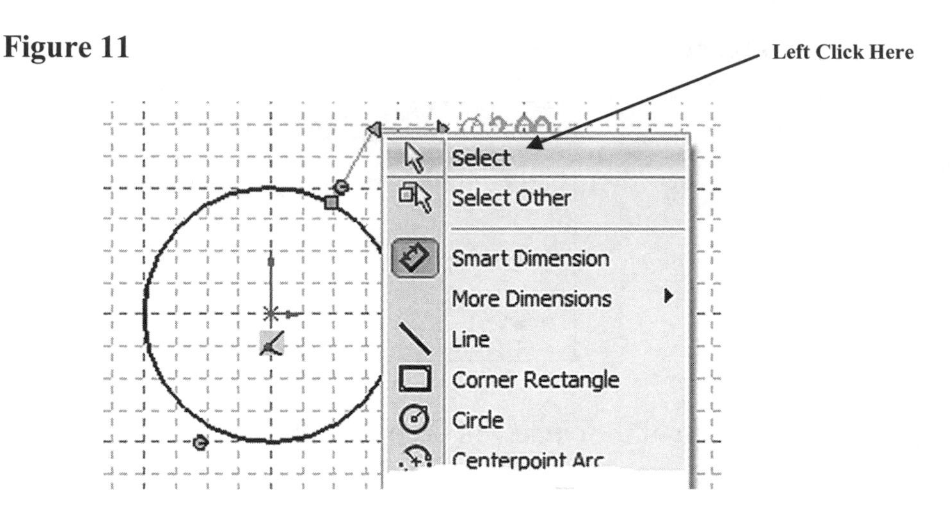

14. To view the entire drawing move the cursor to the middle portion of the screen and left click on the "Zoom To Fit" icon as shown in Figure 12.

Figure 12

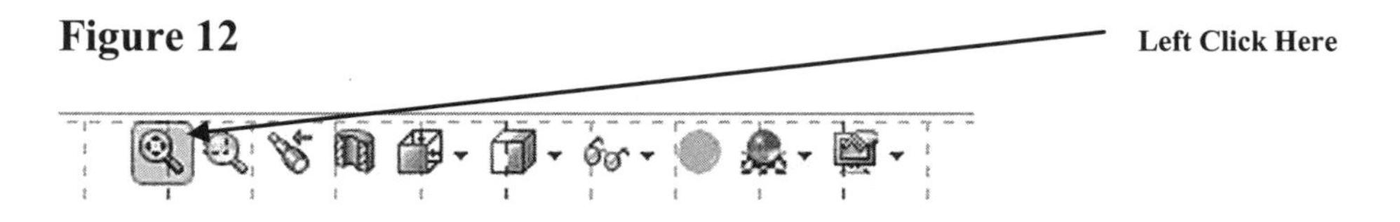

15. The drawing will "fill up" the entire screen. Zooming in and out can also be done by scrolling the scroll wheel to achieve the desired view. If for some reason you need to return to the previous view, left click on the Previous View icon as shown in Figure 13.

Figure 13

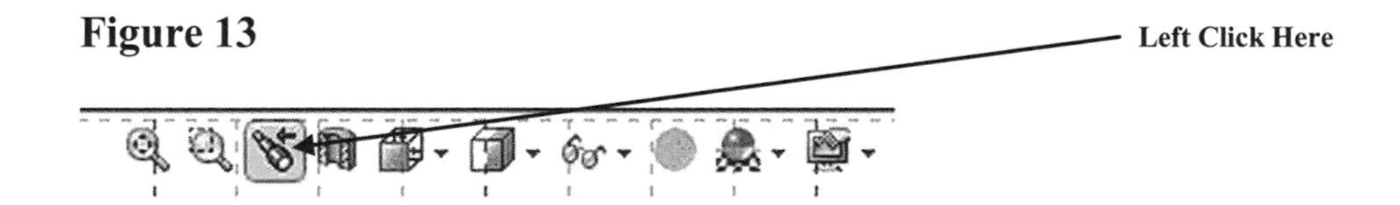

16. After the sketch is complete it is time to extrude the sketch into a solid. Right click anywhere on the drawing. A pop up menu will appear. Left click on **Select** as shown in Figure 14.

Figure 14

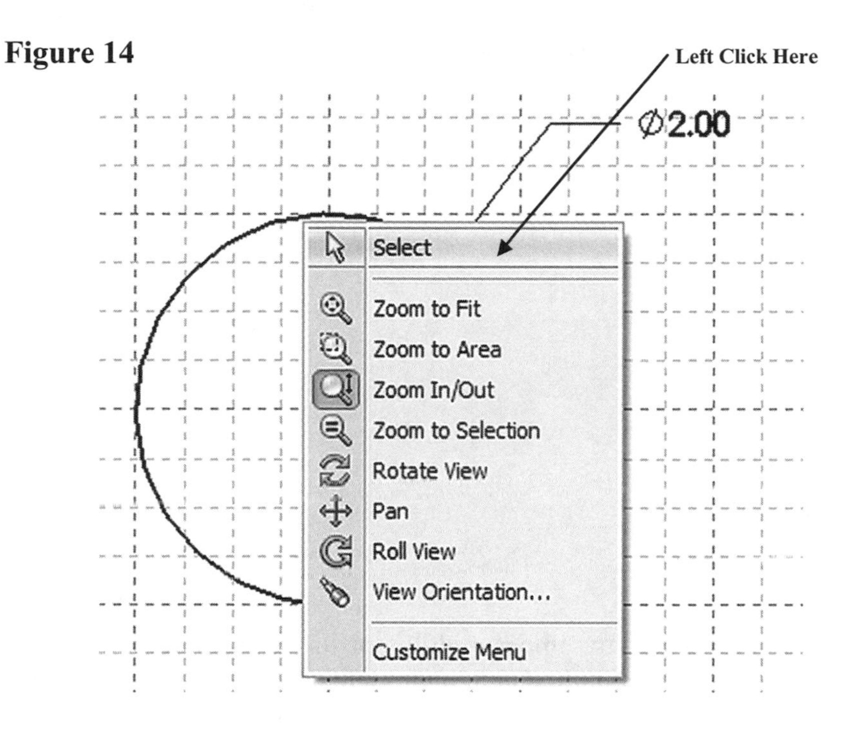

17. After you have verified that no commands are active, right click anywhere on the sketch. A pop up menu will appear. Left click on the **Exit Sketch** icon as shown in Figure 15. If the Features tools did not appear, left click on the **Features** tab at the upper left portion of the screen.

Figure 15

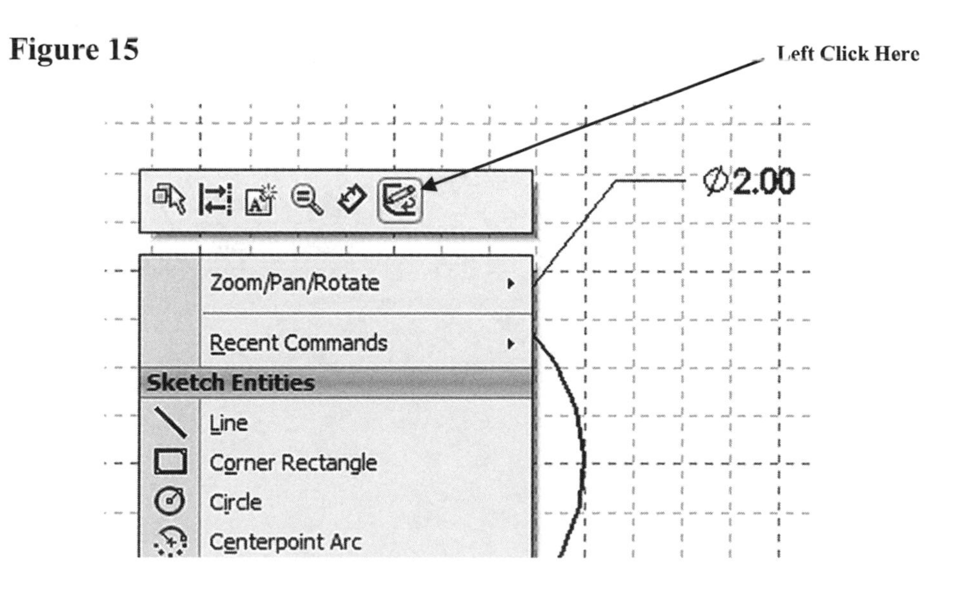

18. SolidWorks is now out of the Sketch commands and into the Features commands. Notice that the commands at the top of the screen and now different. To work in the Features commands a sketch must be present and have no opens (non-connected lines). If there are any opens in the sketch an error message will appear. You screen should look similar to Figure 16.

Figure 16

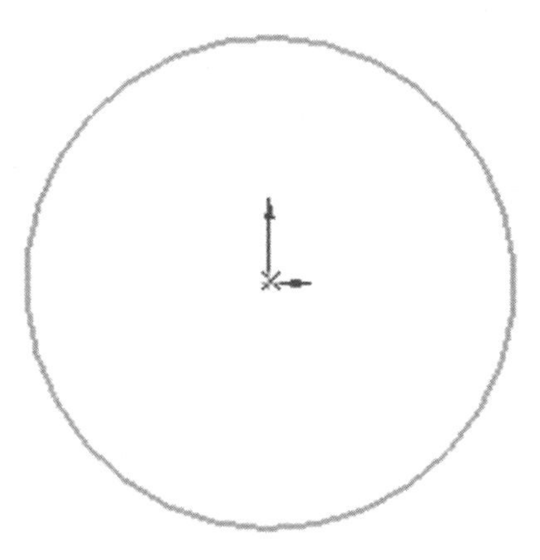

19. Move the cursor to the upper middle portion of the screen and left click on the drop down arrow next to the "Views Orientation" icon. A drop down menu will appear. Left click on **Trimetric** as shown in Figure 17.

Figure 17

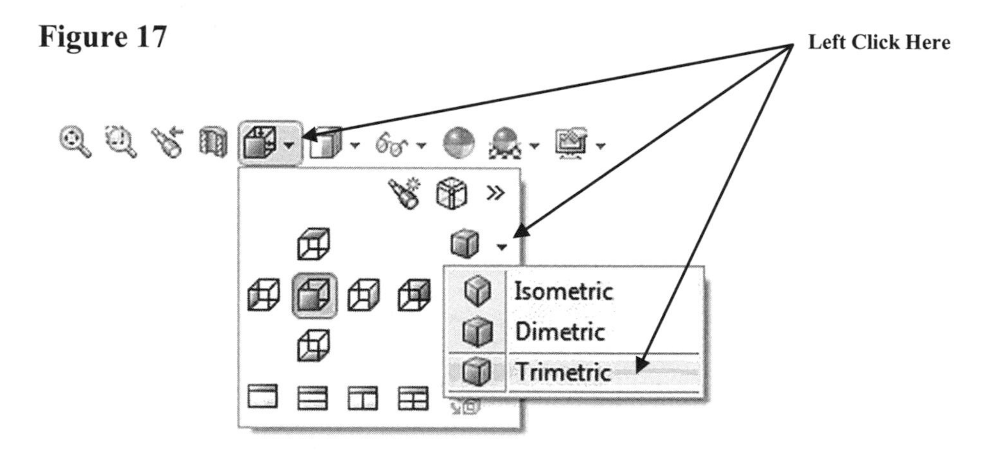

20. The view will become trimetric as shown in Figure 18.

Figure 18

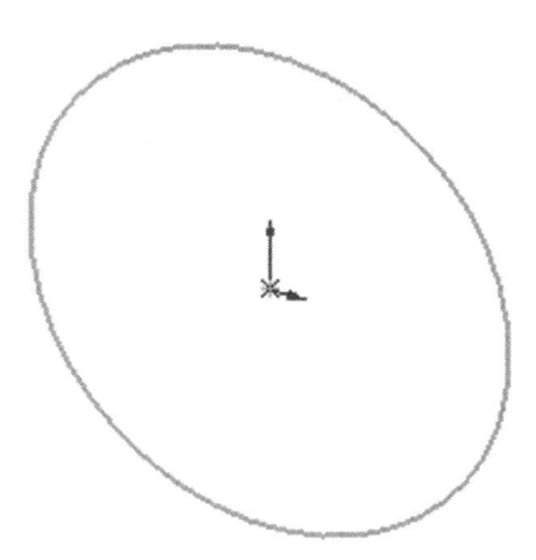

21. Move the cursor to the upper left portion of the screen and left click on **Extruded Boss/Base.** If the **Extruded Boss/Base** command does not appear, left click on the **Features** tab as shown in Figure 19.

Figure 19

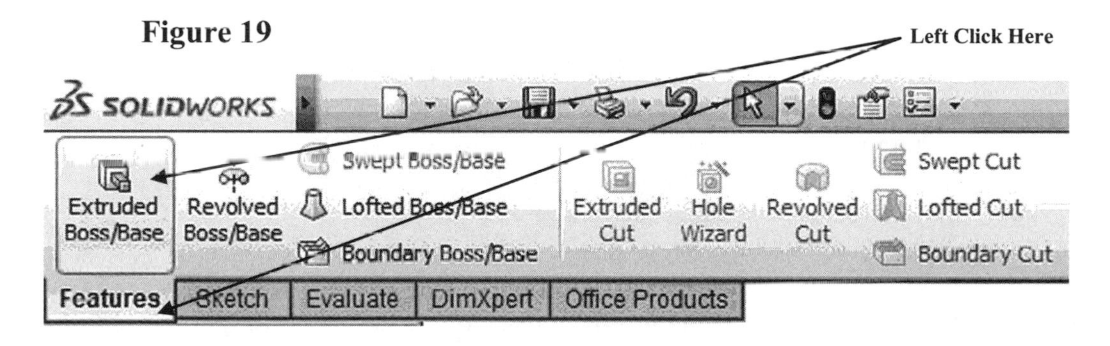

22. A preview of the extrusion will be displayed. <u>If a preview is not displayed, left click anywhere on the edge of the circle</u>. If SolidWorks gave you an error message, there are opens (non-connected lines) somewhere on the sketch. Check each intersection for opens by using the **Extend** and **Trim** commands.

23. Move the cursor to the upper left portion of the screen and enter **.25** next to D1 as shown in Figure 20.

Figure 20

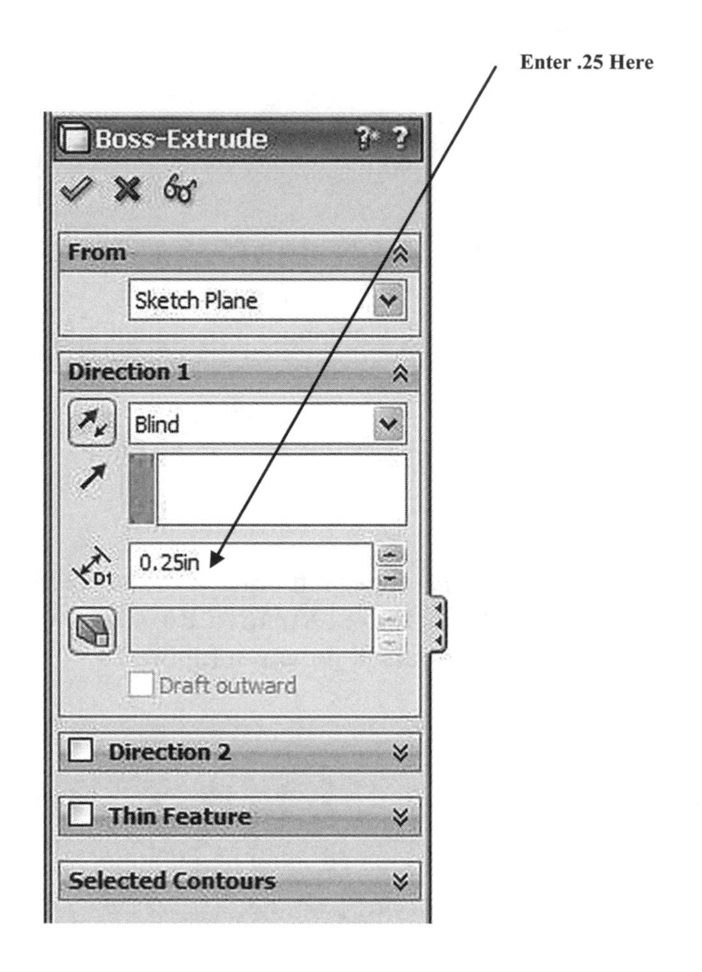

24. Move the cursor to the upper left portion of the screen and left click on the green checkmark as shown in Figure 21.

Figure 21

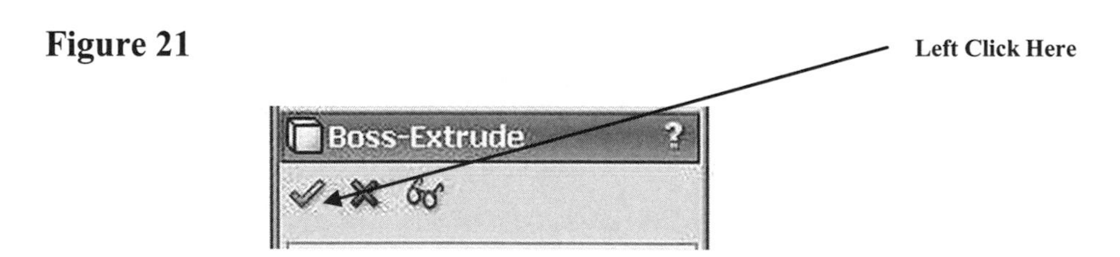

25. Your screen should look similar to Figure 22.

Figure 22

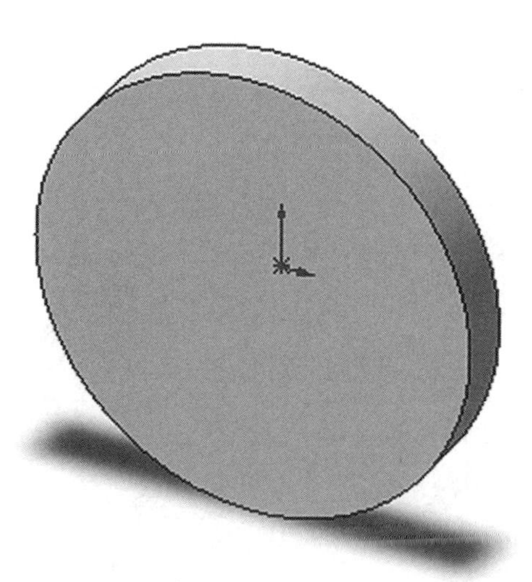

26. Move the cursor to the front surface causing the edges of the surface to turn red. Right click on the surface once. A pop up menu will appear. Left click on the **Insert Sketch** icon as shown in Figure 23.

Figure 23

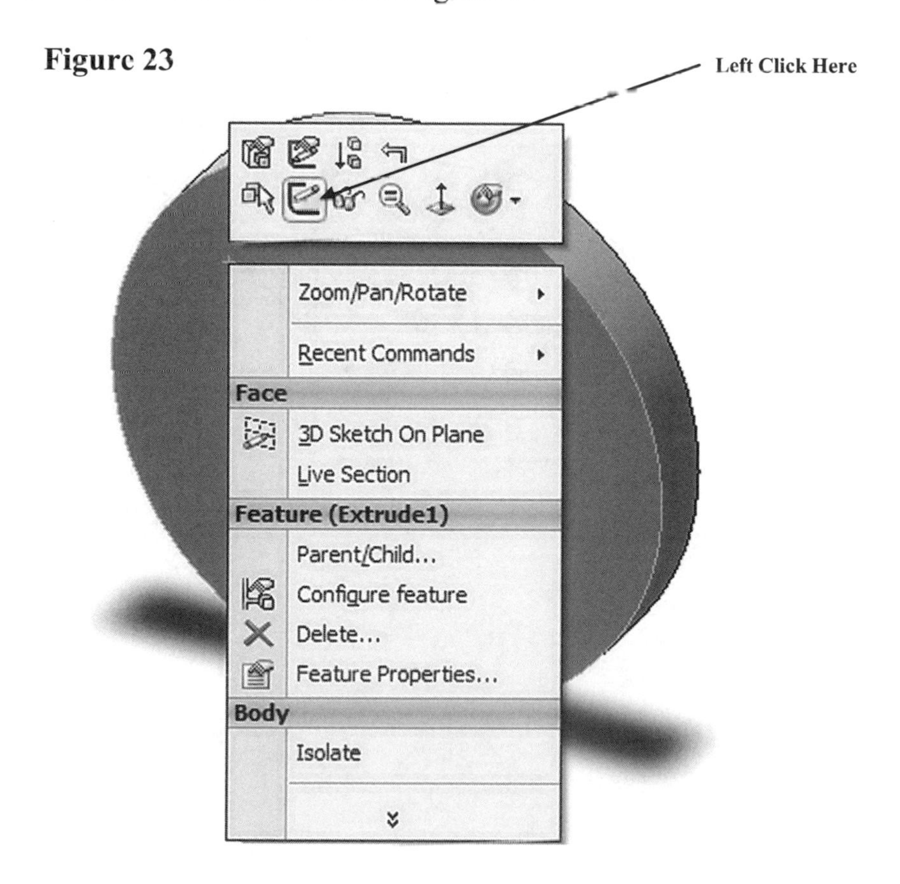

27. SolidWorks will start a new sketch on the selected surface as shown in Figure 24.

Figure 24

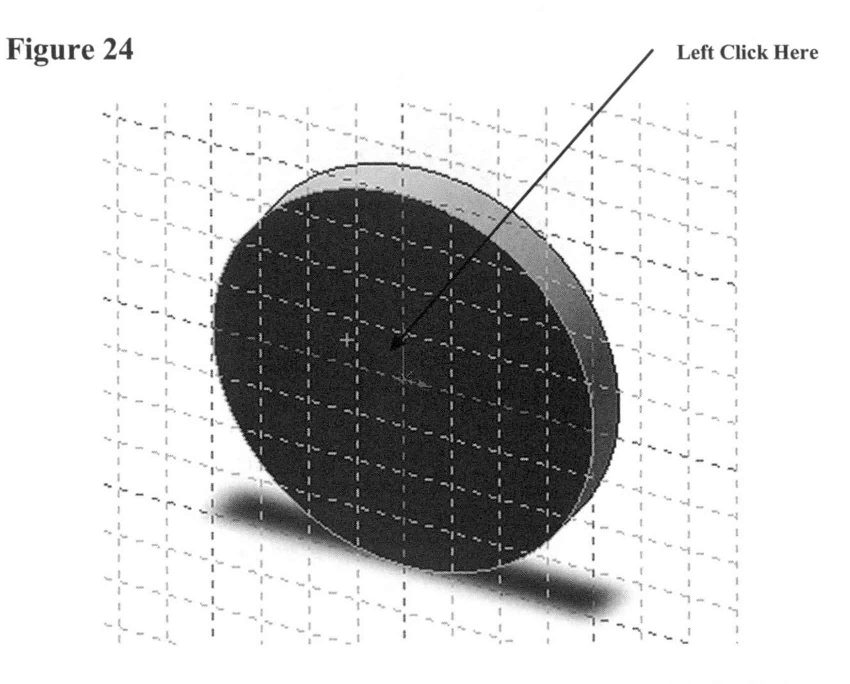

28. Move the cursor to the upper right portion of the screen and left click on the drop down arrow next to the "Standard Views" icon. A drop down menu will appear. Left click on the **Front** icon. You can also left click on the **Normal To** icon as shown in Figure 25.

Figure 25

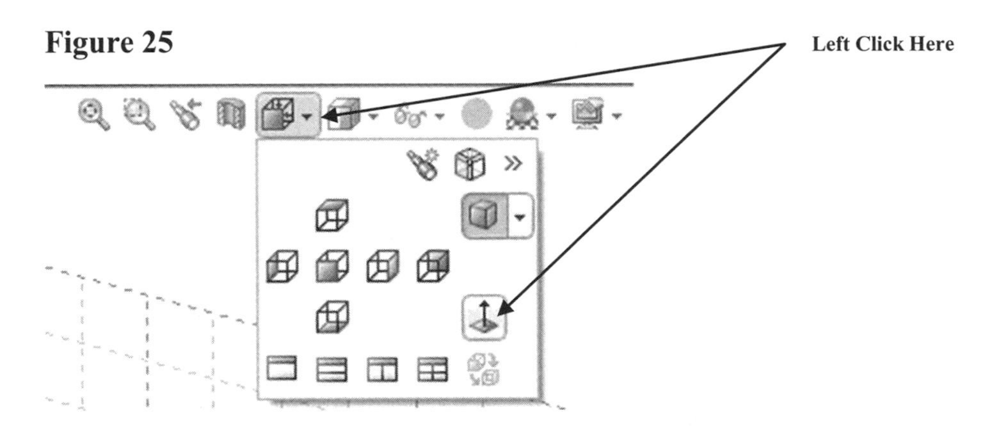

29. The part will rotate providing a perpendicular view of the surface as shown in Figure 26.

Figure 26

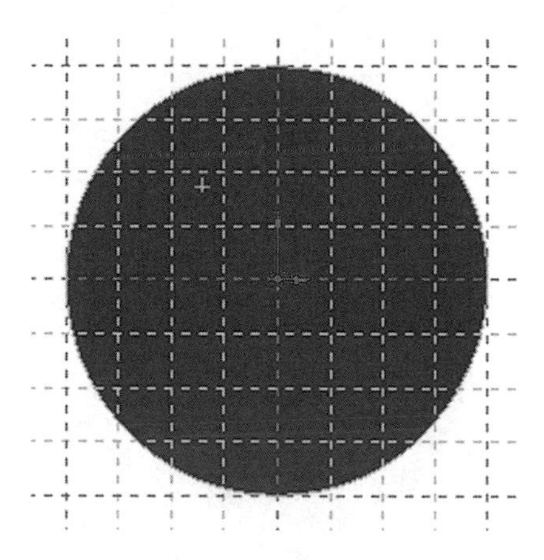

30. Move the cursor to the upper middle portion of the screen and left click on **Circle** as shown in Figure 27.

Figure 27

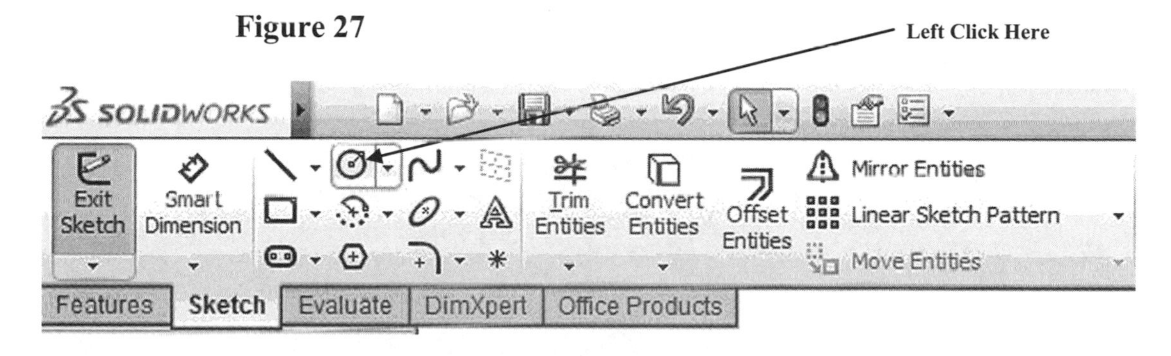

31. Move the cursor to the center of the circle. A red dot will appear in the center as shown in Figure 28.

Figure 28

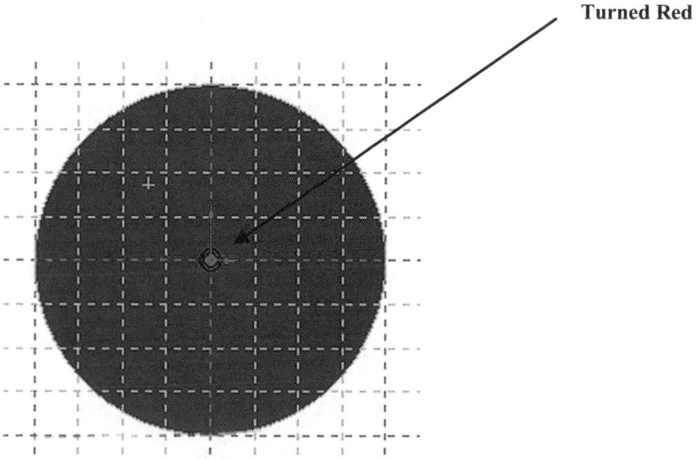

Turned Red

32. After the red dot appears, move the cursor straight up causing a blue dashed line to appear. Left click once as shown in Figure 29.

Figure 29

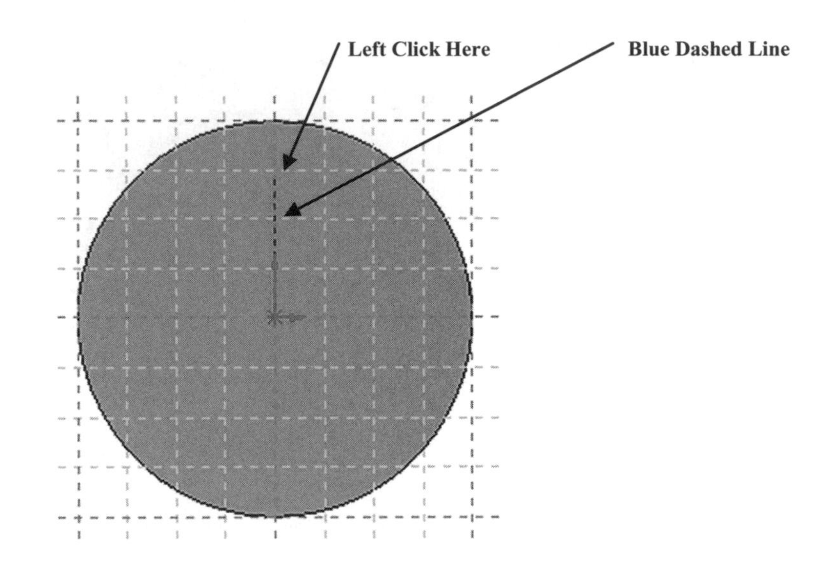

33. Move the cursor to the right side. A circle will form. Let click once as shown in Figure 30.

Figure 30

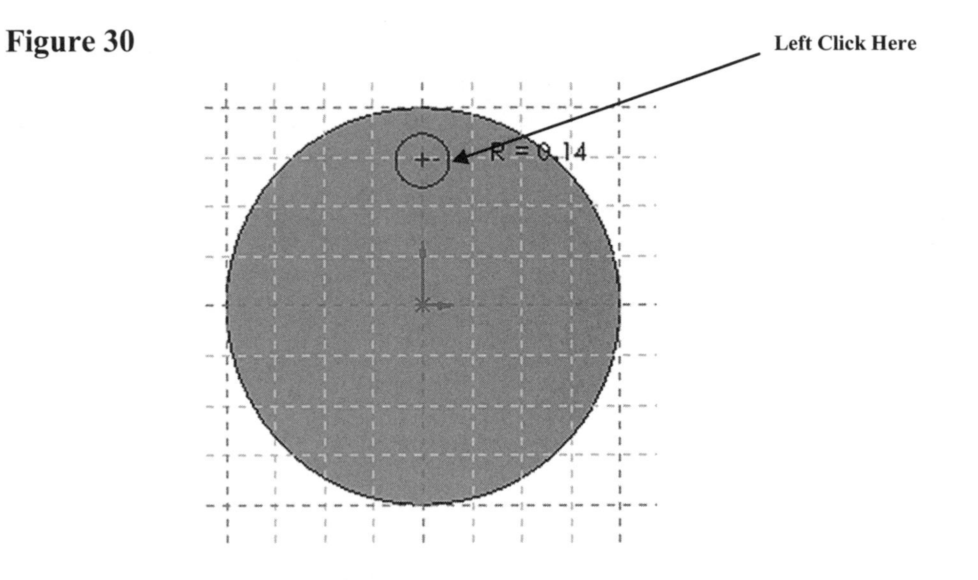

34. Right click anywhere on the drawing. A pop up menu will appear. Left click on **Select** as shown in Figure 31.

Figure 31

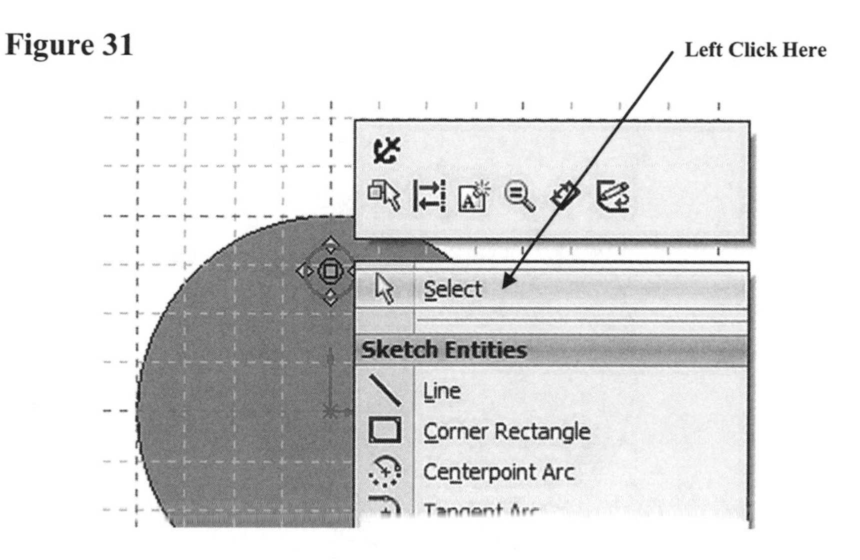

35. Move the cursor to the upper left portion of the screen and left click on **Smart Dimension** as shown in Figure 32.

Figure 32

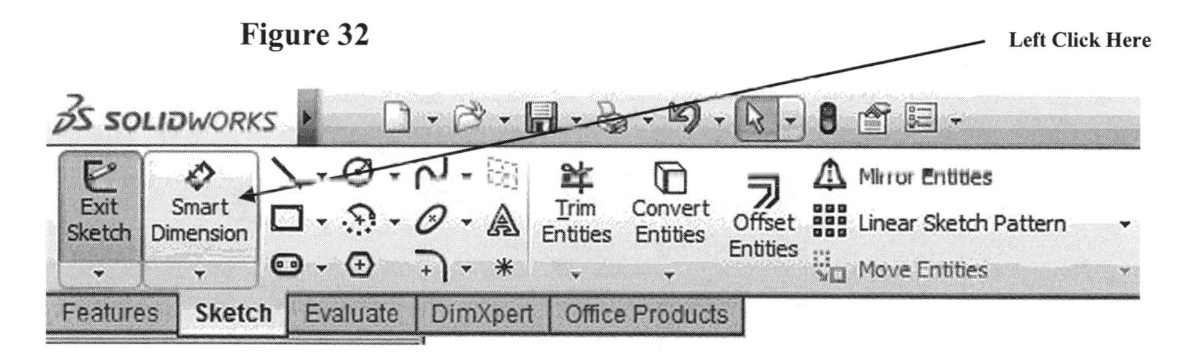

36. Move the cursor over the edge (not center) of the circle until it turns red as shown in Figure 33. Select the circle by left clicking anywhere on the edge. The dimension will be attached to the cursor.

Figure 33

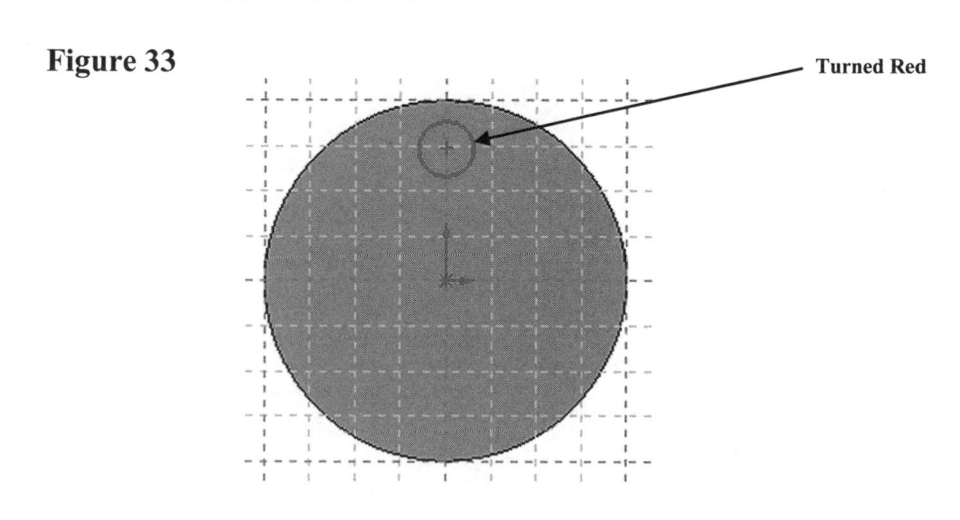

37. Move the cursor to where the dimension will be placed and left click once as shown in Figure 34.

Figure 34

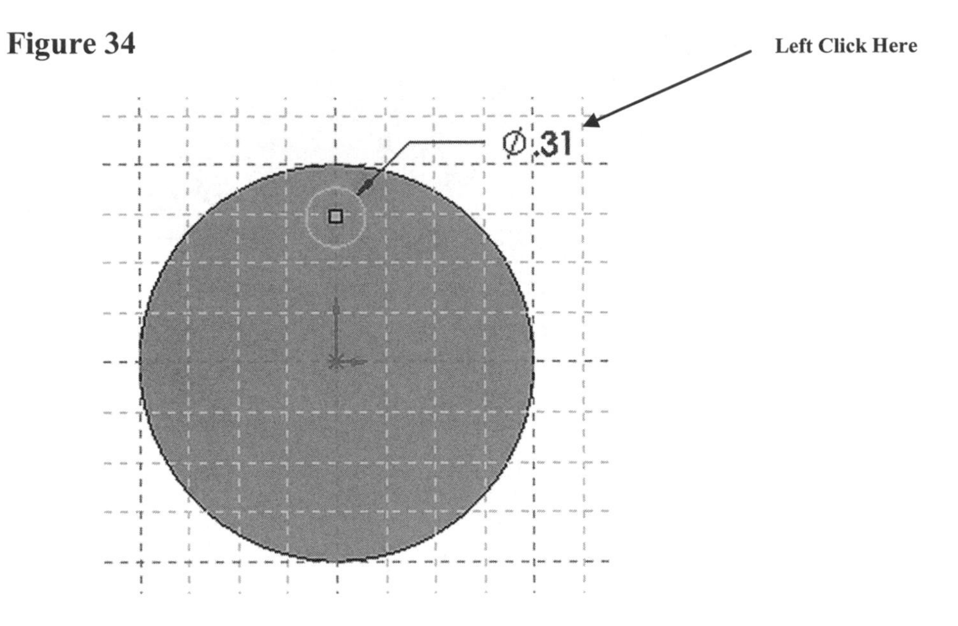

38. The Modify dialog box will appear as shown in Figure 35.

Figure 35

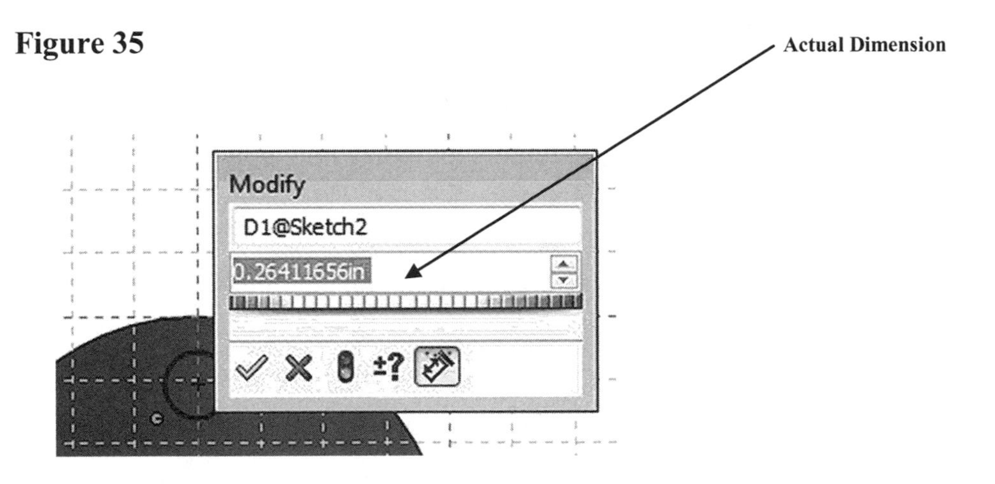

39. To edit the dimension, type **.375** in the Modify dialog box (while the current dimension is highlighted) and left click on the green check mark as shown in Figure 36.

Figure 36

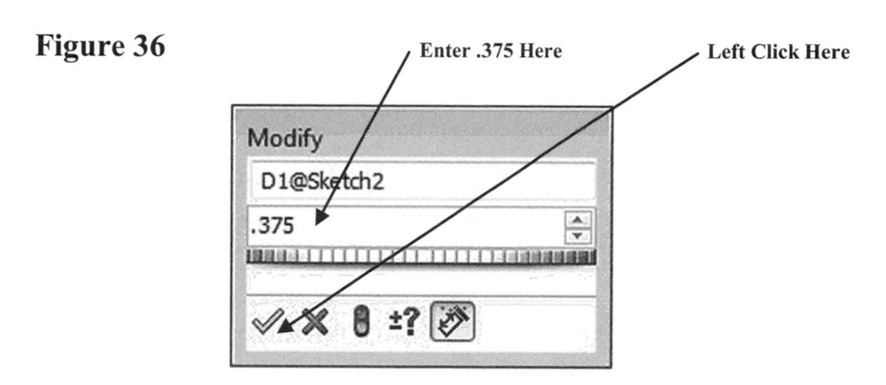

40. The dimension of the circle will become .375 inches as shown in Figure 37. Use the Zoom icons to zoom out if necessary.

Figure 37

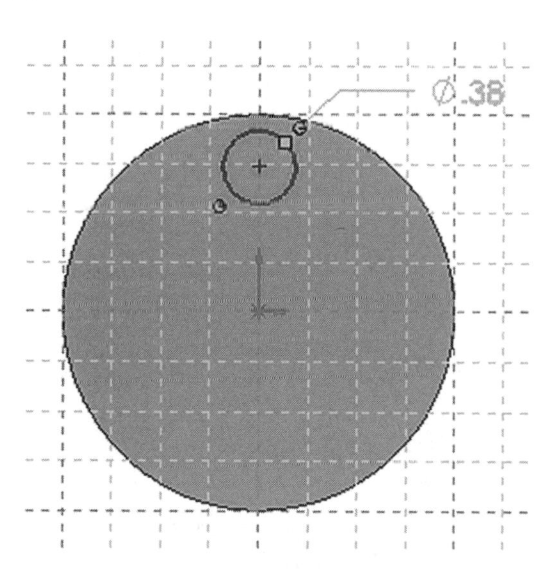

41. Right click anywhere around the drawing. A pop up menu will appear. Left click on **Select** as shown in Figure 38.

Figure 38

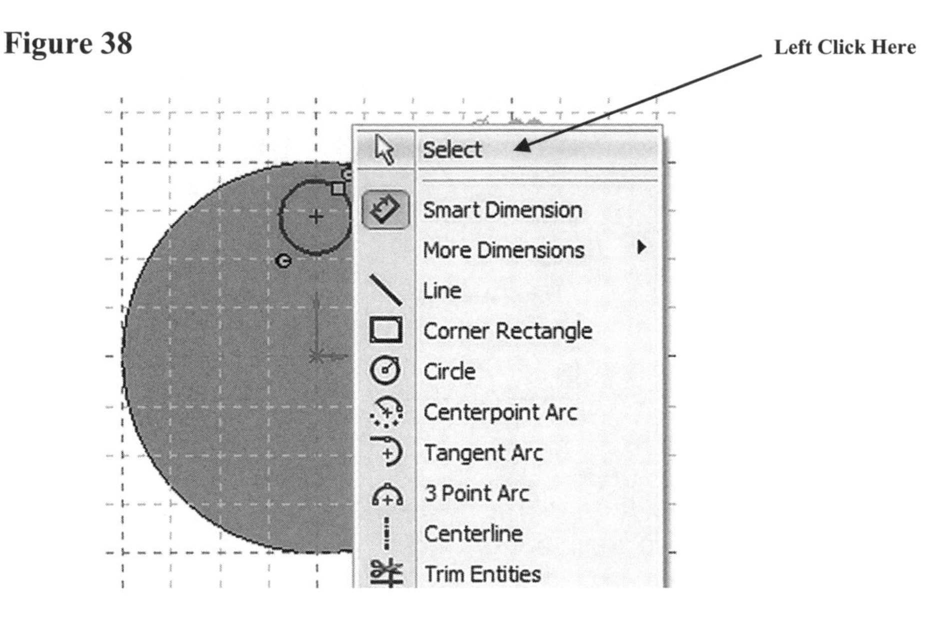

42. After you have verified that no commands are active, right click anywhere on the sketch. A pop up menu will appear. Left click on the **Exit Sketch** icon as shown in Figure 39.

Figure 39

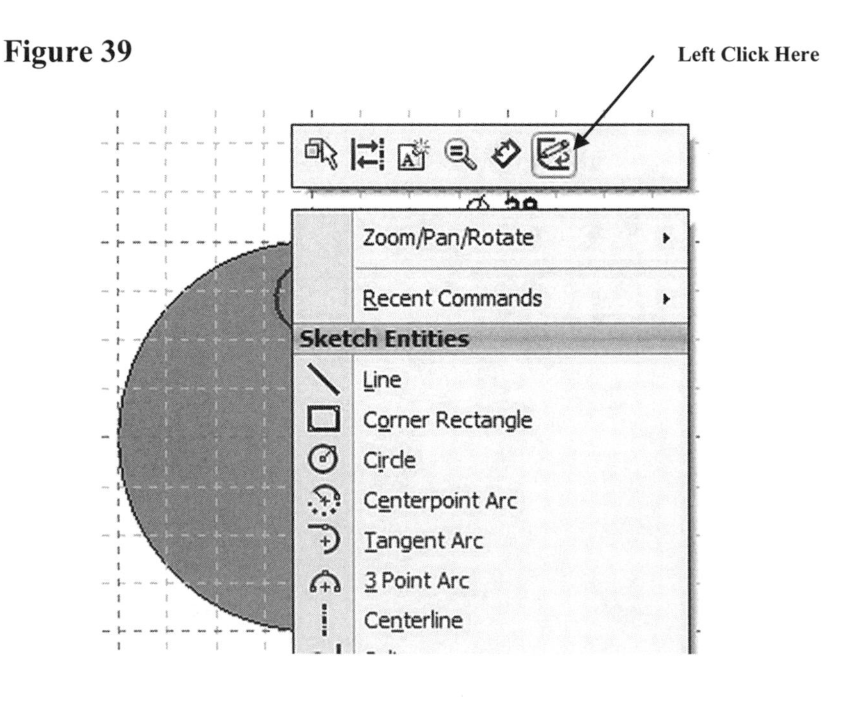

43. SolidWorks is now out of the Sketch commands and into the Features commands. Notice that the commands at the top of the screen are now different. To work in the Features commands a sketch must be present and have no opens (non-connected lines). If there are any opens in the sketch an error message will appear. Your screen should look similar to Figure 40.

Figure 40

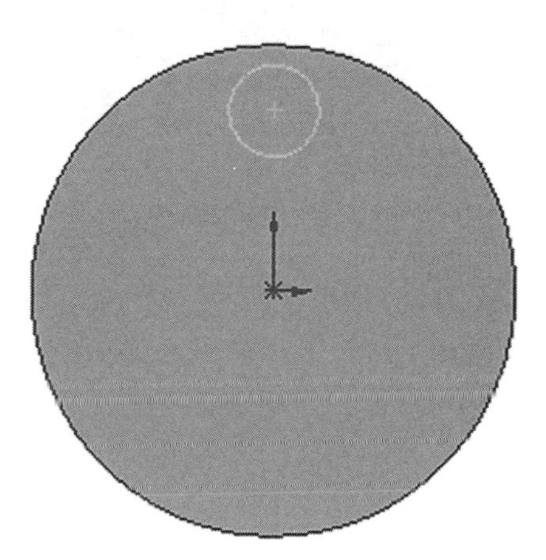

44. Move the cursor to the upper right portion of the screen and left click on the drop down arrow next to the "Views Orientation" icon. A drop down menu will appear. Left click on **Trimetric** as shown in Figure 41.

Figure 41

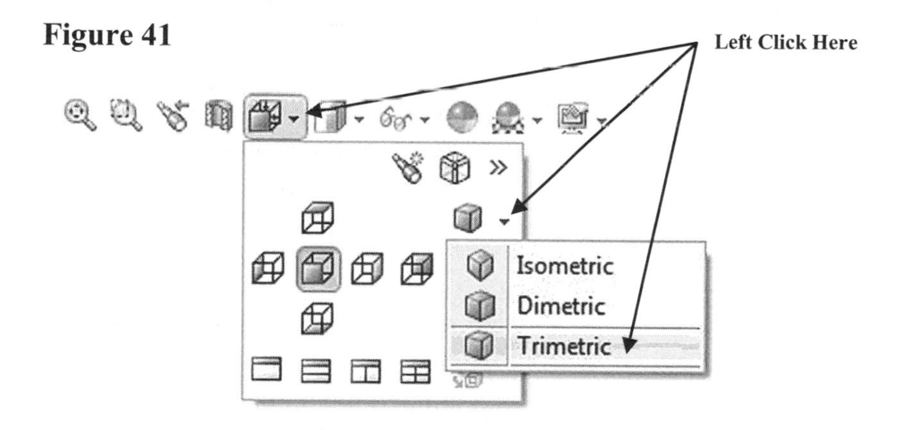

45. The view will become trimetric as shown in Figure 42.

Figure 42

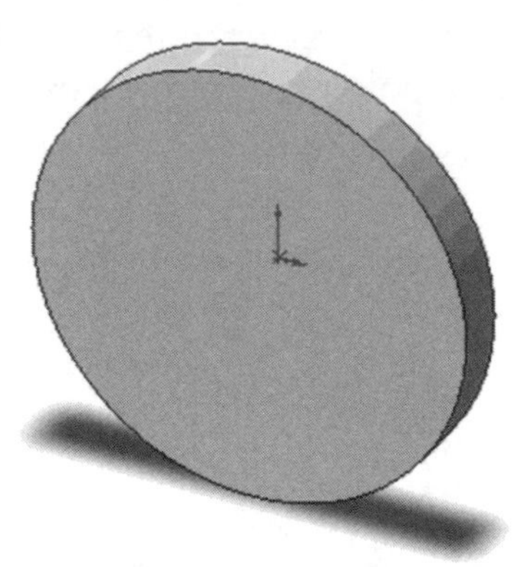

46. Move the cursor to the upper middle portion of the screen and left click on **Extruded Cut** as shown in Figure 43.

Figure 43

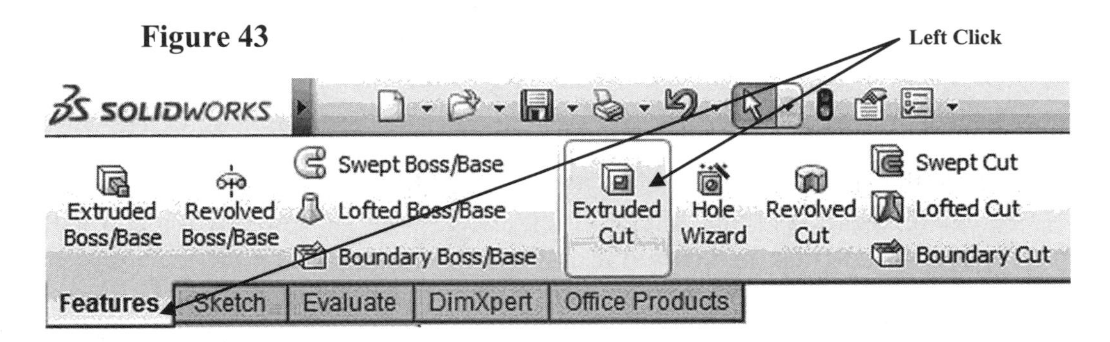

47. A preview of the cut extrusion will be displayed. If a preview is not displayed left click on the edge of the circle. SolidWorks gave you an error message, there are opens (non-connected lines) somewhere on the sketch. Check each intersection for opens by using the Extend and Trim commands. Your screen should look similar to Figure 44.

Figure 44

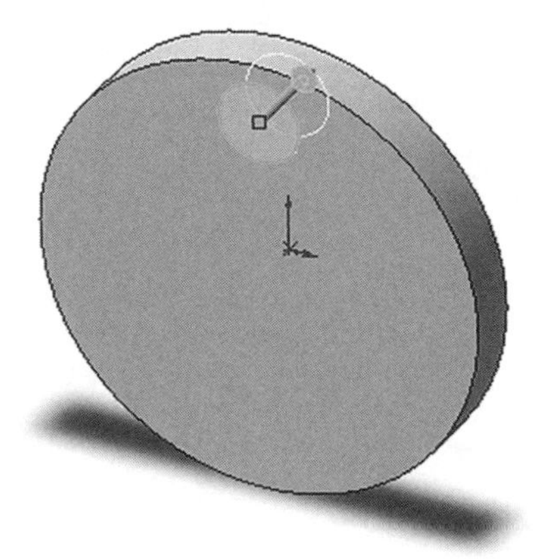

48. Enter **.25** next to D1 as shown in Figure 45.

Figure 45

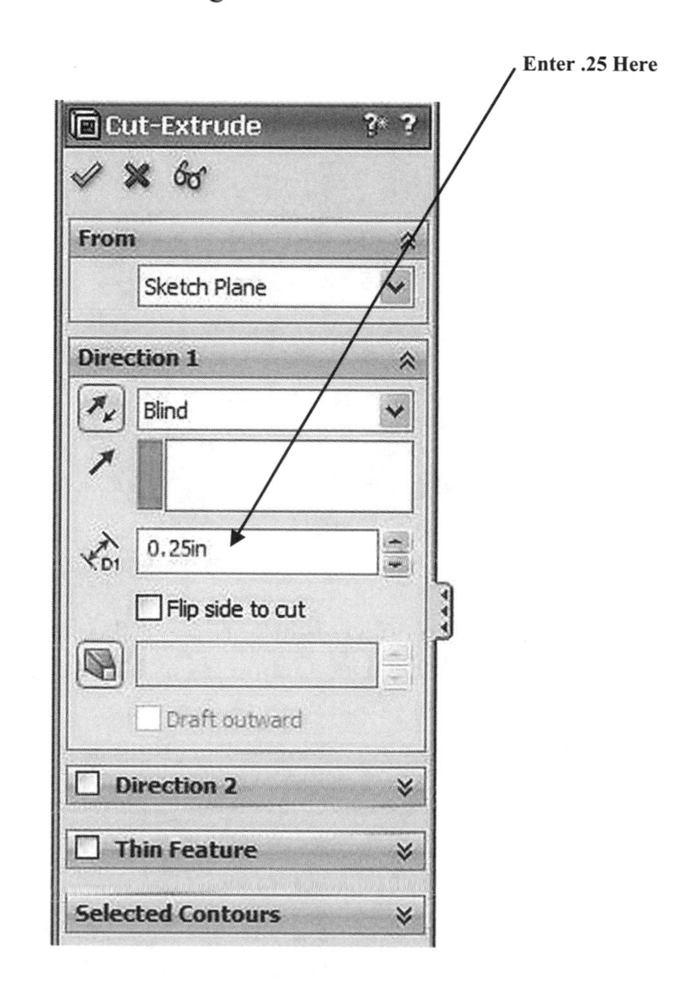

49. Move the cursor to the upper left portion of the screen and left click on the green checkmark as shown in Figure 46.

Figure 46

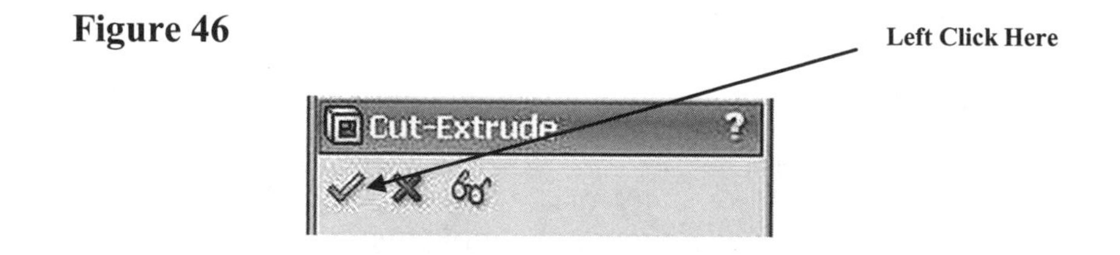

50. Your screen should look similar to Figure 47. You may have to use the zoom out command to view the part.

Figure 47

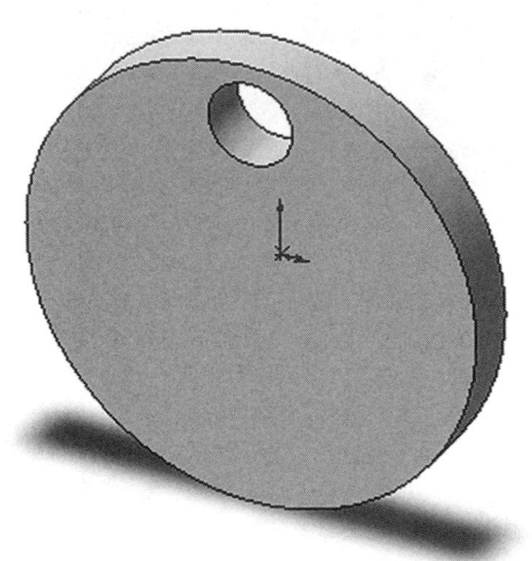

51. Move the cursor to the upper right portion of the screen and left click on the double arrow under Reference Geometry as shown in Figure 48.

Figure 48

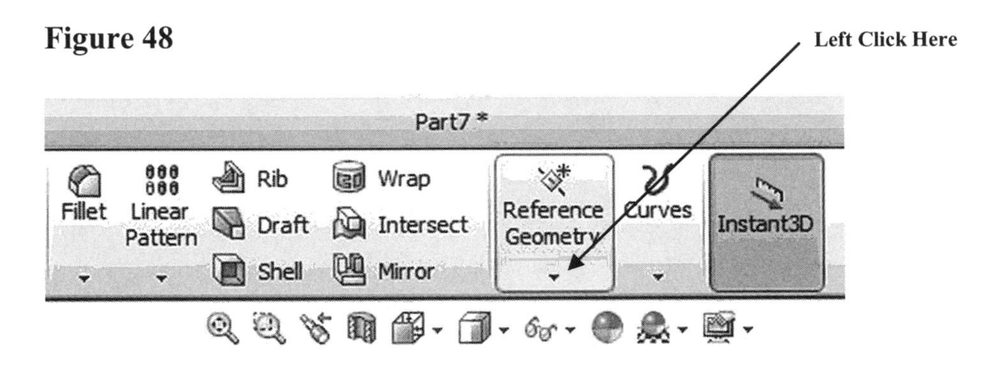

52. A drop down menu will appear. Left click on **Axis** as shown in Figure 49.

Figure 49

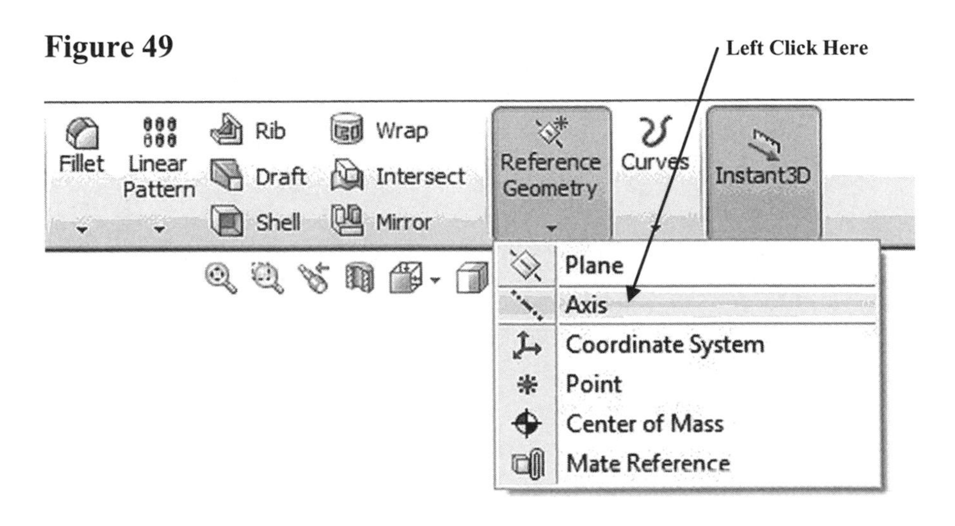

53. Move the cursor to the side of the part causing both edges to turn red. Left click once as shown in Figure 50.

Figure 50

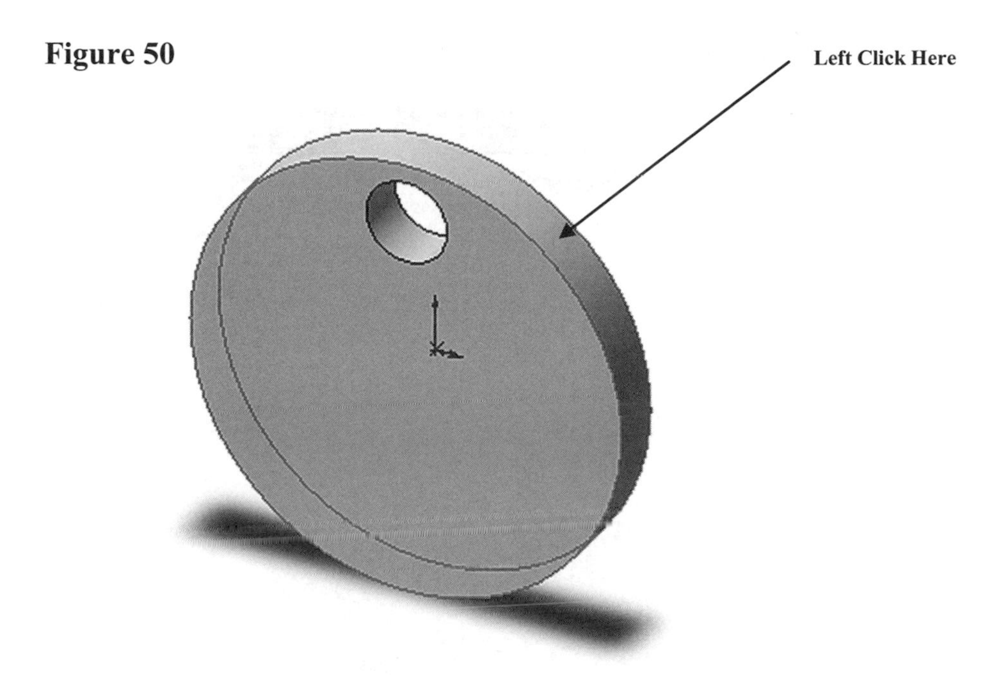

54. SolidWorks will project a center axis as shown in Figure 51. This axis will be used when creating a circular pattern.

Figure 51

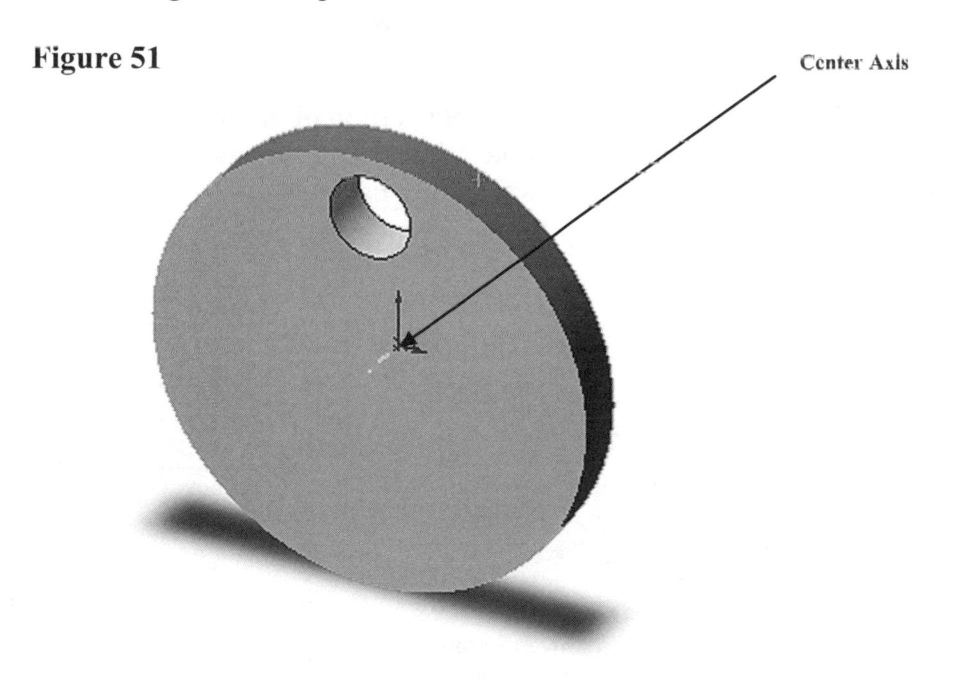

55. Move the cursor to the middle portion of the screen and left click on the green checkmark as shown in Figure 52.

Figure 52

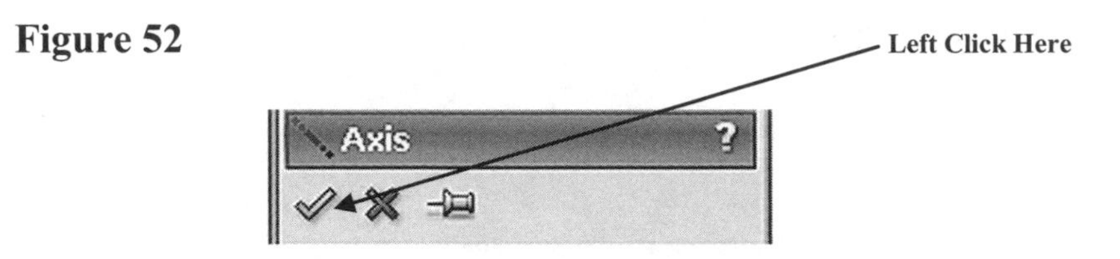

56. The axis will be displayed as shown in Figure 53.

Figure 53

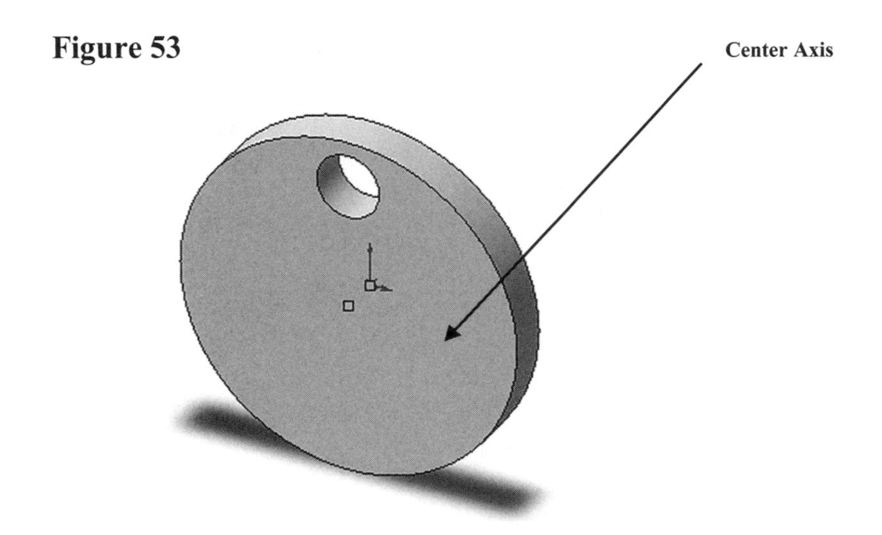

57. Move the cursor to the upper middle portion of the screen and left click on the drop down arrow under Linear Pattern. A drop down menu will appear. Left click on **Circular Pattern** as shown in Figure 54.

Figure 54

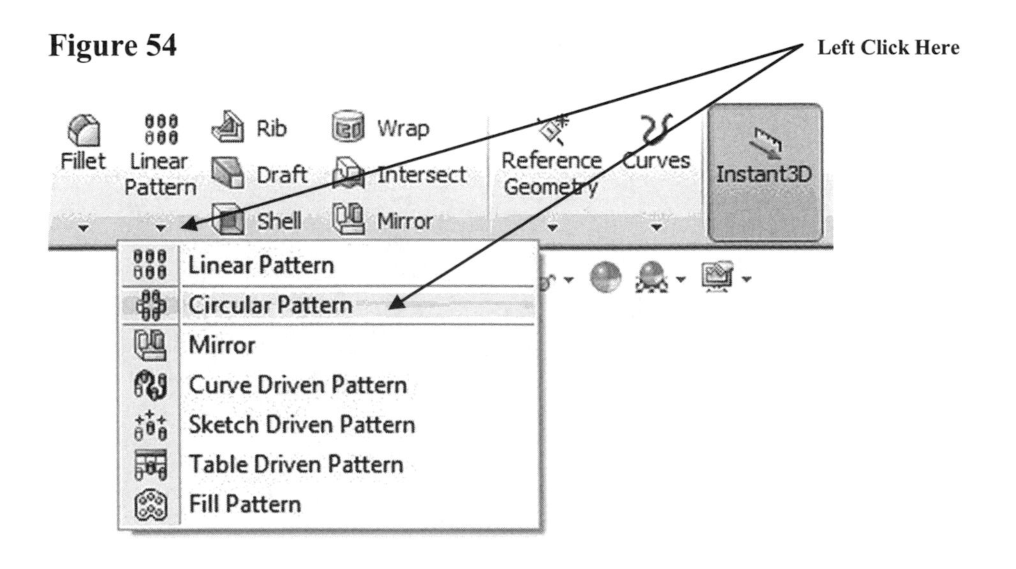

58. Move the cursor to the left portion of the screen and enter **3** for the number of holes and left click in the box next to the text "Equal spacing" as shown in Figure 55.

Figure 55

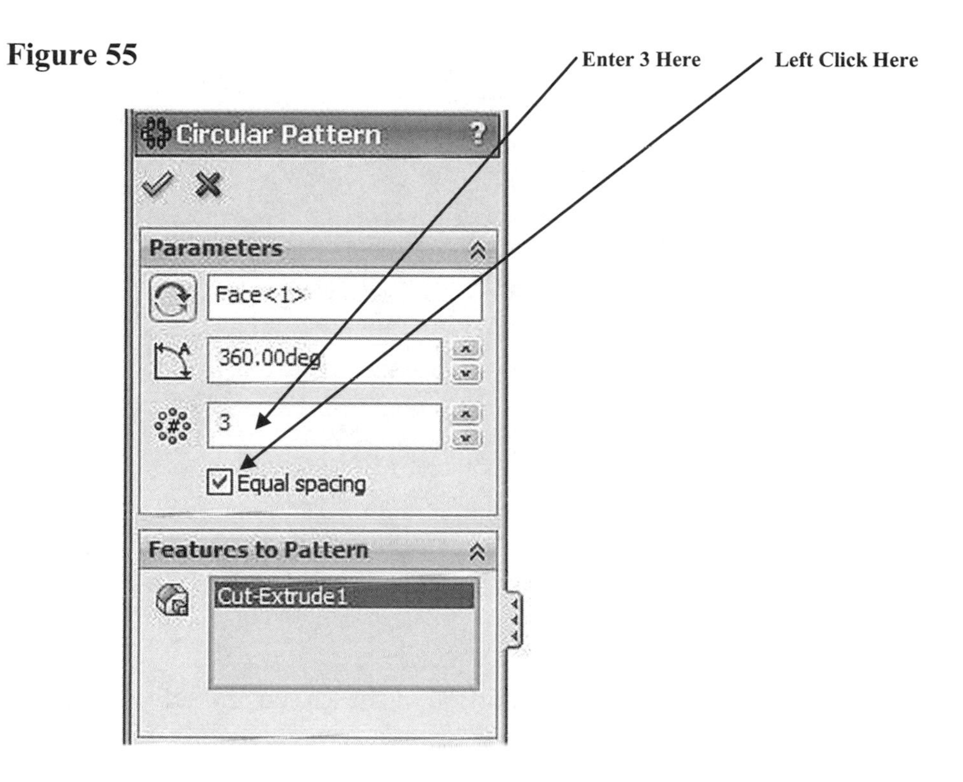

59. Move the cursor to the center of the circle (hole) as shown in Figure 56. The edges will become red lines. SolidWorks will only find the circle (hole) if the view is trimetric or isometric.

Figure 56

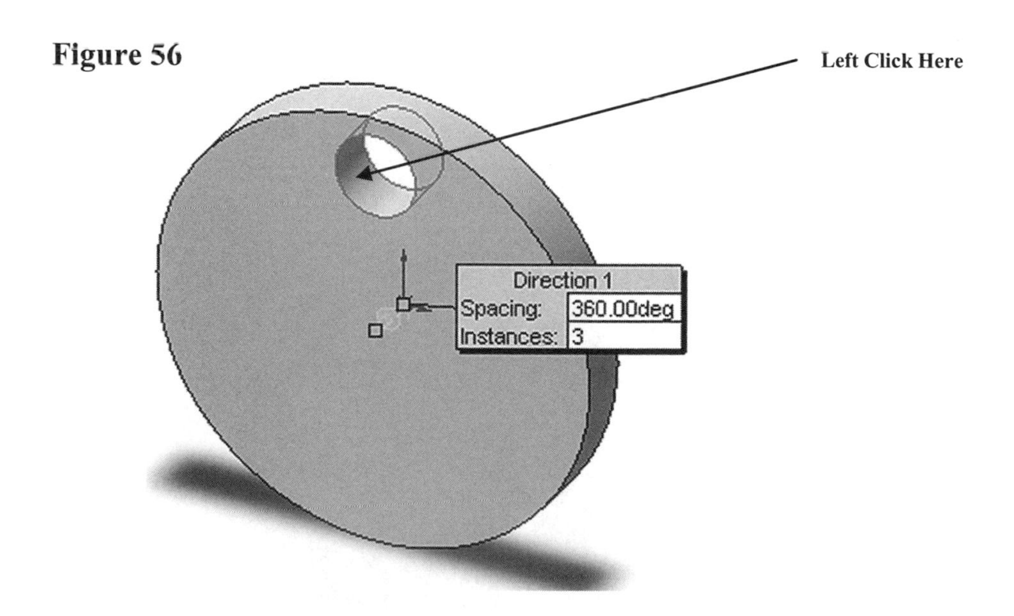

60. Left click inside the circle (hole). The inside of the hole will turn green. A preview of the holes will be displayed as shown in Figure 57.

Figure 57

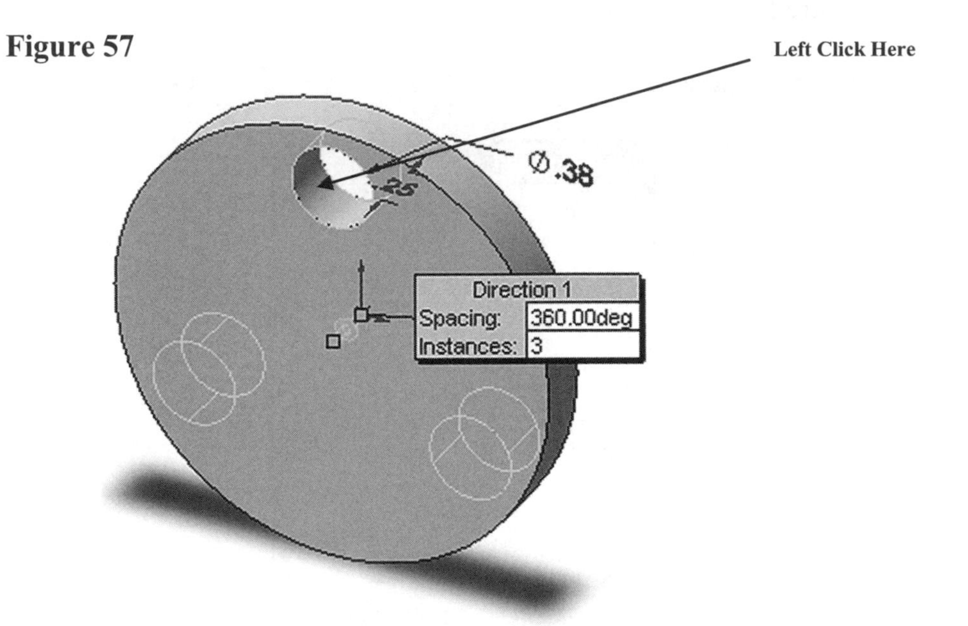

61. Move the cursor to the left portion of the screen and left click on the green checkmark as shown in Figure 58.

Figure 58

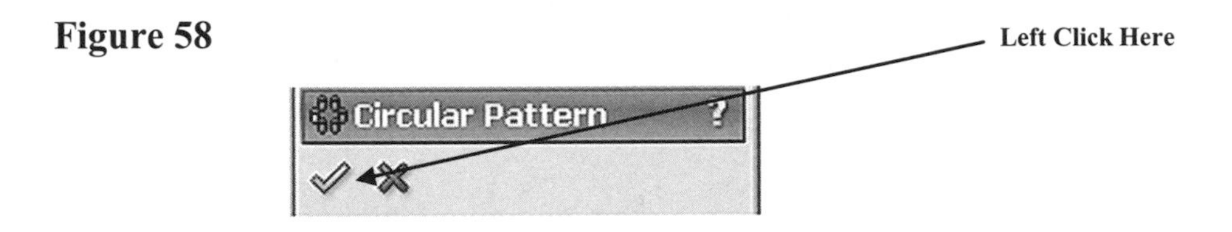

62. Your screen should look similar to Figure 59.

Figure 59

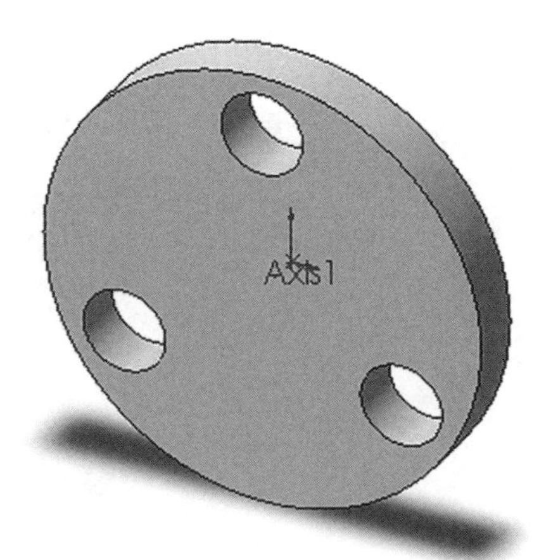

63. Move the cursor to the upper middle portion of the screen and left click the drop down arrow under Fillet. A drop menu will appear. Left click on **Chamfer** as shown in Figure 60.

Figure 60

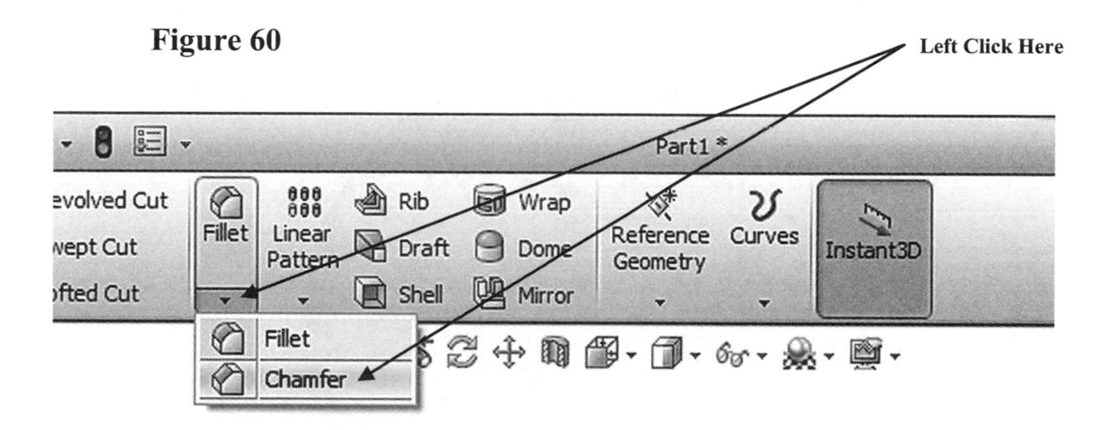

64. Enter **.0625** for the distance as shown in Figure 61. Left click next to Full Preview. Press **Enter** on the keyboard.

Figure 61

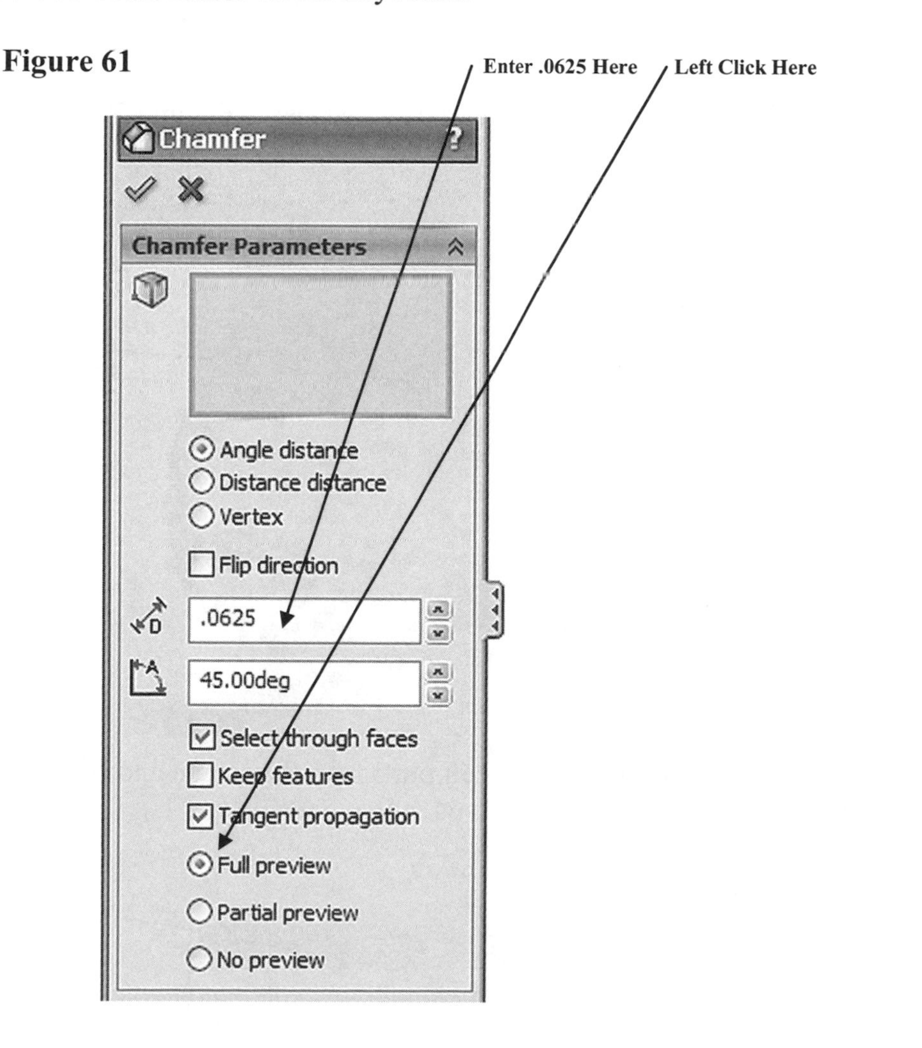

65. Move the cursor over the edge of the part causing it to turn red as shown in Figure 62. Left click once.

Figure 62

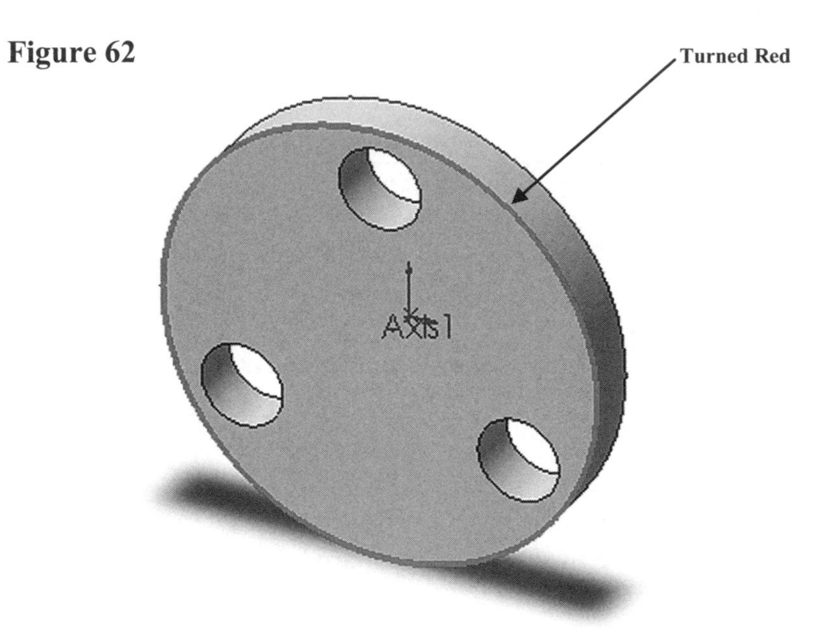

66. A preview of the chamfer will be displayed as in Figure 63.

Figure 63

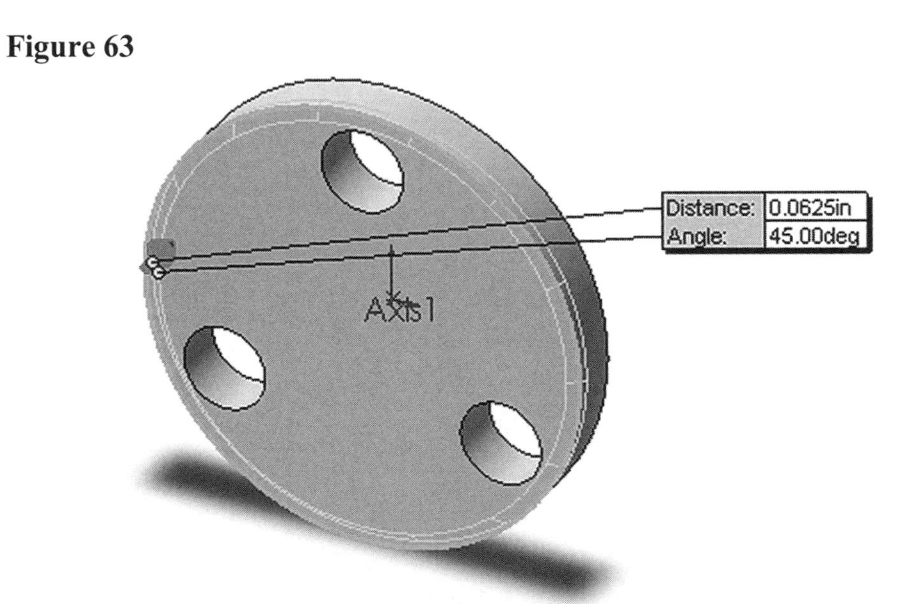

67. Move the cursor to the upper left portion of the screen and left click on the green checkmark as shown in Figure 64.

Figure 64

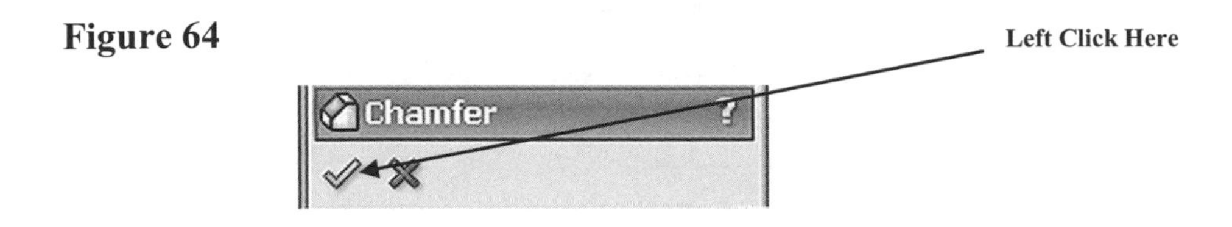

68. Your screen should look similar to Figure 65.

Figure 65

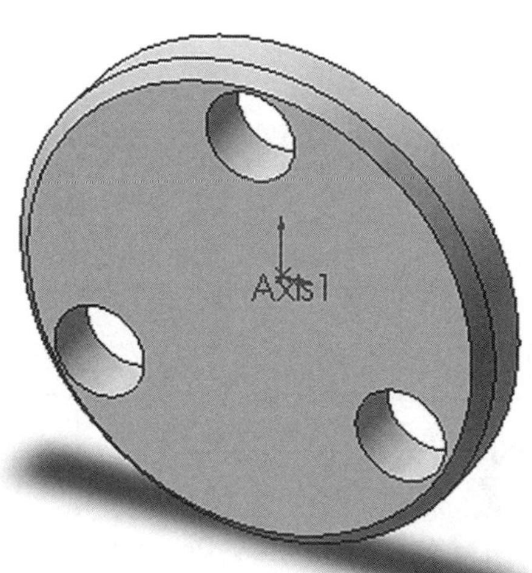

69. If for some reason a change needs to be made to this part, it can be accomplished by either editing a sketch or feature located in the Part Tree at the middle left corner of the screen as shown in Figure 66.

Figure 66

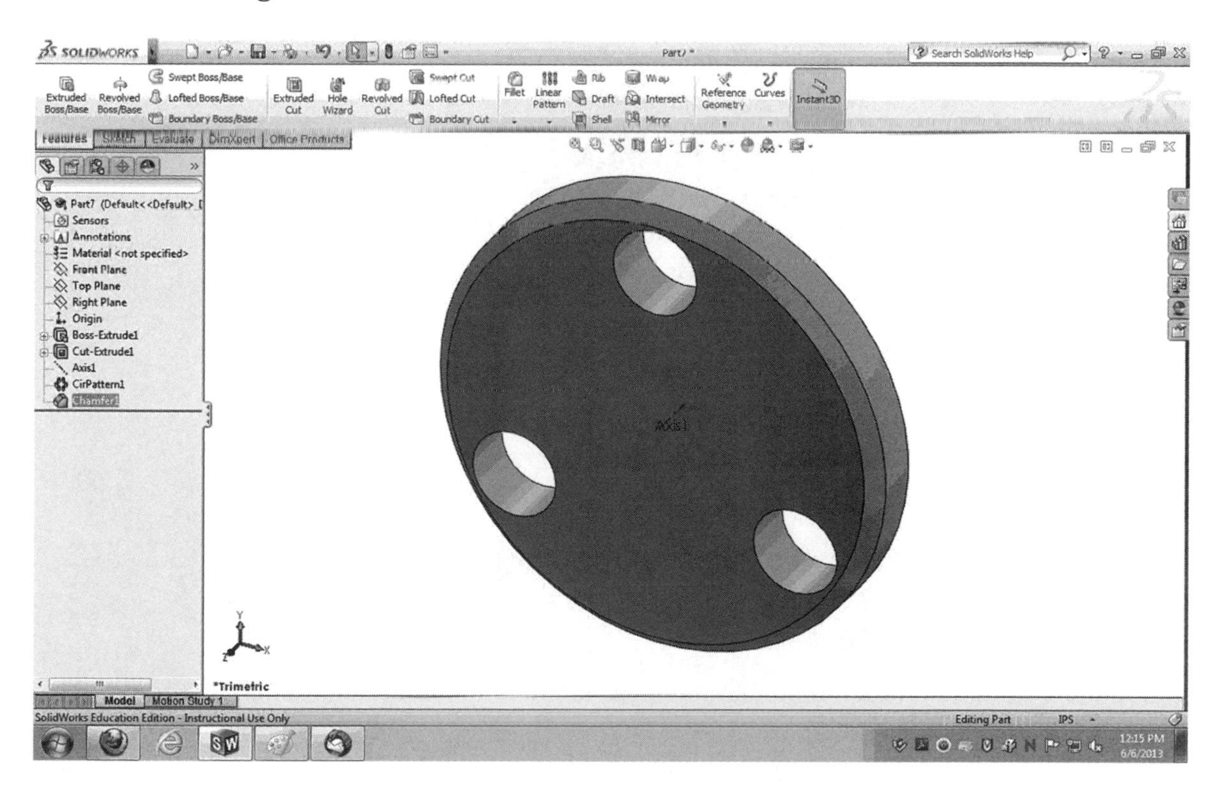

70. The Part Tree is shown in Figure 67. Left click on each of the "plus" signs in the part tree. The tree will expand showing more details for part construction.

Figure 67

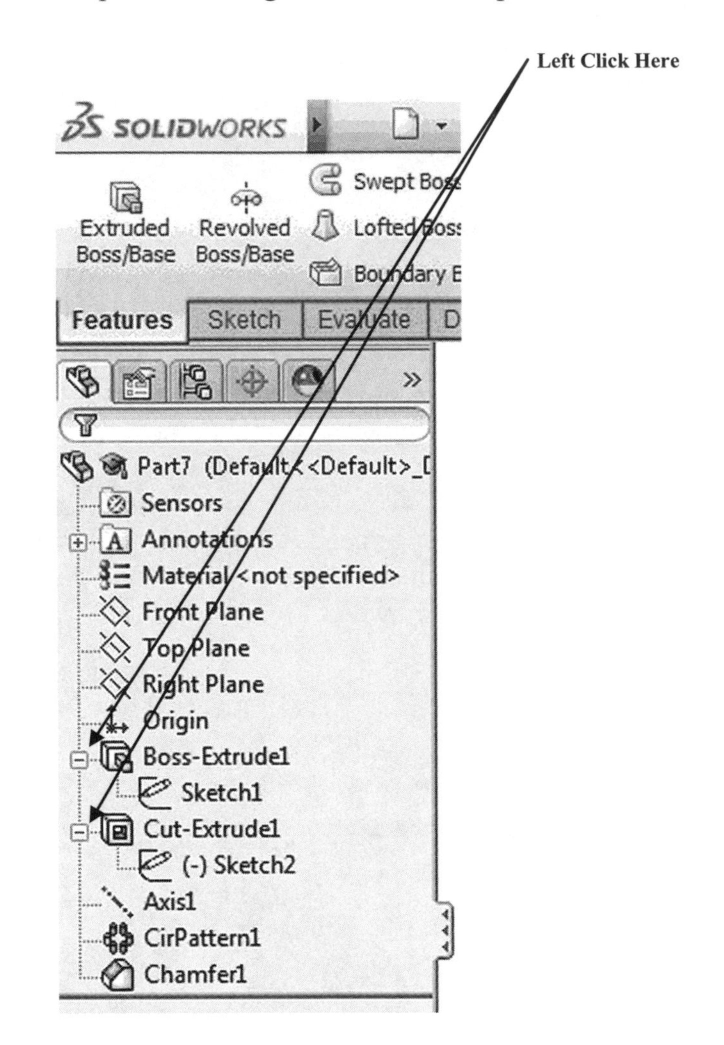

71. Move the cursor over the text “Sketch1”. The text will become highlighted as shown in Figure 68.

Figure 68

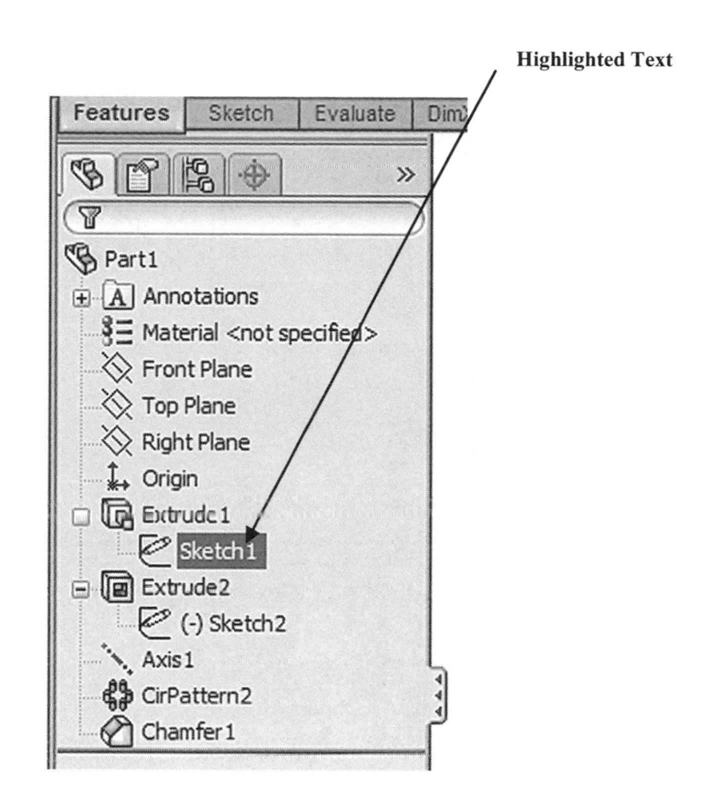

72. The original sketch will also appear as shown in Figure 69.

Figure 69

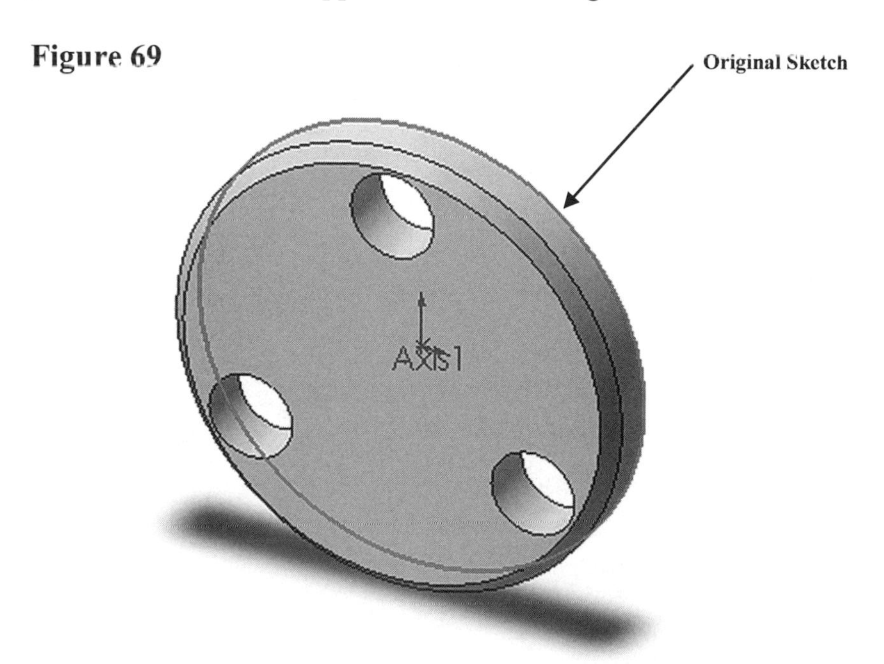

73. Right click on **Sketch1**. A pop up menu will appear. Left click on the **Edit Sketch** icon as shown in Figure 70.

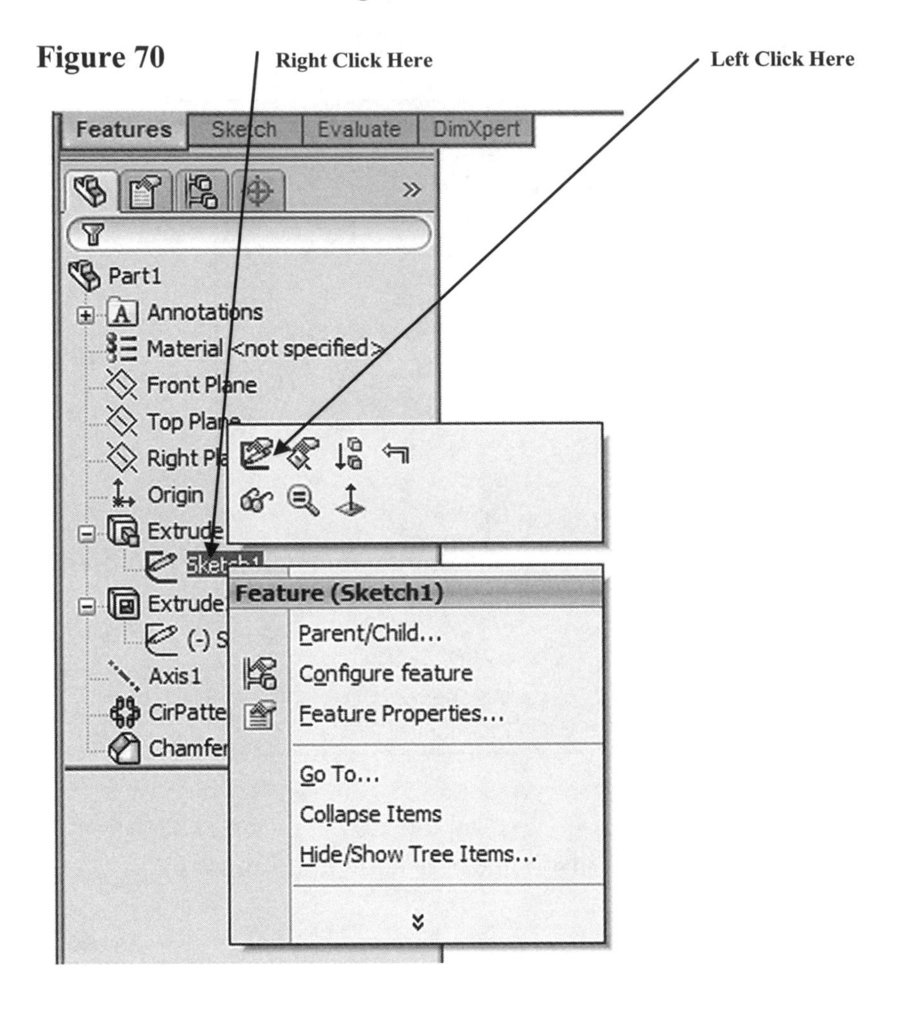

Figure 70

74. The original sketch will appear as shown in Figure 71.

Figure 71

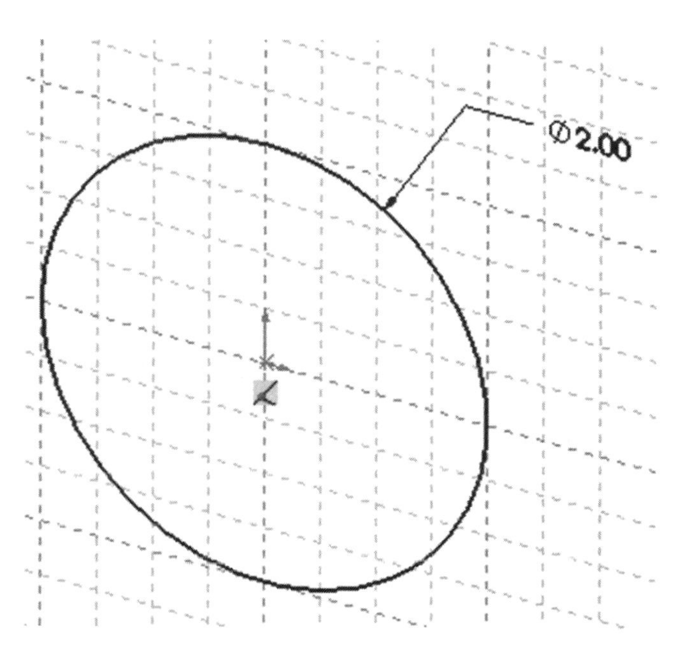

75. Move the cursor to the upper middle portion of the screen and left click on the drop down arrow next to the "Views Orientation" icon. A drop down menu will appear. Left click on **Front** or **Normal To** as shown in Figure 72.

Figure 72

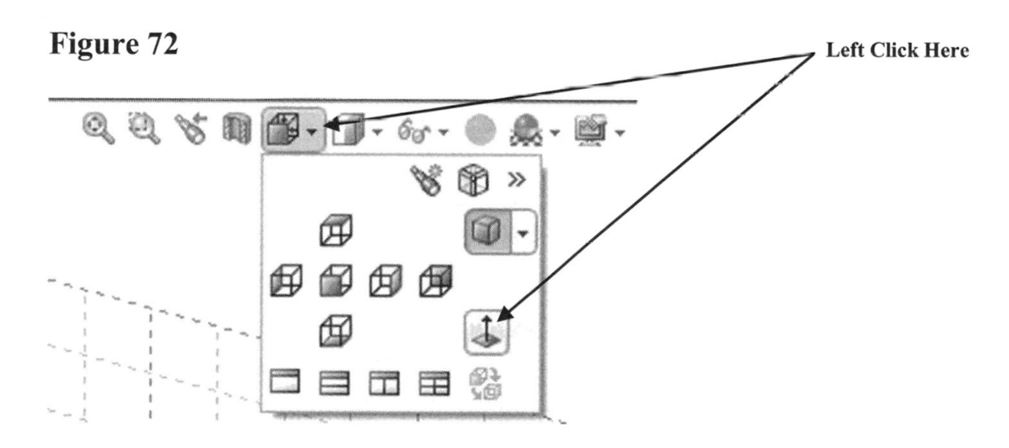

76. A perpendicular view of the sketch will be displayed. The sketch will look similar to when the sketch was first constructed. Your screen should look similar to Figure 73.

Figure 73

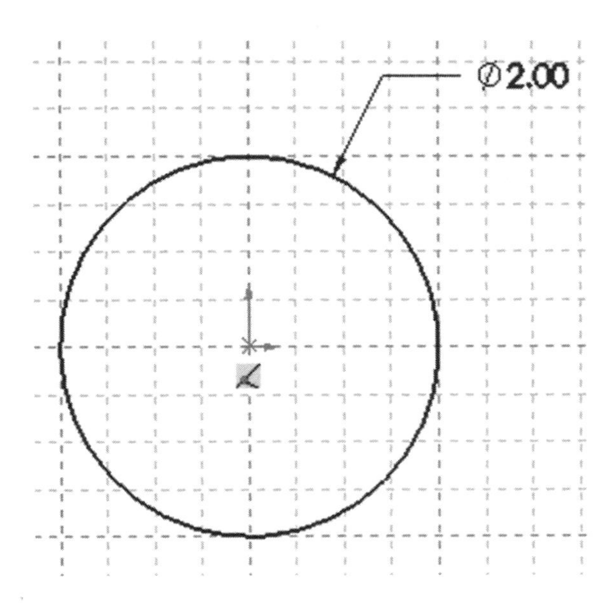

77. Start by modifying the diameter of the part. Double click on the overall dimension. The Modify dialog box will appear as shown in Figure 74.

Figure 74

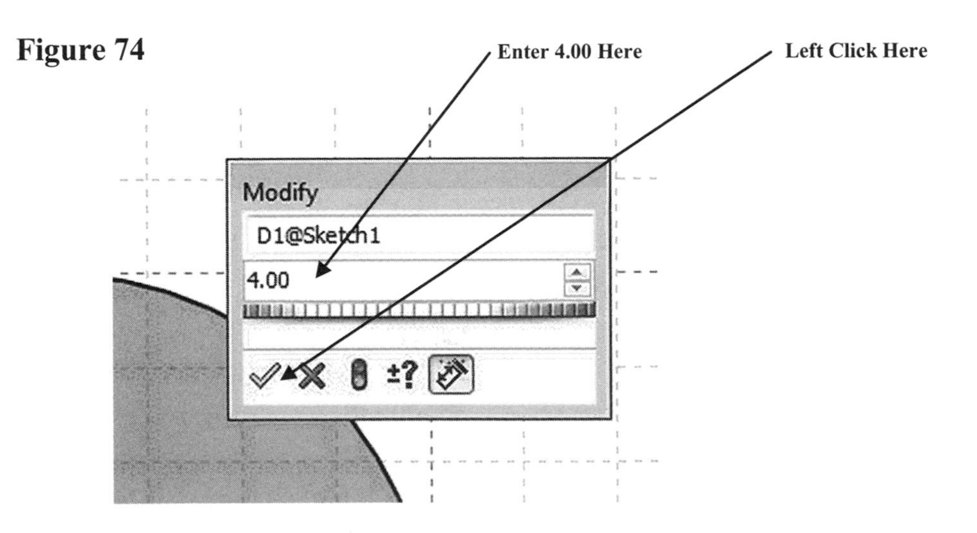

78. Enter **4.00** and left click on the green checkmark as shown in Figure 74.

79. The diameter of the part will increase to 4.00 as shown in Figure 75.

Figure 75

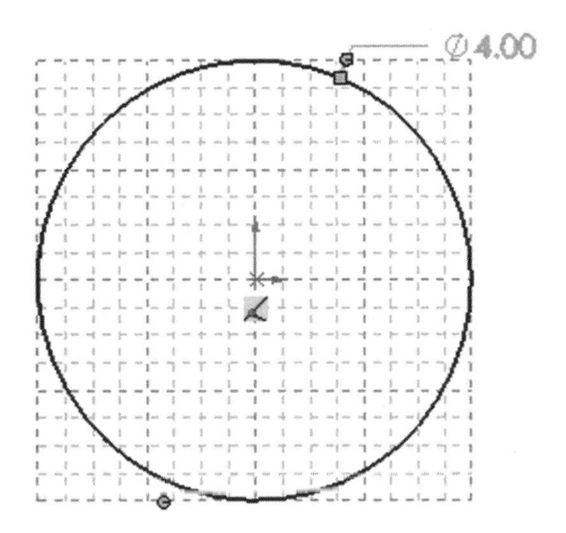

80. Move the cursor to the upper middle portion of the screen and left click on the green checkmark as shown in Figure 76.

Figure 76

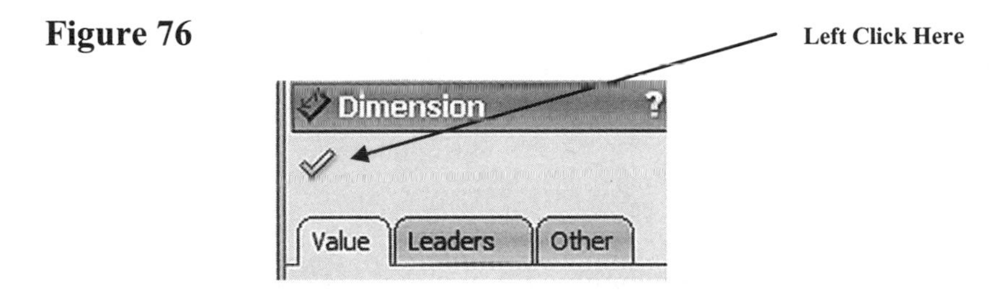

81. Right click anywhere on the screen. A pop up menu will appear. Left click on the **Exit Sketch** icon as shown in Figure 77.

Figure 77

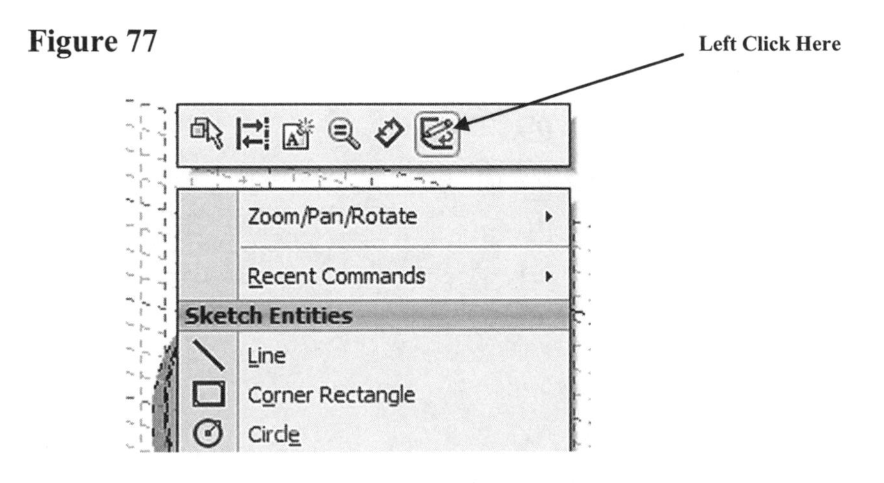

82. SolidWorks will automatically update the part to reflect the changes made in the sketch. It will not be necessary to extrude the part again. Your screen should look similar to Figure 78.

Figure 78

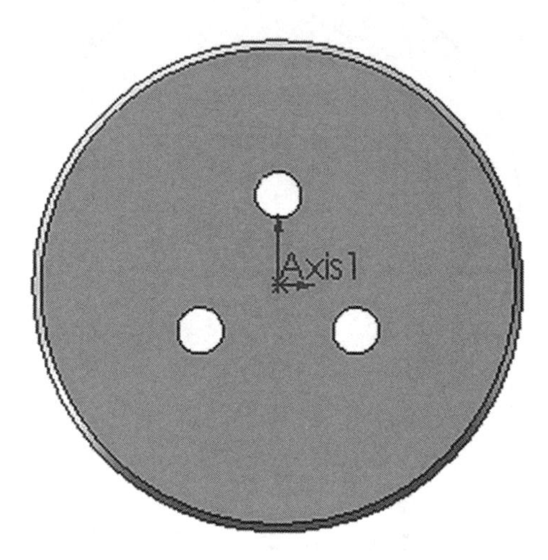

83. Move the cursor to the upper middle portion of the screen and left click on the drop down arrow next to the "View Orientation " icon. A drop down menu will appear. Left click on the **Trimetric** icon as shown in Figure 79.

Figure 79

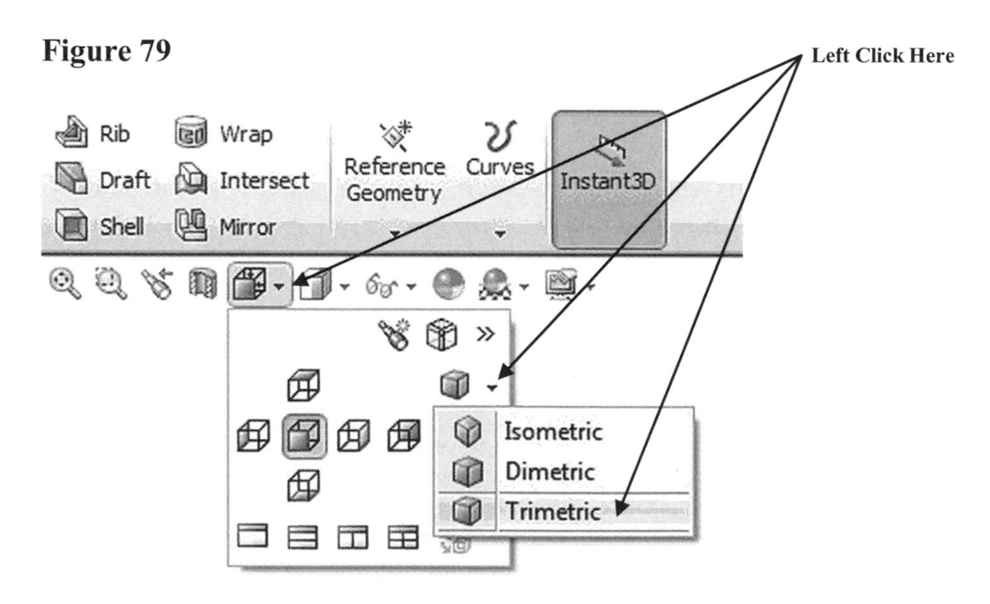

84. Your screen should look similar to Figure 80.

Figure 80

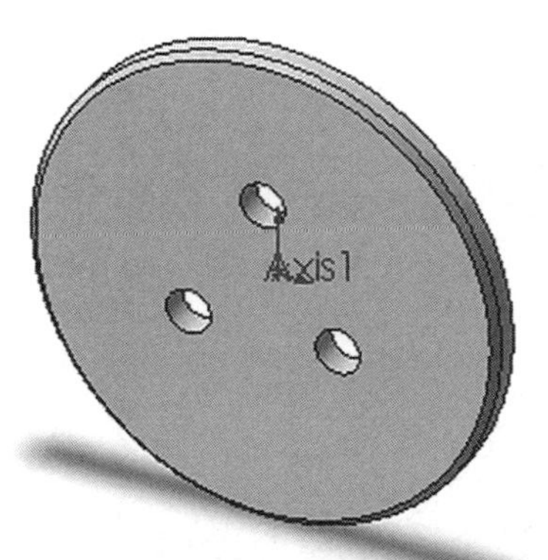

85. Move the cursor to the middle left portion of the screen where the part tree is located. Right click once on **Boss-Extrude1/Extrude1 (2009 and earlier versions of SolidWorks display "Extrude1" rather than "Boss-Extrude1"**. A pop up menu will appear. Left click on the **Edit Feature** icon as shown in Figure 81.

Figure 81

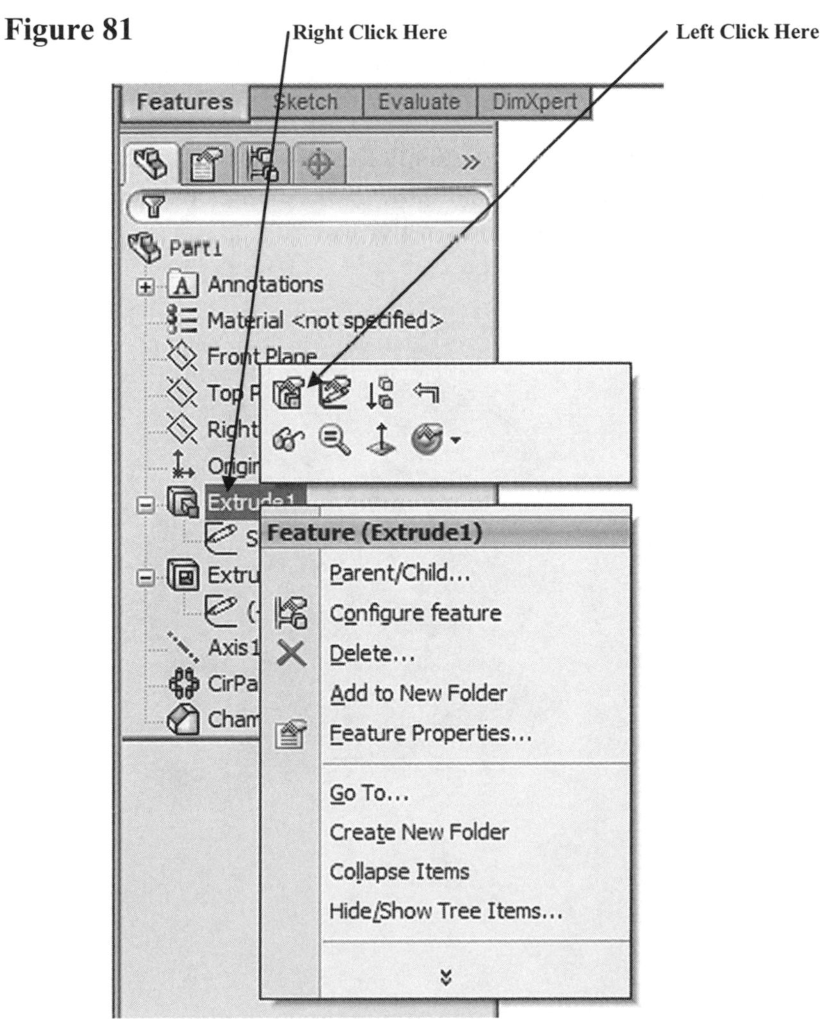

86. Move the cursor to the middle left portion of the screen. The Extrude1 dialog box will appear. Enter **.750** for the extrusion distance and left click on the green checkmark as shown in Figure 82.

Figure 82

Enter .750 Here

Left Click Here

Boss-Extrude1

From

Sketch Plane

Direction 1

Blind

0.75in

87. SolidWorks will automatically update the part without the need to repeat any of the steps that created the original part. Notice that the holes are no longer "thru" holes. Your screen should look similar to Figure 83.

Figure 83

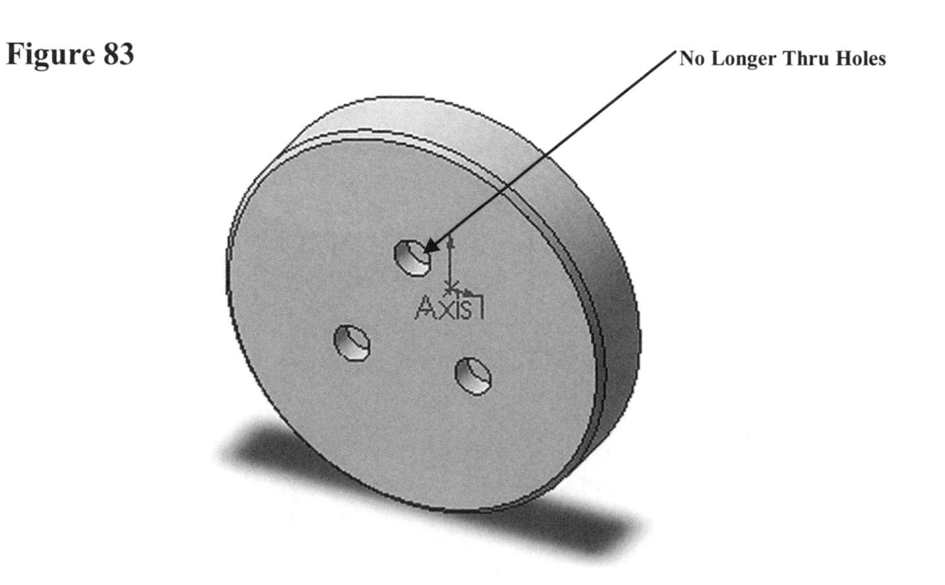

88. Move the cursor to the middle left portion of the screen where the part tree is located. Right click once on **Sketch2**. A pop up menu will appear. Left click on the **Edit Sketch** icon as shown in Figure 84.

Figure 84

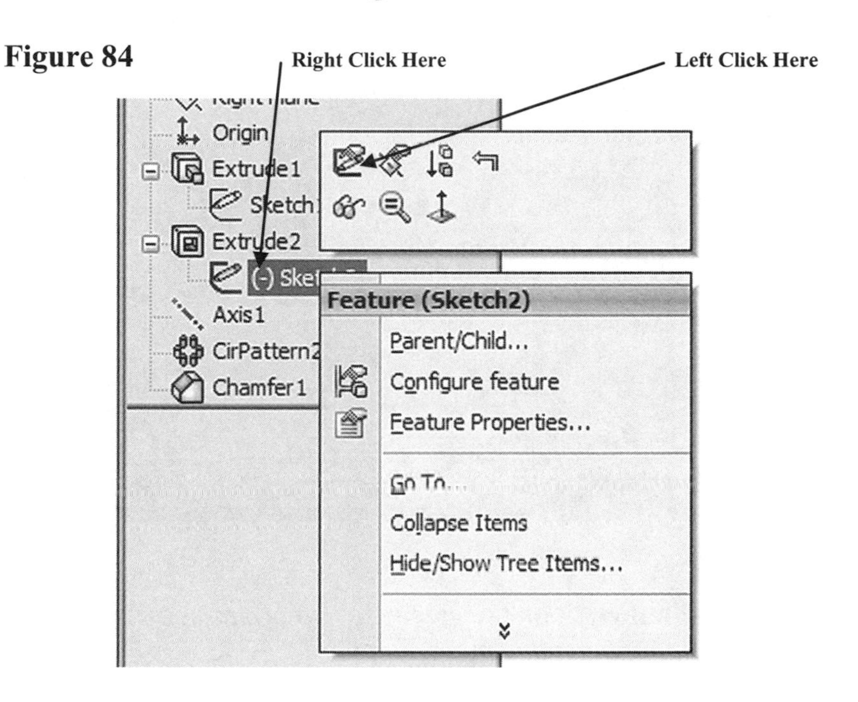

89. The original sketch will appear as shown in Figure 85.

Figure 85

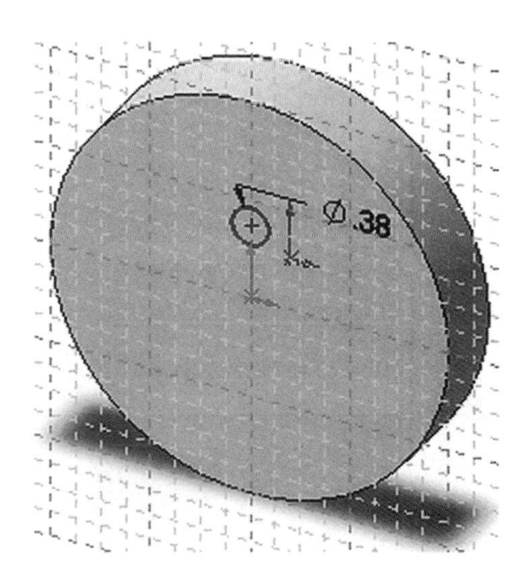

90. Double click on the hole diameter dimension. The Modify dialog box will appear as shown in Figure 86.

Figure 86

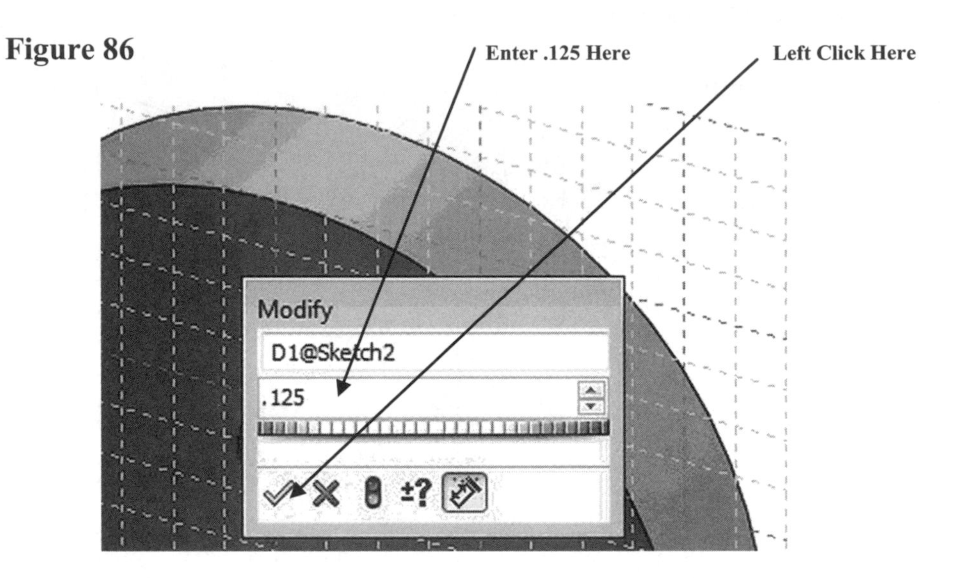

91. Enter **.125** and left click on the green checkmark as shown in Figure 86.

92. The diameter of the holes will be reduced to .125 as shown in Figure 87.

Figure 87

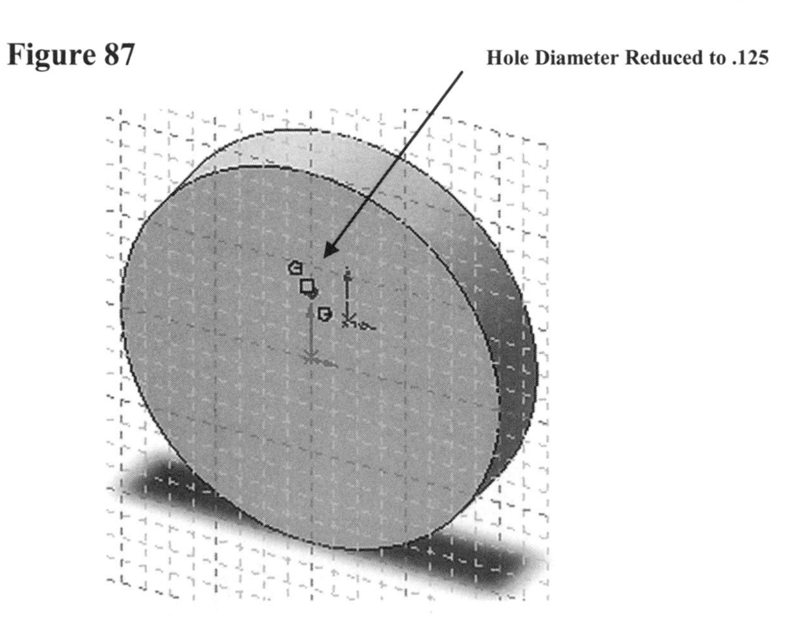

93. Move the cursor to the upper left portion of the screen and left click on the green checkmark as shown in Figure 88.

Figure 88

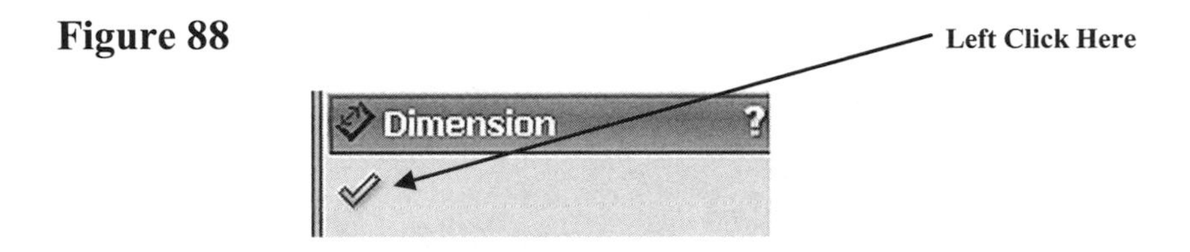

94. Right click anywhere around the drawing. A pop up menu will appear. Left click on the **Exit Sketch** icon as shown in Figure 89.

Figure 89

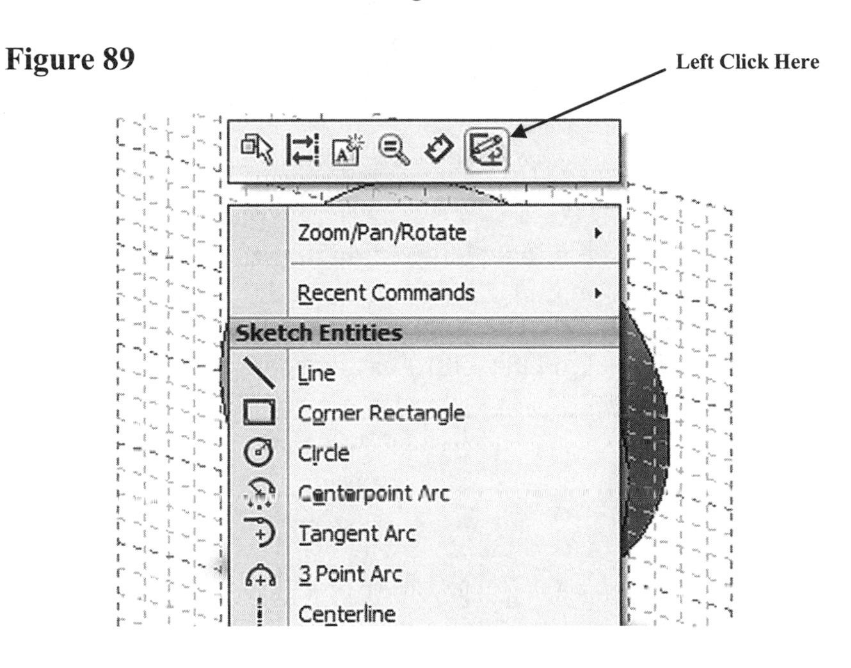

95. SolidWorks will automatically update the part as shown in Figure 90.

Figure 90

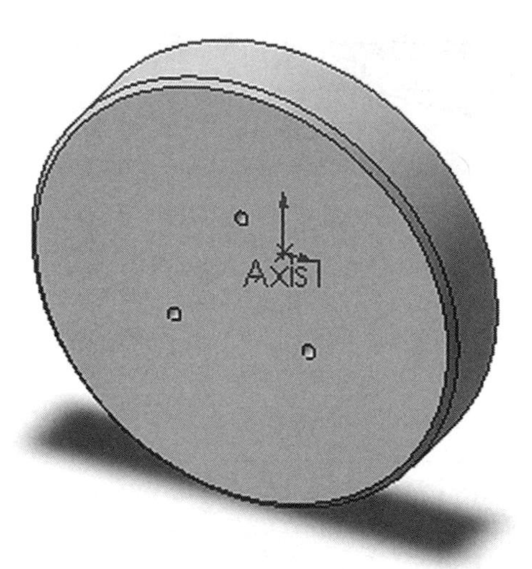

96. Move the cursor to the middle left portion of the screen where the part tree is located. Right click on **Cut-Extrude1/Extrude2 (2009 and earlier versions of SolidWorks display "Extrude2" rather than "Cut-Extrude1"**. A pop up menu will appear. Left click on the **Edit Feature** icon as shown in Figure 91.

Figure 91

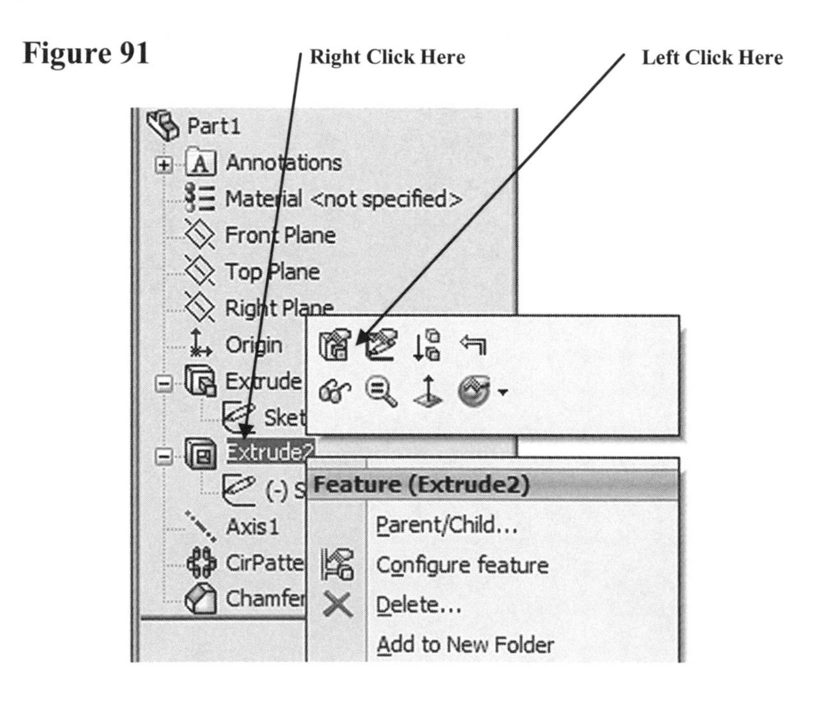

97. Enter **.750** of the extrusion distance as shown in Figure 92.

Figure 92

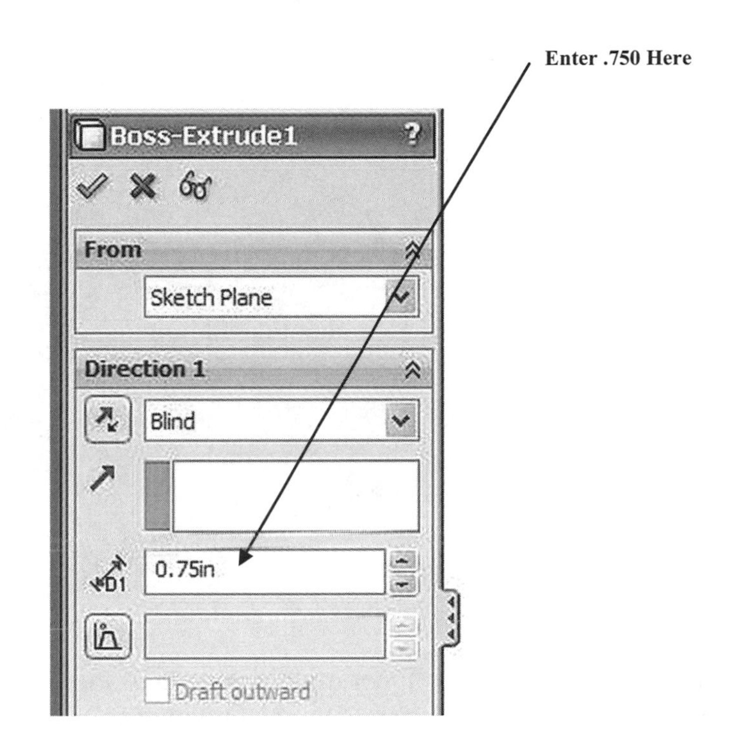

98. Move the cursor to the upper left portion of the screen and left click on the green checkmark as shown in Figure 93.

Figure 93

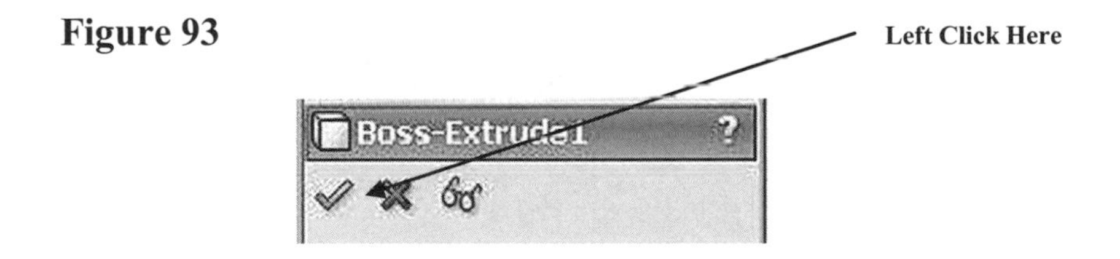

99. SolidWorks will automatically update the part. Notice that the holes are now "thru" holes as shown in Figure 94.

Figure 94

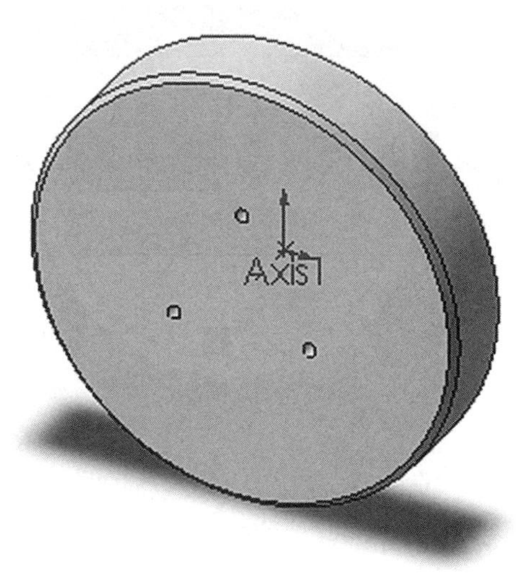

100. Move the cursor to the upper middle portion of the screen and left click on the drop down arrow next to the "Views Orientation" icon. A drop down menu will appear. Left click on the **Front** icon as shown in Figure 95.

Figure 95

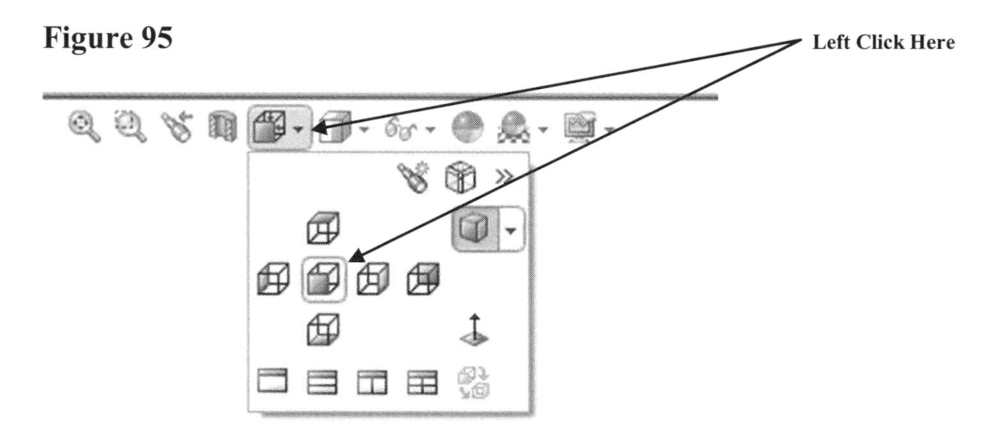

101. A perpendicular view of the part will be displayed. Verify the holes are actually thru holes as shown in Figure 96.

Figure 96

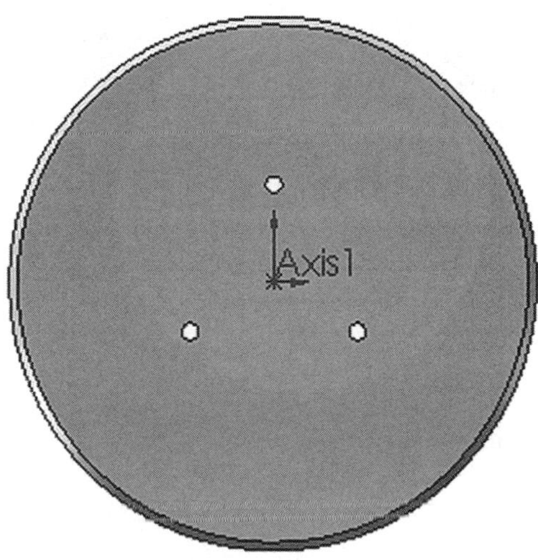

102. Move the cursor to the middle left portion of the screen where the part tree is located. Right click on **CirPattern1**. A pop up menu will appear. Left click on the **Edit Feature** icon as shown in Figure 97.

Figure 97

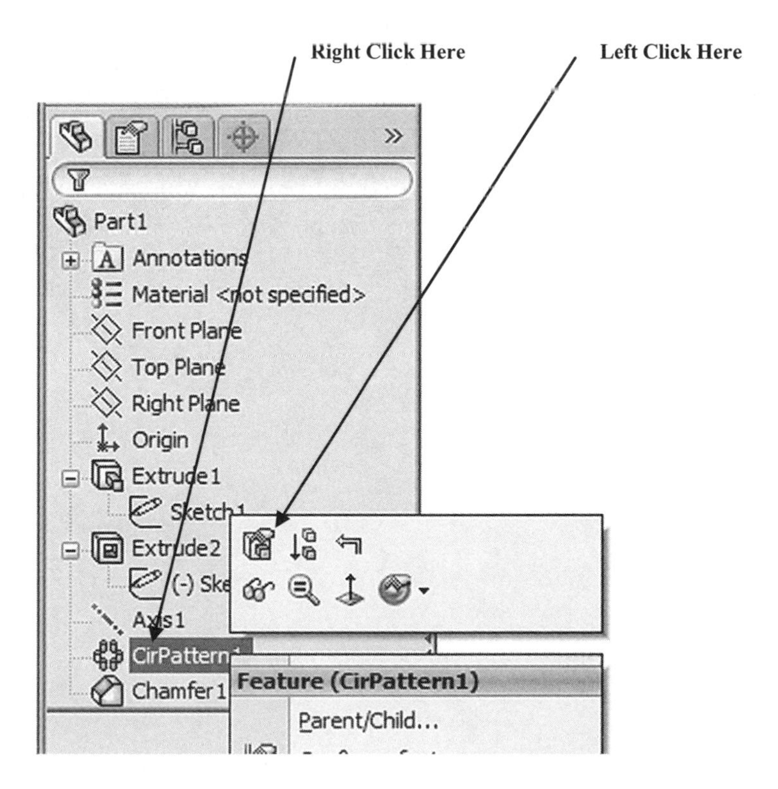

103. Enter **6** for the number of holes and left click next to Equal Spacing as shown in Figure 98. Press the **Enter** key on the keyboard.

Figure 98

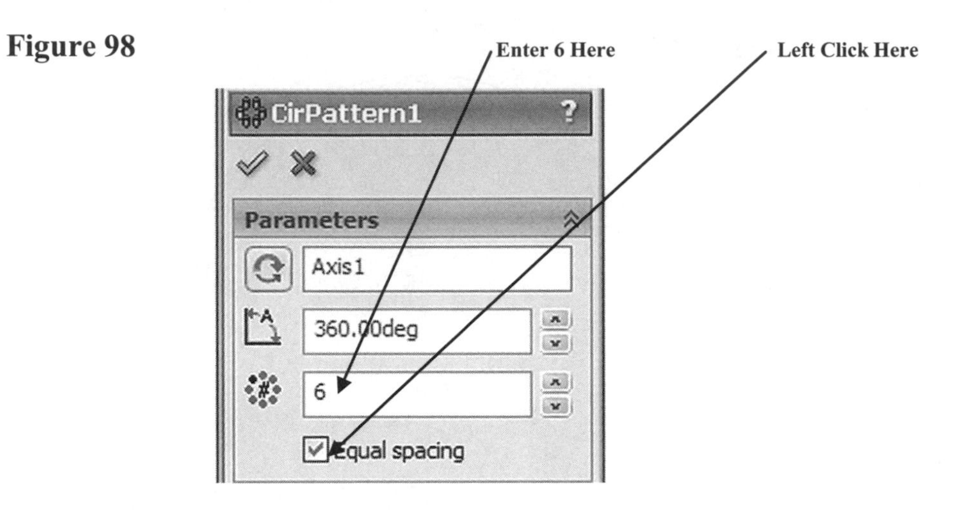

104. A preview will be displayed as shown in Figure 99.

Figure 99

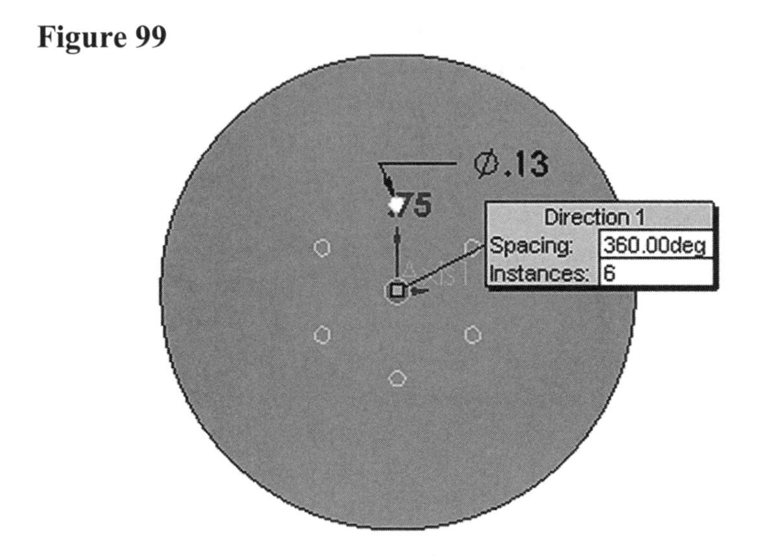

105. Left click on the green checkmark as shown in Figure 100.

Figure 100

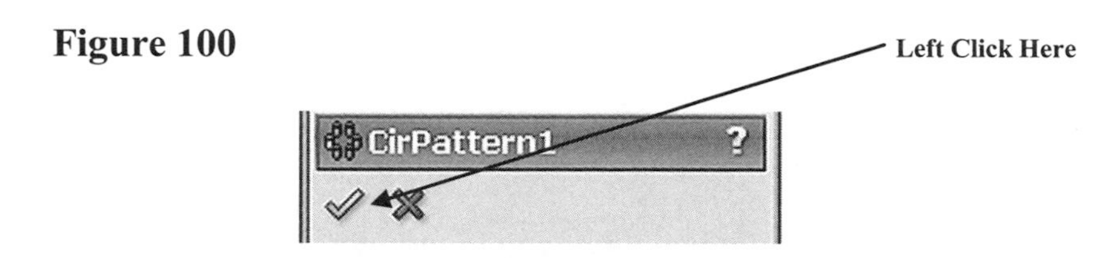

106. Your screen should look similar to Figure 101.

Figure 101

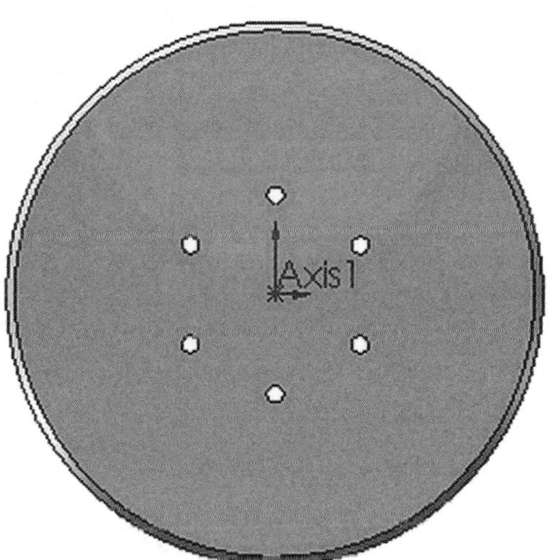

107. Move the cursor to the upper middle portion of the screen and left click on the drop down arrow next to the "Views Orientation " icon. A drop down menu will appear. Left click on the **Trimetric** icon as shown in Figure 102.

Figure 102

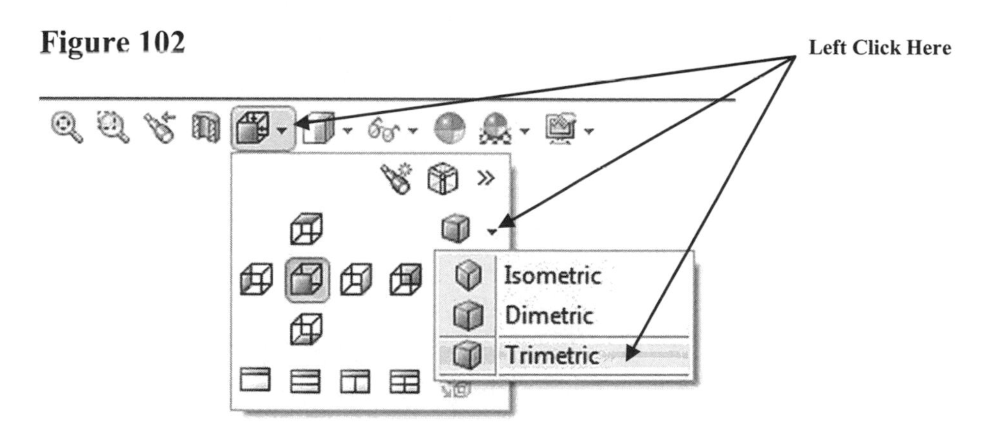

108. The part will be displayed in trimetric as shown in Figure 103.

Figure 103

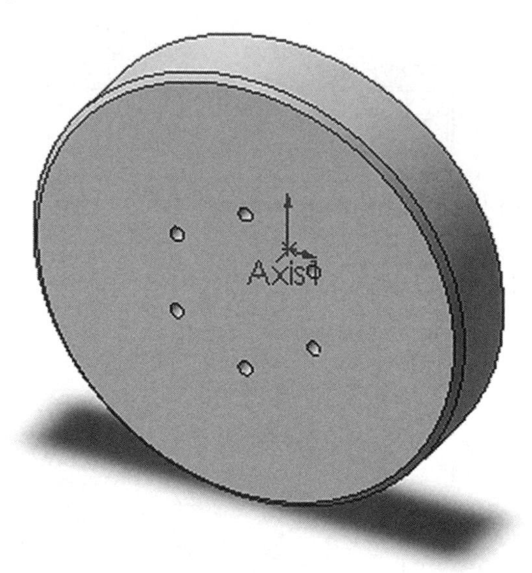

109. Move the cursor to the middle left portion of the screen where the part tree is located. Right click on **Chamfer1.** A pop up menu will appear. Left click on the **Edit Feature** icon as shown in Figure 104.

Figure 104

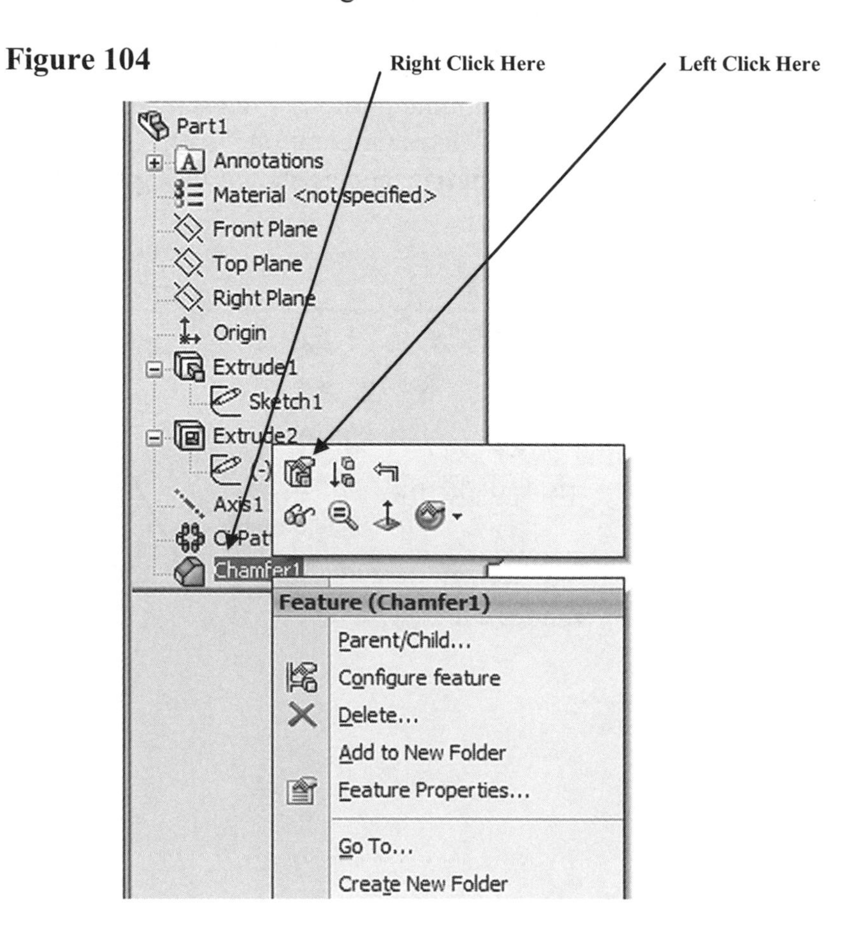

110. Enter **.25** for the Distance a shown in Figure 105.

Figure 105

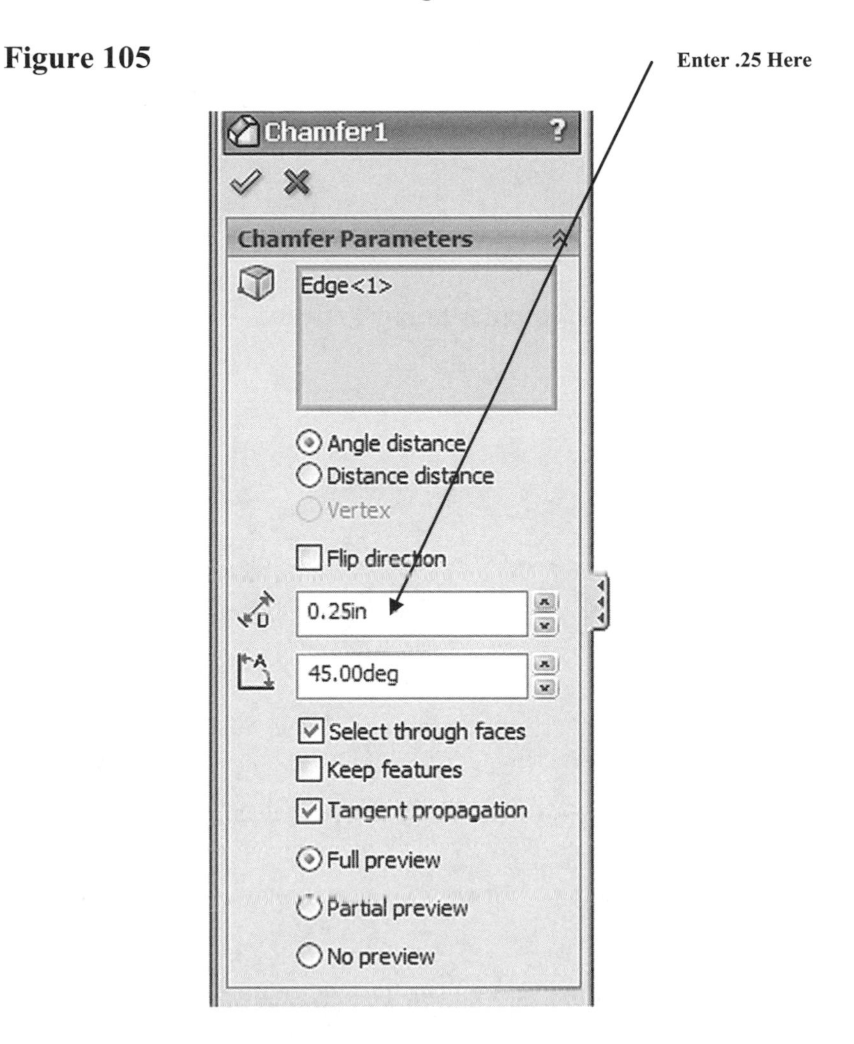

111. A preview of the chamfer will be displayed as shown in Figure 106.

Figure 106

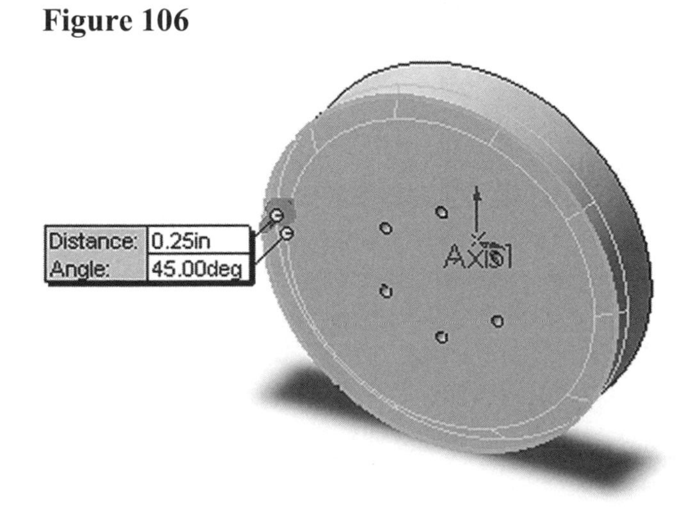

112. Left click on the green checkmark as shown in Figure 107.

Figure 107

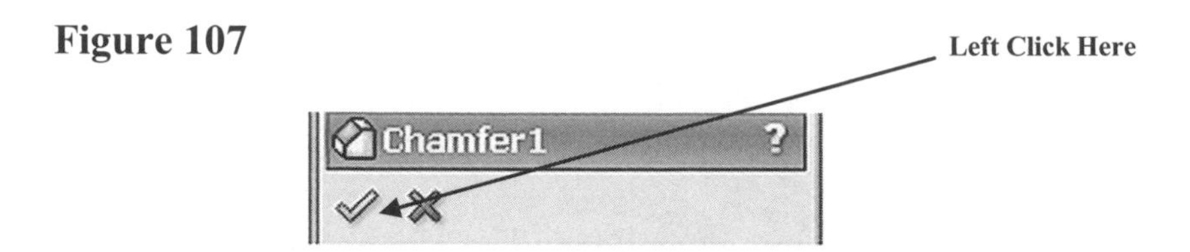

113. Your screen should look similar to Figure 108

Figure 108

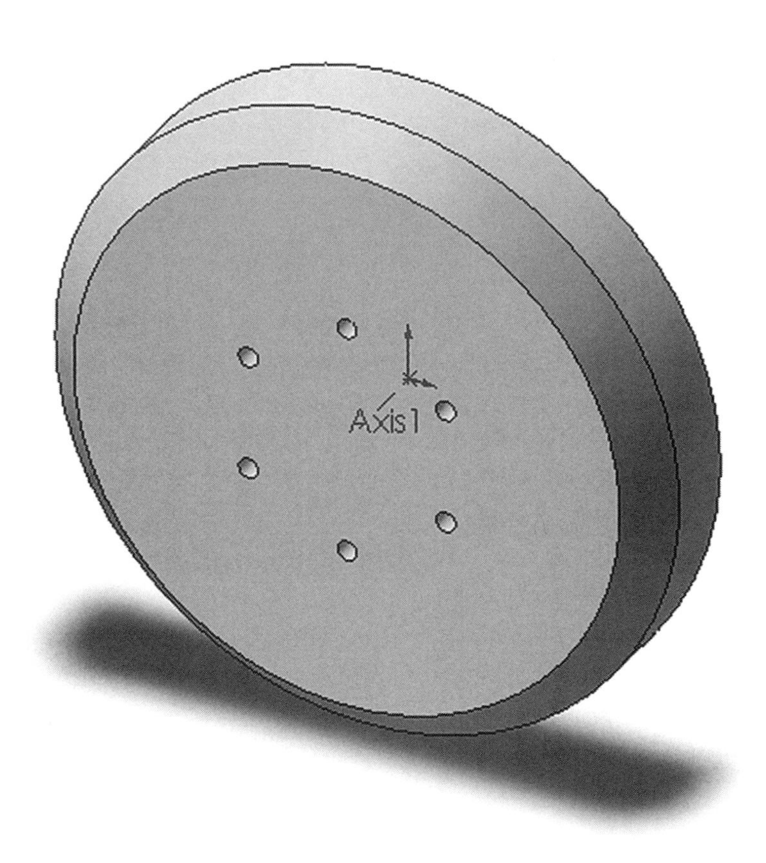

114. Move the cursor to the middle left portion of the screen where the part tree is located. Right click on **CirPattern1**. A pop up menu will appear. Left click on the **Suppress** icon as shown in Figure 109.

Figure 109

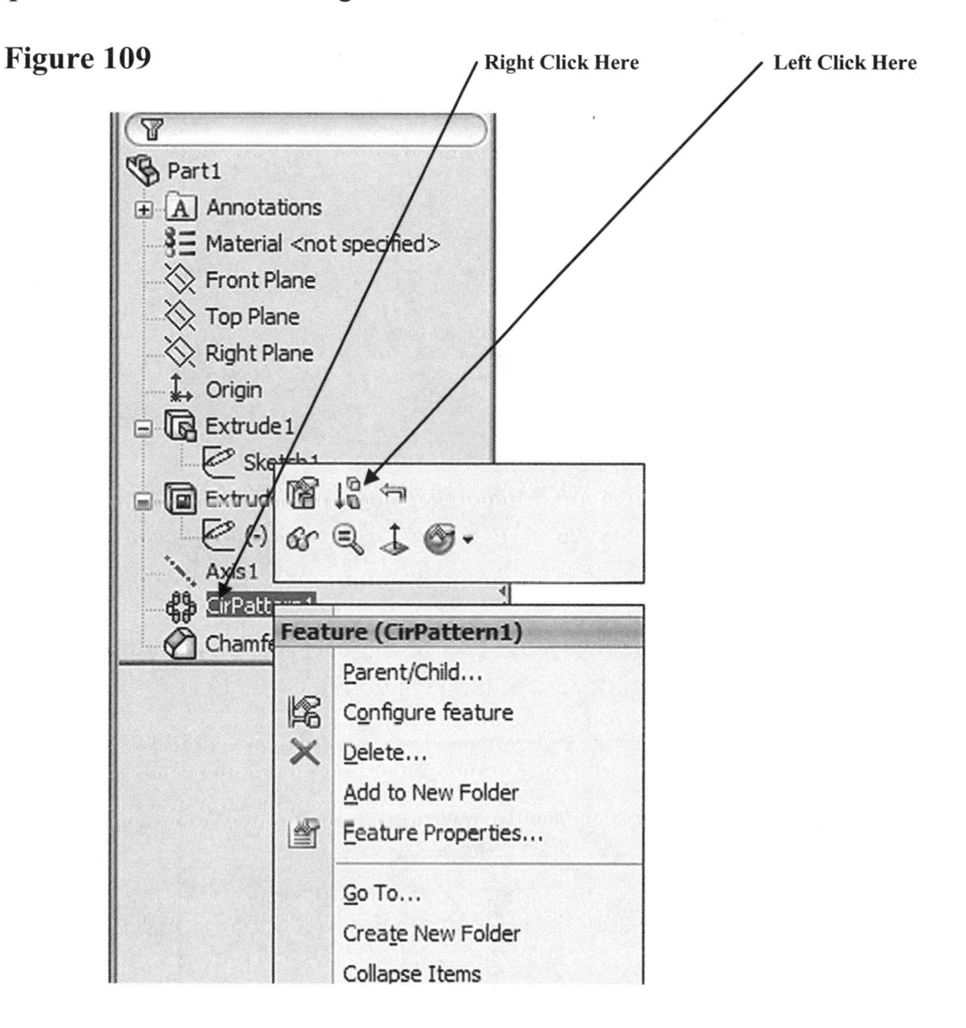

115. All holes created using the Circular Pattern command will be suppressed except for the original hole as shown in Figure 110.

Figure 110

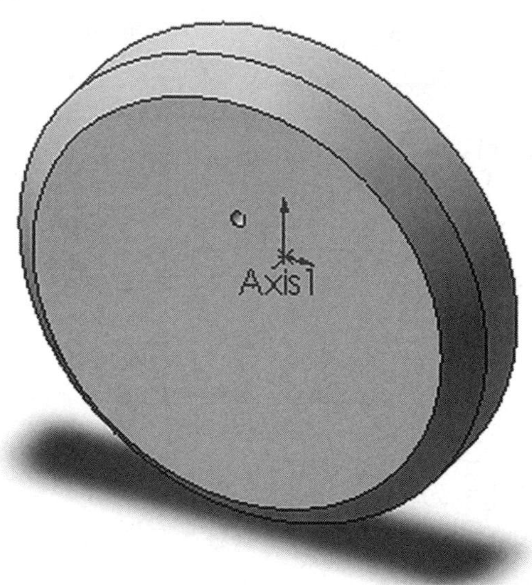

116. SolidWorks will gray out the text in the part tree as shown in Figure 111. Use the same steps to "unsuppress" the CirPattern1.

Figure 111

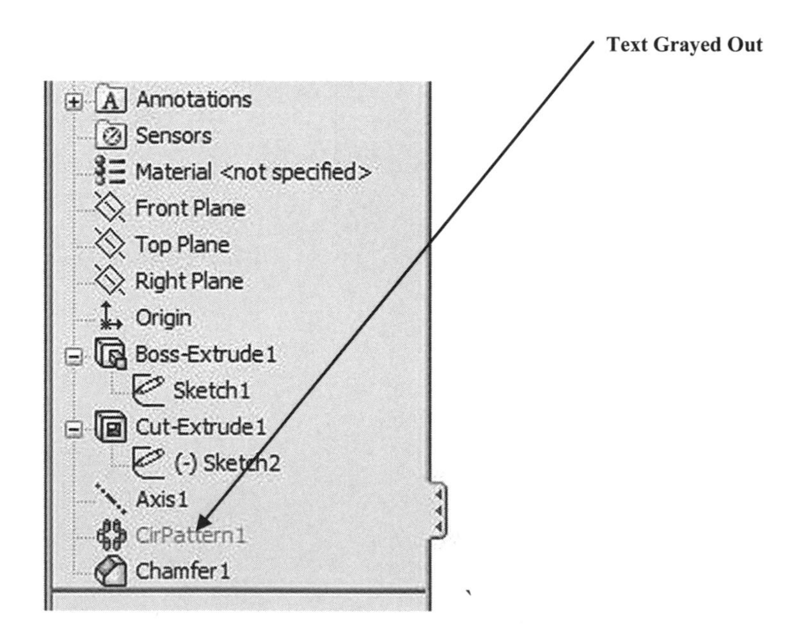

117. The names of all branches in the part tree can also be edited. Left click once on the text **Boss-Extude1/Extrude1**. The text will become highlighted as shown in Figure 112. After the text is highlighted, left click once. The text can now be edited.

Figure 112

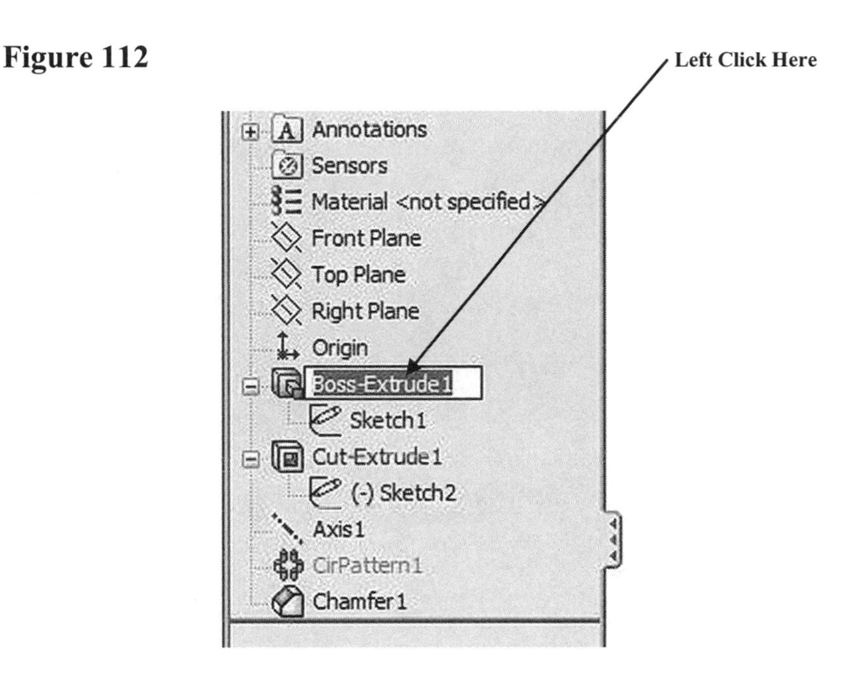

118. Enter the text **Base Extrusion** as shown in Figure 113. Press **Enter** on the keyboard. Text for each of the individual operations can be edited if desired.

Figure 113

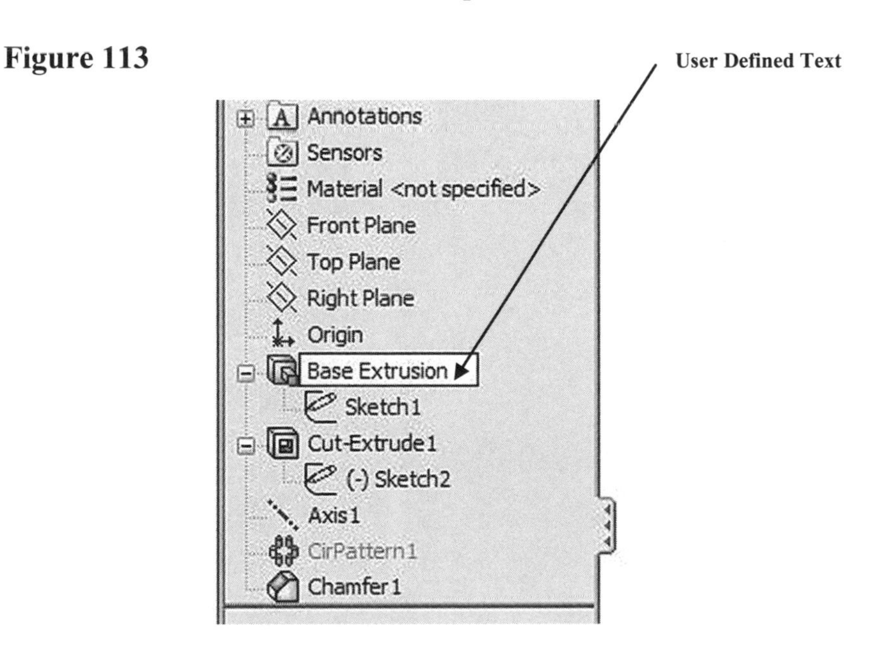

119. Notice that the final design looks significantly different than the original design. The new part was redesigned by modifying the existing part as shown in Figure 114.

Figure 114

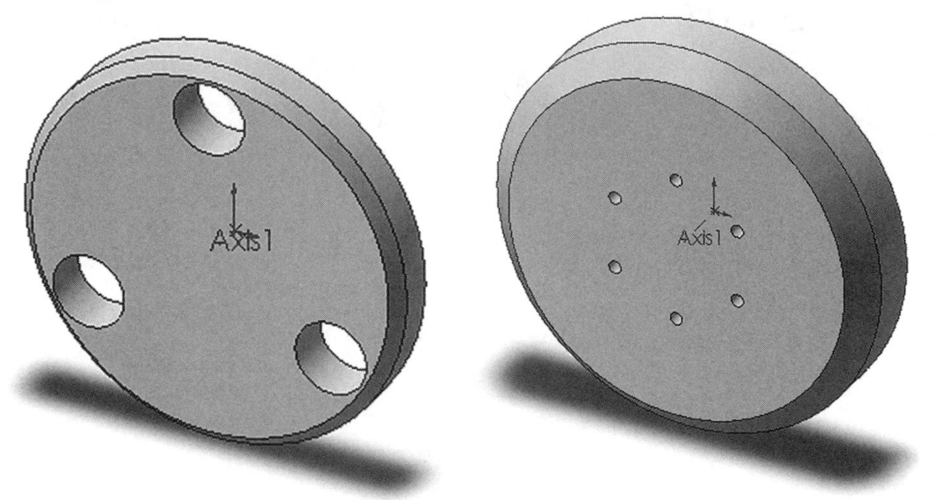

Drawing Activities

Use these problems from Chapters 1 and 2 to create redesigned parts.

Problem 1

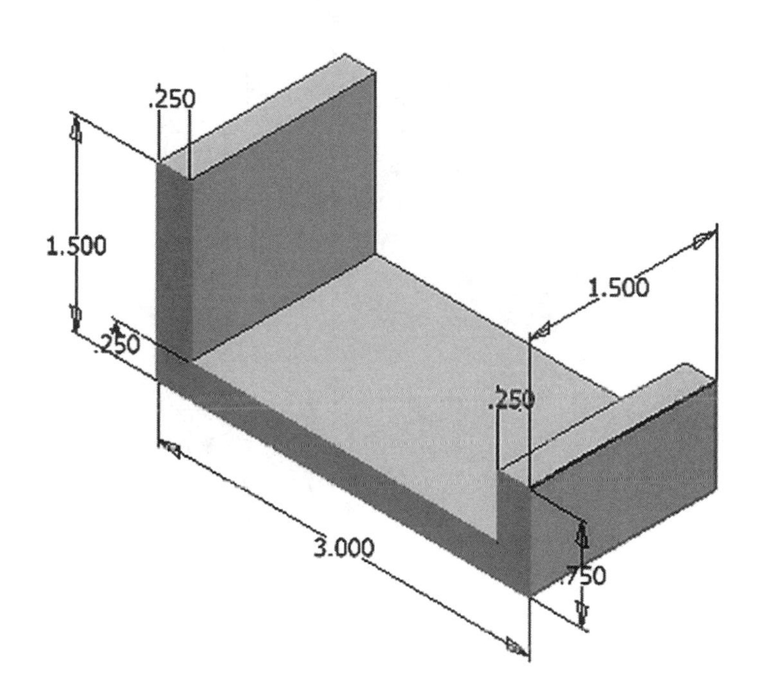

Problem 2

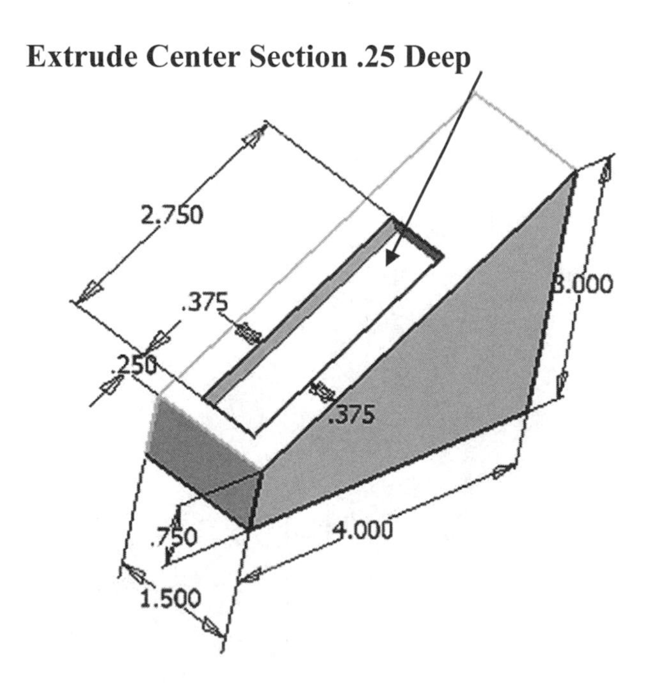

Problem 3

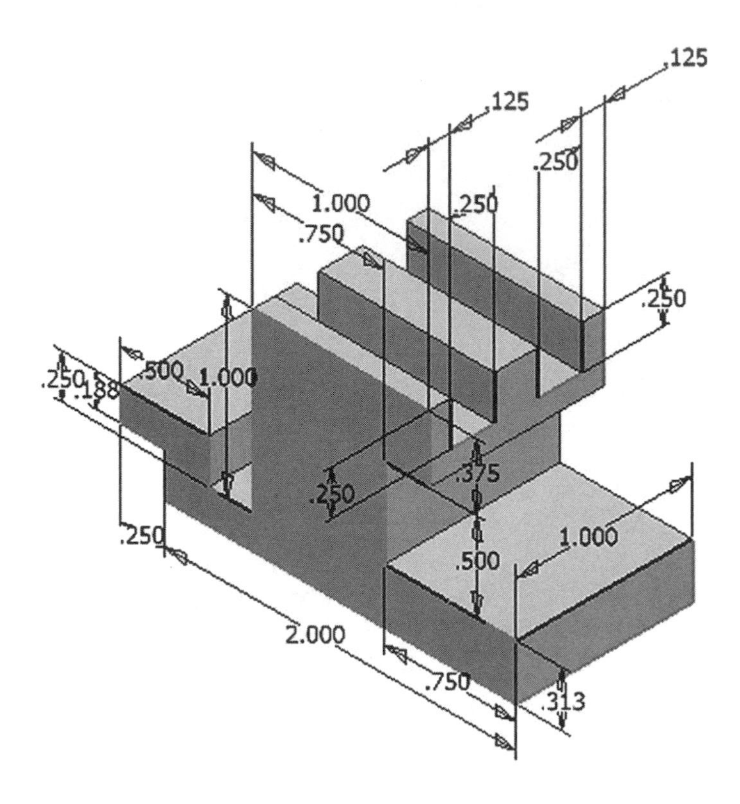

Problem 4

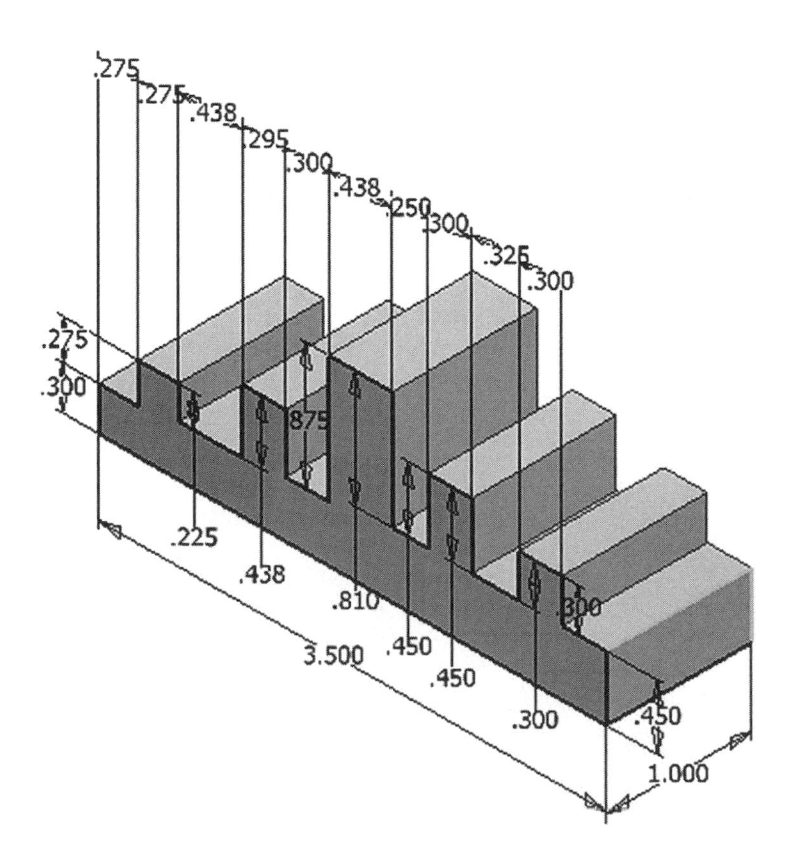

Problem 5

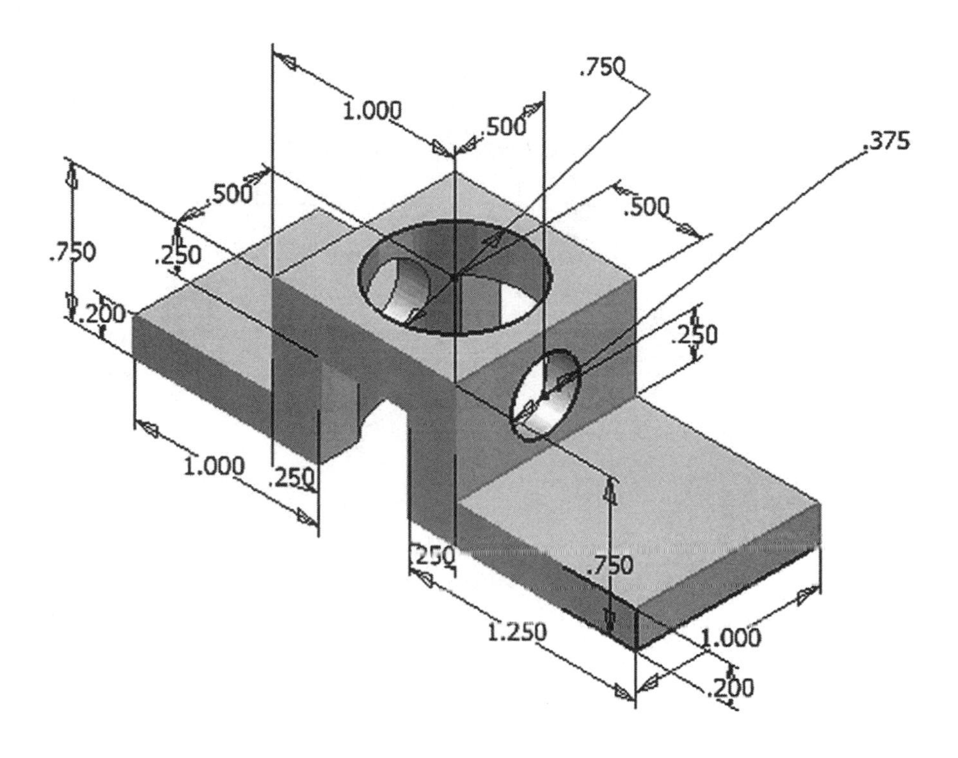

Problem 6

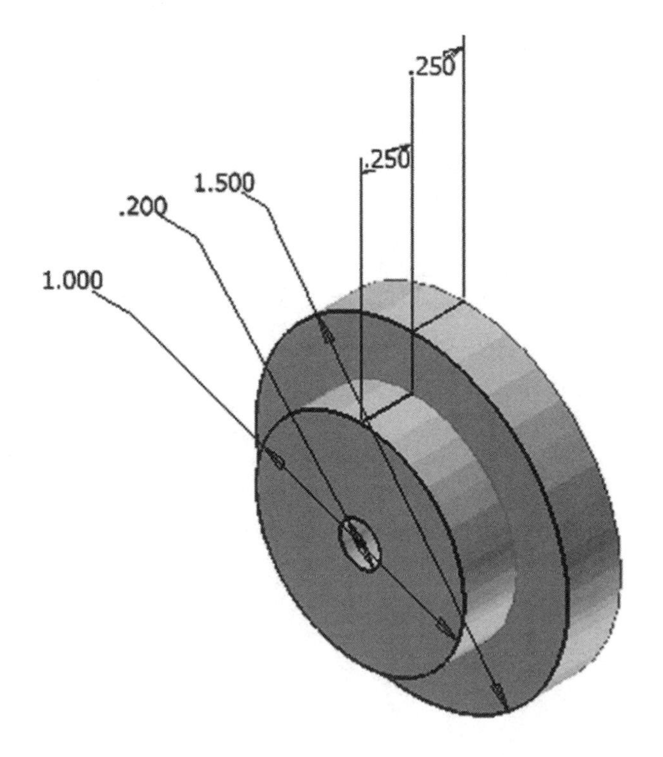

Problem 7

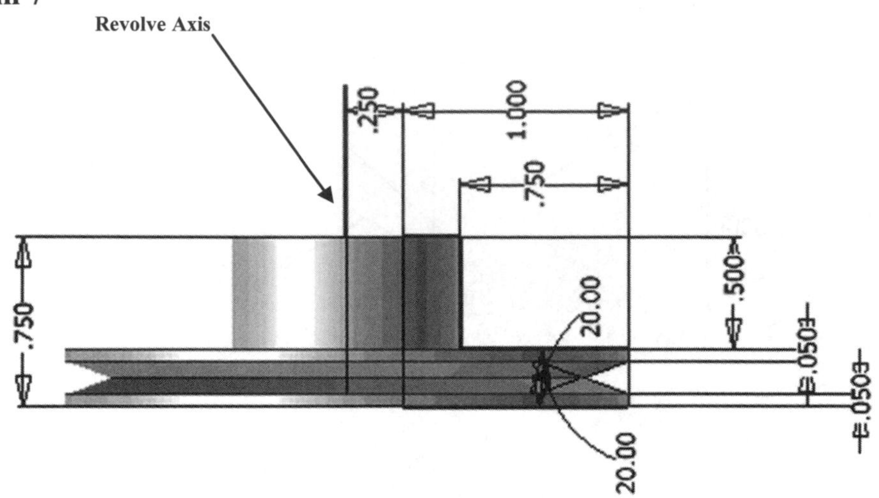

Problem 8

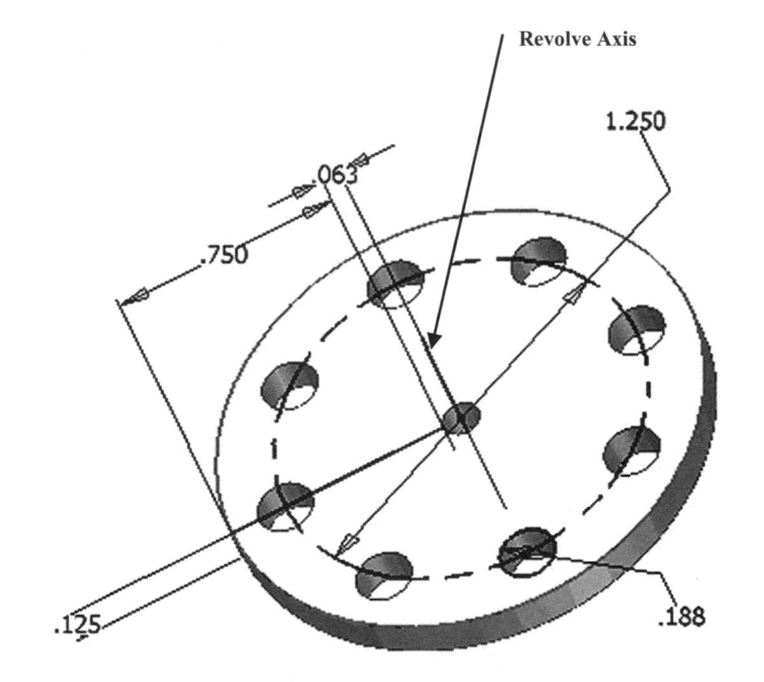

Chapter 6 Advanced Design Procedures

Objectives:

- Design multiple sketch parts
- Learn to use the Front, Top, and Right Planes
- Learn to use the Shell command
- Learn to use the Wireframe viewing command

Chapter 6 includes instruction on how to design the parts shown below.

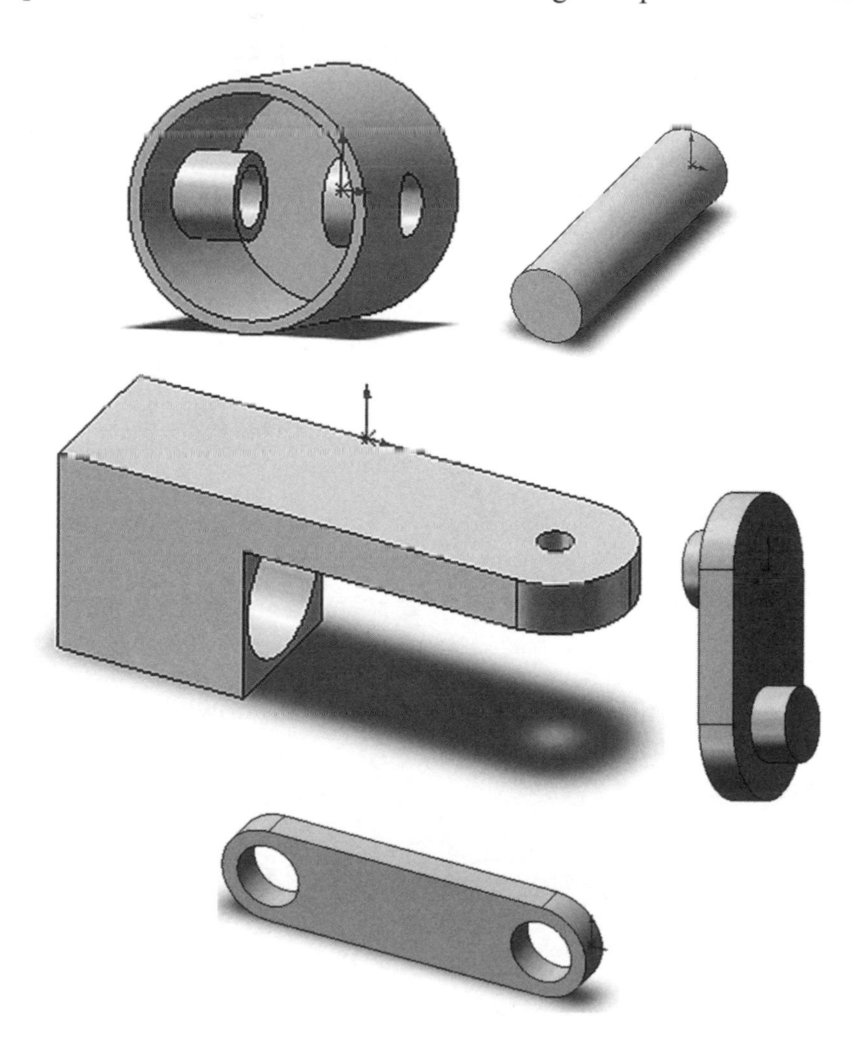

1. Start SolidWorks by referring to "Chapter 1 Getting Started".

2. After SolidWorks is running, begin a new sketch.

3. Move the cursor to the upper middle portion of the screen and left click on **Circle** as shown in Figure 1.

Figure 1

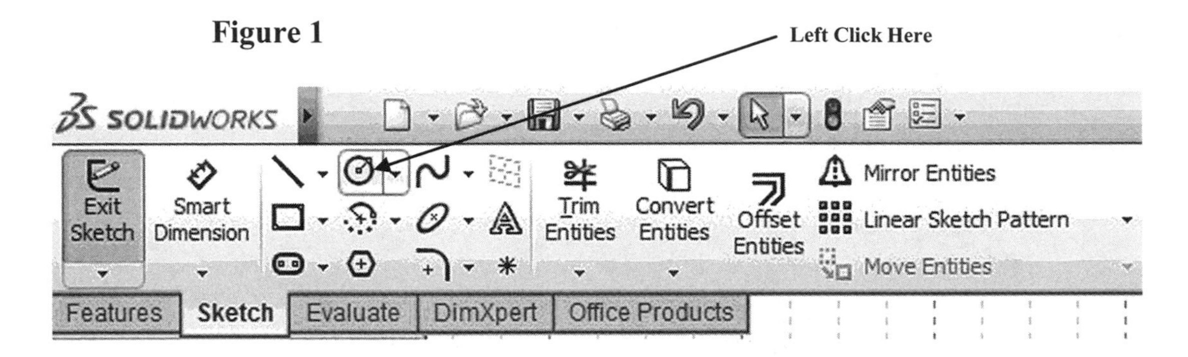

4. Move the cursor to the center of the screen and left click once. This will be the center of the circle as shown in Figure 2.

Figure 2

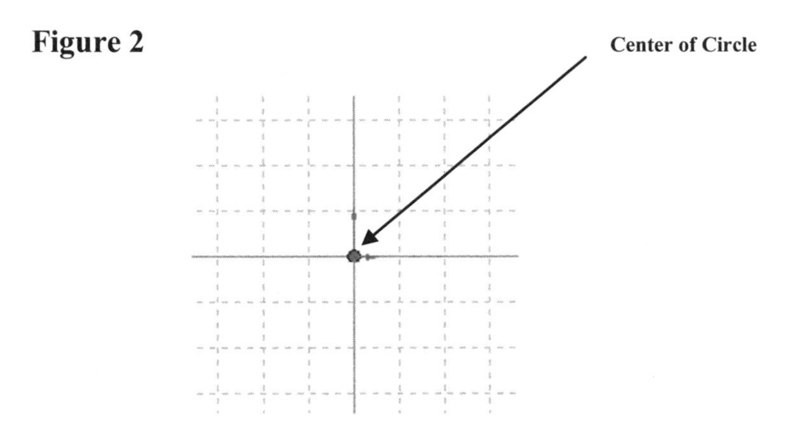

5. Move the cursor to the right and left click once as shown in Figure 3.

Figure 3

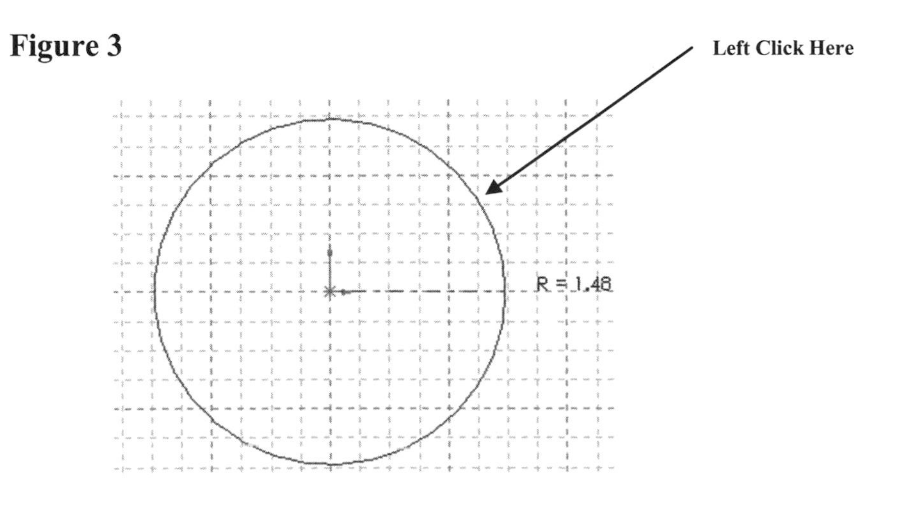

6. Right click anywhere around the drawing. A pop up menu will appear. Left click on **Select** as shown in Figure 4.

Figure 4

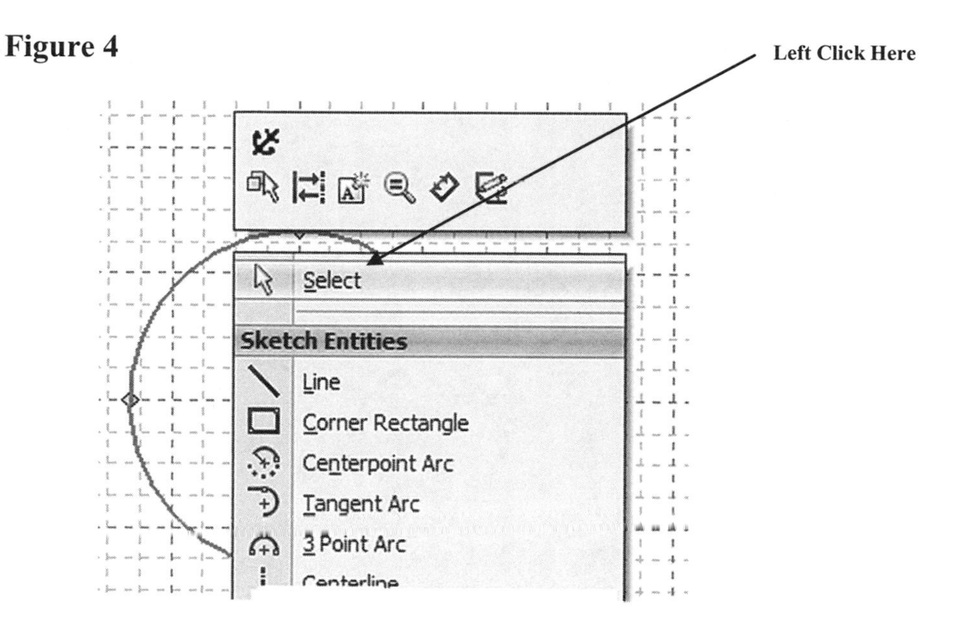

7. Move the cursor to the upper left portion of the screen and left click on **Smart Dimension** as shown in Figure 5.

Figure 5

8. After selecting **Smart Dimension** move the cursor over the edge of the circle causing it to turn red as shown in Figure 6. Left click once. The dimension will be attached to the cursor.

Figure 6

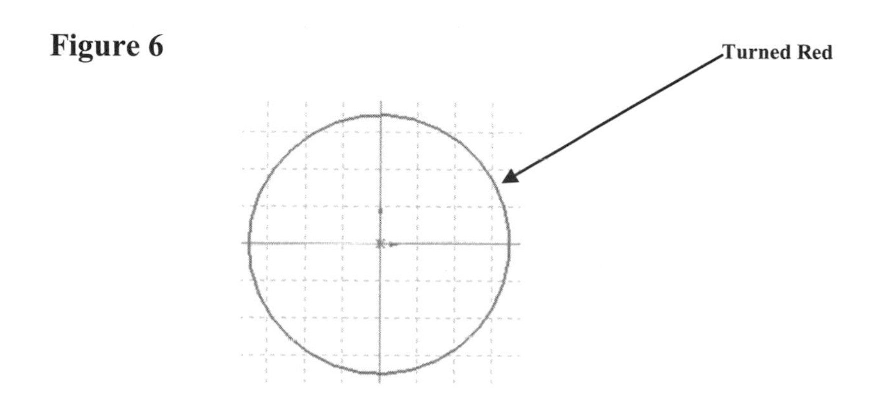

9. Move the cursor to where the dimension will be placed and left click once as shown in Figure 7.

Figure 7

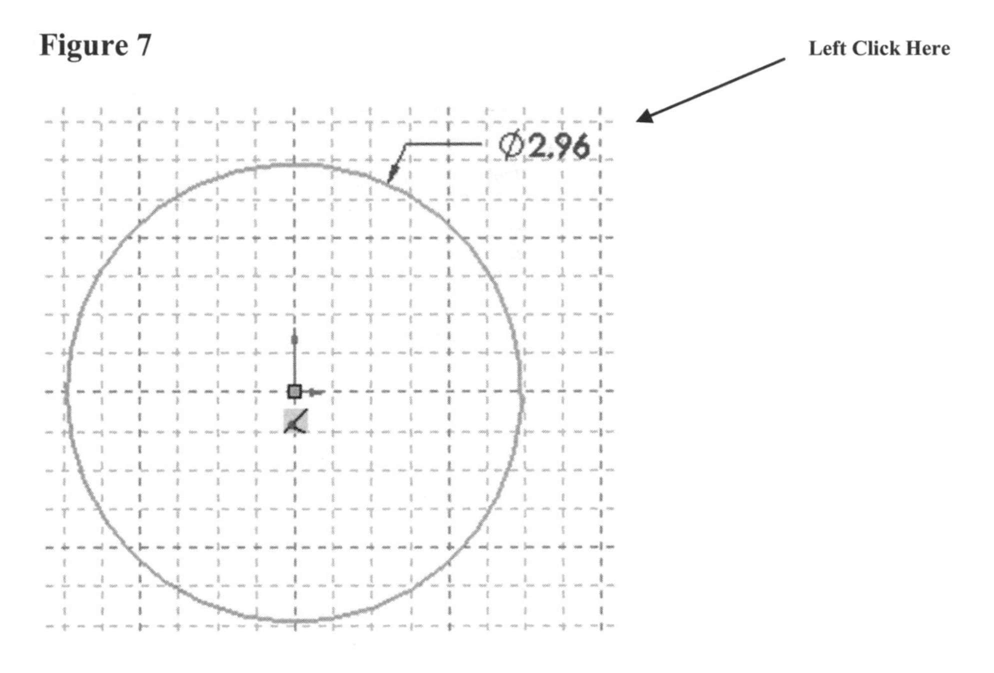

10. The Modify dialog box will appear as shown in Figure 8.

Figure 8

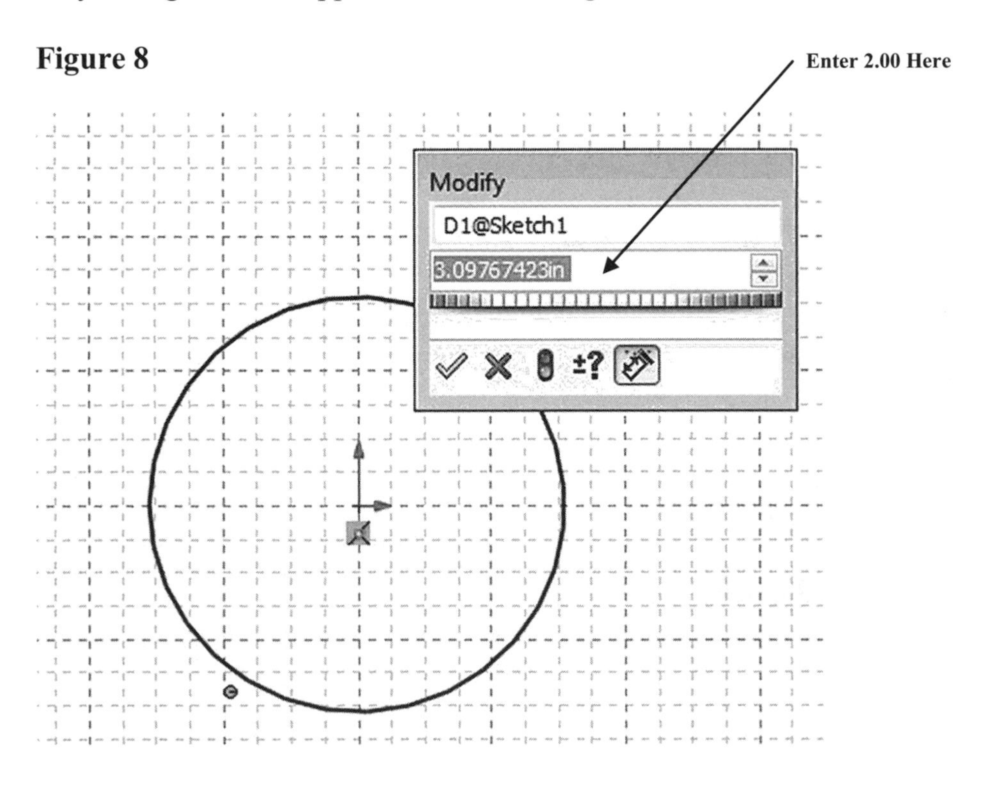

11. To edit the dimension, enter **2.00** in the Modify dialog box (while the current dimension is highlighted). Left click on the green checkmark as shown in Figure 9.

Figure 9

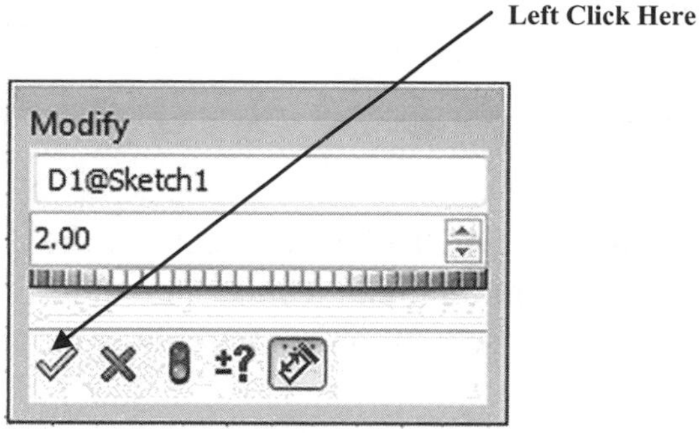

Left Click Here

Modify

D1@Sketch1

2.00

12. The dimension of the circle will become 2.00 inches as shown in Figure 10.

Figure 10

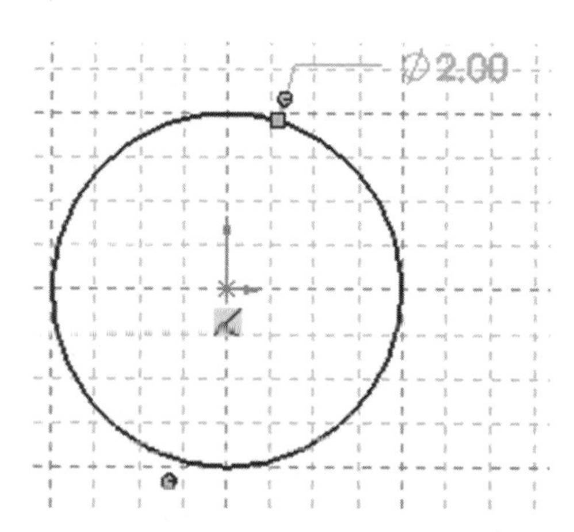

13. Move the cursor to the upper left portion of the screen and left click on the green checkmark as shown in Figure 11.

Figure 11

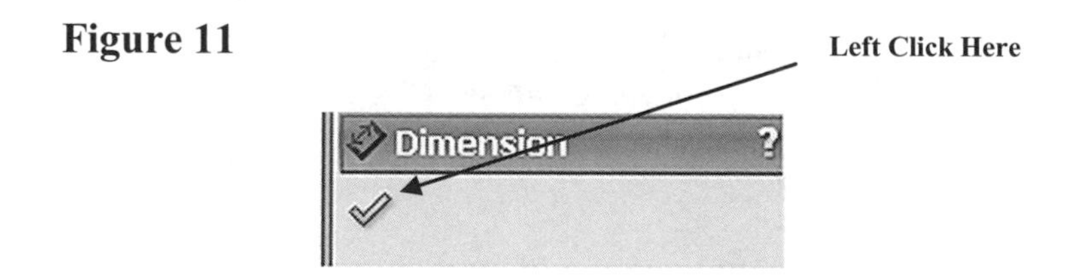

14. In order to view the entire drawing it may be necessary to move the cursor to the middle portion of the screen and left click once on the "Zoom To Fit" icon as shown in Figure 12.

Figure 12

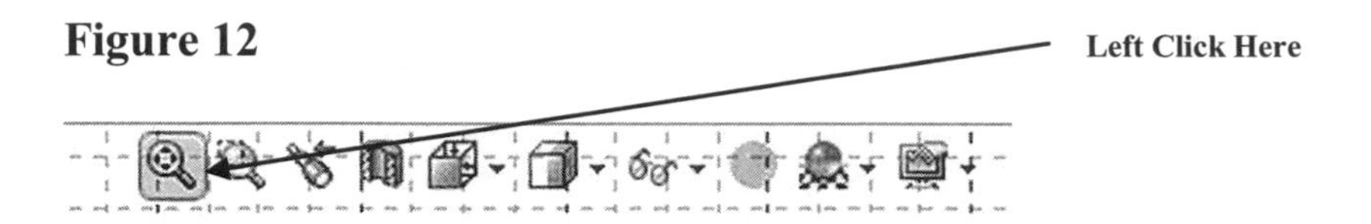

15. The drawing will "fill up" the entire screen. If the drawing is too large or small, left click on the "Zoom to Area" icon as shown in Figure 13. After selecting the icon, hold the left mouse button down and drag a box around the desired area.

Figure 13

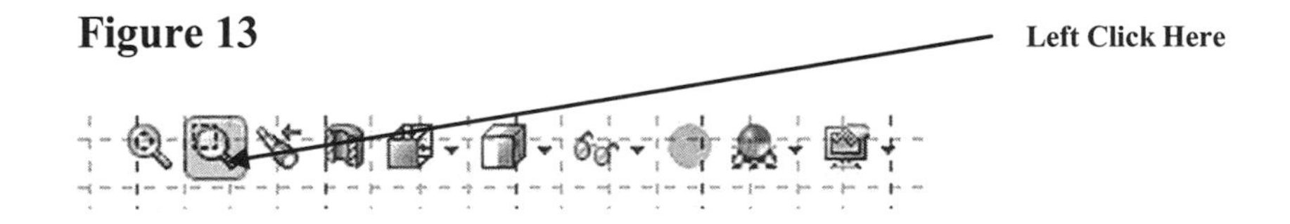

16. Right click anywhere on the sketch. A pop up menu will appear. Left click on **Exit Sketch** as shown in Figure 14.

Figure 14

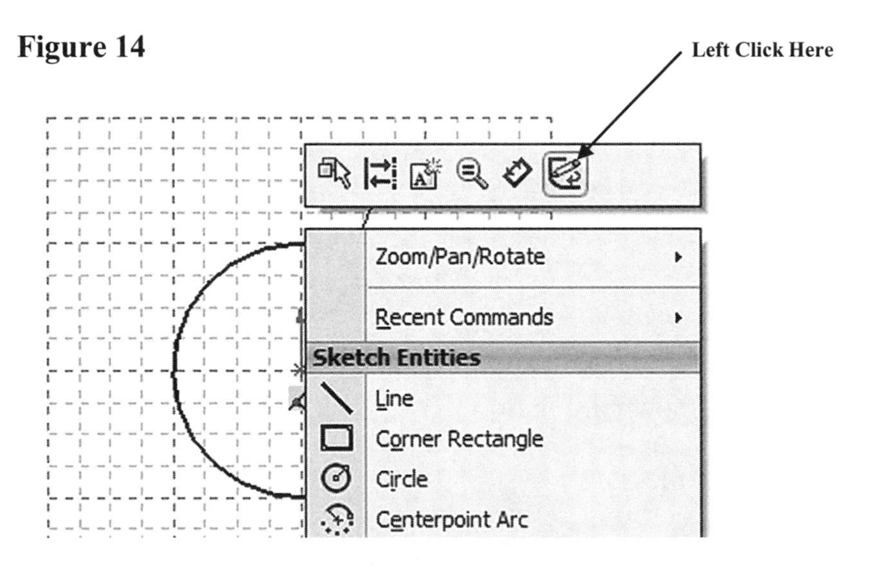

17. SolidWorks is now out of the Sketch commands and into the Features commands. Notice that the commands at the top of the screen are now different. To work in the Features commands a sketch must be present and have no opens (non-connected lines). If there are any opens in the sketch an error message will appear. Your screen should look similar to Figure 15.

Figure 15

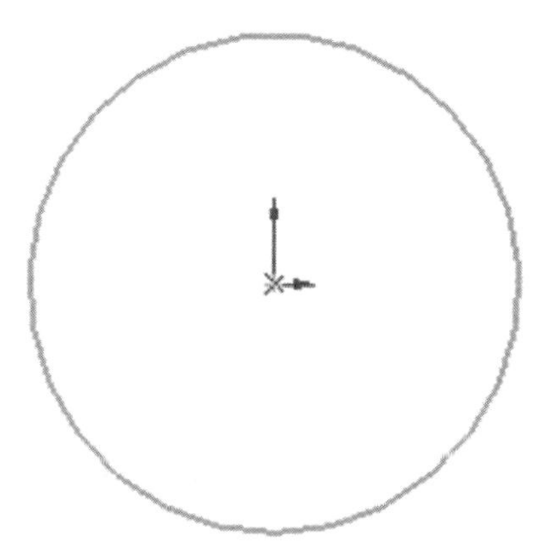

18. Move the cursor to the upper right portion of the screen and left click on the drop down arrow next to the "View Orientation" icon. A drop down menu will appear. Left click on **Trimetric** as shown in Figure 16.

Figure 16

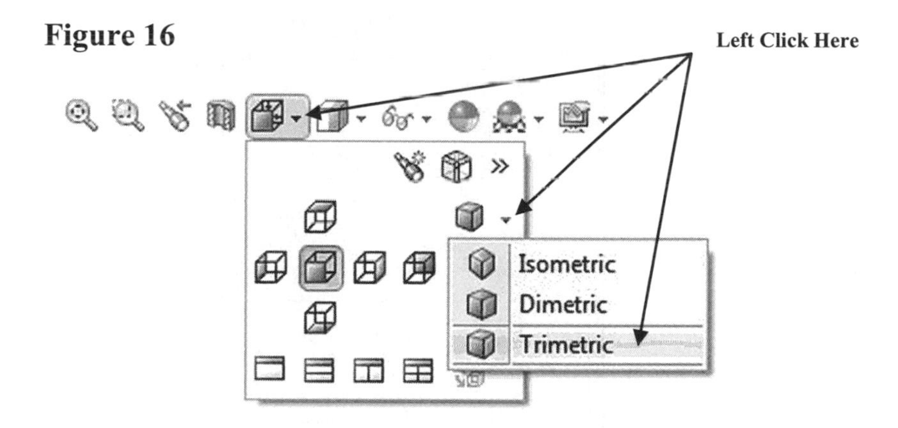

19. The view will become trimetric as shown in Figure 17.

Figure 17

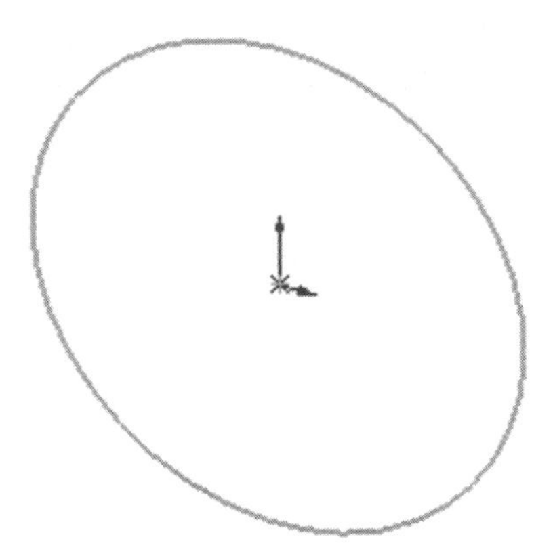

20. Move the cursor to the upper left portion of the screen and left click on **Extruded Boss/Base**. If the Extruded Boss/Base icon did not appear, left click on the Features tab at the upper left portion of the screen. A preview of the extrusion will be displayed.

21. Enter **2.00** for D1 and left click on the green checkmark as shown in Figure 18.

Figure 18

Enter 2.00 Here **Left Click Here**

Boss-Extrude1

From

Sketch Plane

Direction 1

Blind

D1 2.00

Draft outward

22. Left click anywhere around the part. Your screen should look similar to Figure 19.

Figure 19

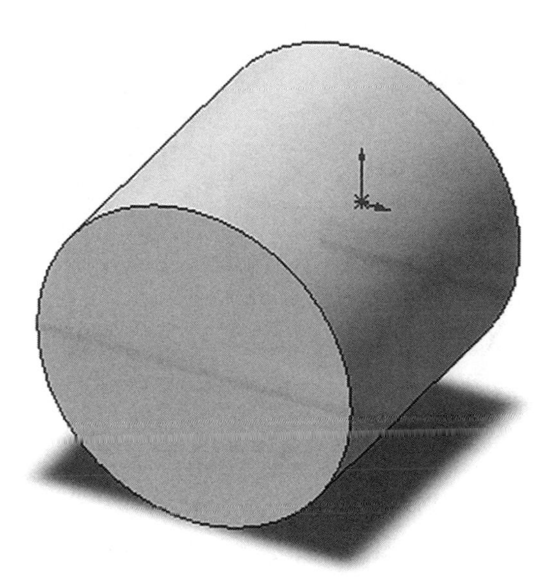

23. Move the cursor to the upper left portion of the screen causing the text "Right Plane" to become highlighted as shown in Figure 20.

Figure 20

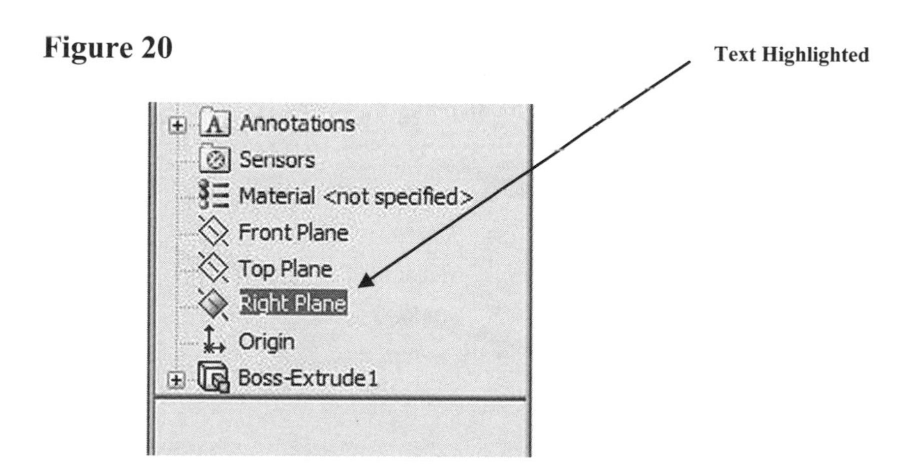

24. The Right Plane will become visible as shown in Figure 21.

Figure 21

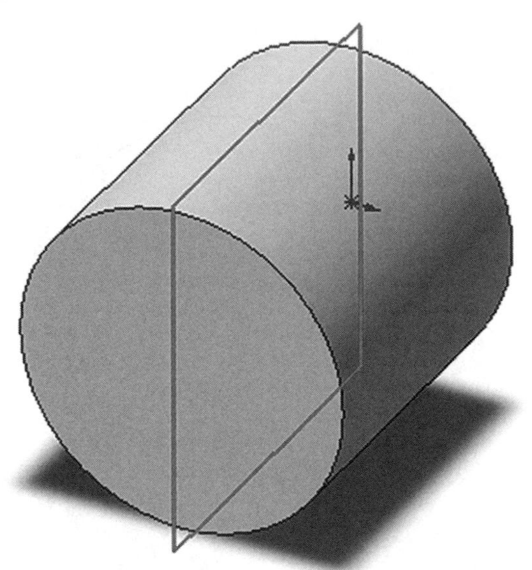

25. Right click on the **Right Plane** text. A pop up menu will appear. Left click on the **Insert Sketch** icon as shown in Figure 22.

Figure 22

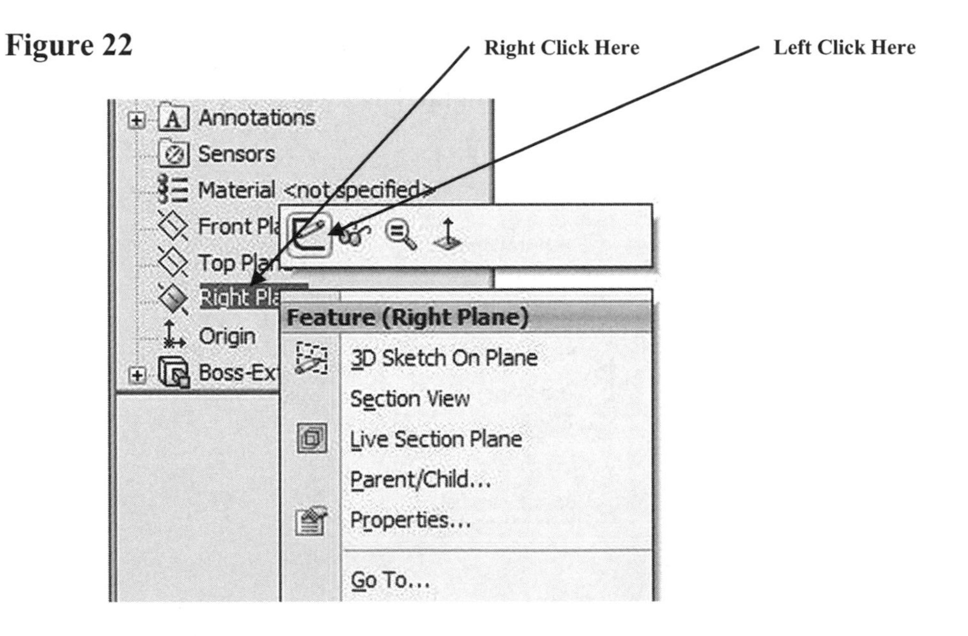

26. Your screen should look similar to Figure 23.

Figure 23

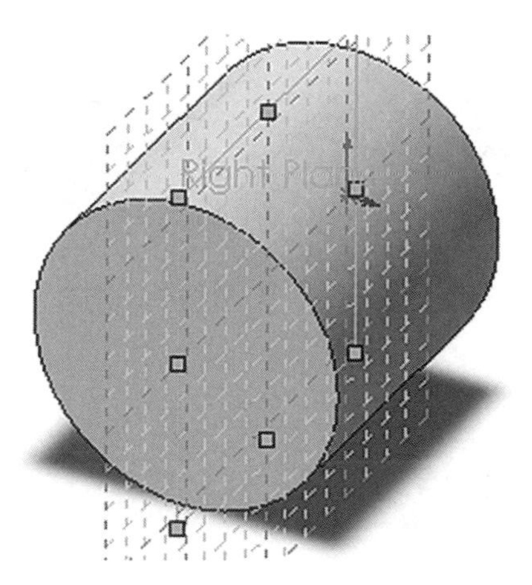

27. Move the cursor to the upper middle portion of the screen and left click on the "Wire Frame Display" icon as shown in Figure 24.

Figure 24

Left Click Here

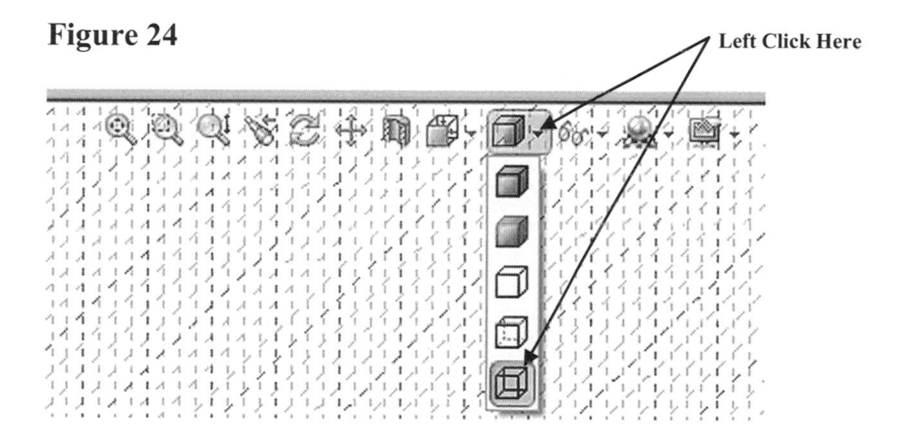

28. Your screen should look similar to Figure 25.

Figure 25

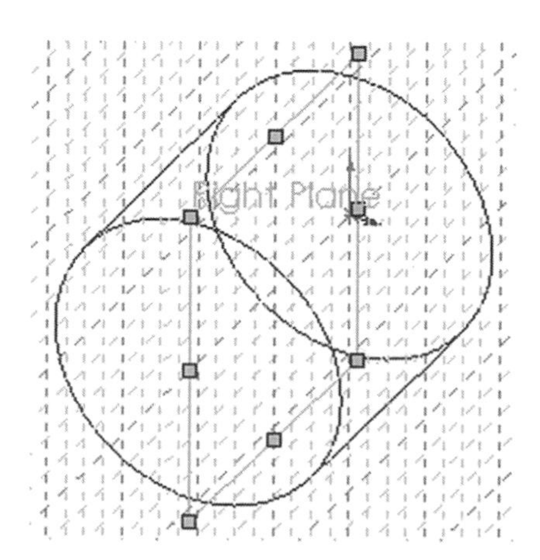

29. Move the cursor to the upper right portion of the screen and left click on the drop down arrow next to the "View Orientation" icon. A drop down menu will appear. Left click on **Right** as shown in Figure 26.

Figure 26

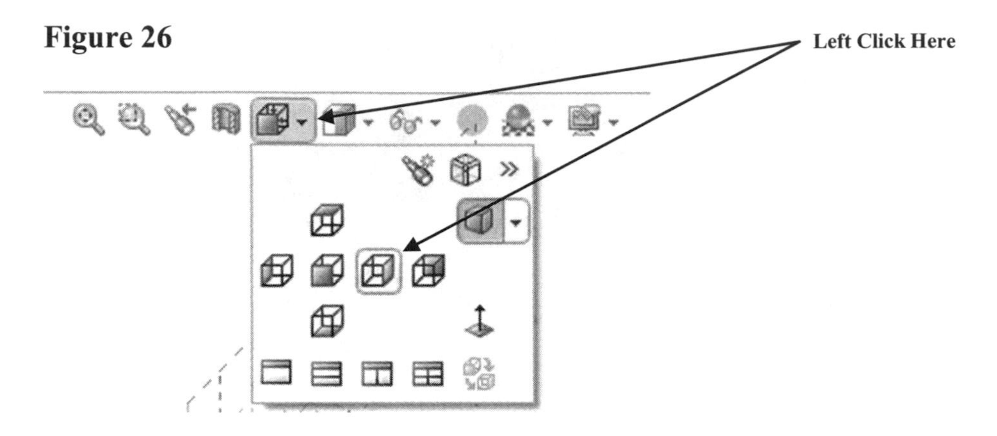

30. The Right Plane will rotate to a perpendicular view as shown in Figure 27.

Figure 27

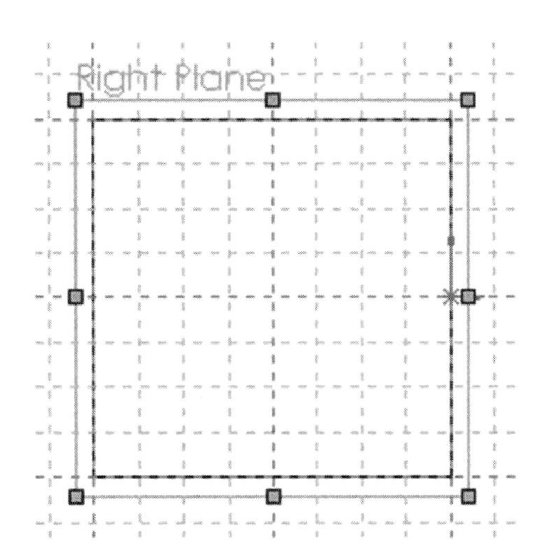

31. The lines that make up the part may not be visible. Move the cursor around where they were located. The lines will appear in red. Move the cursor over the top line causing it to turn red and left click once as shown in Figure 28.

Figure 28

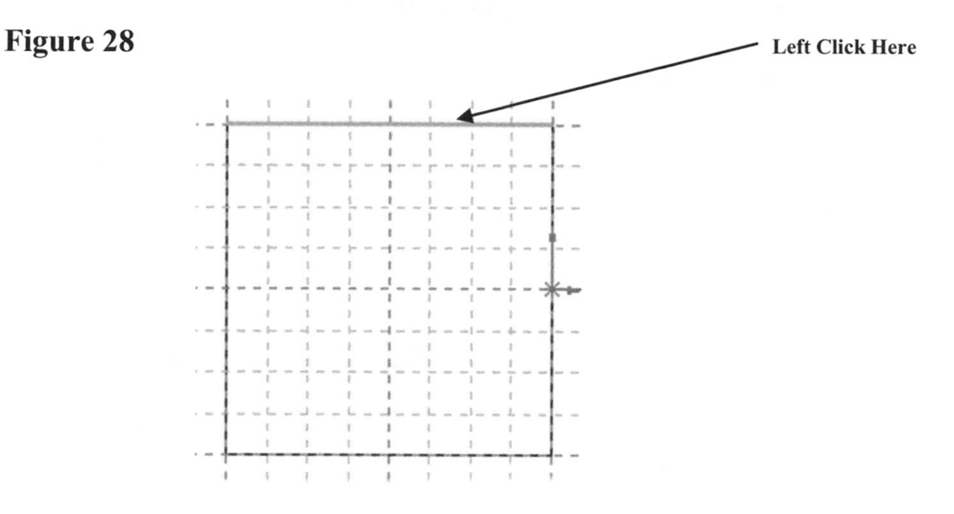

32. While the line is still green move the cursor to the upper middle portion of the screen and left click on **Convert Entities** as shown in Figure 29. The line will become visible.

Figure 29

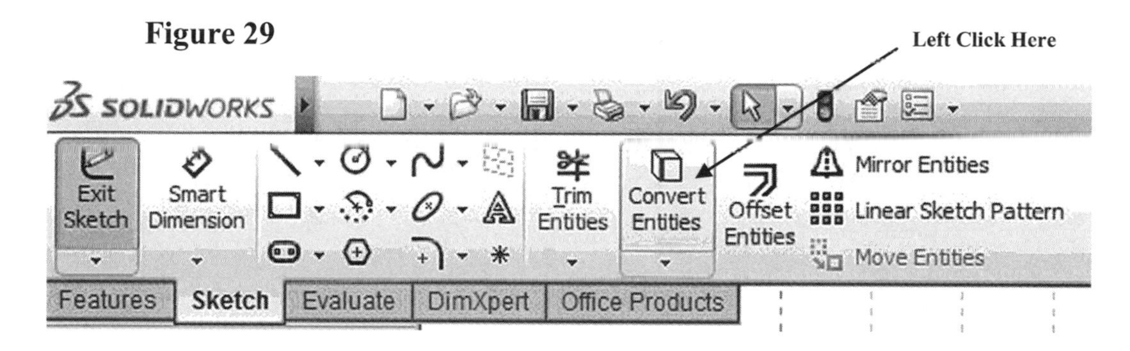

33. Move the cursor over the left vertical line causing it to turn red. Left click once as shown in Figure 30.

Figure 30

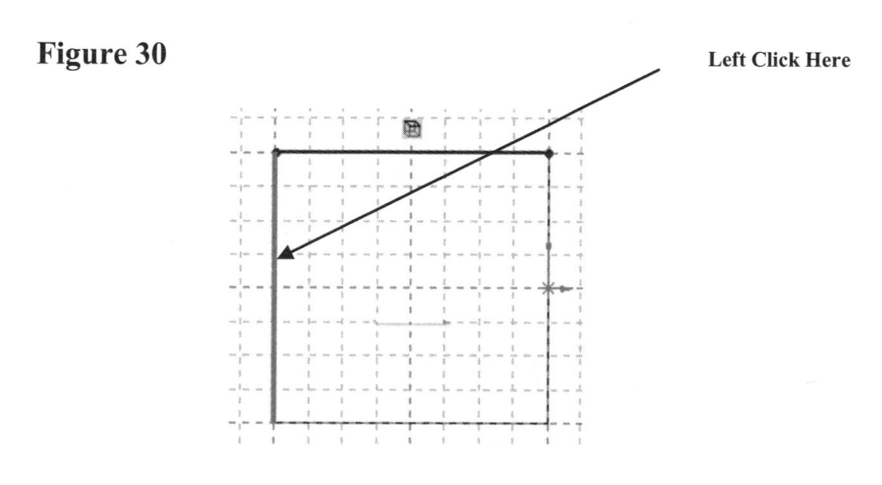

34. While the line is still green move the cursor to the upper middle portion of the screen and left click on **Convert Entities** as shown in Figure 31. The line will become visible on the new plane. You may want to slightly rotate the view isometric to see the line projected on to the new plane.

Figure 31

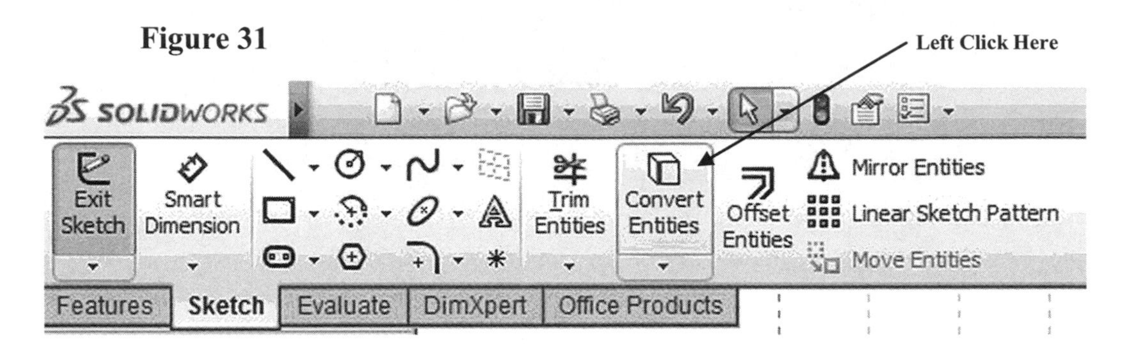

35. Move the cursor over the lower line causing it to turn red. Left click once as shown in Figure 32.

Figure 32

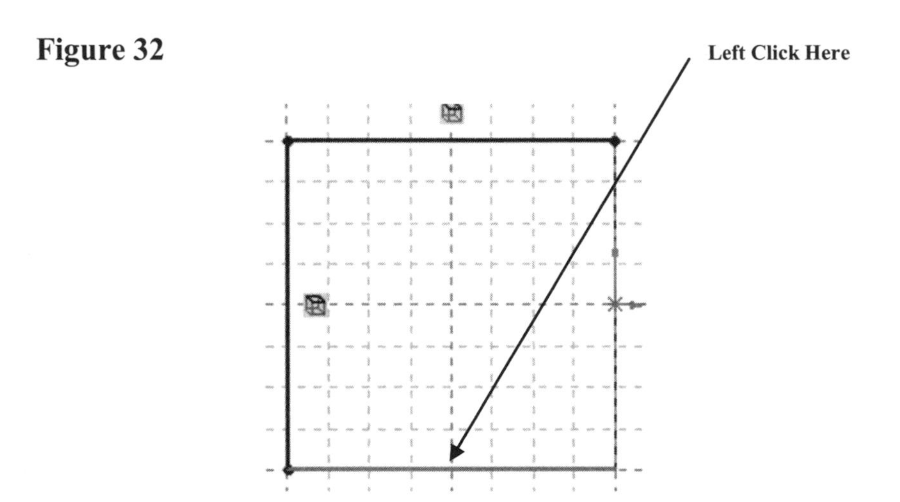

36. While the line is still green move the cursor to the upper middle portion of the screen and left click on **Convert Entities** as shown in Figure 33. The line will become visible.

Figure 33

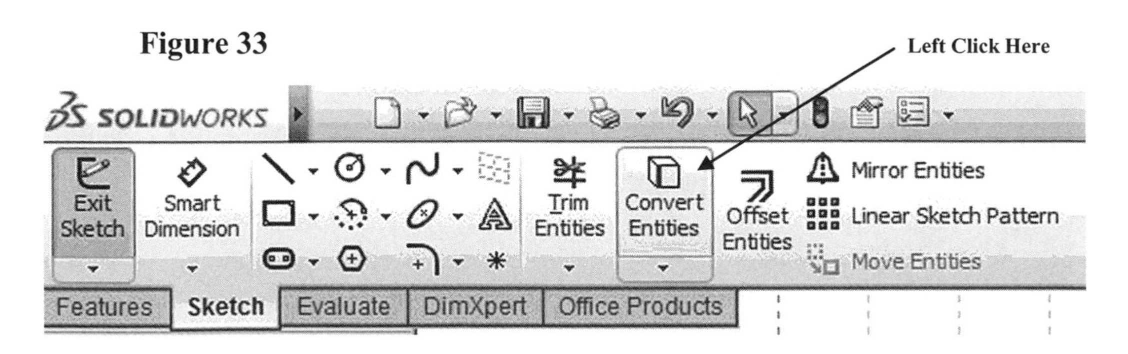

37. Move the cursor over the right vertical line causing it to turn red. Left click once as shown in Figure 34.

Figure 34

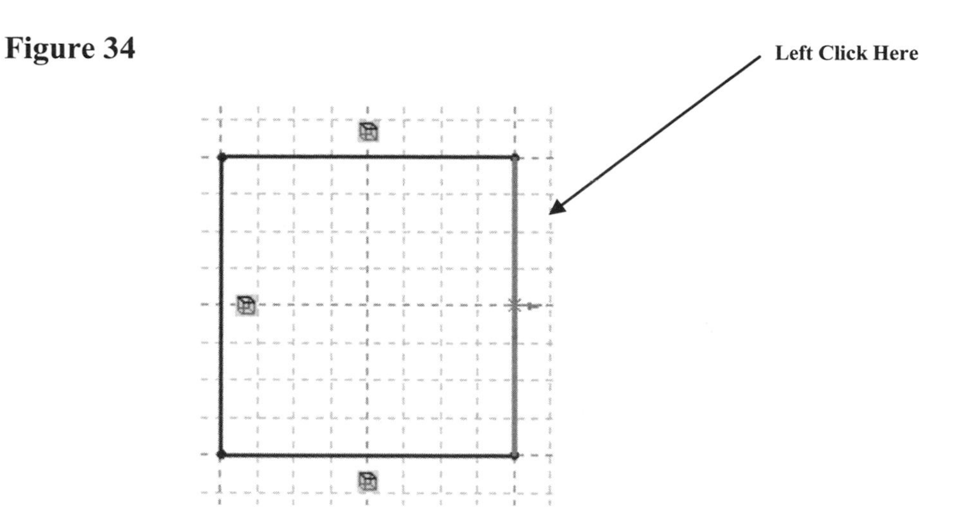

38. While the line is still green move the cursor to the upper right portion of the screen and left click on **Convert Entities** as shown in Figure 35. The line will become visible.

Figure 35

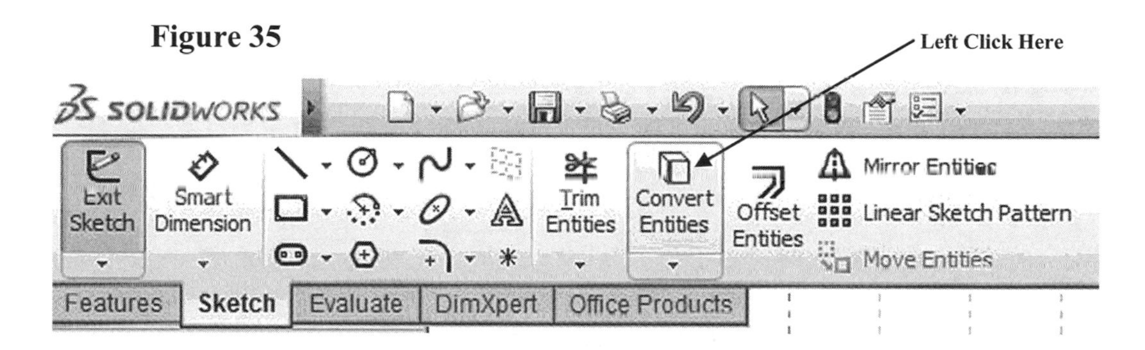

39. Your screen should look similar to Figure 36.

Figure 36

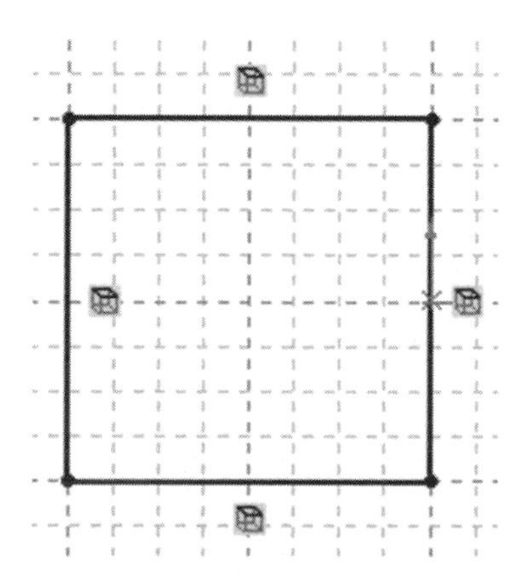

40. Move the cursor to the upper left portion of the screen and left click on **Line** as shown in Figure 37.

Figure 37

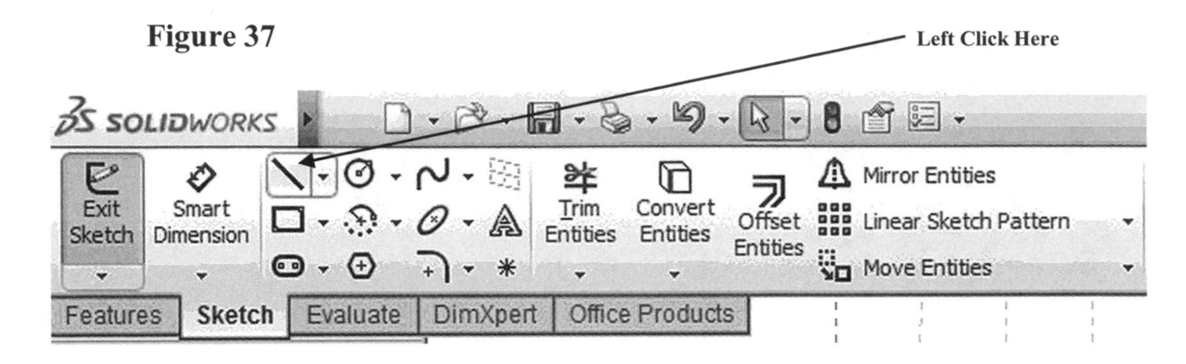

41. Move the cursor to the midpoint of the upper line causing a red dot to appear. Left click as shown in Figure 38.

Figure 38

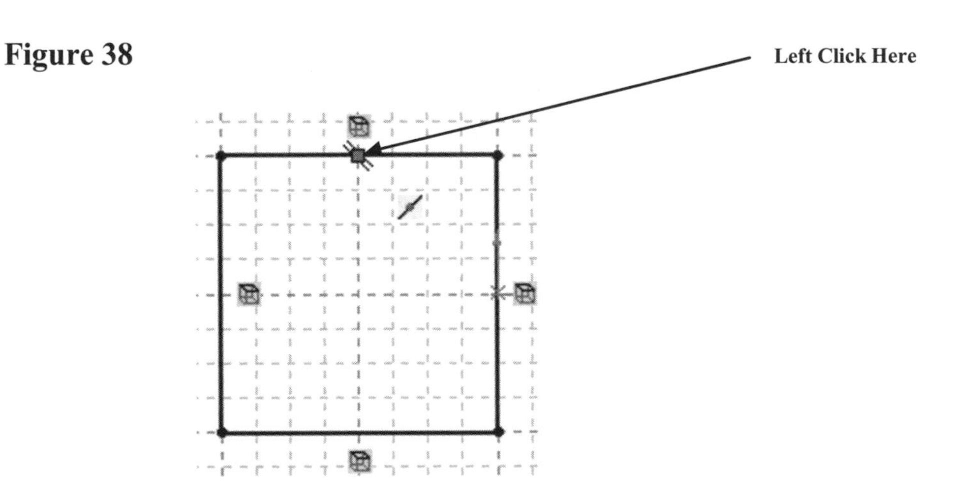

42. Move the cursor down to the lower line causing a red dot to appear. Left click as shown in Figure 39.

Figure 39

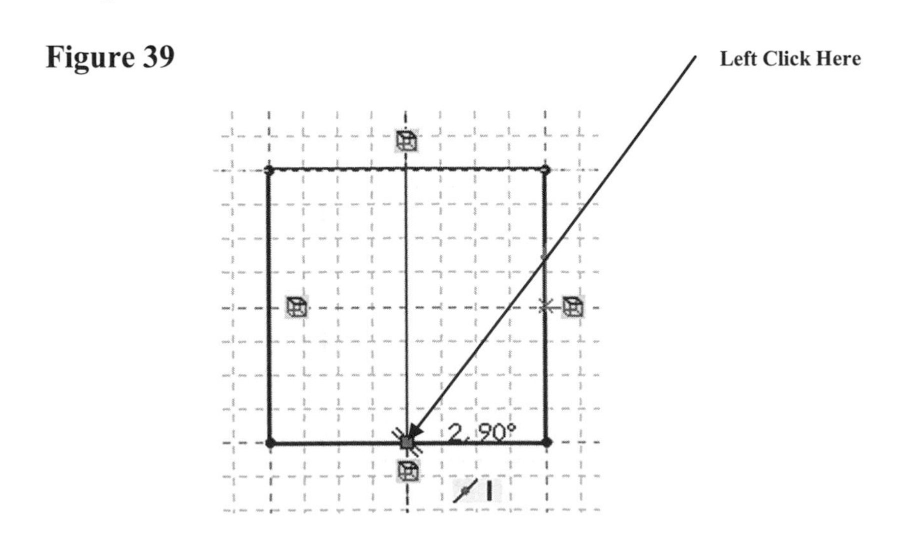

43. Right click anywhere around the drawing. A pop up menu will appear. Left click on **Select** as shown in Figure 40.

Figure 40

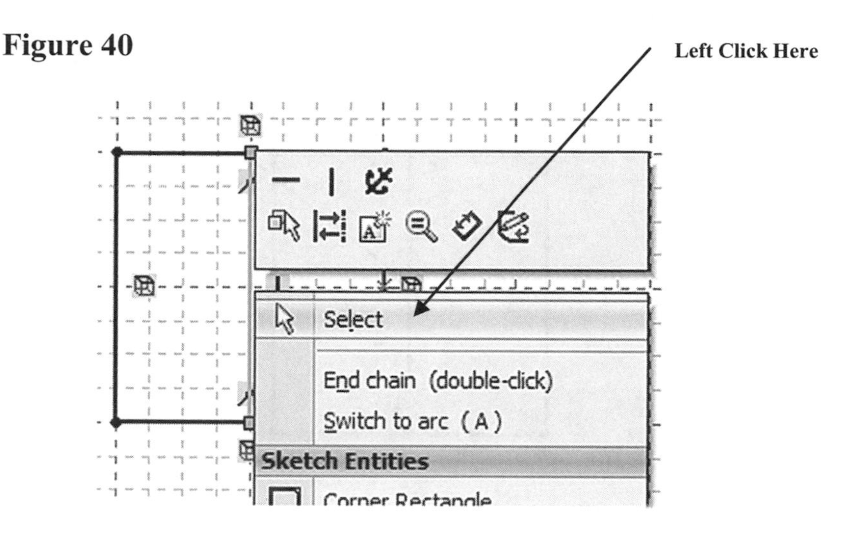

44. Your screen should look similar to Figure 41.

Figure 41

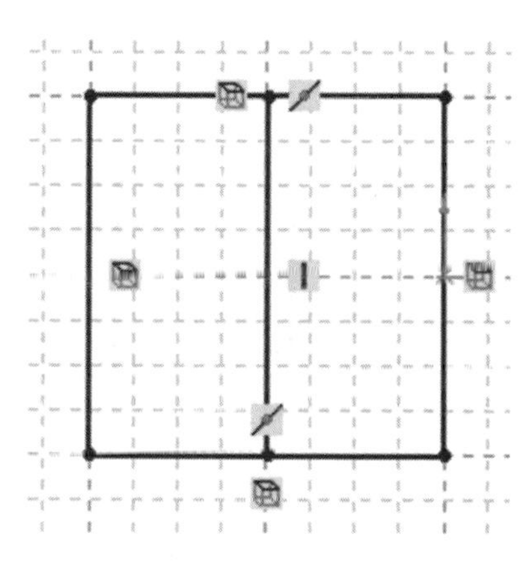

45. Move the cursor to the upper left portion of the screen and left click on **Circle** as shown in Figure 42.

Figure 42

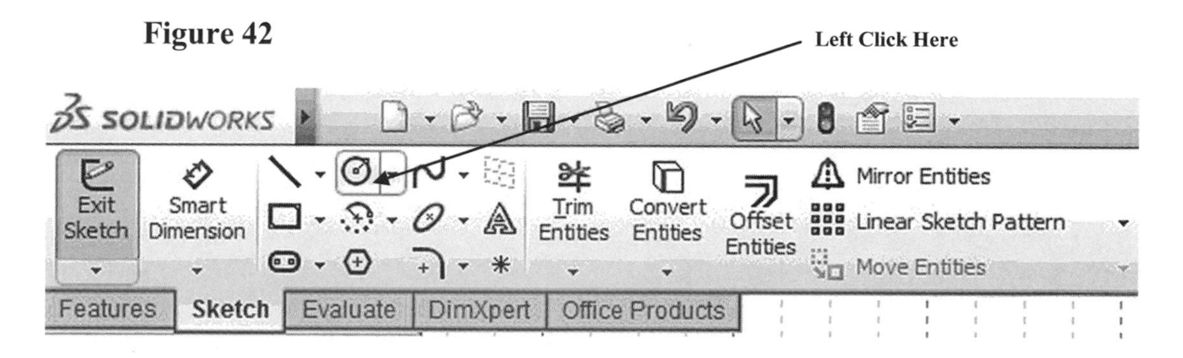

46. Move the cursor to the midpoint of the center line. A yellow dot will appear. Left click as shown in Figure 43.

Figure 43

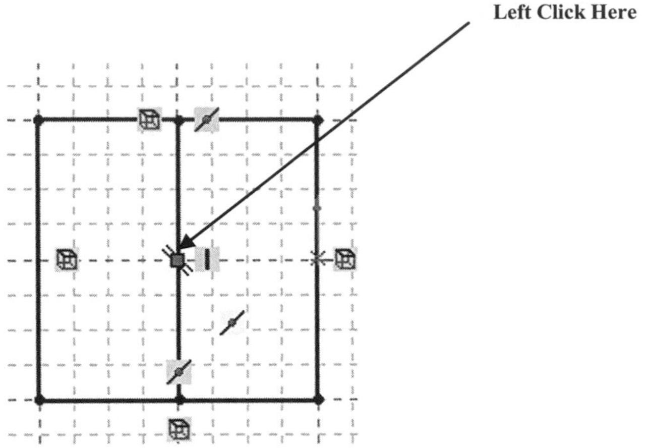

Left Click Here

47. Move the cursor out to the side and left click as shown in Figure 44.

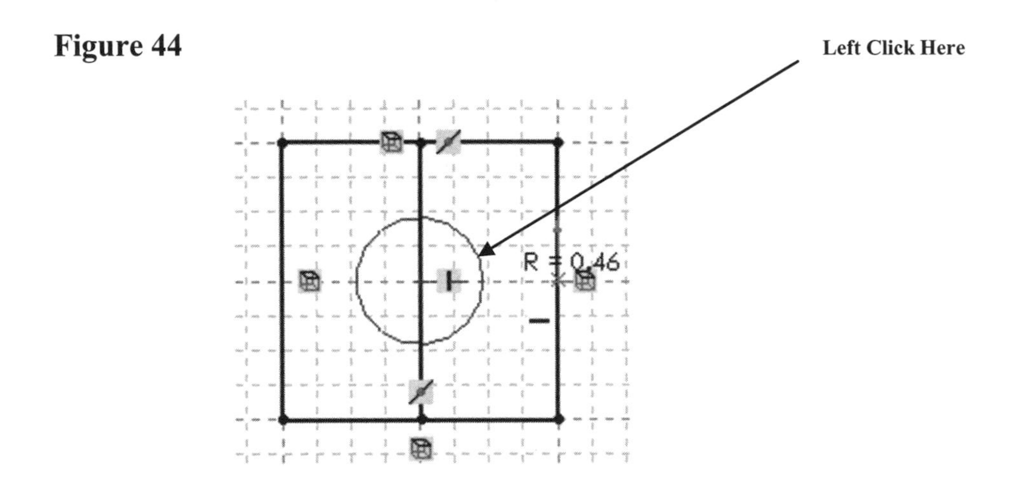

Figure 44

48. Right click anywhere around the drawing. A pop up menu will appear. Left click on **Select** as shown in Figure 45.

Figure 45

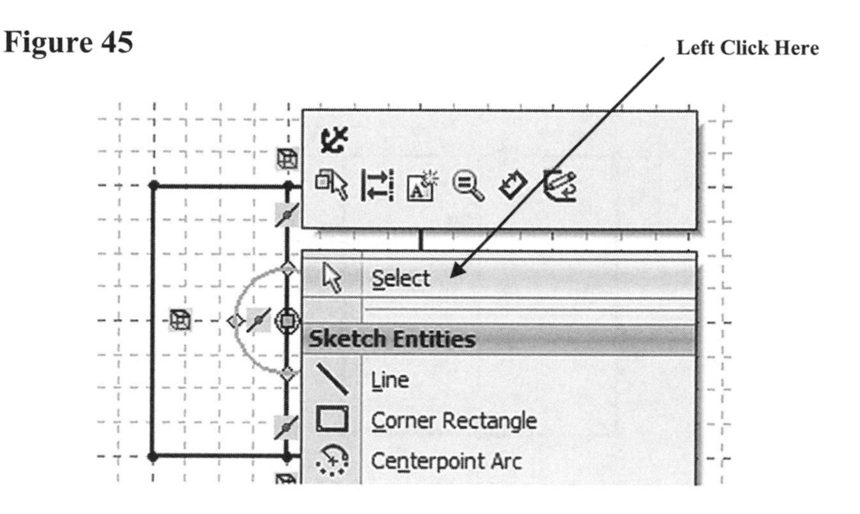

49. Move the cursor to the upper left portion of the screen and left click on **Smart Dimension** as shown in Figure 46.

Figure 46

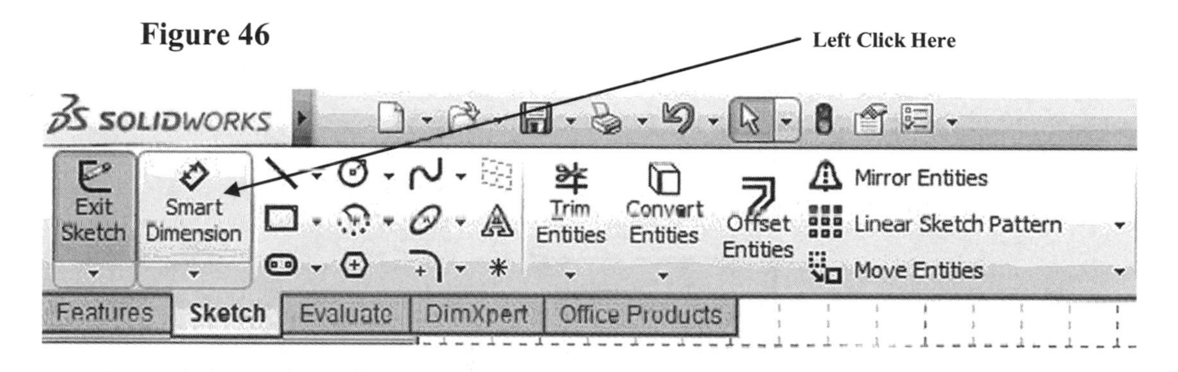

50. Move the cursor over the edge (not center) of the circle until it turns red. Left click once as shown in Figure 47.

Figure 47

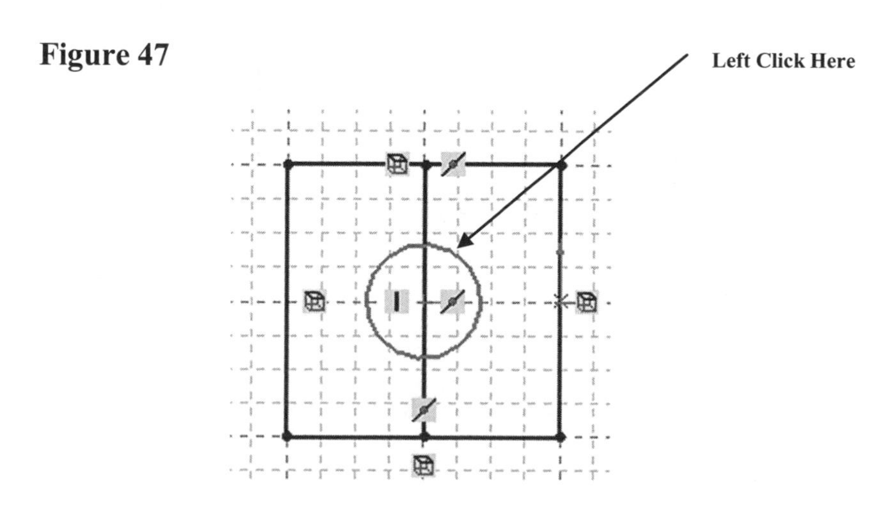

51. Move the cursor around. The dimension of the line will appear as shown in Figure 48. The dimension is attached to the cursor.

Figure 48

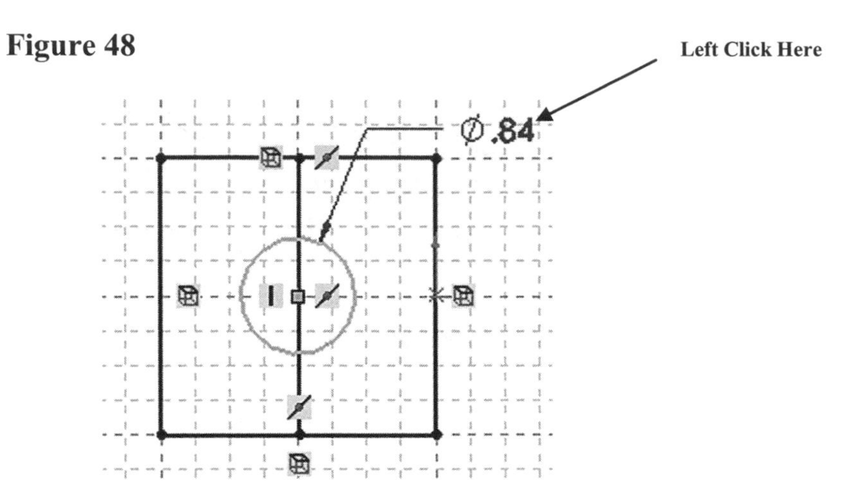

52. Move the cursor to where the dimension will be placed and left click once. The Modify dialog box will appear as shown in Figure 49.

Figure 49

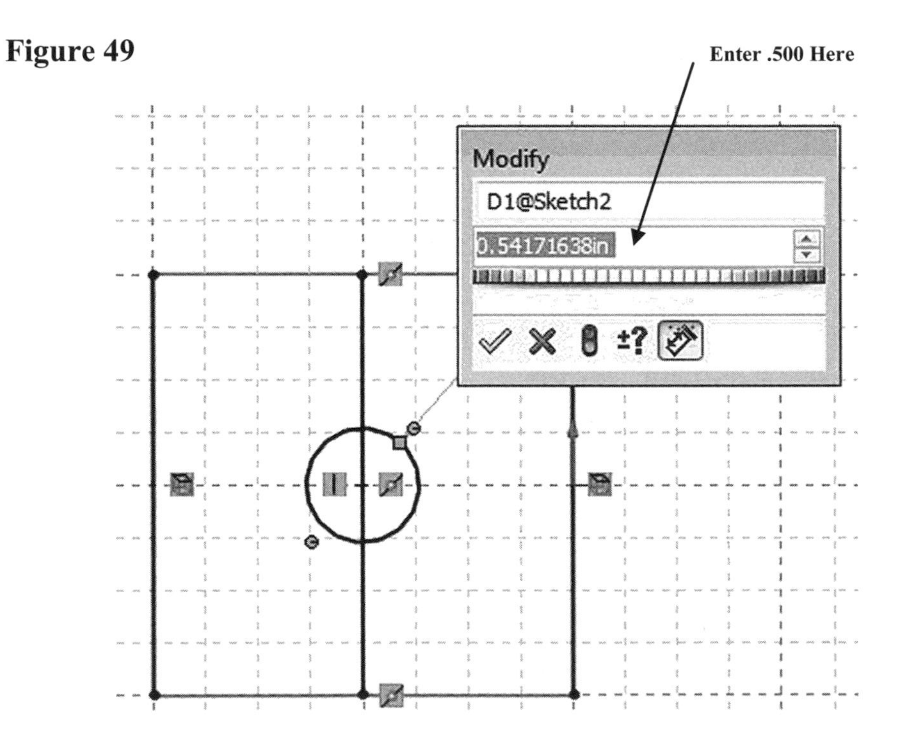

53. To edit the dimension, enter **.500** in the Modify dialog box (while the current dimension is highlighted) and left click on the green checkmark as shown in Figure 50.

Figure 50

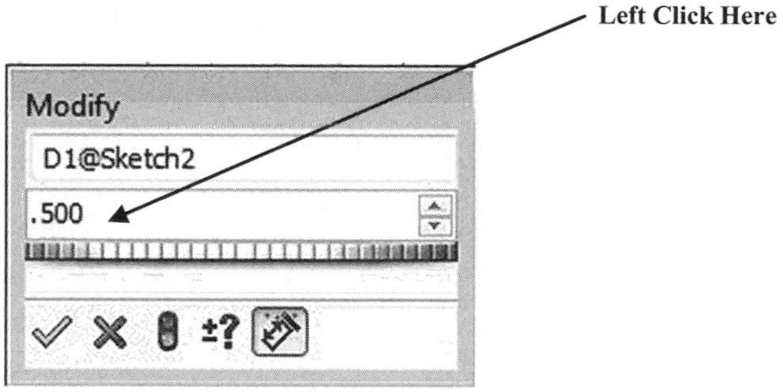

Left Click Here

Modify

D1@Sketch2

.500

54. Your screen should look similar to Figure 51.

Figure 51

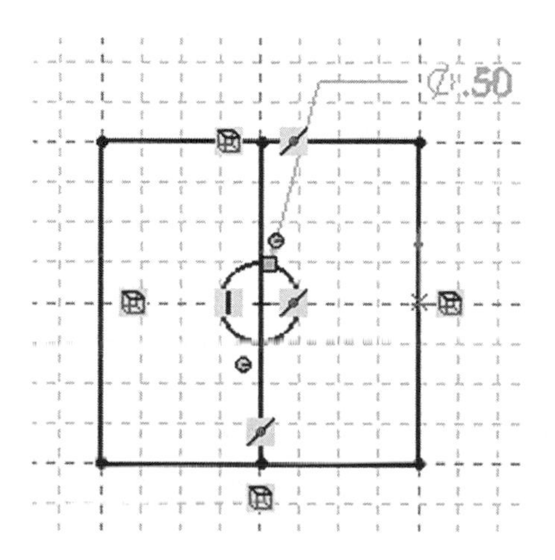

55. Right click anywhere around the drawing. A pop up menu will appear. Left click on **Select** as shown in Figure 52.

Figure 52

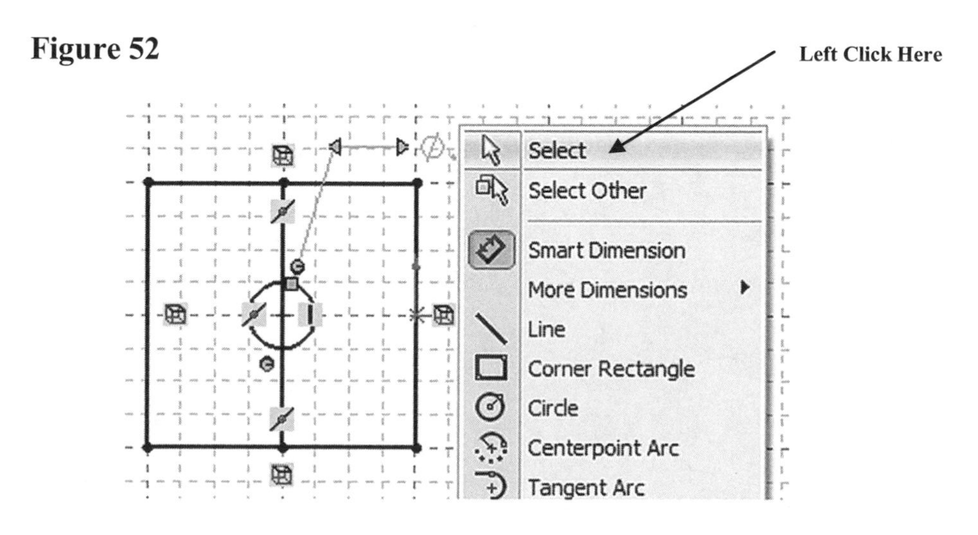

56. You will need to delete each line that was "converted" (5 total) to the new plane. Move the cursor over each line causing it to turn red. Right click on the line. A pop up menu will appear. Left click on **Delete** as shown in Figure 53.

Figure 53

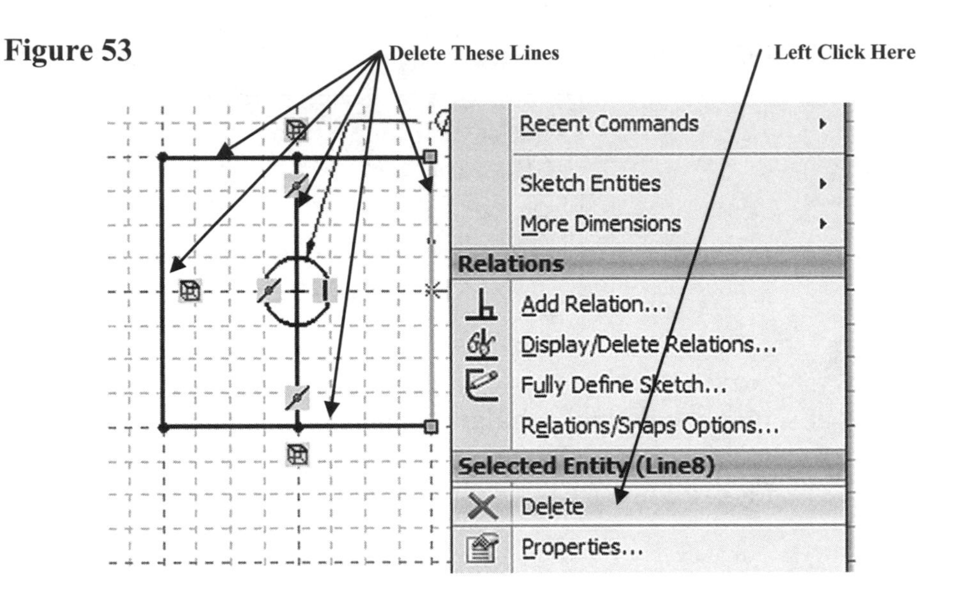

57. Your screen should look similar to Figure 54.

Figure 54

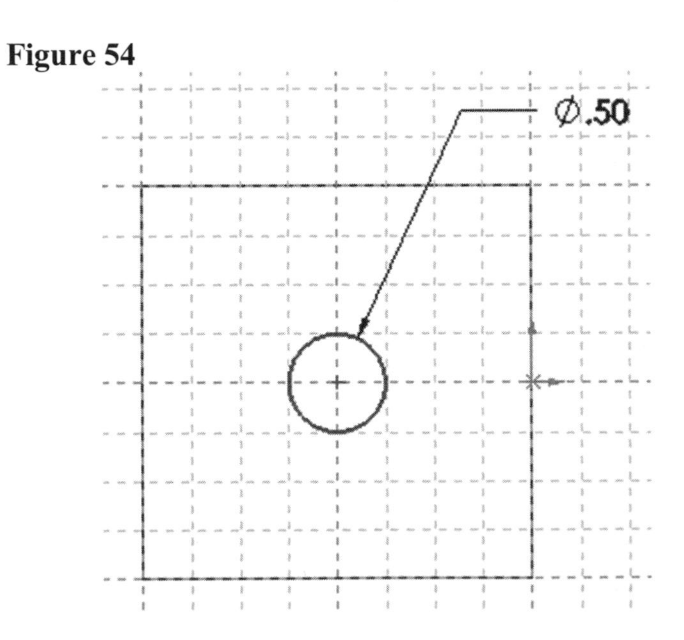

58. Move the cursor to the upper right portion of the screen and left click on the "Shaded with Edges" icon as shown in Figure 55.

Figure 55

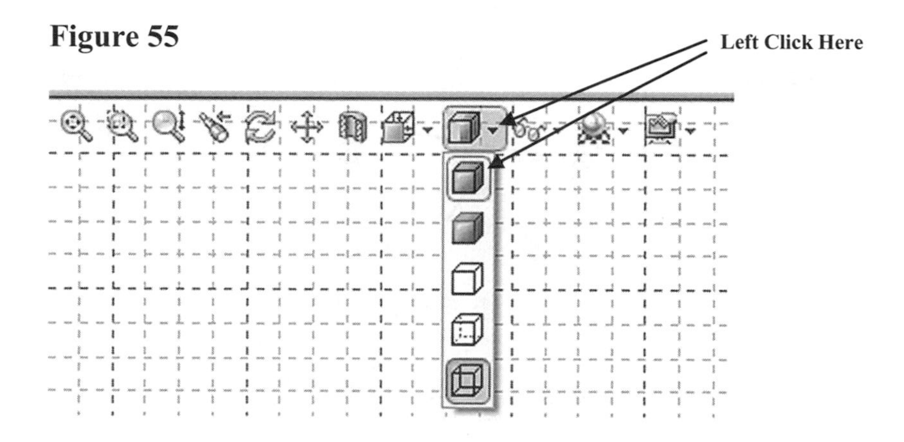

59. Your screen should look similar to Figure 56.

Figure 56

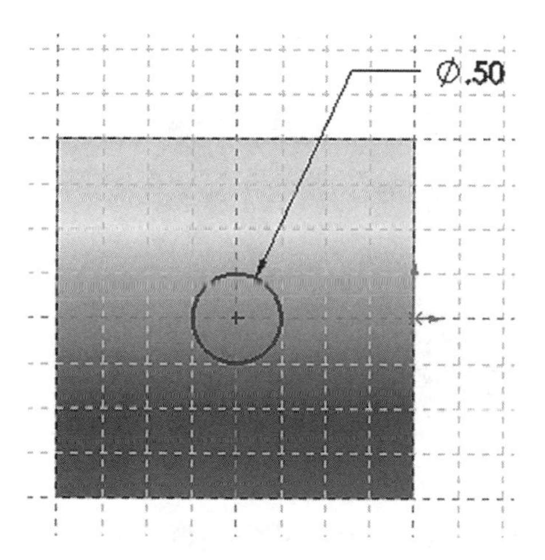

60. After you have verified that no commands are active, right click anywhere on the sketch. A pop up menu will appear. Left click on **Exit Sketch** as shown in Figure 57.

Figure 57

Left Click Here

Zoom/Pan/Rotate

Recent Commands

Sketch Entities

Line

61. Move the cursor to the upper right portion of the screen and left click on the drop down arrow next to the "View Orientation" icon. A drop down menu will appear. Left click on **Trimetric** as shown in Figure 58.

Figure 58

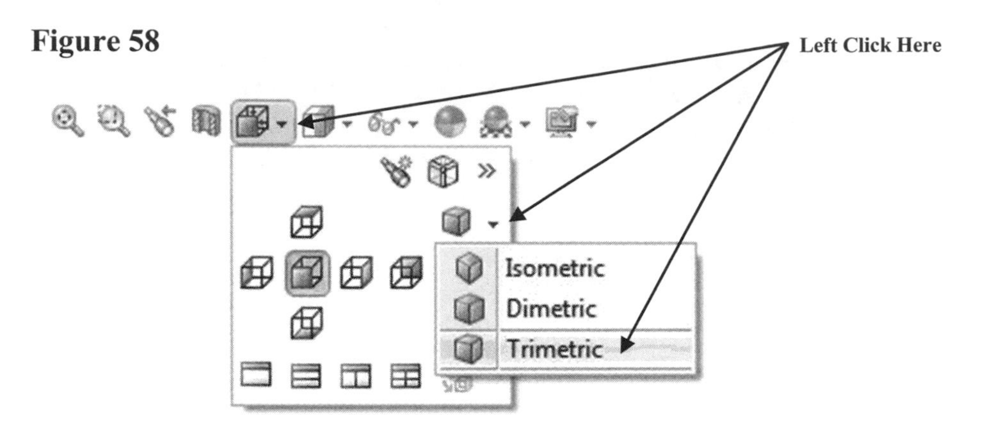

62. Your screen should look similar to Figure 59.

Figure 59

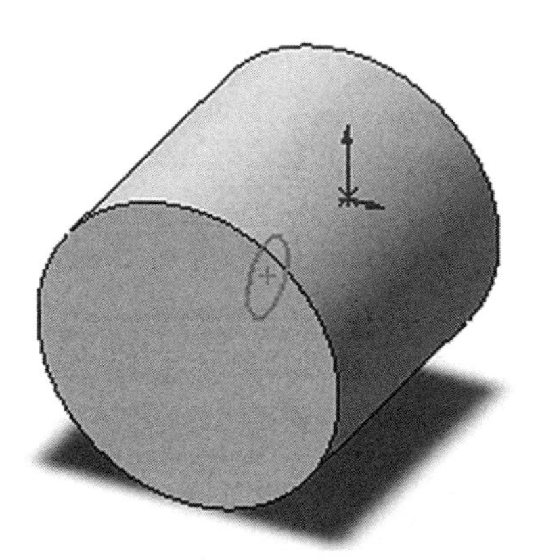

63. Move the cursor to the upper left portion of the screen and left click on **Extruded Cut** as shown in Figure 60. The Extrude dialog box will appear. Move the cursor to the edge of the small circle causing it to turn red. Left click once.

Figure 60

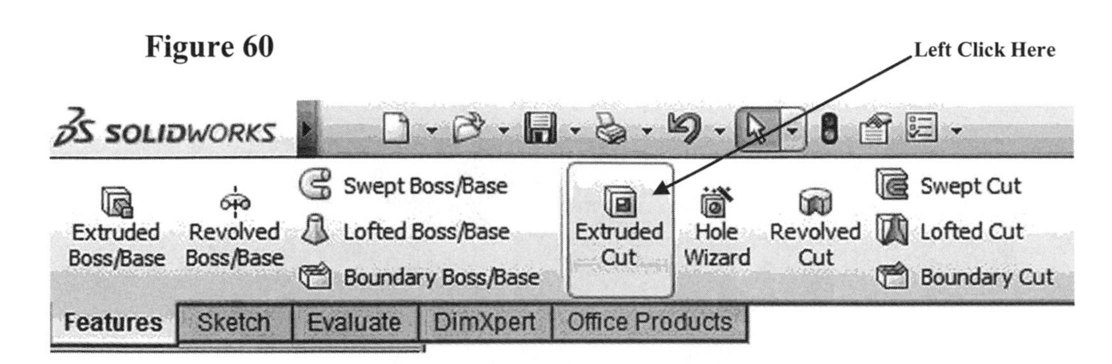

64. Move the cursor to the middle left portion of the screen. Enter **1.00** for D1. Left click in the box next to the text "Direction 2". Enter **1.00** for D2 as shown in Figure 61.

Figure 61

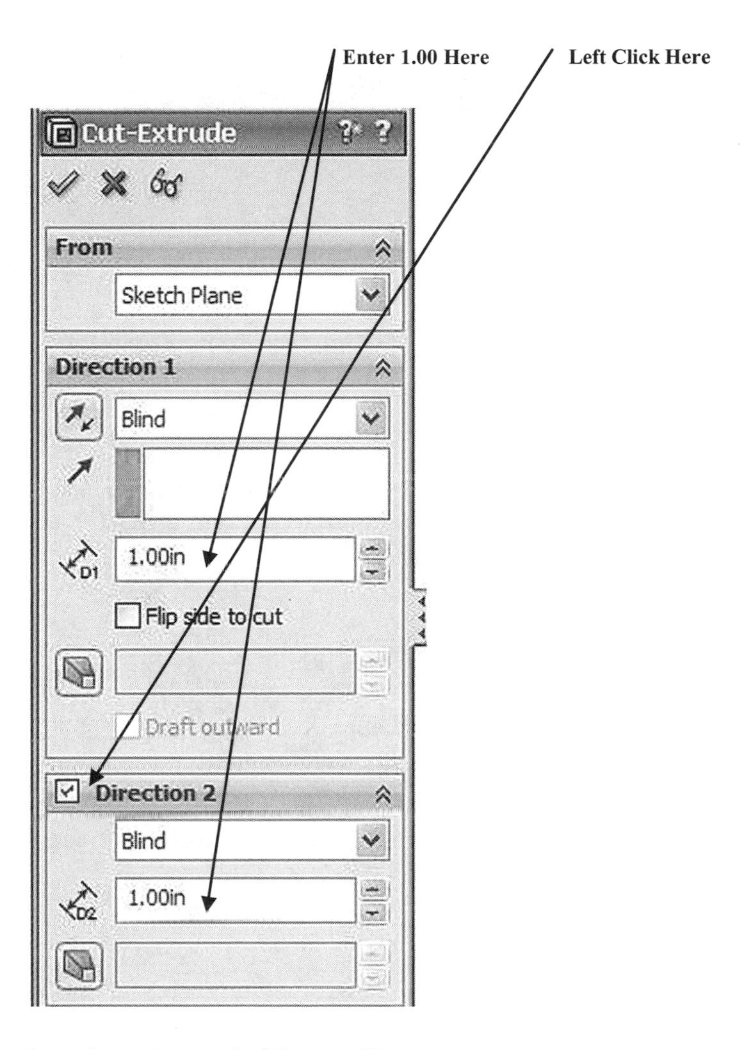

65. A preview will be displayed as shown in Figure 62.

Figure 62

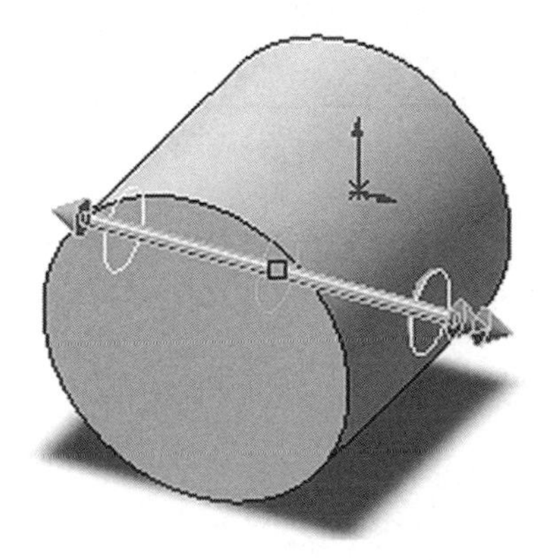

66. Move the cursor to the upper left portion of the screen and left click on the green checkmark as shown in Figure 63.

Figure 63

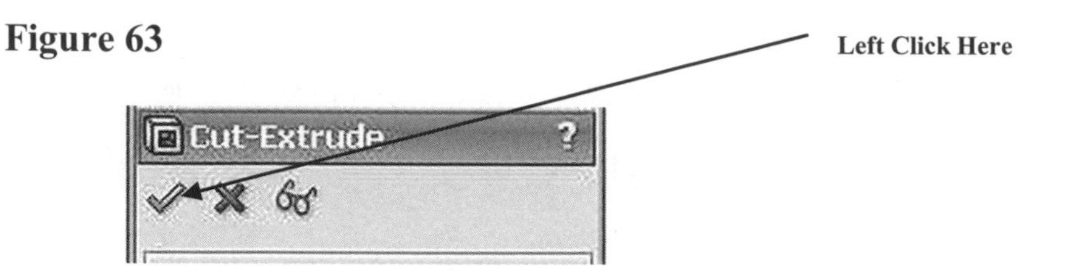

67. Your screen should look similar to Figure 64.

Figure 64

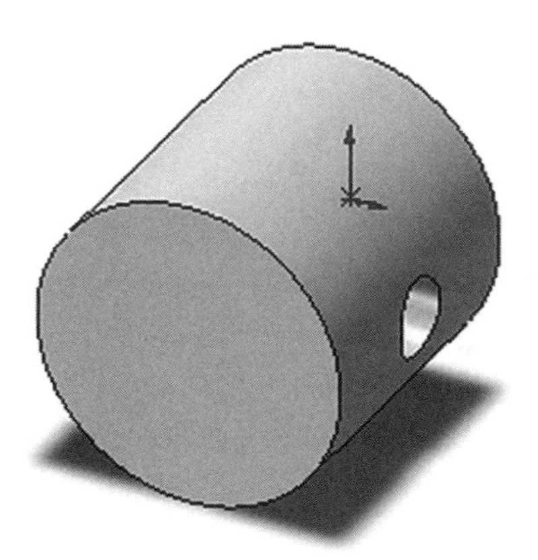

68. Move the cursor to the upper middle portion of the screen and left click on **Shell** as shown in Figure 65.

Figure 65

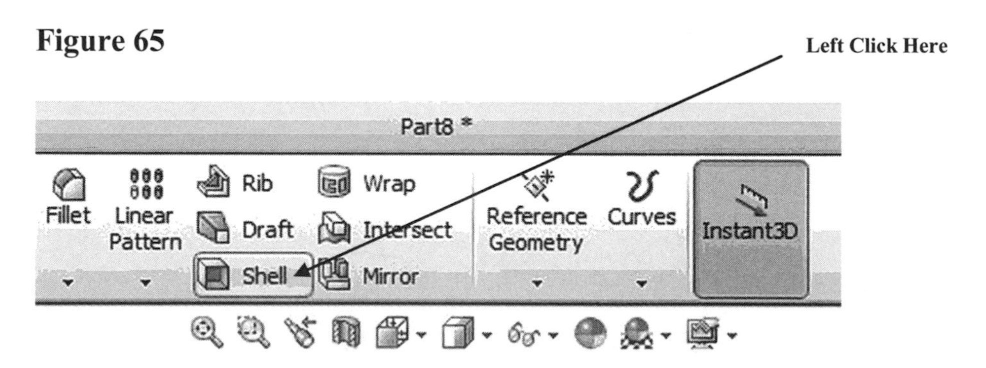

69. Left click on the lower surface as shown in Figure 66.

Figure 66

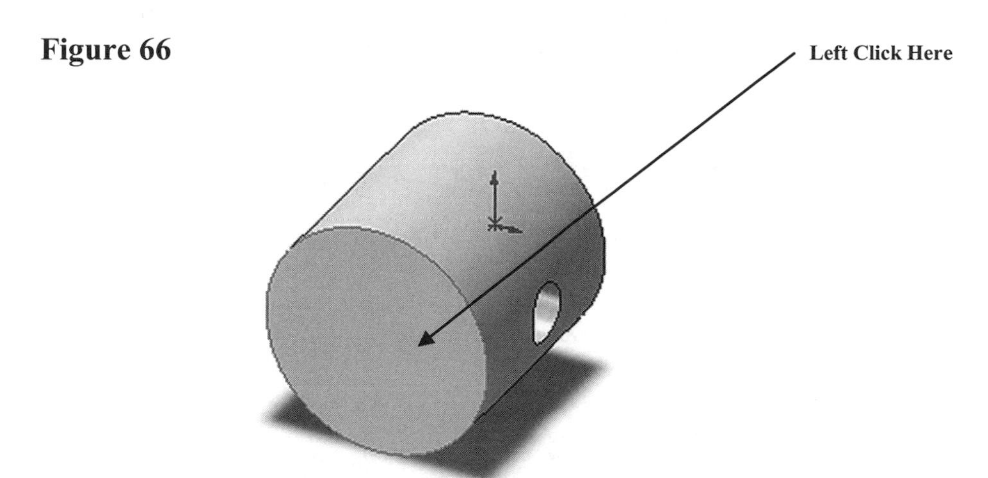

70. Enter **.10** next to D1. Left click on the green checkmark as shown in Figure 67.

Figure 67

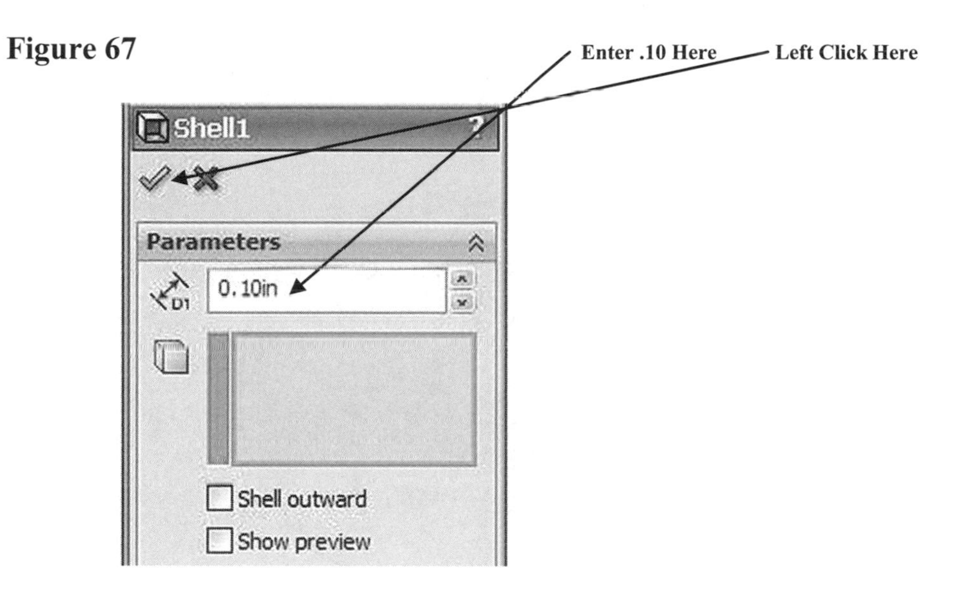

71. Your screen should look similar to Figure 68.

Figure 68

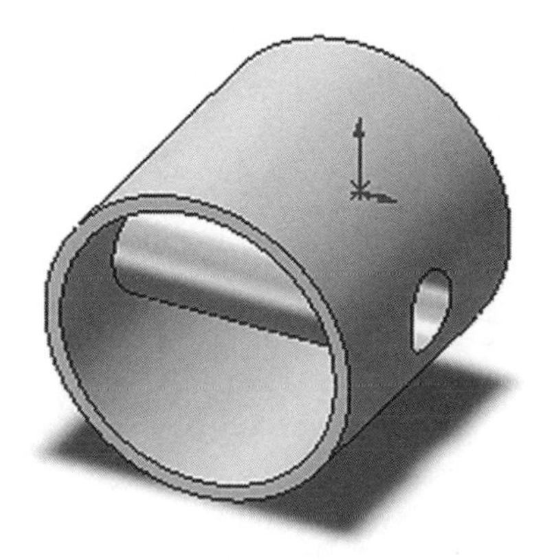

72. Move the cursor to the upper right portion of the screen and left click on the drop down arrow next to the "View Orientation" icon. A drop down menu will appear. Left click on **Front** as shown in Figure 69.

Figure 69

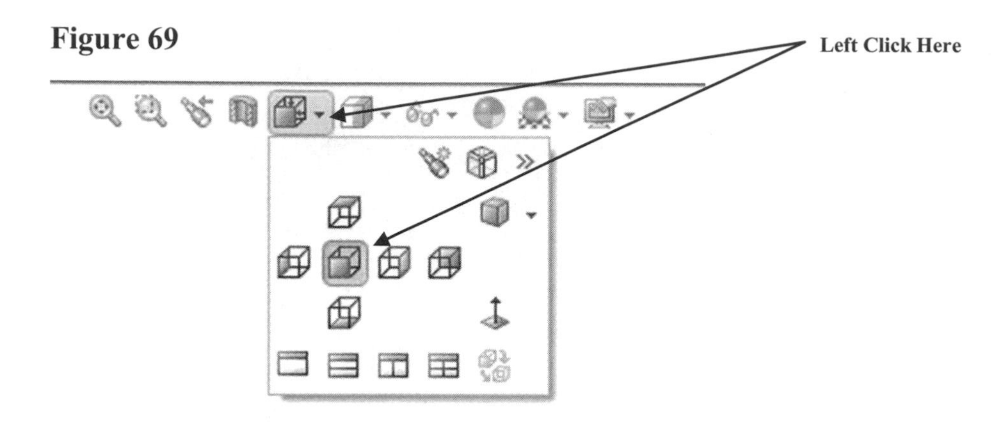

73. The part will rotate to provide a perpendicular view of the inside as shown in Figure 70.

Figure 70

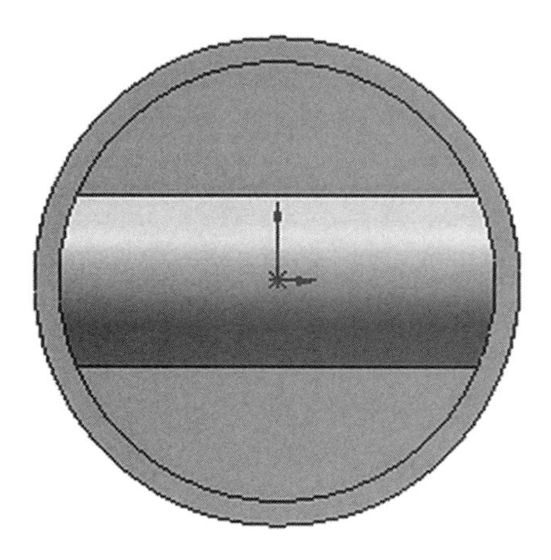

74. Move the cursor over the same surface causing both the inside and outside lines to turn red. You may have to zoom in for SolidWorks to find both lines at the same time. Both lines must be red at the same time. After both lines are red at the same time, right click on the surface as shown in Figure 71. The surface will turn green.

Figure 71

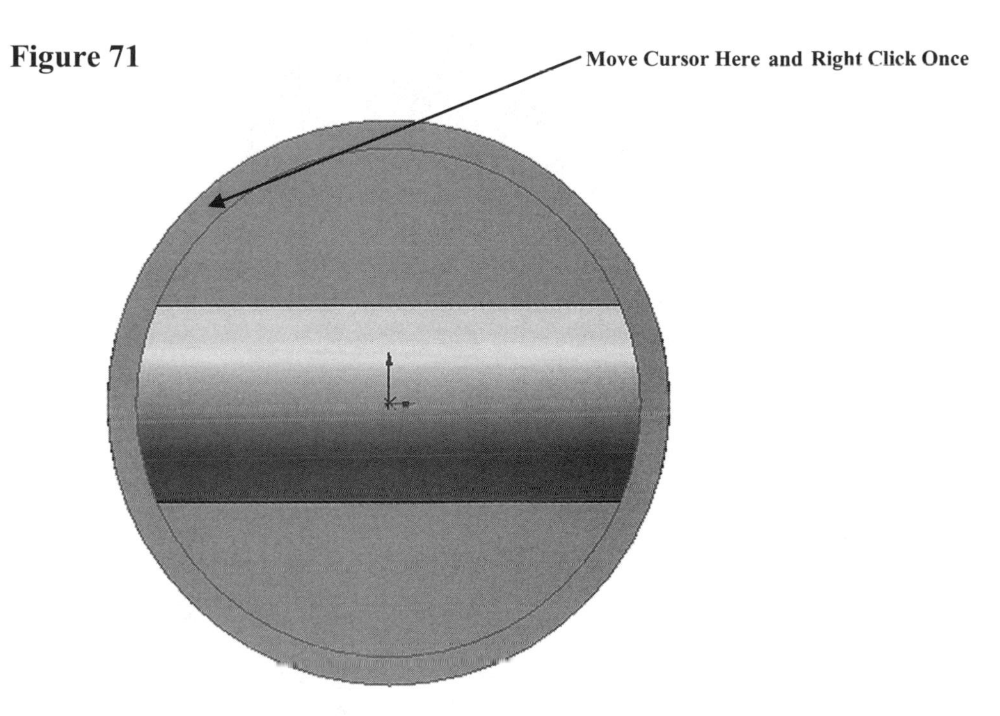

75. A pop up menu will appear. Left click on **Insert Sketch** as shown in Figure 72.

Figure 72

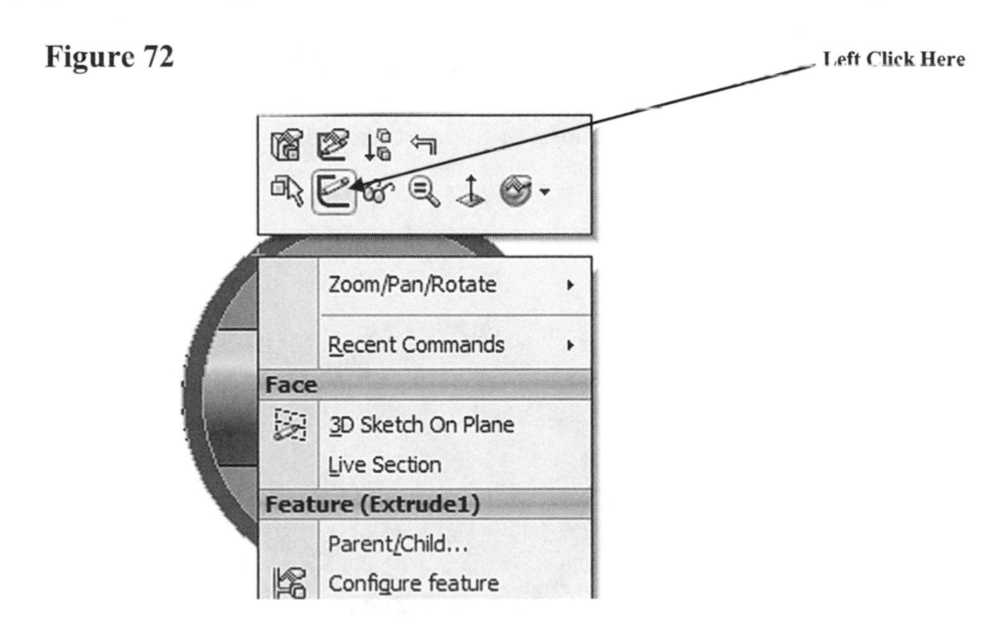

76. A new sketch will appear on the selected surface. Your screen should look similar to Figure 73.

Figure 73

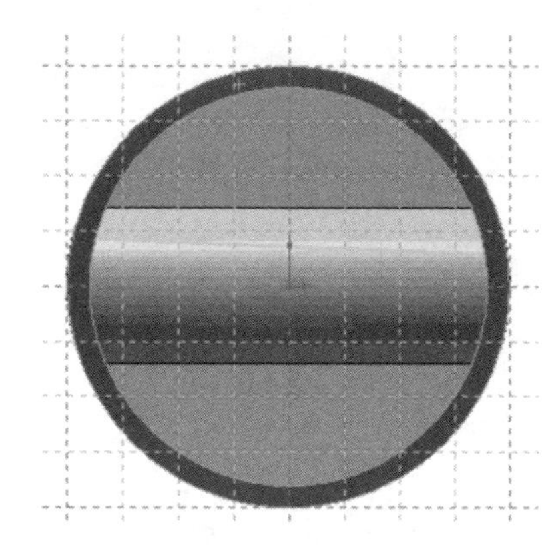

77. Move the cursor to the upper left portion of the screen and left click on **Line** as shown in Figure 74.

Figure 74

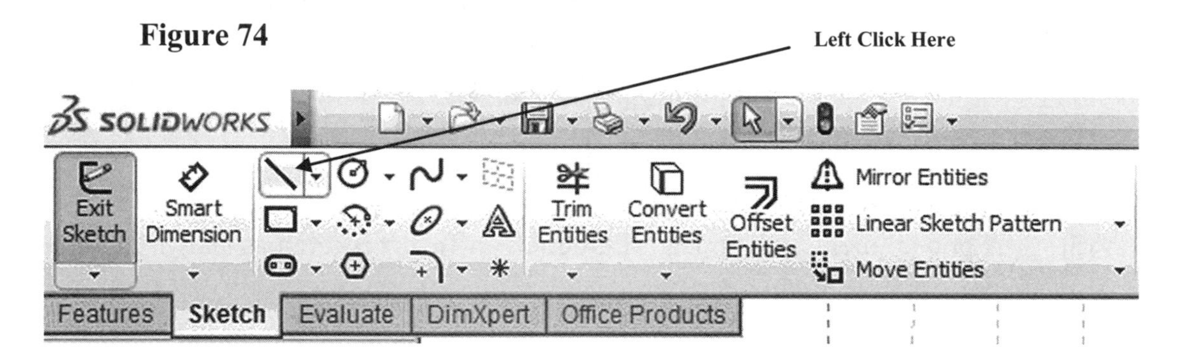

78. Move the cursor to the center of the part causing a red dot to appear. Left click as shown in Figure 75.

Figure 75

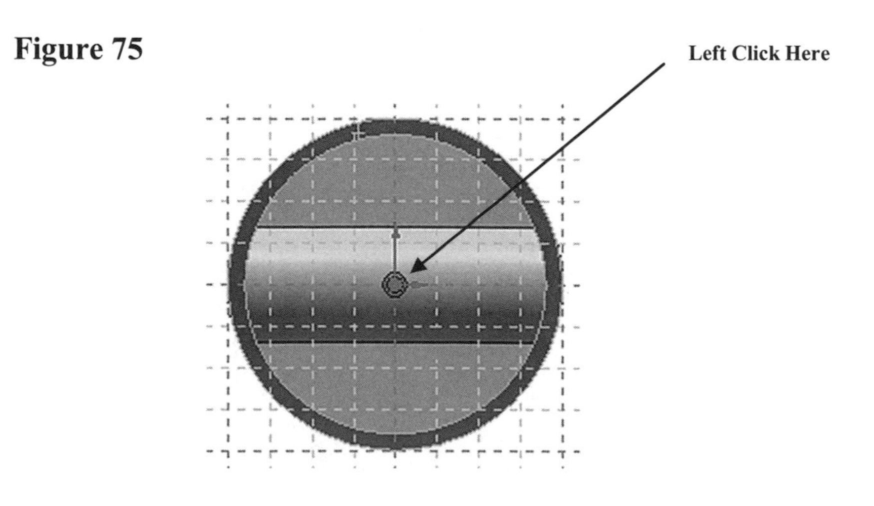

79. Move the cursor upward and left click as shown in Figure 76.

Figure 76

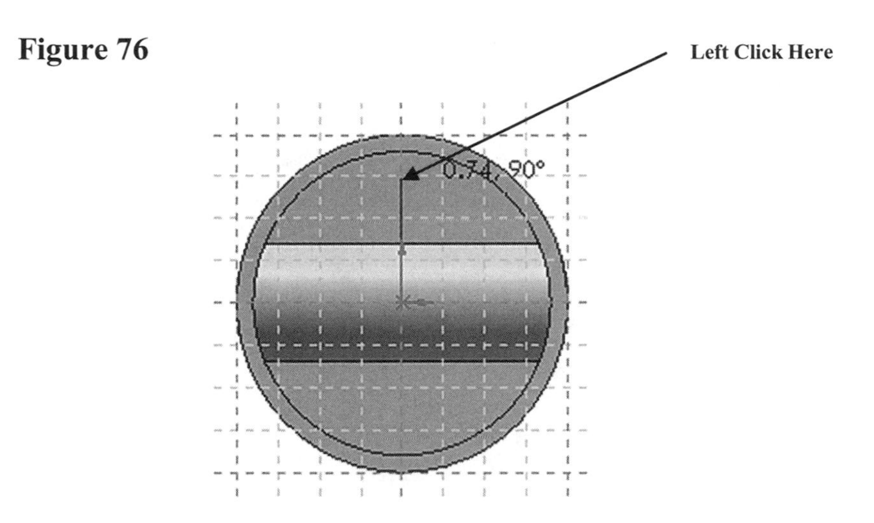

80. Right click anywhere around the drawing. A pop up menu will appear. Left click on **Select** as shown in Figure 77.

Figure 77

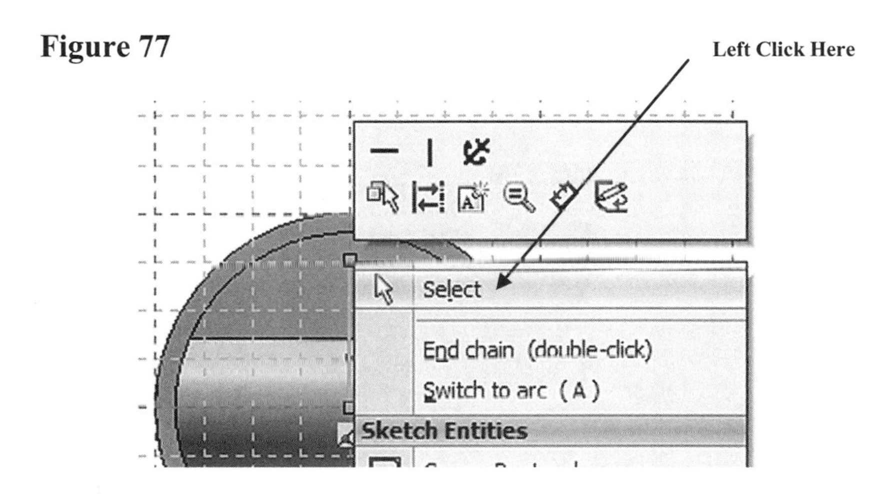

81. Move the cursor to the upper left portion of the screen and left click on **Line** as shown in Figure 78.

Figure 78

82. Move the cursor to the center of the part causing a red dot to appear. Left click as shown in Figure 79.

Figure 79

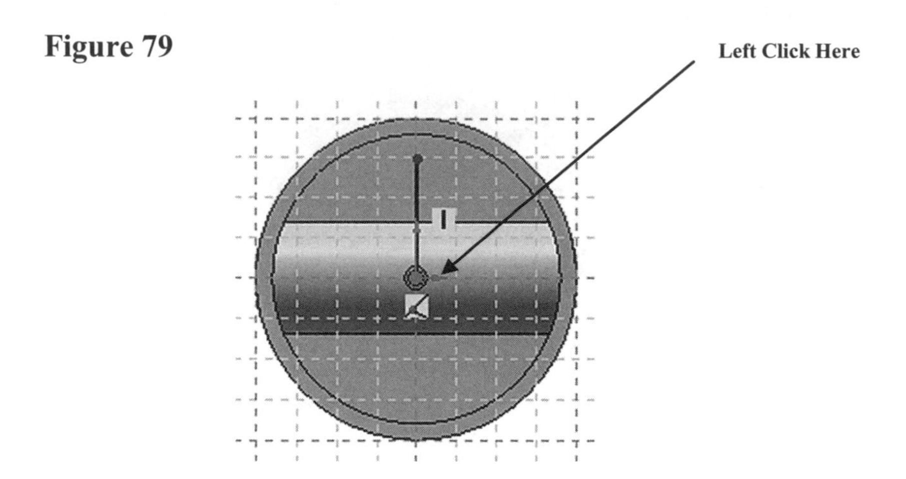

83. Move the cursor to the left and left click as shown in Figure 80.

Figure 80

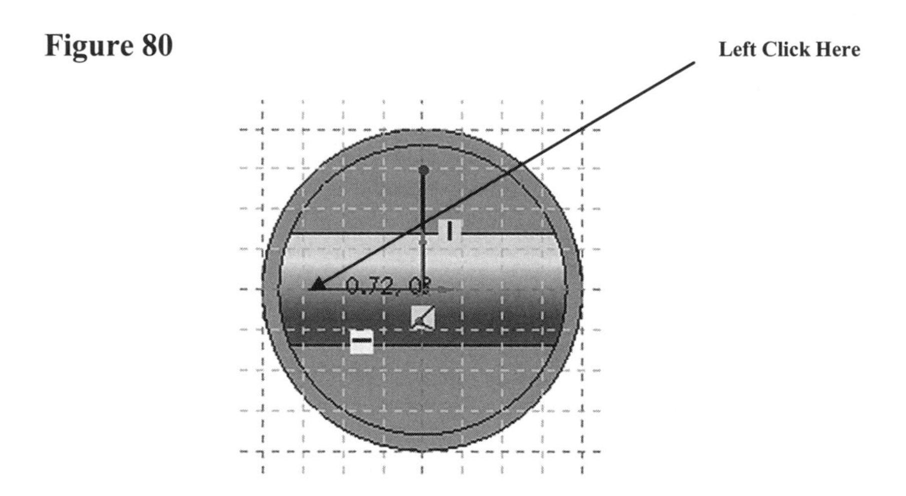

84. Right click anywhere around the drawing. A pop up menu will appear. Left click on **Select** as shown in Figure 81.

Figure 81

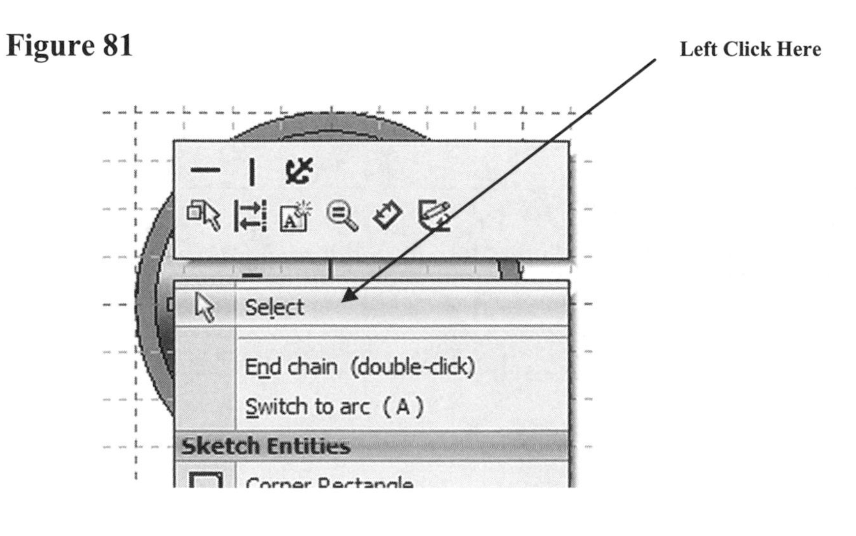

85. Your screen should look similar to Figure 82.

Figure 82

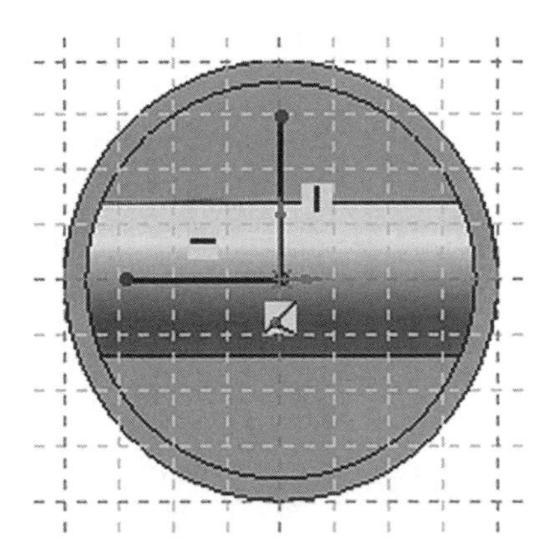

86. Move the cursor to the upper left portion of the screen and left click on **Linc** as shown in Figure 83.

Figure 83

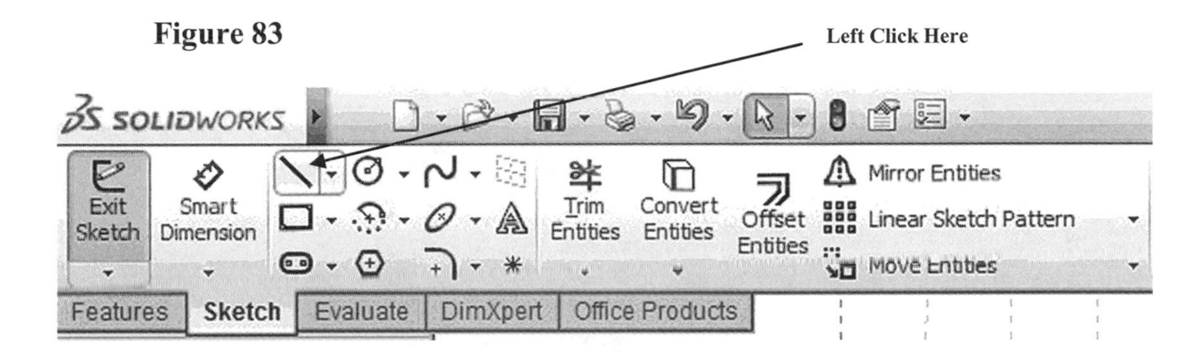

87. Move the cursor to the position shown in Figure 84 and left click once.

Figure 84

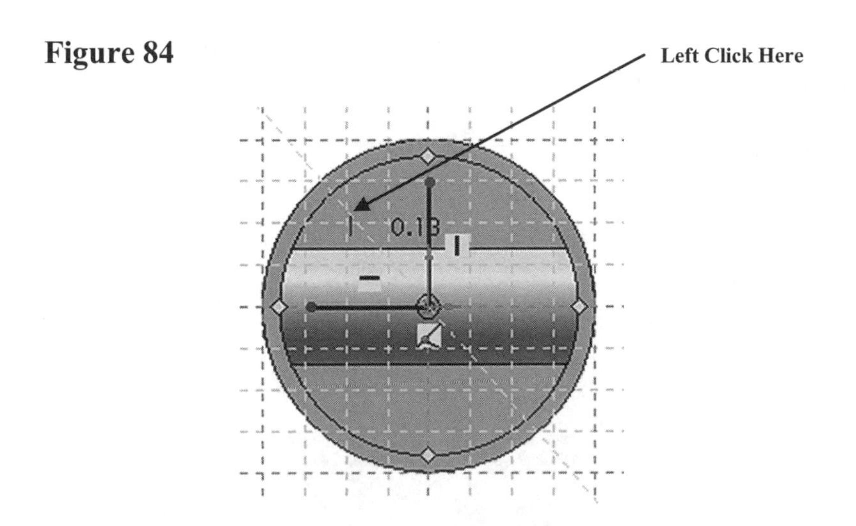

88. Move the cursor downward and left click as shown in Figure 85.

Figure 85

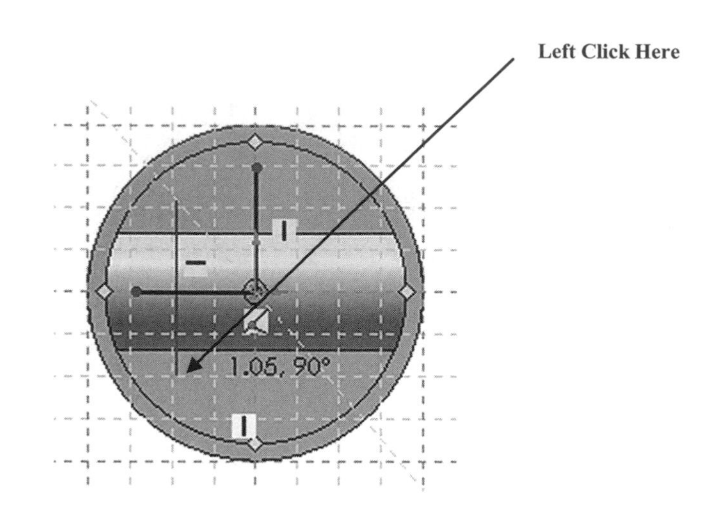

89. Move the cursor to the right and left click as shown in Figure 86.

Figure 86

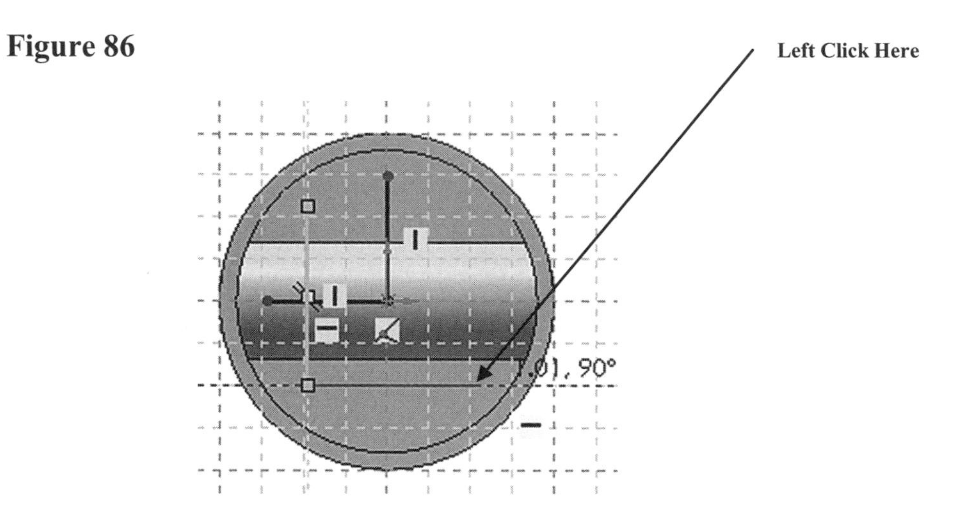

90. Move the cursor upward. Ensure that dots appear from the original starting point. Left click as shown in Figure 87.

Figure 87

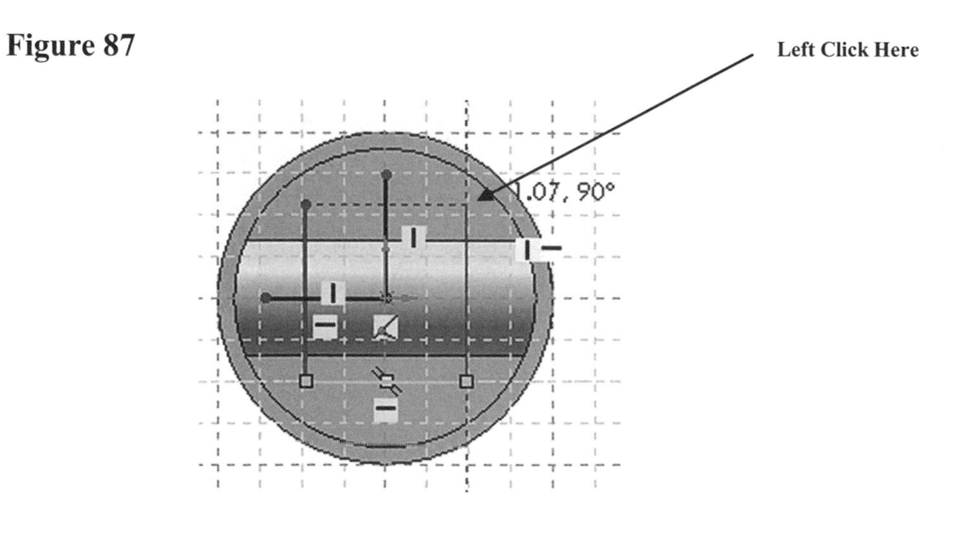

91. Move the cursor to the left and left click as shown in Figure 88.

Figure 88

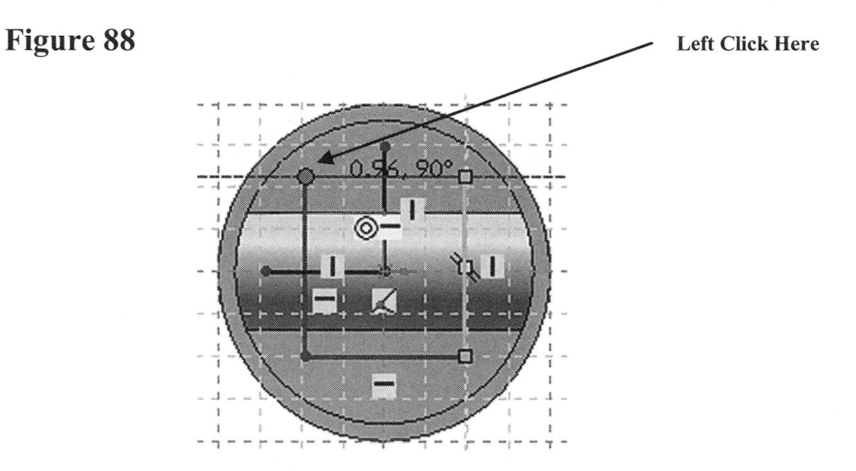

92. Right click anywhere around the drawing. A pop up menu will appear. Left click on **Select** as shown in Figure 89.

Figure 89

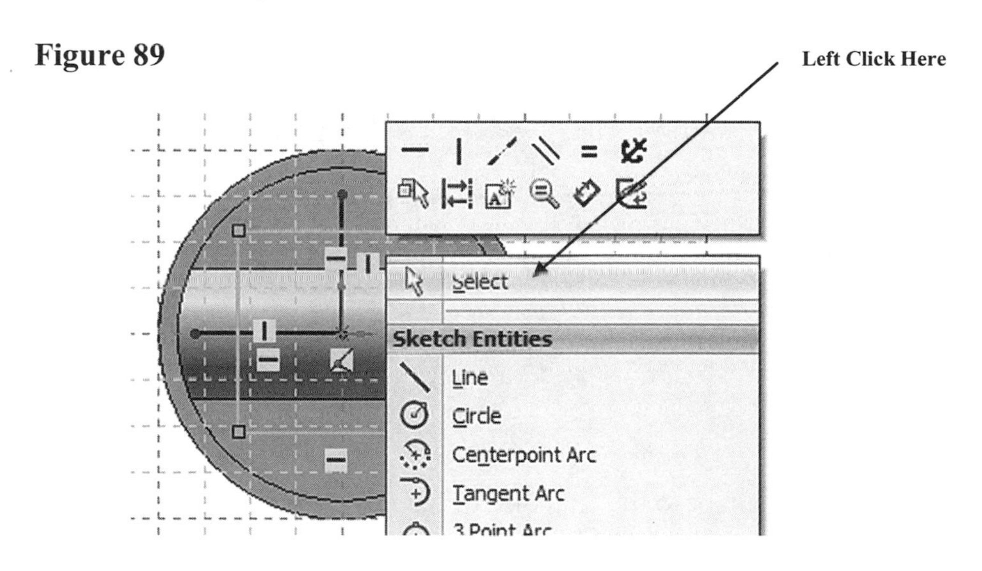

93. Your screen should look similar to Figure 90.

Figure 90

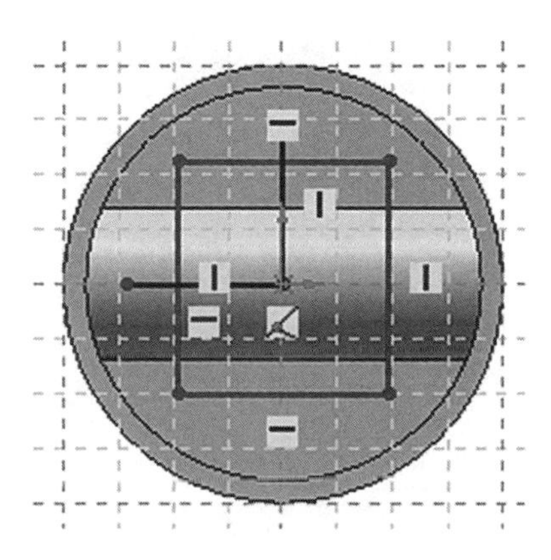

94. Move the cursor to the upper left portion of the screen and left click on **Smart Dimension** as shown in Figure 91.

Figure 91

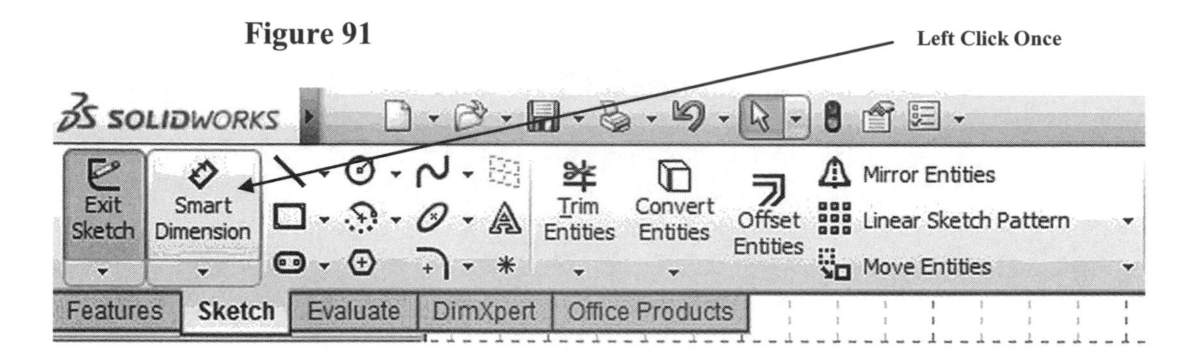

95. After selecting **Smart Dimension** move the cursor over the lower endpoint of the vertical line coming out of the center of the part. A red dot will appear. Left click on the endpoint as shown in Figure 92.

Figure 92

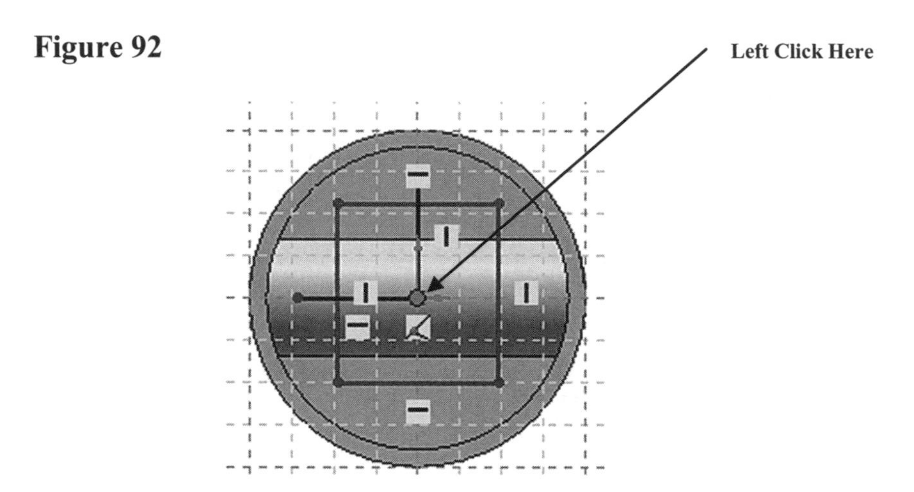

96. Move the cursor to the endpoint of the far left line. A red dot will appear. Left click once as shown in Figure 93.

Figure 93

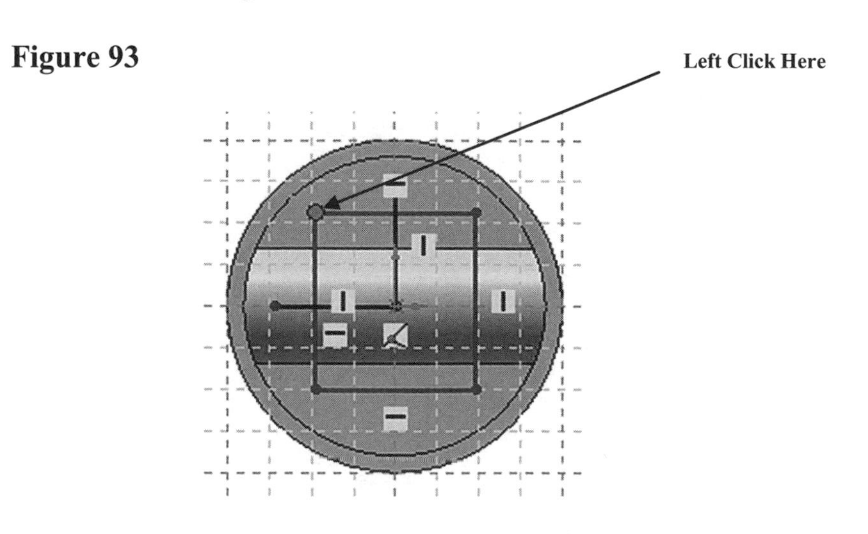

97. Move the cursor upward. The dimension of the line will appear as shown in Figure 94. The dimension is attached to the cursor.

Figure 94

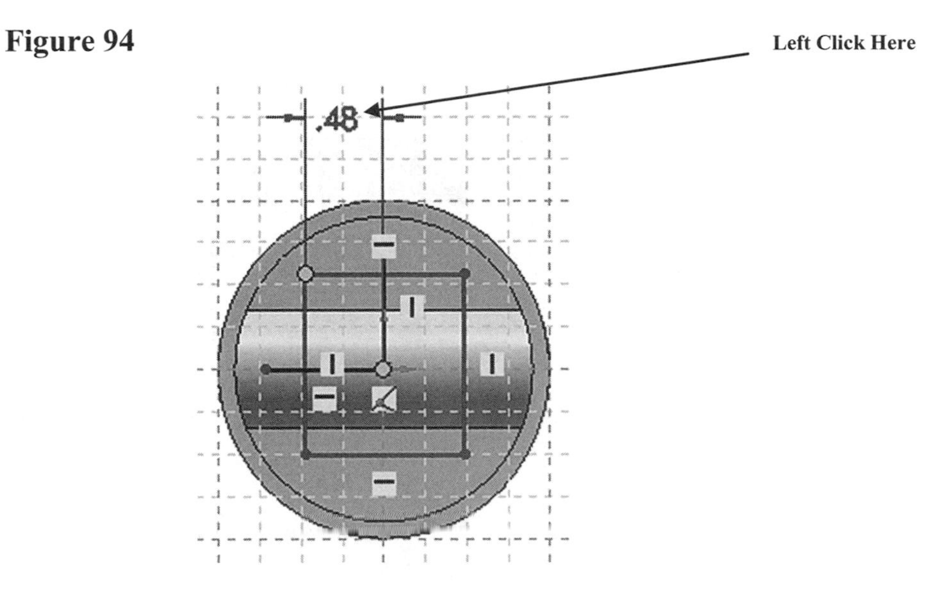

98. Move the cursor to where the dimension will be placed and left click once. The Modify dialog box will appear as shown in Figure 95.

Figure 95

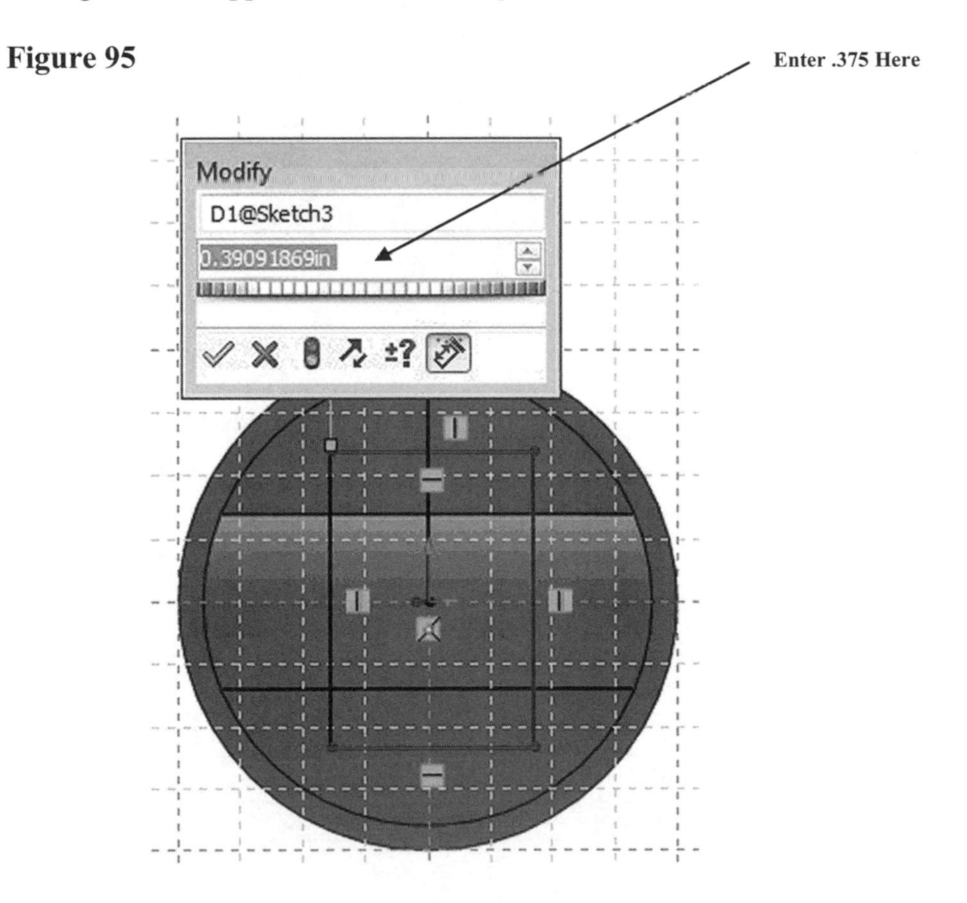

99. To edit the dimension, enter **.375** in the Modify dialog box (while the current dimension is highlighted). Left click on the green checkmark as shown in Figure 96.

Figure 96

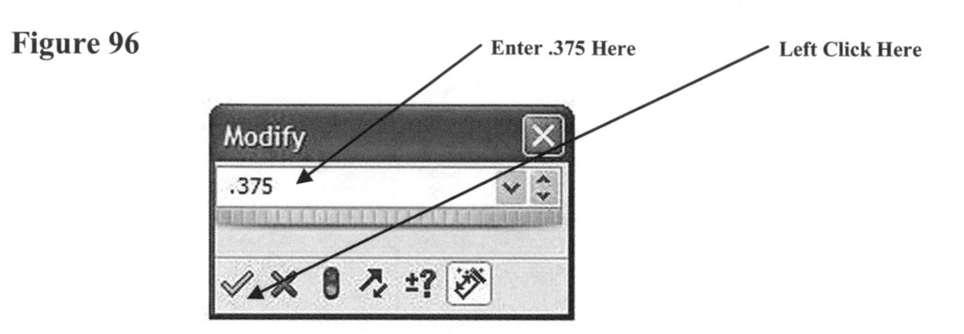

100. The dimension of the line will become .375 inches as shown in Figure 97.

Figure 97

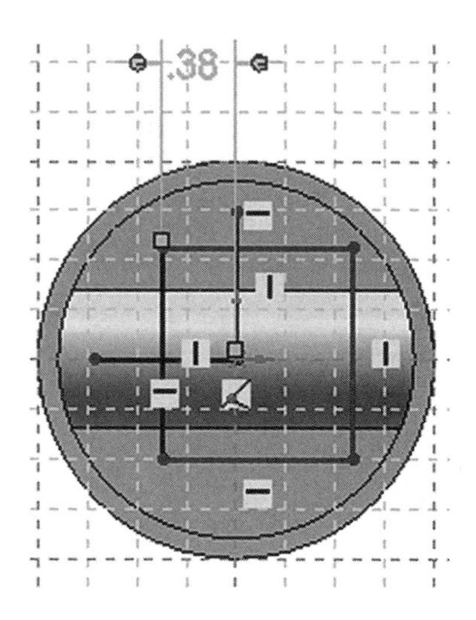

101. Move the cursor to the upper left portion of the screen and left click on **Smart Dimension** as shown in Figure 98.

Figure 98

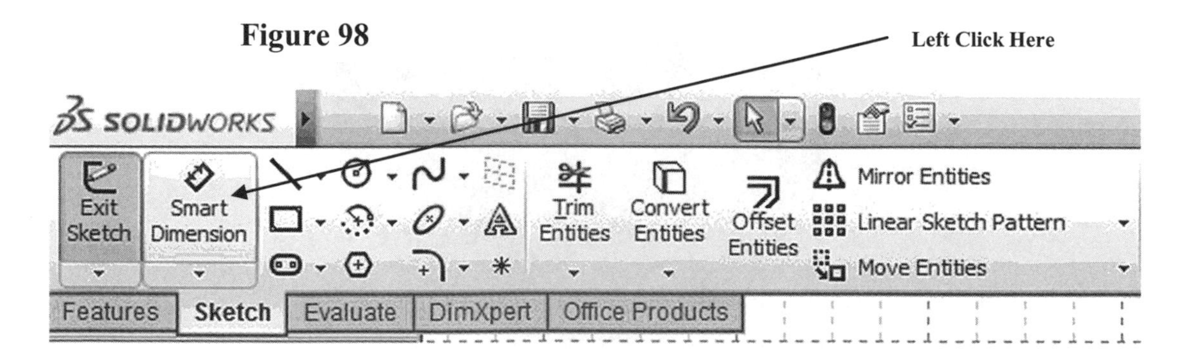

102. Move the cursor over the end point of the horizontal line coming out of the center. Left click on the endpoint as shown in Figure 99.

Figure 99

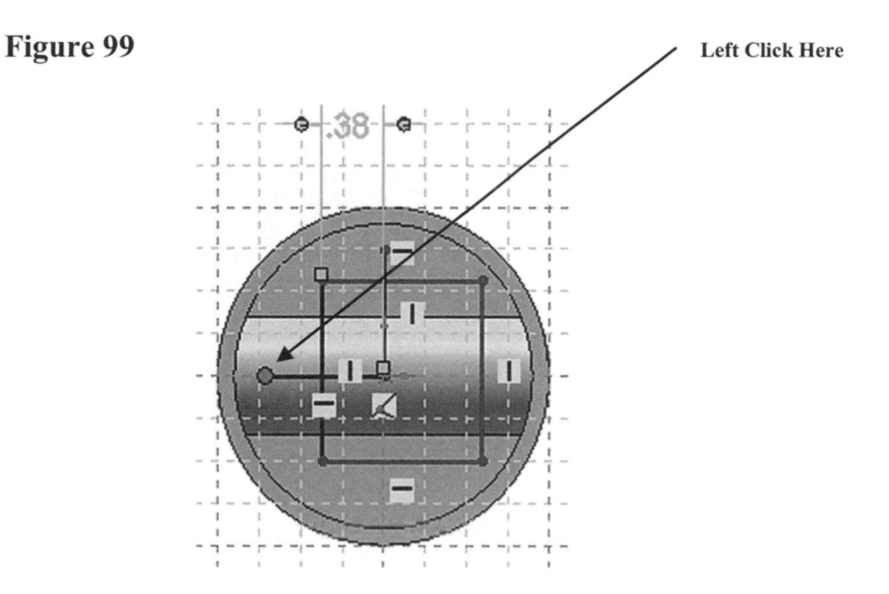

103. Move the cursor to the endpoint of the upper line and left click once as shown in Figure 100.

Figure 100

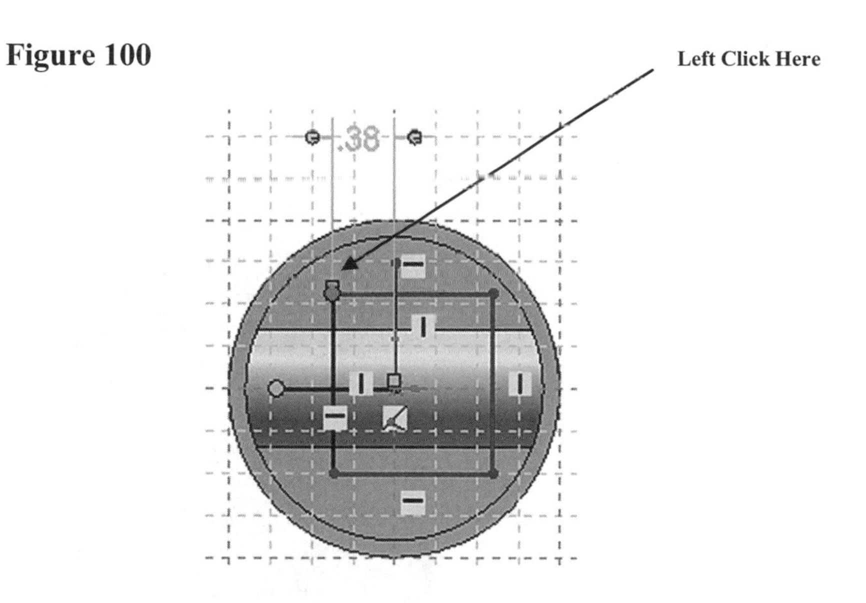

104. Move the cursor out to the side. The dimension of the line will appear as shown in Figure 101. The dimension is attached to the cursor.

Figure 101

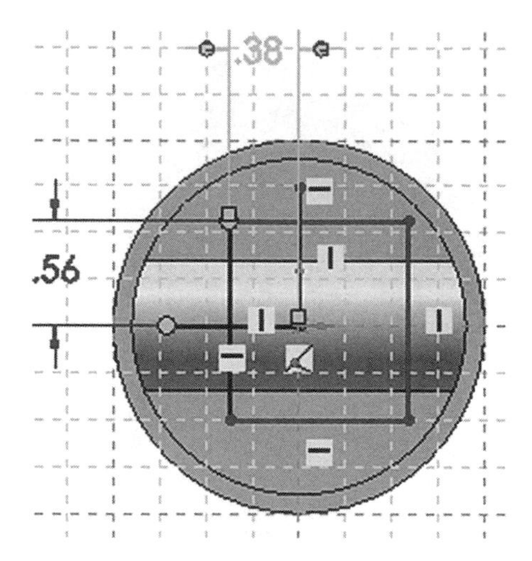

105. Move the cursor to where the dimension will be placed and left click once. The Modify dialog box will appear as shown in Figure 102.

Figure 102

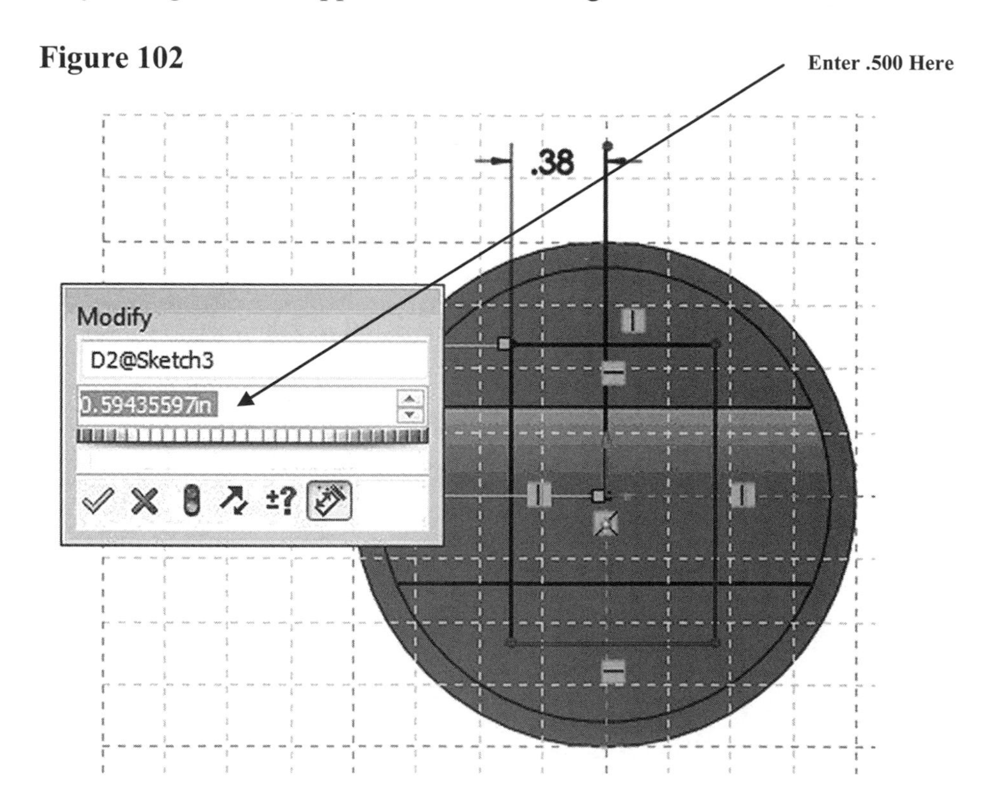

106. To edit the dimension, enter **.500** in the Modify dialog box (while the current dimension is highlighted). Left click on the green checkmark as shown in Figure 103.

Figure 103

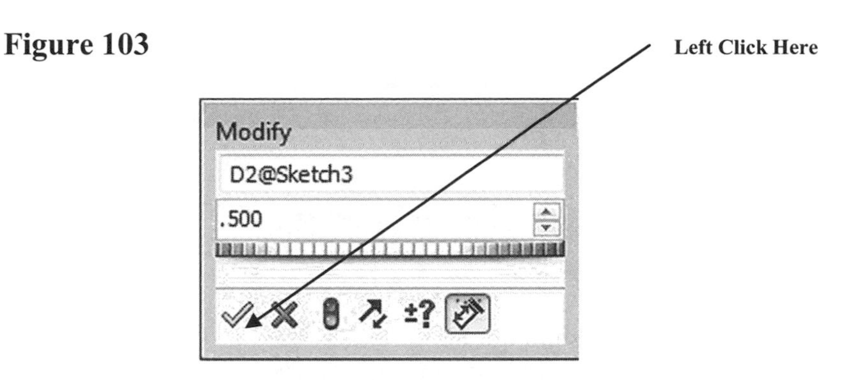

107. The dimension of the line will become .500 inches as shown in Figure 104.

Figure 104

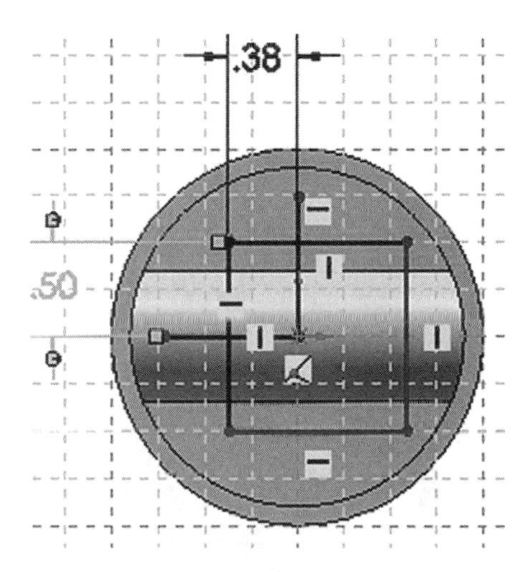

108. Move the cursor to the upper left portion of the screen and left click on **Smart Dimension** as shown in Figure 105.

Figure 105

109. Move the cursor over the endpoint of the horizontal line coming out of the center of the part. Left click on the endpoint as shown in Figure 106.

Figure 106

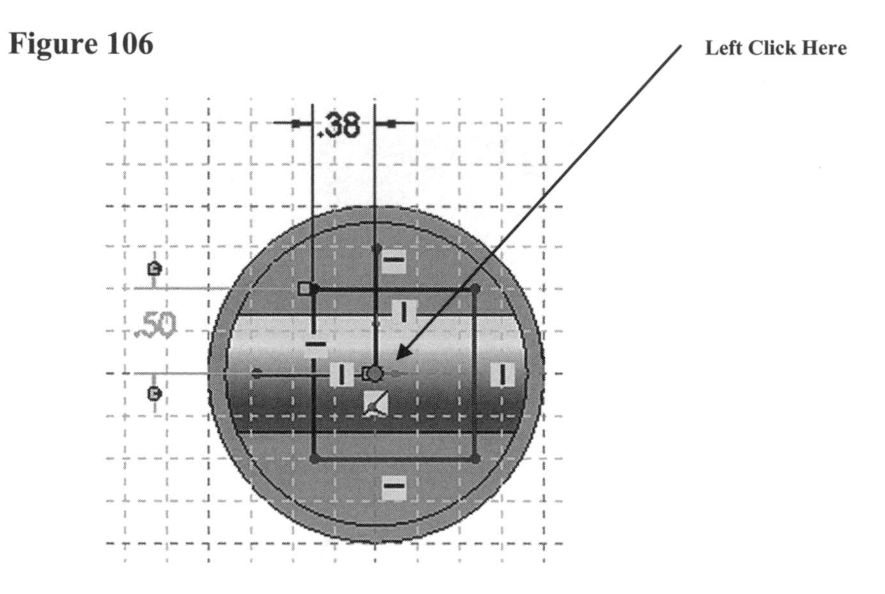

110. Move the cursor to the endpoint of the lower line and left click once as shown in Figure 107.

Figure 107

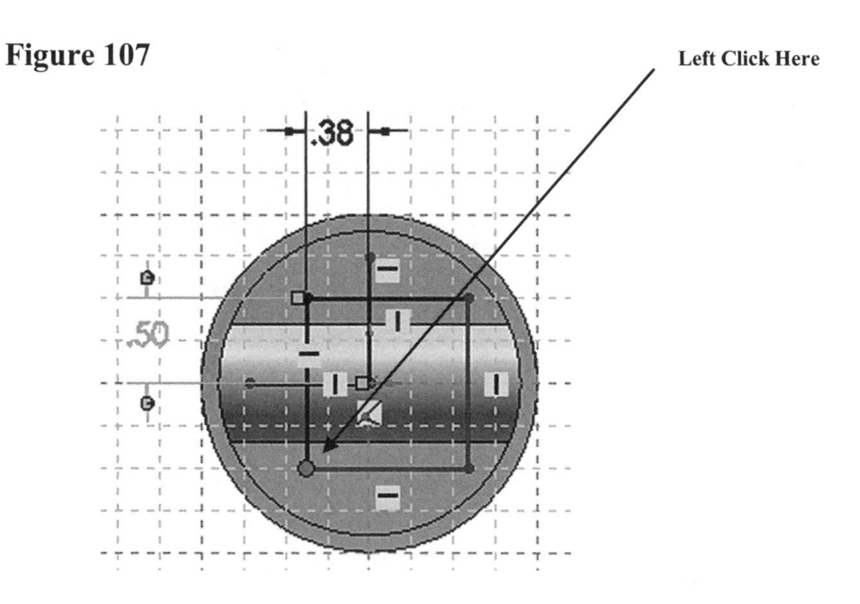

111. Move the cursor out to the side. The dimension of the line will appear as shown in Figure 108. The dimension is attached to the cursor.

Figure 108

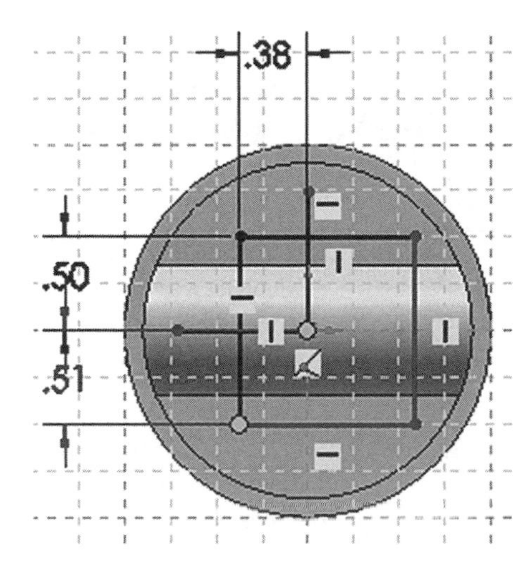

112. Move the cursor to where the dimension will be placed and left click once. The Modify dialog box will appear as shown in Figure 109.

Figure 109

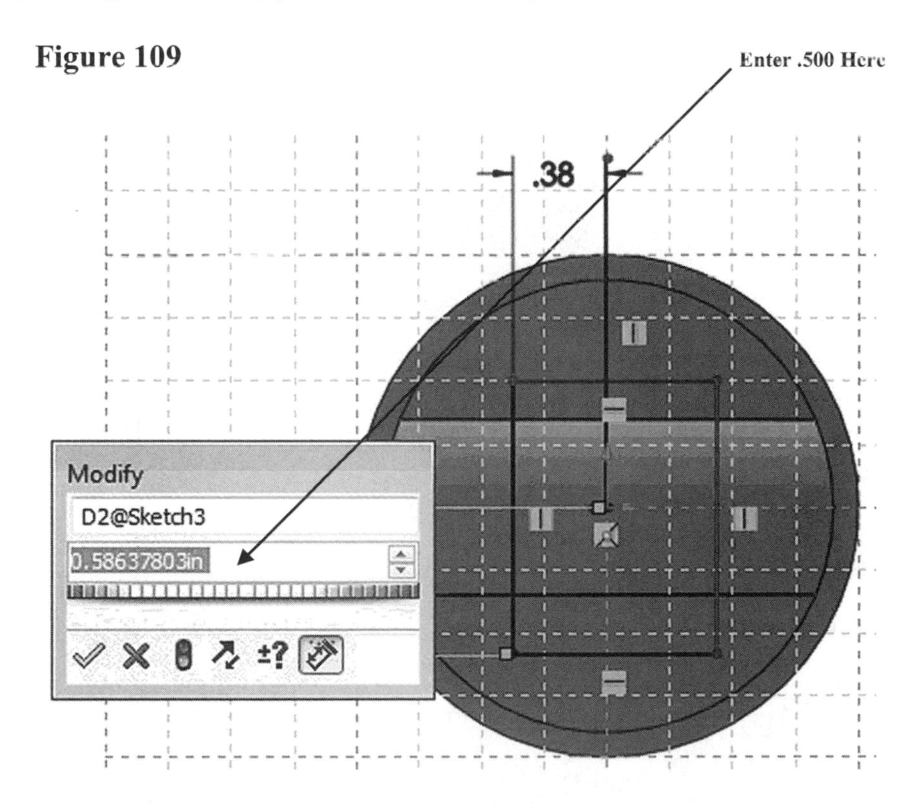

113. To edit the dimension, enter **.500** in the Modify dialog box (while the current dimension is highlighted). Left click on the green checkmark as shown in Figure 110.

Figure 110

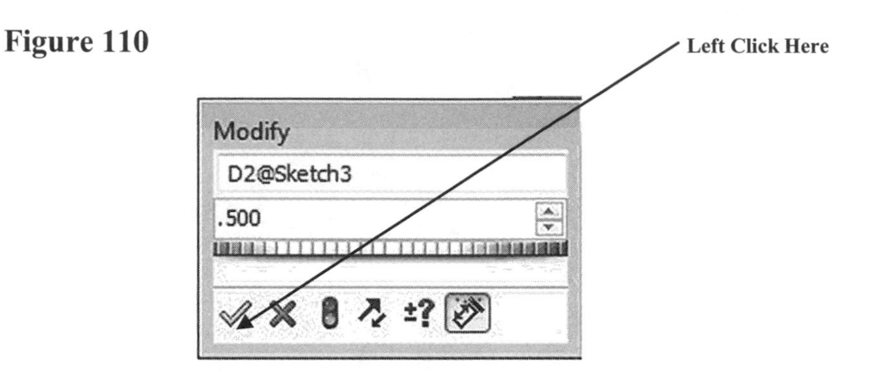

114. The dimension of the line will become .500 inches as shown in Figure 111.

Figure 111

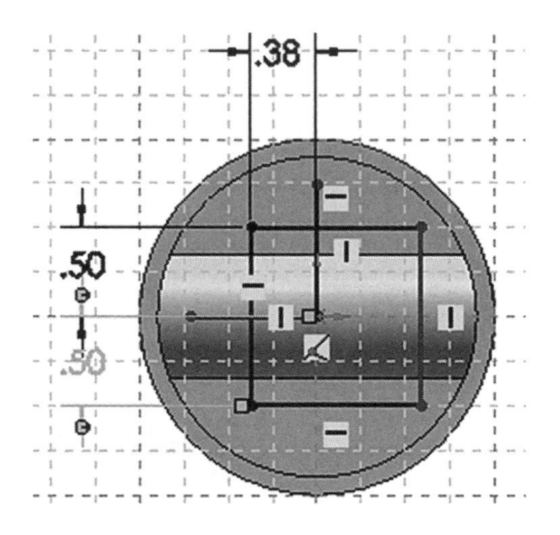

115. Move the cursor to the upper left portion of the screen and left click on **Smart Dimension** as shown in Figure 112.

Figure 112

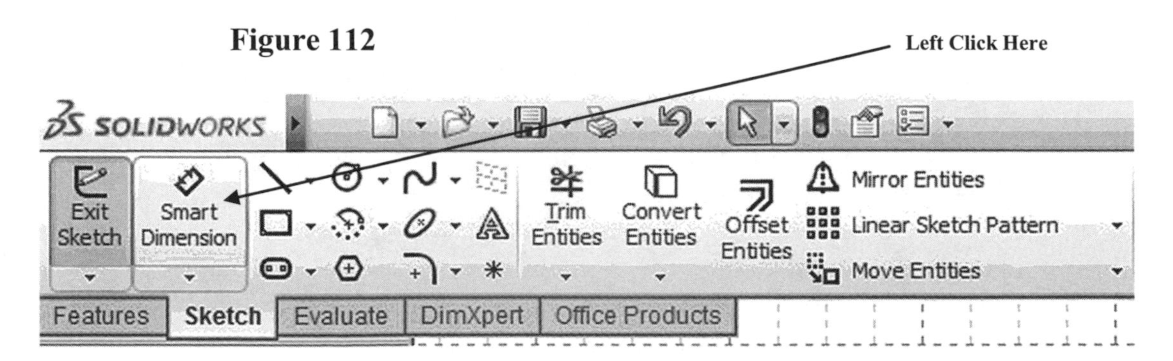

116. After selecting **Smart Dimension** move the cursor to the upper endpoint of the vertical line coming out of the center of the part. Left click as shown in Figure 113.

Figure 113

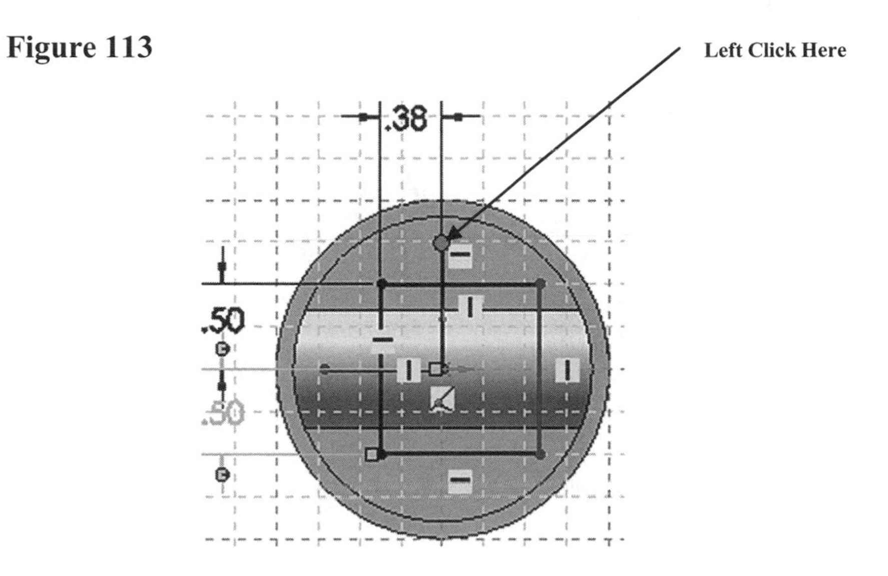

117. Move the cursor to the endpoint of the right vertical line and left click once as shown in Figure 114.

Figure 114

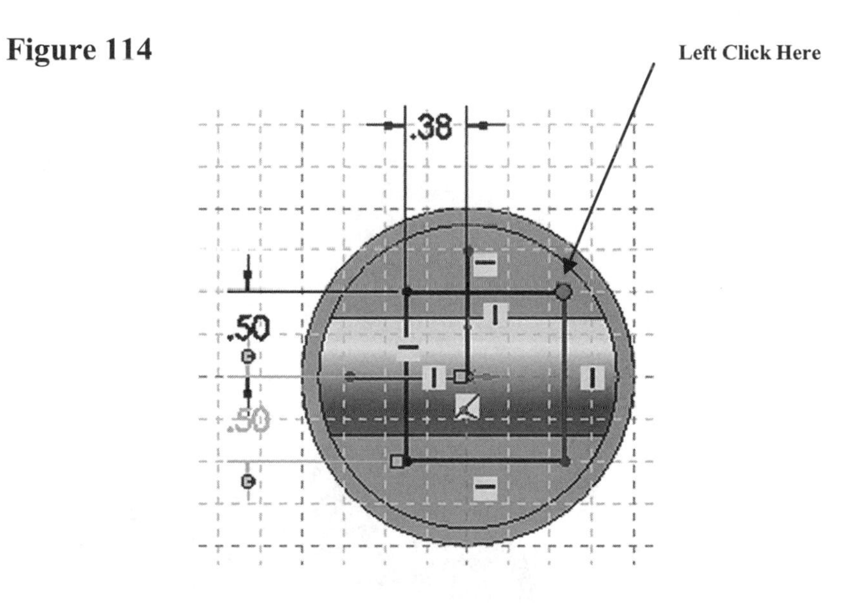

118. Move the cursor out to the side. The dimension of the line will appear as shown in Figure 115. The dimension is attached to the cursor.

Figure 115

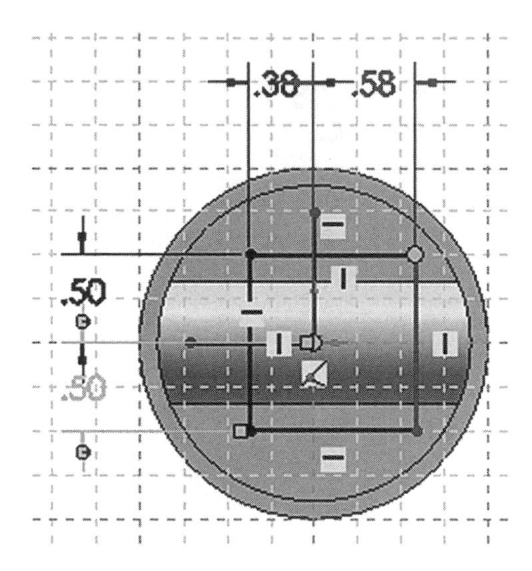

119. Move the cursor to where the dimension will be placed and left click once. The Modify dialog box will appear as shown in Figure 116.

Figure 116

Enter .375 Here

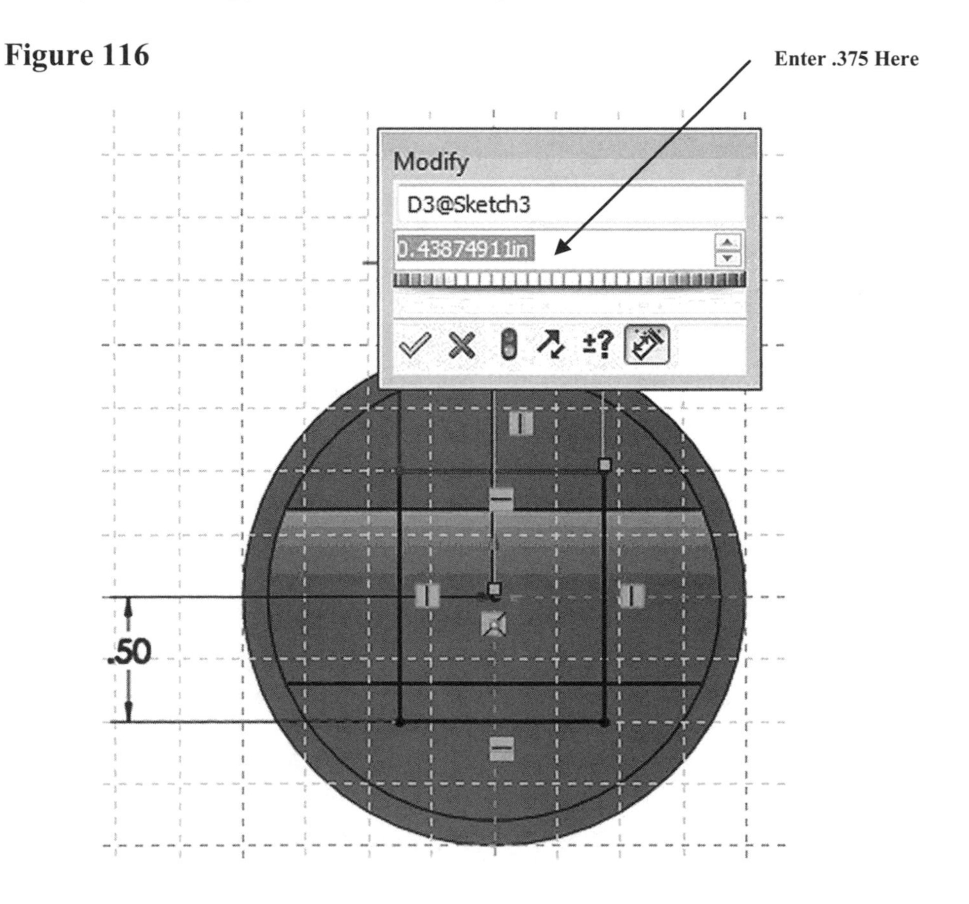

120. To edit the dimension, enter **.375** in the Modify dialog box (while the current dimension is highlighted) and left click on the green checkmark as shown in Figure 117.

Figure 117

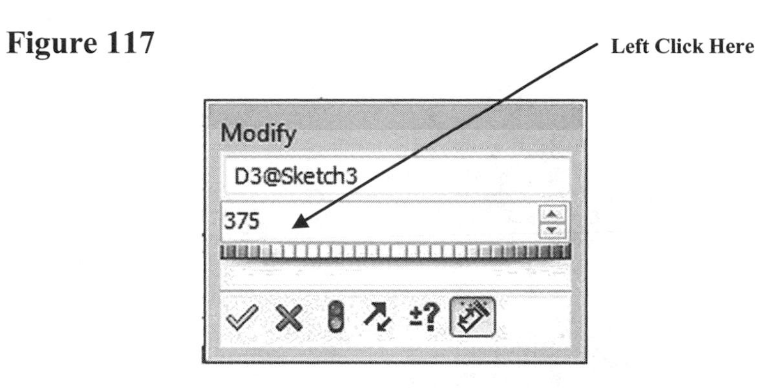

121. The dimension of the line will become .375 inches as shown in Figure 118.

Figurc 118

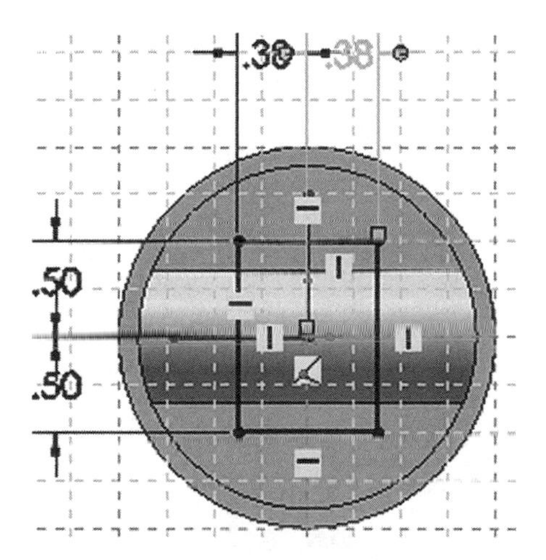

122. Right click anywhere around the drawing. A pop up menu will appear. Left click on **Select** as shown in Figure 119.

Figure 119

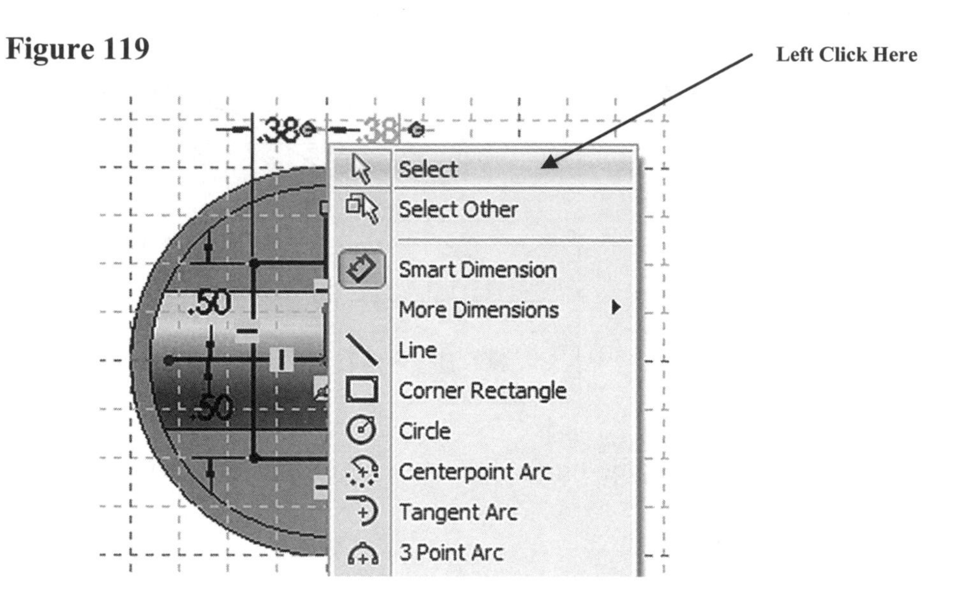

123. Move the cursor over the vertical line coming out of the center of the part causing it to turn red as shown in Figure 120.

Figure 120

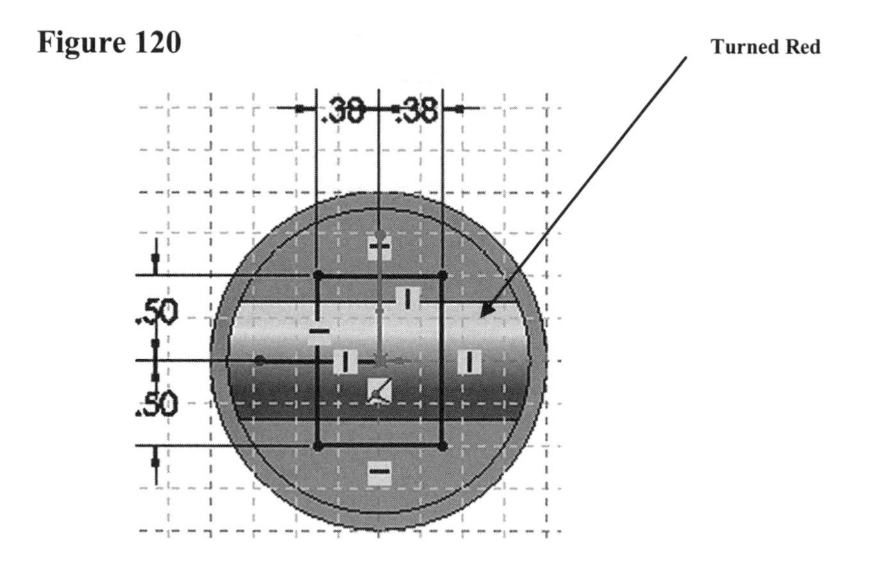

124. After the line has turned red, right click once. A pop up menu will appear. Left click on **Delete** as shown in Figure 121.

Figure 121

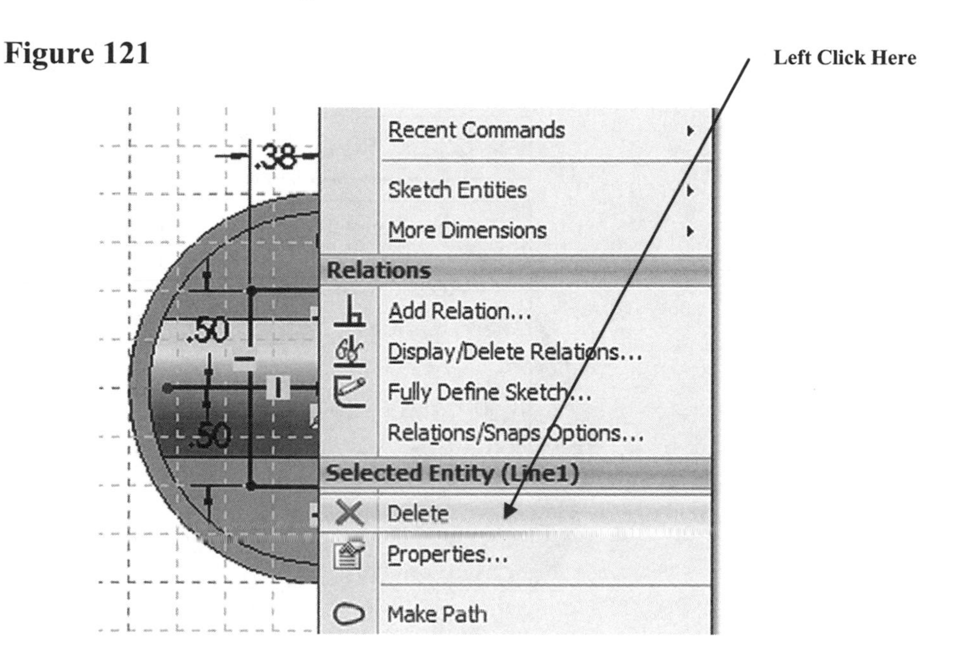

125. Use the same steps to delete the horizontal line coming out of the center of the part as shown in Figure 122. If the dimensions disappear do not be concerned.

Figure 122

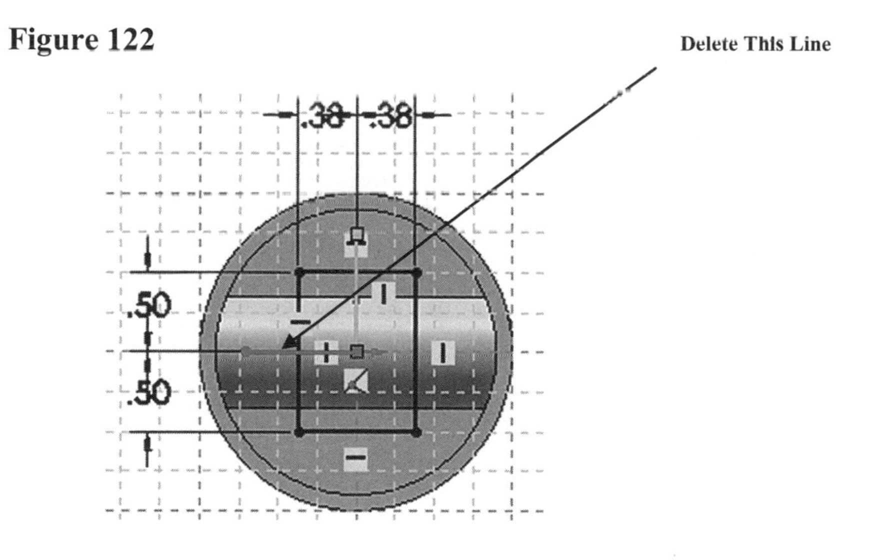

126. Right click anywhere around the drawing. A pop up menu will appear. Left click on **Exit Sketch** as shown in Figure 123.

Figure 123

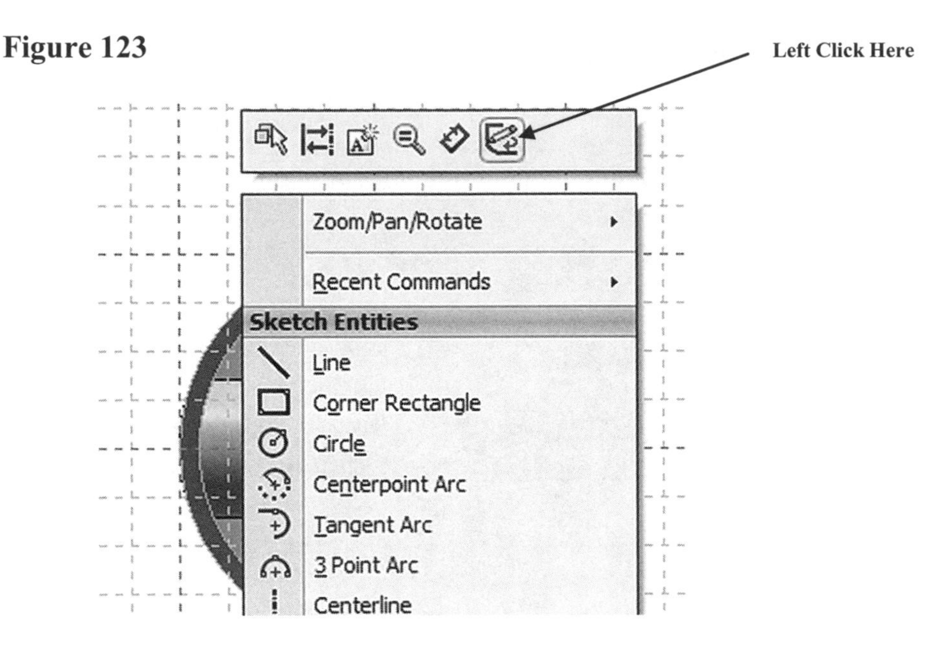

127. Move the cursor to the upper middle portion of the screen and left click on the drop down arrow next to the "View Orientation" icon. A drop down menu will appear. Left click on **Trimetric** as shown in Figure 124.

Figure 124

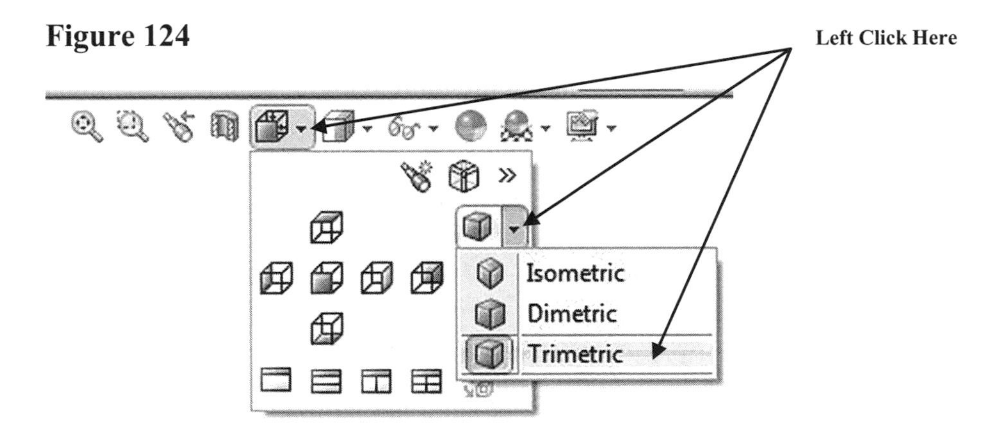

128. Your screen should look similar to Figure 125.

Figure 125

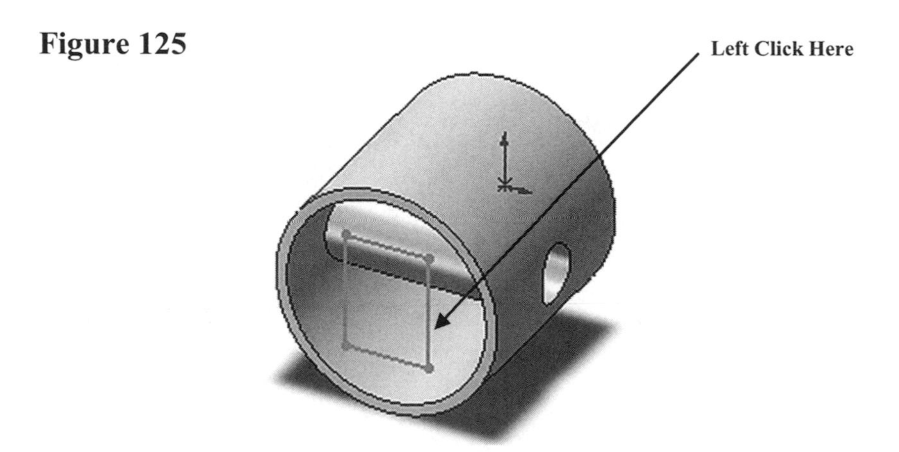

129. Move the cursor to the upper left portion of the screen and left click on **Extruded Cut** as shown in Figure 126. If SolidWorks does not provide a preview of the extruded cut, left click on any of the lines that form the box.

Figure 126

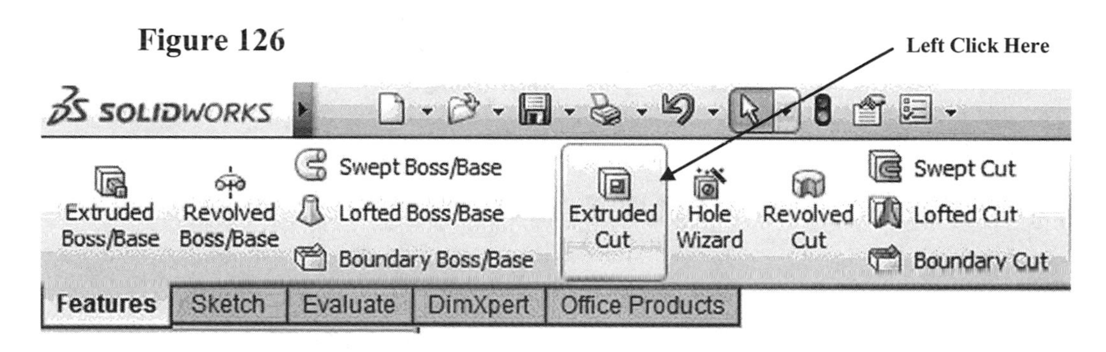

130. Enter **1.875** for D1 as shown in Figure 127.

Figure 127

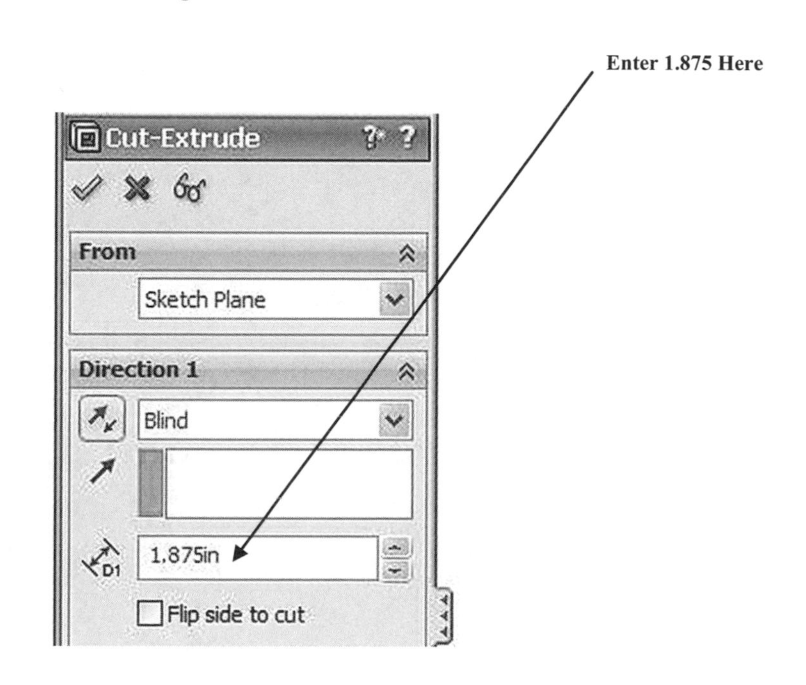

131. Left click on the green checkmark as shown in Figure 128.

Figure 128

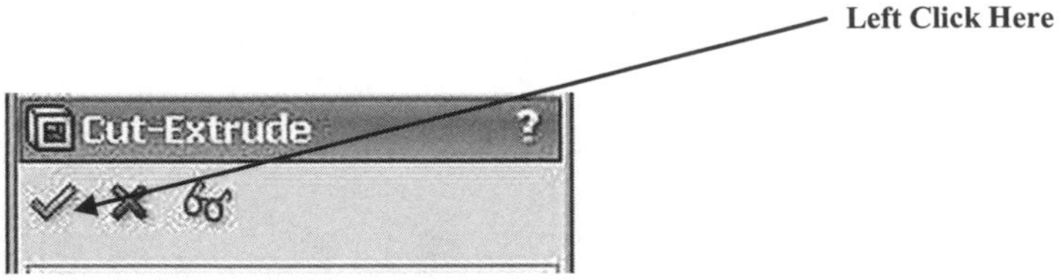

Left Click Here

Cut-Extrude

132. Your screen should look similar to Figure 129. Use the Rotate command to roll the part around to view the inside.

Figure 129

133. Save the part as Piston1.SLDPRT where it can be easily retired later.

134. Begin a new drawing as described in Chapter 1.

135. Draw a circle in the center of the grid as shown in Figure 130.

Figure 130

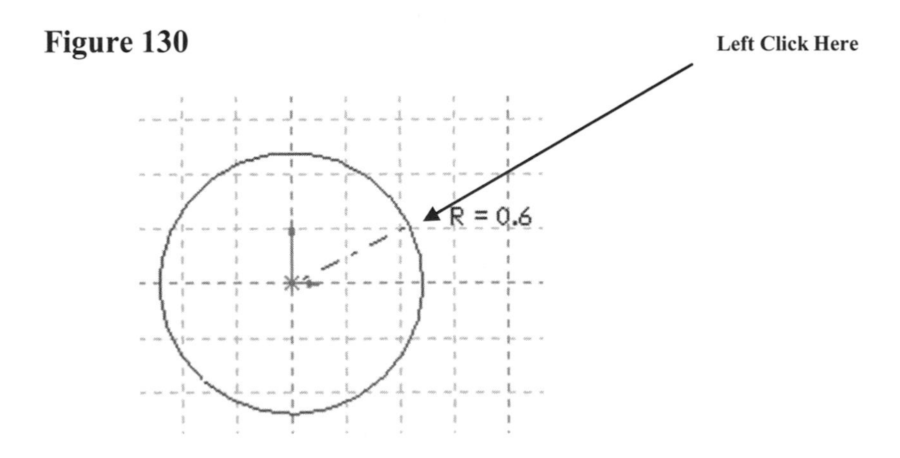

136. Use the **Smart Dimension** command to dimension the circle to **.500** inches as shown in Figure 131.

Figure 131

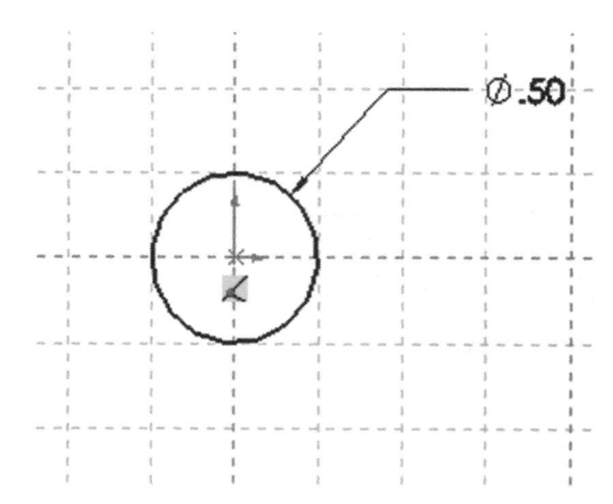

137. Exit the Sketch commands and Extrude the circle to a length of **1.875** inches as shown in Figure 132.

Figure 132

138. Your screen should look similar to Figure 133.

Figure 133

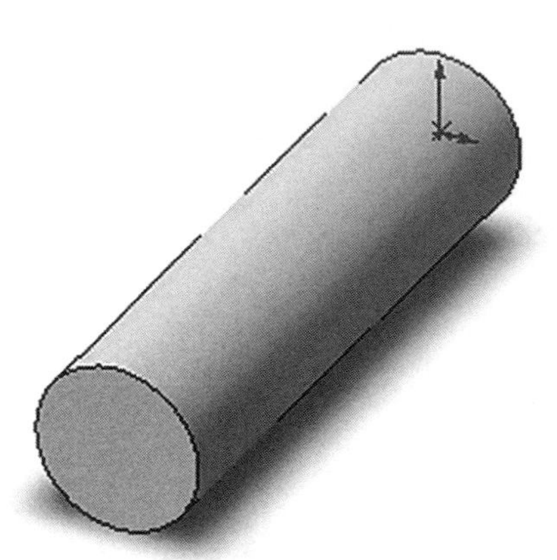

139. Save the part as Wristpin1.SLDPRT where it can be easily retrieved later.

140. Begin a new sketch as described in Chapter 1.

141. Complete the sketch shown in Figure 134.

Figure 134

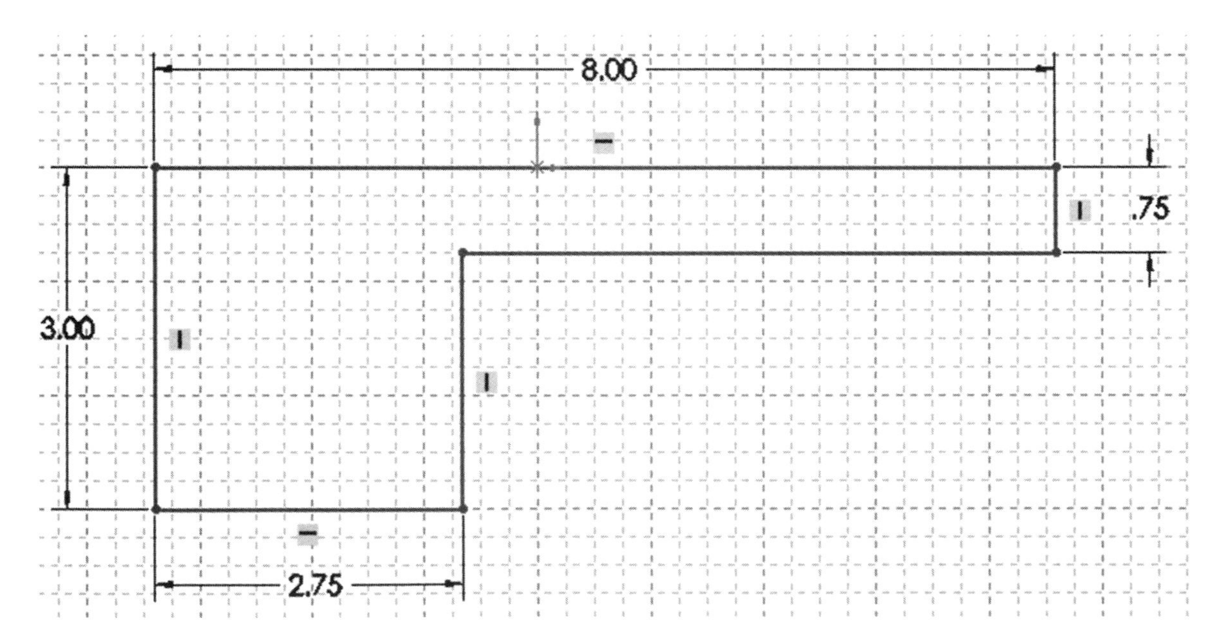

142. Exit the Sketch commands and change the view to Trimetric. A trimetric view will be displayed as shown in Figure 135.

Figure 135

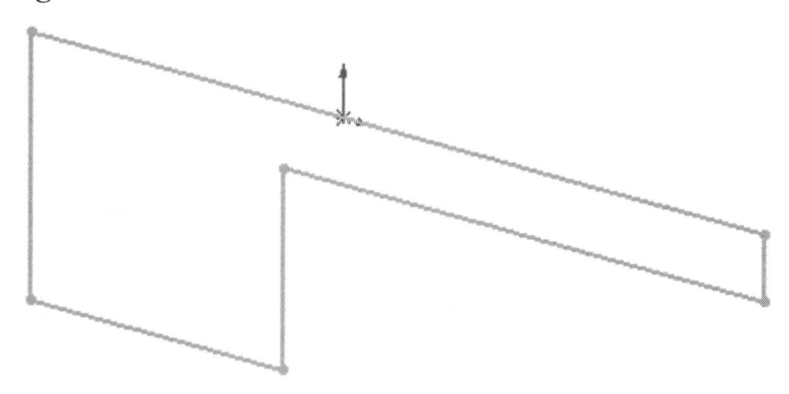

143. Extrude the sketch to a distance of **2.25** inches. Your screen should look similar to Figure 136.

Figure 136

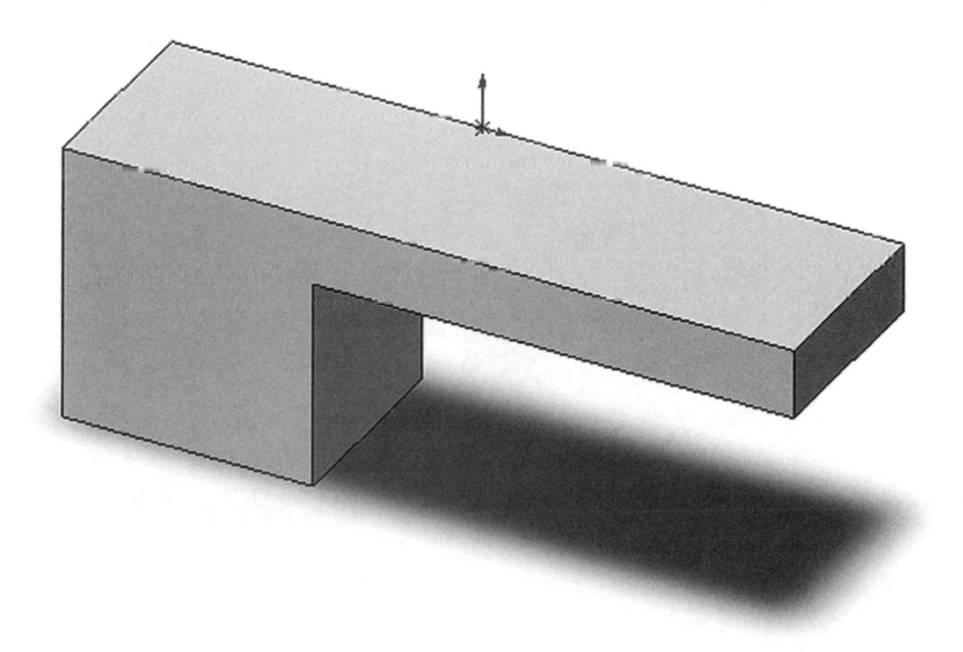

144. Use the **Fillet** command to create **1.125** inch fillets on the front portion of the part as shown in Figure 137.

Figure 137

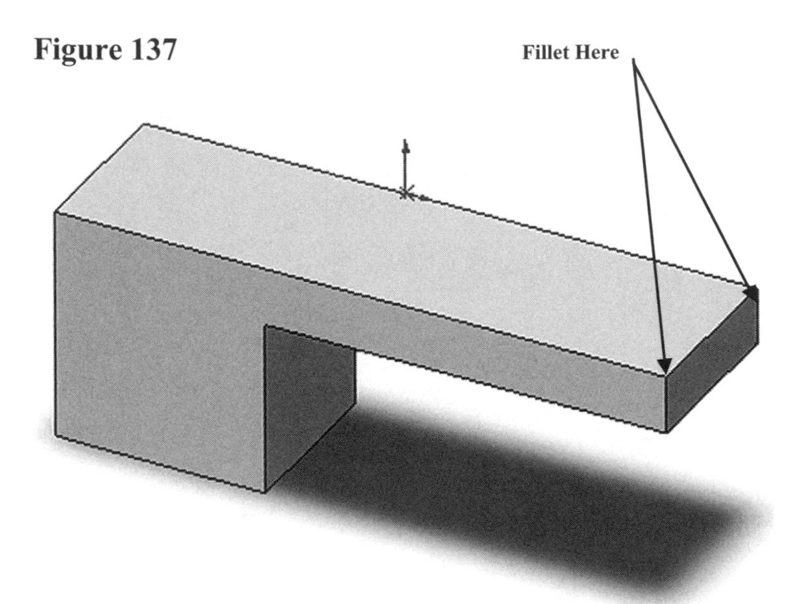

145. Your screen should look similar to Figure 138.

Figure 138

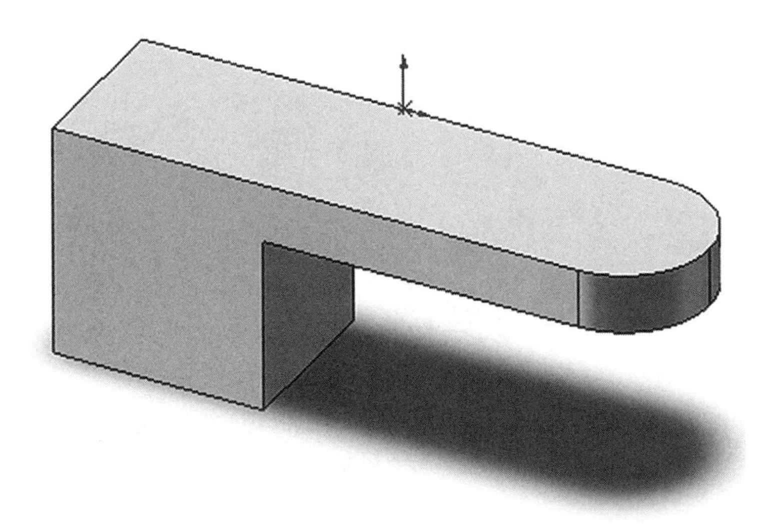

146. Move the cursor to the surface shown in Figure 139 causing it to turn red.

Figure 139

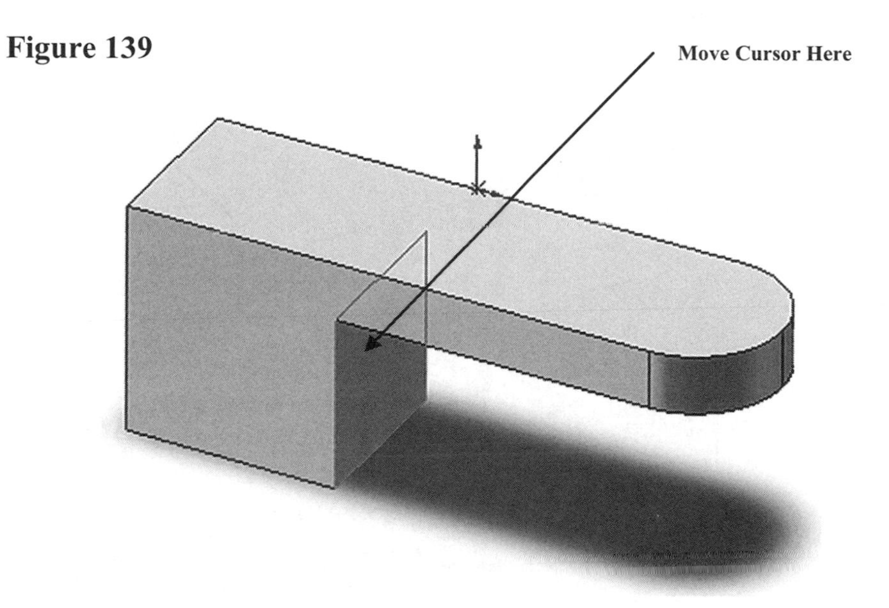

147. Left click once as shown in Figure 140. The surface will turn green.

Figure 140

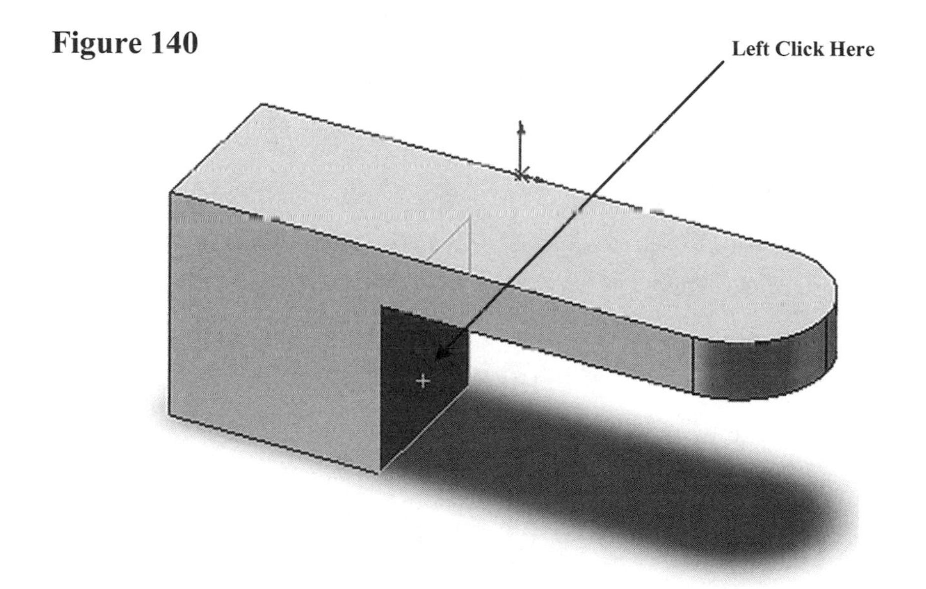

148. Right click on the surface. A pop up menu will appear. Left click on **Insert Sketch** as shown in Figure 141.

Figure 141

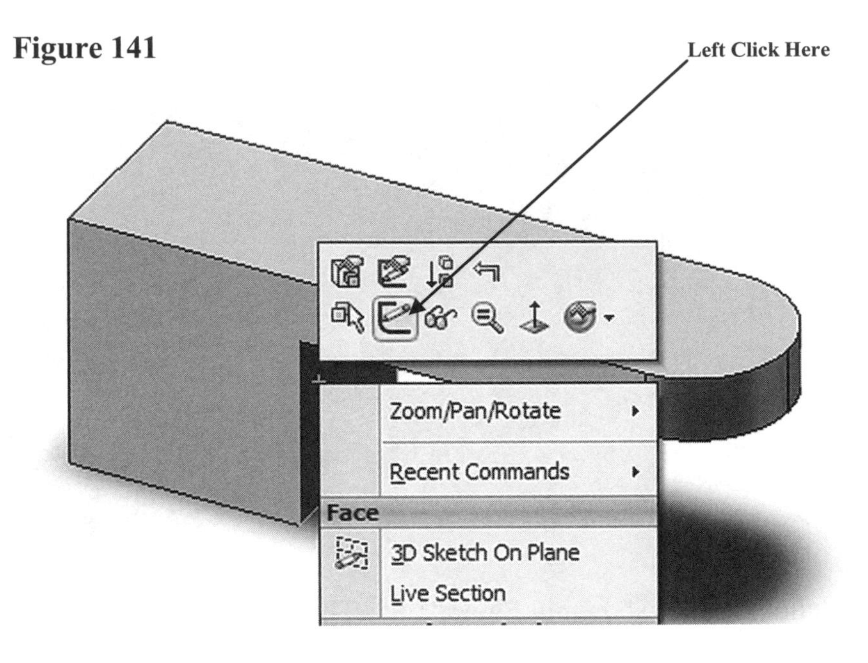

149. Move the cursor to the upper middle portion of the screen and left click on the drop down arrow next to the "Views Orientation" icon. A drop down menu will appear. Left click on **Normal To** as shown in Figure 142.

Figure 142

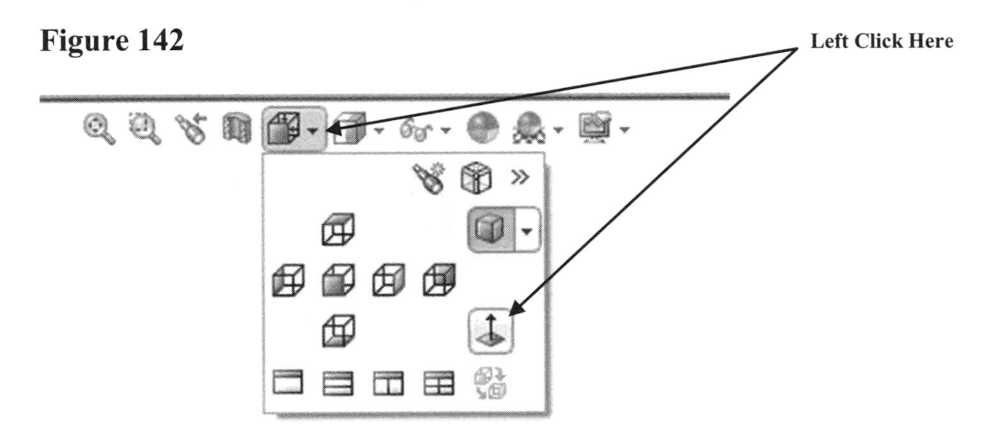

150. A perpendicular view of the surface will be displayed as shown in Figure 143.

Figure 143

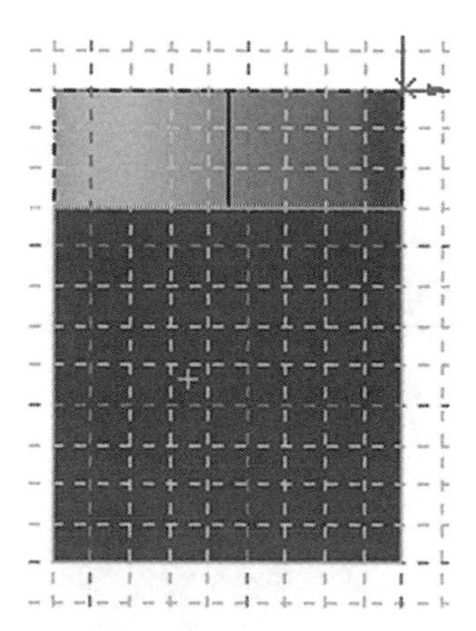

151. Create a sketch on the selected surface as shown in Figure 144.

Figure 144

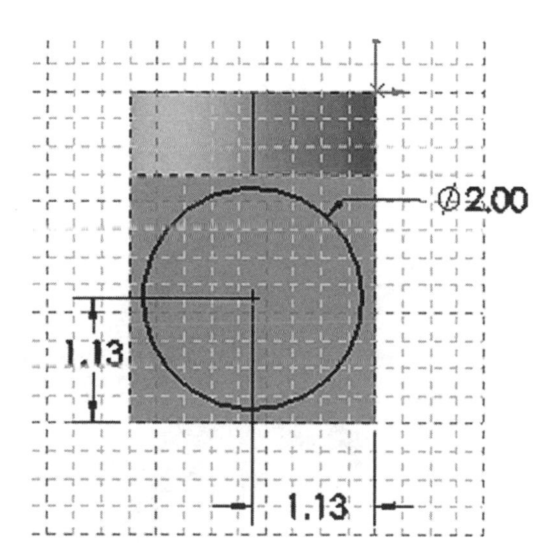

152. Exit out of the Sketch commands and change the view to trimetric as shown in Figure 145.

Figure 145

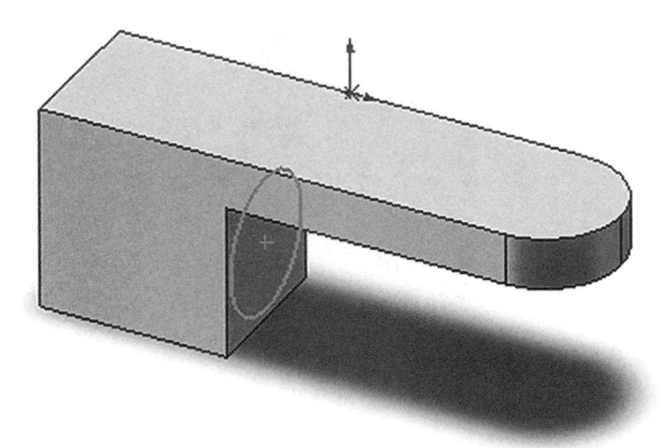

153. Use the **Extruded Cut** command to extrude or "cut" out (clear through the part) the circle that was just completed. Your screen should look similar to Figure 146.

Figure 146

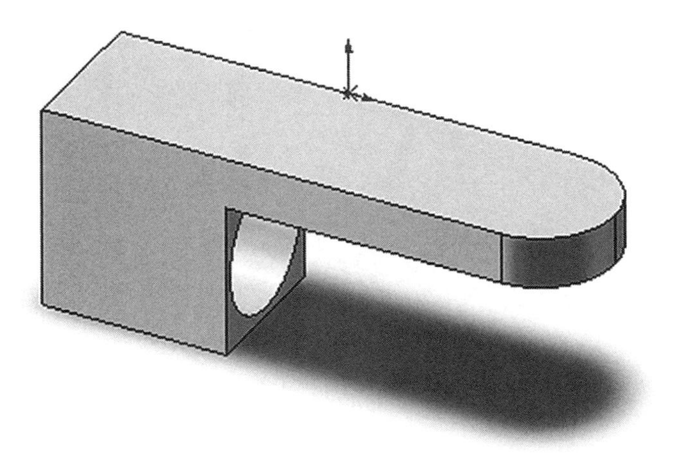

154. Left click on the top portion of the part causing it to turn green as shown in Figure 147.

Figure 147

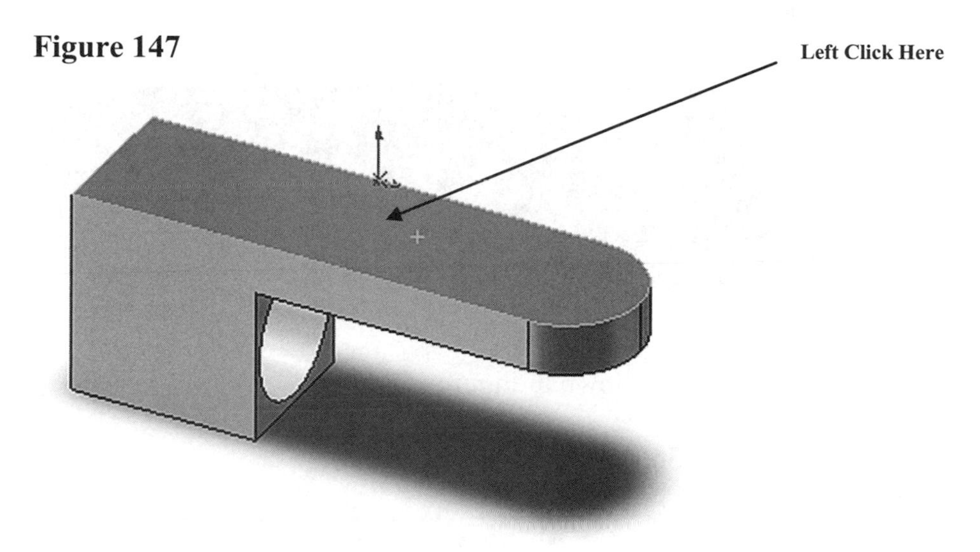

155. Move the cursor to the upper middle portion of the screen and left click on the drop down arrow next to the "View Orientation" icon. A drop down menu will appear. Left click on **Normal To** as shown in Figure 148.

Figure 148

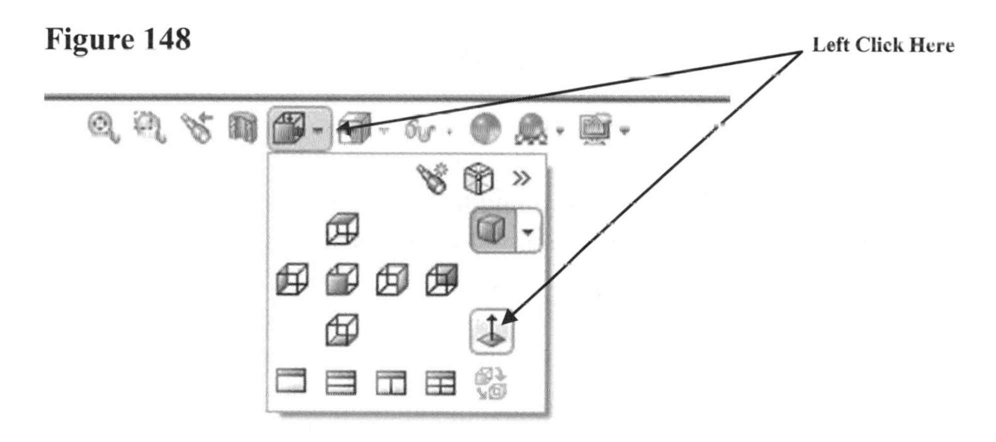

156. A perpendicular view will be displayed as shown in Figure 149.

Figure 149

157. Draw the sketch as shown in Figure 150.

Figure 150

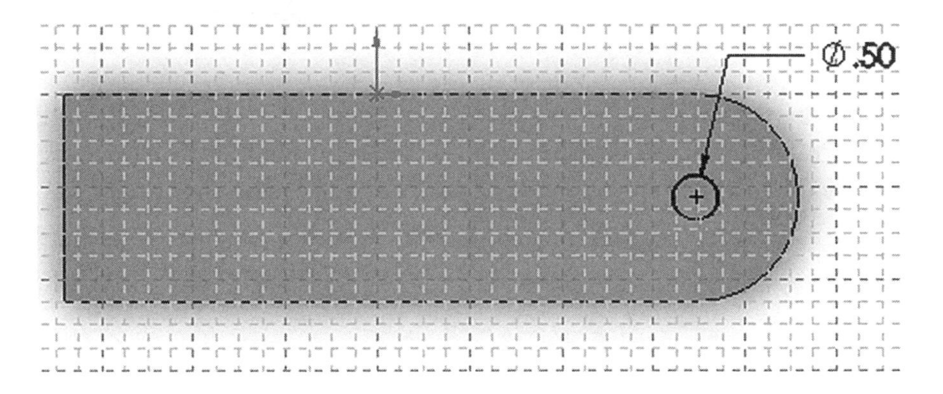

158. Use the **Extruded Cut** command to extrude or "cut" out the circle that was just completed. Your screen should look similar to Figure 151.

Figure 151

159. Change the view to trimetric. Your screen should look similar to Figure 152.

Figure 152

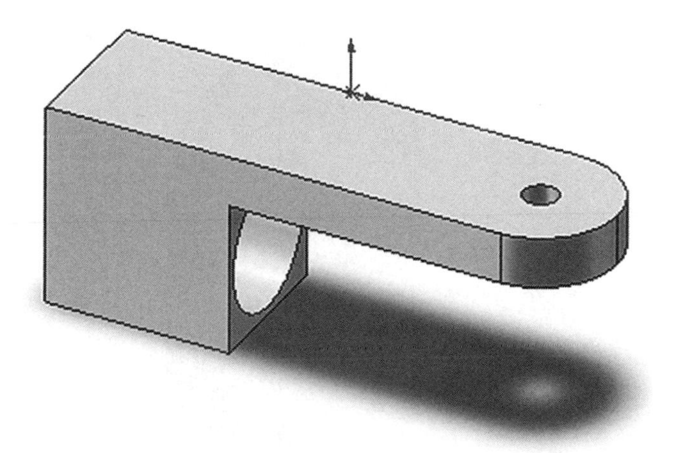

160. Save the part as Pistoncase1.SLDPRT where it can be easily retrieved later.

161. Begin a new drawing as described in Chapter 1.

162. Complete the sketch shown in Figure 153.

Figure 153

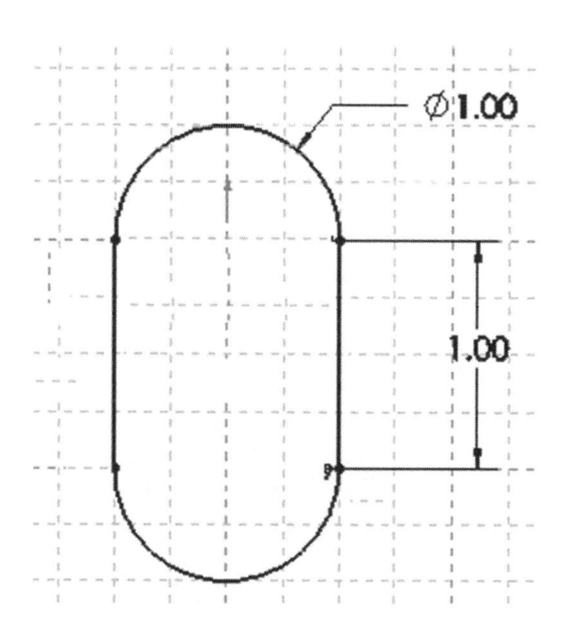

163. Extrude the sketch into a solid with a thickness of **.25** as shown in Figure 154.

Figure 154

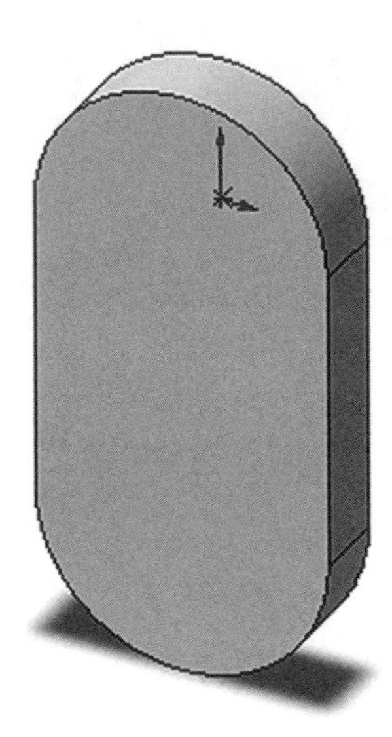

164. Complete the following sketch. Use the center of the outside fillet radius as the center of the circle as shown in Figure 155.

Figure 155

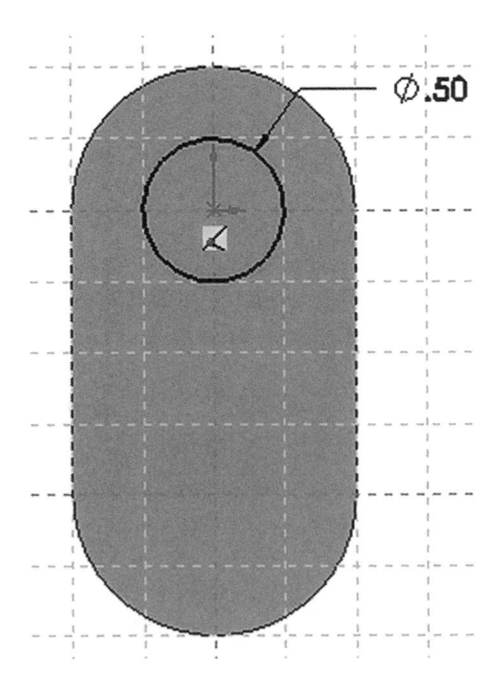

165. Extrude the sketch into a solid with a thickness of **.25** as shown in Figure 156.

Figure 156

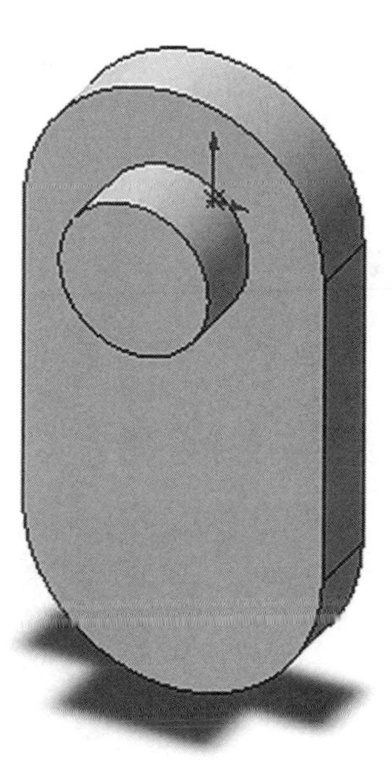

166. Use the rotate command and roll the part around to gain access to the opposite side as shown in Figure 157.

Figure 157

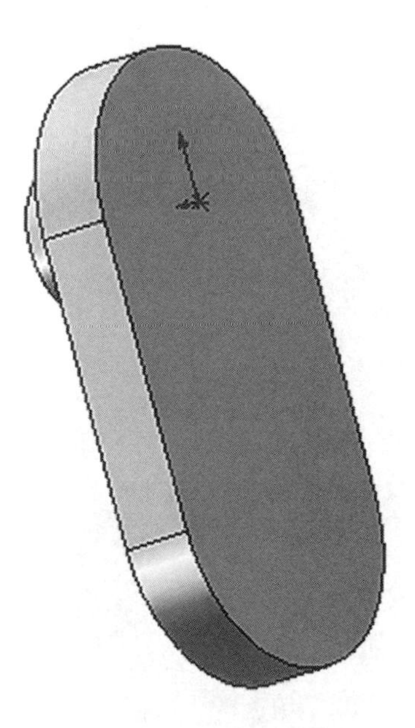

167. Begin a new sketch on the opposite side as shown in Figure 158.

Figure 158

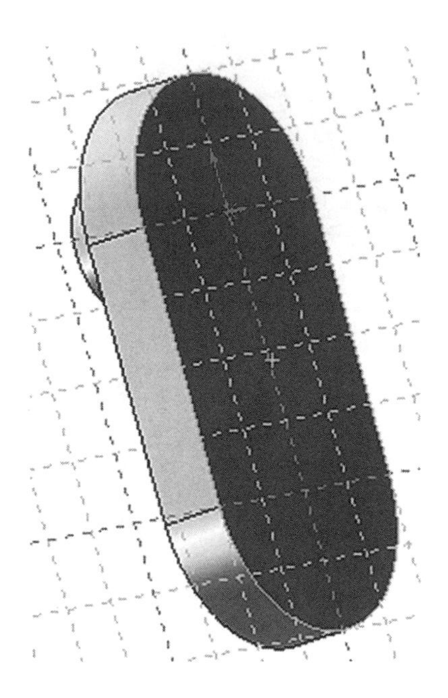

168. Use the **Normal To** command to gain a perpendicular view as shown in Figure 159.

Figure 159

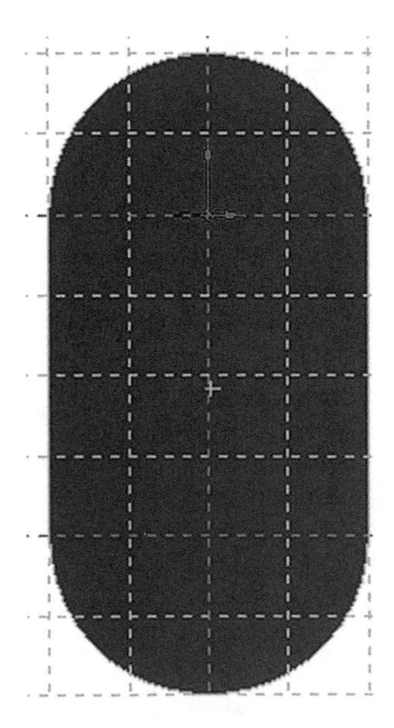

169. Complete the following sketch as shown in Figure 160.

Figure 160

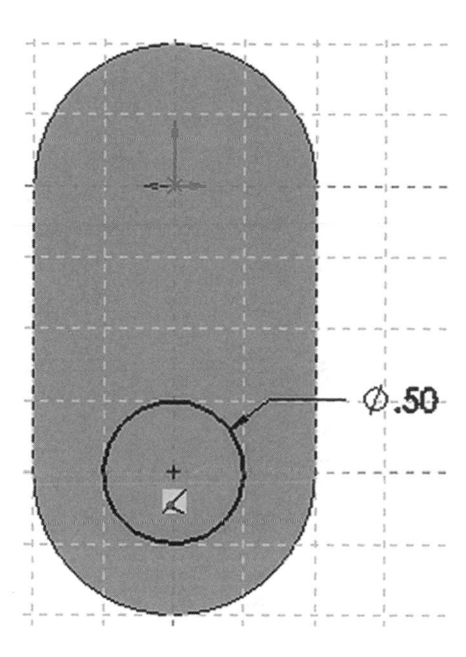

170. Extrude the sketch into a solid with a thickness of **.25** as shown in Figure 161.

Figure 161

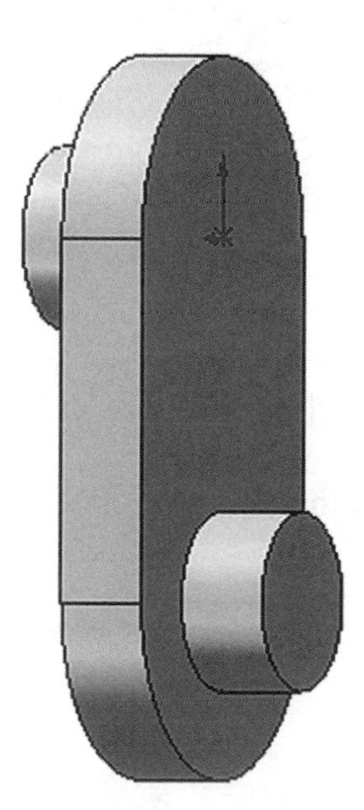

171. Save the part as Crankshaft1.SLDPRT where it can be easily retrieved later.

172. Begin a new drawing as described in Chapter 1.

173. Complete the sketch shown in Figure 162. Extrude the sketch into a solid with a thickness of **.25**.

Figure 162

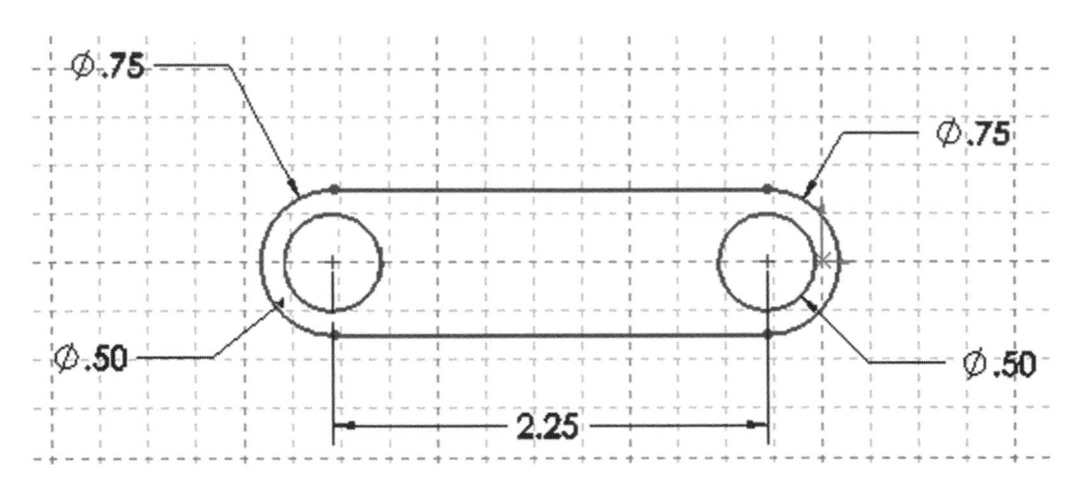

174. Your screen should look similar to Figure 163.

Figure 163

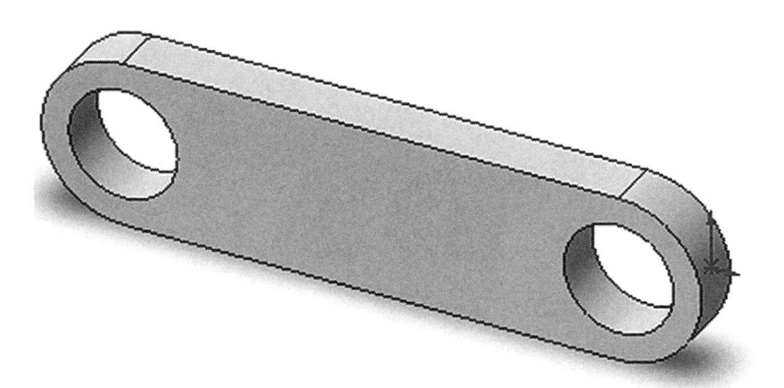

175. Save the part as Conrod1.SLDPRT where it can be easily retrieved later.

176. All of these parts will be used in the next chapter.

Chapter 7 Introduction to Assembly View Procedures

Objectives:

- Learn to import existing solid models into an Assembly
- Learn to constrain all parts in an Assembly
- Learn to edit/modify parts while in an Assembly
- Learn to use the Motion Study command in an Assembly

Chapter 7 includes instruction on how to construct the assembly shown below.

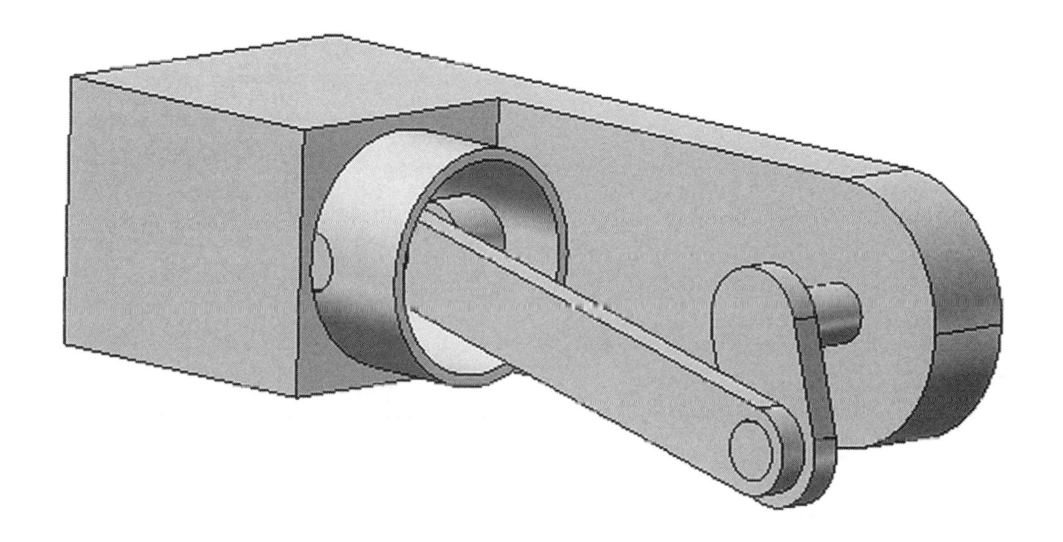

1. Start SolidWorks by referring to "Chapter 1 Getting Started".

2. After SolidWorks is running, begin an assembly drawing. Move the cursor to the upper left corner of the screen and left on **New**. The New SolidWorks Document dialog box will appear. Left click on **Assembly** as shown in Figure 1.

Figure 1

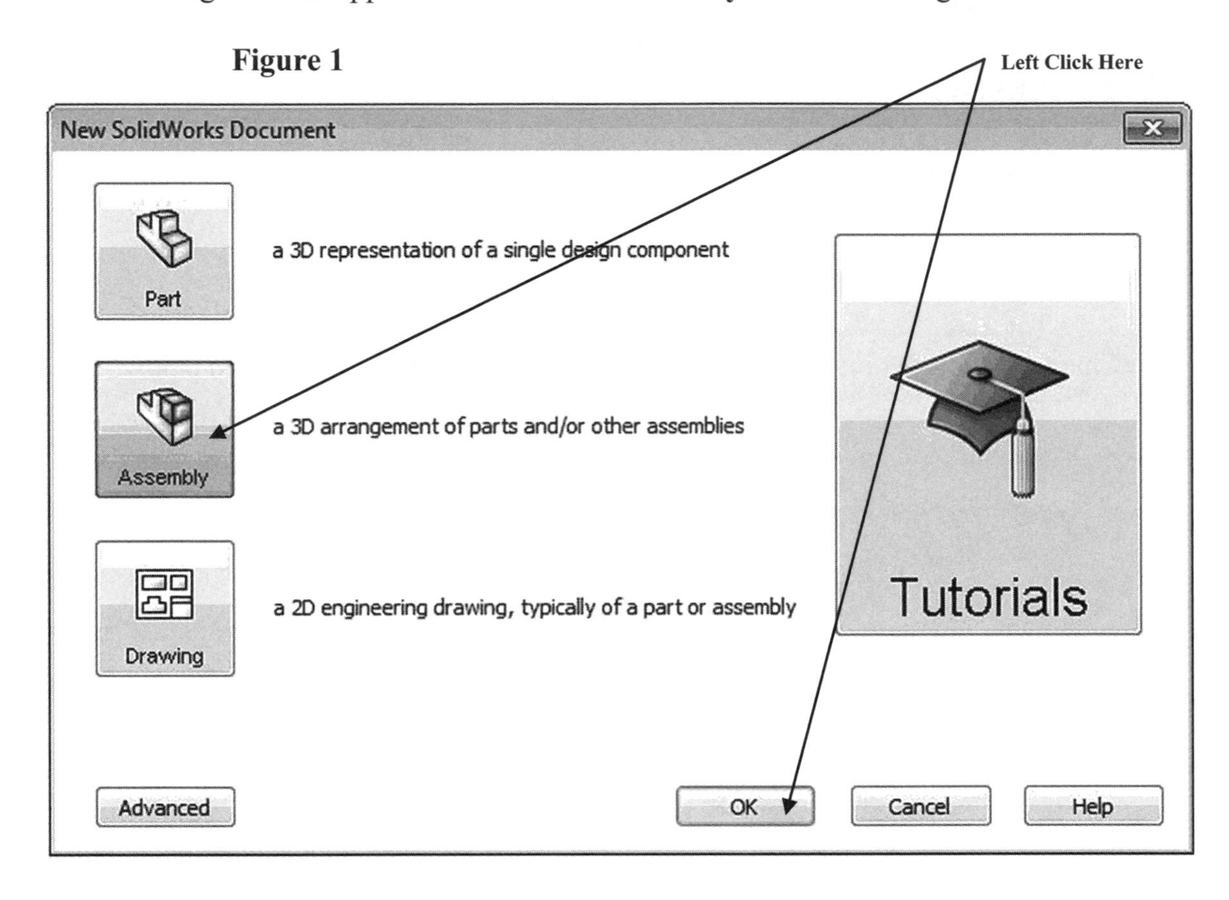

3. Left click on **OK**.

4. The Assembly feature of SolidWorks will open. Left click on the **Assembly** tab as shown. Your screen should look similar to Figure 2.

Figure 2

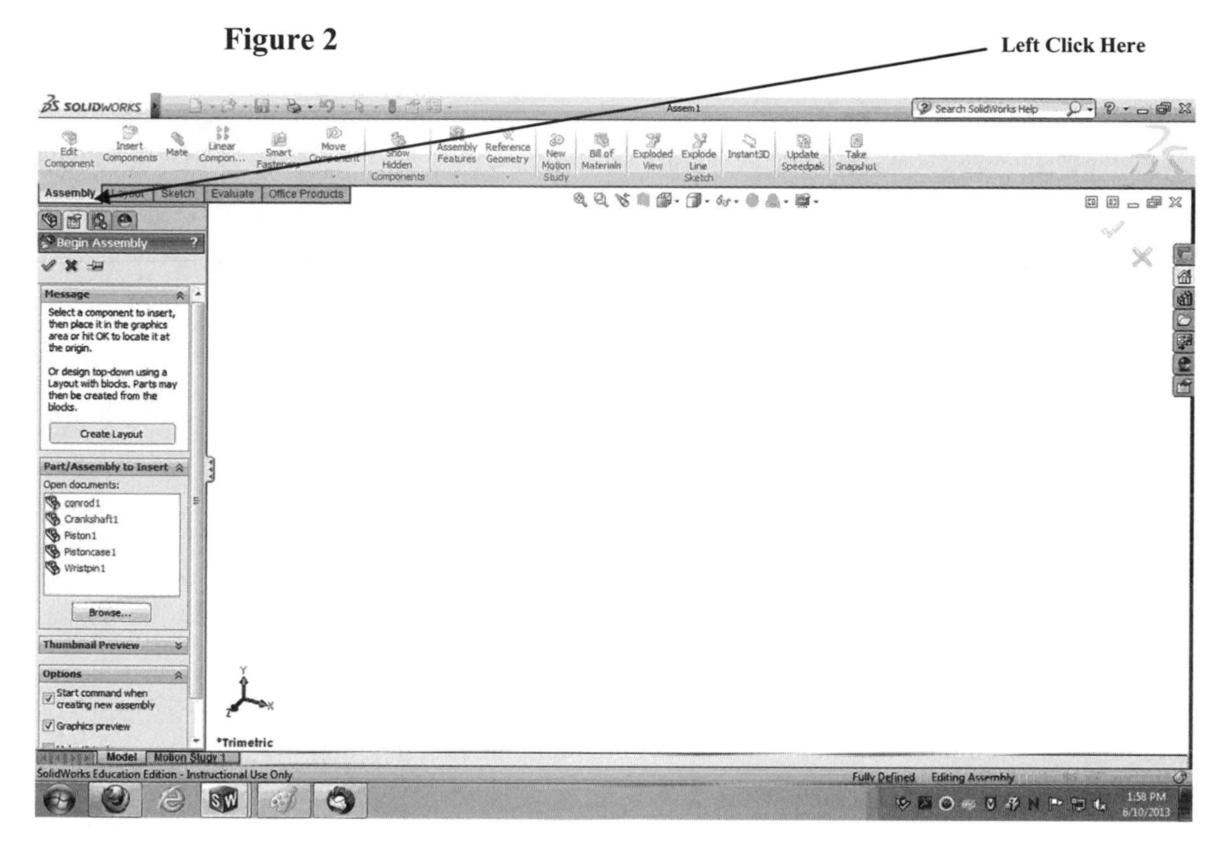

5. Move the cursor to the lower left portion of the screen and left click **Browse** as shown in Figure 3.

Figure 3

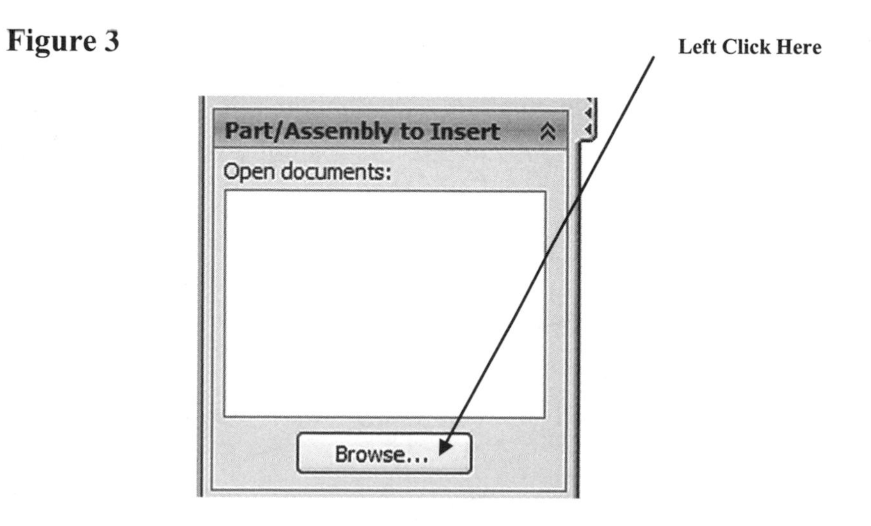

6. The Open dialog box will appear. Locate the Pistoncase1.SLDPART file and left click on **Open** as shown in Figure 4.

Figure 4

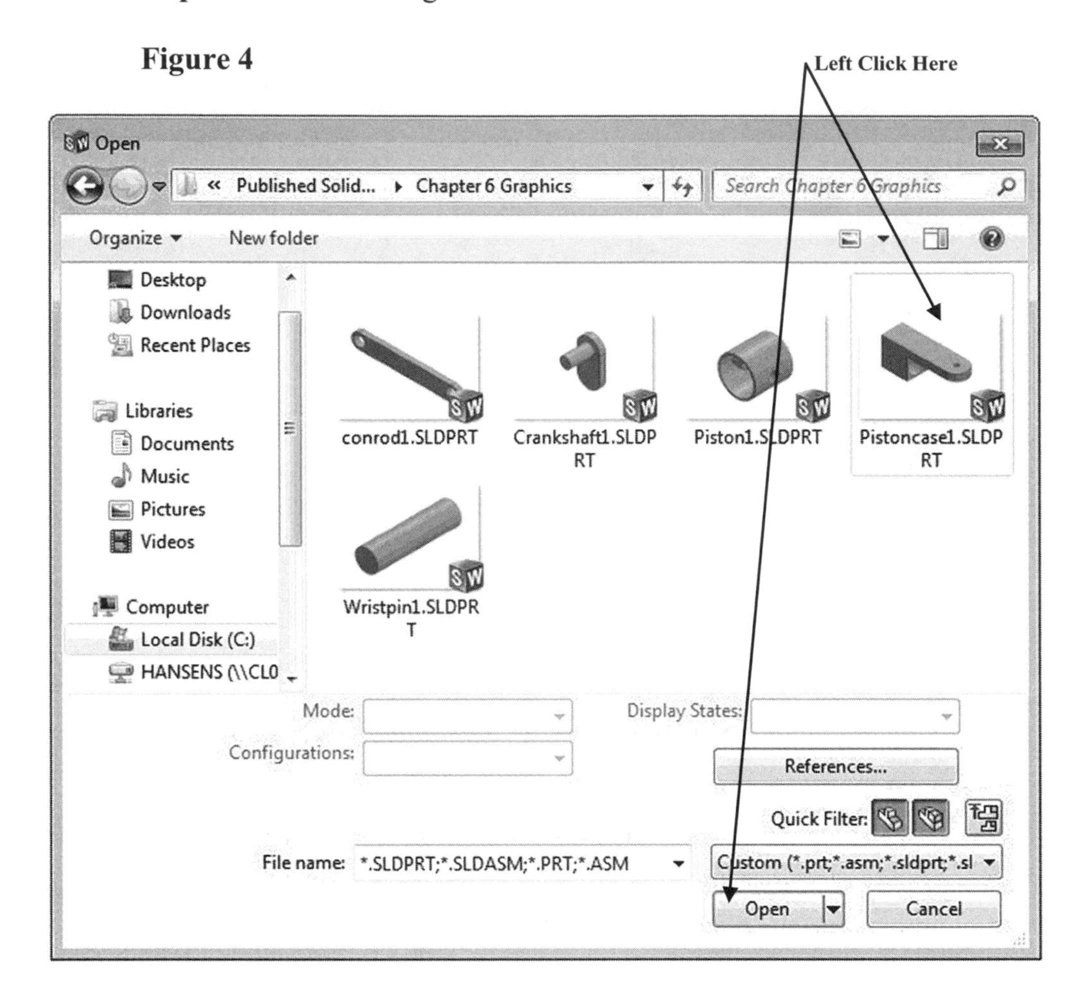

7. SolidWorks will place one piston case in the drawing space. The piston case will be attached to the cursor. The piston case will appear as shown in Figure 5. If the part does not appear attached to the cursor simply left click anywhere in the work area.

Figure 5

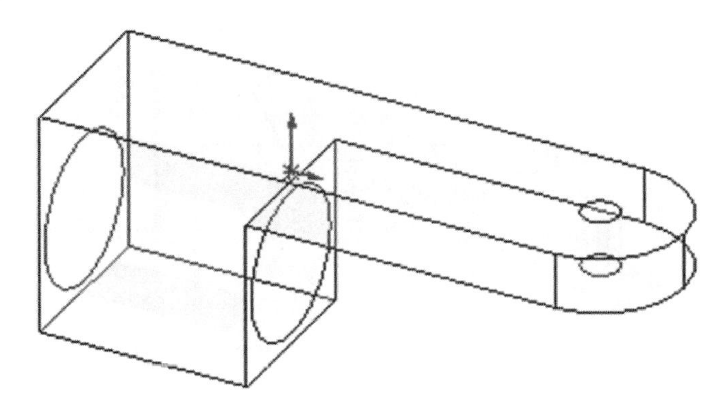

8. Left click once. The piston case will be placed in the assembly drawing. Your screen should look similar to Figure 6.

Figure 6

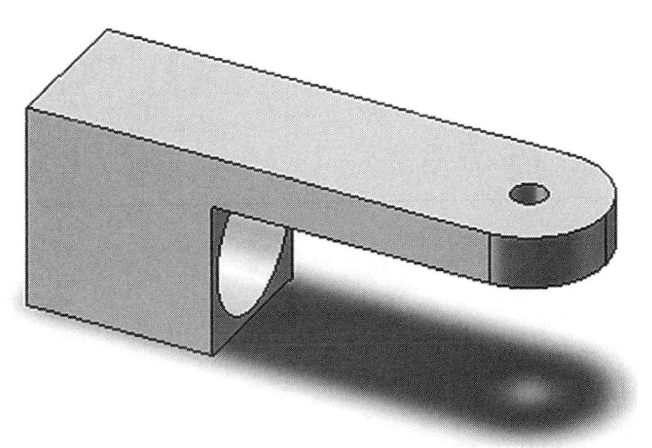

9. Move the cursor to the upper left portion of the screen and left click on **Insert Components**. If the Assembly tools are not visible, left click on the **Assembly** tab as shown in Figure 7.

Figure 7

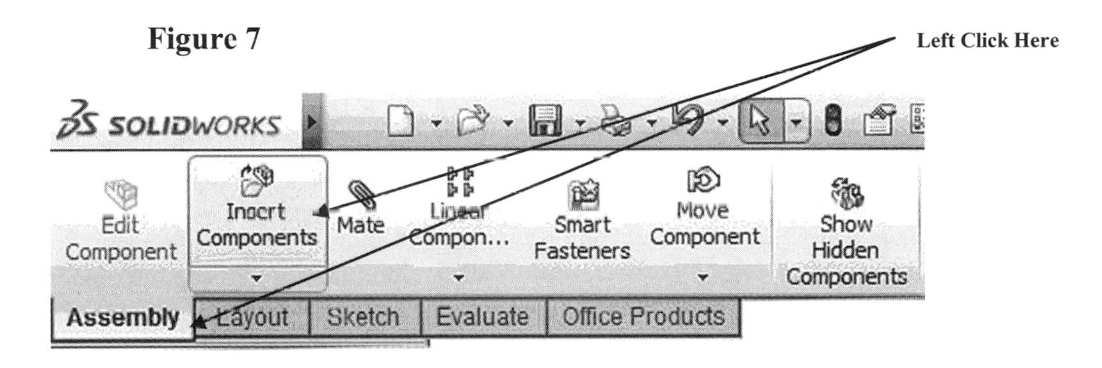

10. Move the cursor to the lower left portion of the screen and left click on **Browse** as shown in Figure 8.

Figure 8

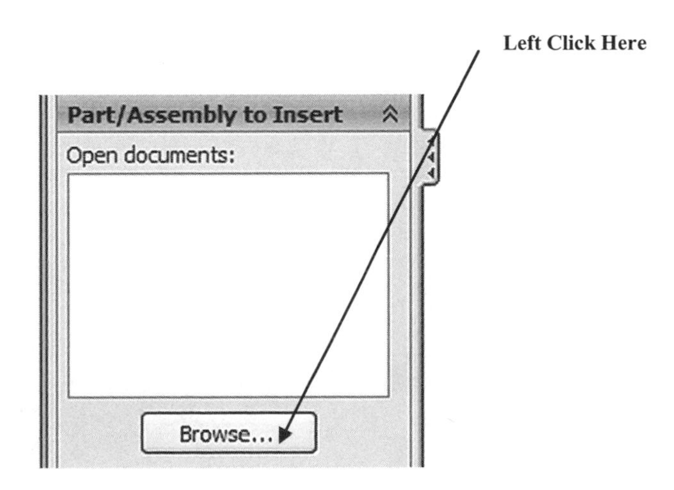

11. The Open dialog box will appear. Locate the Piston1.SLDPRT file and left click on **Open** as shown in Figure 9.

Figure 9

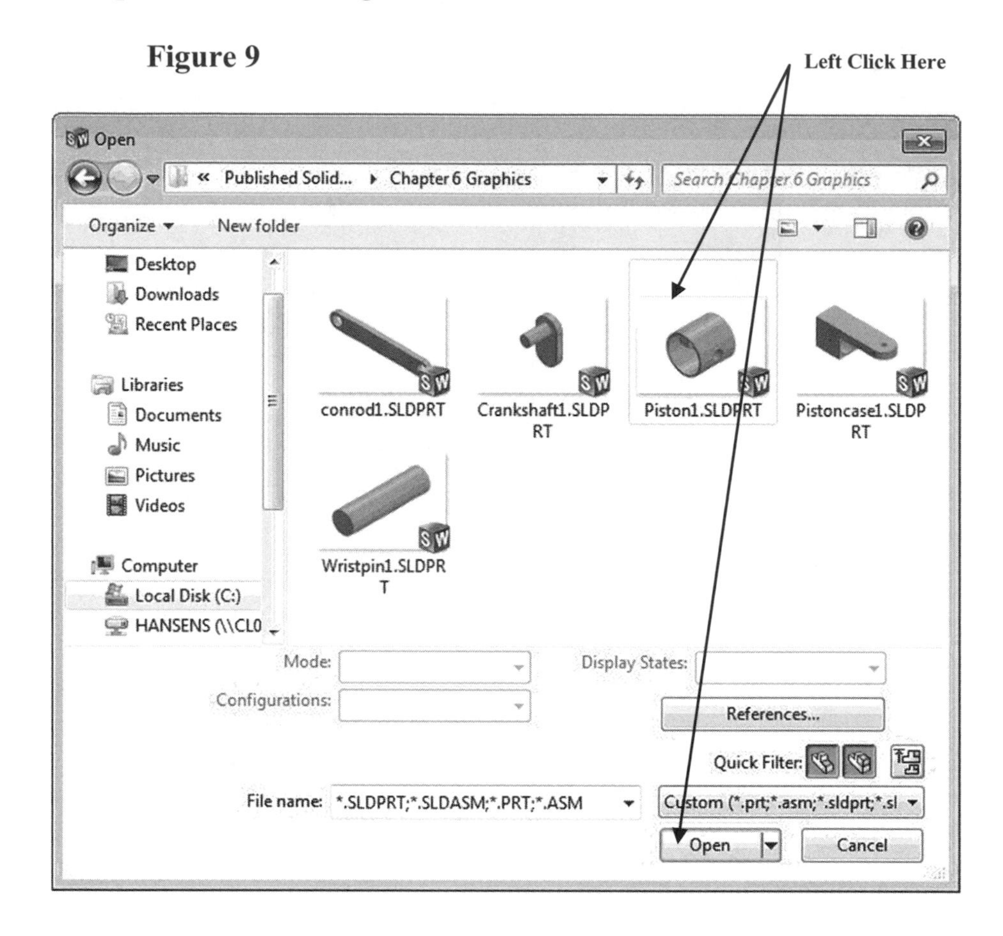

12. The piston will be attached to the cursor. Place the piston anywhere near the piston case and left click once. Your screen should look similar to Figure 10.

Figure 10

13. Move the cursor to the upper left portion of the screen and left click on **Insert Components** as shown in Figure 11.

Figure 11

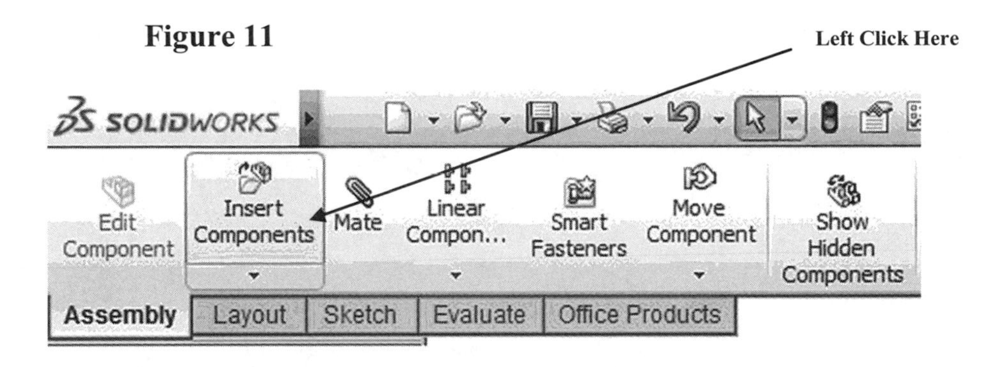

14. Move the cursor to the lower left portion of the screen and left click on **Browse** as shown in Figure 12.

Figure 12

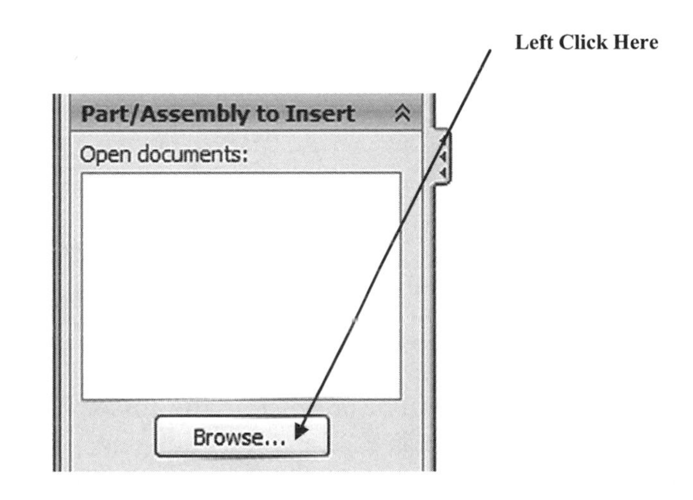

15. The Open dialog box will appear. Locate the Conrod1.SLDPRT file and left click on **Open** as shown in Figure 13.

Figure 13

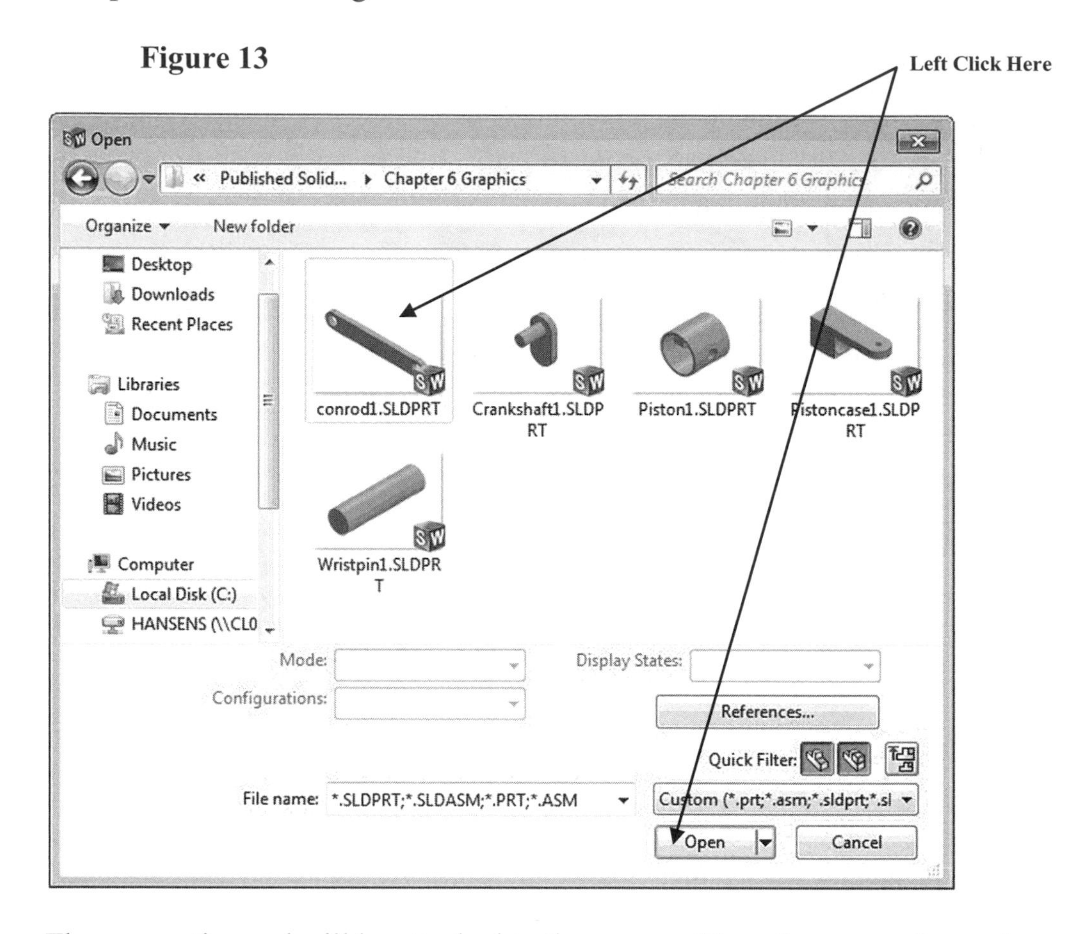

16. The connecting rod will be attached to the cursor. Place the connecting rod anywhere near the piston case and left click once. On the keyboard press the **Esc** button on the key board once. Your screen should look similar to Figure 14.

Figure 14

17. Move the cursor to the upper left portion of the screen and left click on **Insert Components** as shown in Figure 15.

Figure 15

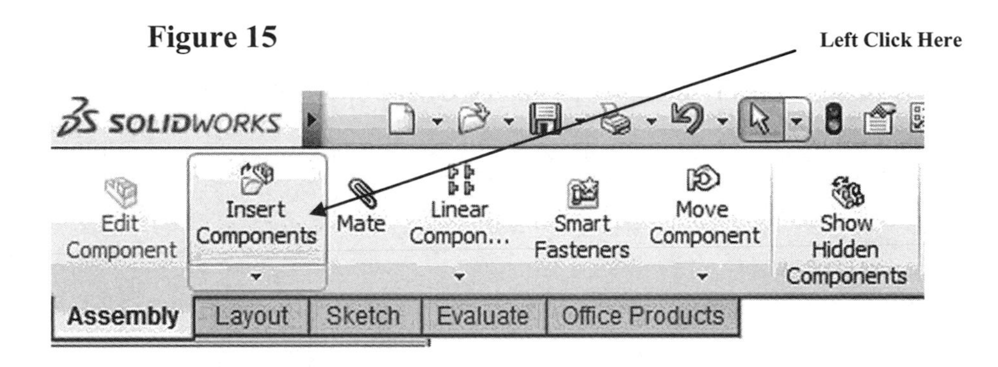

18. Move the cursor to the lower left portion of the screen and left click on **Browse** as shown in Figure 16.

Figure 16

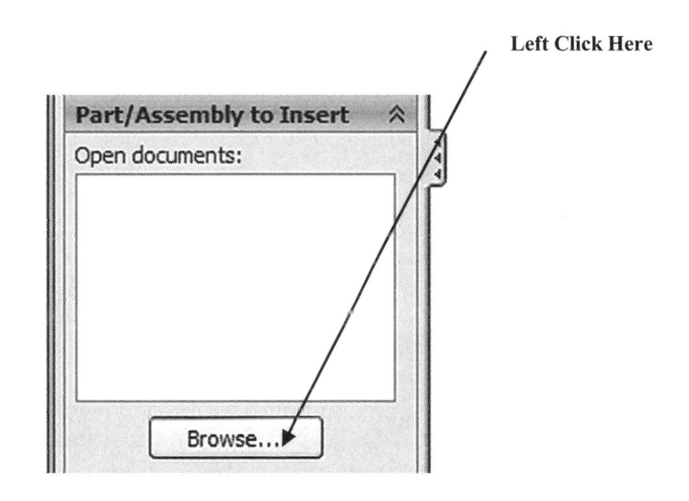

19. The Open dialog box will appear. Locate the crankshaft1.SLDPRT file and left click on **Open** as shown in Figure 17.

Figure 17

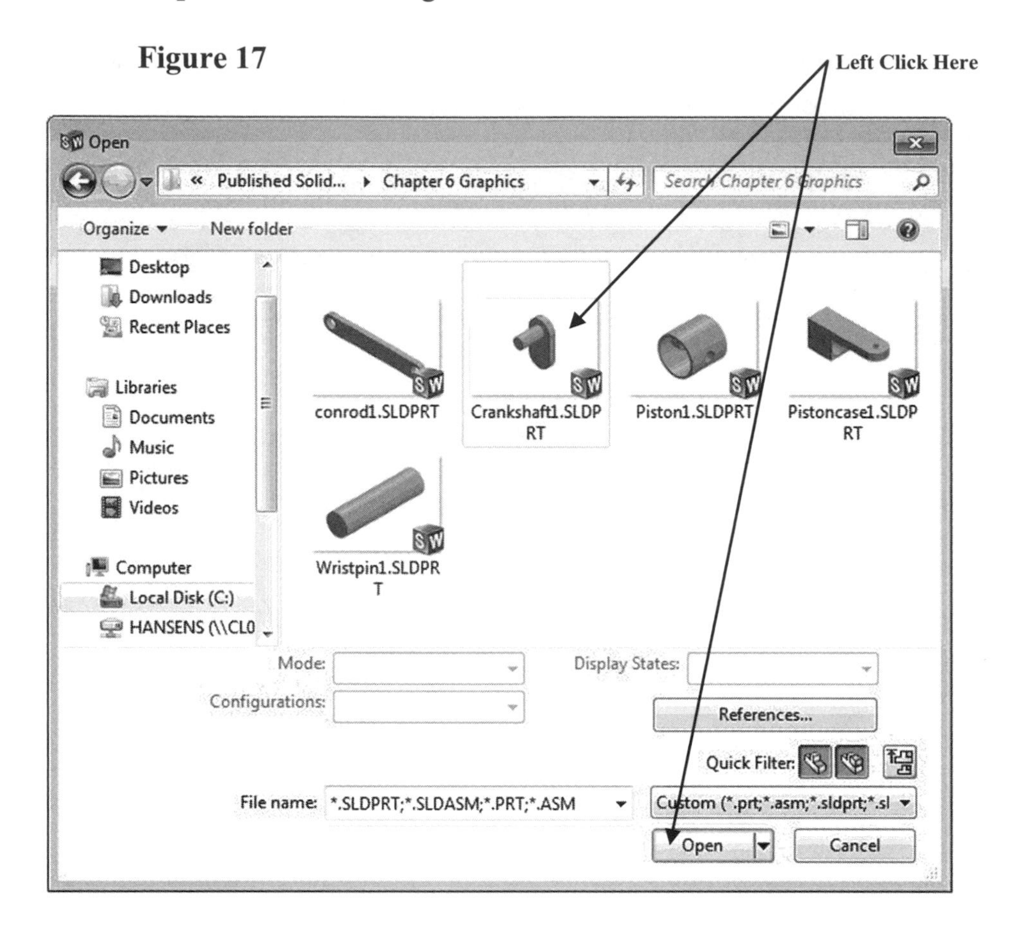

20. The crankshaft will be attached to the cursor. Place the crankshaft anywhere near the piston case and left click once. Your screen should look similar to Figure 18.

Figure 18

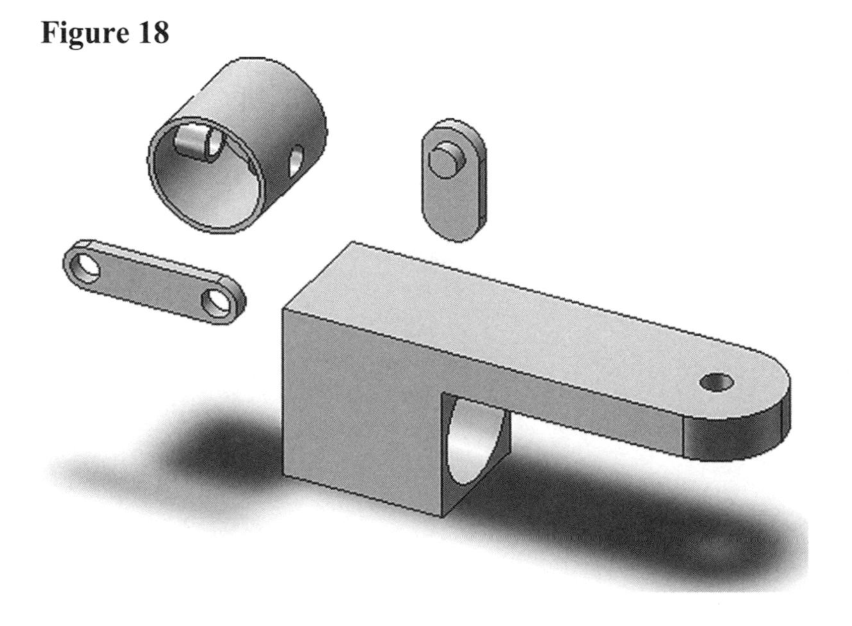

21. Move the cursor to the upper left portion of the screen and left click on **Insert Components** as shown in Figure 19.

Figure 19

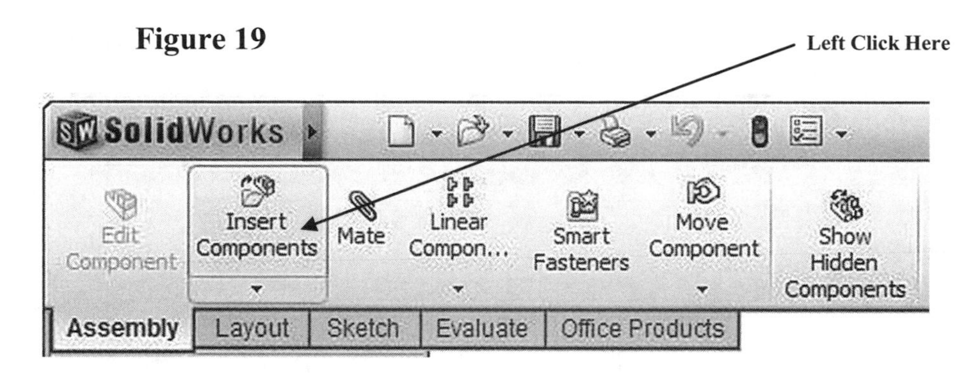

22. Move the cursor to the lower left portion of the screen and left click on **Browse** as shown in Figure 20.

Figure 20

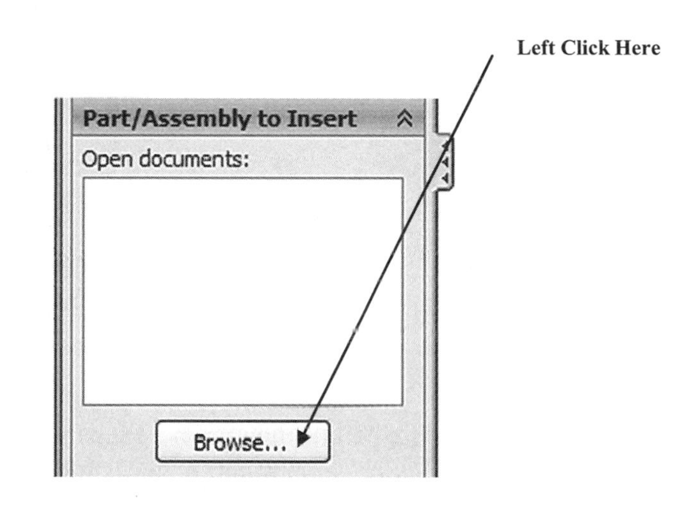

23. The Open dialog box will appear. Locate the Wristpin1.SLDPRT file and left click on **Open** as shown in Figure 21.

Figure 21

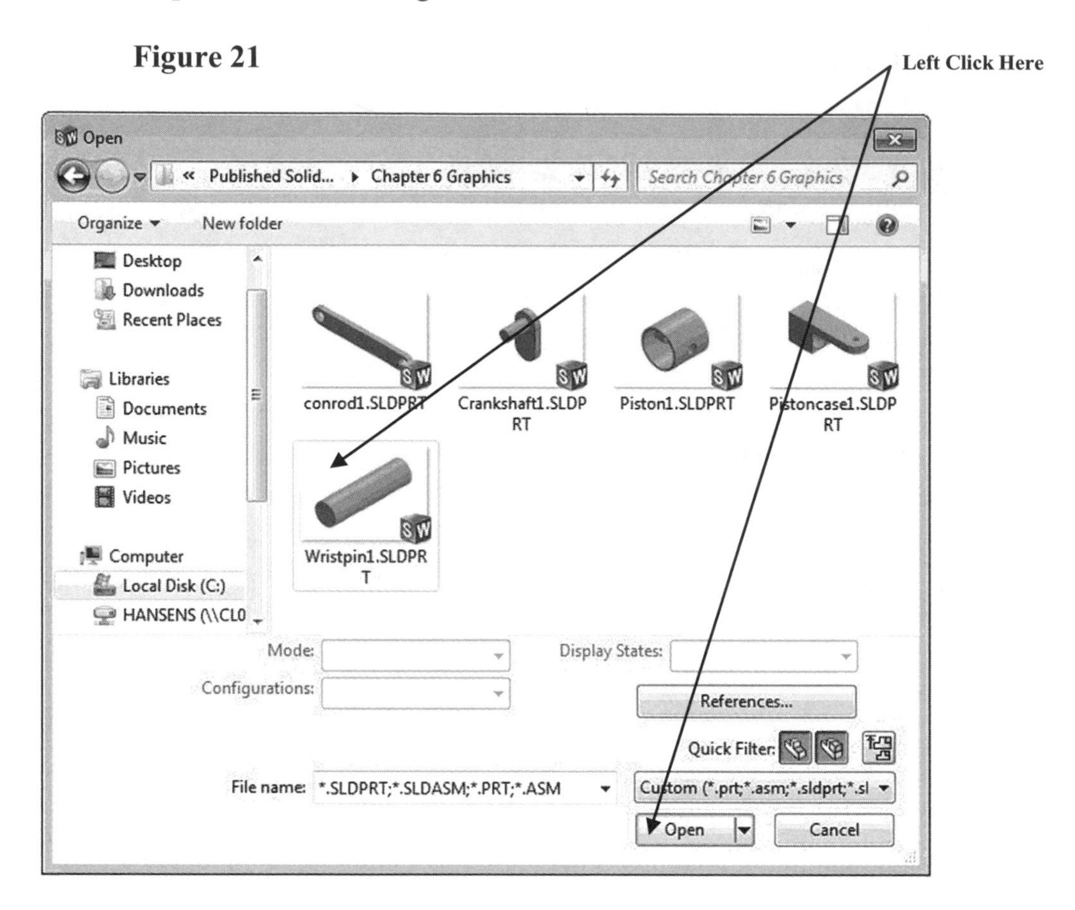

24. The wristpin will be attached to the cursor. Place the wristpin anywhere near the piston case and left click once. Your screen should look similar to Figure 22.

Figure 22

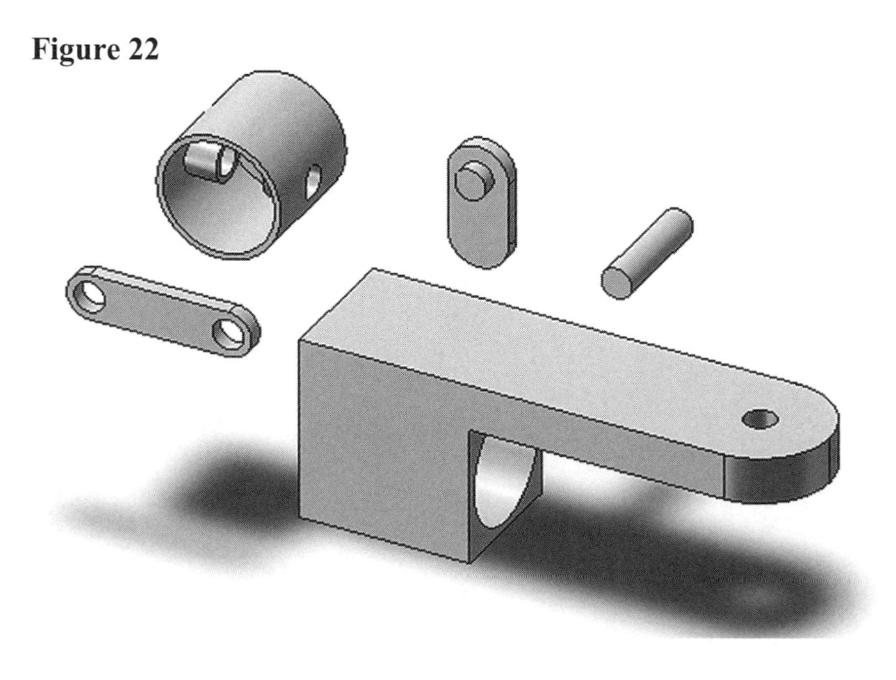

25. The first part inserted into the assembly becomes "fixed" meaning it cannot be moved. All other parts can be moved by left clicking (holding the left mouse button down) and dragging the part to the desired location.

26. Use the **Rotate** command (or press the mouse wheel down) to rotate the entire parts group to the position shown in Figure 23.

Figure 23

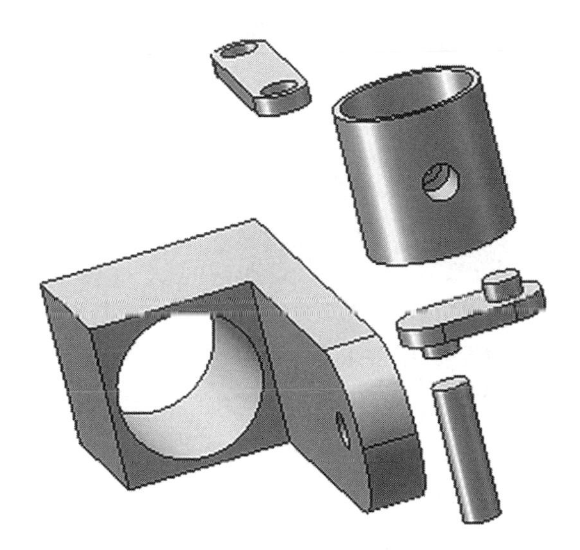

27. Move the cursor to the upper middle portion of the screen and left click on **Rotate Component** as shown in Figure 24.

Figure 24

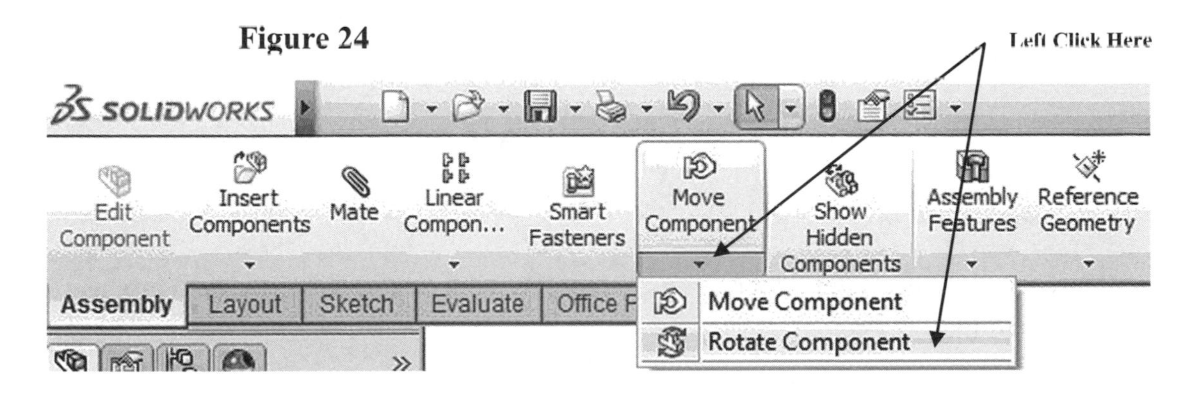

28. Left click (holding the left mouse button down) and rotate the piston to a horizontal position as shown in Figure 25.

Figure 25

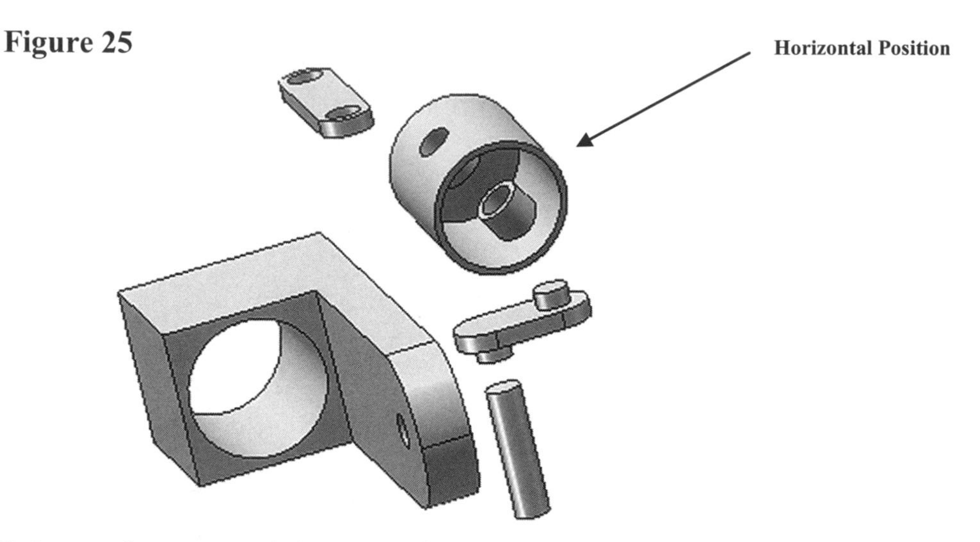

29. Right click anywhere around the part. A pop up menu will appear. Left click on **Select** as shown in Figure 26.

Figure 26

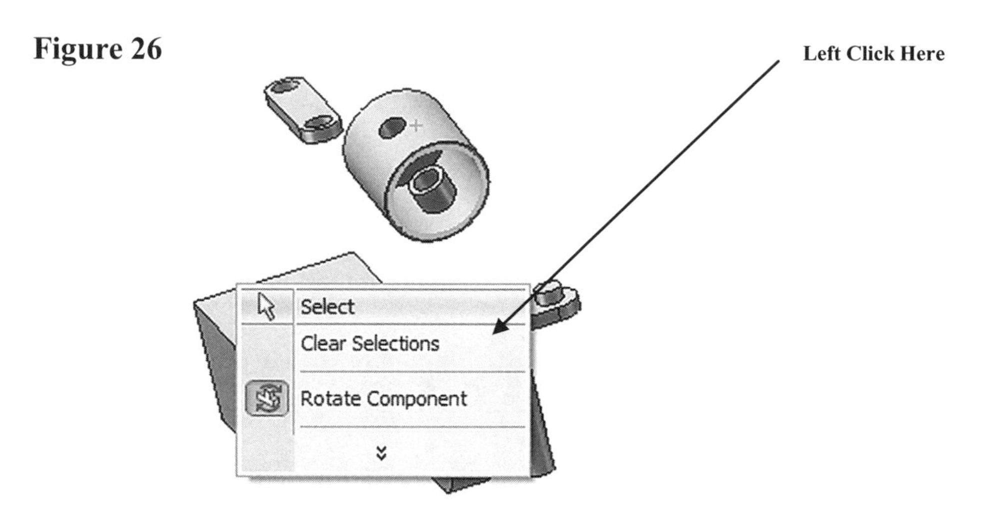

30. Move the cursor to the upper left portion of the screen and left click on **Mate** as shown in Figure 27.

Figure 27

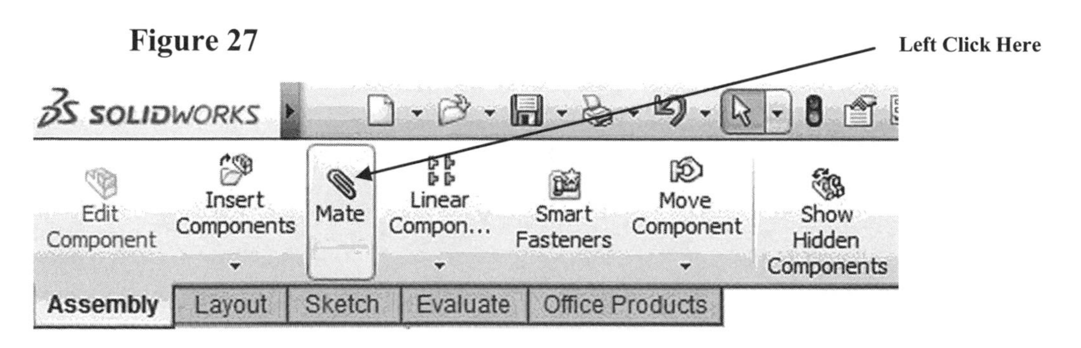

31. Move the cursor to the piston causing the OUTSIDE edges of the part to turn red. Left click once. The part will turn green as shown in Figure 28.

Figure 28

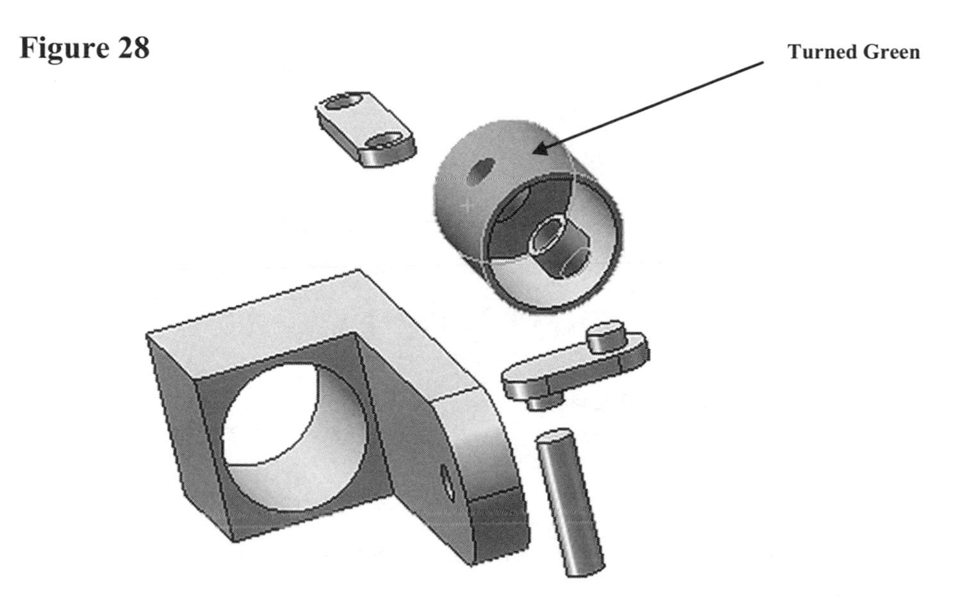

32. Move the cursor to the piston case causing the cylinder edges to turn red. Left click once. The cylinder will turn green as shown in Figure 29.

Figure 29

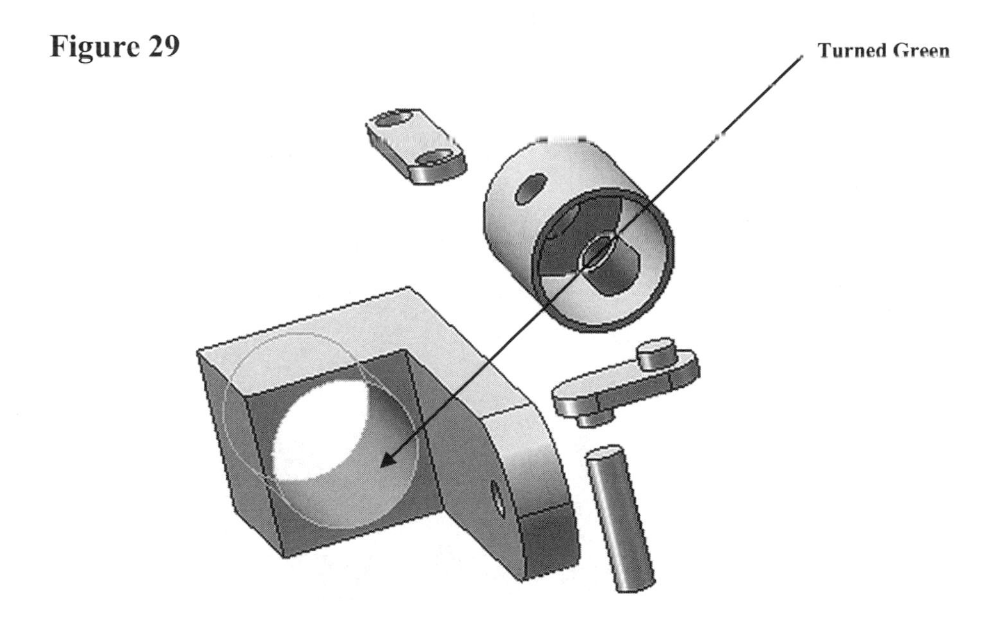

33. The centers of the piston and the cylinder will be aligned. Your screen should look similar to Figure 30.

Figure 30

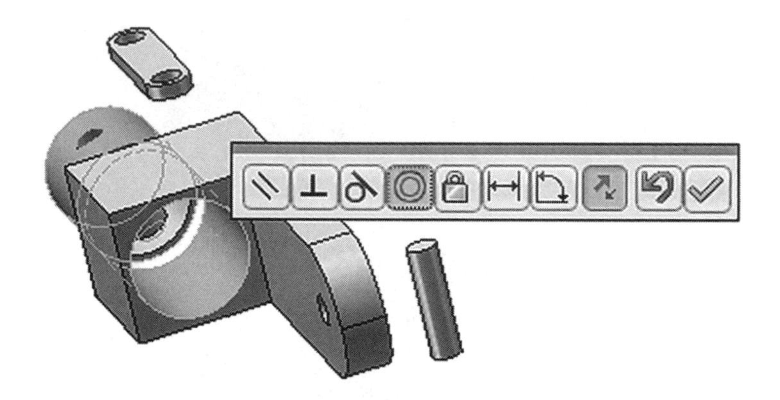

34. Left click on the green checkmark beneath the text "Concentric1"as shown in Figure 31.

Figure 31

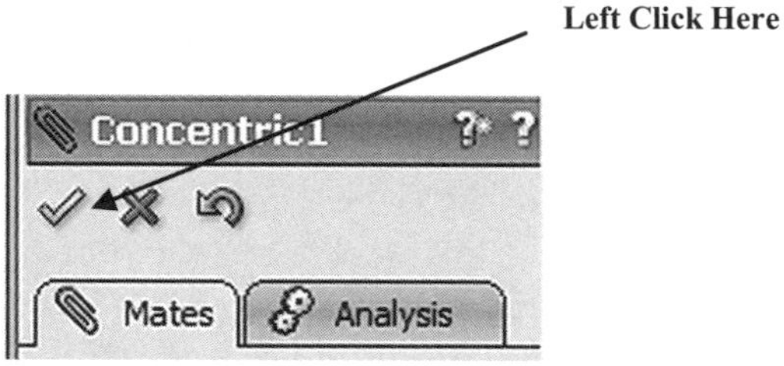

Left Click Here

35. Left click on the green checkmark beneath the text "Mate" as shown in Figure 32.

Figure 32

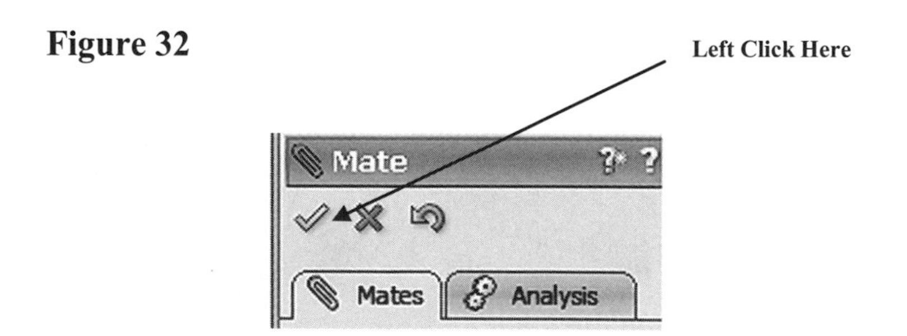

36. Left click (holding the left mouse button down) and drag the piston down into the cylinder as shown in Figure 33.

Figure 33

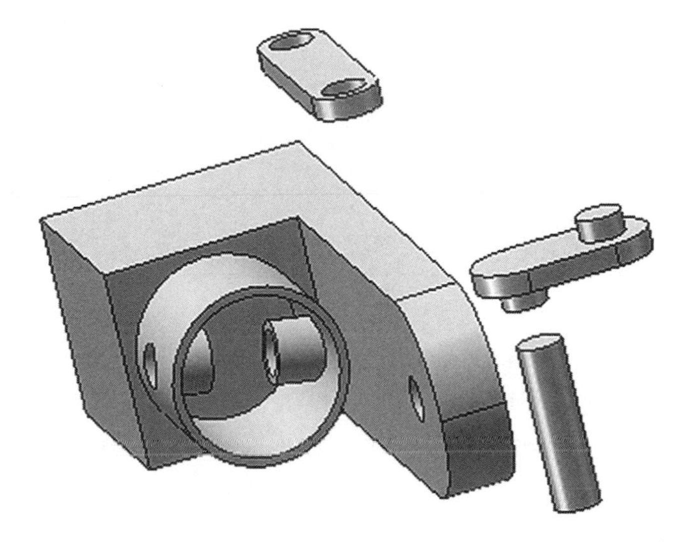

37. Move the cursor to the lower left portion of the piston. Left click (holding the left mouse button down) and slide the piston downward, out below the bore as shown in Figure 34.

Figure 34

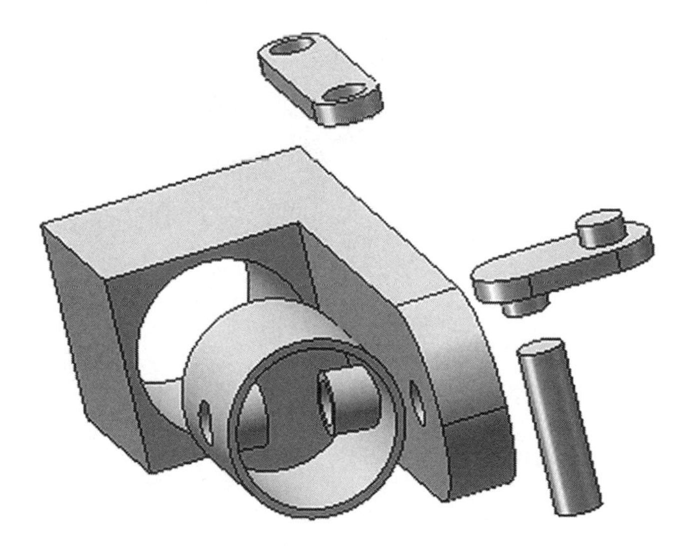

38. Move the cursor to the upper left portion of the screen and left click on **Mate** as shown in Figure 35.

Figure 35

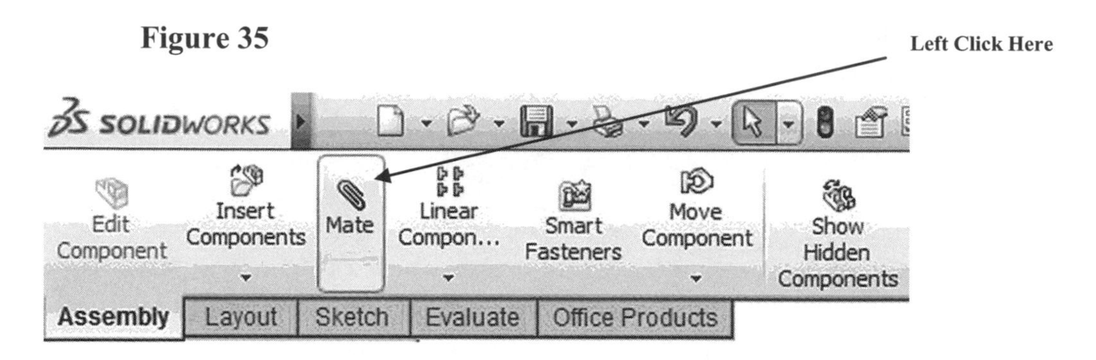

39. Move the cursor to the wristpin hole on the piston causing the inside edge of the hole to turn red. Left click once as shown in Figure 36.

Figure 36

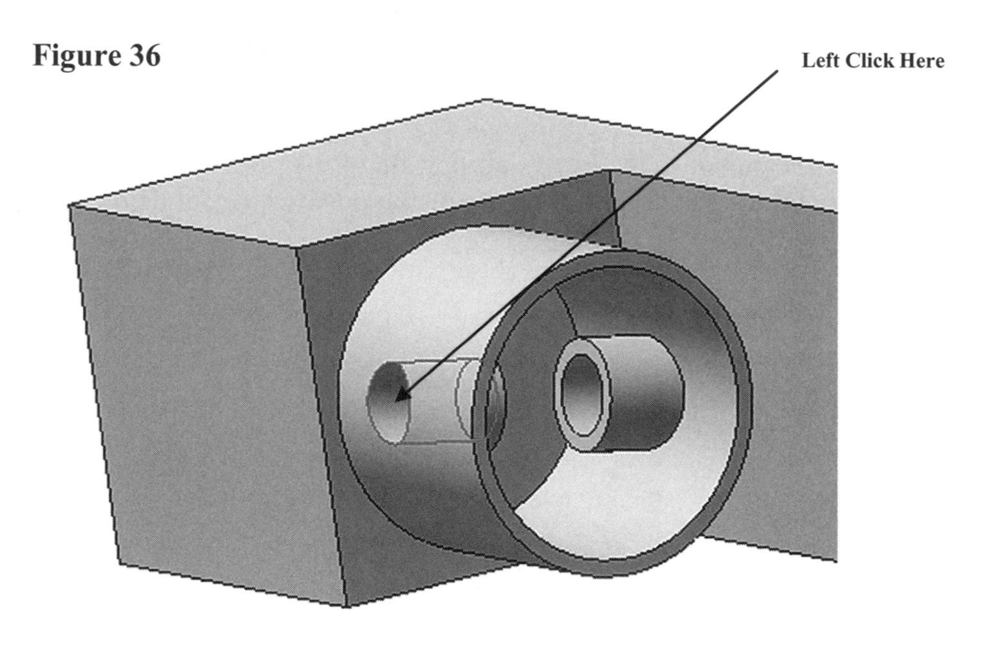

40. Move the cursor to the upper portion of the connecting rod causing the inside edges of the hole to turn red. You may have to drag the connecting rod to a position where the end holes are accessible. Left click as shown in Figure 37.

Figure 37

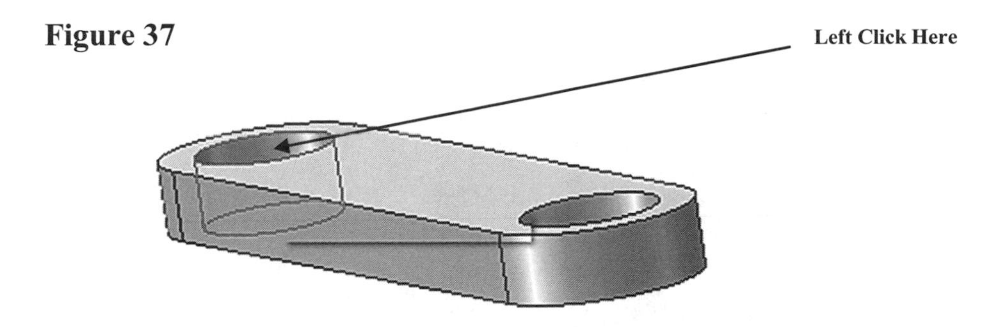

41. The connecting rod hole and the center of the wristpin hole in the piston will be aligned as shown in Figure 38.

Figure 38

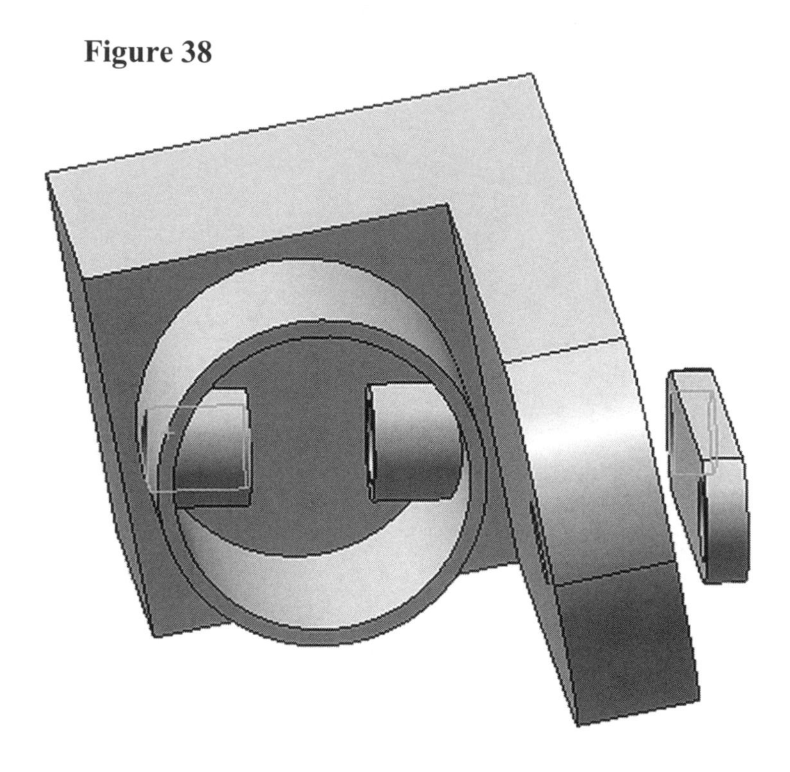

42. Left click on the green checkmark beneath the text "Concentric2" as shown in Figure 39.

Figure 39

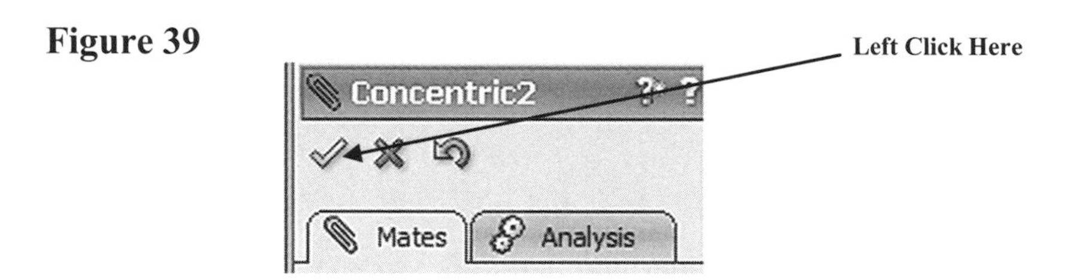

43. Left click on the green checkmark beneath the text "Mate" as shown in Figure 40.

Figure 40

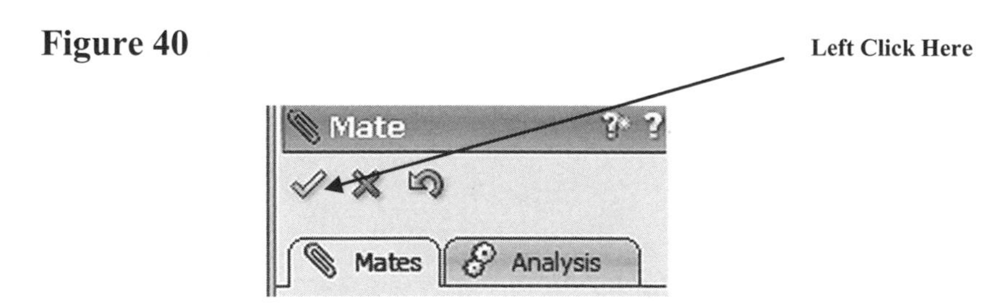

44. Use the rotate command to rotate the entire assembly around to gain access to the underside of the piston. Move the connecting rod to the location shown in Figure 41. This will take some skill. If the piston moves during this process, use the cursor to move the piston to the original location.

Figure 41

45. Move the cursor to the upper left portion of the screen and left click on **Mate** as shown in Figure 42.

Figure 42

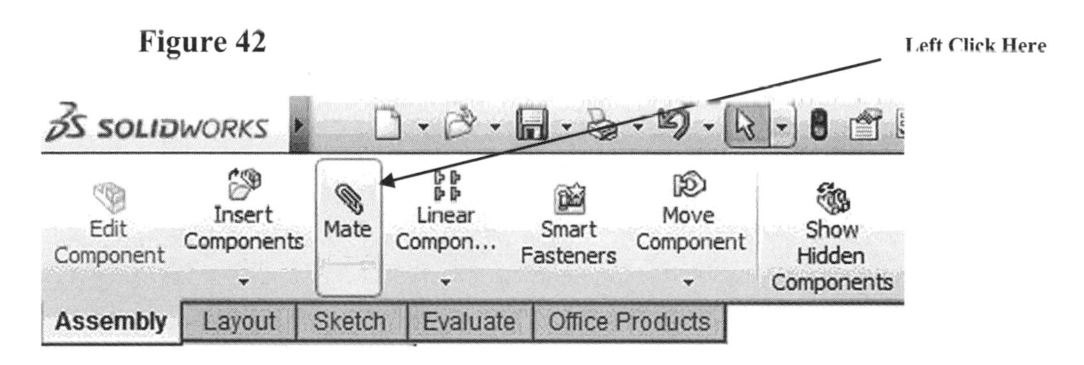

46. Move the cursor to the left side of the connecting rod causing the edges of the connecting rod to turn red. Left click as shown in Figure 43. You may have to zoom in so that SolidWorks will find the proper surface.

Figure 43

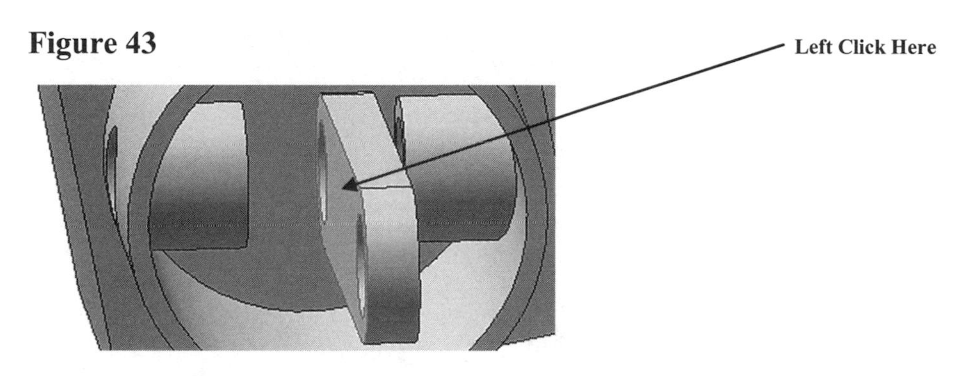

47. Press the mouse wheel down to rotate the assembly around to gain access to the surface opposite the previously selected surface as shown in Figure 44. Select the second mating surface (while both edges are red) by left clicking as shown in Figure 44.

Figure 44

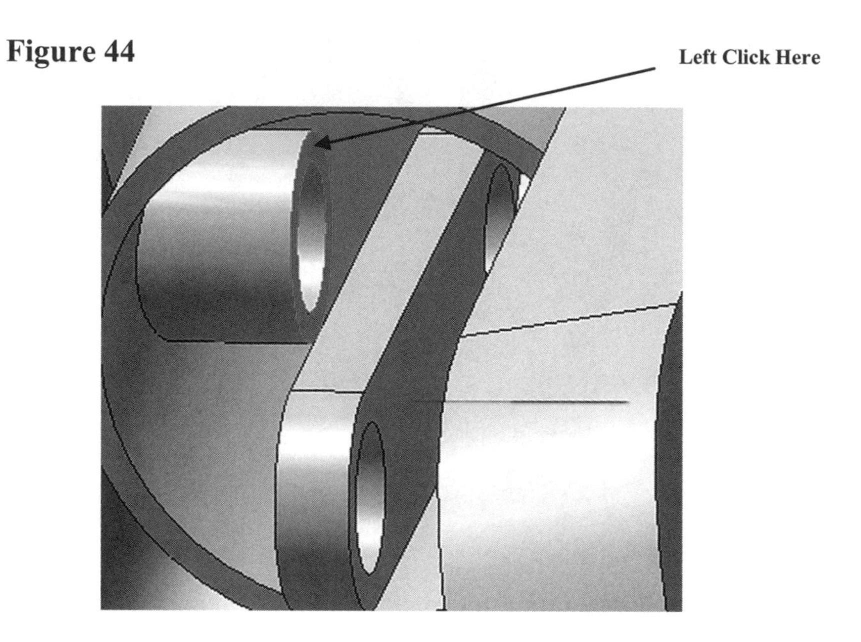

48. Left click on the Distance icon and enter **.25** as shown in Figure 45.

Figure 45

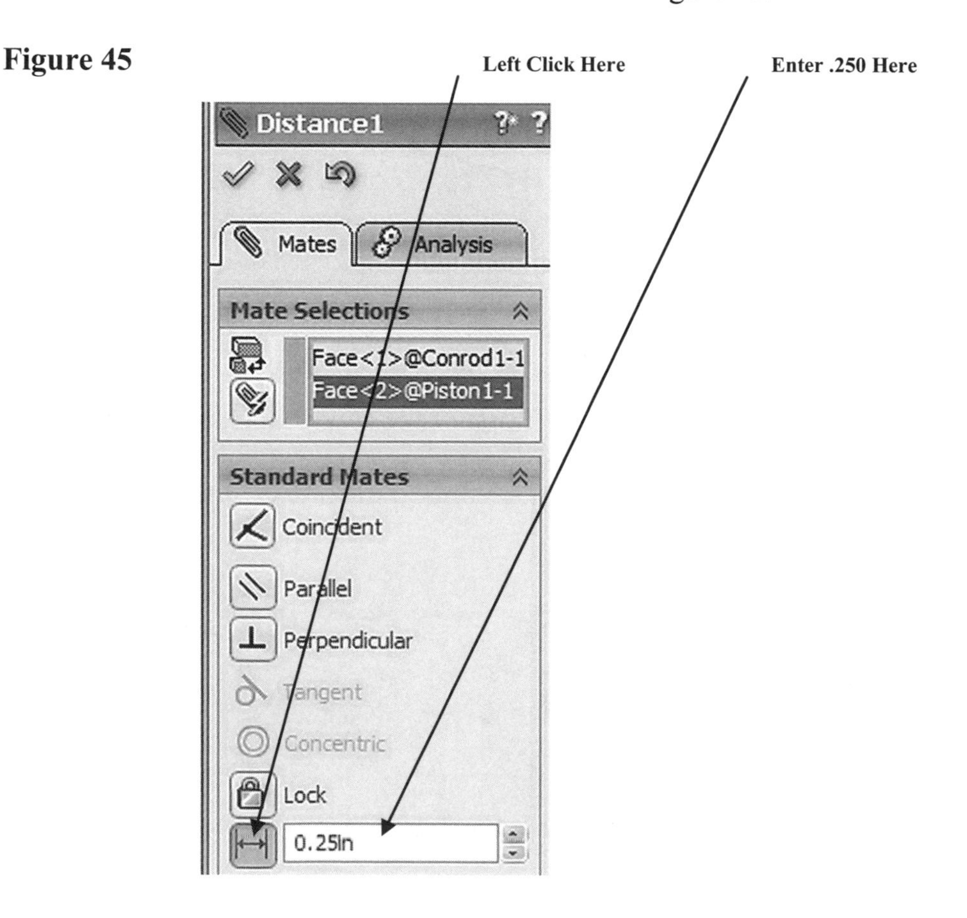

49. Left click on the green checkmark beneath the text "Distance1" as shown in Figure 46.

Figure 46

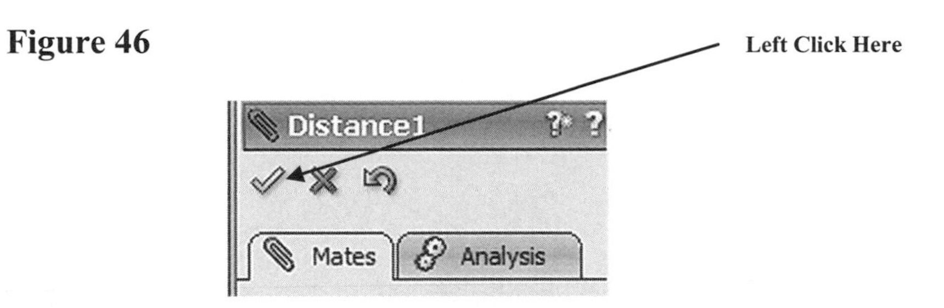

50. Left click on the green checkmark beneath the text "Mate" as shown in Figure 47.

Figure 47

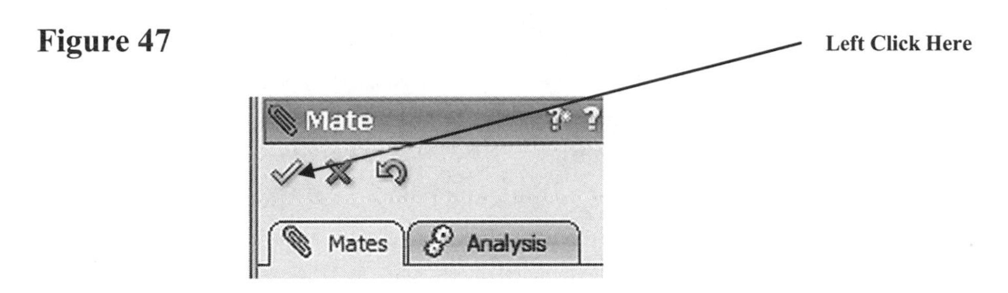

51. The connecting rod should be centered in the piston. Your screen should look similar to Figure 48.

Figure 48

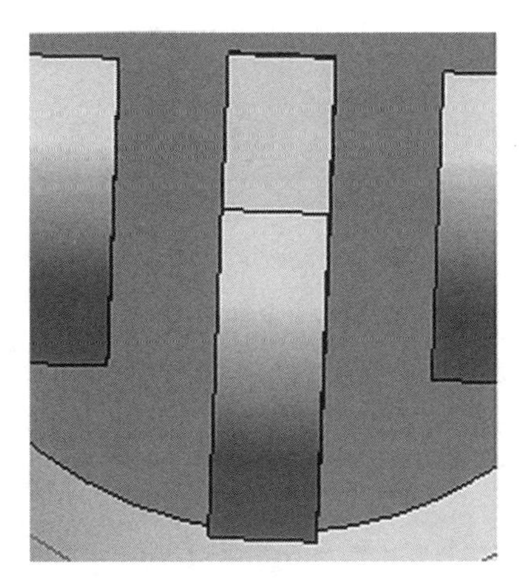

52. Rotate the entire assembly to the position shown in Figure 49.

Figure 49

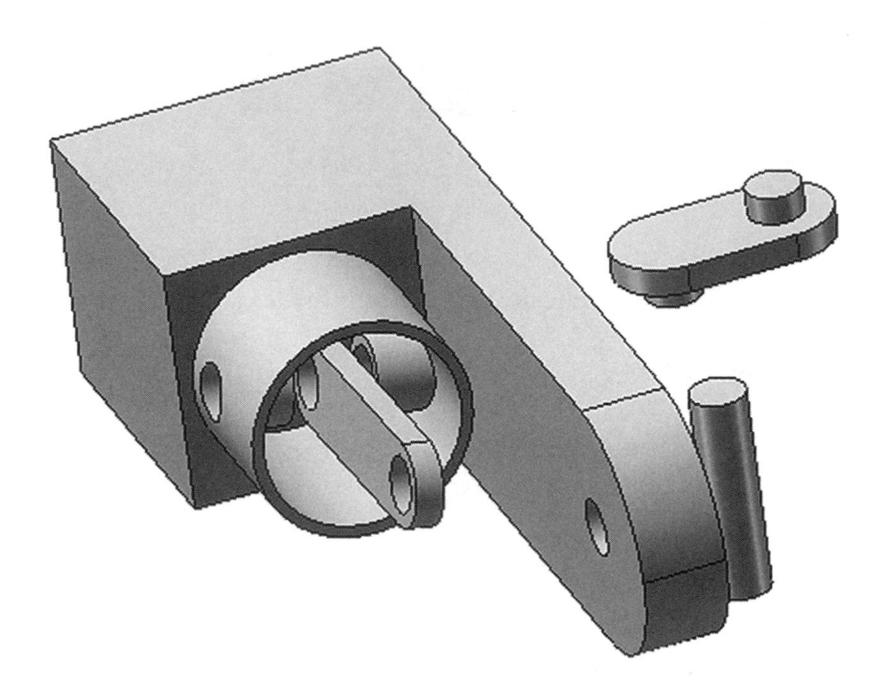

53. Move the cursor to the upper left portion of the screen and left click on **Mate** as shown in Figure 50.

Figure 50

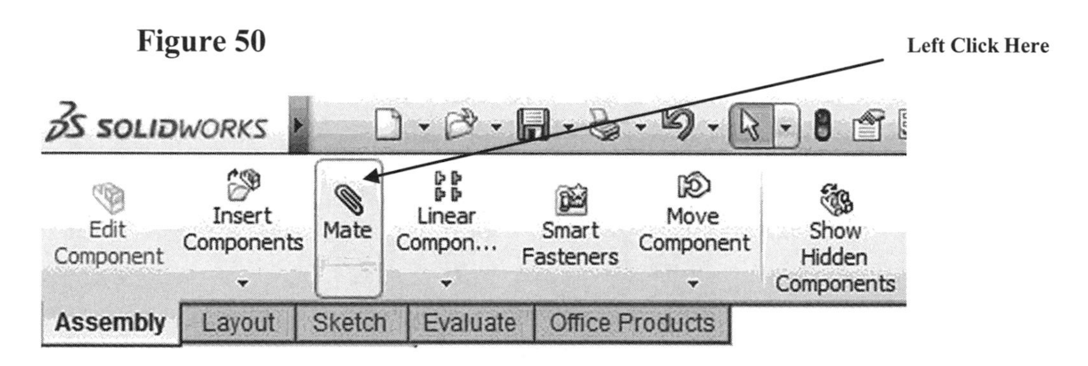

54. Move the cursor to the wristpin causing the edges of the wristpin to turn red. Left click once as shown in Figure 51.

Figure 51

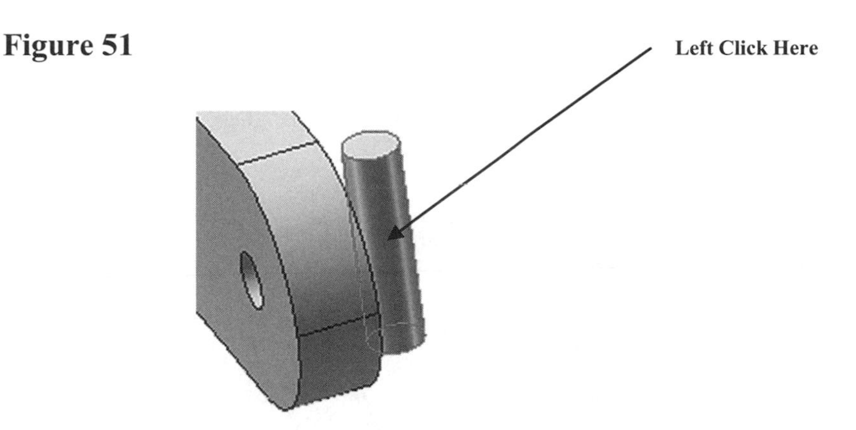

55. Move the cursor to the wristpin hole causing the edges of the hole to turn red. Left click once as shown in Figure 52.

Figure 52

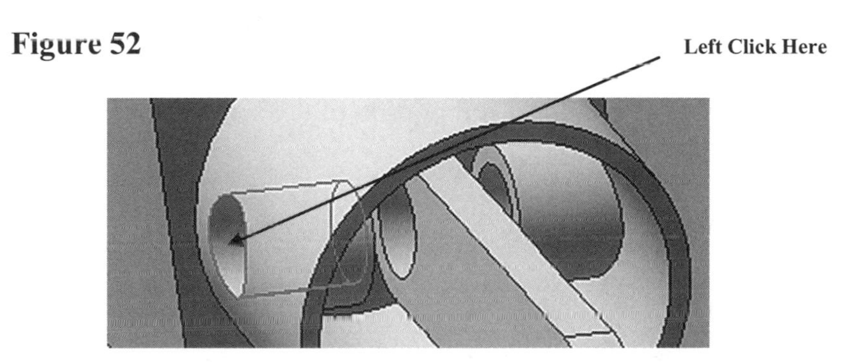

56. Left click on the green checkmark beneath the text "Concentric3" as shown in Figure 53.

Figure 53

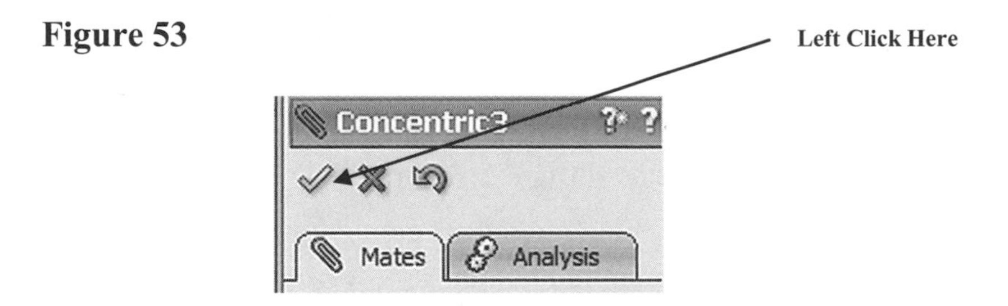

57. Left click on the green checkmark beneath the text "Mate" as shown in Figure 54.

Figure 54

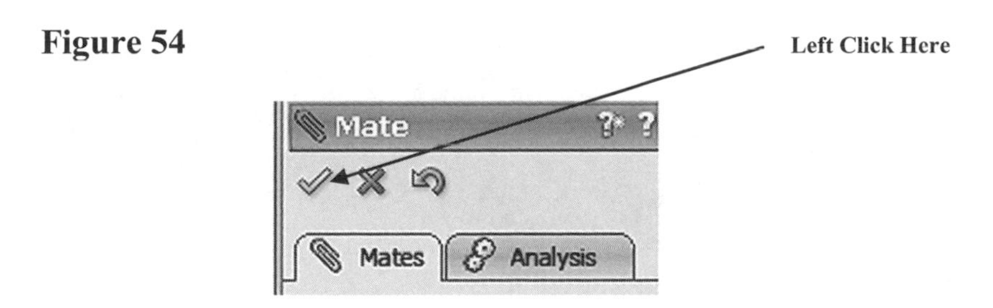

58. The wristpin will be placed in line with the wristpin hole as shown in Figure 55.

Figure 55

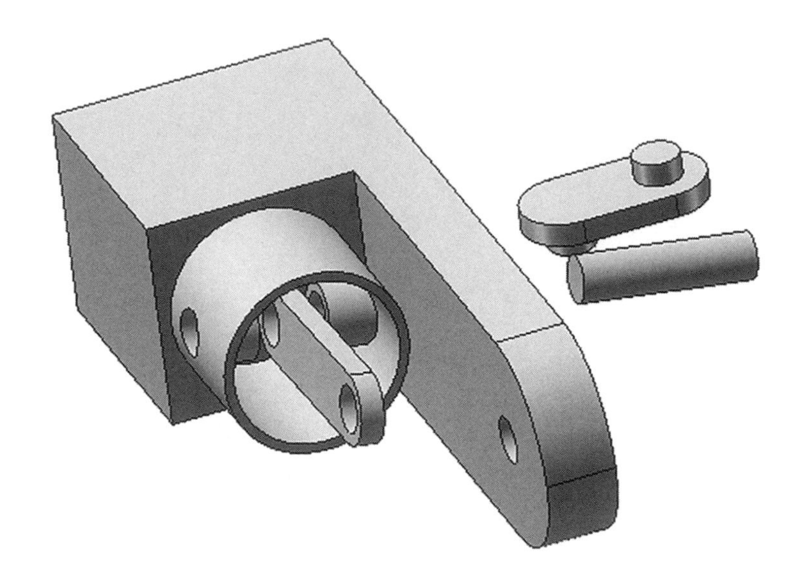

59. Move the cursor to the upper left portion of the screen and left click on **Mate** as shown in Figure 56.

Figure 56

Left Click Here

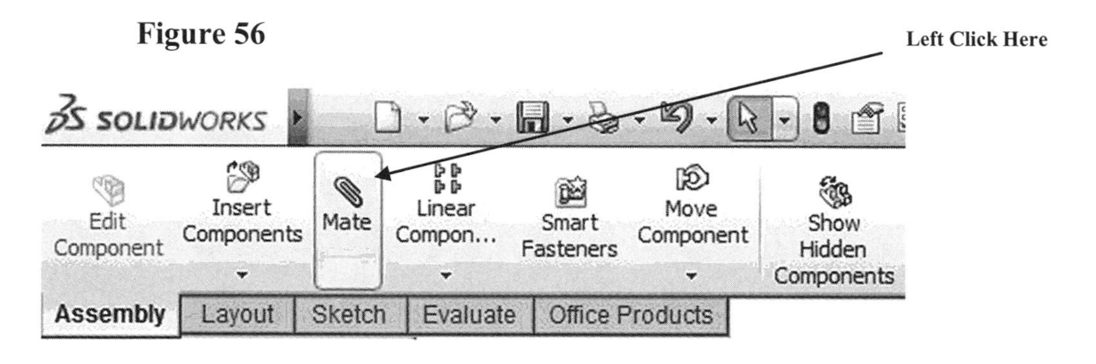

60. Move the cursor to the side of the wristpin causing the edges of the wristpin to turn red. Left click once as shown in Figure 57.

Figure 57

Left Click Here

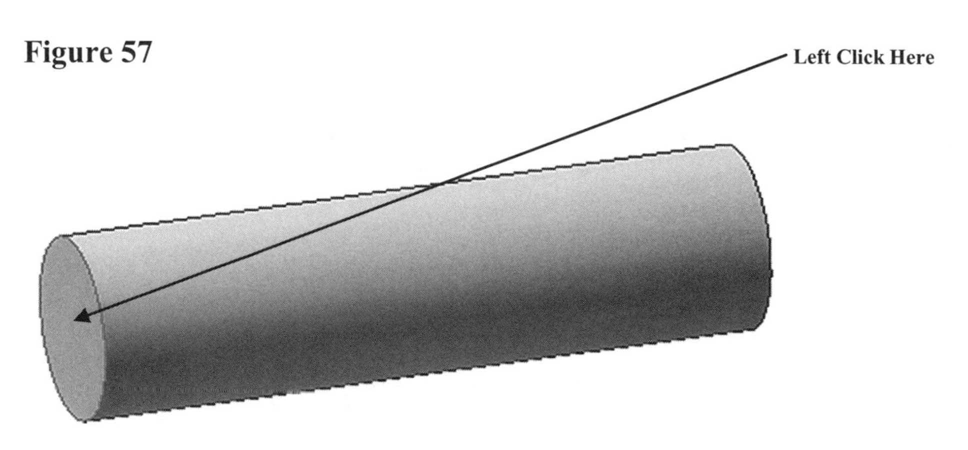

61. Move the cursor to the side of the connecting rod causing the edges of the connecting rod to turn red. Left click once as shown in Figure 58.

Figure 58

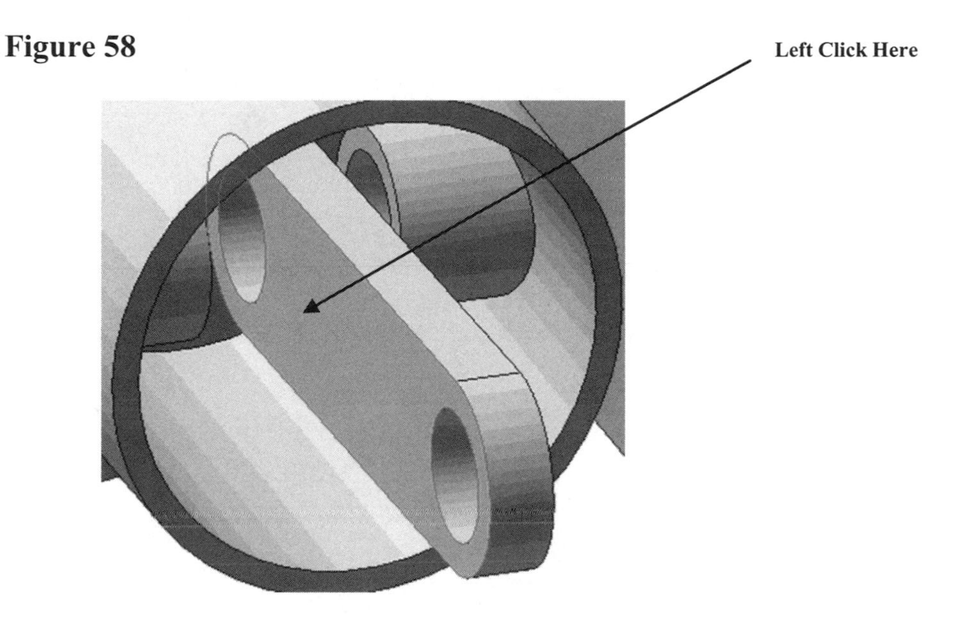

62. Left click on the Distance icon and enter **.7825**. Left click next to "Flip dimension" as shown in Figure 59.

Figure 59

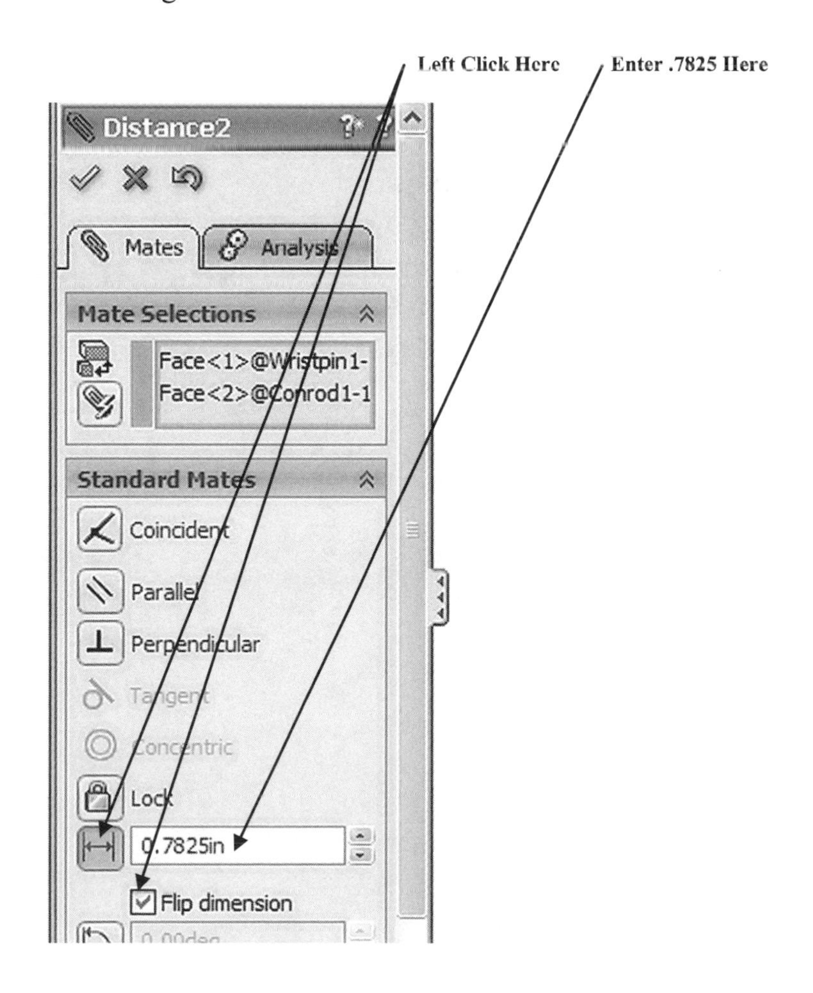

63. Left click on the green checkmark beneath the text "Distance2" as shown in Figure 60.

Figure 60

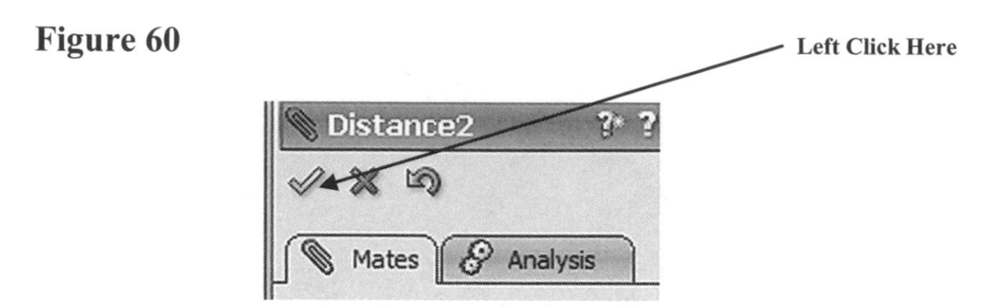

64. Left click on the green checkmark beneath the text "Mate" as shown in Figure 61.

Figure 61

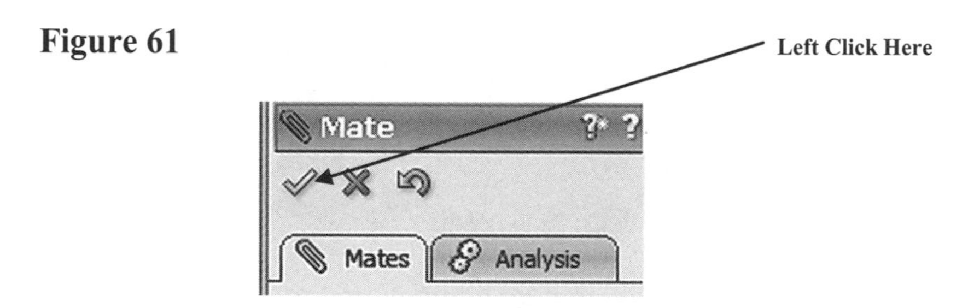

65. Your screen should look similar to Figure 62.

Figure 62

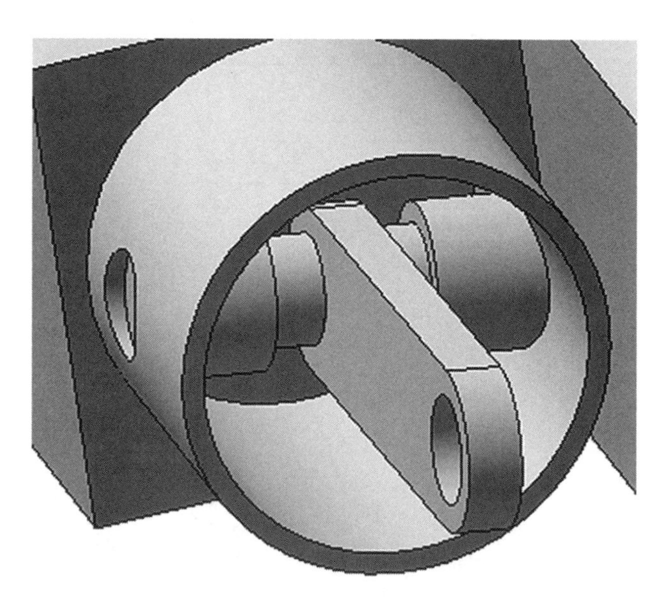

66. Move the cursor to the upper middle portion of the screen and left click on **Move Component** as shown in Figure 63.

Figure 63

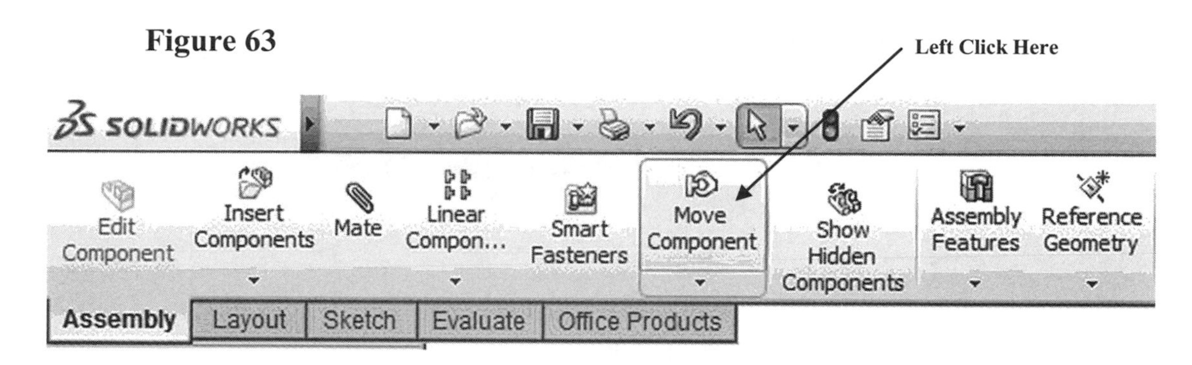

67. Left click again (holding the left mouse button down) and drag the crankshaft to where it can be rotated as shown in Figure 64.

Figure 64

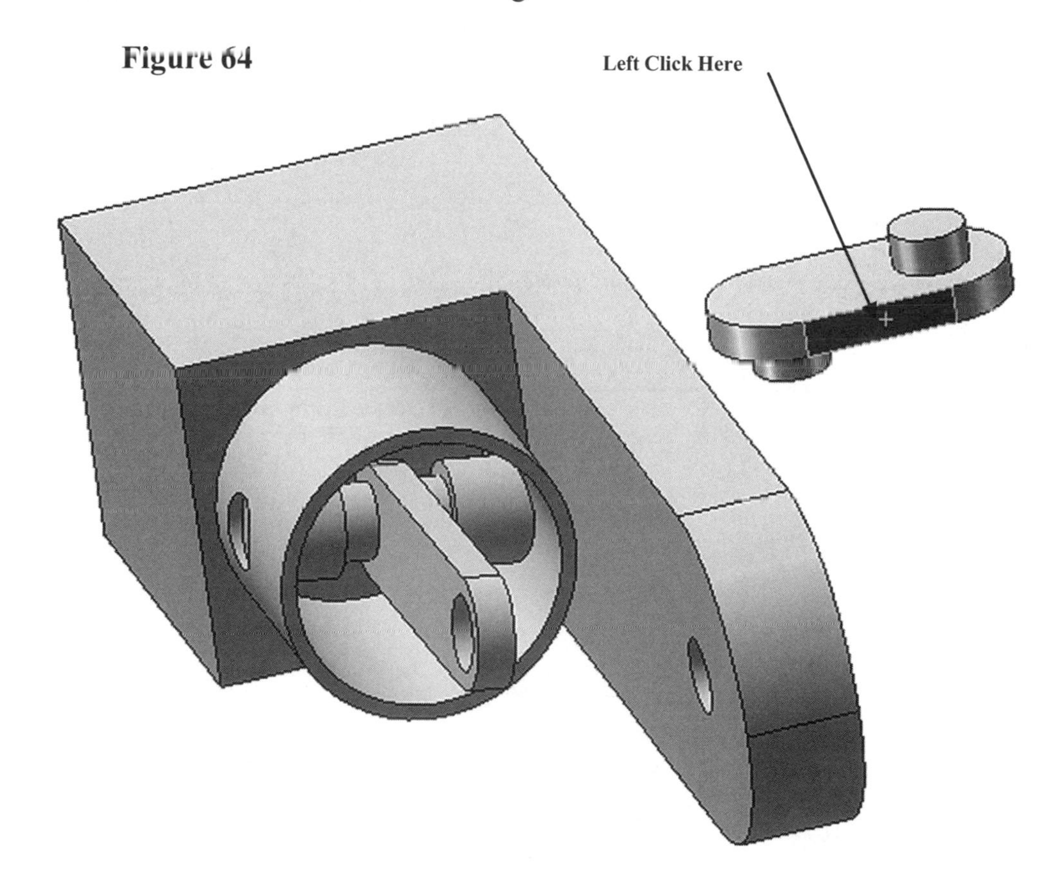

68. Right click anywhere around the parts. A pop up menu will appear. Left click on **Select** as shown in Figure 65.

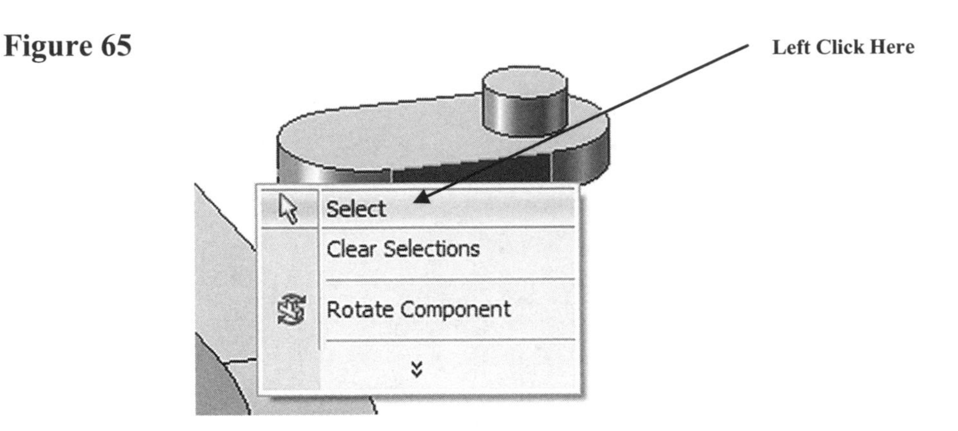

Figure 65

69. Move the cursor to the upper middle portion of the screen and left click on **Rotate Component** as shown in Figure 66.

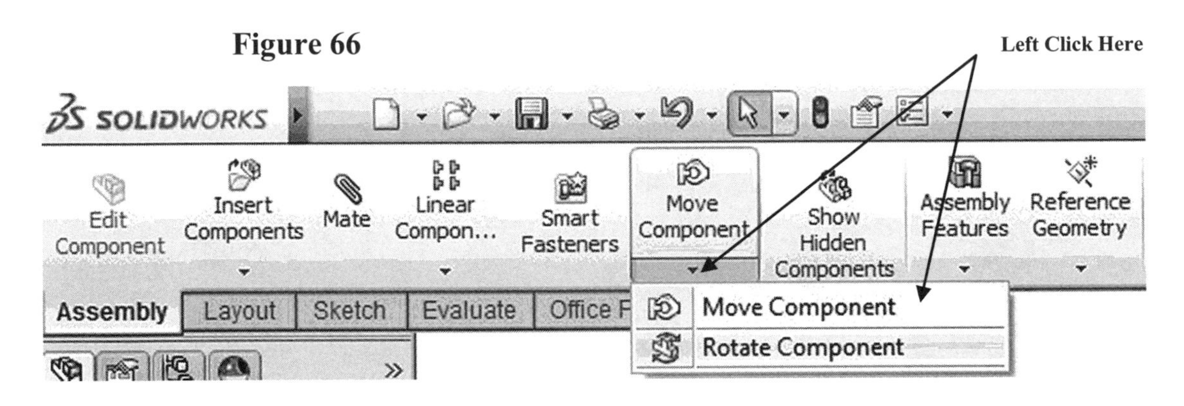

Figure 66

70. Left click once. Left click again (holding the left mouse button down) and rotate the connecting rod upward as shown in Figure 67.

Figure 67

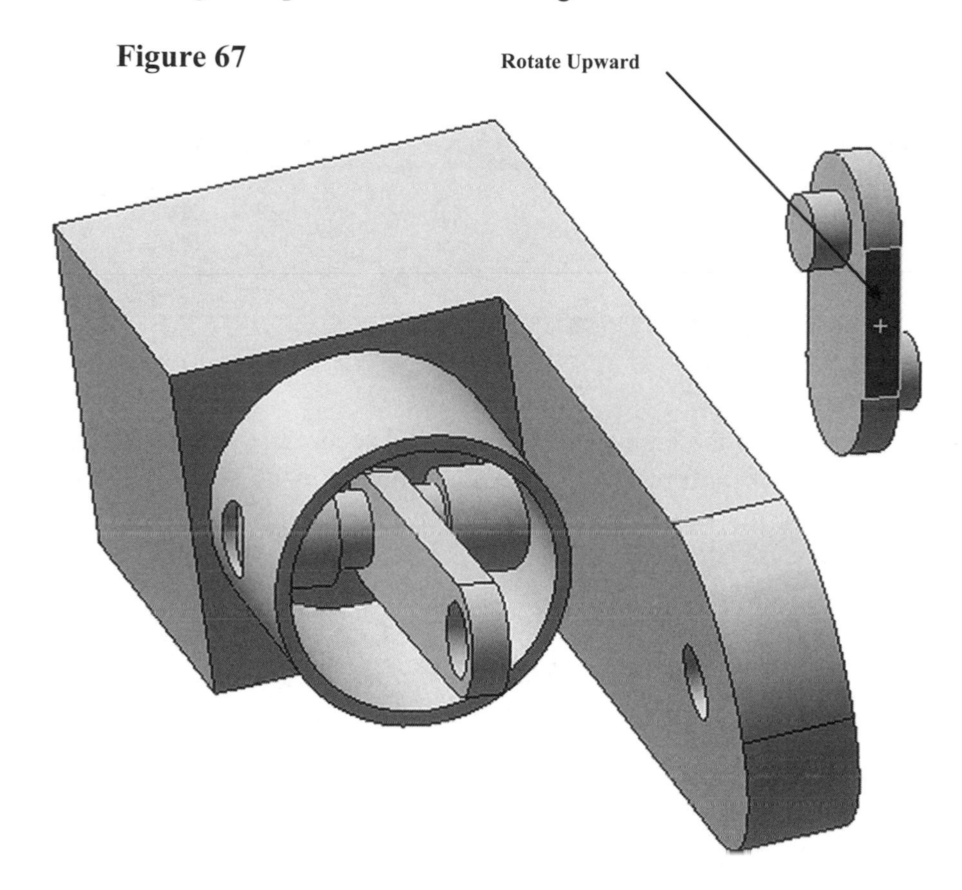

71. Right click anywhere around the parts. A pop up menu will appear. Left click on **Select** as shown in Figure 68.

Figure 68

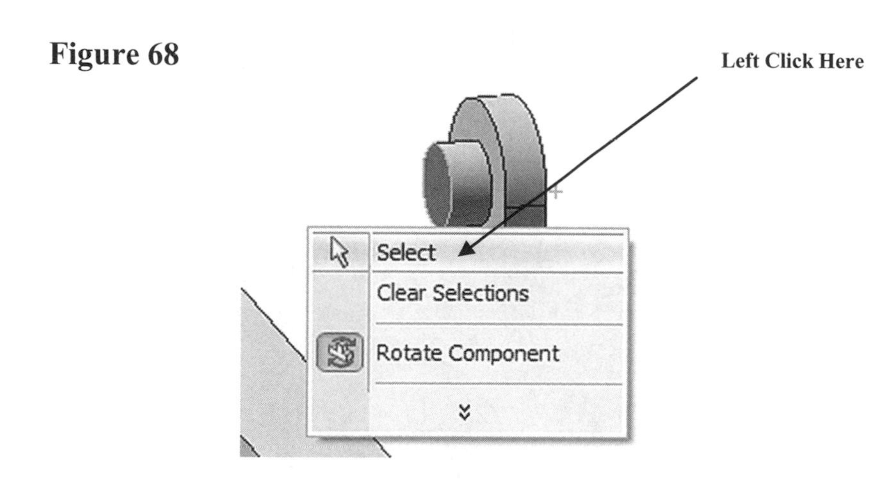

72. Move the cursor to the upper left portion of the screen and left click on **Mate** as shown in Figure 69.

Figure 69

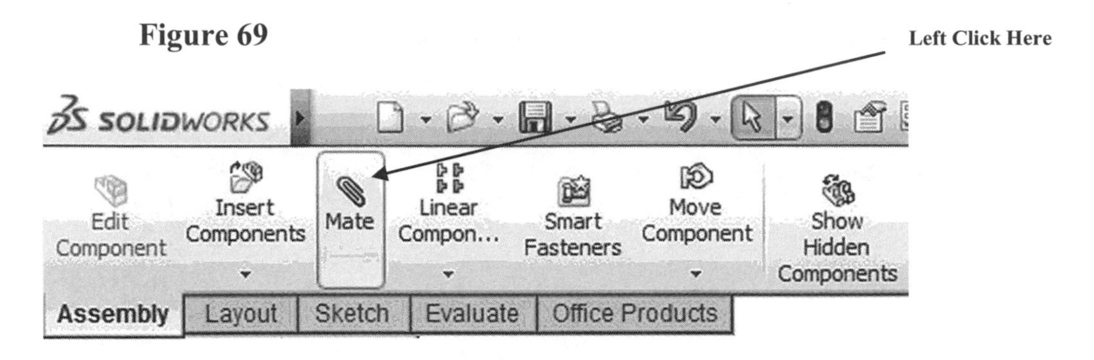

73. Move the cursor to the crankshaft pin that will secure the connecting rod. The edges of the crankshaft pin will turn red. Left click once as shown in Figure 70.

Figure 70

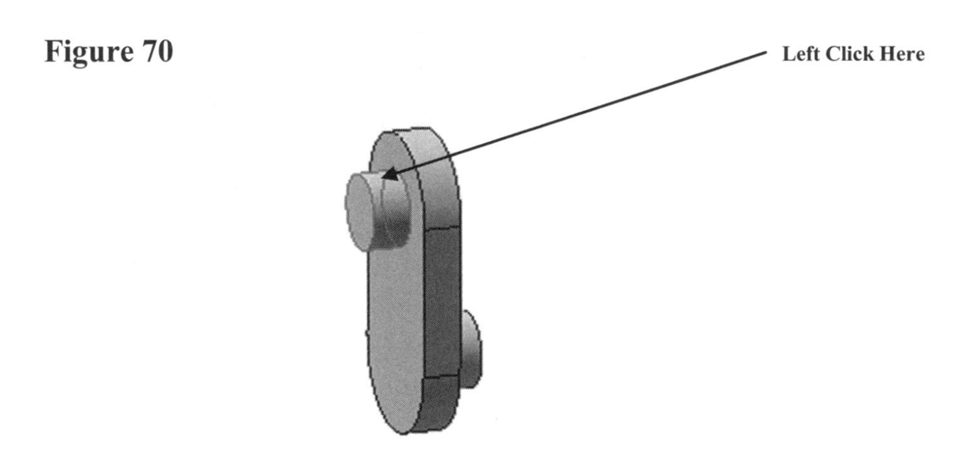

74. Move the cursor to the connecting rod hole that will be secured to the crankshaft. The edges of the connecting rod hole will turn red. Left click once as shown in Figure 71.

Figure 71

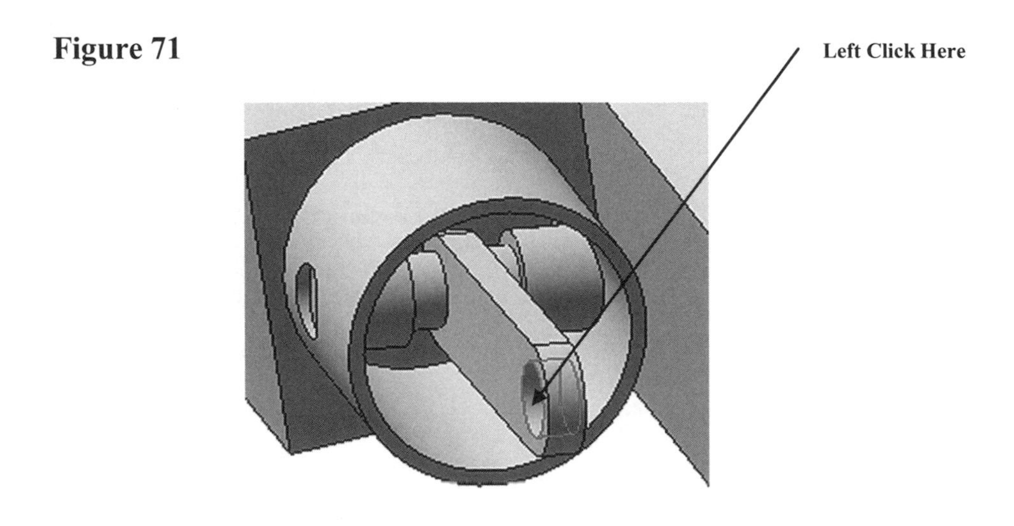

75. Left click on the green checkmark beneath the text "Concentric4" as shown in Figure 72.

Figure 72

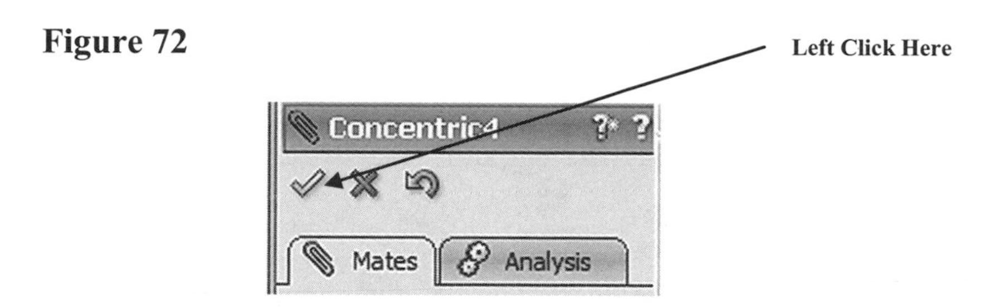

76. Left click on the green checkmark beneath the text "Mate" as shown in Figure 73.

Figure 73

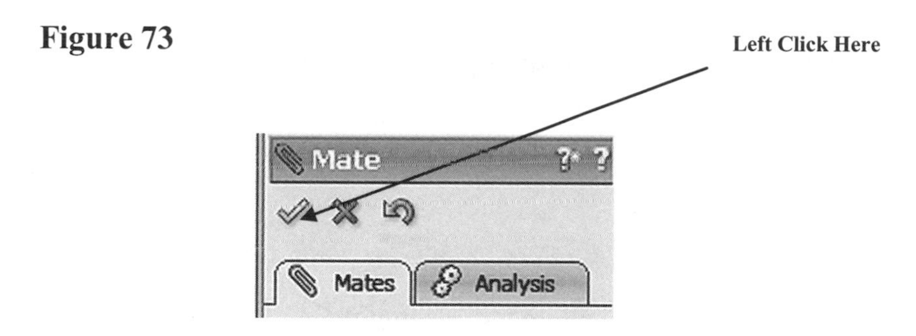

77. The center of the crankshaft and the center of the connecting rod are now aligned.

78. Left click on the crankshaft (holding the left mouse button down) and drag it to the connecting rod as shown in Figure 74.

Figure 74

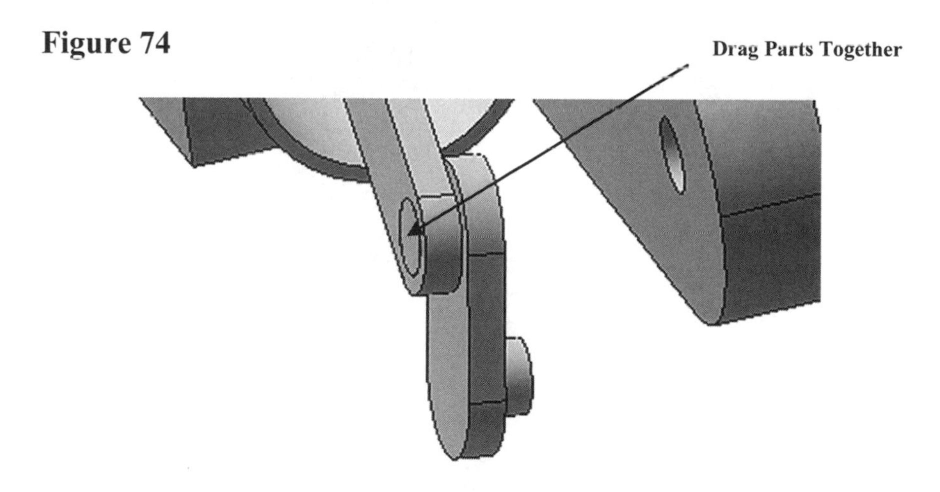

79. Move the cursor to the upper left portion of the screen and left click on **Mate** as shown in Figure 75.

Figure 75

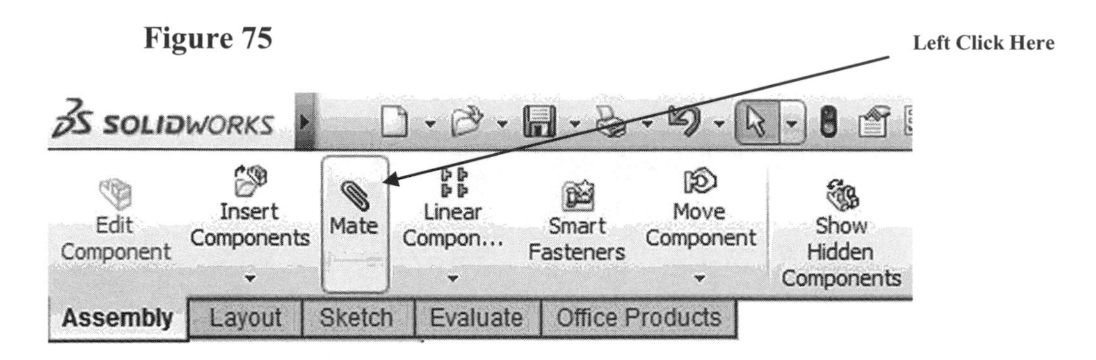

80. Move the cursor to the crankshaft pin causing the edge of the pin to turn red. Left click as shown in Figure 76.

Figure 76

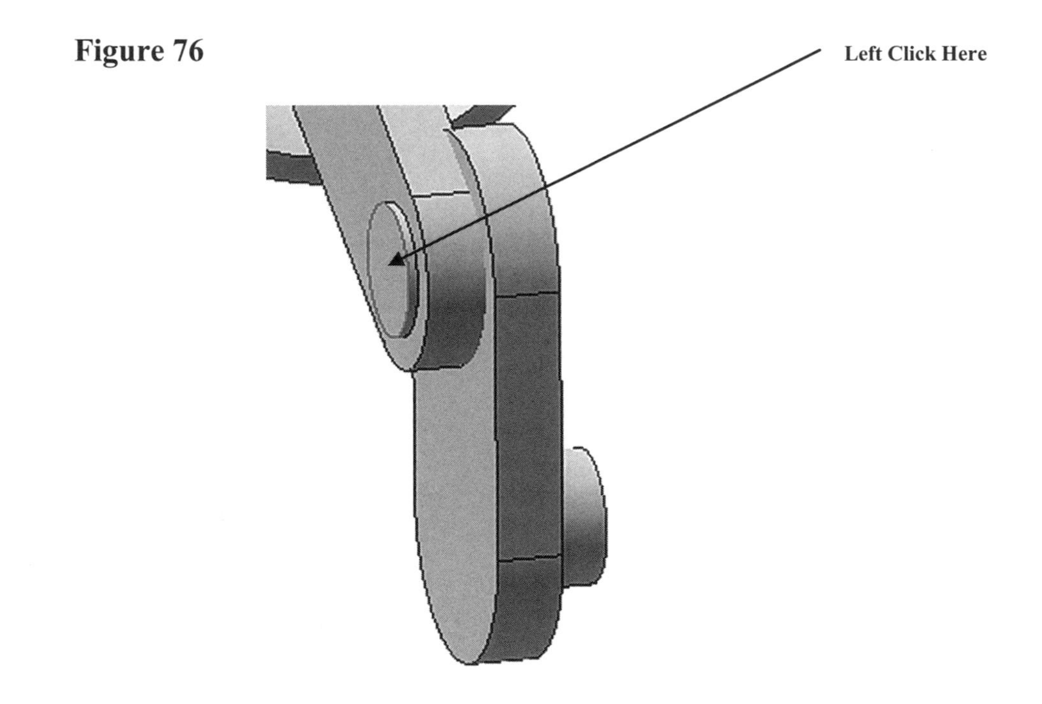

81. Move the cursor to the connecting rod side. The edge of the connecting rod will turn red. Left click as shown in Figure 77.

Figure 77

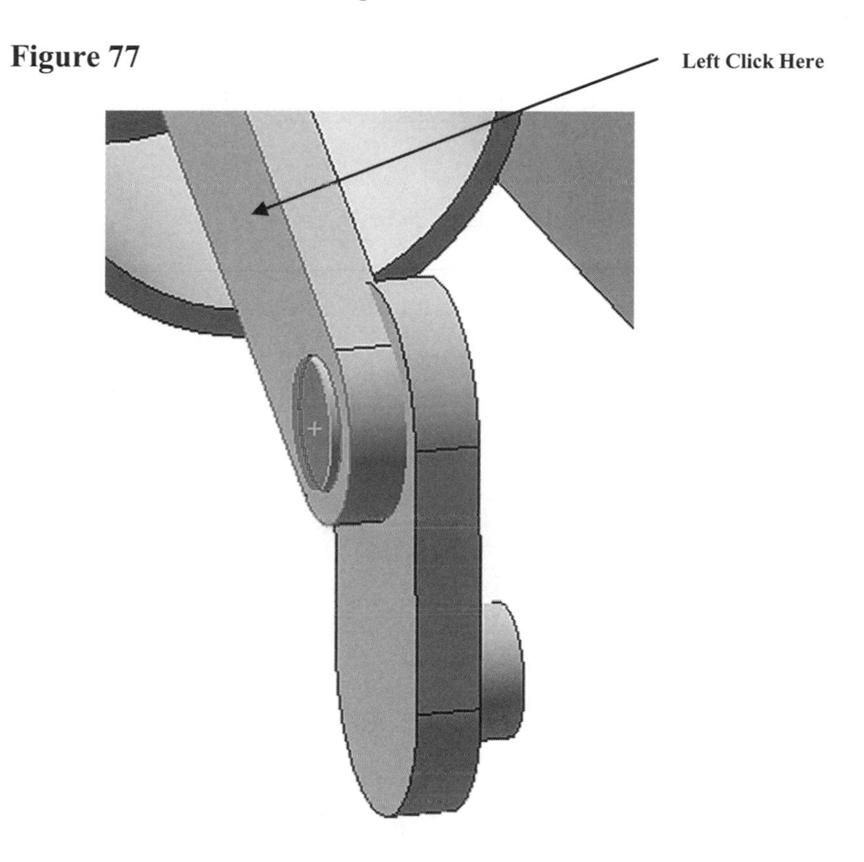

82. Left click on the green checkmark beneath the text "Coincident3" as shown in Figure 78.

Figure 78

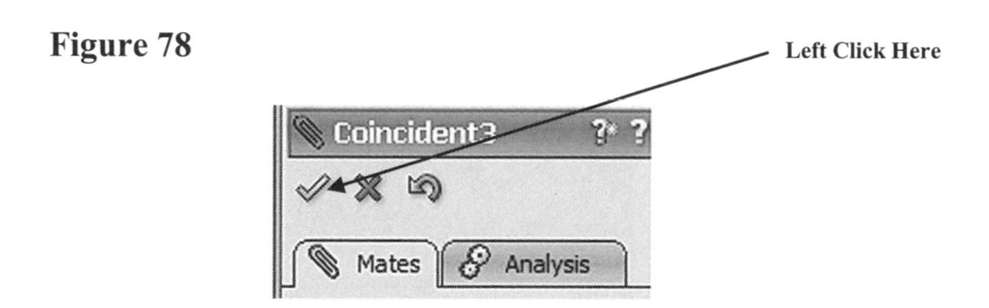

83. Left click on the green checkmark beneath the text "Mate" as shown in Figure 79.

Figure 79

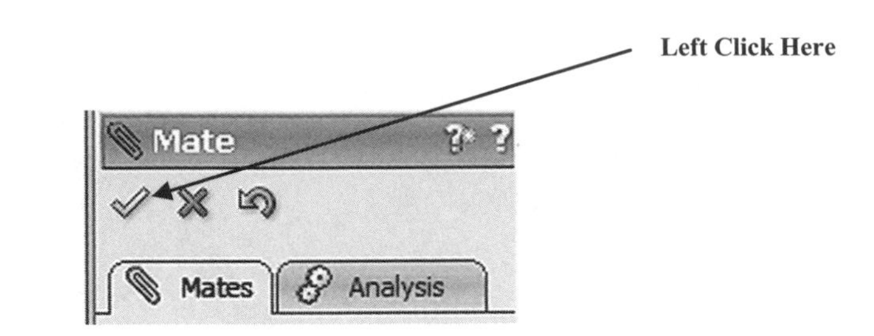

84. SolidWorks will place the connecting rod and crankshaft together as shown in Figure 80.

Figure 80

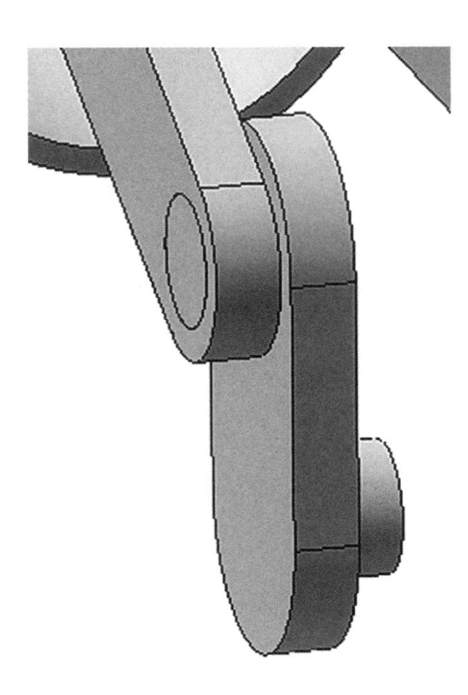

85. Use the rotate command and roll the assembly around to gain access to the opposite side as shown in Figure 81.

Figure 81

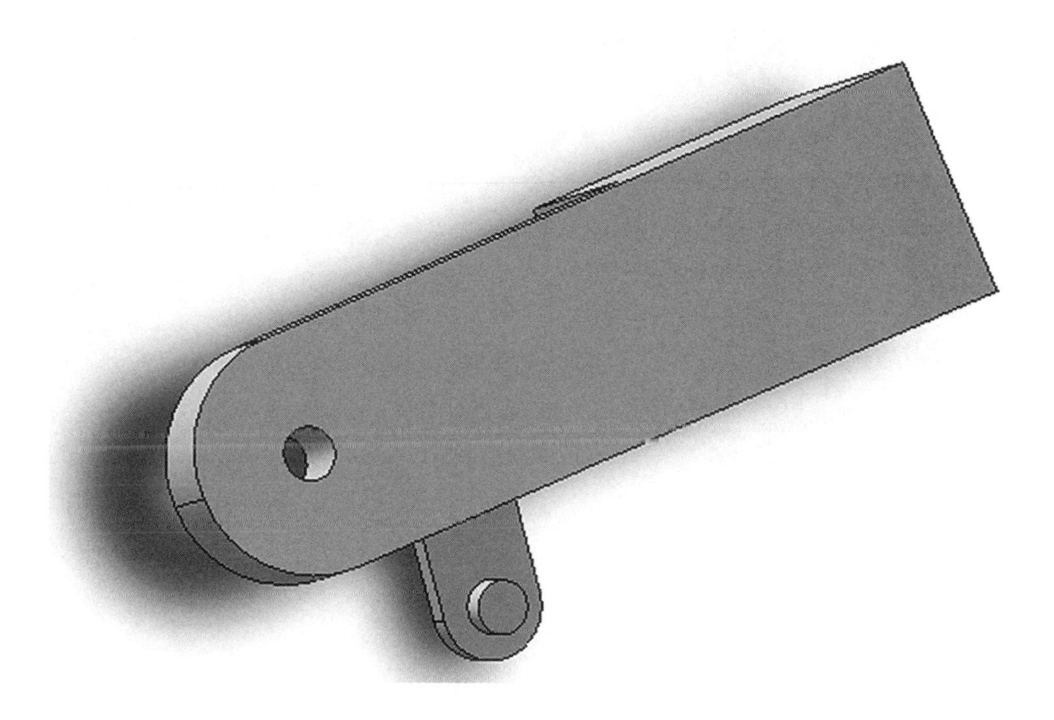

86. Move the cursor to the upper left portion of the screen and left click on **Mate** as shown in Figure 82.

Figure 82

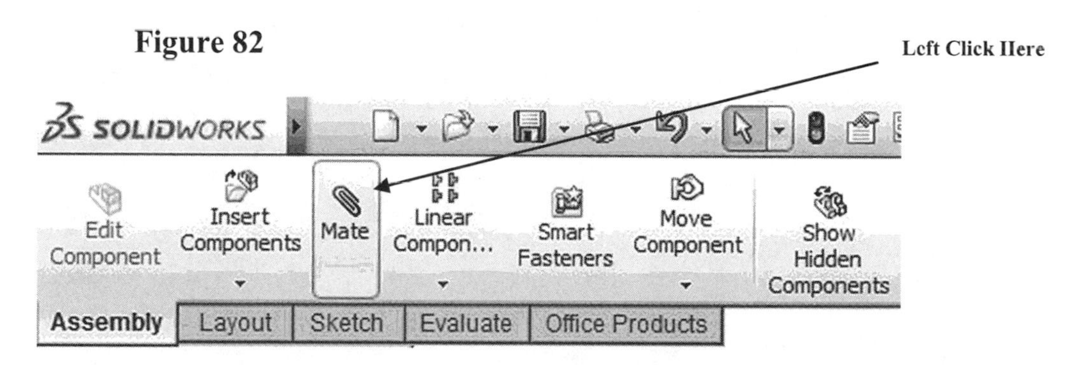

87. Move the cursor to the crankshaft pin that will be secured in the piston case. The edges of the pin will turn red. Left click once as shown in Figure 83.

Figure 83

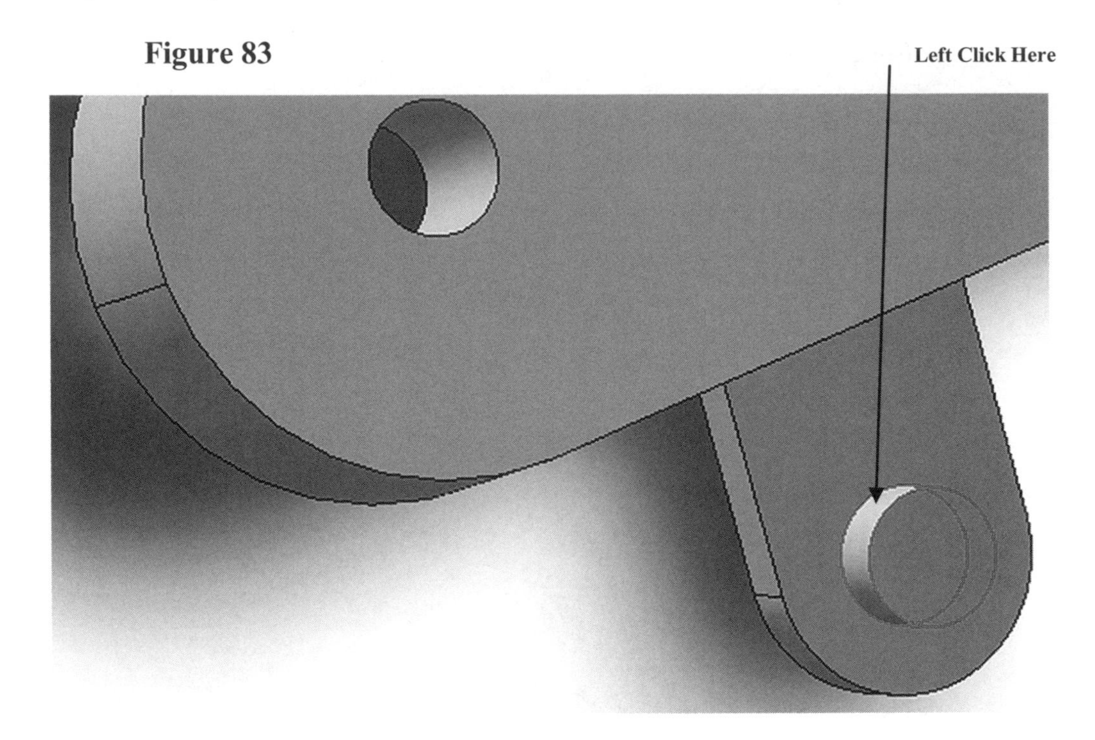

88. Move the cursor to the piston case hole that will secure the crankshaft. The edges of the hole will turn red. Left click once as shown in Figure 84.

Figure 84

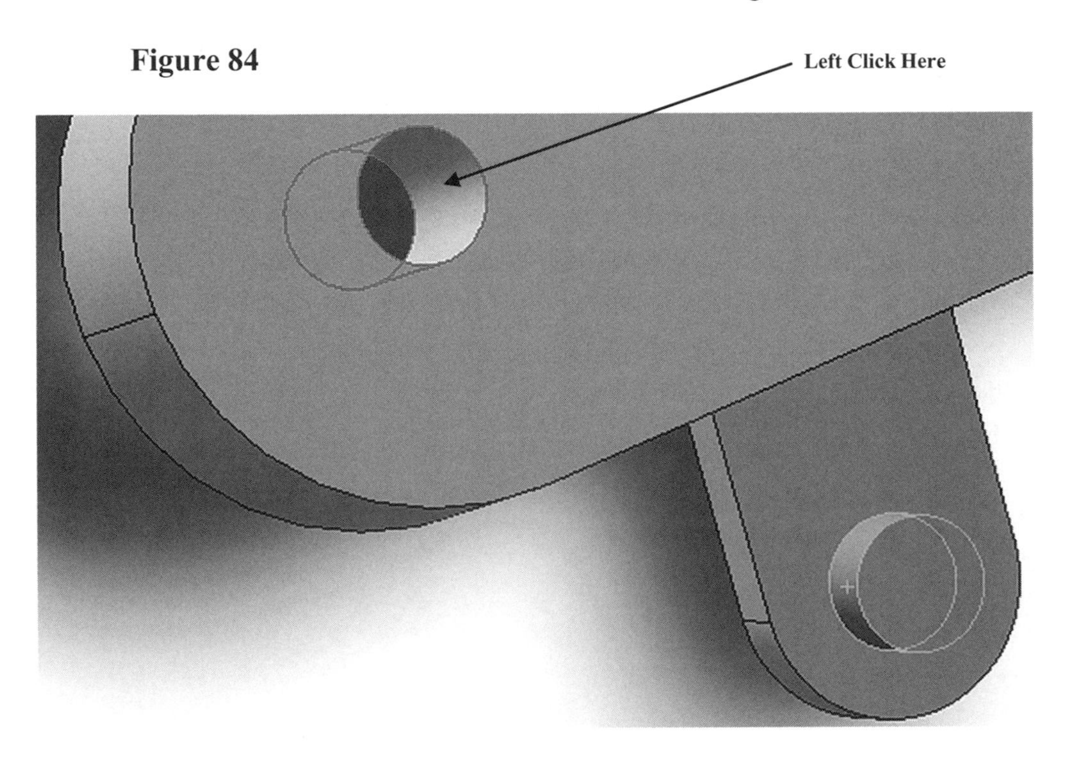

89. SolidWorks will place the crankshaft pin into the piston case as shown in Figure 85.

Figure 85

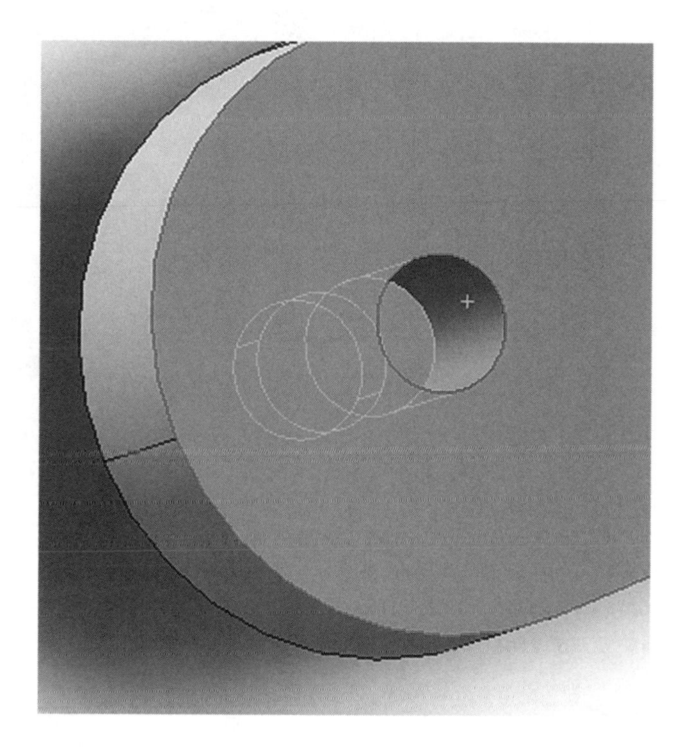

90. Left click on the green checkmark beneath the text "Concentric5" as shown in Figure 86.

Figure 86

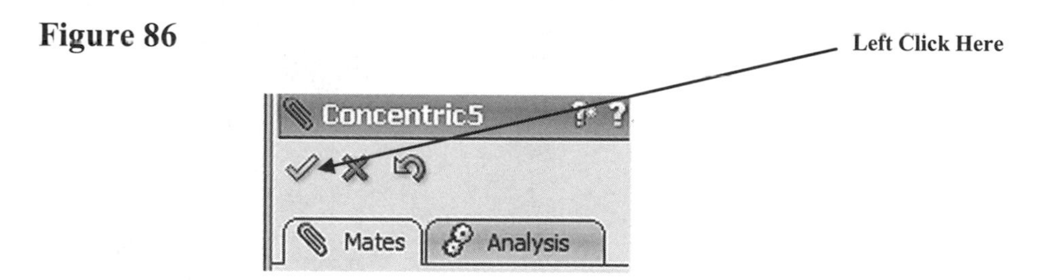

91. Left click on the green checkmark beneath the text "Mate" as shown in Figure 87.

Figure 87

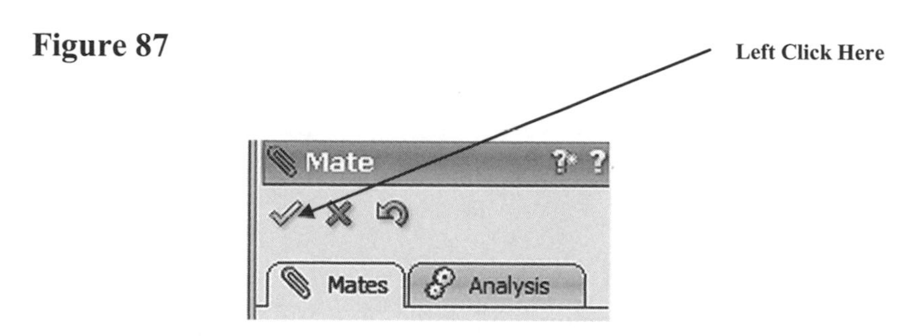

92. Your screen should look similar to Figure 88.

Figure 88

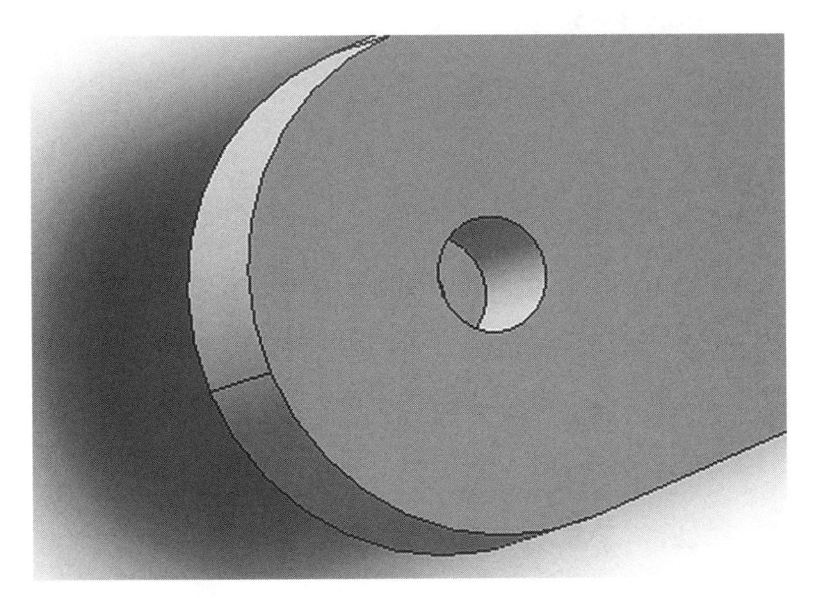

93. Rotate the parts around to view the assembly. Your screen should look similar to Figure 89. Save the file at this time.

Figure 89

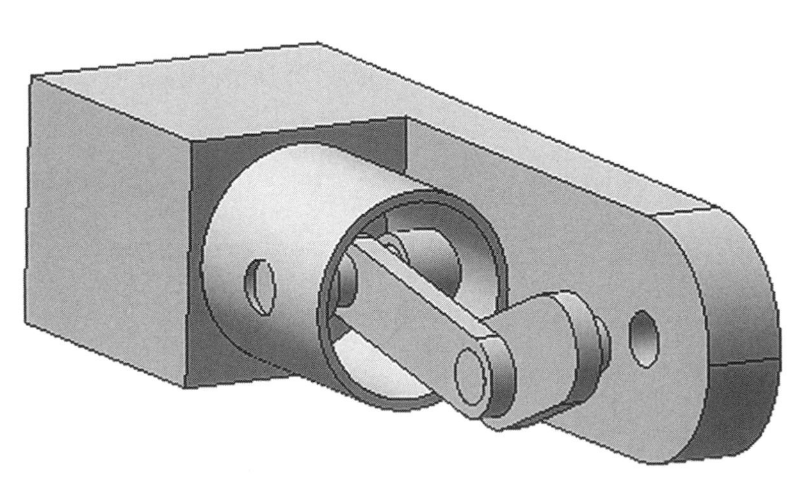

94. The length of the connecting rod must be modified. Move the cursor over the connecting rod as shown in Figure 90. The edges will turn red.

Figure 90

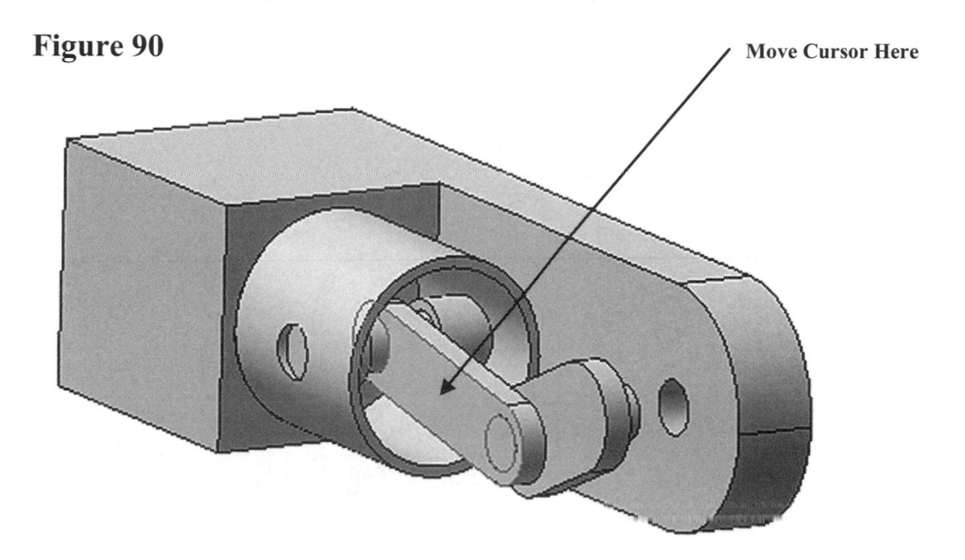

95. Left click once. The surface of the connecting rod will turn green. Right click once. A pop up menu will appear. Left click on **Edit Sketch** as shown in Figure 91.

Figure 91

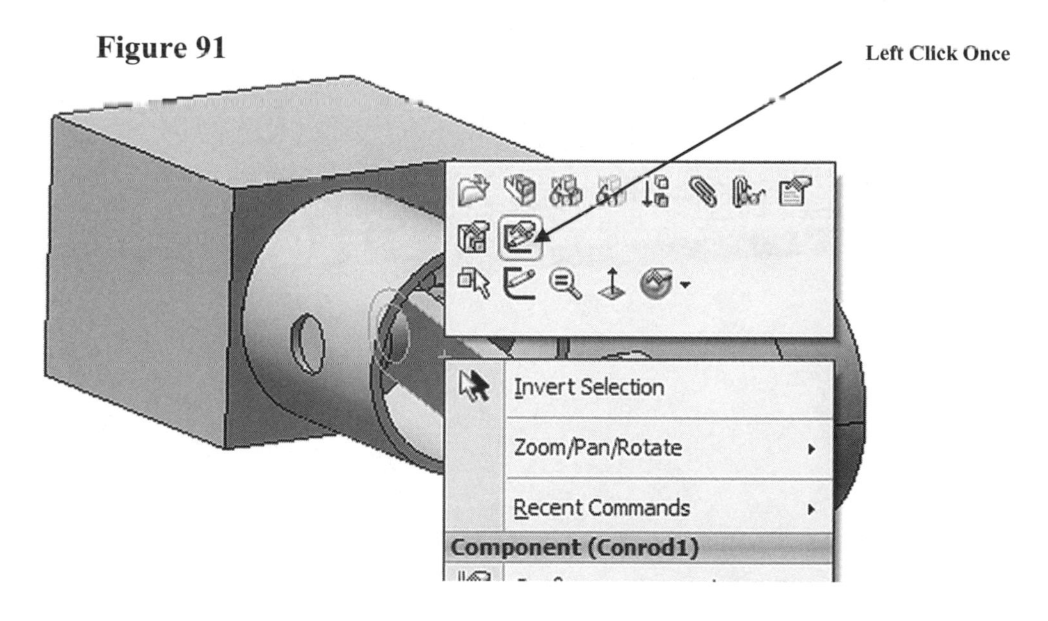

96. The sketch used to create the connecting rod will be displayed as shown in Figure 92. Use the Rotate command to obtain a better view of the connecting rod.

Figure 92

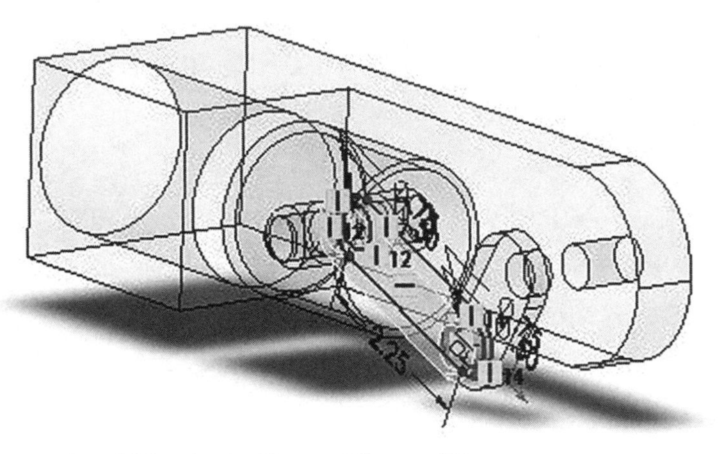

97. Your screen should look similar to Figure 93.

Figure 93

98. Move the cursor over the 2.25 dimension. The dimension will turn red. Double click. The Modify dialog box will appear as shown in Figure 94.

Figure 94

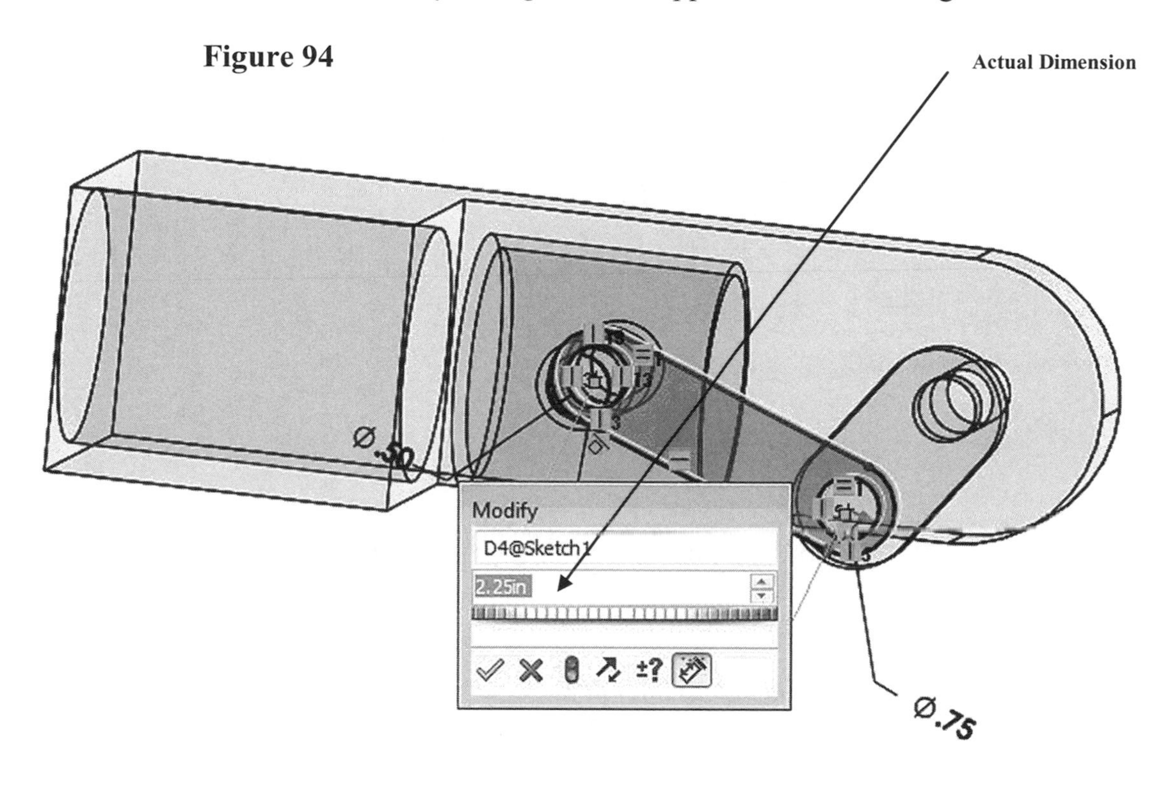

99. While the text is still highlighted, enter **4.5** and left click on the green checkmark as shown in Figure 95.

Figure 95

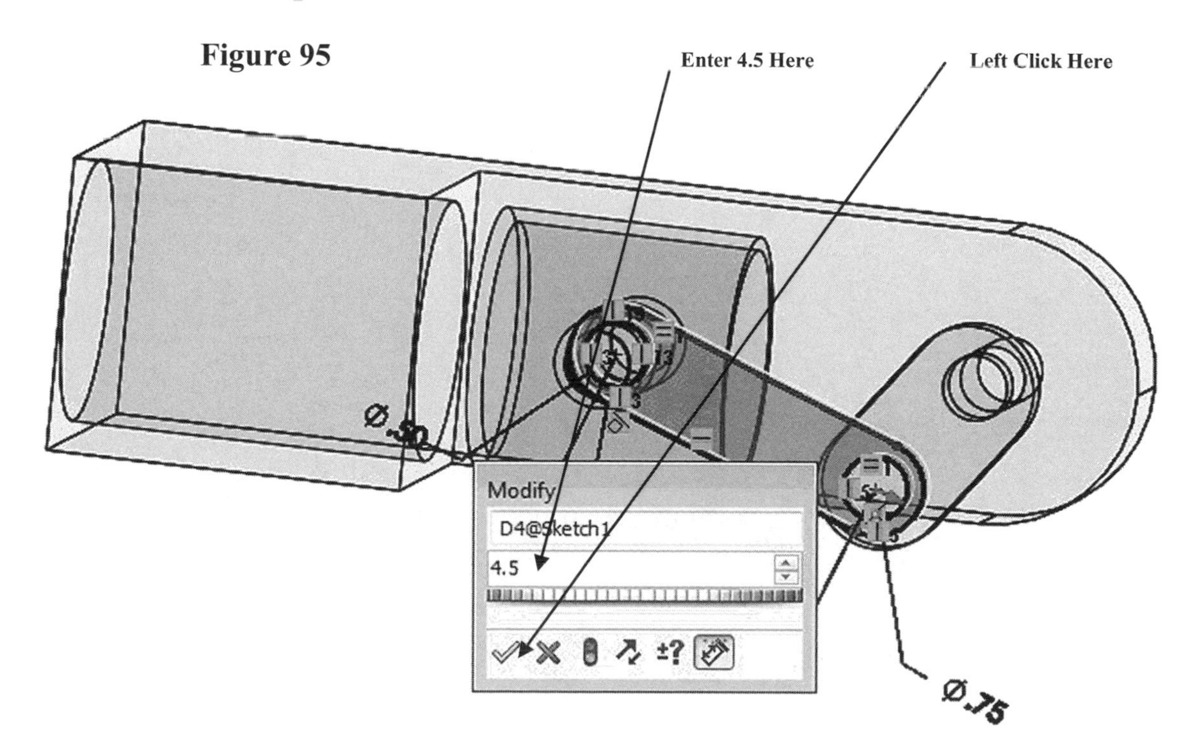

100. The length of the connecting rod will become 4.50 inches as shown in Figure 96.

Figure 96

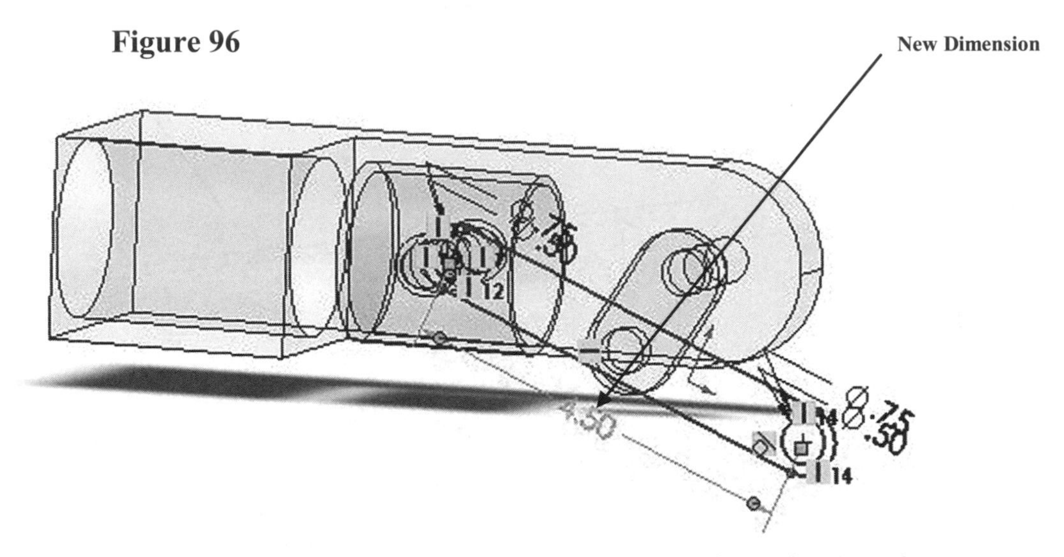

101. Left click on the green checkmark beneath the text "Dimension" as shown as shown in Figure 97.

Figure 97

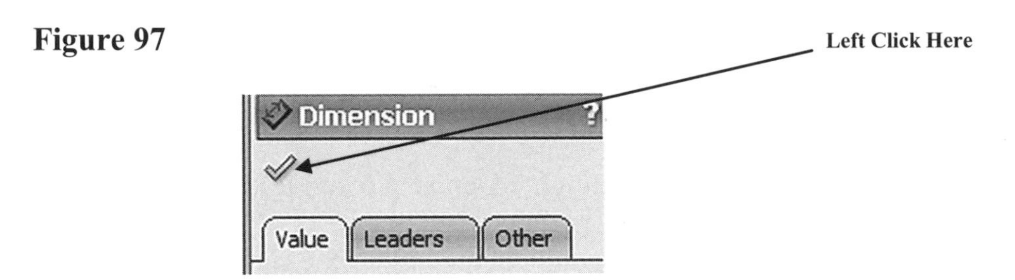

102. Right click anywhere around the parts. A pop up menu will appear. Left click on **Exit Sketch** as shown in Figure 98.

Figure 98

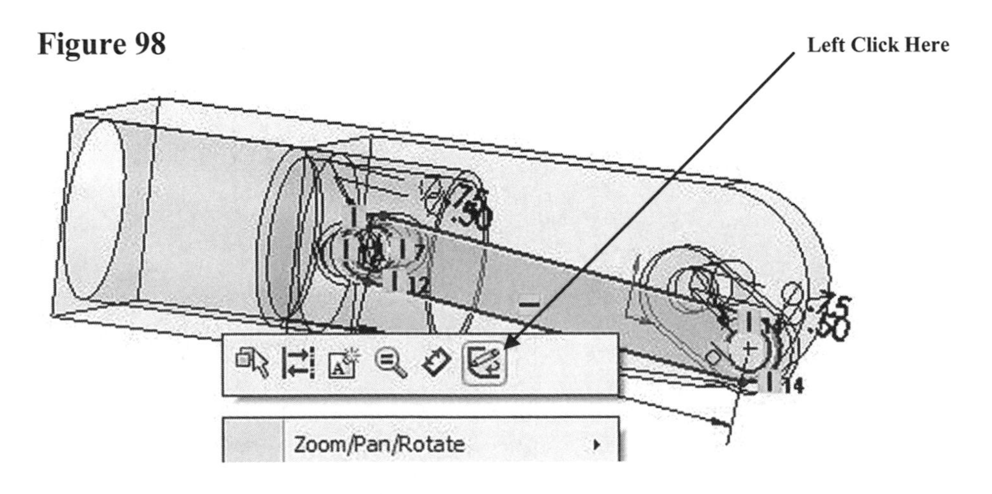

103. Right click anywhere around the parts. A pop up menu will appear. Left click on **Edit Assembly: Assem1** as shown in Figure 99.

Figure 99

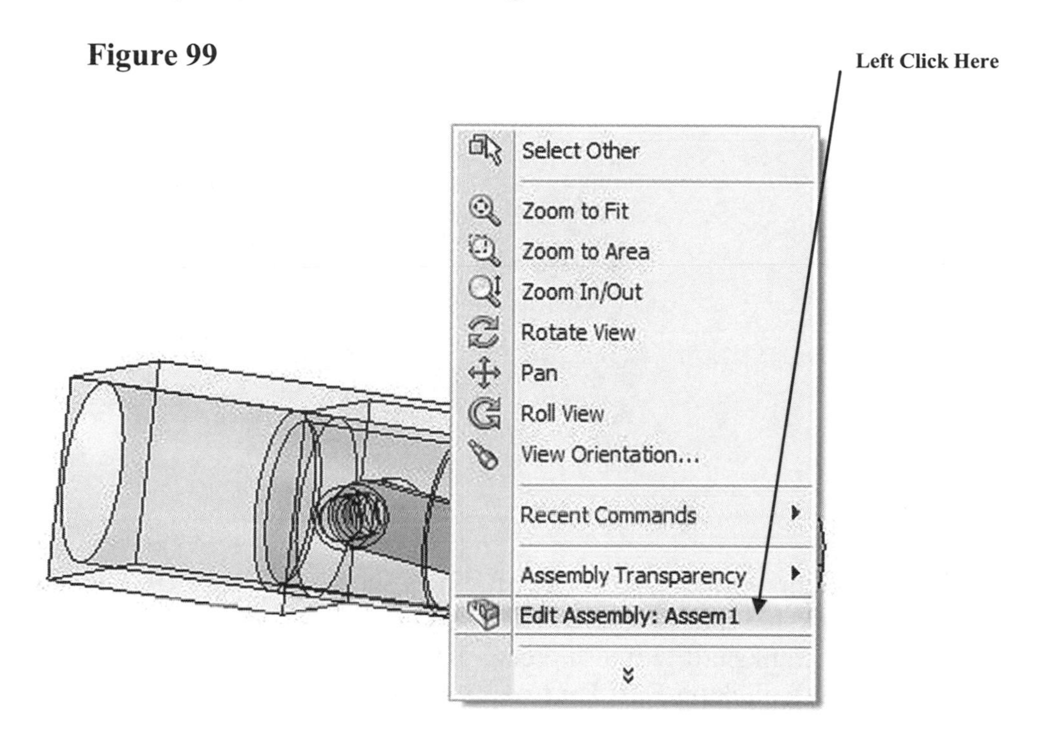

104. The changes made to the connecting rod will be displayed. Your screen should look similar to Figure 100.

Figure 100

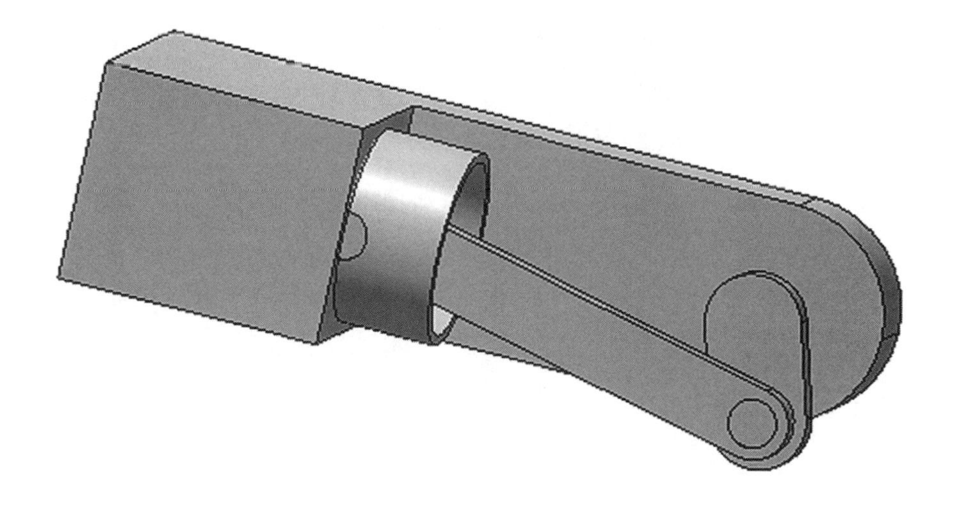

105. The length of the crankshaft pin also must be modified. Rotate the assembly to gain access to the backside of the crankshaft as shown in Figure 101.

Figure 101

Move Cursor Here

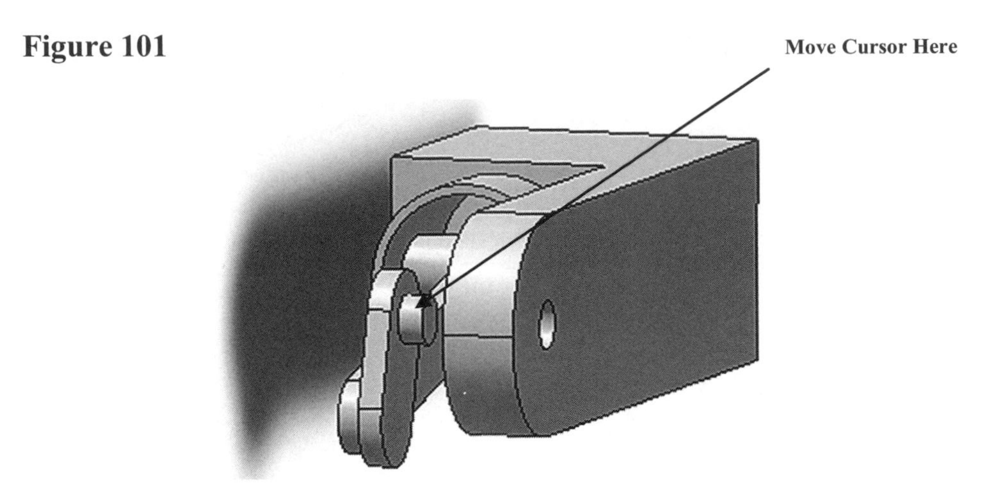

106. Move the cursor over the crankshaft pin as shown in Figure 101. Ensure that the entire crankshaft pin is highlighted. Left click on the crankshaft pin. The entire pin of the crankshaft will turn green. Right click once. A pop up menu will appear. Left click on **Edit Feature** as shown in Figure 102.

Figure 102

Left Click Here

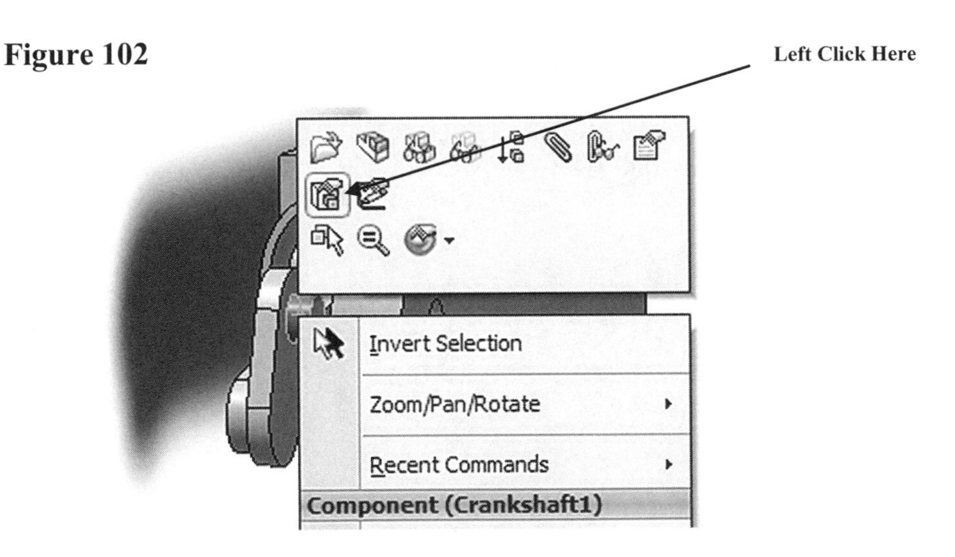

107. The existing extrusion length will be displayed as shown in Figure 103.

Figure 103

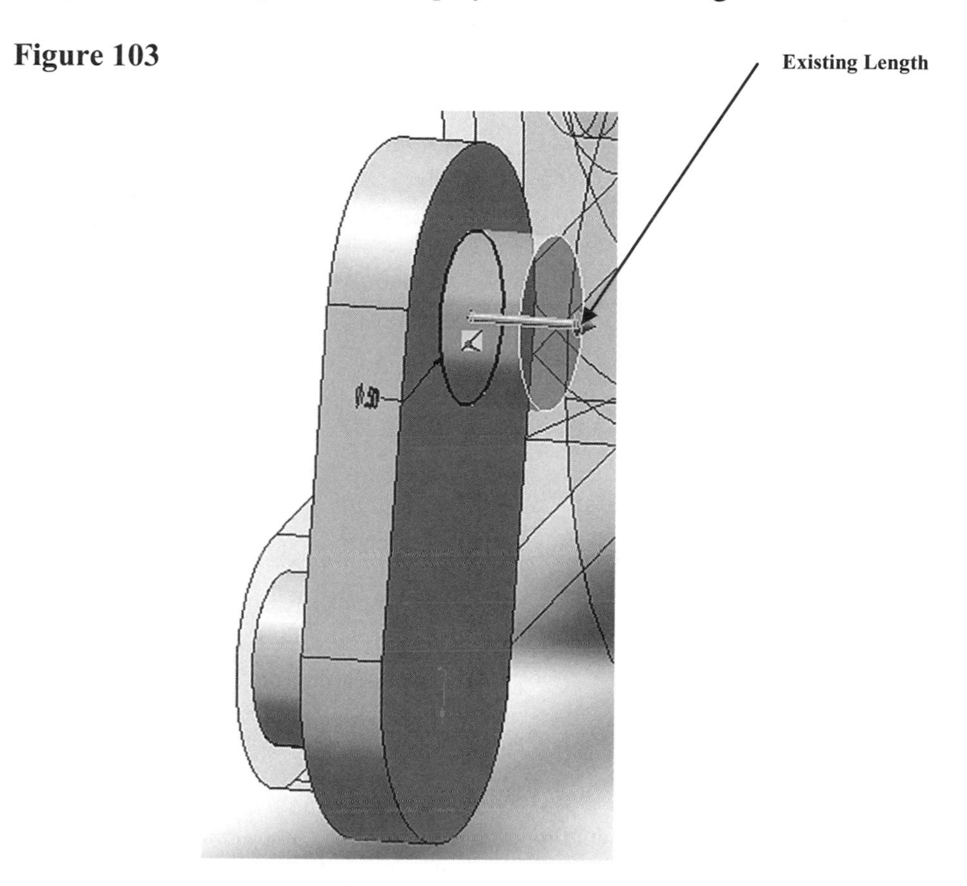

108. Enter **2.00** for the extrusion distance. A preview of the extrusion will be displayed. Left click on the green checkmark as shown in Figure 104.

Figure 104

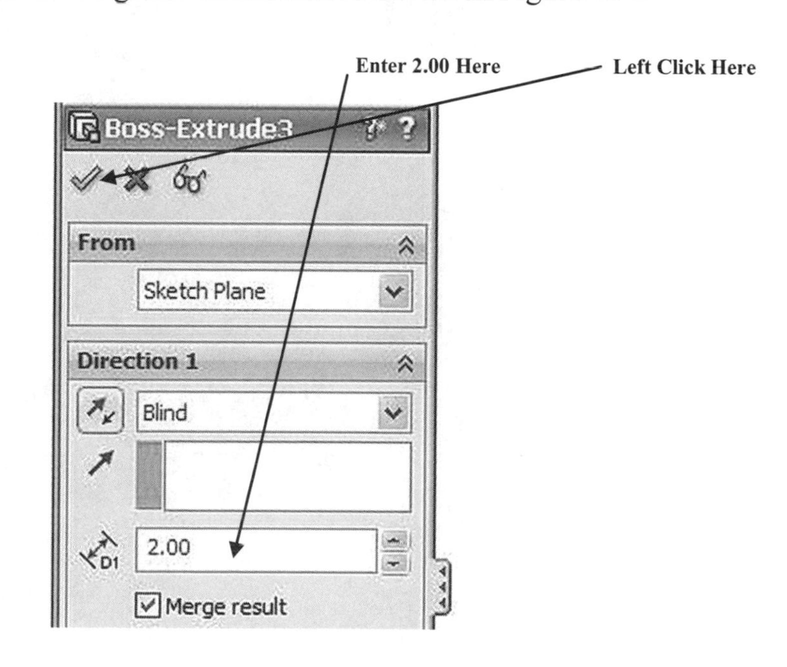

109. Right click anywhere around the parts. A pop up menu will appear. Left click on **Edit Assembly: Assem1** as shown in Figure 105.

Figure 105

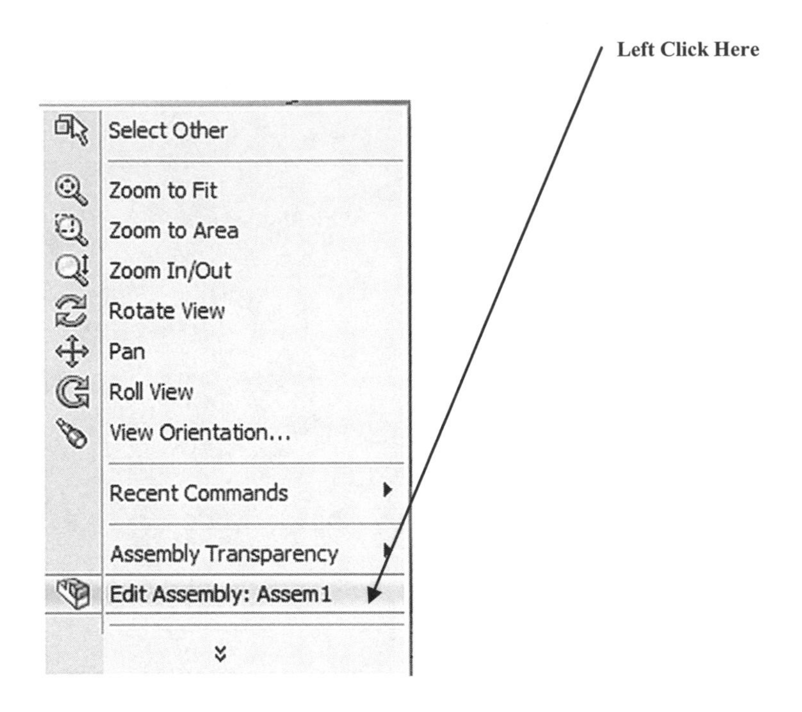

110. The changes made to the crankshaft will be displayed. Your screen should look similar to Figure 106.

Figure 106

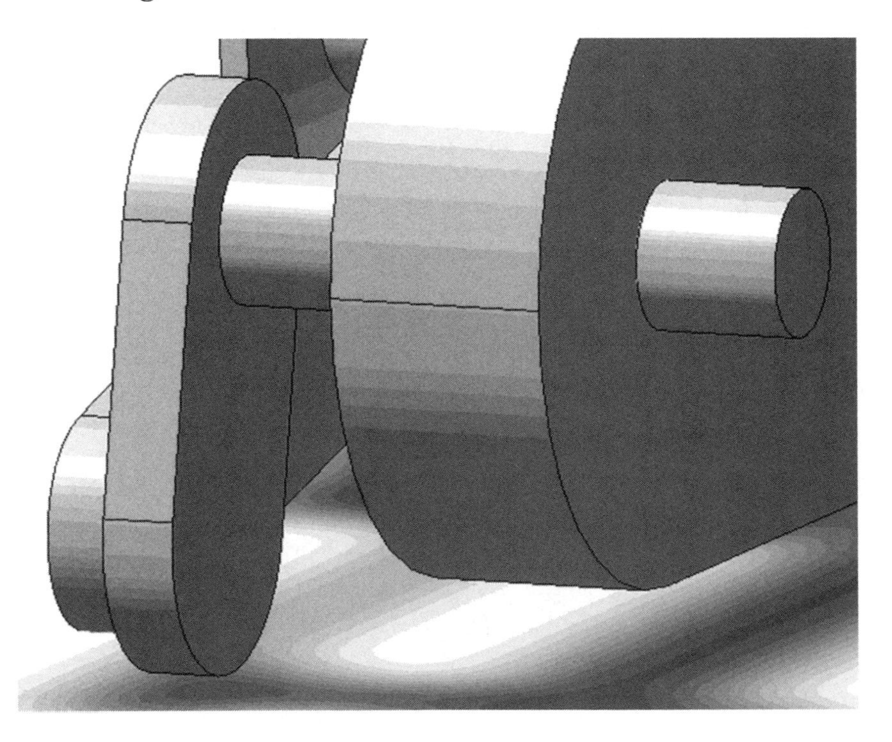

111. Rotate the part around as shown in Figure 107.

Figure 107

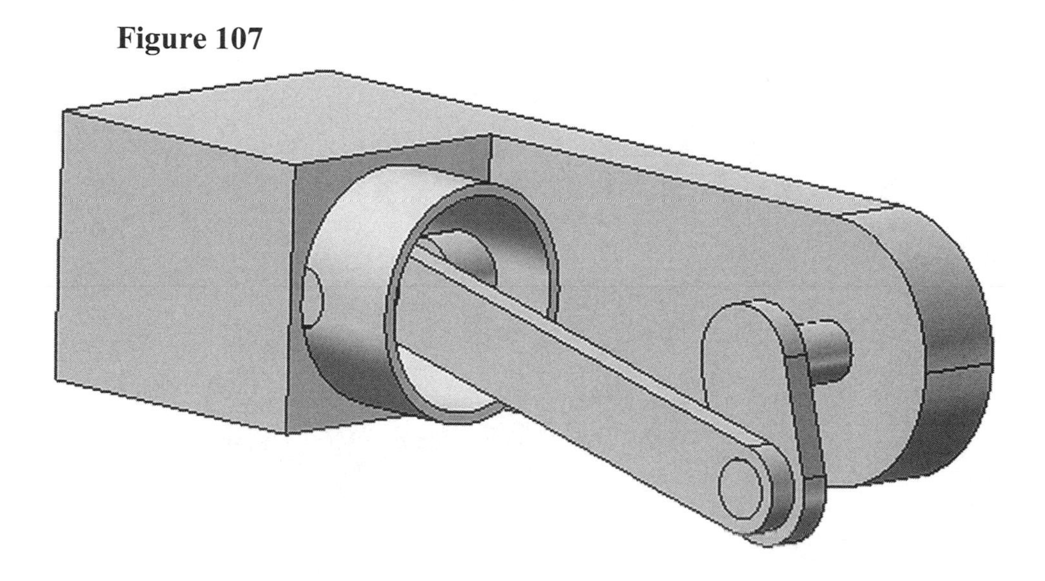

112. Move the cursor to the lower left portion of the screen and left click on **Motion Study 1** as shown in Figure 108.

Figure 108

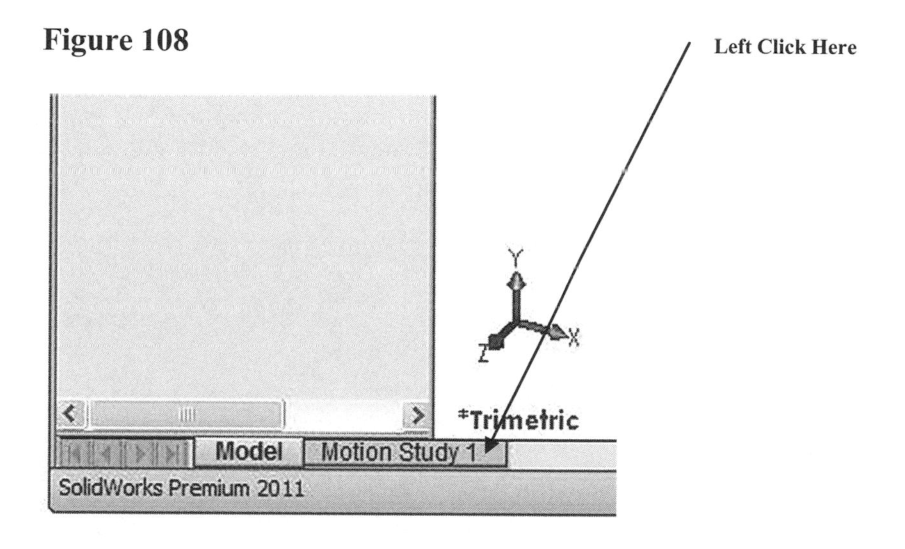

113. The Study Motion dialog box will appear. Your screen should look similar to Figure 109.

Figure 109

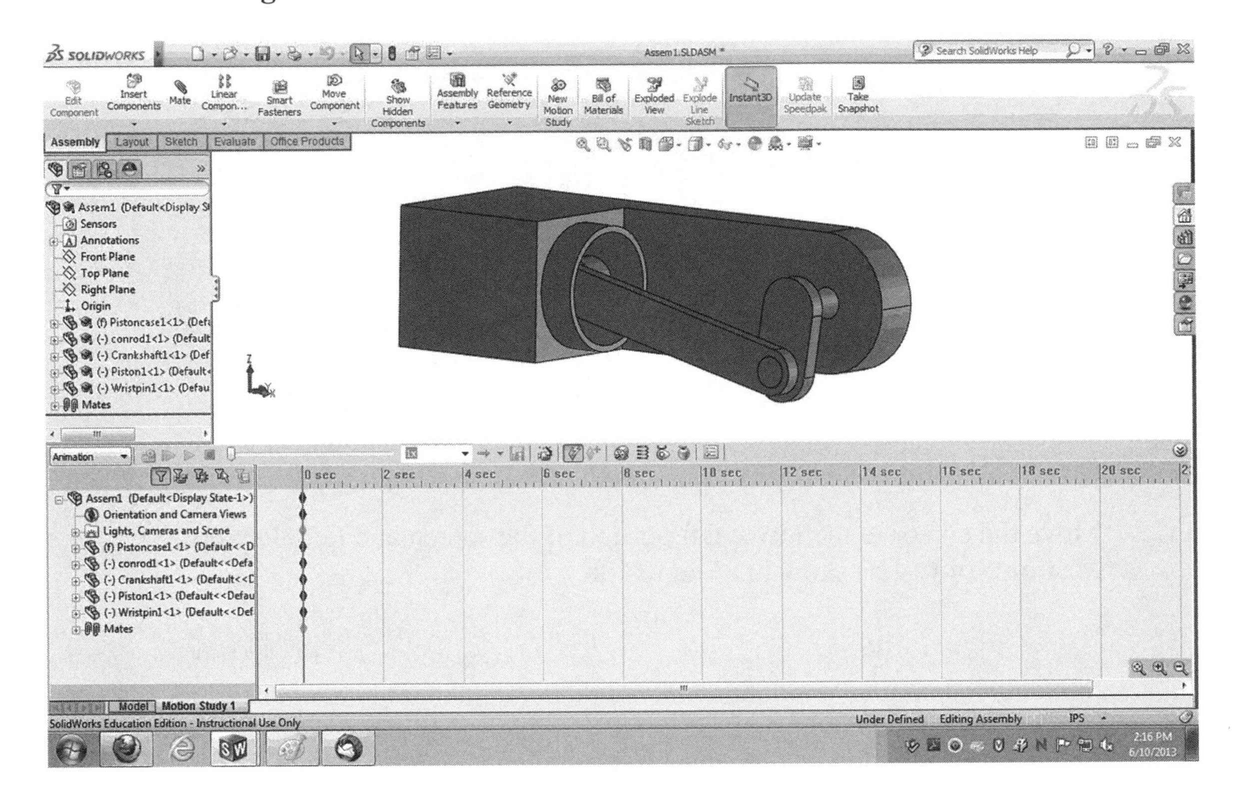

114. Move the cursor to the middle right portion of the screen and left click on "Motor" icon as shown in Figure 110.

Figure 110

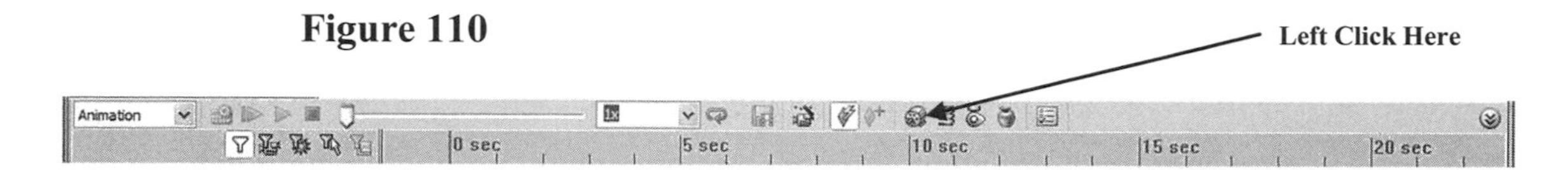

115. Move the cursor to the crankshaft pin that protrudes through the piston case. The crankshaft pin will turn red as shown in Figure 111.

Figure 111

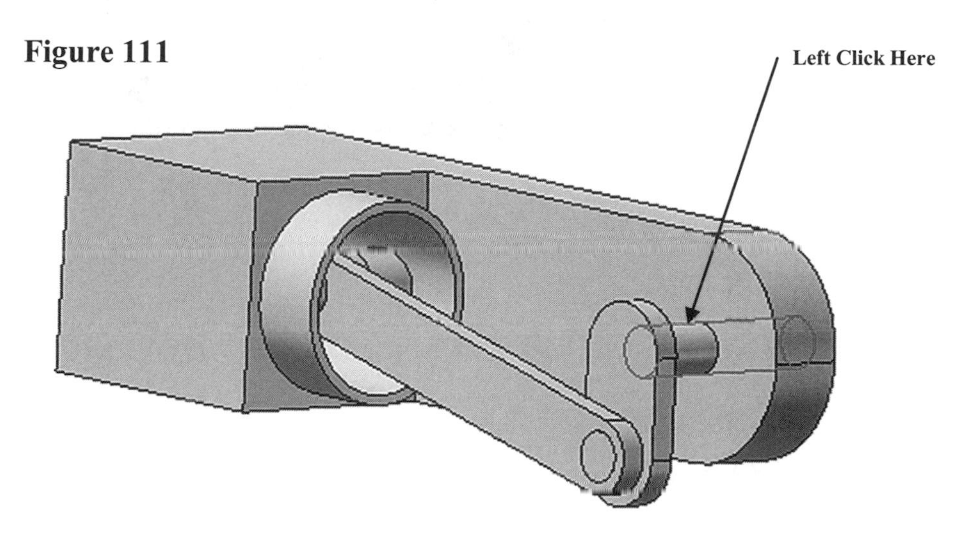

116. Left click once. A red arrow will be displayed showing the direction of motion as shown in Figure 112.

Figure 112

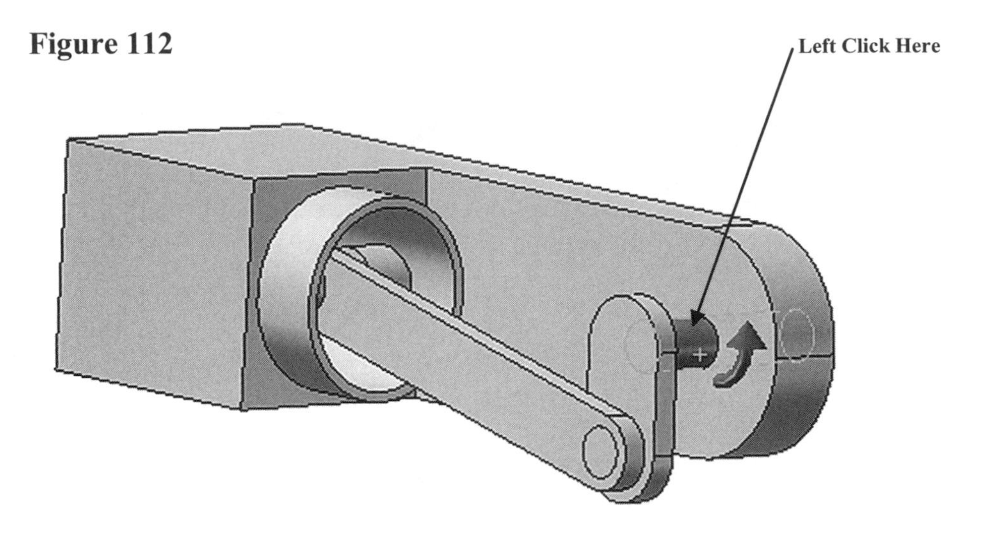

117. Left click on the green checkmark beneath the text "Motor" as shown in Figure 113.

Figure 113

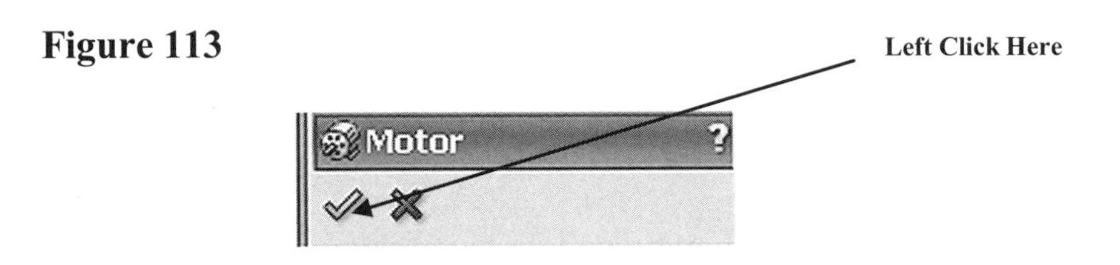

118. Move the cursor to the lower left portion of the screen and left click on the "Calculate" icon as shown in Figure 114.

Figure 114

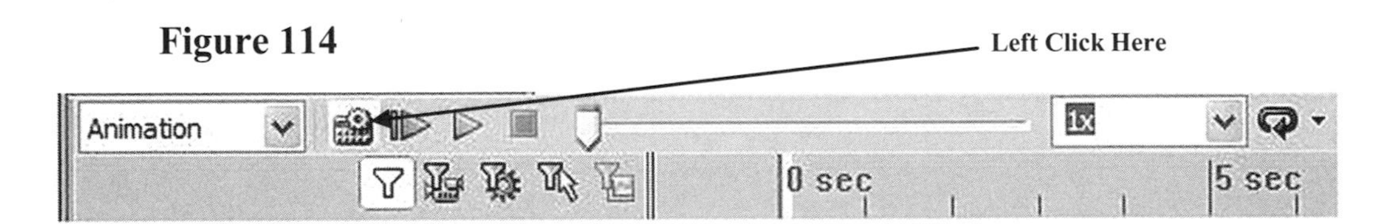

119. SolidWorks will begin animating the part. If SolidWorks detects any part collision or interference, the animation will stop playing.

120. Move the cursor to the lower left portion of the screen and left click on the "Play" icon as shown in Figure 115.

Figure 115

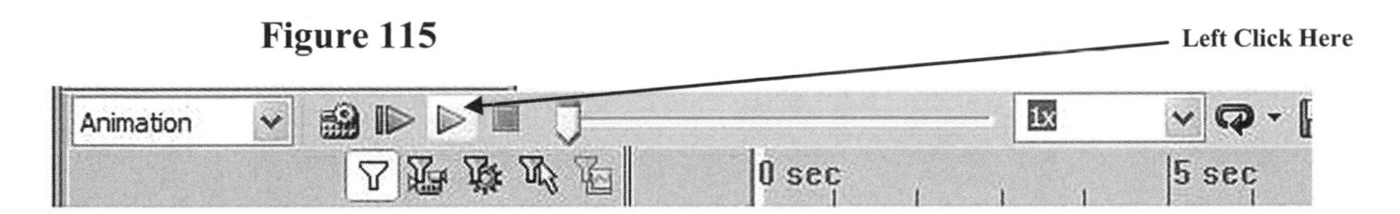

121. The Animation can be controlled by increasing or decreasing the speed or looping. Left click on the Motion Study Properties icon as shown in Figure 116. The Motion Study Properties panel will appear at the left (not shown). Increase the frames per second to 30 to smooth out the animation. Left click on the minimize icon to enlarge the animation to full screen.

Figure 116

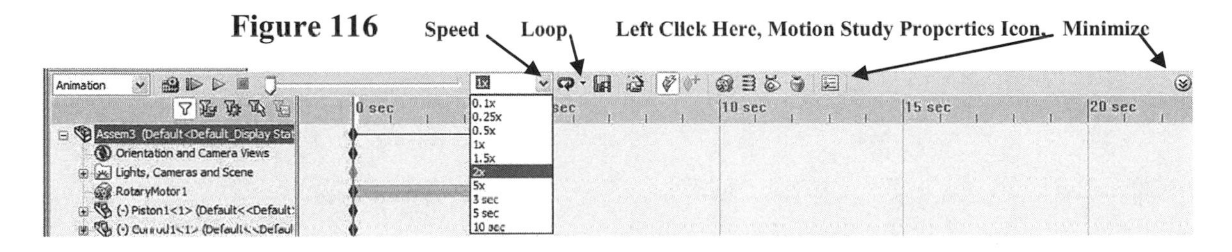

122. Save the assembly where it can easily be retrieved. When closing SolidWorks, a dialog box will appear indicating that models making up the assembly have been modified. Left click on **Yes** as shown in Figure 117.

Figure 117

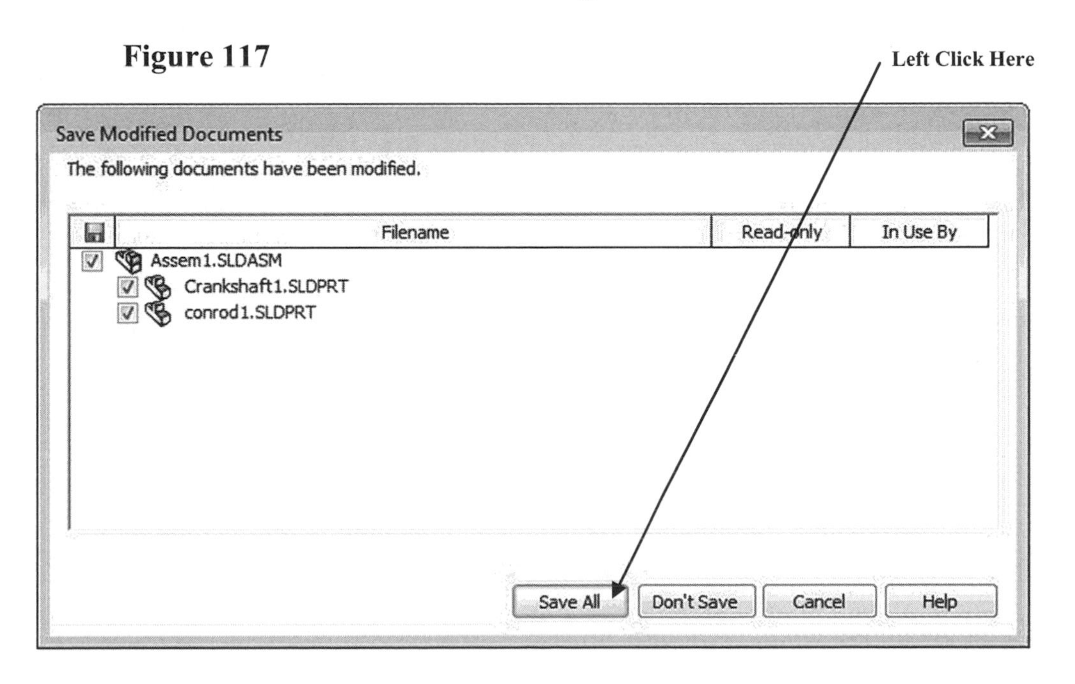

Chapter 8 Introduction to Advanced Commands

Objectives:

- Learn to use the Swept Boss/Base command
- Learn to use the Lofted/Boss Base command
- Learn to use the Plane command

Chapter 8 includes instruction on how to design the parts shown below.

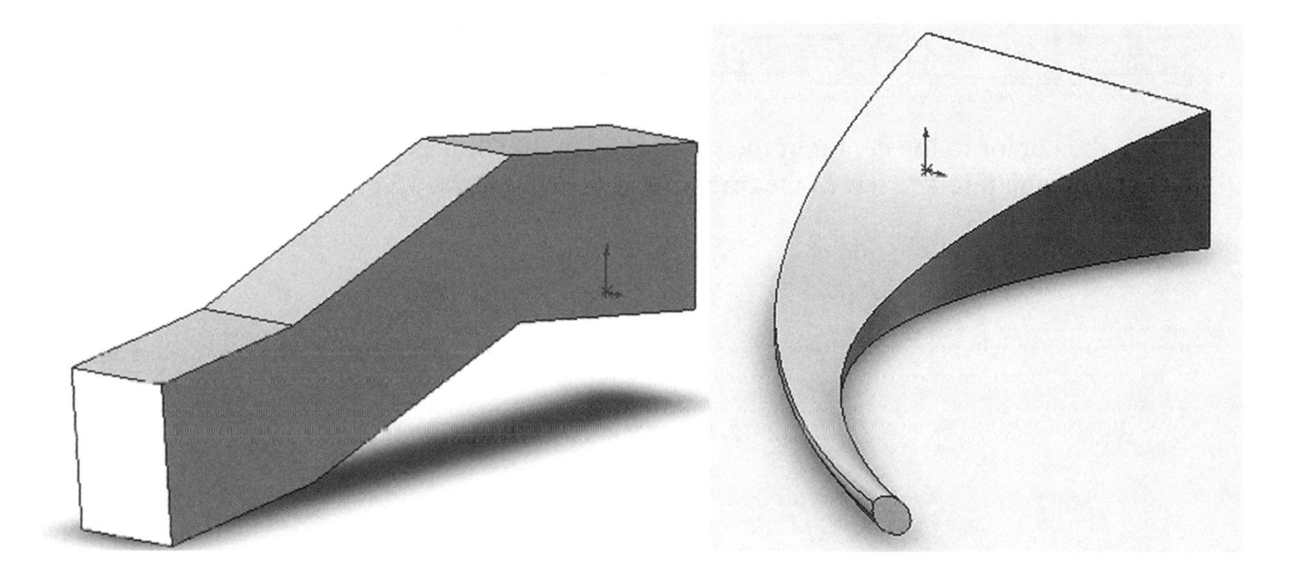

1. Start SolidWorks by referring to "Chapter 1 Getting Started".

2. After SolidWorks is running, begin a New Sketch.

3. Move the cursor to the upper left portion of the screen and left click on **Line** as shown in Figure 1.

Figure 1

Left Click Here

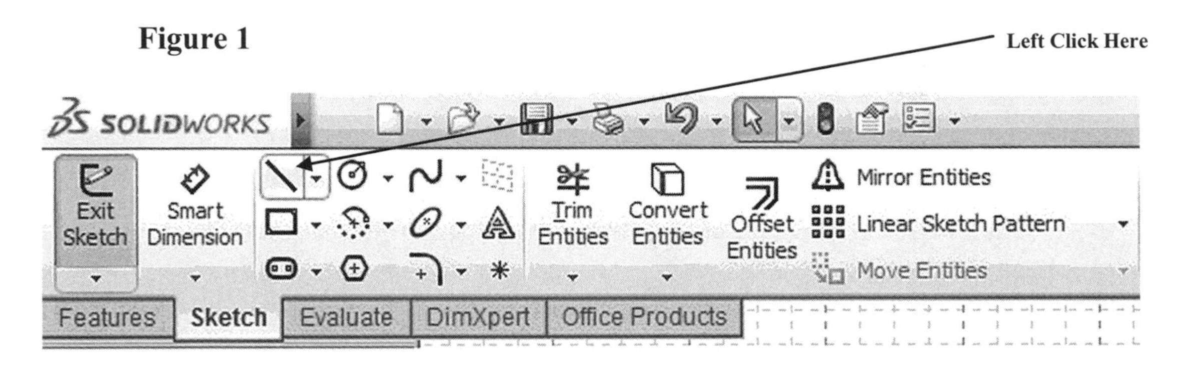

4. Move the cursor to the center of the screen and left click once. Ensure that the red dot appears on the intersection of the origin as shown in Figure 2.

Figure 2

Left Click Here

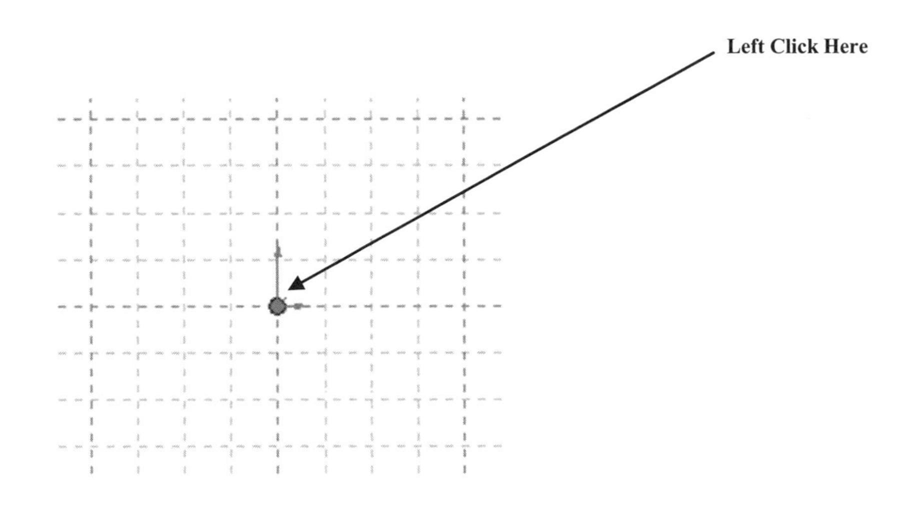

5. Move the cursor to the left portion of the screen and left click once as shown in Figure 3. Right click anywhere on the screen. A pop up menu will appear. Left click on **Select**.

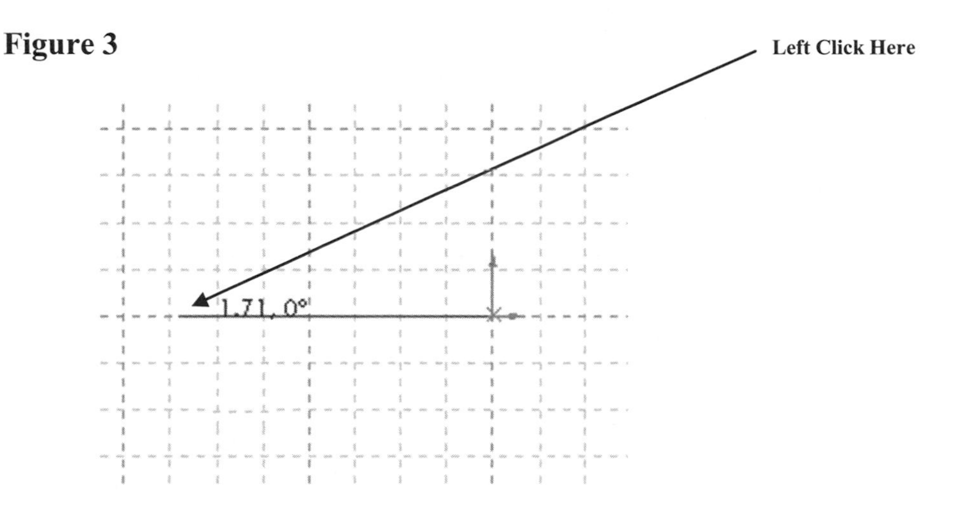

Figure 3

6. Move the cursor to the upper left portion of the screen and left click on **Line** as shown in Figure 4.

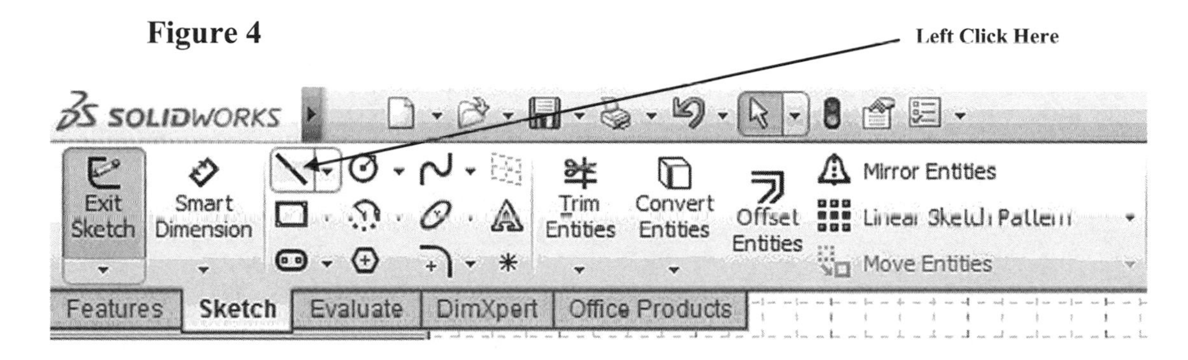

Figure 4

7. Move the cursor to the center of the grid and left click once as shown in Figure 5. Move the cursor to the left and left click (draw a short line going off to the left). Move the cursor to the upper left portion of the screen and left click on **Rectangle**. Complete the sketch shown in Figure 5. Be sure to start the rectangle at the origin. After the sketch has been dimensioned, delete the horizontal line that was drawn first. Right click anywhere on the screen. A pop up menu will appear. Left click on **Select**.

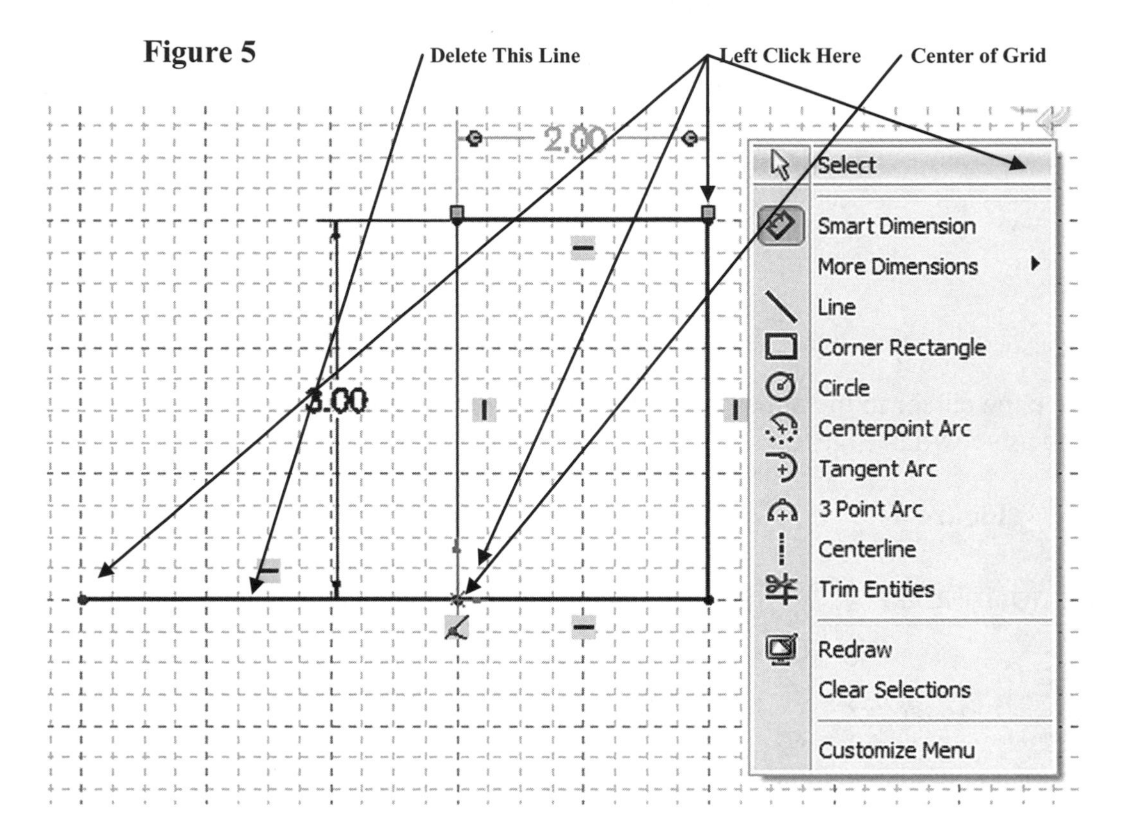

8. Right click anywhere on the screen. A pop up menu will appear. Left click **Exit Sketch** as shown in Figure 6.

Figure 6

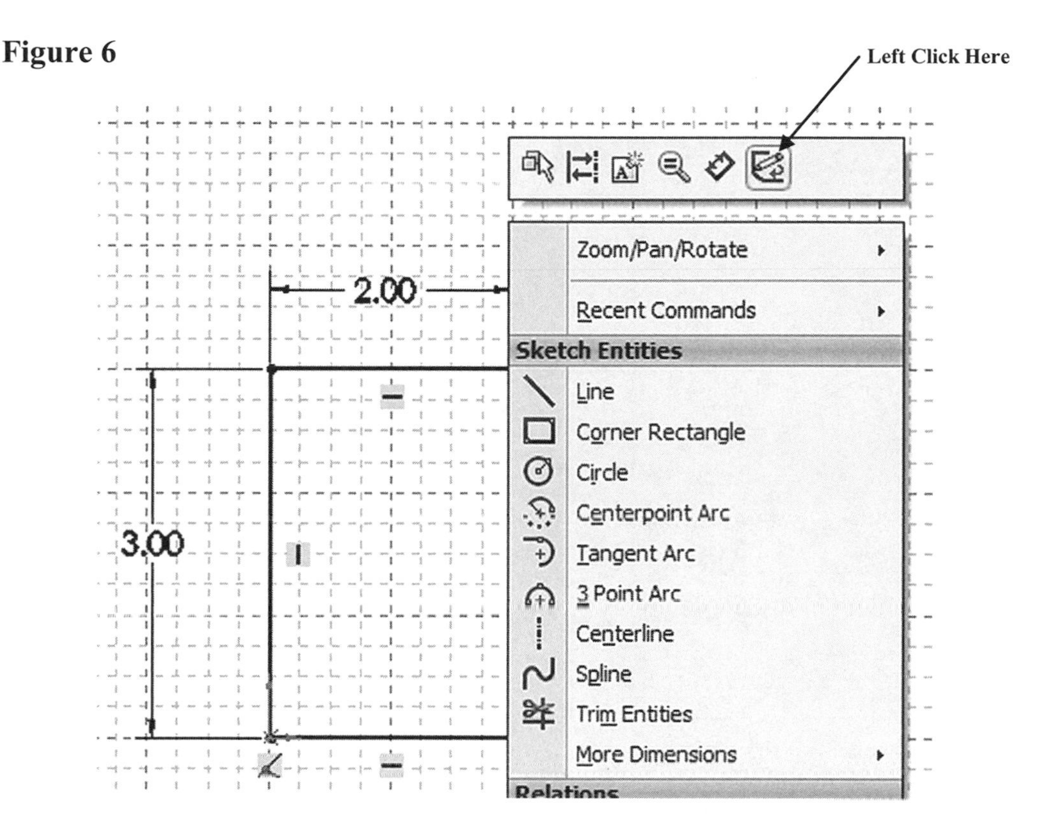

9. Move the cursor to the upper right portion of the screen and left click on the drop down arrow next to the Standard Views icon. A drop down menu will appear. Left click on **Trimetric** as shown in Figure 7.

Figure 7

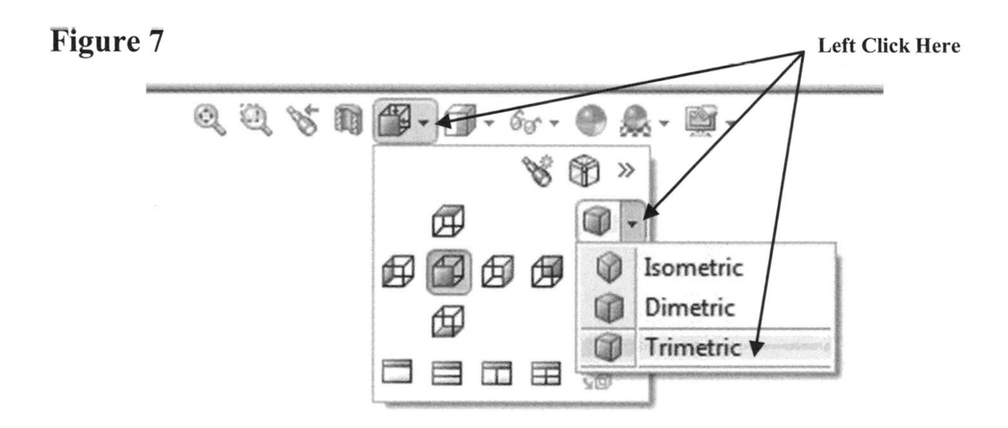

10. Your screen should look similar to Figure 8.

Figure 8

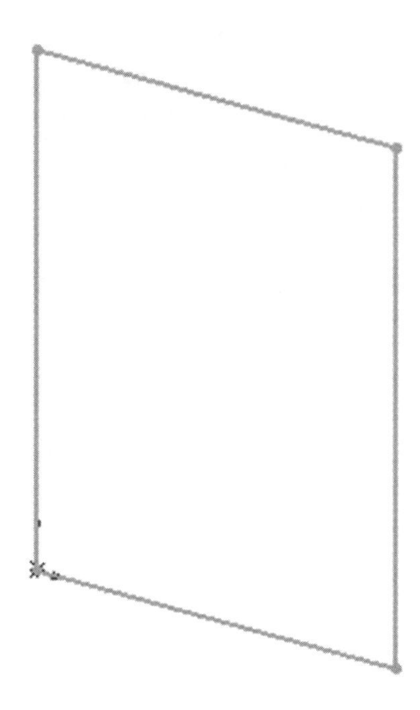

11. Move the cursor over the text "Right Plane" causing the text to become highlighted. Notice the right plane is highlighted. Right click once. A pop up menu will appear. Left click on **Sketch** as shown in Figure 9.

Figure 9

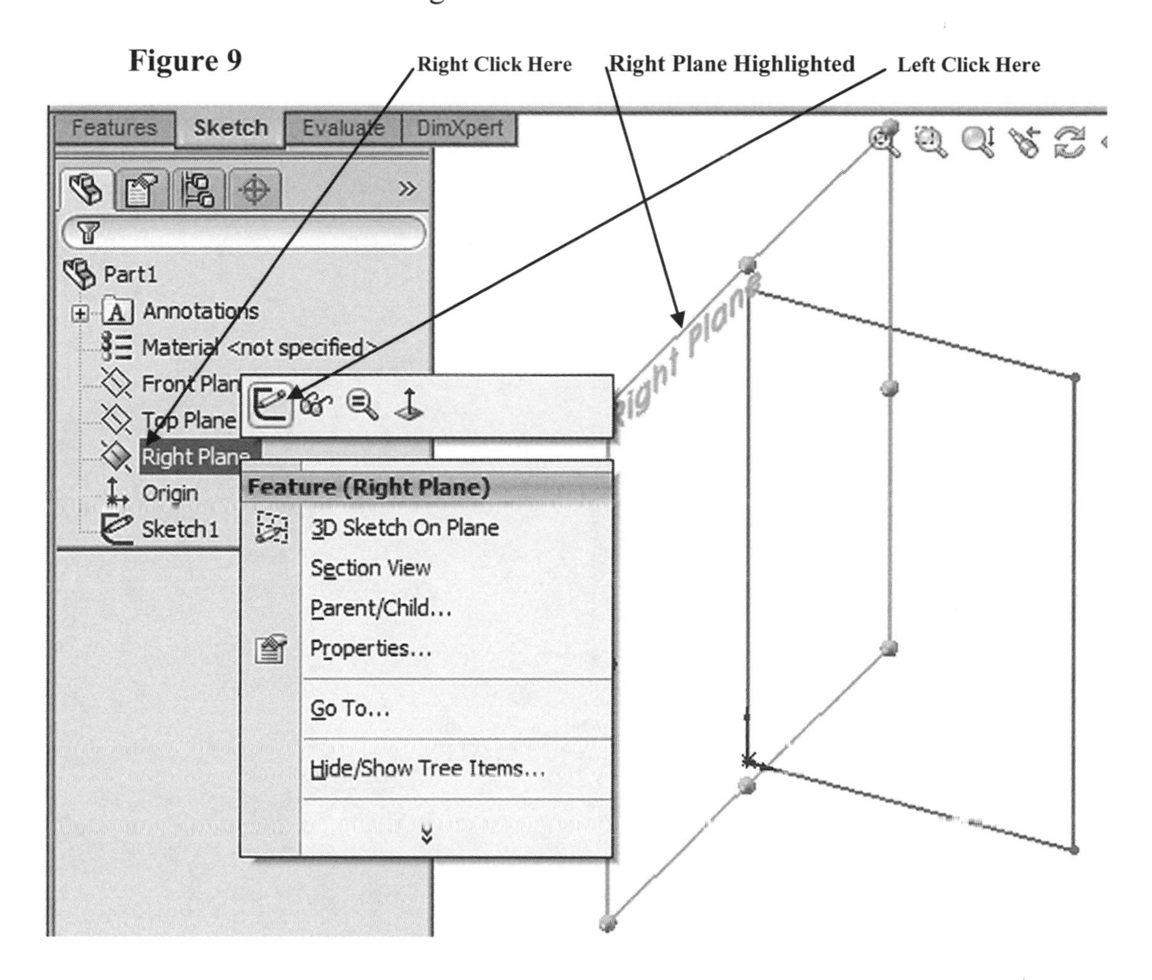

12. Your screen should look similar to Figure 10.

Figure 10

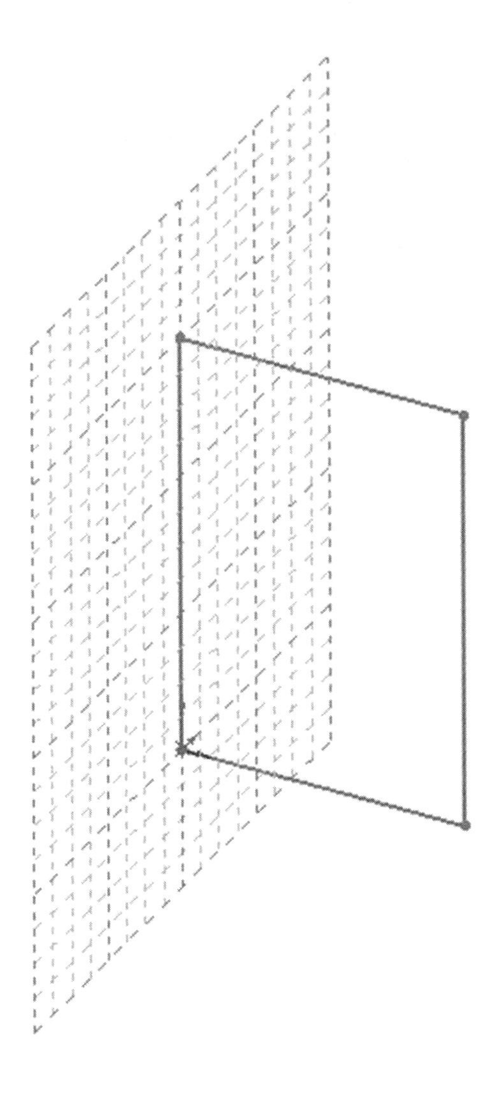

13. Complete the following sketch. Use the Rotate command to provide a better view of the sweep line. The angle of the lines can be estimated. **The sketch lines must intersect with the corner of the 2 inch by 3 inch box as shown in Figure 11**. To dimension on an angle keep the cursor close to the line to be dimensions (while the dimension is still attached to the cursor) then left click once. Press the **Esc** key on the keyboard after completing the dimensioning.

Figure 11

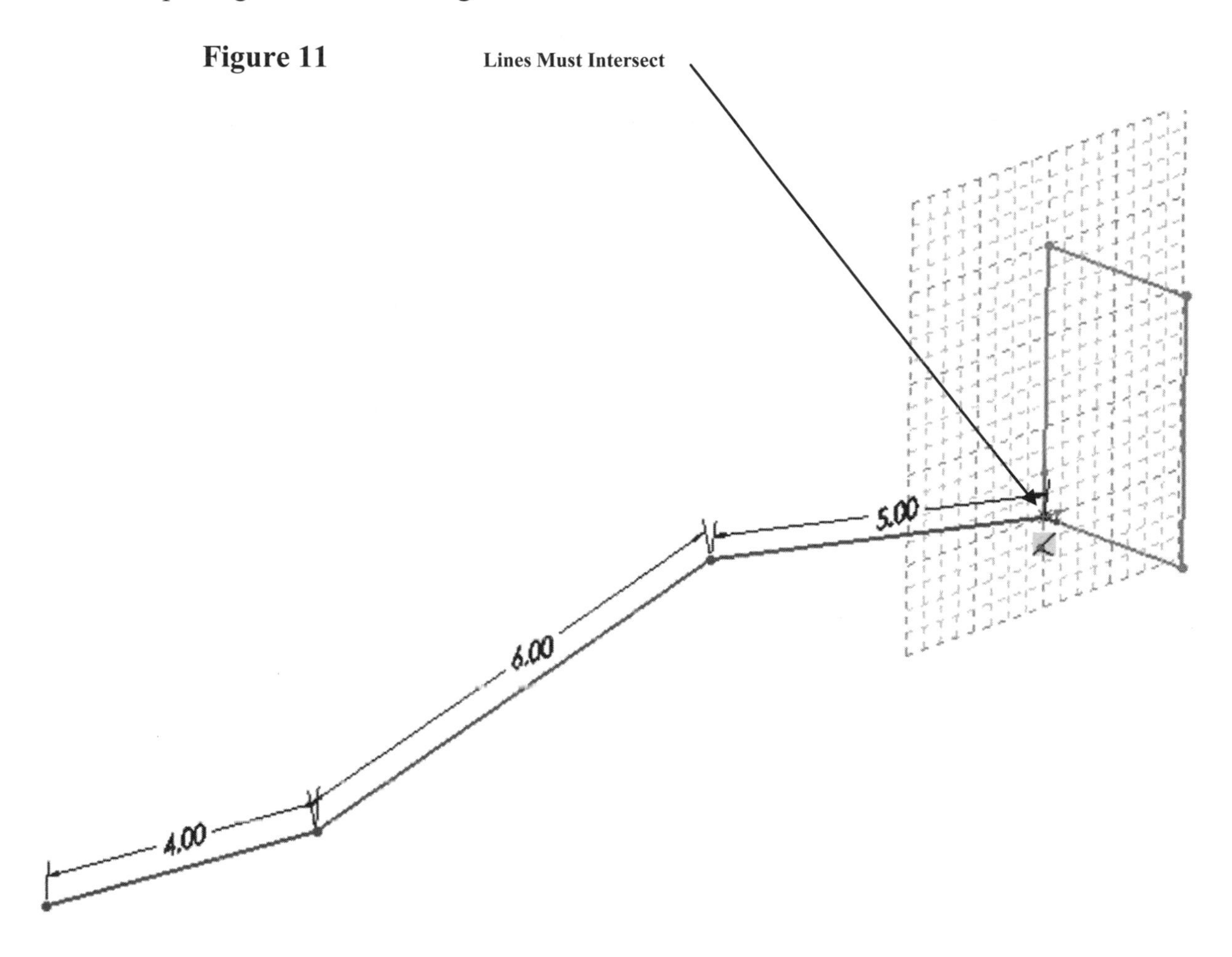

14. Right click anywhere on the screen. A pop up menu will appear. Left click on **Exit Sketch** as shown in Figure 12.

Figure 12

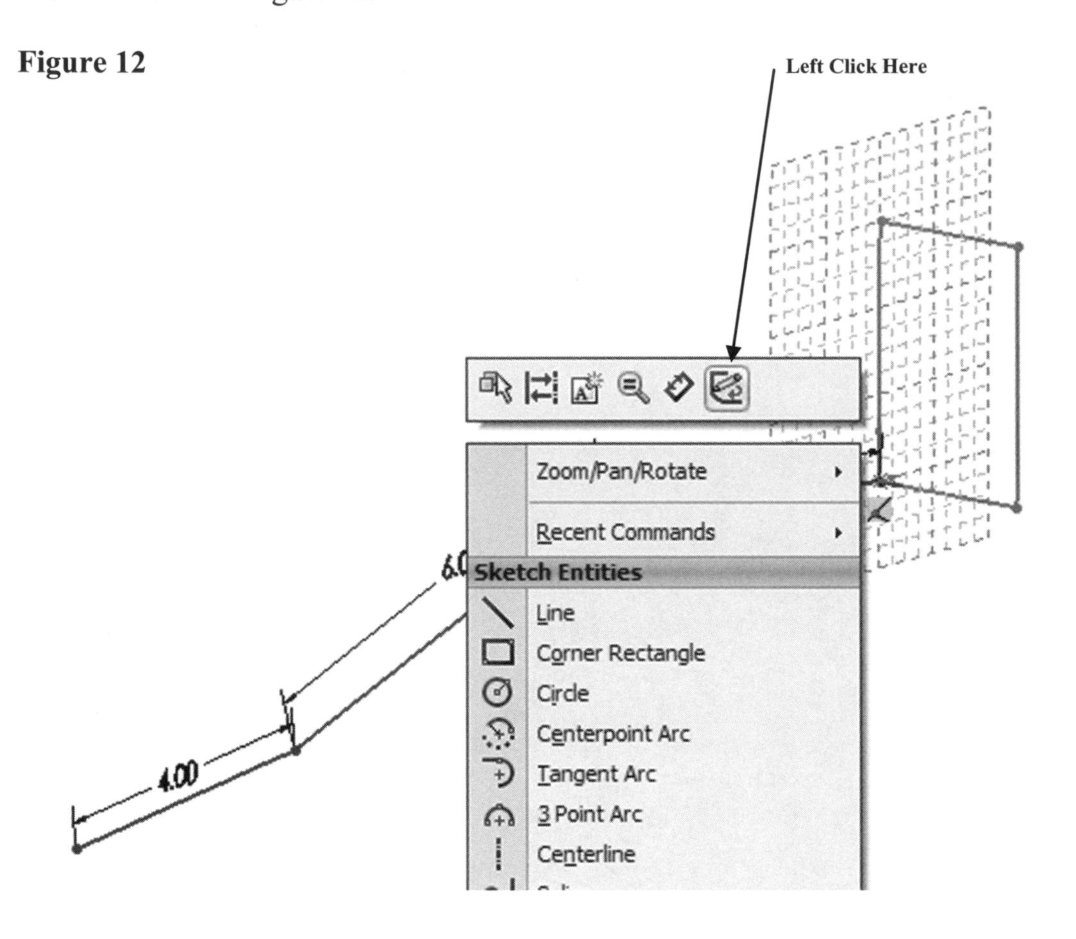

15. Your screen should look similar to Figure 13.

Figure 13

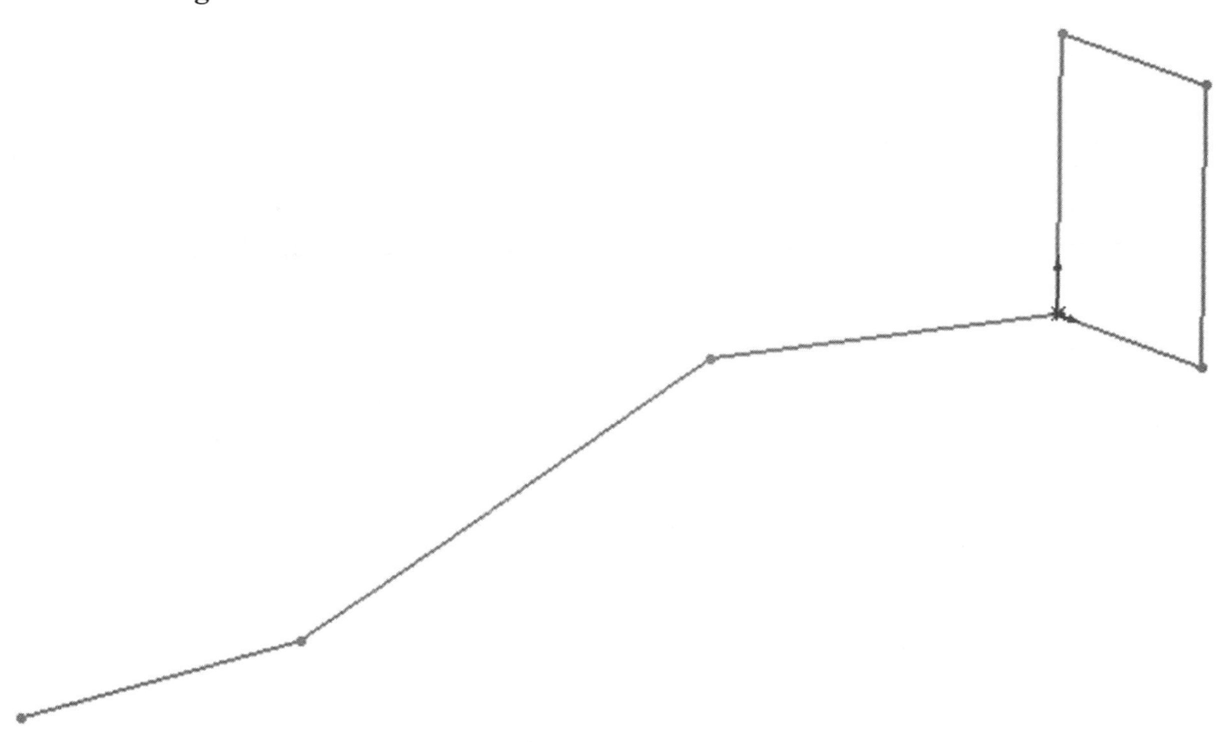

16. Move the cursor to the upper left portion of the screen and left click on **Swept Boss/Base** as shown in Figure 14.

Figure 14

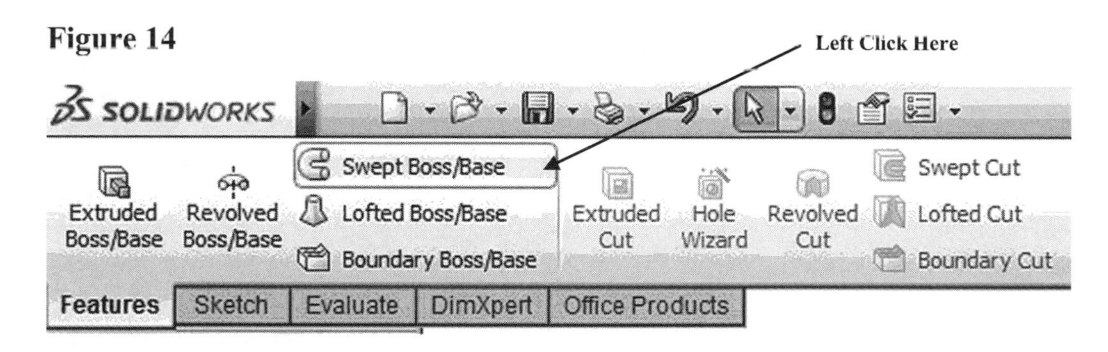

17. Left click on one of the lines that make up the rectangle as shown in Figure 15.

Figure 15

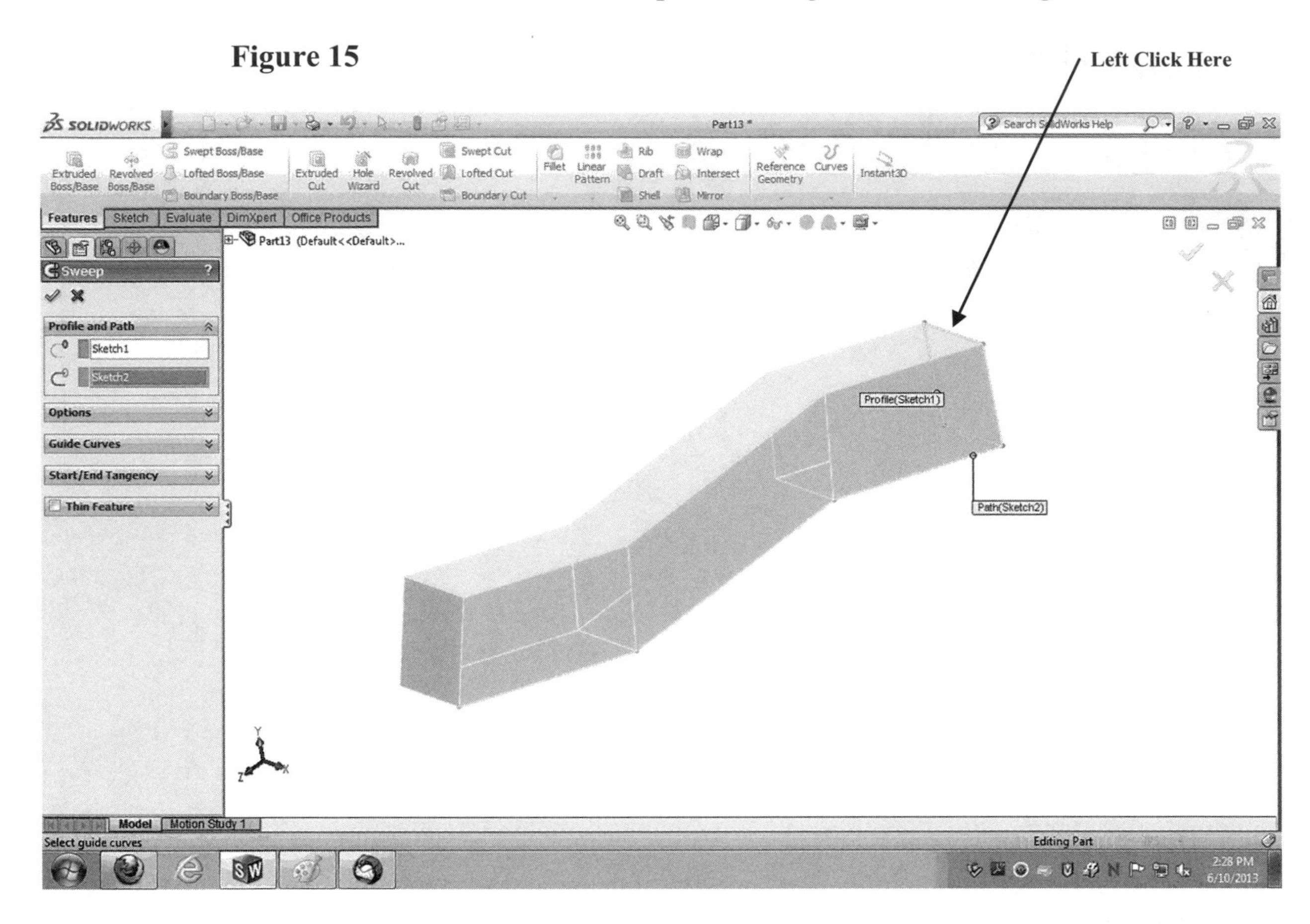

18. SolidWorks will provide a preview of the sweep. Left click on the green check mark as shown in Figure 16.

Figure 16

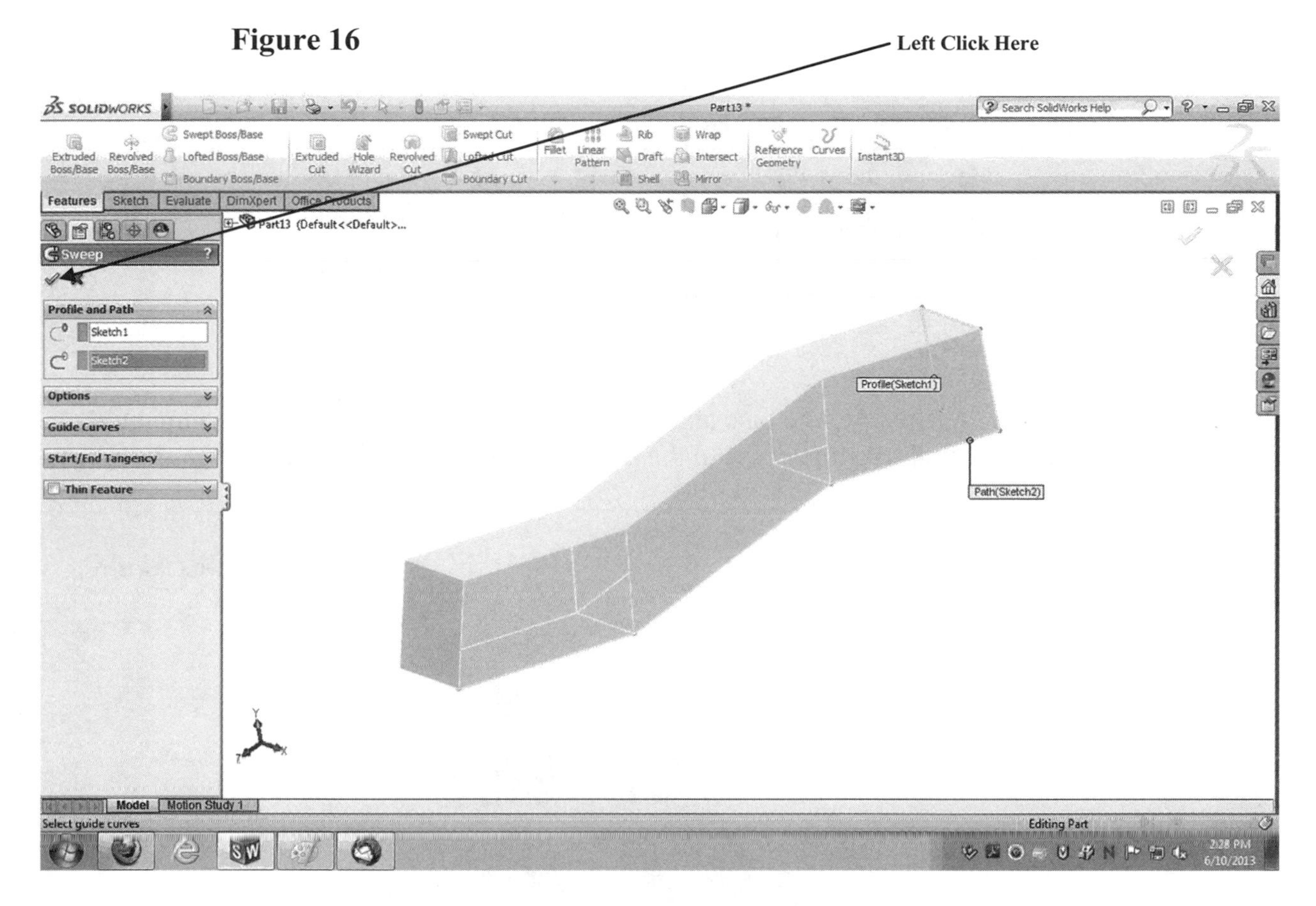

19. Your screen should look similar to Figure 17.

Figure 17

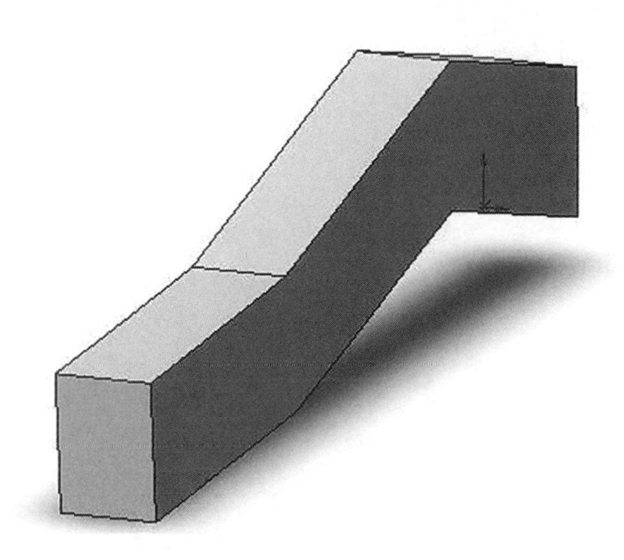

20. Move the cursor to the upper middle portion of the screen and left click on **Shell** as shown in Figure 18.

Figure 18

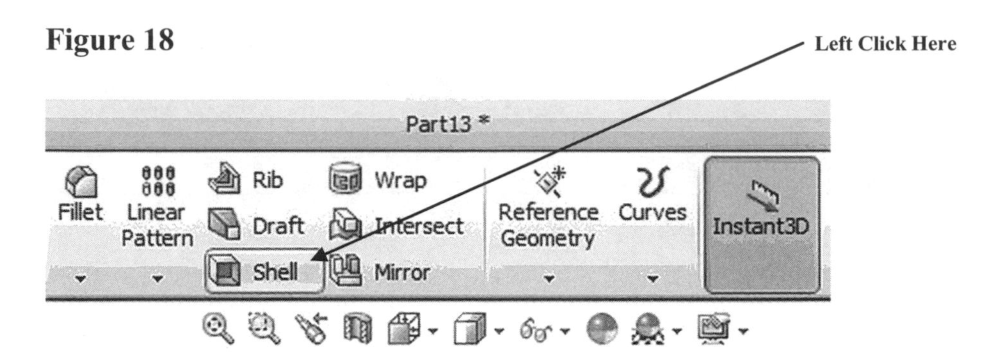

21. Move the cursor to the left side face and left click once. Using the **Rotate** command, rotate the part around to gain access to the right side face and left click once as shown in Figure 19.

Figure 19

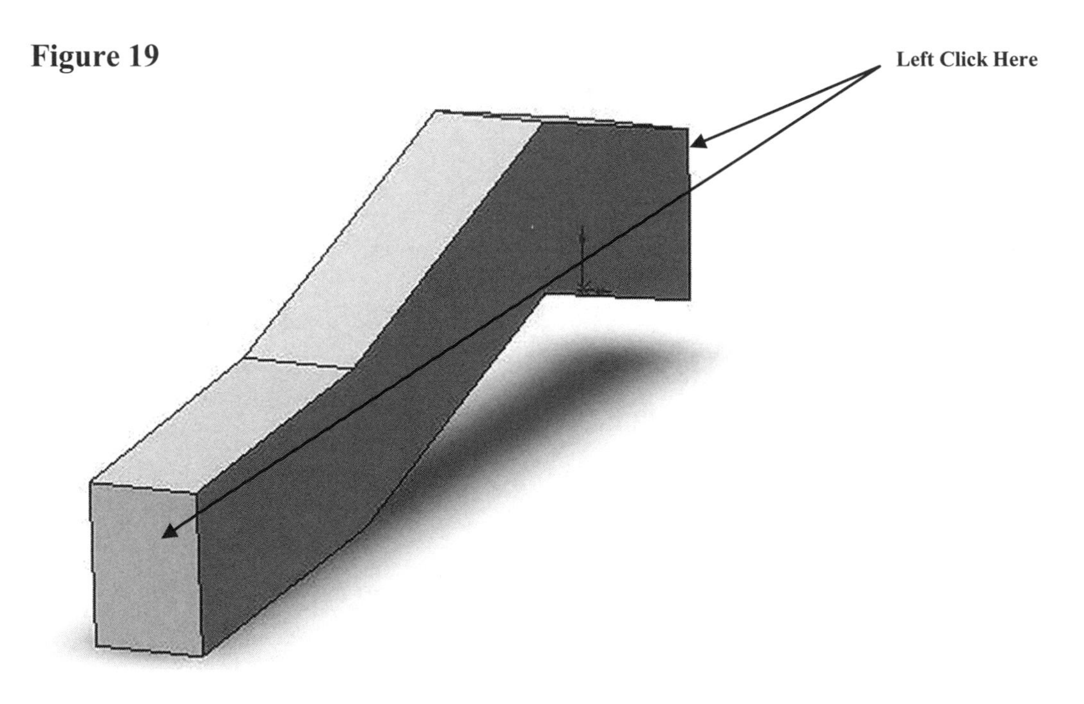

22. Left click on the green check mark as shown in Figure 20.

Figure 20

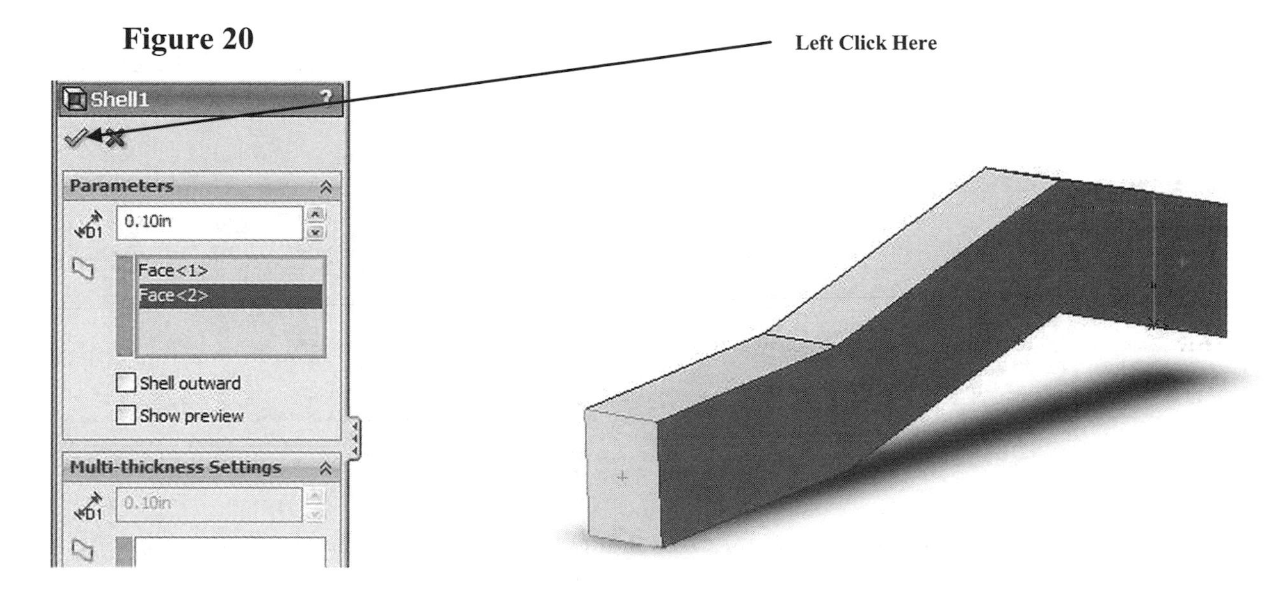

23. Use the **Rotate** command to access the ends of the model. The model should be open on both ends similar to a piece of rectangular tubing. Your screen should look similar to Figure 21.

Figure 21

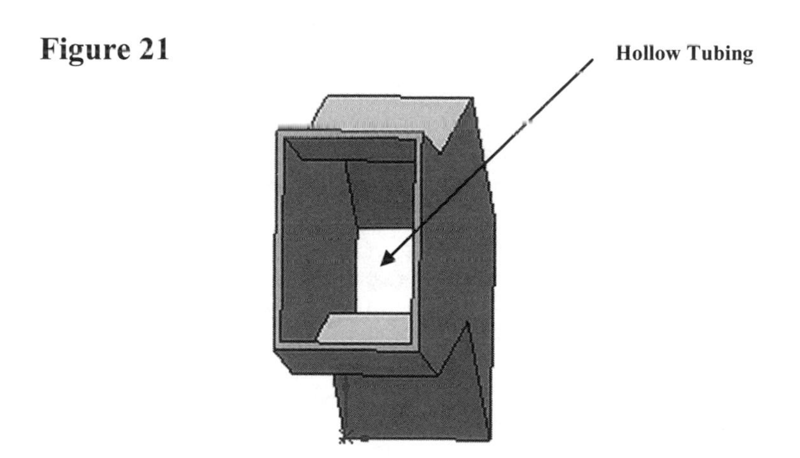

24. Begin a New Sketch.

25. Move the cursor to the upper left portion of the screen and left click on **Rectangle** as shown in Figure 22.

Figure 22

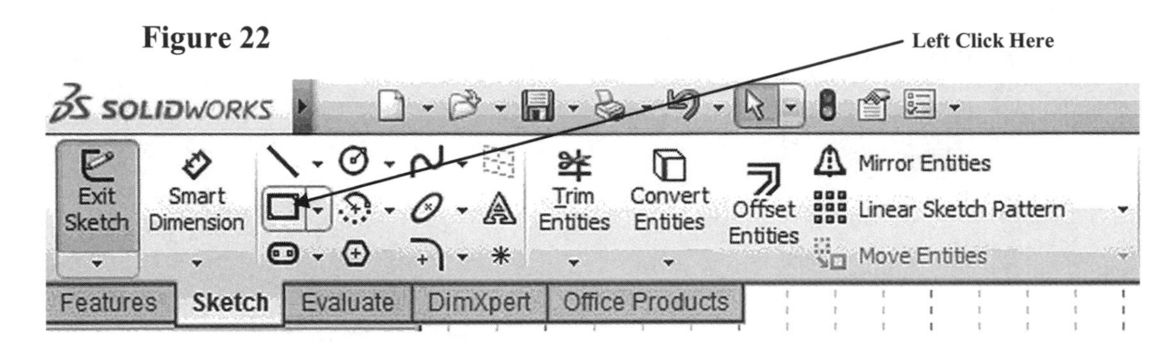

26. Move the cursor to the center of the screen and left click once. Ensure that the red dot appears on the intersection of the origin as shown in Figure 23.

Figure 23

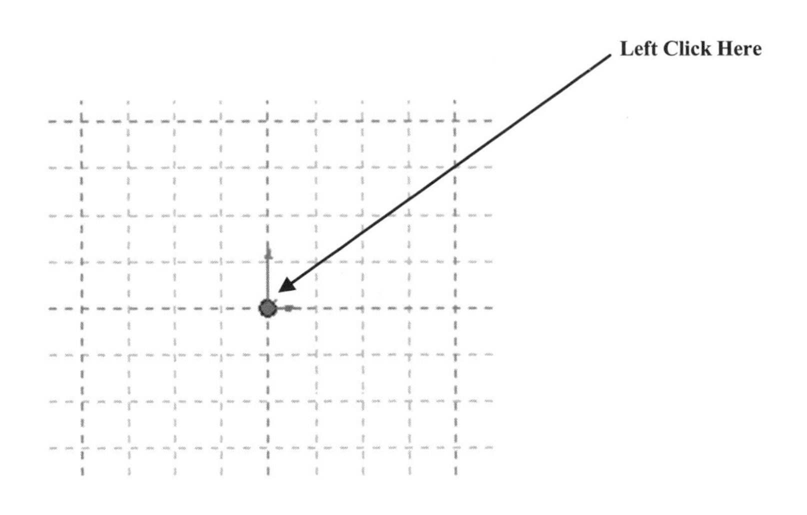

27. Complete the sketch as shown in Figure 24. Right click anywhere on the screen. A pop up menu will appear. Left click on **Select** as shown in Figure 24.

Figure 24

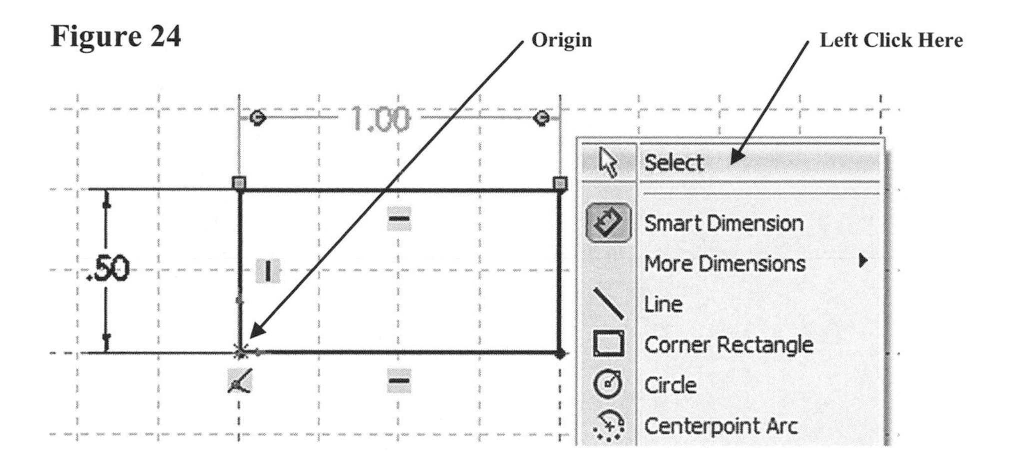

28. Right click anywhere on the screen. A pop up menu will appear. Left click on **Exit Sketch** as shown in Figure 25.

Figure 25

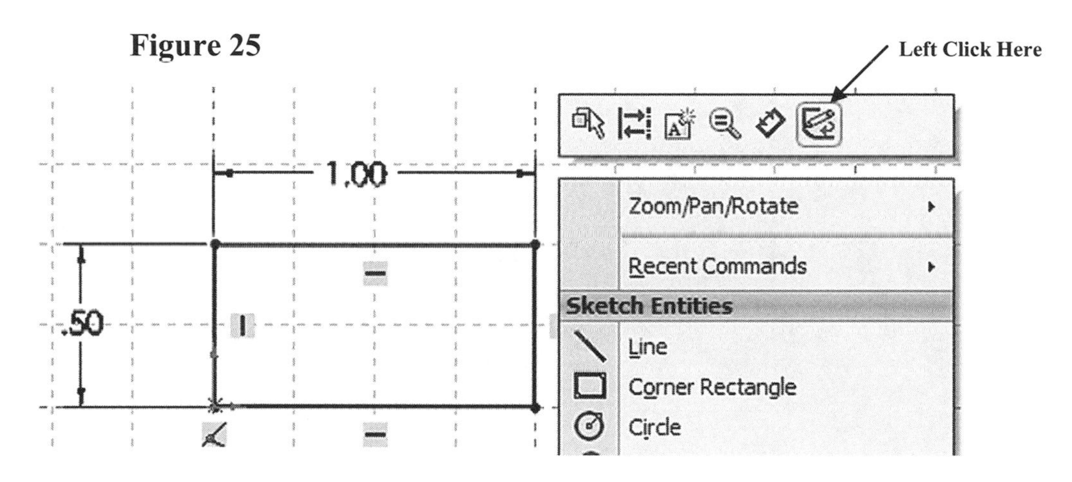

29. Move the cursor to the upper middle portion of the screen and left click on the drop down arrow to the right of the Views Orientation icon. A drop down menu will appear. Left click on **Trimetric** as shown in Figure 26.

Figure 26

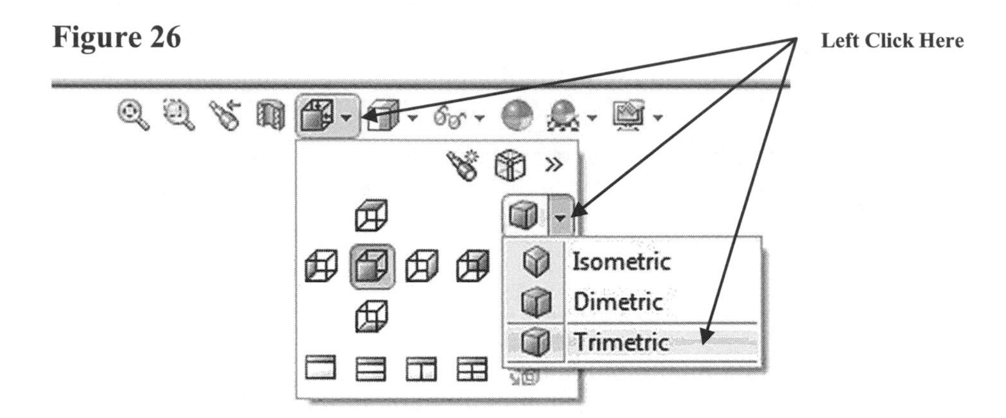

30. Use the **Extruded Boss/Base** command to extrude the sketch to a thickness of .001 of an inch as shown in Figure 27.

Figure 27

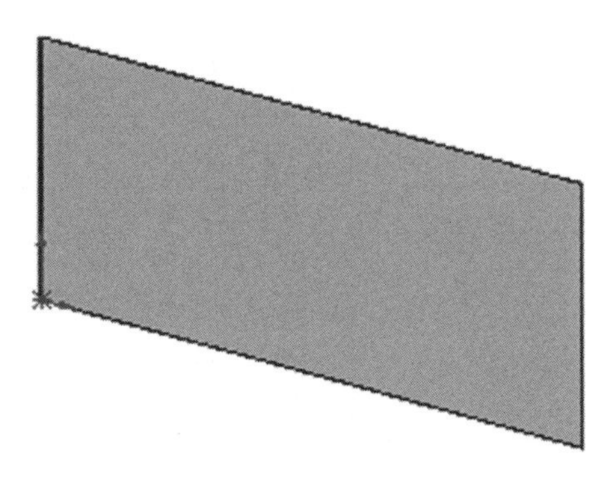

31. Move the cursor to the upper middle portion of the screen and left click on the drop down arrow underneath Reference Geometry. A drop down menu will appear. Left click on **Plane** as shown in Figure 28.

Figure 28

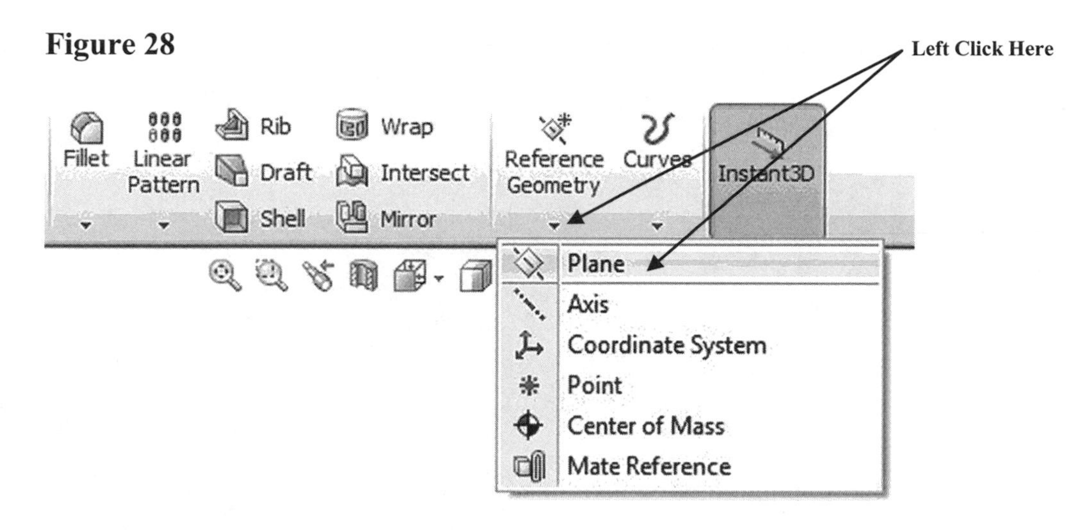

32. Left click on the surface of the solid. SolidWorks will provide a preview of the new plane. Left click on the Distance icon and enter 1.00. Left click on the green check mark as shown in Figure 29.

Figure 29

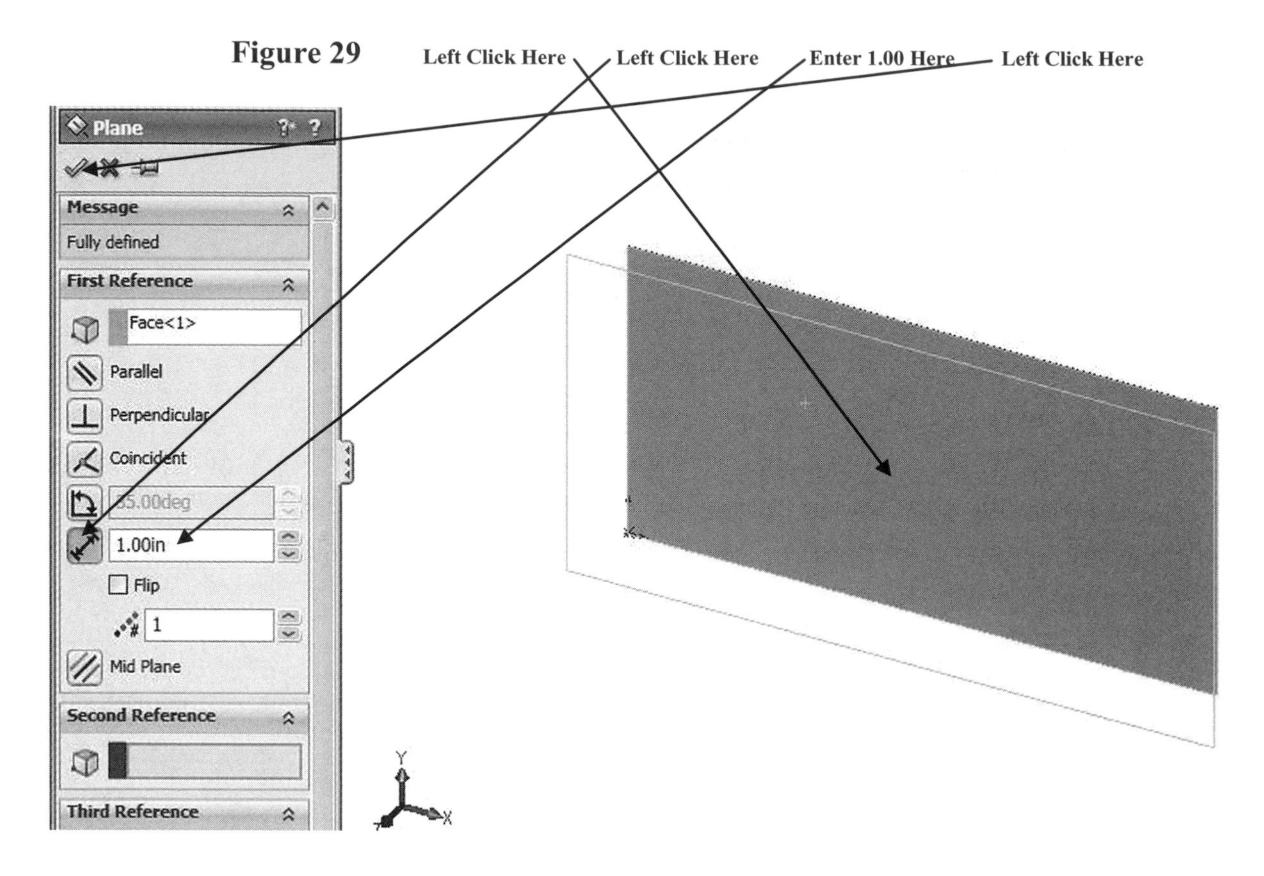

33. Your screen should look similar to Figure 30.

Figure 30

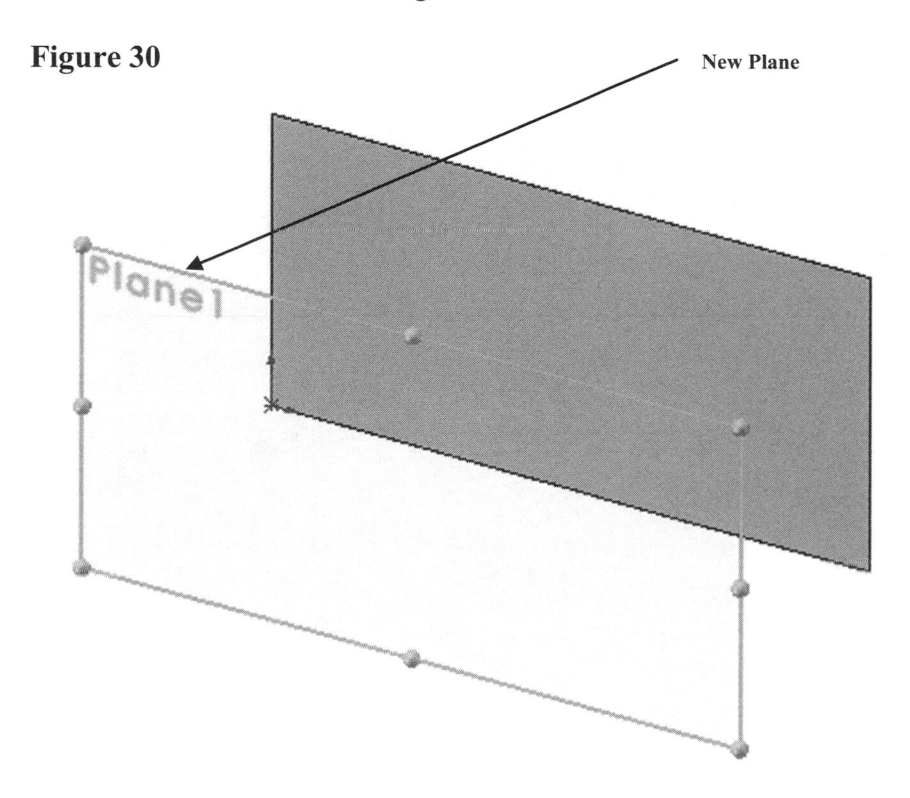

34. Move the cursor to the edge of the newly created plane and right click. A pop up menu will appear. Left click on **Sketch** as shown in Figure 31.

Figure 31

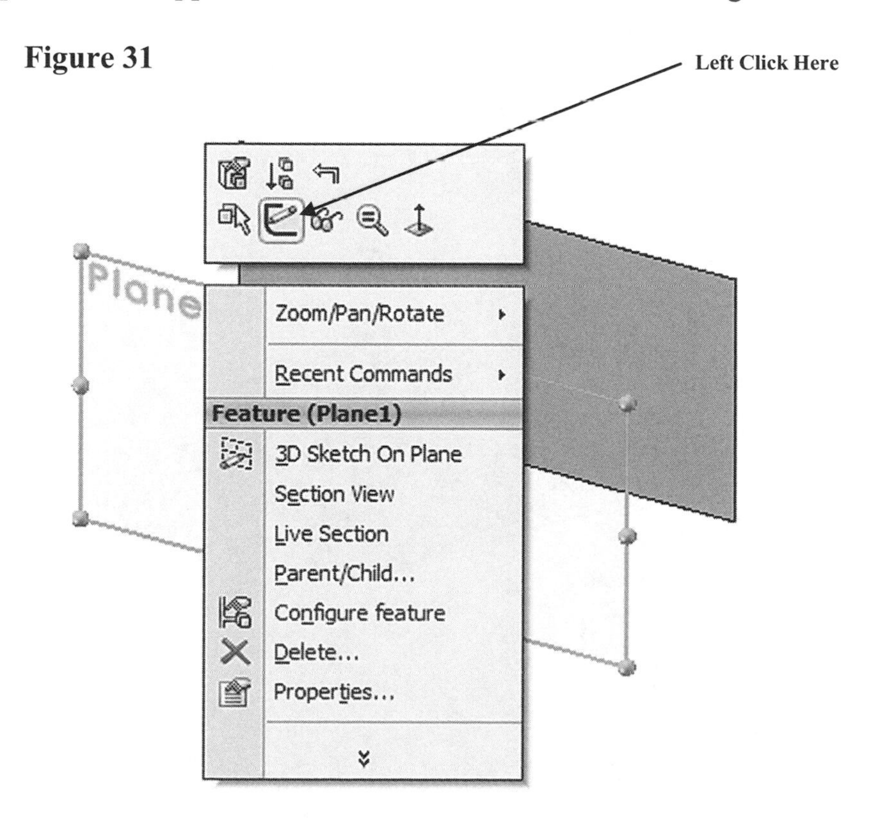

35. A new sketch will appear on the newly created plane as shown in Figure 32.

Figure 32

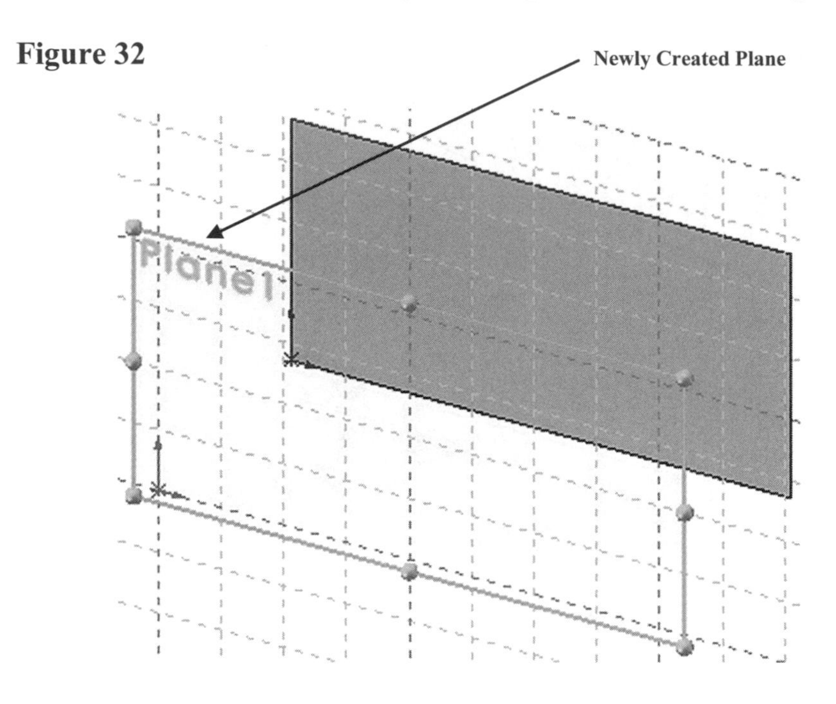

36. Draw a small rectangle in the corner of the sketch as shown in Figure 33. Estimate size and distance.

Figure 33

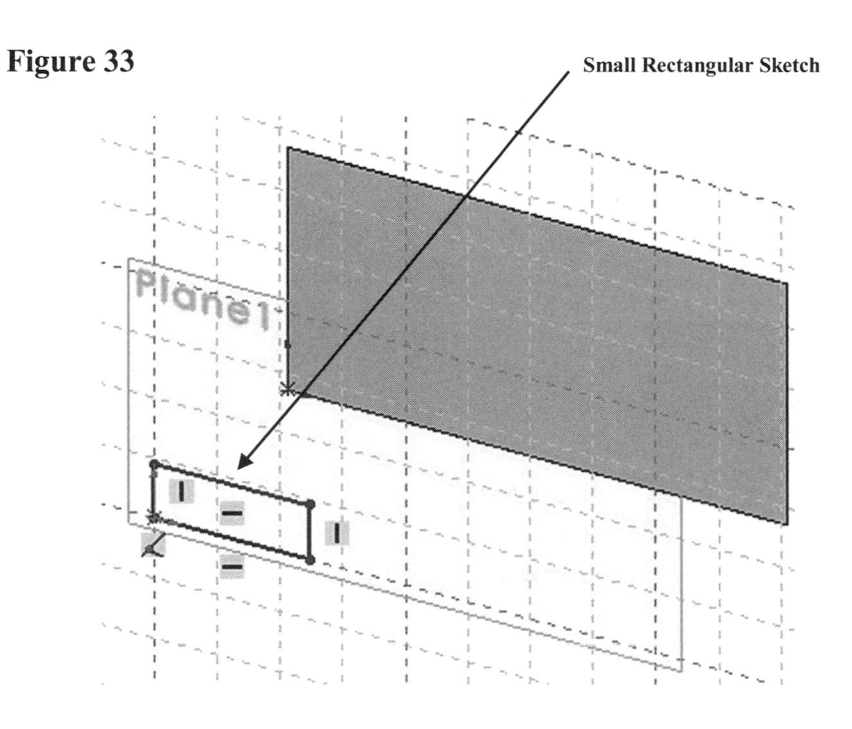

37. Once the sketch is complete, right click anywhere around the part. A pop up menu will appear. Left click on **Exit Sketch** as shown in Figure 34.

Figure 34

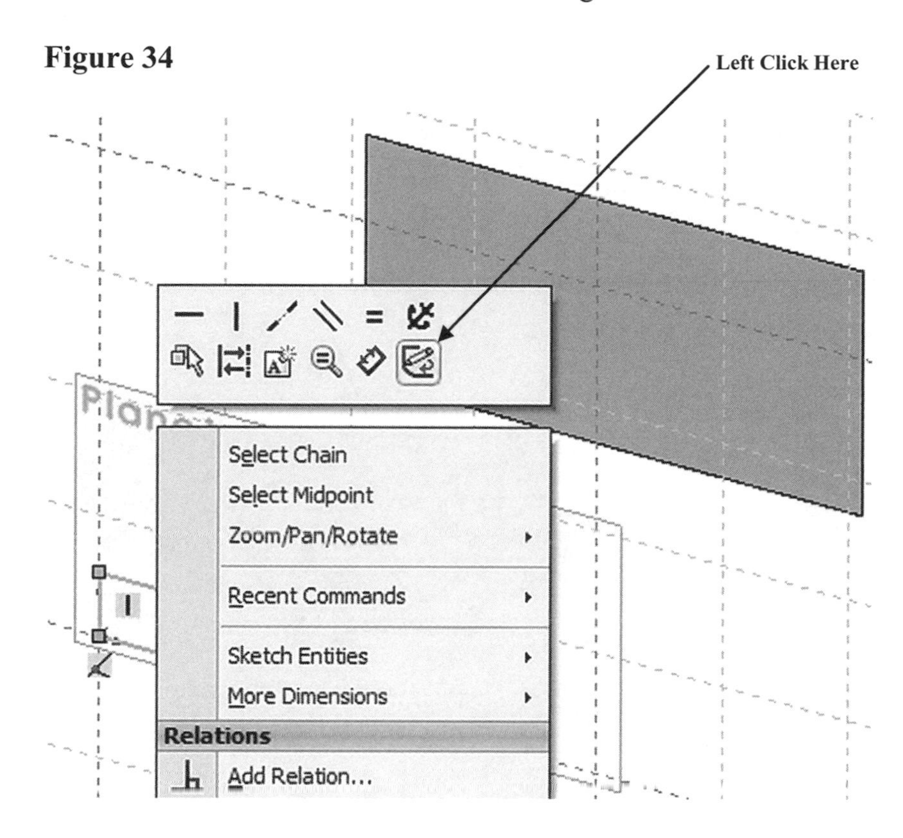

38. Move the cursor to the upper middle portion of the screen and left click on the drop down arrow underneath Reference Geometry. A drop down menu will appear. Left click on **Plane** as shown in Figure 35.

Figure 35

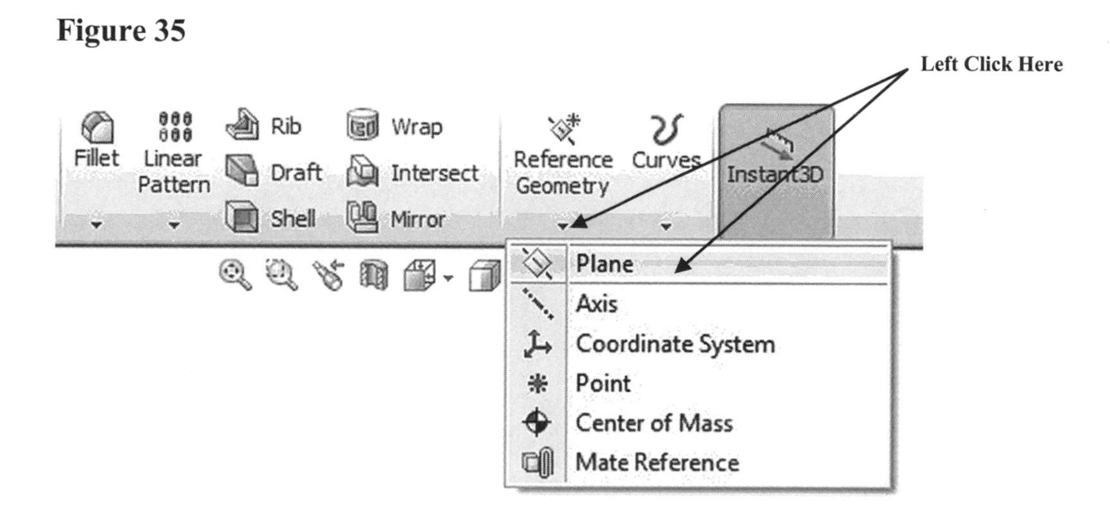

39. Left click on the previously create workplane. SolidWorks will place another plane 1.00 inch in front of the previous plane. If another plane does not appear, left click on the edge of Plane 1. Left click on the green check mark. Your screen should look similar to Figure 36.

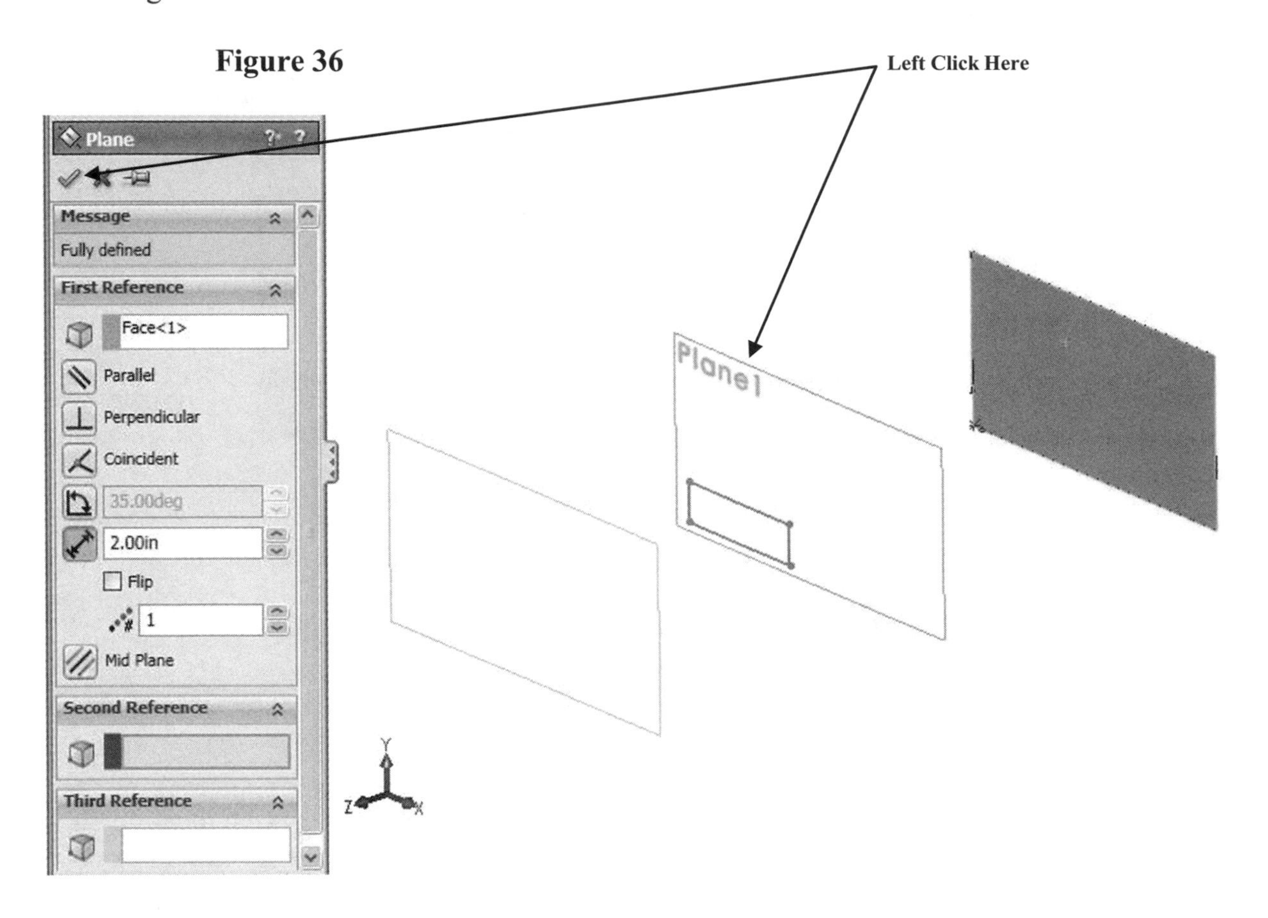

Figure 36

40. Your screen should look similar to Figure 37.

Figure 37

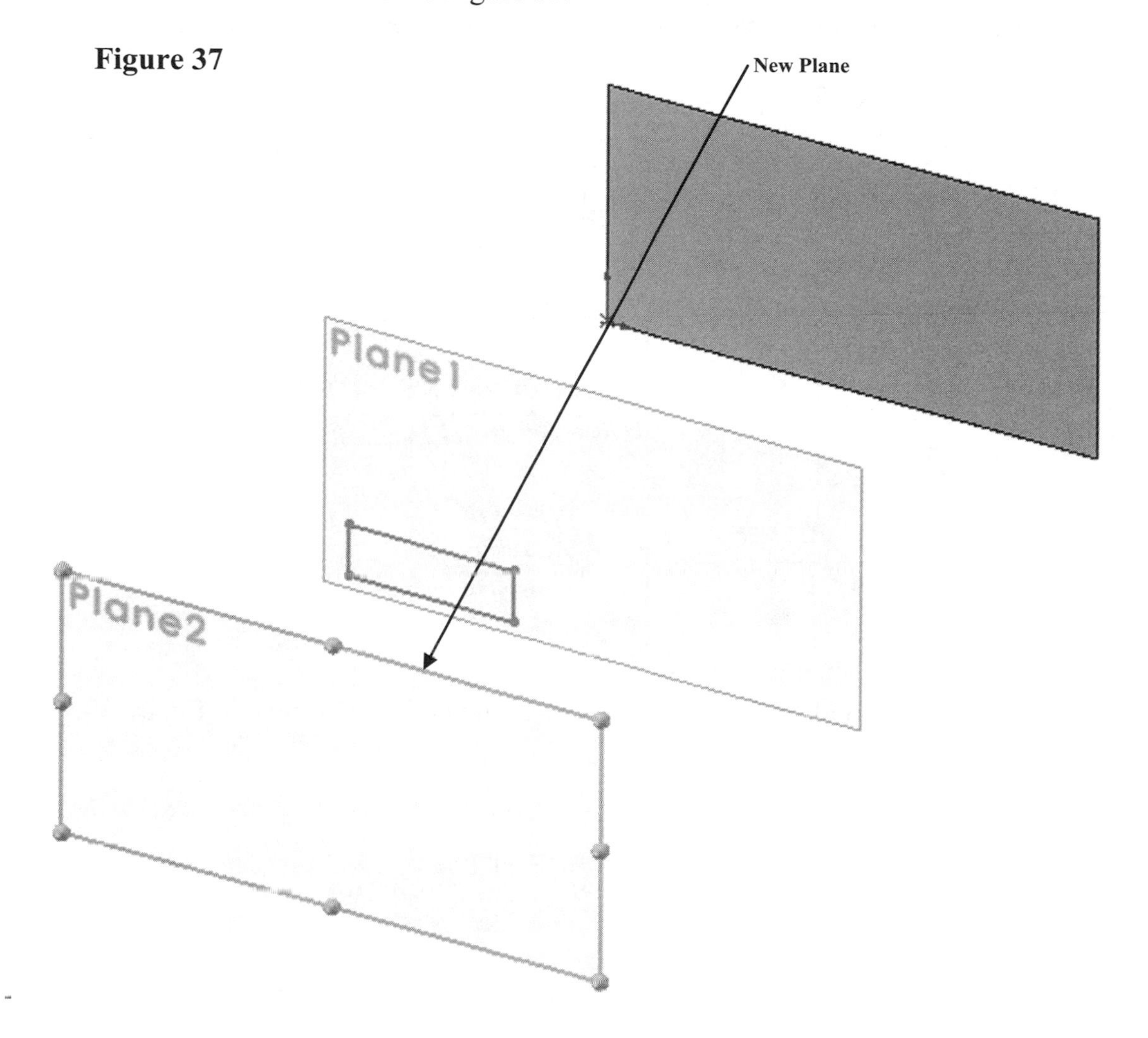

41 Move the cursor to the edge of the newly created plane and right click. A pop up menu will appear. Left click on **Sketch** as shown in Figure 38.

Figure 38

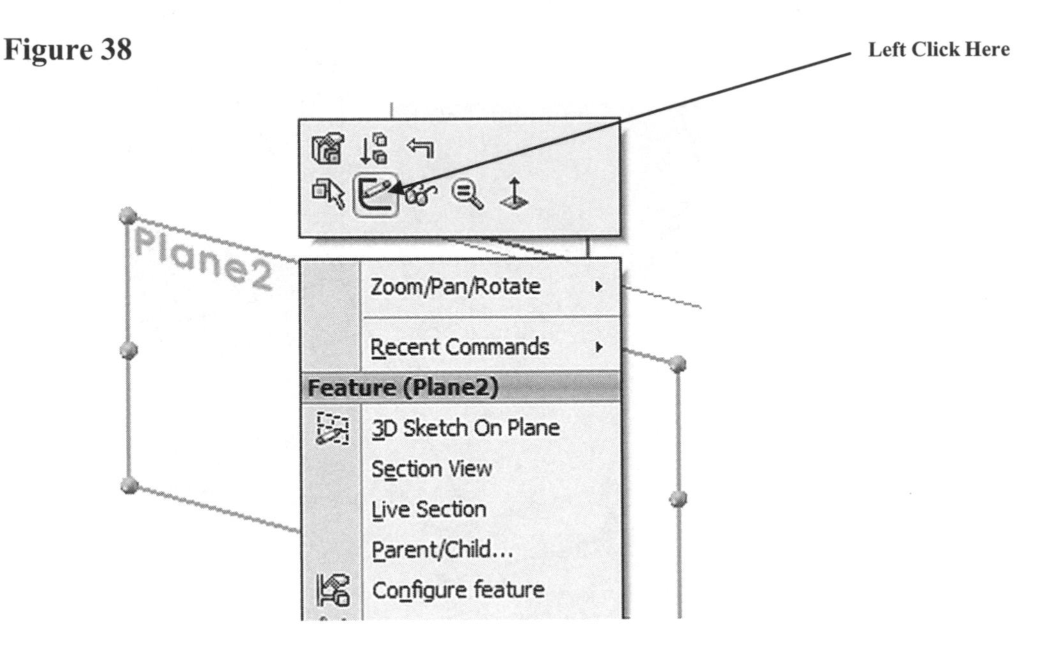

42. A new sketch will appear on the newly created plane as shown in Figure 39.

Figure 39

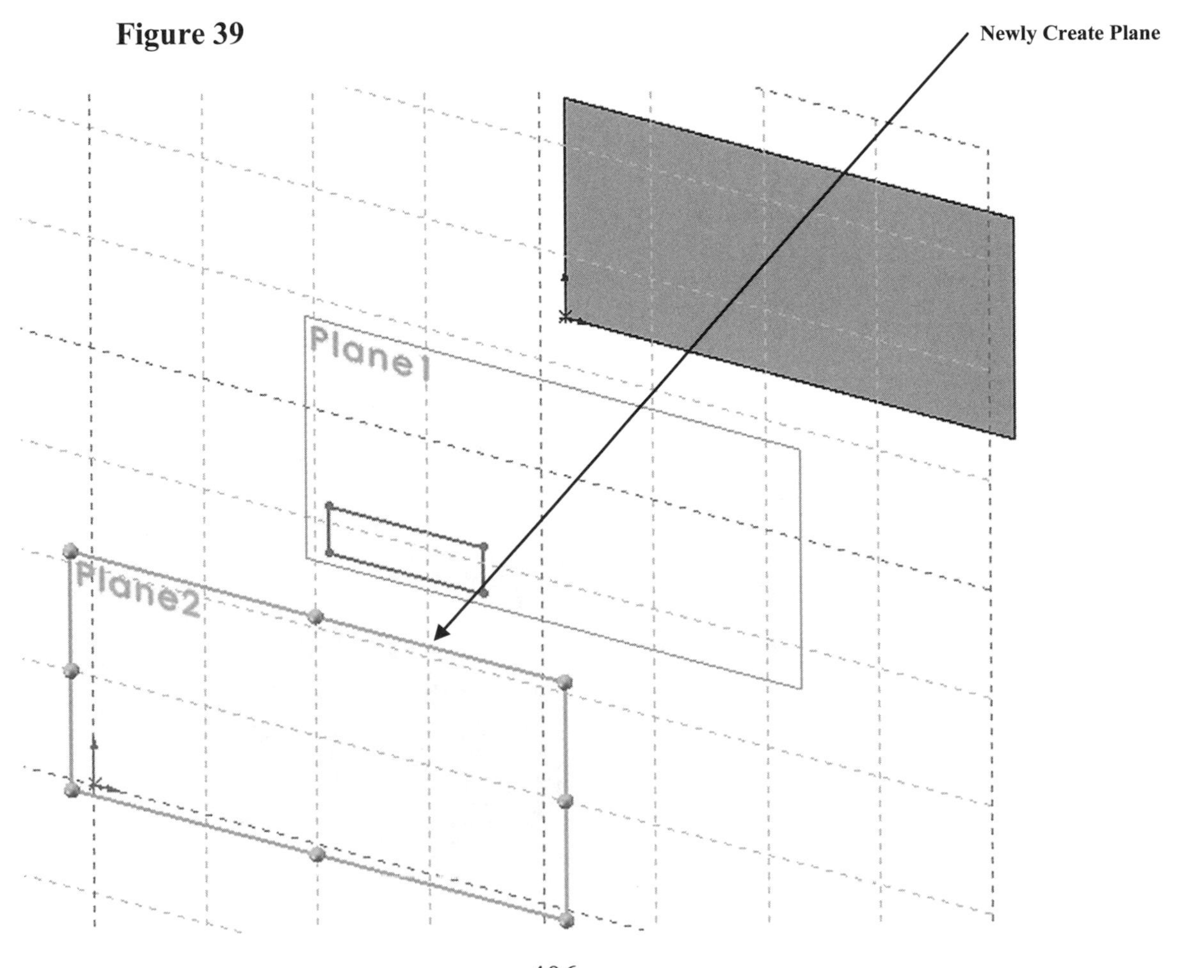

43. Draw a small circle in the corner of the sketch as shown in Figure 40. Estimate size and distance.

Figure 40

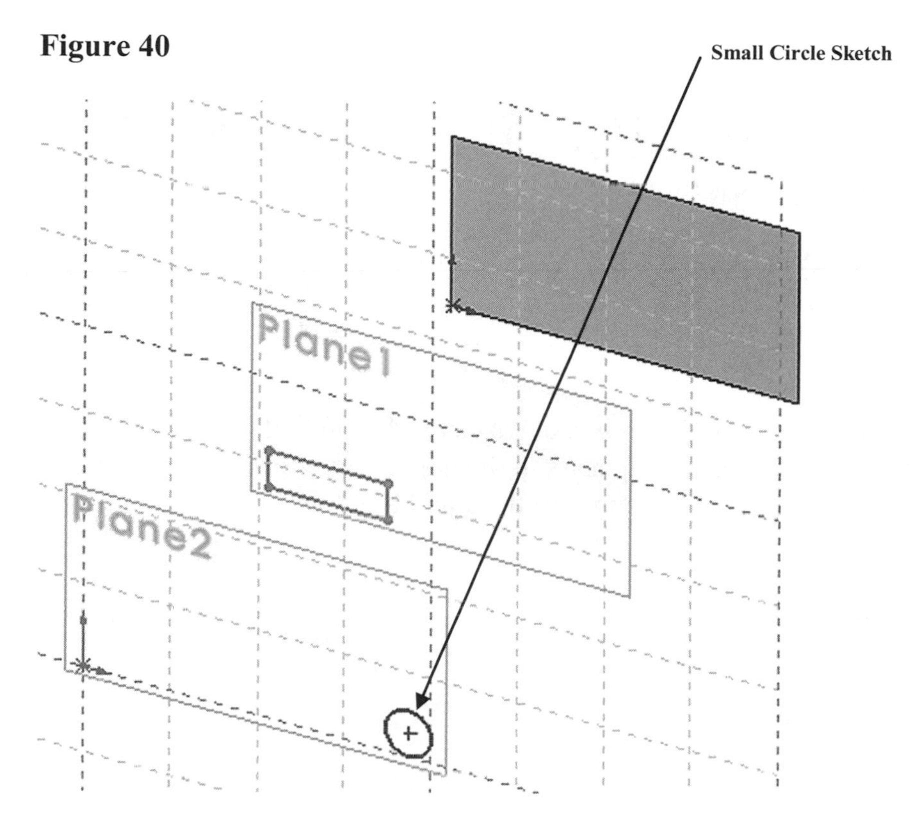

44. Once the sketch is complete, right click anywhere around the part. A pop up menu will appear. Left click on **Exit Sketch** as shown in Figure 41.

Figure 41

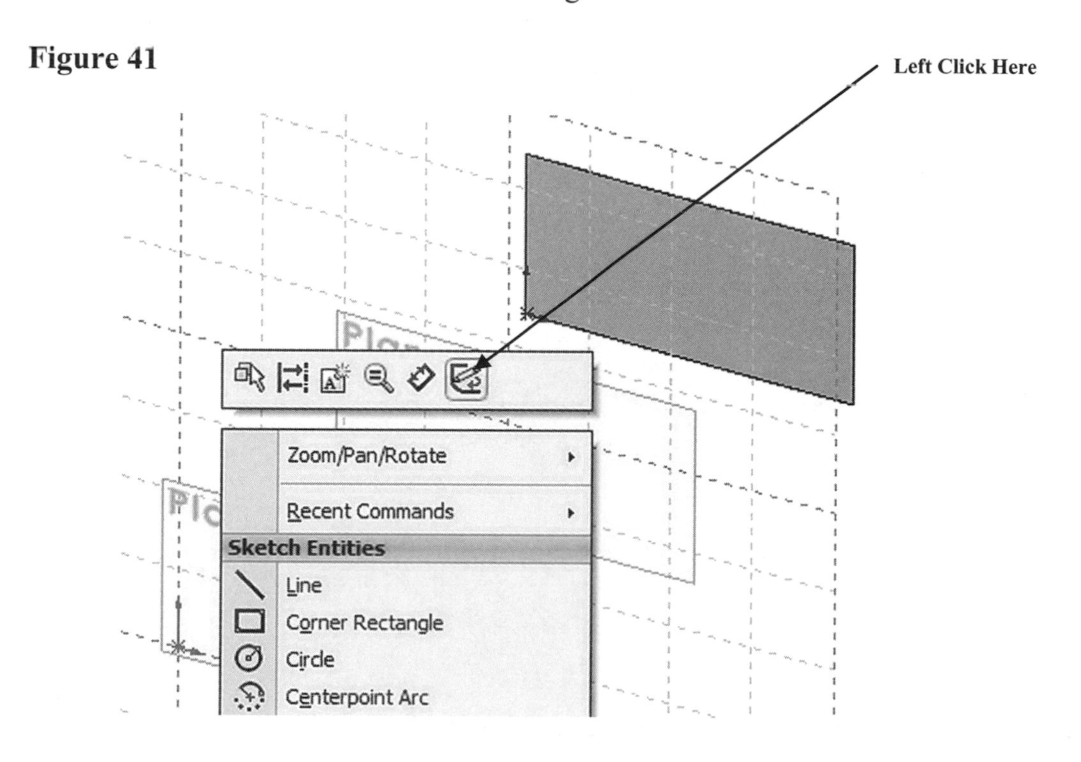

45. Move the cursor to the upper left portion of the screen and left click on **Lofted Boss/Base** as shown in Figure 42.

Figure 42

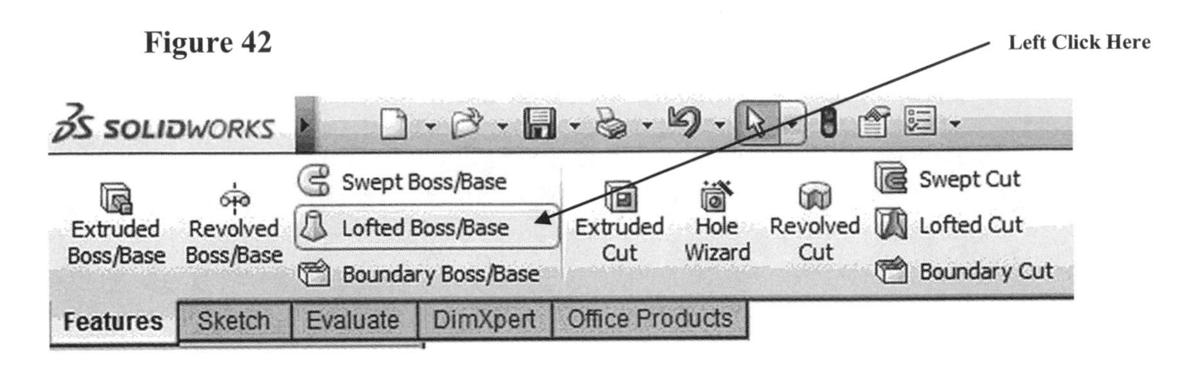

46. Move the cursor to the first box that was extruded and left click on the face. Because the Profile banner appears in the upper corner of the first sketch, move the cursor to the upper left corner of the small box (second sketch in the same location) and left click.

Figure 43

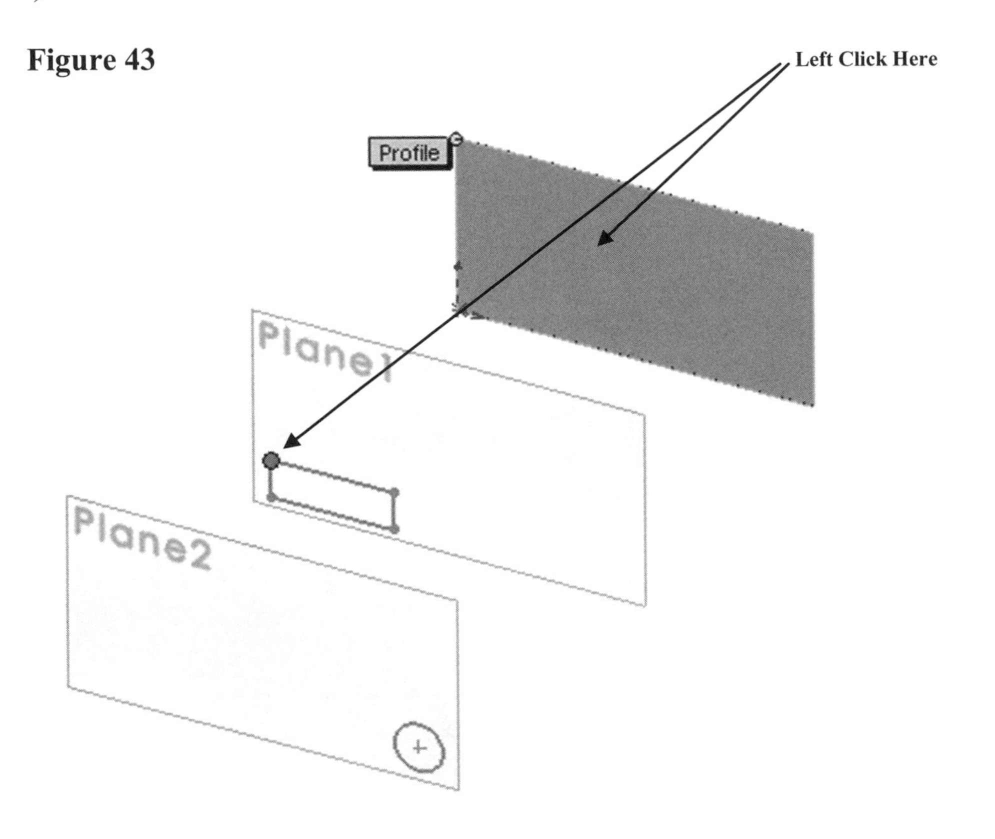

47. Your screen should look similar to Figure 44.

Figure 44

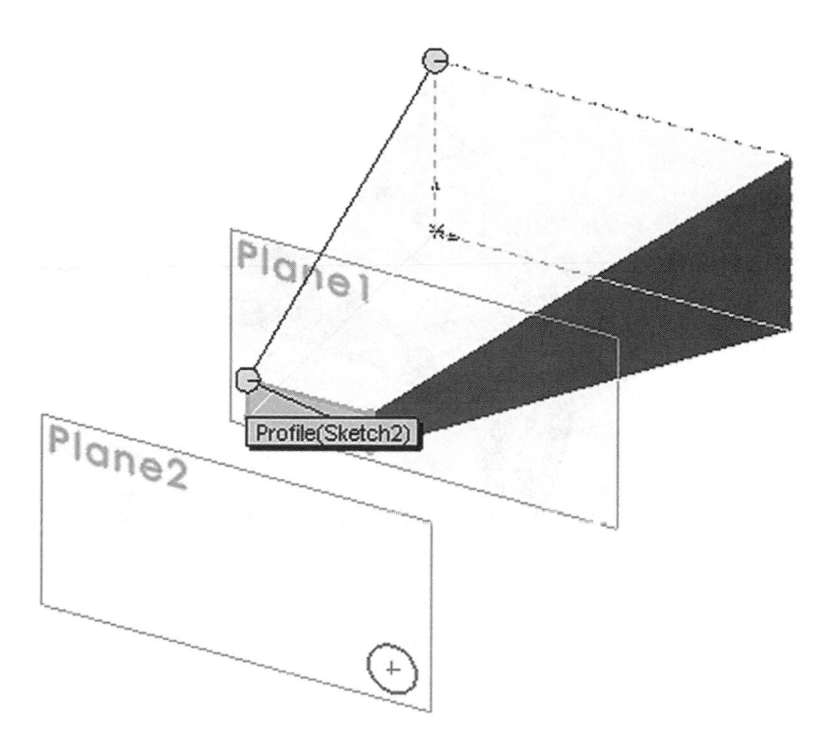

48. Move the cursor to the edge of the circle and left click as shown in Figure 45.

Figure 45

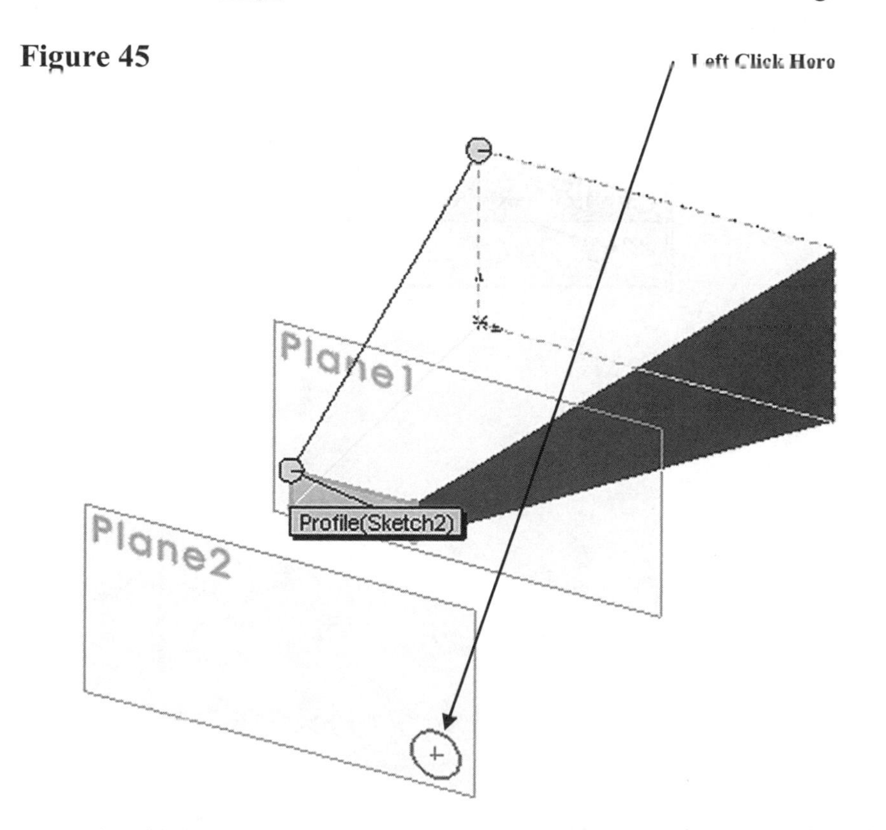

49. Your screen should look similar to Figure 46.

Figure 46

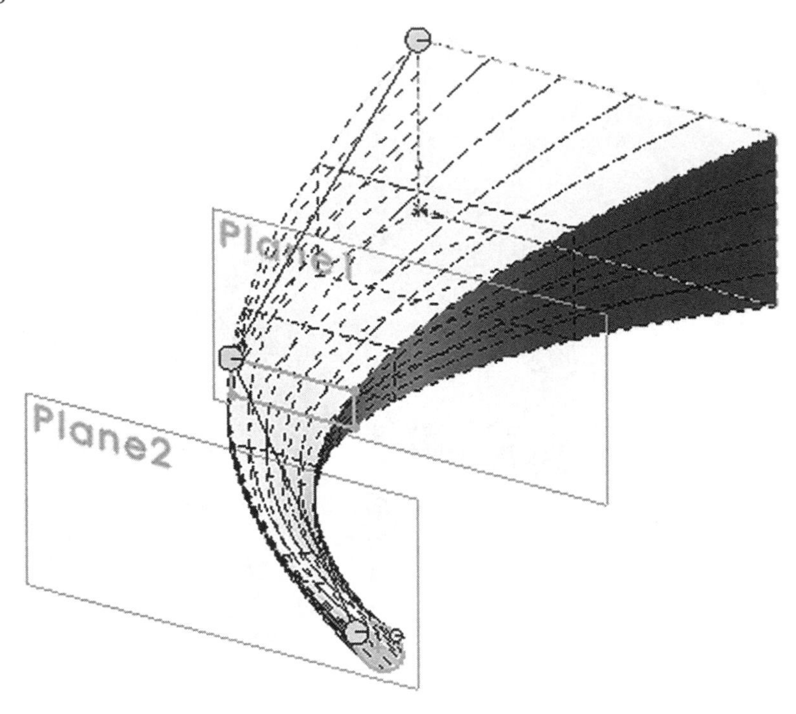

50. Left click on the green check mark as shown in Figure 47. Press the **Esc** key several times.

Figure 47

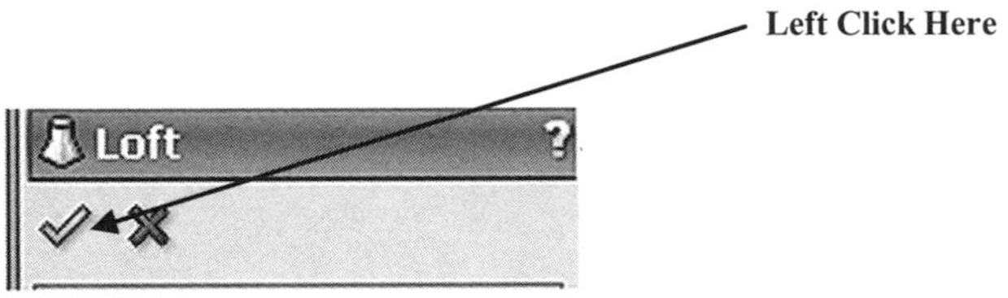

Left Click Here

Loft

51. Your screen should look similar to Figure 48.

Figure 48

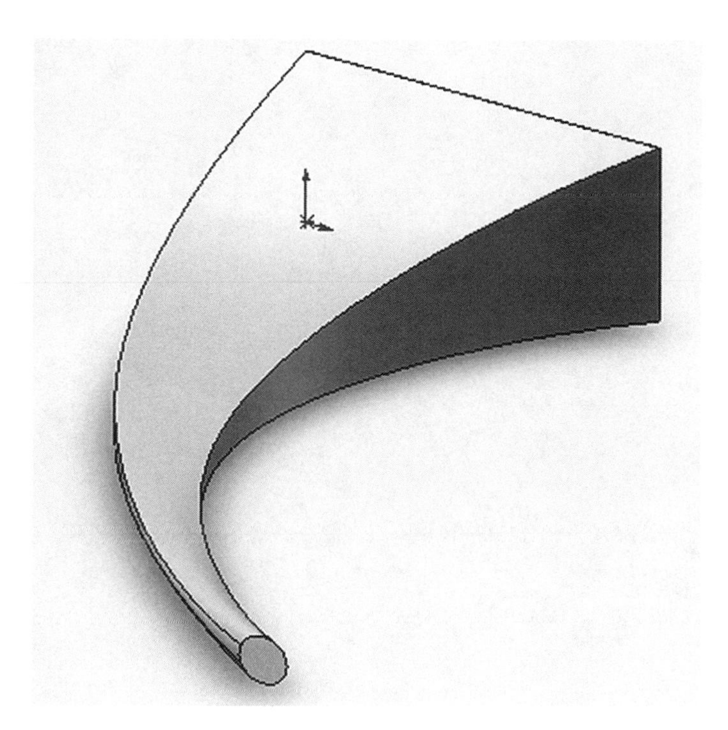

Chapter 9 Introduction to SimulationXpress

Objectives:

- Learn to create simple parts used to run SimulationXpress
- Learn to interpret the results of an analysis

Chapter 9 includes instruction on how to run the analysis shown below.

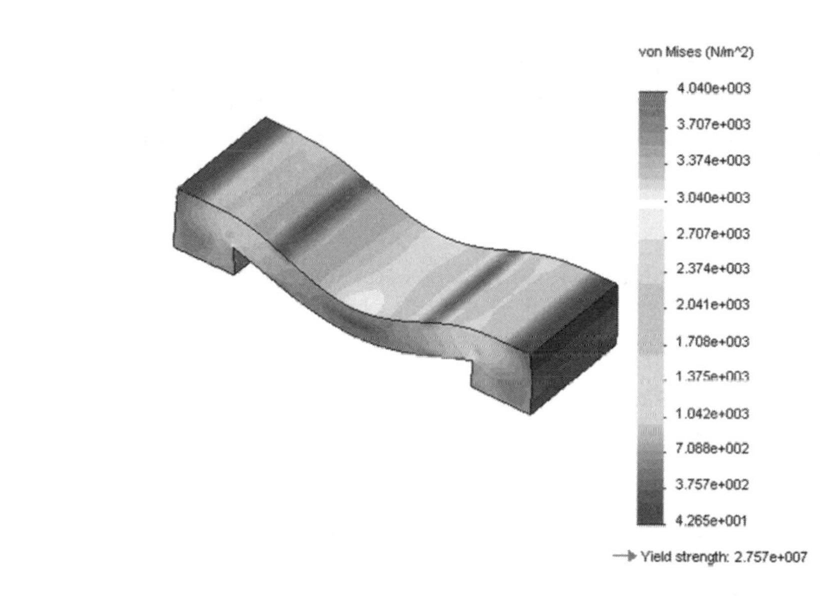

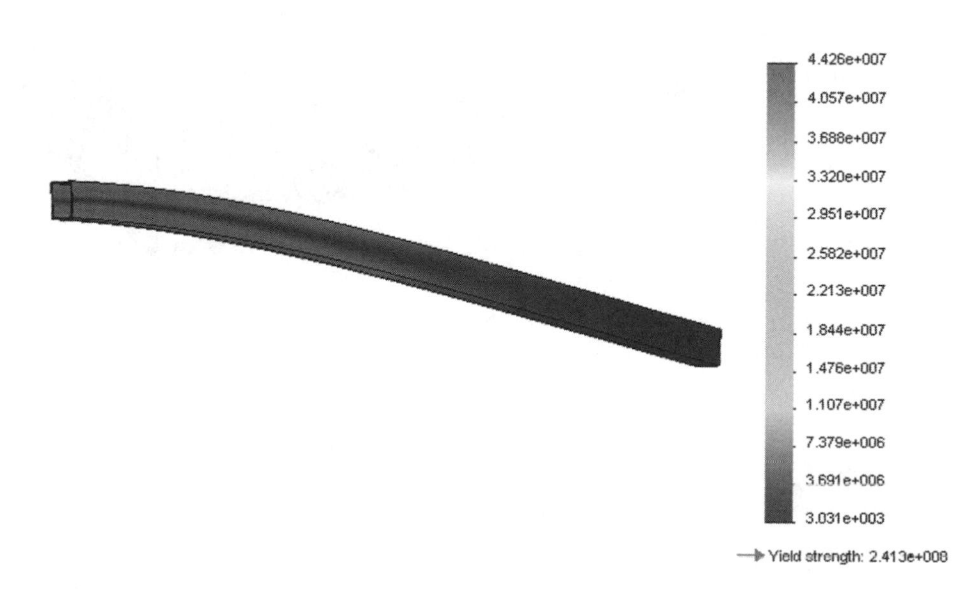

1. Start SolidWorks by referring to "Chapter 1 Getting Started".

2. Begin a new Sketch.

3. Complete the sketch as shown in Figure 1.

Figure 1

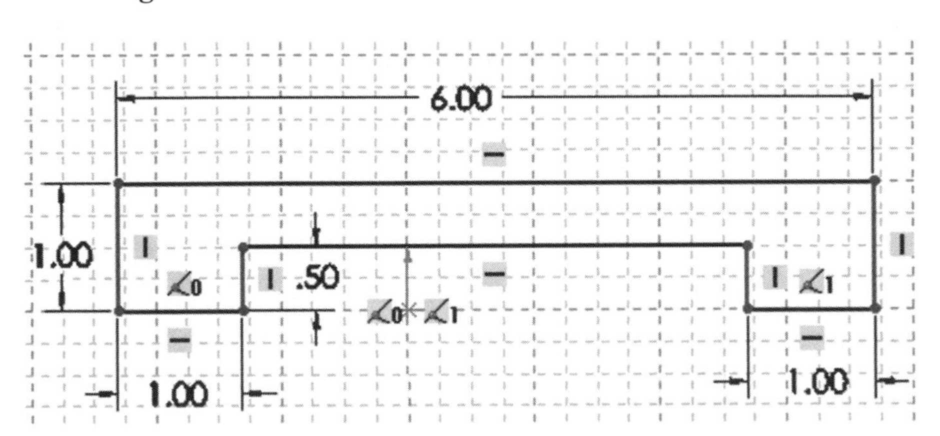

4. Extrude the sketch a distance of 2 inches as shown in Figure 2.

Figure 2

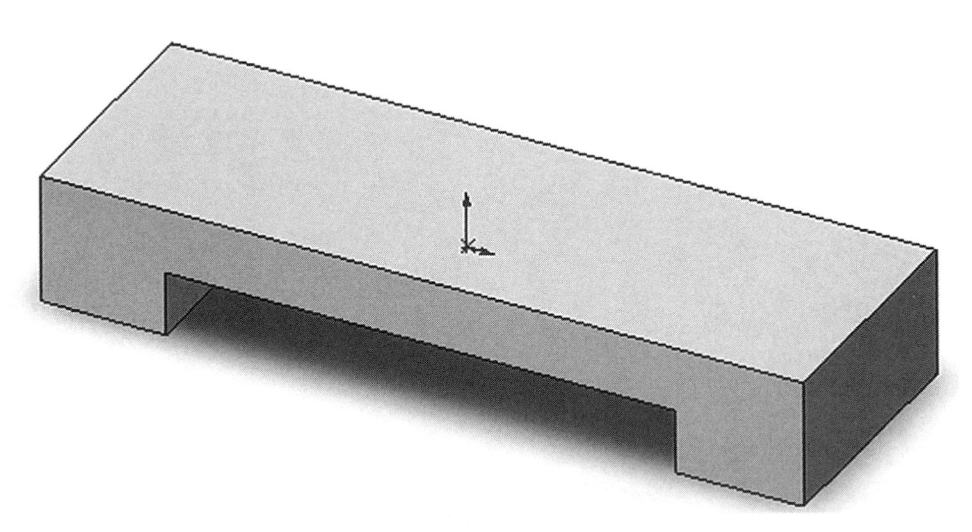

5. Save the part where it can easily be retrieved later.

6. Move the cursor to the upper left portion of the screen and left click on the arrow to the right of SolidWorks. A fly out menu will appear. Left click on **Tools**. A drop down menu will appear. Left click on **SimulationXpress** as shown in Figure 3.

Figure 3

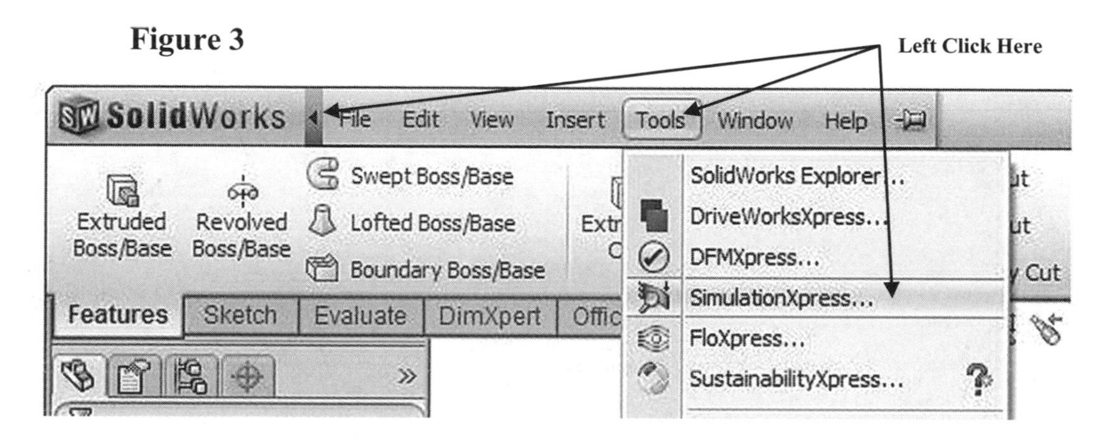

7. The SimulationXpress dialog box will appear. Left click on **Next** as shown in Figure 4.

Figure 4

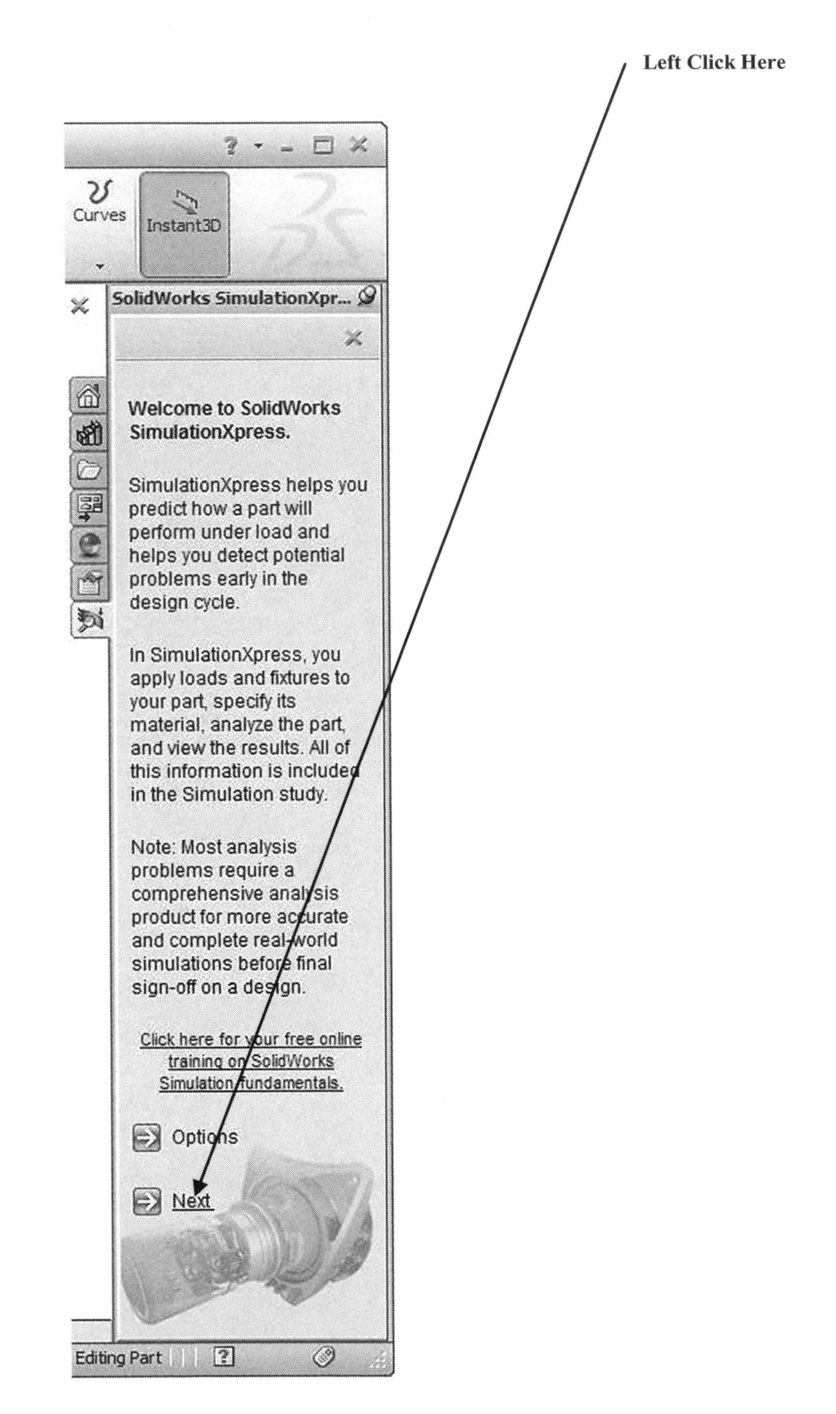

8. Left click on **Add a Fixture** as shown in Figure 5.

Figure 5

Left Click Here

SolidWorks SimulationXpr...

1 Fixtures
2 Loads
3 Material
4 Run
5 Results
6 Optimize

Apply fixtures to keep the part from moving when loads are applied.

Warning: Faces with fixtures are treated as perfectly rigid. This can cause unrealistic results in the vicinity of the fixture. Examples:

Fixed Holes
Fixed vs. Supported
Fixed vs. Attached Parts

Note: More flexible fixture types are available in SolidWorks Simulation Professional.

Add a fixture

Back Start Over

9. Use the **Rotate** command to rotate the part upward to gain access to the bottom as shown in Figure 6.

Figure 6

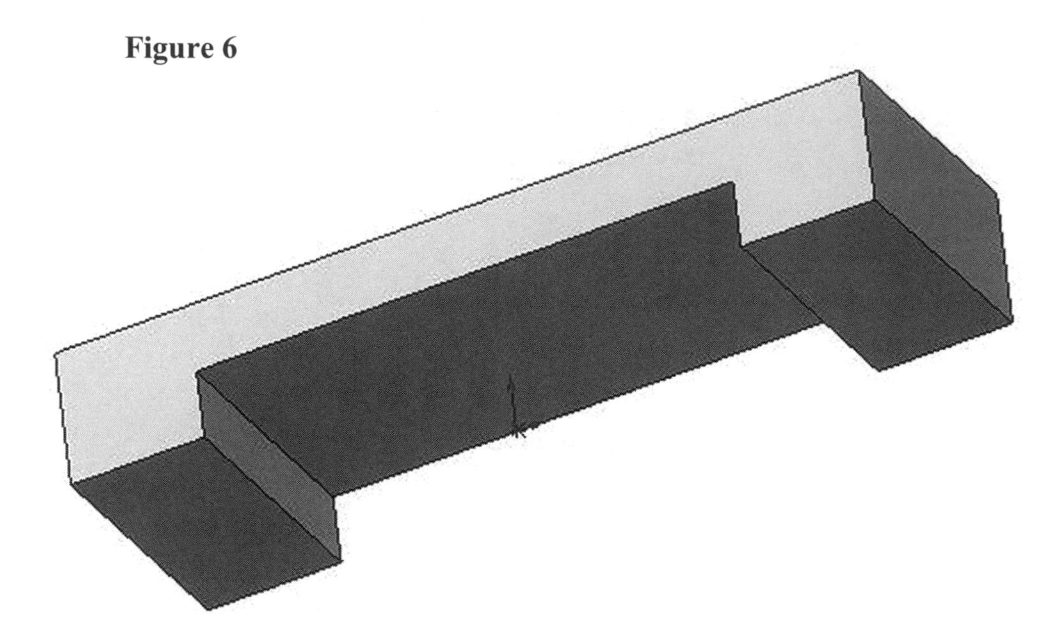

10. Left click on the bottom surfaces as shown in Figure 7.

Figure 7

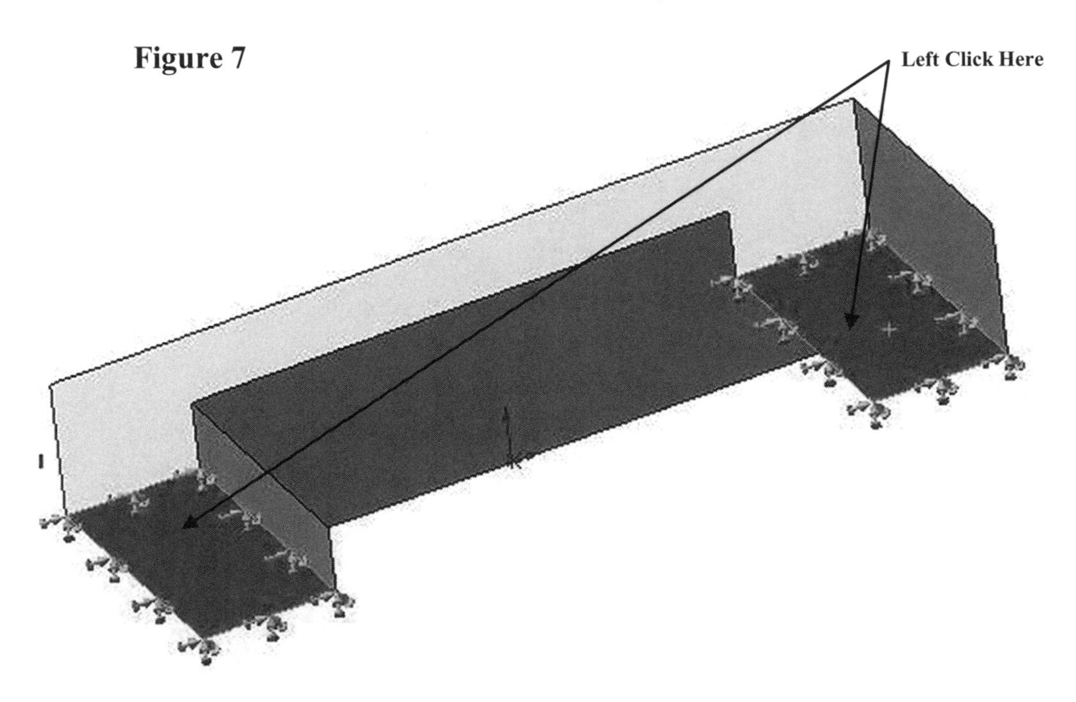

11. Move the cursor to the upper left portion of the screen and left click on the green check mark as shown in Figure 8.

Figure 8

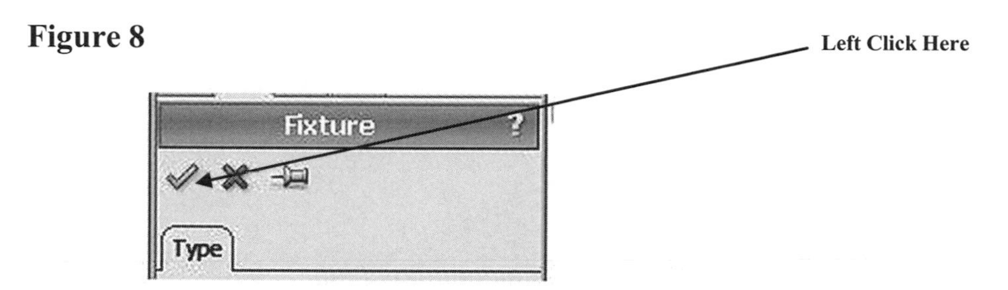

12. Move the cursor to the upper right portion of the screen and left click on **Loads** as shown in Figure 9.

Figure 9

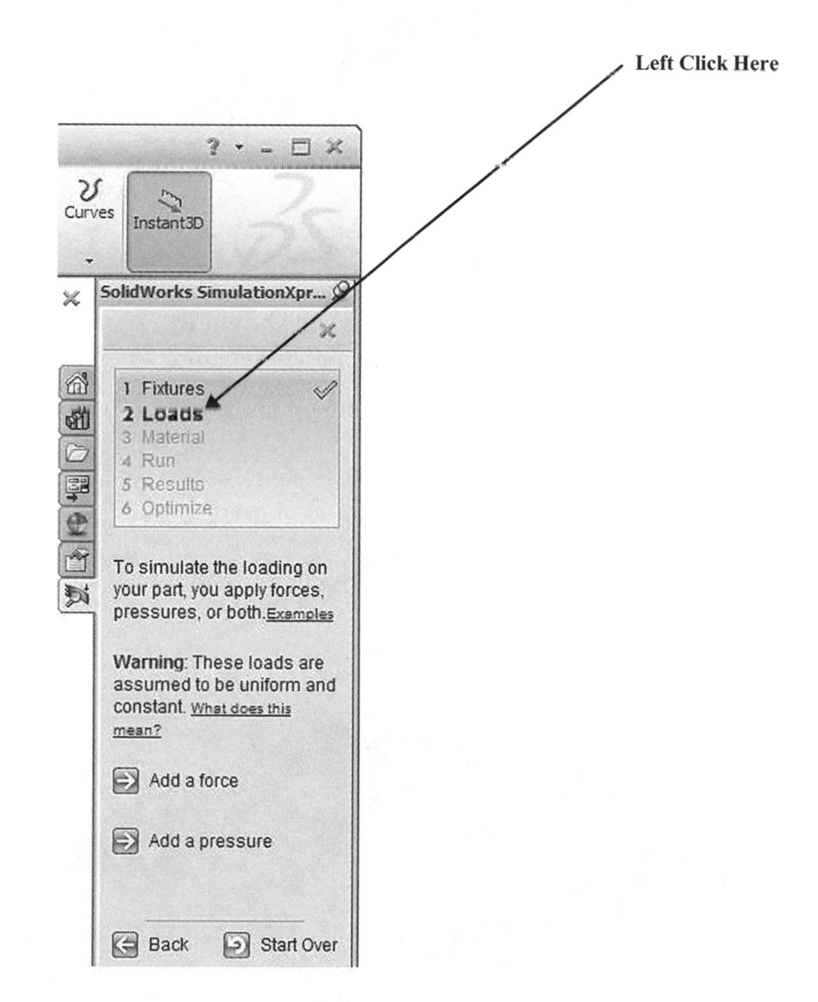

13. Use the **Rotate** command to rotate the part downward to gain access to the top as shown in Figure 10.

Figure 10

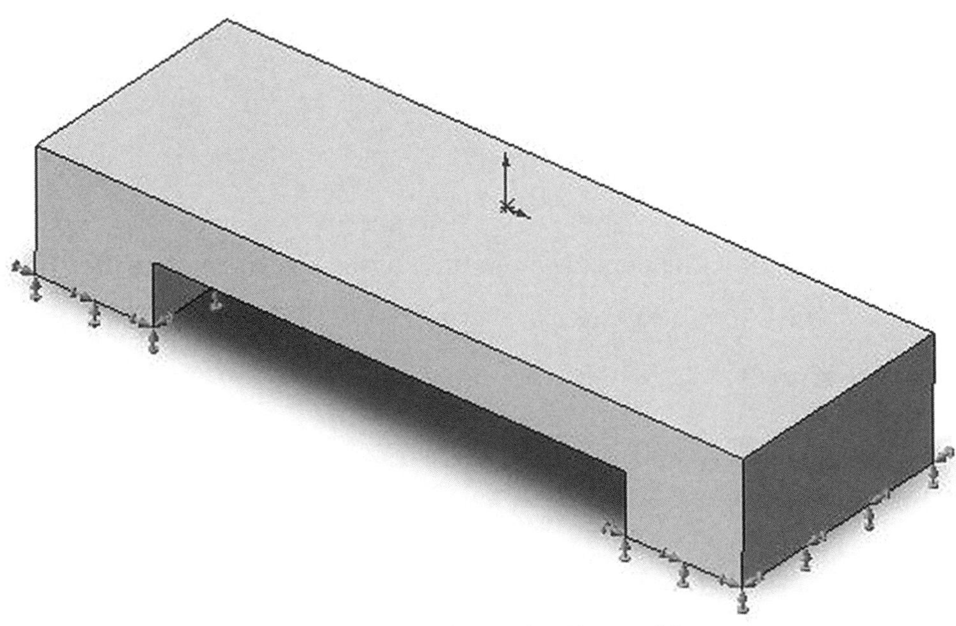

14. Left click on the top of the part as shown in Figure 11.

Figure 11

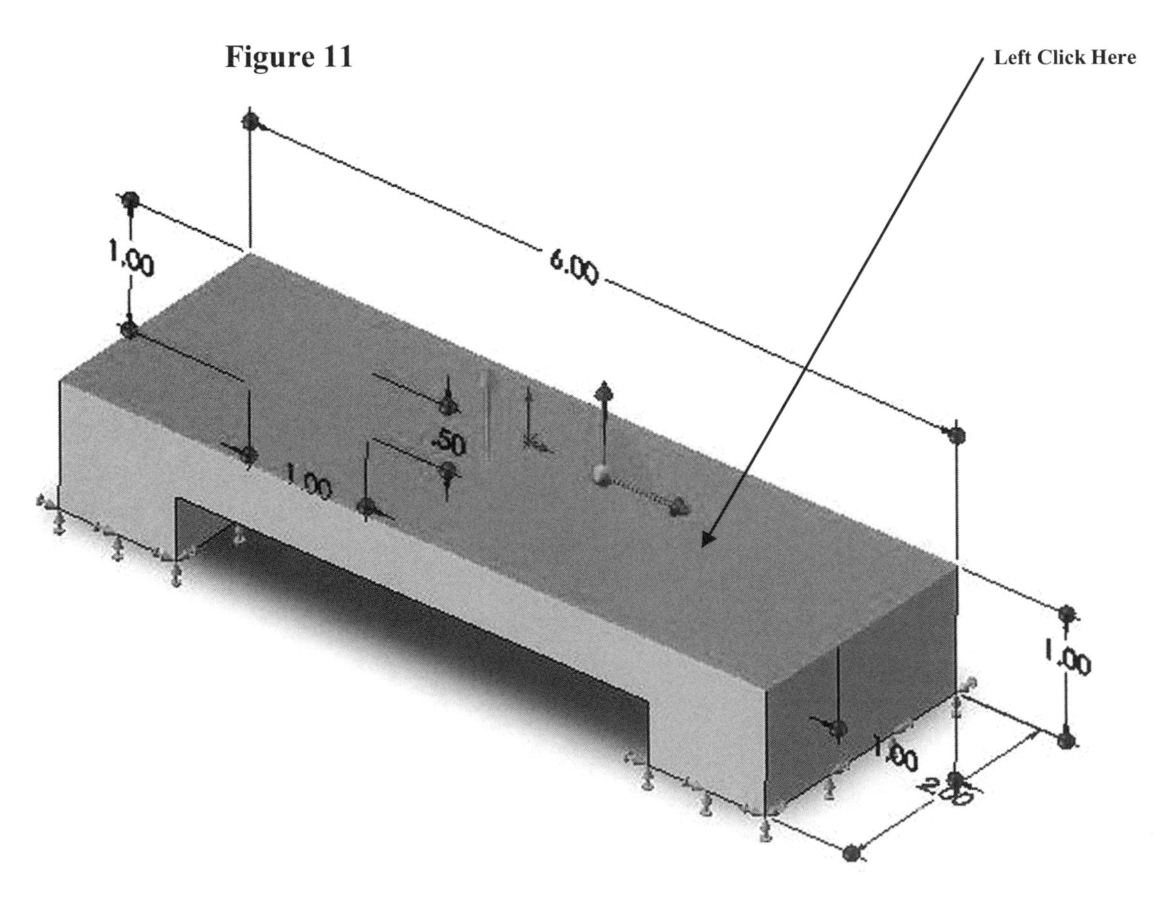

15. Move the cursor to the lower right portion of the screen and left click on **Add a force** as shown in Figure 12.

Figure 12

Left Click Here

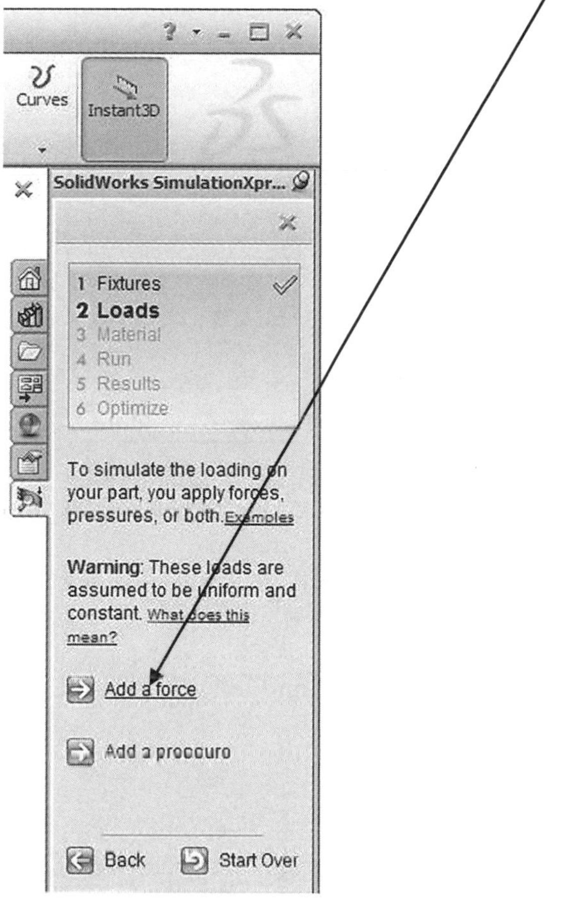

16. Move the cursor to the middle left portion of the screen and left click on the drop down arrow and change the units to **English (IPS)** as shown in Figure 13.

Figure 13

Left Click Here

17. Enter **100** for the load value and left click on the green check mark as shown in Figure 14

Figure 14

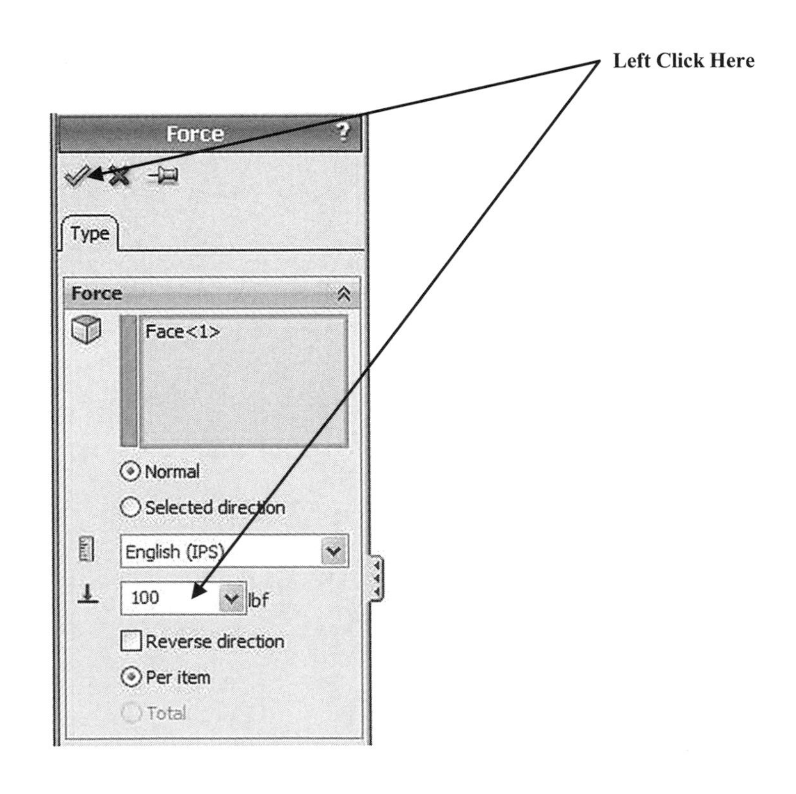

18. Move the cursor to the upper right portion of the screen and left click on **Material.** Left click on **<u>Choose Material</u>** as shown in Figure 15

Figure 15

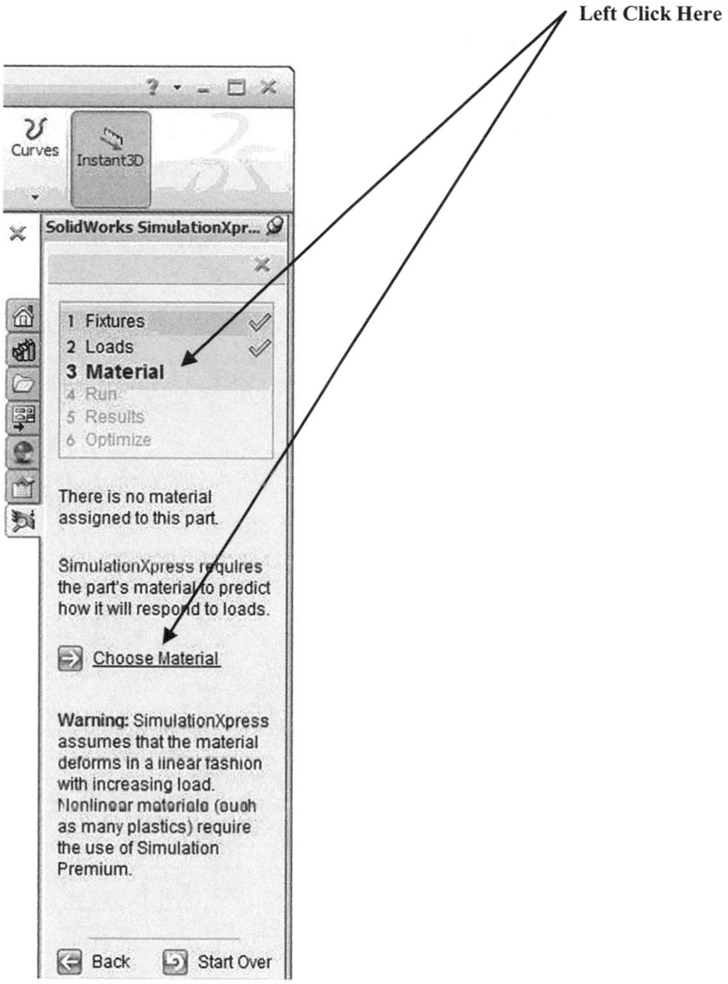
Left Click Here
Curves
Instant3D
SolidWorks SimulationXpr...
1 Fixtures
2 Loads
3 Material
4 Run
5 Results
6 Optimize
There is no material assigned to this part.
SimulationXpress requires the part's material to predict how it will respond to loads.
Choose Material
Warning: SimulationXpress assumes that the material deforms in a linear fashion with increasing load. Nonlinear materials (such as many plastics) require the use of Simulation Premium.
Back
Start Over

19. The Material dialog box will appear. Scroll down to **Aluminum Alloys** and select **1060 Alloy.** Left click on **Apply.** Left click on **Close** as shown in Figure 16.

Figure 16

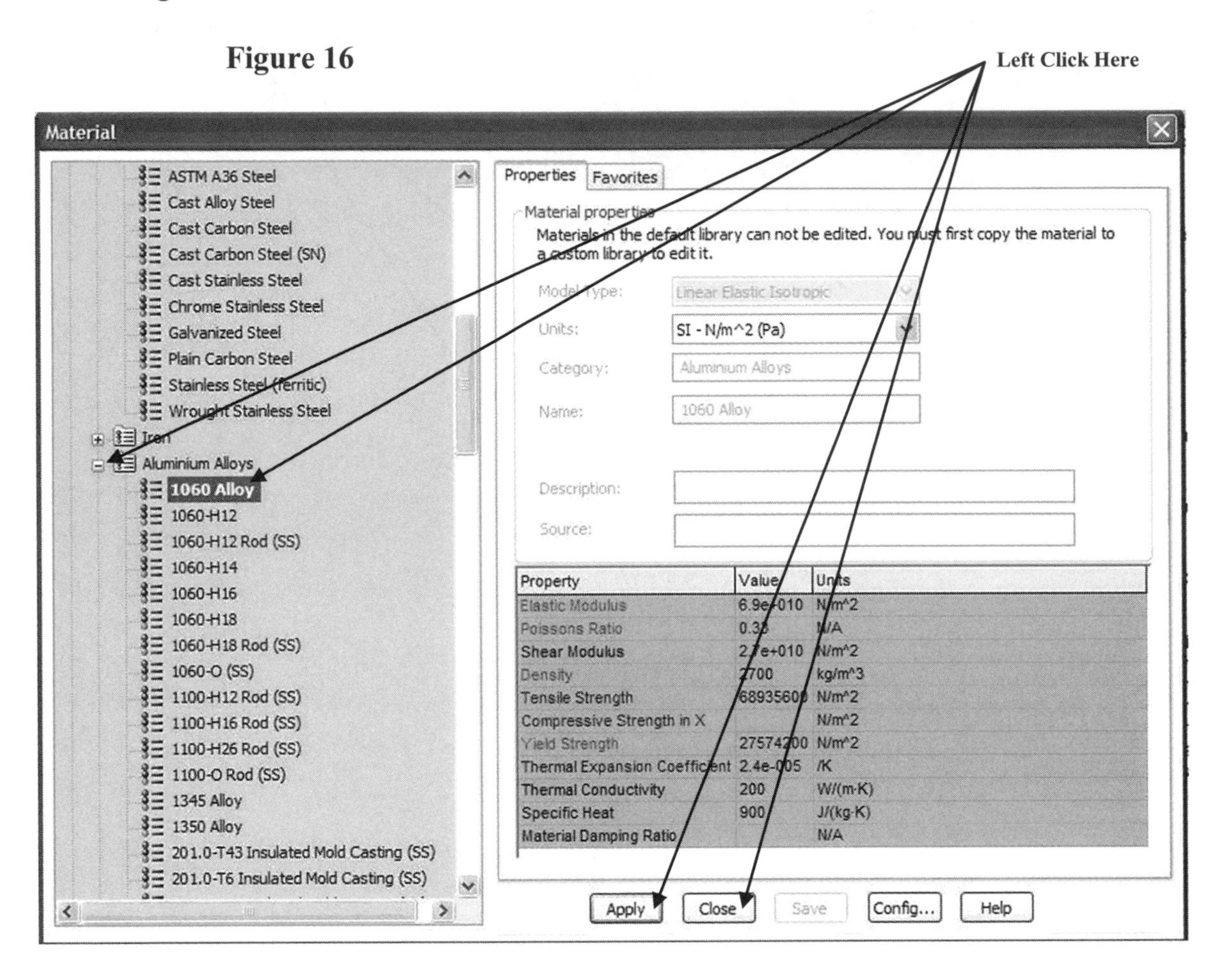

20. Move the cursor to the upper right portion of the screen and left click on **Run.** Left click on **<u>Run Simulation</u>** as shown in Figure 17.

Figure 17

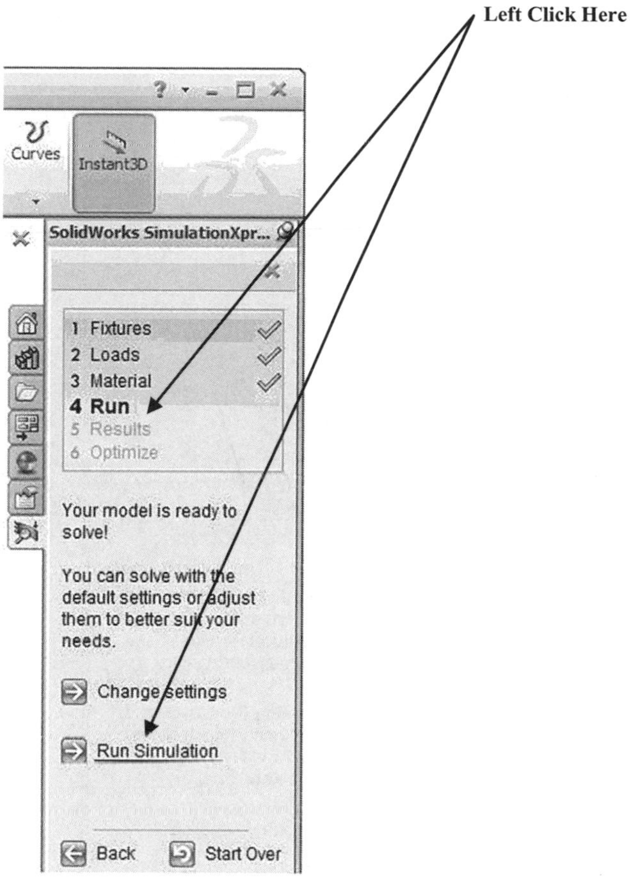

Left Click Here

21. The model will go through the simulation process. Notice that SolidWorks has moved to the Results portion of the simulation. Once this is complete, SolidWorks will display the results. Left click on **Yes, continue** as shown in Figure 18.

Figure 18

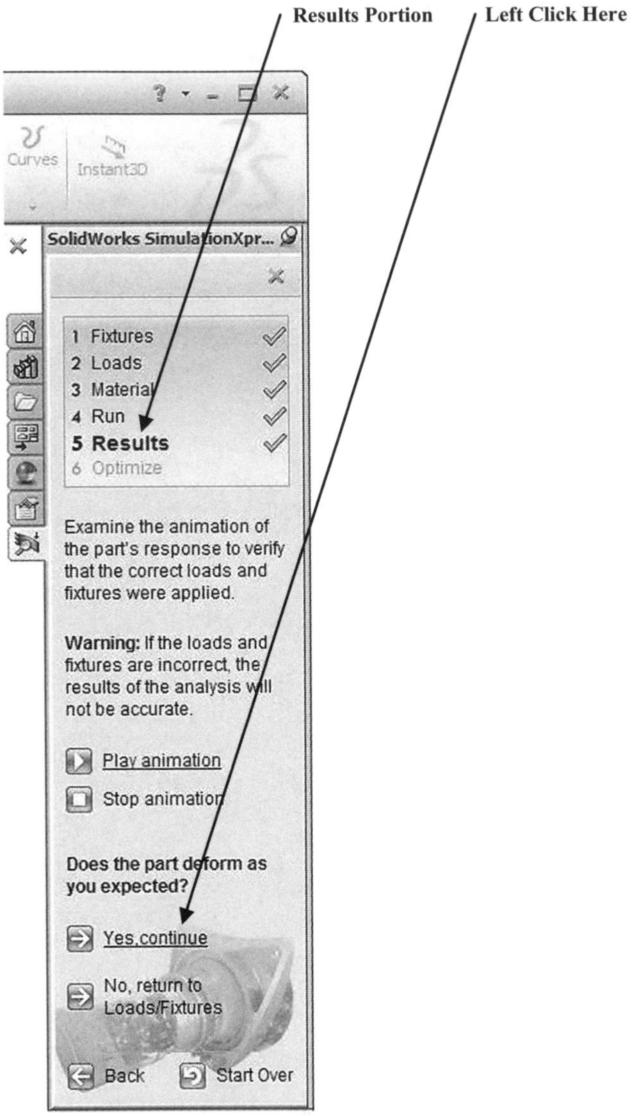

Results Portion

Left Click Here

Curves
Instant3D
SolidWorks SimulationXpr...
1 Fixtures
2 Loads
3 Material
4 Run
5 Results
6 Optimize
Examine the animation of the part's response to verify that the correct loads and fixtures were applied.
Warning: If the loads and fixtures are incorrect, the results of the analysis will not be accurate.
Play animation
Stop animation
Does the part deform as you expected?
Yes,continue
No, return to Loads/Fixtures
Back
Start Over

22. Move the cursor to the upper right portion of the screen and left click on **Show von Mises** stress. SolidWorks will provide an exaggerated display of the simulation during the test. Your screen should look similar to Figure 19. Rotate the part upward to identify any problems areas in the design. Areas of red indicate a potential high stress area. Left click on **Play Animation** to watch the simulation with color representation. When you are finished viewing the animation, move the cursor to the lower right portion of the screen and left click on **Stop Animation**. Move the cursor to the upper right corner of the screen and left click on **Close** as shown in Figure 19.

Figure 19

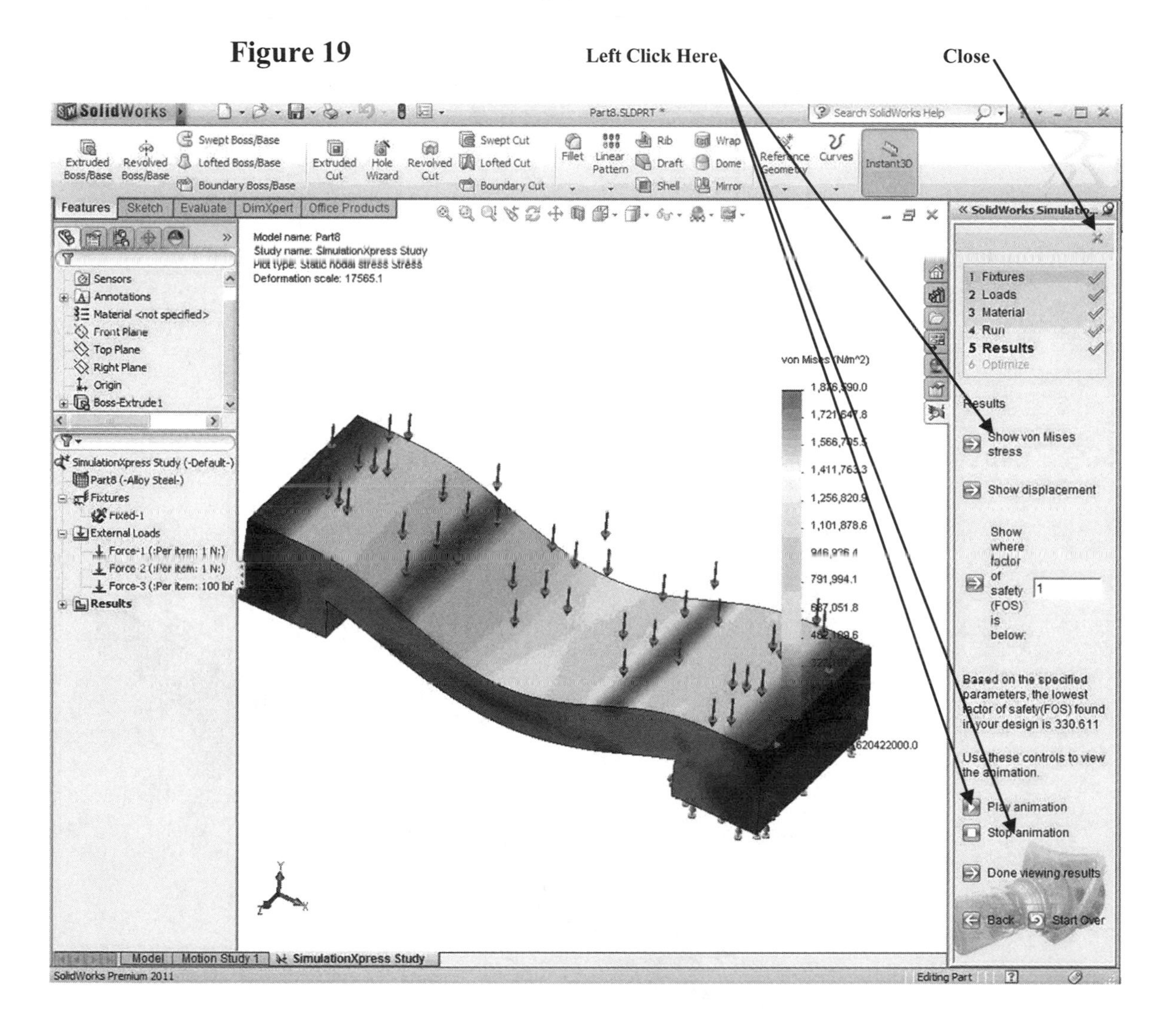

23. Begin a new drawing. Complete the sketch shown in Figure 20.

Figure 20

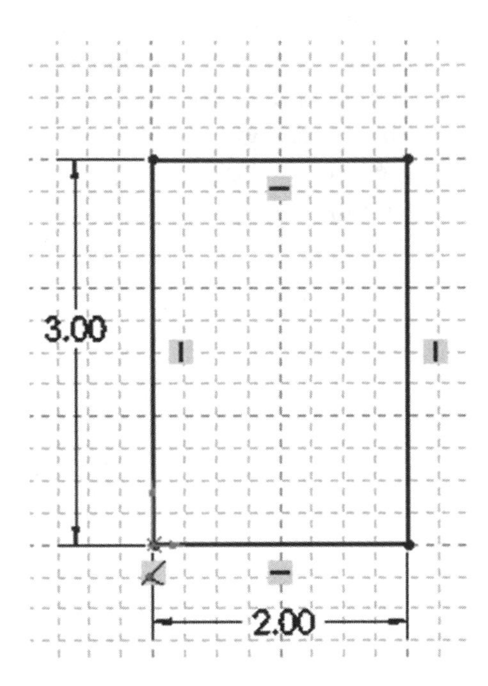

24. Extrude the sketch to a distance of **96** inches as shown in Figure 21.

Figure 21

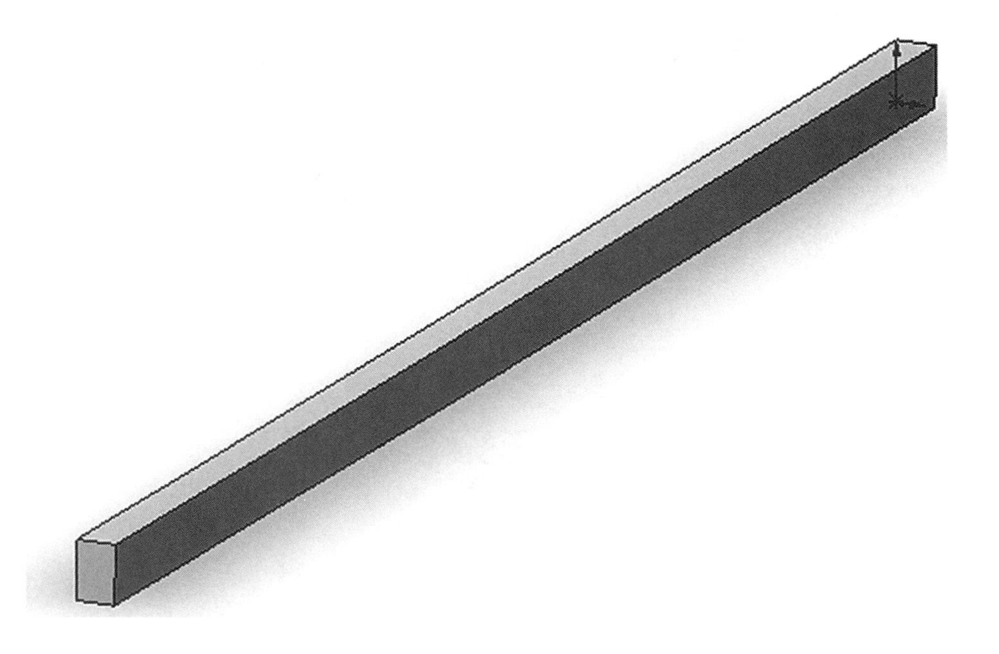

25. Use the **Shell** command to shell the part with a thickness of **.10** as shown in Figure 22.

Figure 22

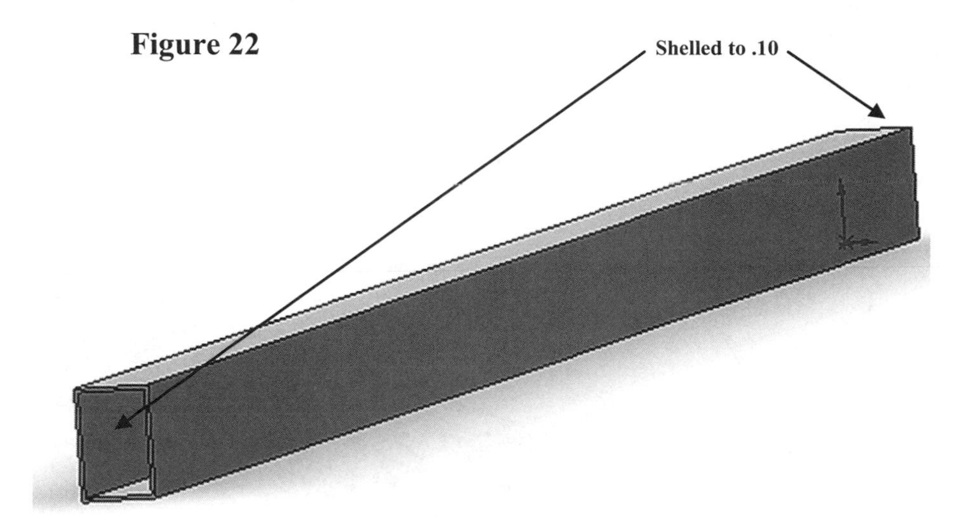

26. Move the cursor to the upper left portion of the screen and left click on the arrow to the right of SolidWorks. A drop down menu will appear. Left click on **SimulationXpress** as shown in Figure 23.

Figure 23

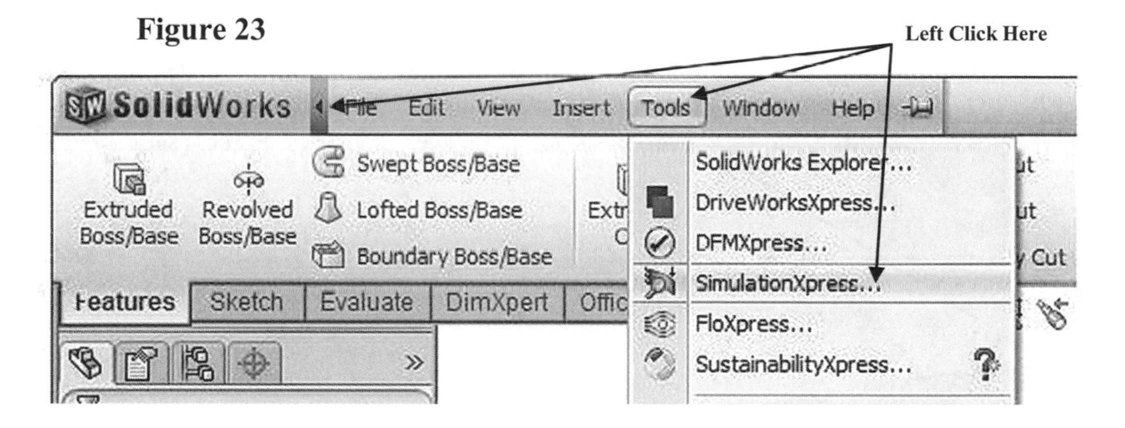

27. Run a simulation on this part (using **Cast Alloy Steel** for material and **100 lbs** for the Load) using the left side as the Restraint surface and the top as the Load surface as shown in Figure 24. This will create a cantilever effect for the simulation. If the simulation requires too much memory to run, reduce the length of the lever and run the simulation again.

Figure 24

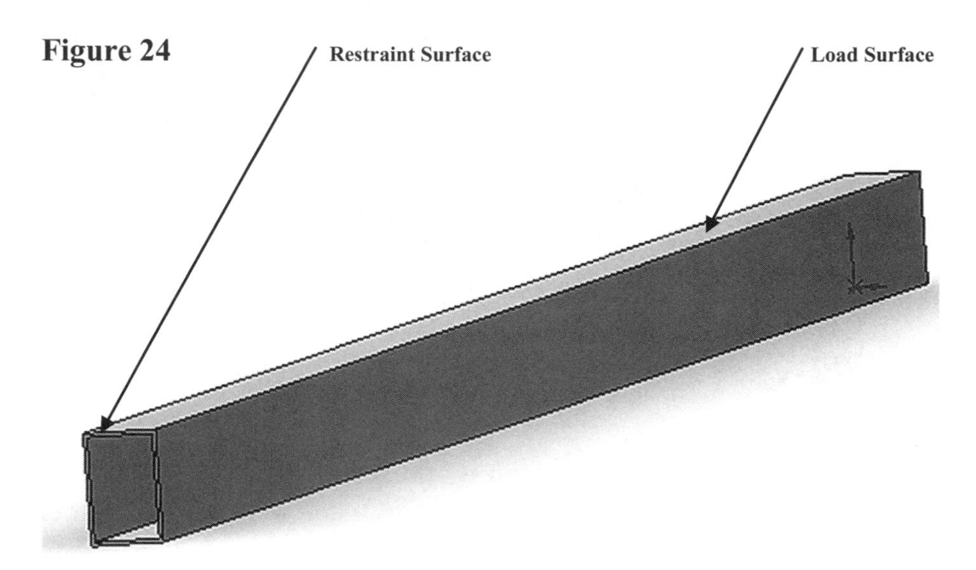

28. After the simulation has been run, your screen should look similar to Figure 25.

Figure 25

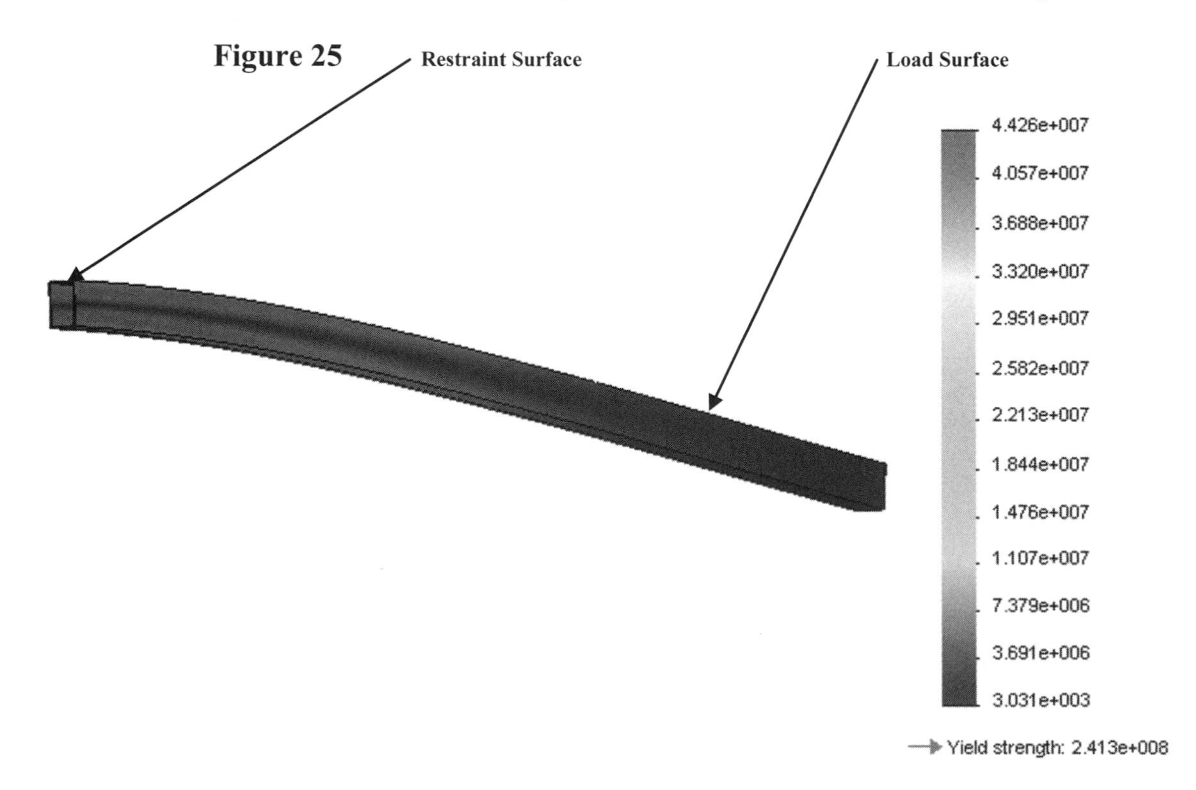

Chapter 10 Introduction to Mechanical Mates

Objectives:

- Learn to create a cam and lifter system of parts
- Learn to use one of the Mechanical Mates options

Chapter 10 includes instruction on how to construct the assembly shown below.

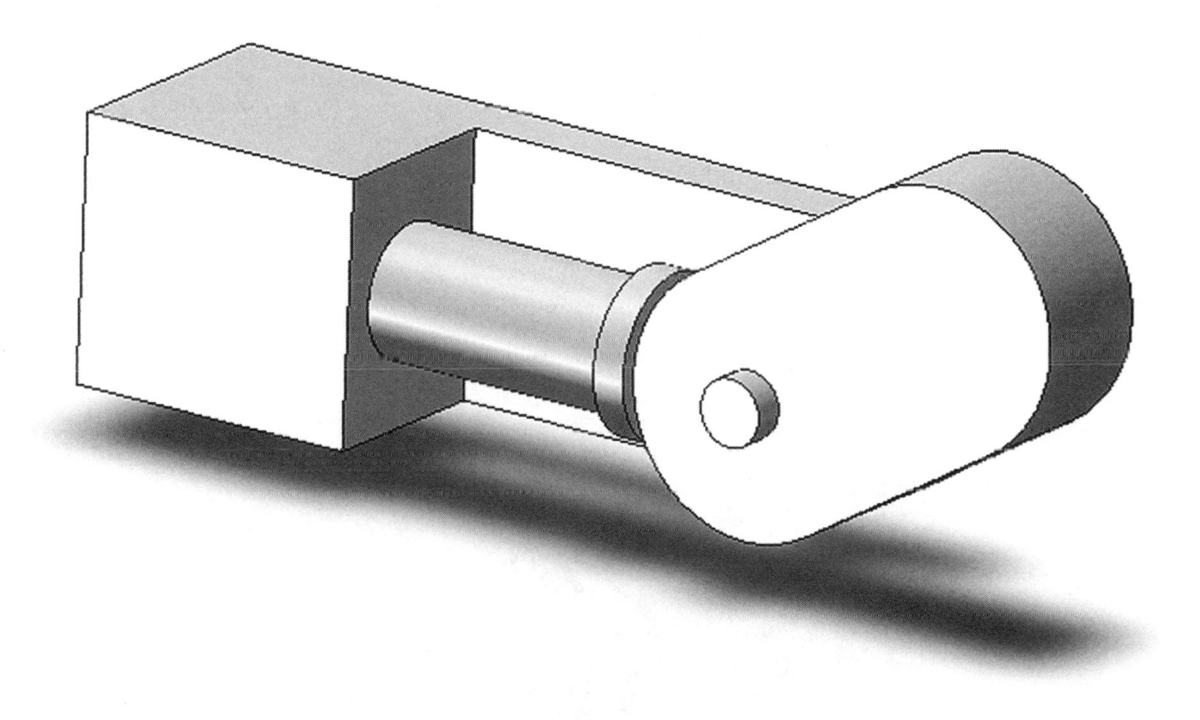

1. Start SolidWorks by referring to “Chapter 1 Getting Started”.

2. After SolidWorks is running, begin a New Sketch.

Figure 1

3. Exit the Sketch area. Extrude the sketch to a thickness of **1.00** in as shown in Figure 2

Figure 2

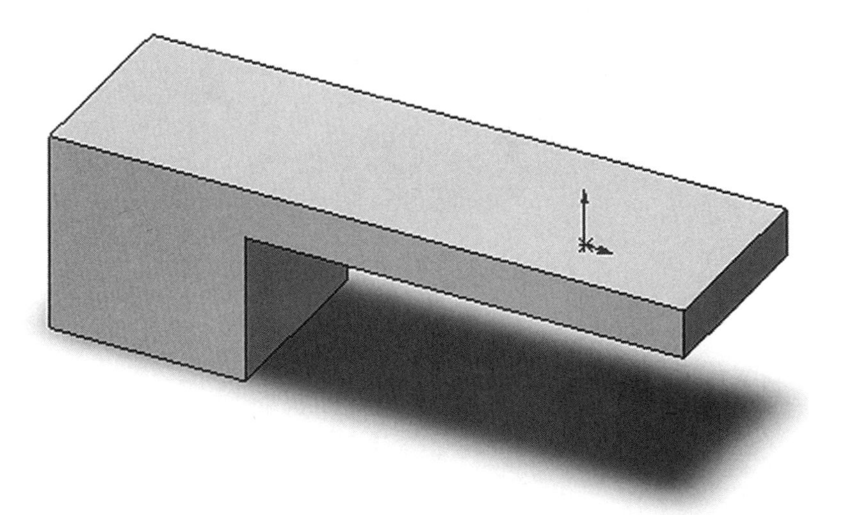

4. Use the **Fillet** command (.5 inch fillets) to radius the lower edge(s) as shown in Figure 3.

Figure 3

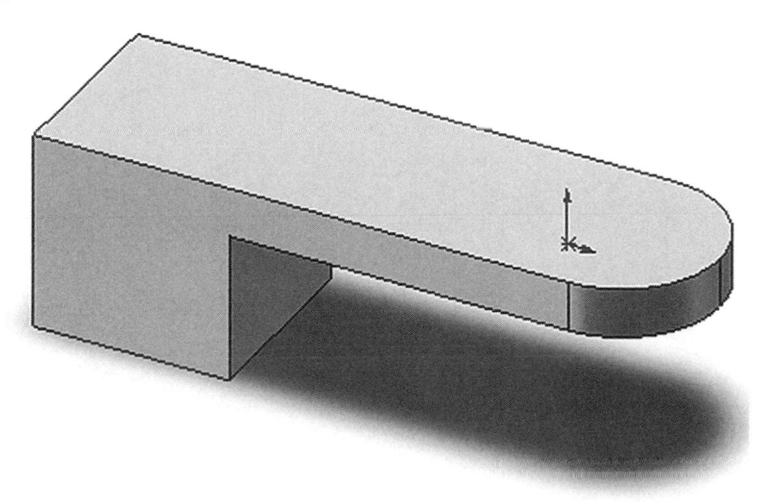

5. Use the **Rotate** command to rotate the part upward as shown in Figure 4.

Figure 4

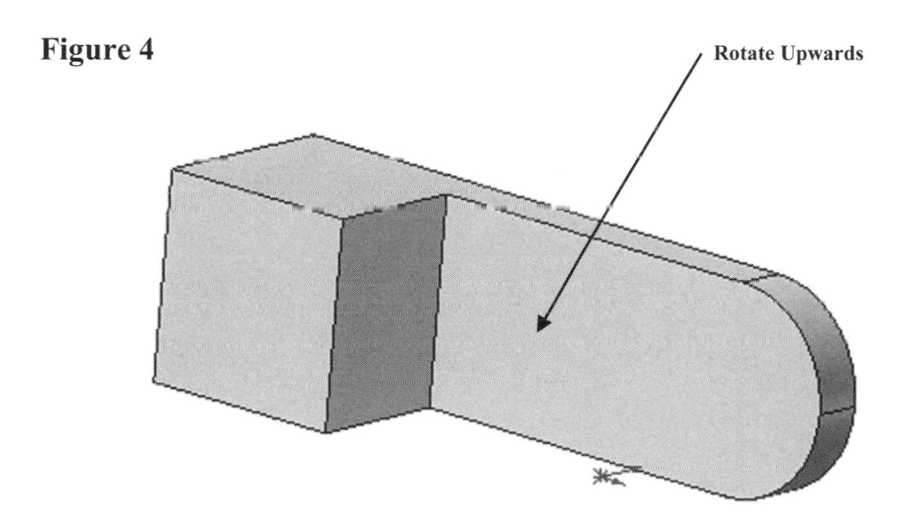

6. Complete the sketch as shown in Figure 5. The center of the circle is located on the corner of the fillet radius.

Figure 5

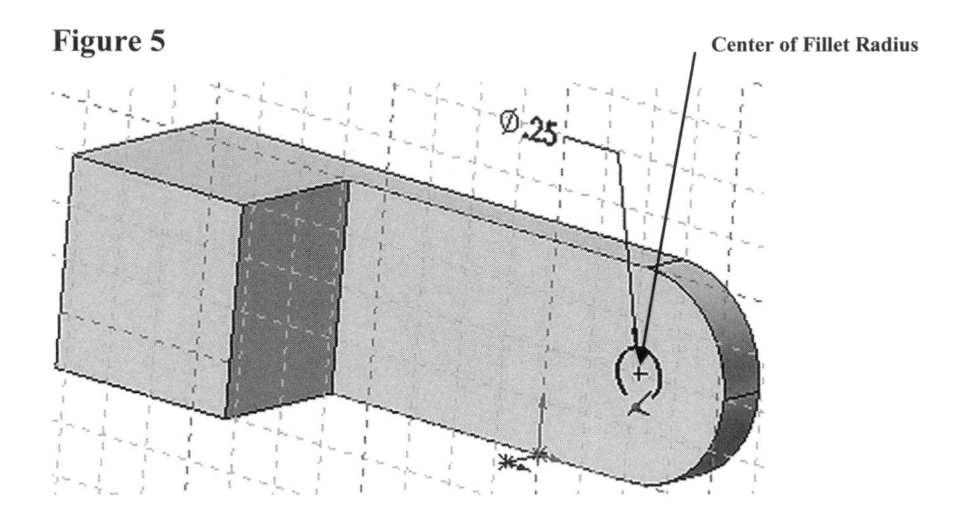

7. Extrude the circle to a distance of **.75** inches as shown in Figure 6.

Figure 6

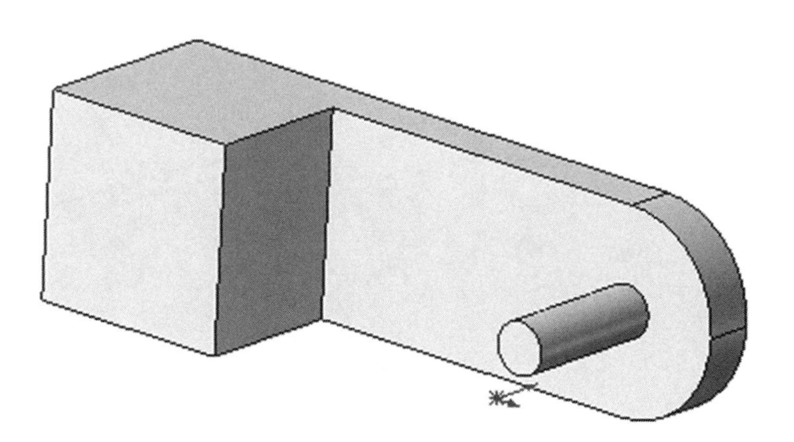

8. Rotate the part around. Complete the sketch shown in Figure 7.

Figure 7

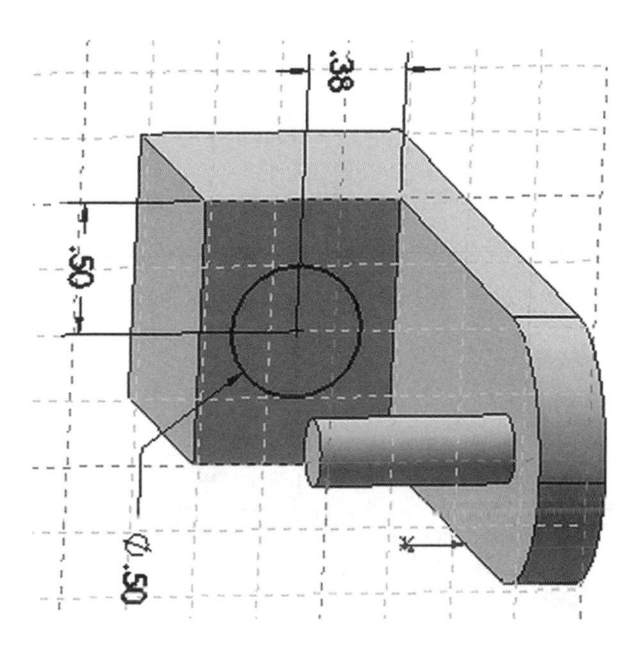

9. Use the **Extruded Cut** option to cut a hole a distance of **1.00 inch** as shown in Figure 8.

Figure 8

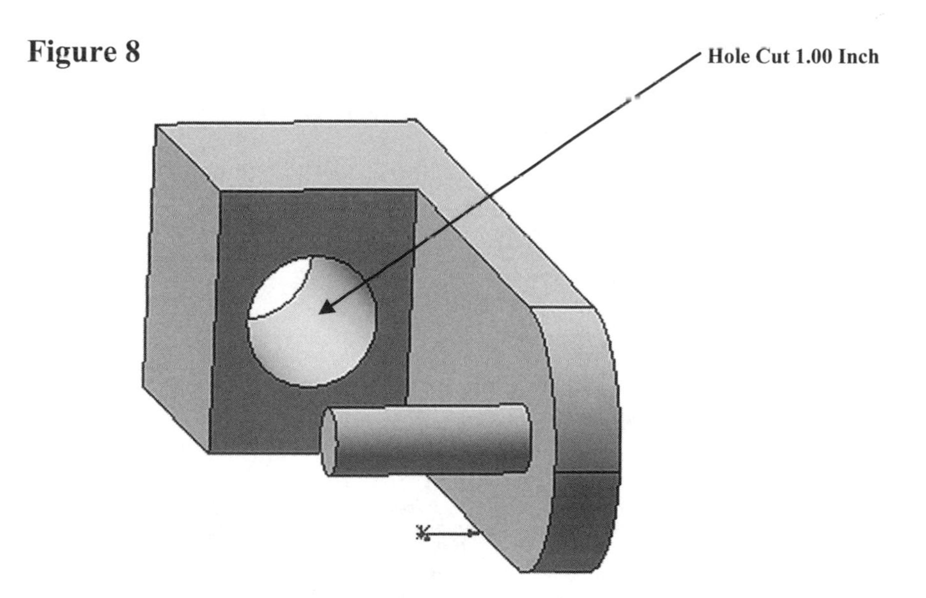

10. Save the part as **Camcase1.SLDPRT** where it can be easily retrieved later.

11. Complete the sketch shown. Extrude the .500 in diameter circle to a distance of 1.00 inch and the .625 in diameter circle to a distance of .125 inches as shown in Figure 9.

Figure 9

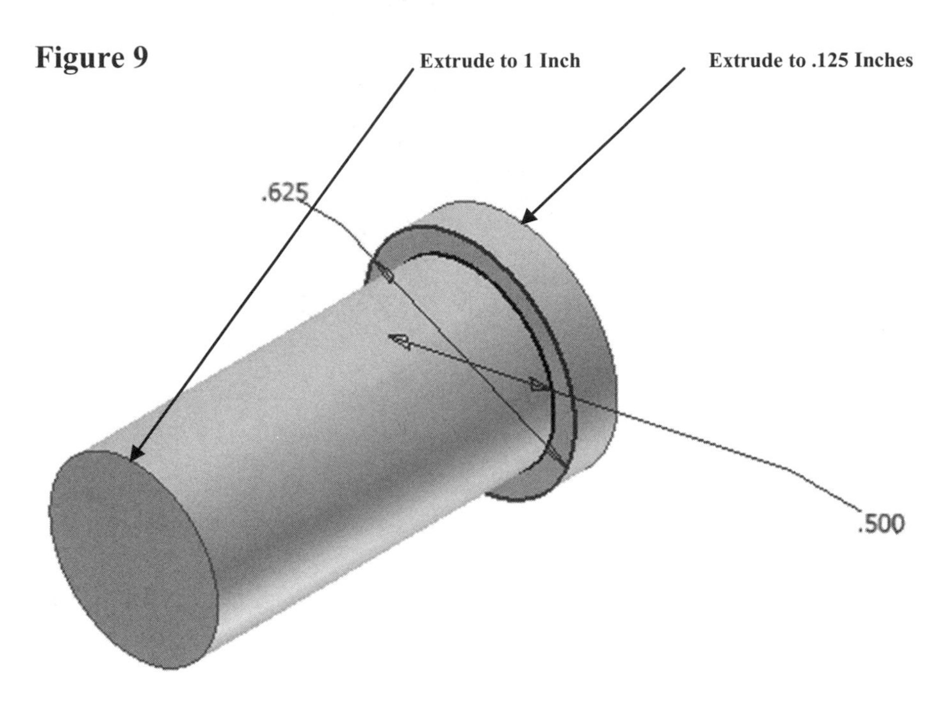

12. The bottom of the part needs to be flat as shown in Figure 10.

Figure 10

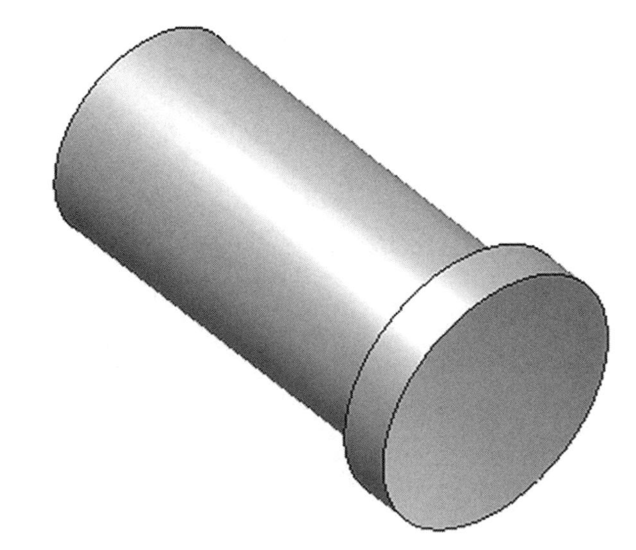

13. Save the part as **Lifter1.SLDPRT** where it can easily be retrieved later.

14. Begin a new drawing. Complete the sketch shown in Figure 11.

Figure 11

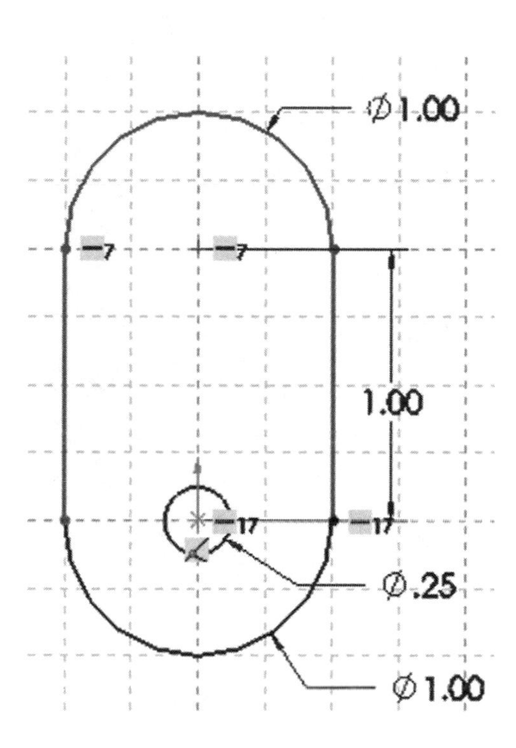

15. Extrude the sketch to a distance of **.500** inches as shown in Figure 12.

Figure 12

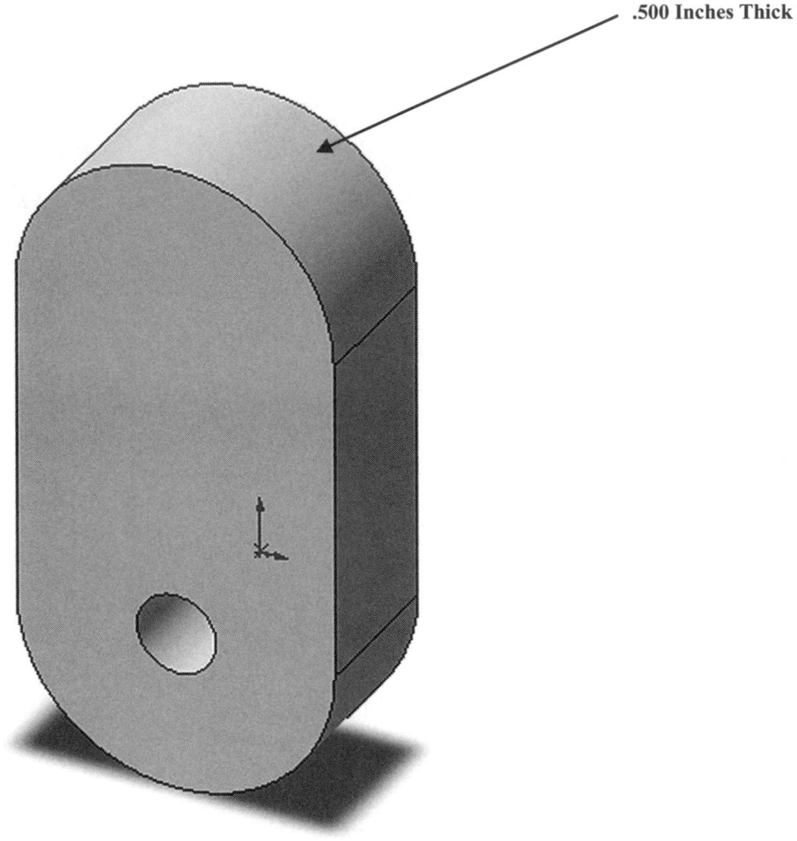

.500 Inches Thick

16. Save the part as **Lobe1.SLDPRT** where it can easily be retrieved later.

17. Start a new Assembly drawing as shown in Chapter 7.

18. Move the cursor to the left middle portion of the screen and left click on **Browse** (not shown). The Open dialog box will appear. Left click on C**amcase1.SLDPRT**. Left click on **Open** as shown in Figure 13.

Figure 13

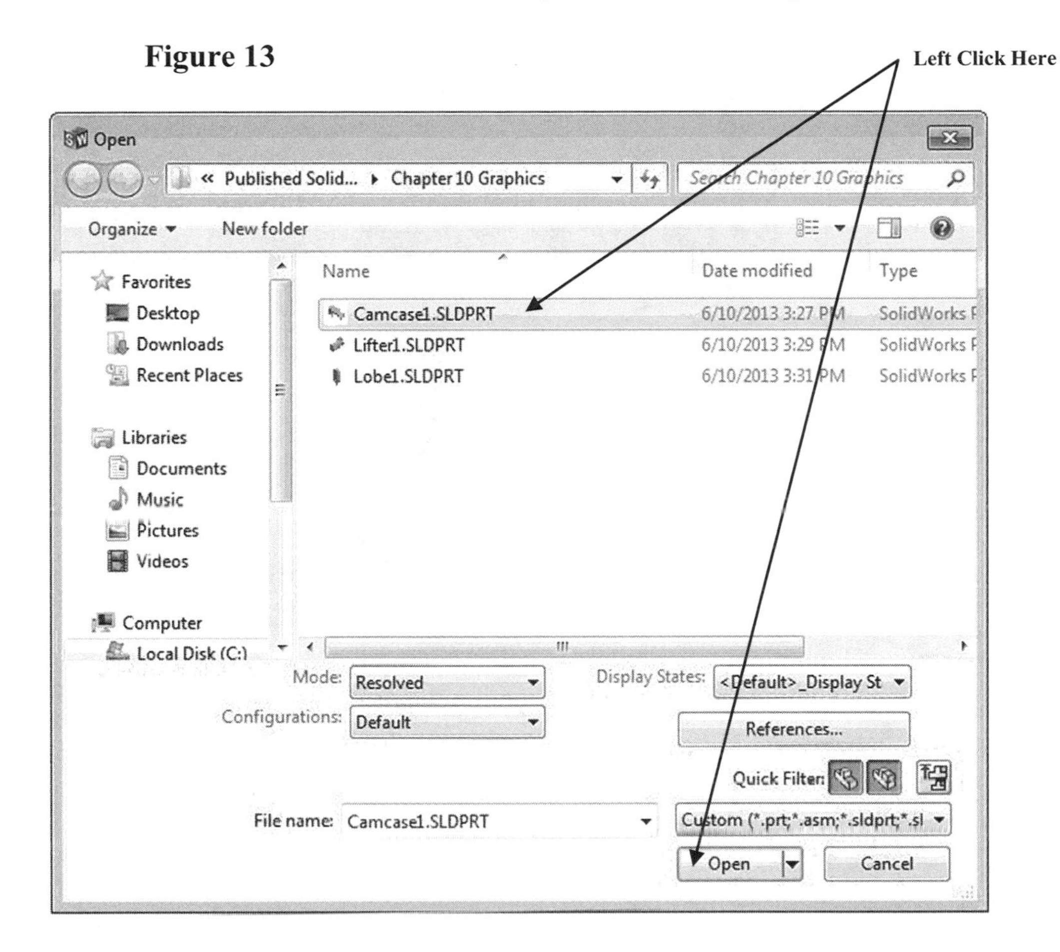

19. SolidWorks will place one cam case in the work area. The cam case will be attached to the cursor. If the part does not appear attached to the cursor simply left click anywhere in the work area.

20. Use the **Insert Components** command to insert the **Lifter1.SLDPRT** and **Lobe1.SLDPRT** parts as shown in Figure 14.

Figure 14

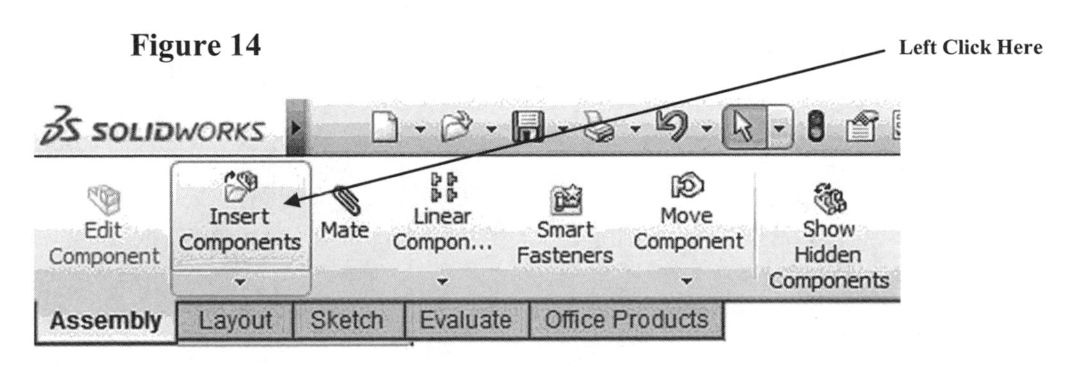

21. Your screen should look similar to Figure 15.

Figure 15

22. Using the **Mate** command, place the lifter in the lifter bore as shown. You may have to change the Mate Alignment, located at the lower left portion of the screen if the lifter is mated upside down as shown in Figure 16.

Figure 16

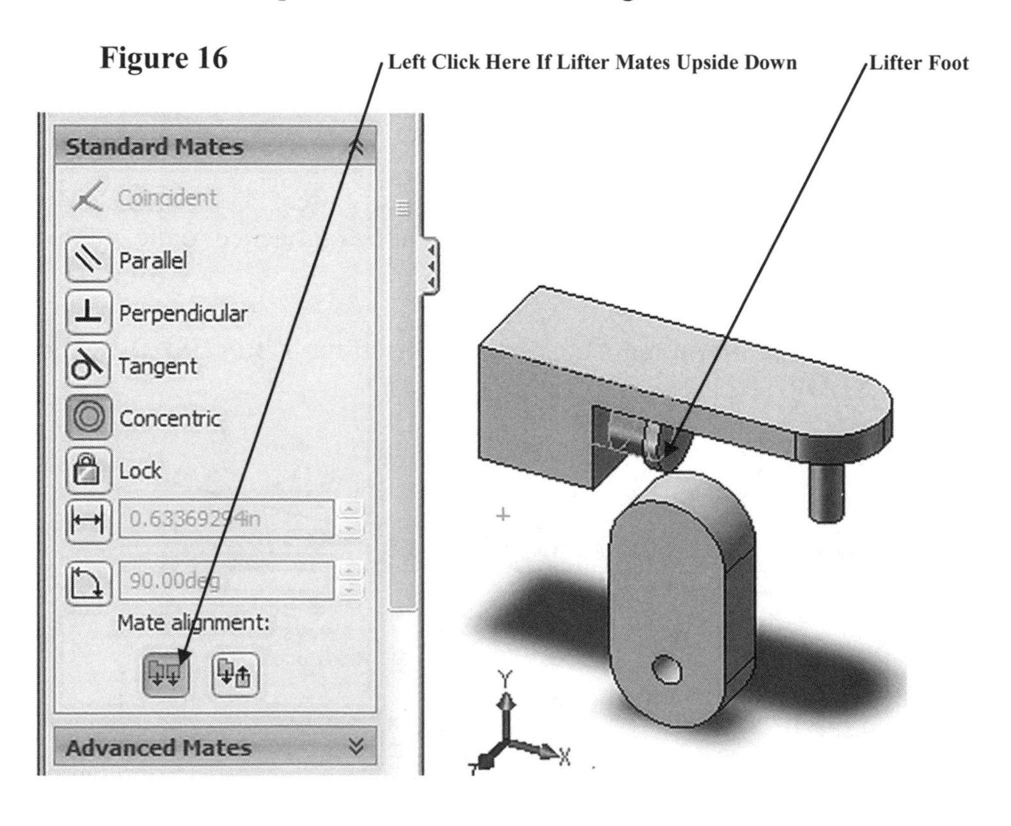

23. Rotate the assembly up as shown in Figure 17.

Figure 17

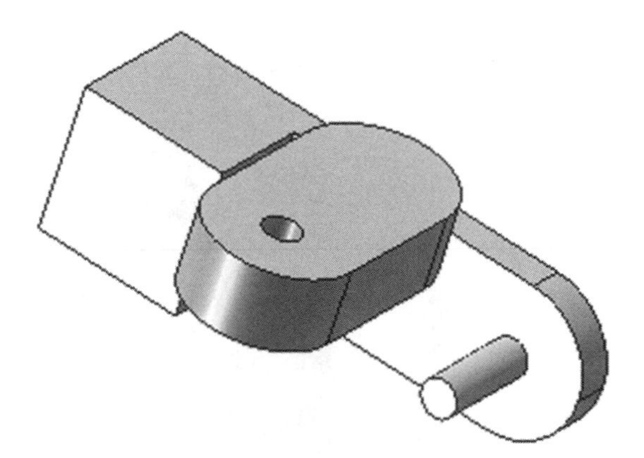

24. Use the **Mate** command to mate the lobe to the shaft as shown in Figure 18.

Figure 18

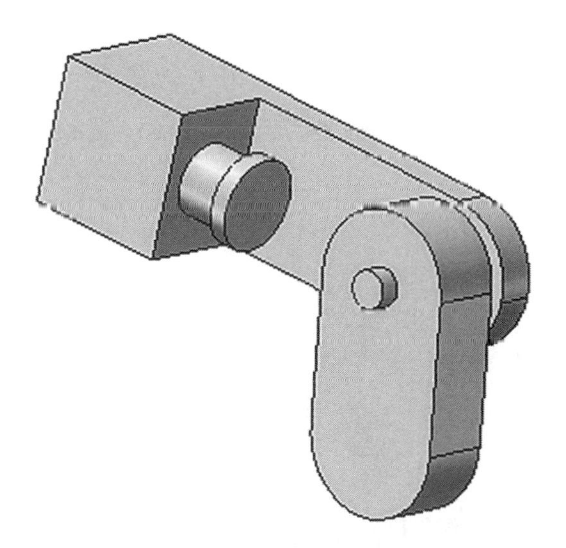

25. Use the **Mate** command to create a **.125 inch** offset between the lobe and the case as shown in Figure 19.

Figure 19

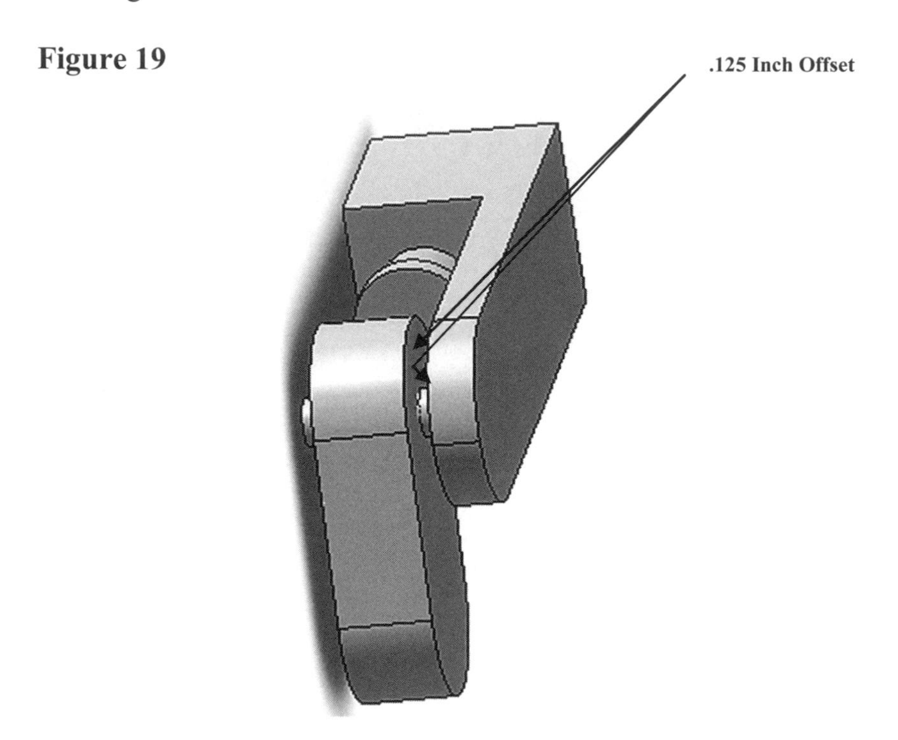

26. Use the **Rotate** command to rotate the assembly around as shown in Figure 20.

Figure 20

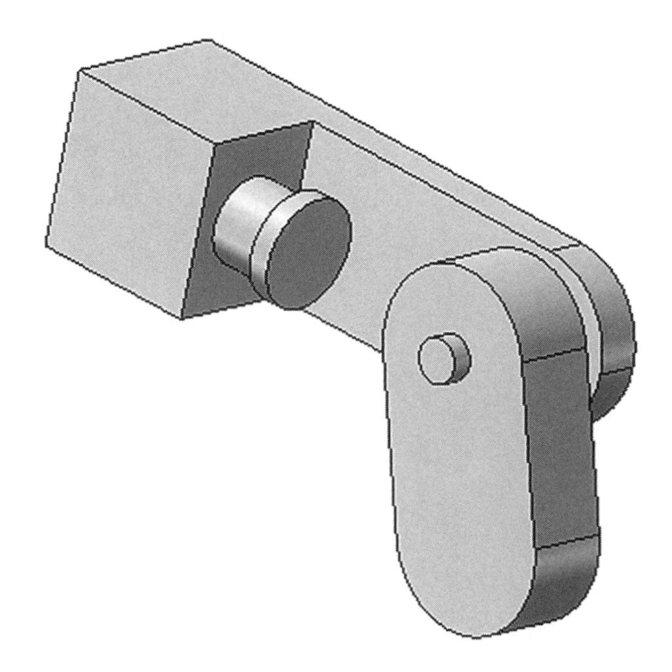

27. Move the cursor to the upper left portion of the screen and left click **Mate** as shown in Figure 21.

Figure 21

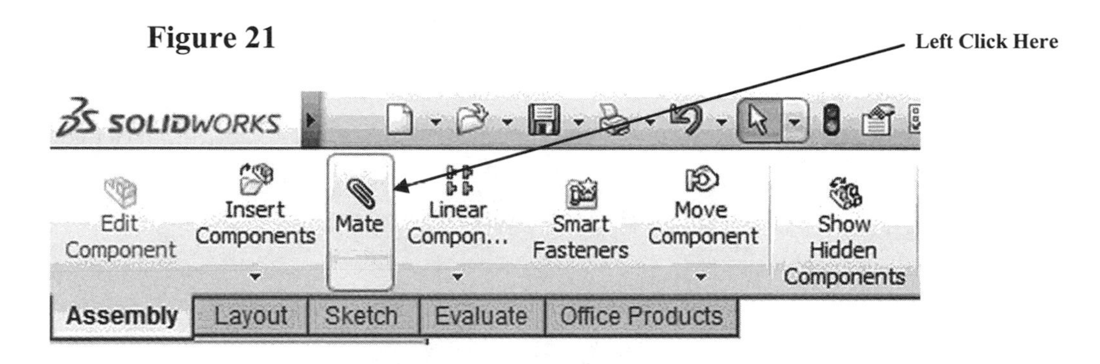

28. Left click on drop down arrows to the right of Mechanical Mates as shown in Figure 22.

Figure 22

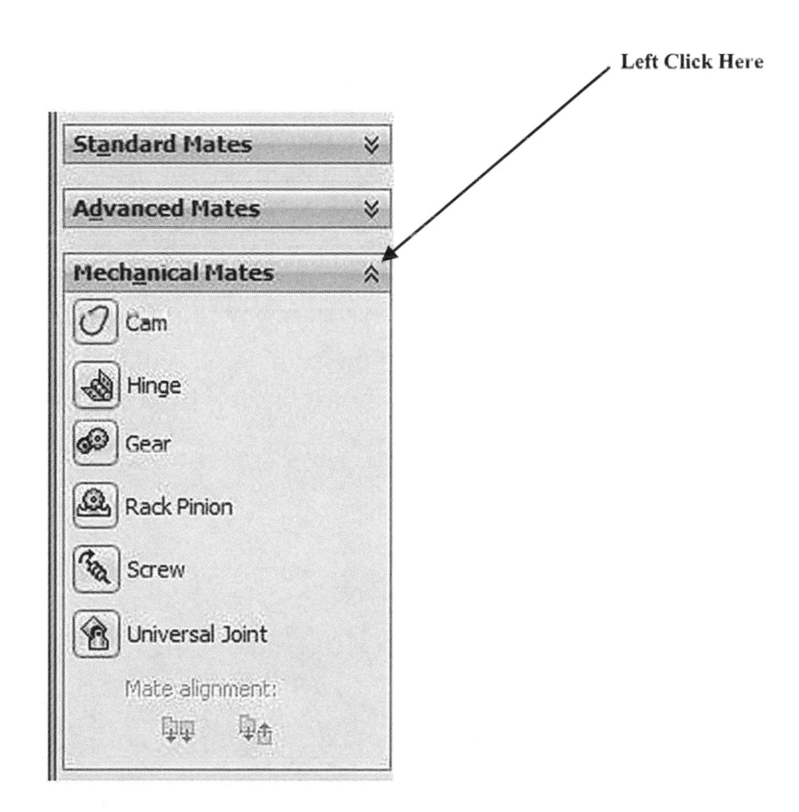

29. The Mechanical Mates drop down box will appear.

30. Left click on **Cam.** The Mate dialog box will appear. Left click on the Analysis tab. This will activate the analysis function. Then, left click on the Mate tab as shown in Figure 23.

Figure 23

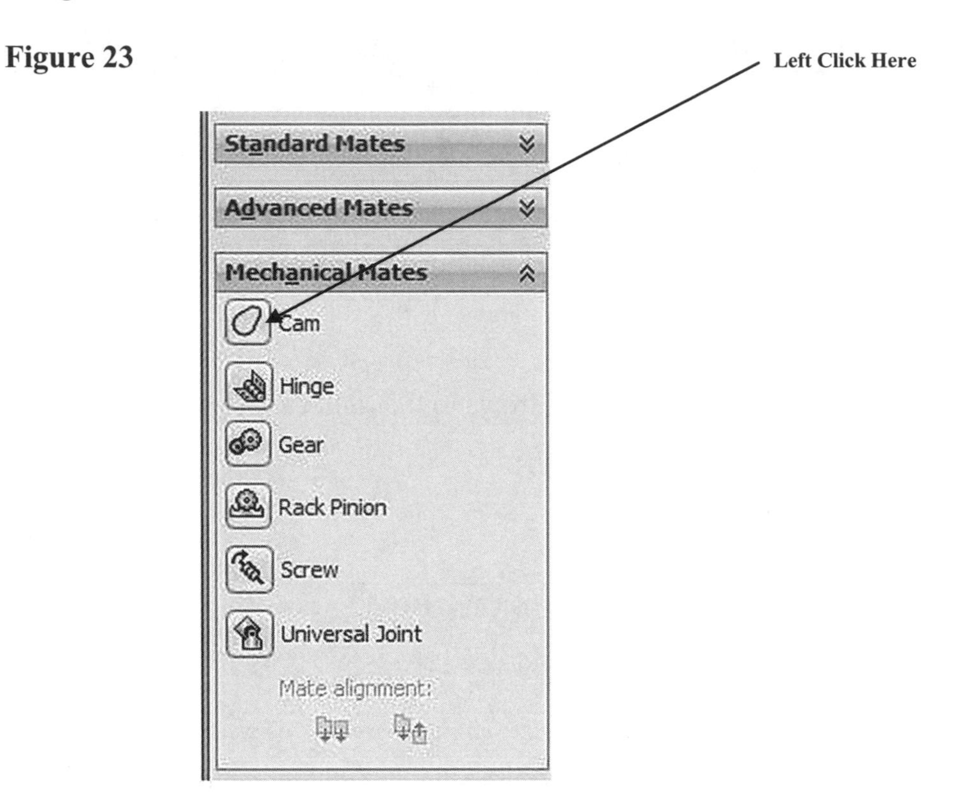

31. Left click on all 4 faces of the cam. You will have to rotate the assembly around (by pressing down on the mouse wheel) to gain access to all 4 faces as shown in Figure 24.

Figure 24

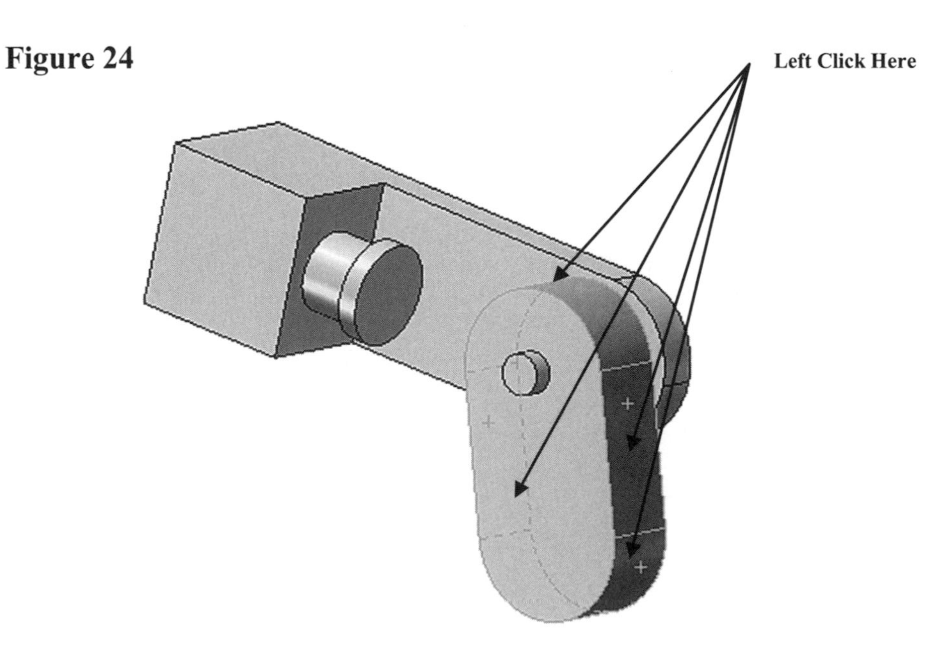

32. Left click in the box under Cam Follower as shown in Figure 25.

Figure 25

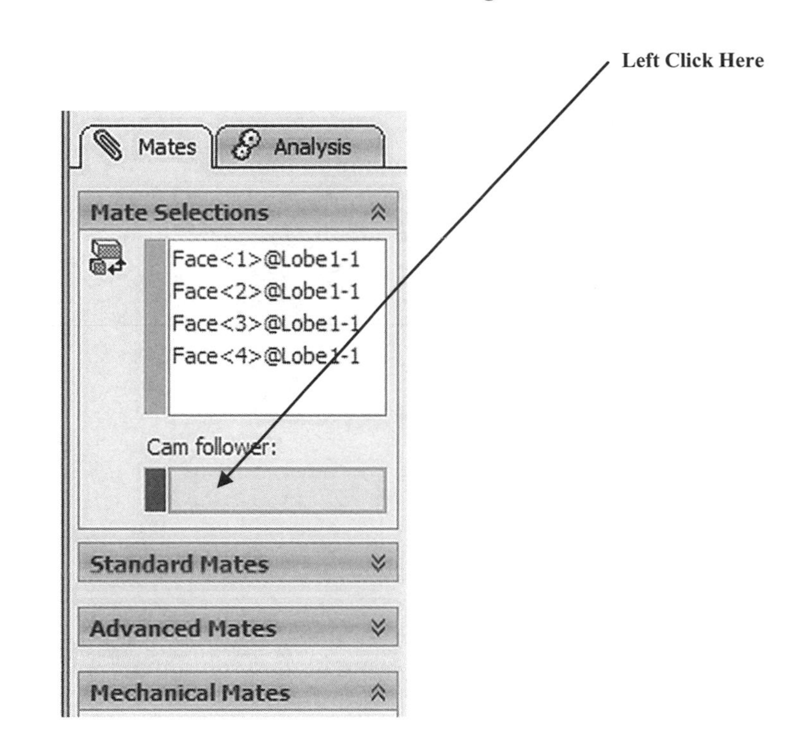

33. Left click on the Lifter foot as shown in Figure 26.

Figure 26

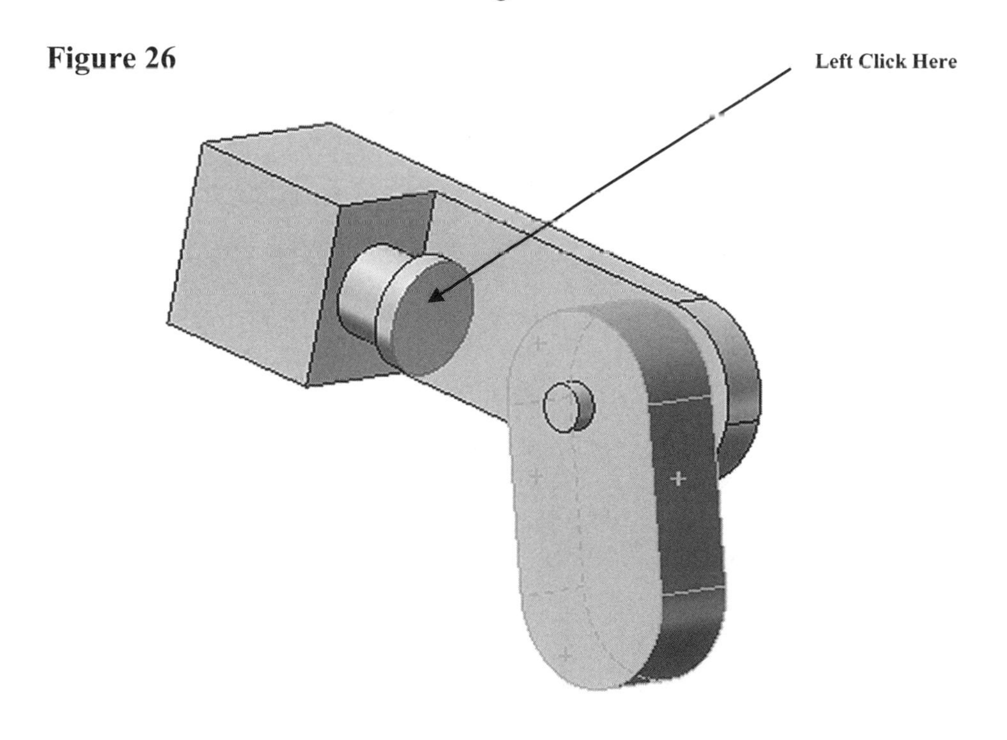

34. SolidWorks will move the lifter down in contact with the cam face as shown in Figure 27.

Figure 27

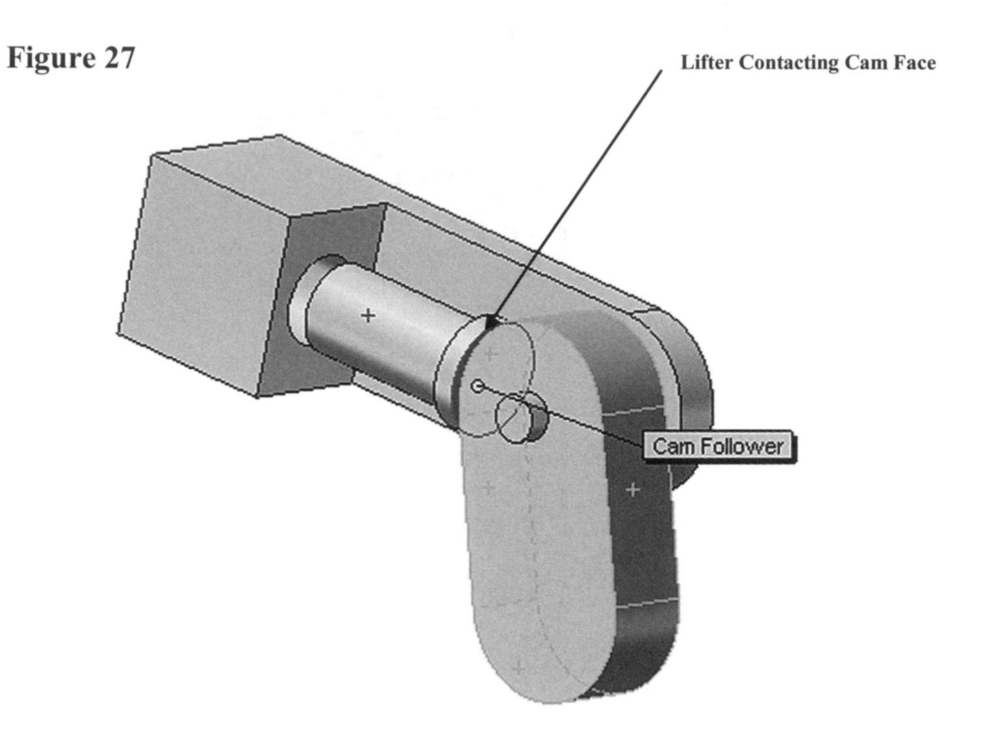

35. Left click on the green check mark as shown in Figure 28.

Figure 28

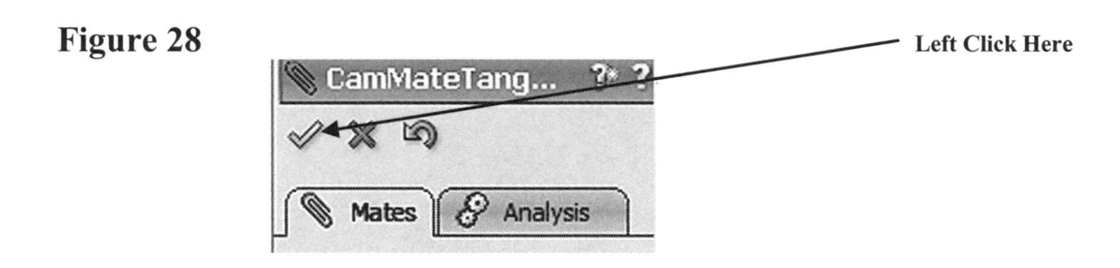

36. An error message will appear indicating that portions of the lobe contain a flat surface. This was done to simplify the exercise. Left click on **OK** as shown in Figure 29.

Figure 29

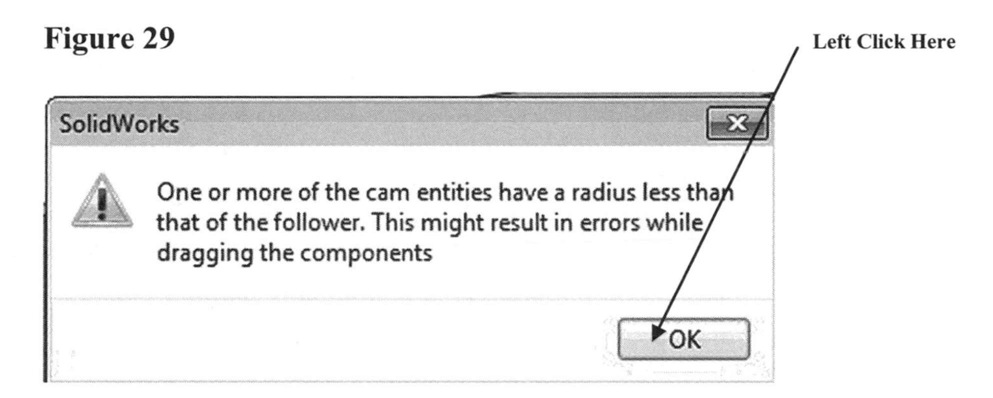

37. Move the cursor to the upper left portion of the screen and left click on the green check mark as shown in Figure 30.

Figure 30

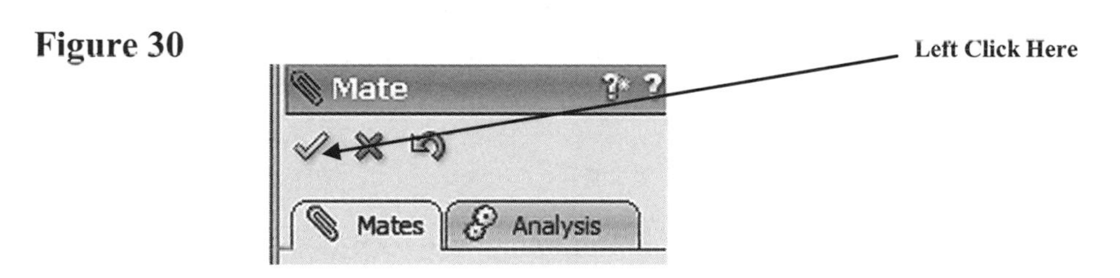

38. Move the cursor to the lower left portion of the screen and left click on **Motion Study 1** as shown in Figure 31.

Figure 31

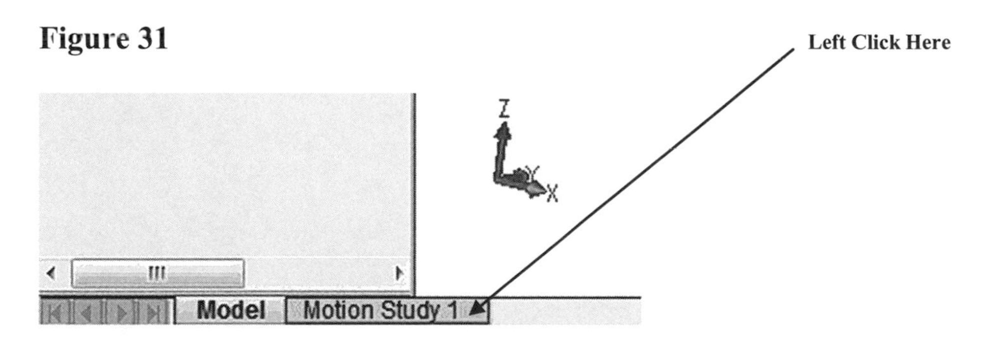

39. Left click on the **Motor** icon as shown in Figure 32.

Figure 32

40. Left click on the lower portion of the lobe (base circle) as shown in Figure 33.

Figure 33

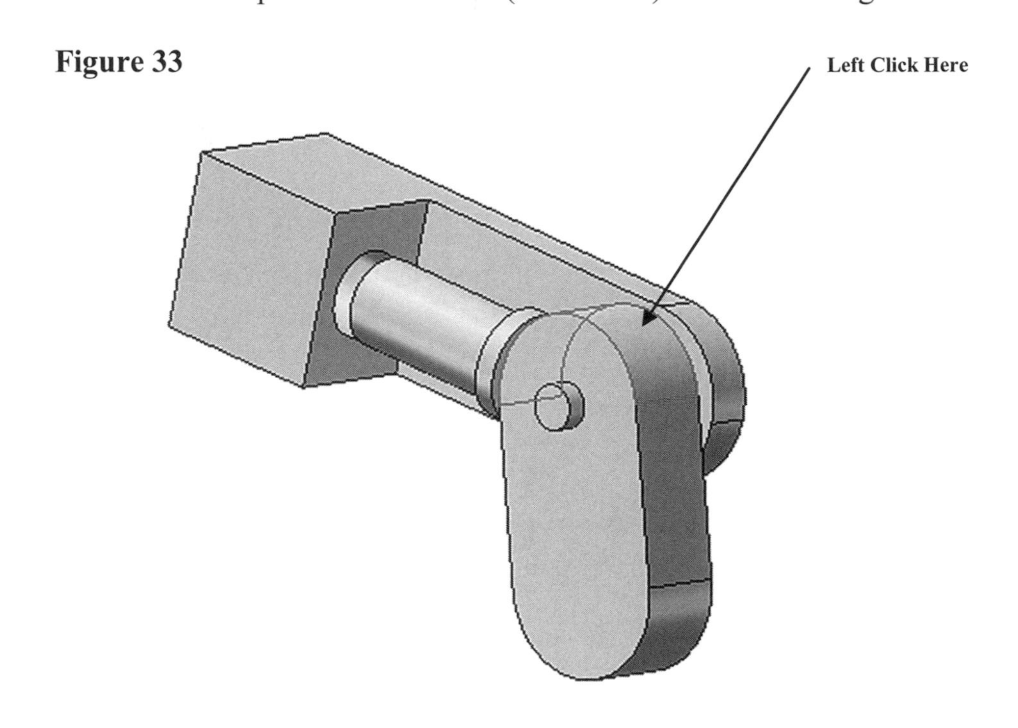

41. SolidWorks will create a motor on the lower portion of the lobe as shown in Figure 34.

Figure 34

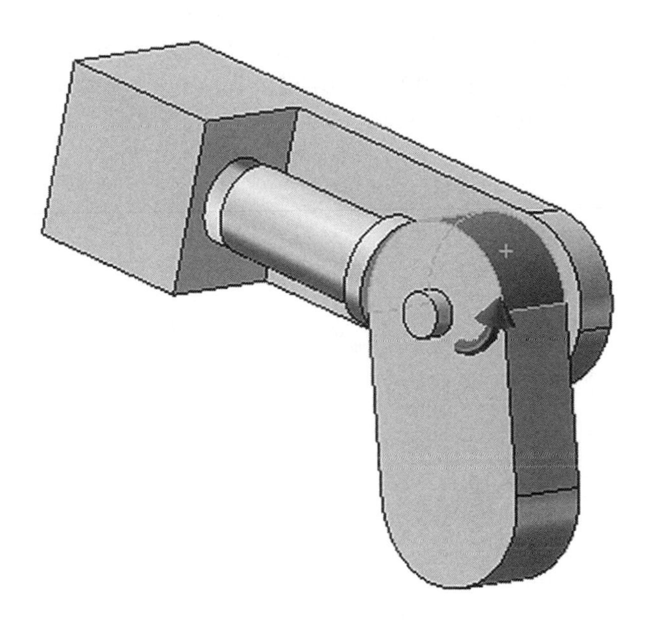

42. Move the cursor to the upper middle portion of the screen and left click on the green check mark as shown in Figure 35.

Figure 35

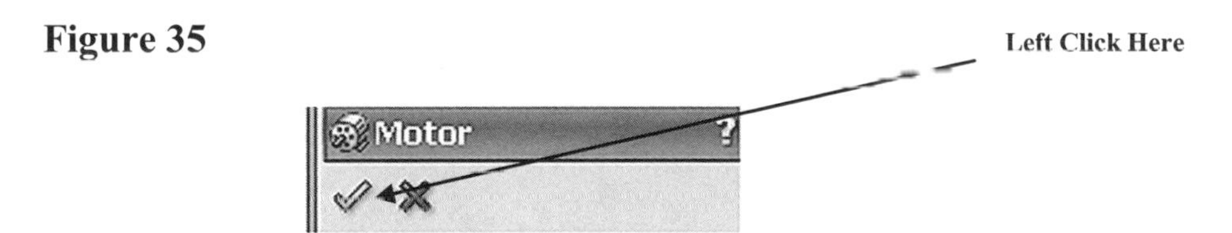

43. Move the cursor to the lower left portion of the screen and left click on the Calculate Motion Study icon. Any time changes are made to the motion study, the Calculate Motion Study icon will need to be selected. Then left click on the Play icon as shown in Figure 36.

Figure 36

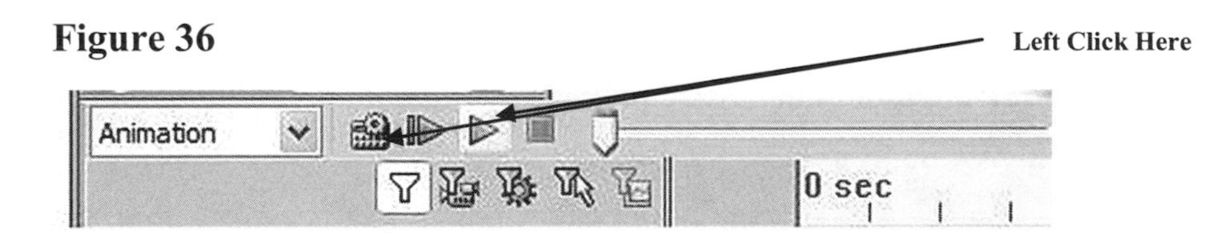

44. Refer to Chapter 7 for more **Play** options. At this time the lifter travels out of the lifter bore. There are 2 ways to correct this. Increase the top portion of the cam case and reduce the height of the lobe. Refer to Chapter 7 on how to edit either the cam case or lobe while in the Assembly area and see the effects it would have on the system of moving parts. You will also notice that the lifter does not stay in contact entirely with the lobe when rotated. This is because the lobe contains 2 flat surfaces which are not typically found on a typical camshaft lobe. The cam lobe constructed for this exercise was simplified for instructional purposes.

Chapter 11 Advanced Work Plane Procedures

Objectives:

- Learn to create points on a solid model
- Learn to create an offset work plane
- Learn to create an offset Extruded cut

Chapter 11 includes instruction on how to create multiple Points on a solid model and use these points to create an offset work plane for use as an offset cut extrusion.

Chapter 11 includes instruction on how to design the part shown below.

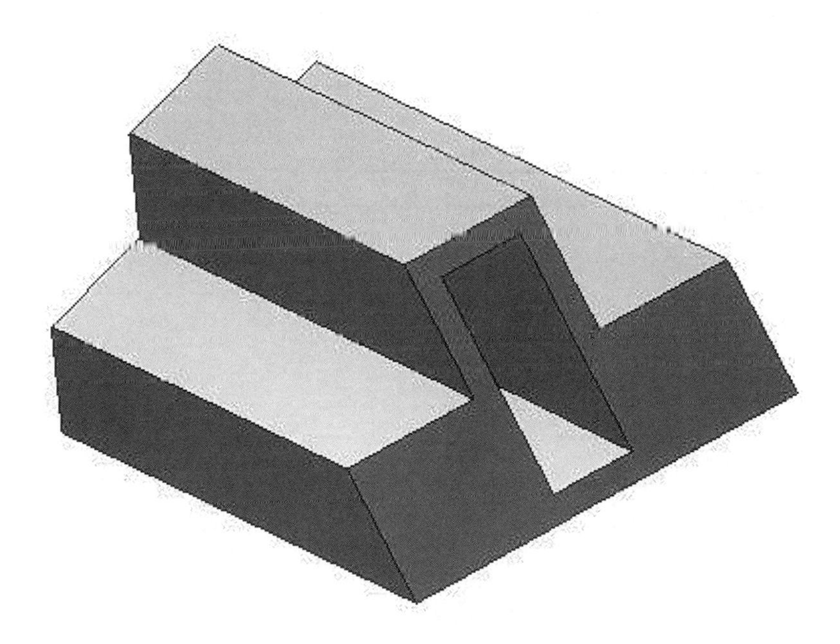

1. Start SolidWorks by referring to "Chapter 1 Getting Started". After SolidWorks is running, begin a New Sketch.

2. Complete the sketch shown in Figure 1.

Figure 1

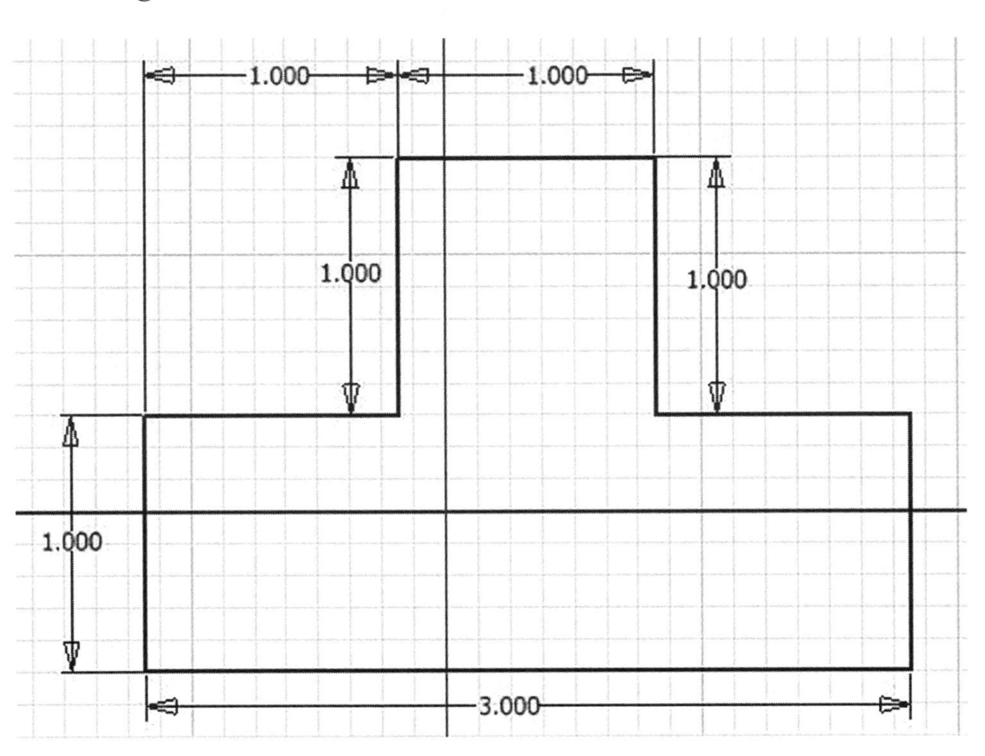

3. Exit the Sketch Panel. Extrude the sketch to a thickness of **4.00** inches as shown in Figure 2.

Figure 2

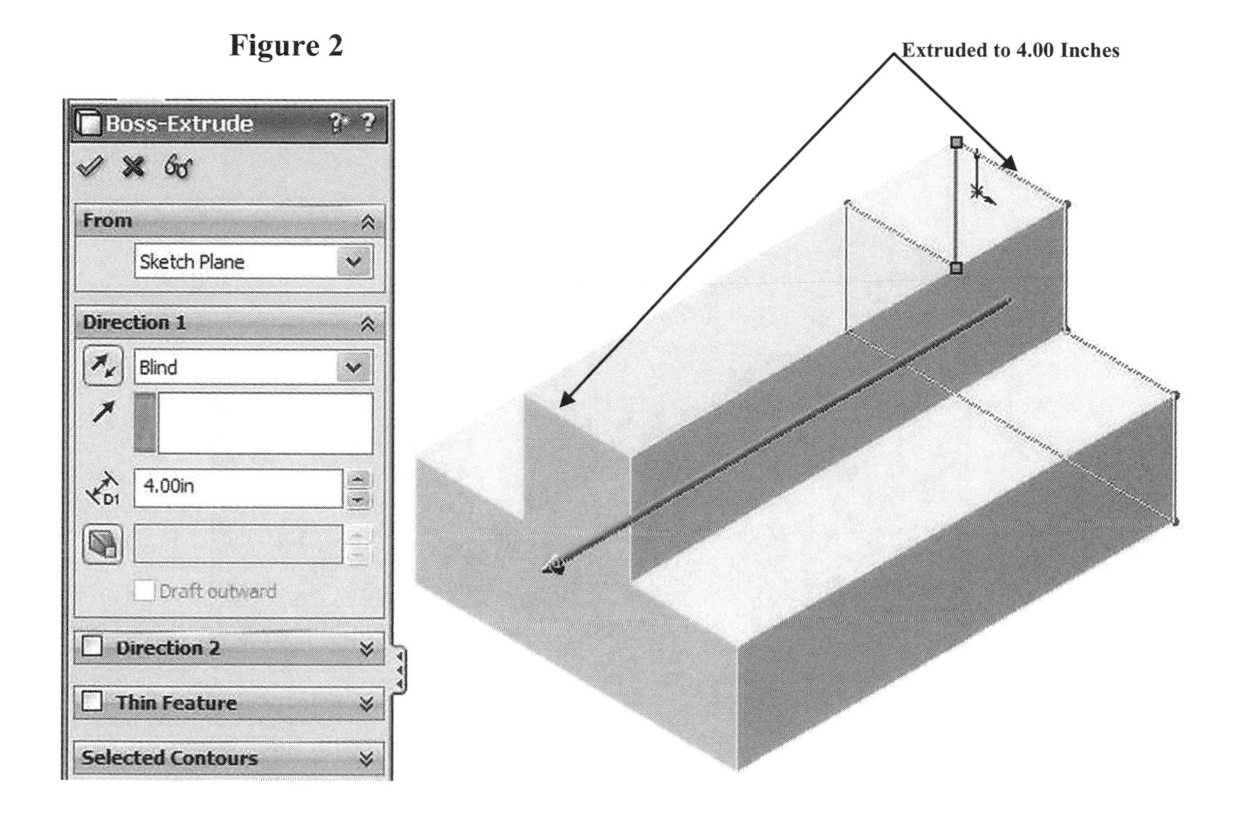

4. Your screen should look similar to Figure 3

Figure 3

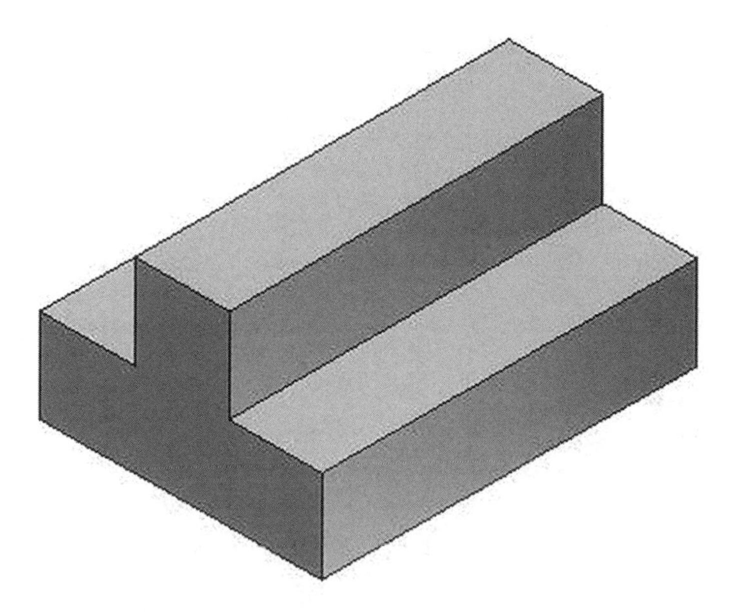

5. Use the **Rotate** command to rotate the part to gain a perpendicular view of the front surface as shown in Figure 4.

Figure 4

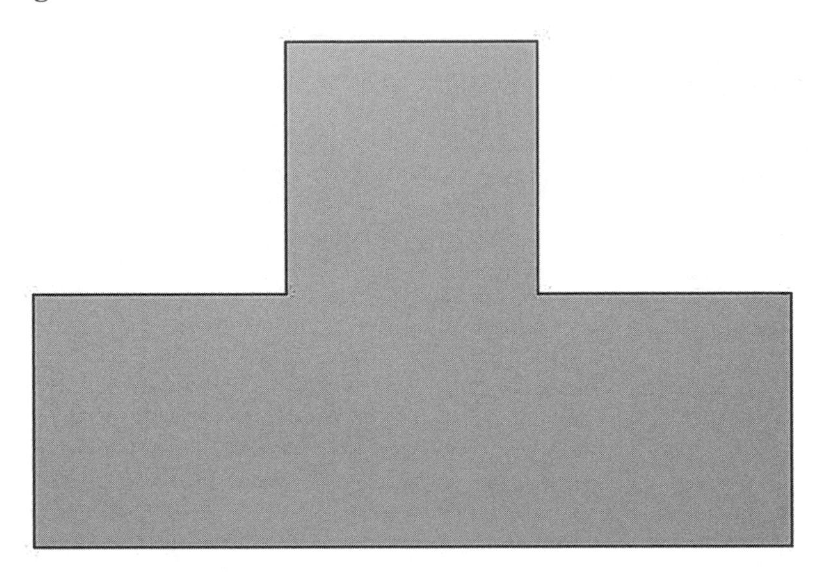

6. Use the **Rectangle** command to complete the sketch as shown in Figure 5. All dimensions are .188.

Figure 5

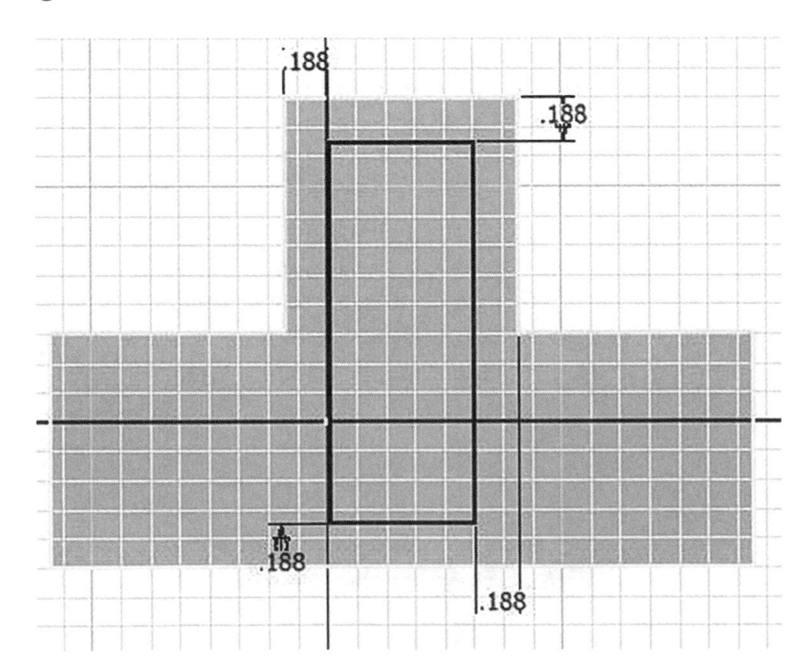

7. Extrude (cut) the rectangle a distance of **4.00** inches creating a square hole as shown in Figure 6.

Figure 6

8. Rotate the part around to gain an isometric (Home View) of the part as shown in Figure 7.

Figure 7

9. Rotate the part around to gain an isometric view of the side of the part as shown in Figure 8.

Figure 8

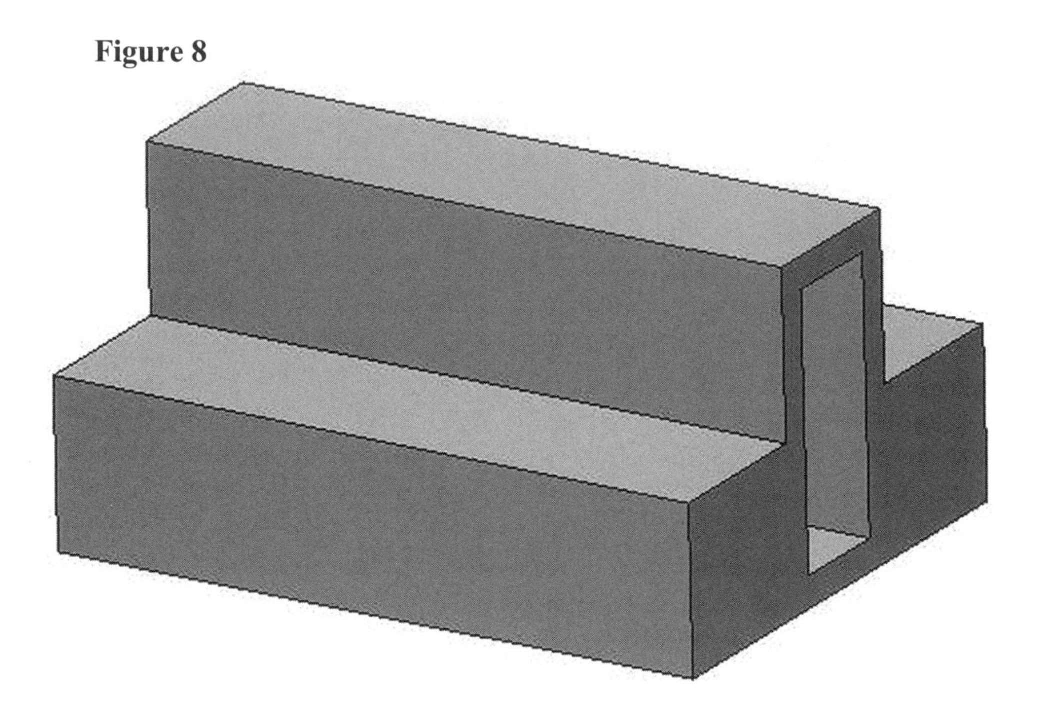

10. Begin a **New Sketch** on the front surface of the part as shown in Figure 9.

Figure 9

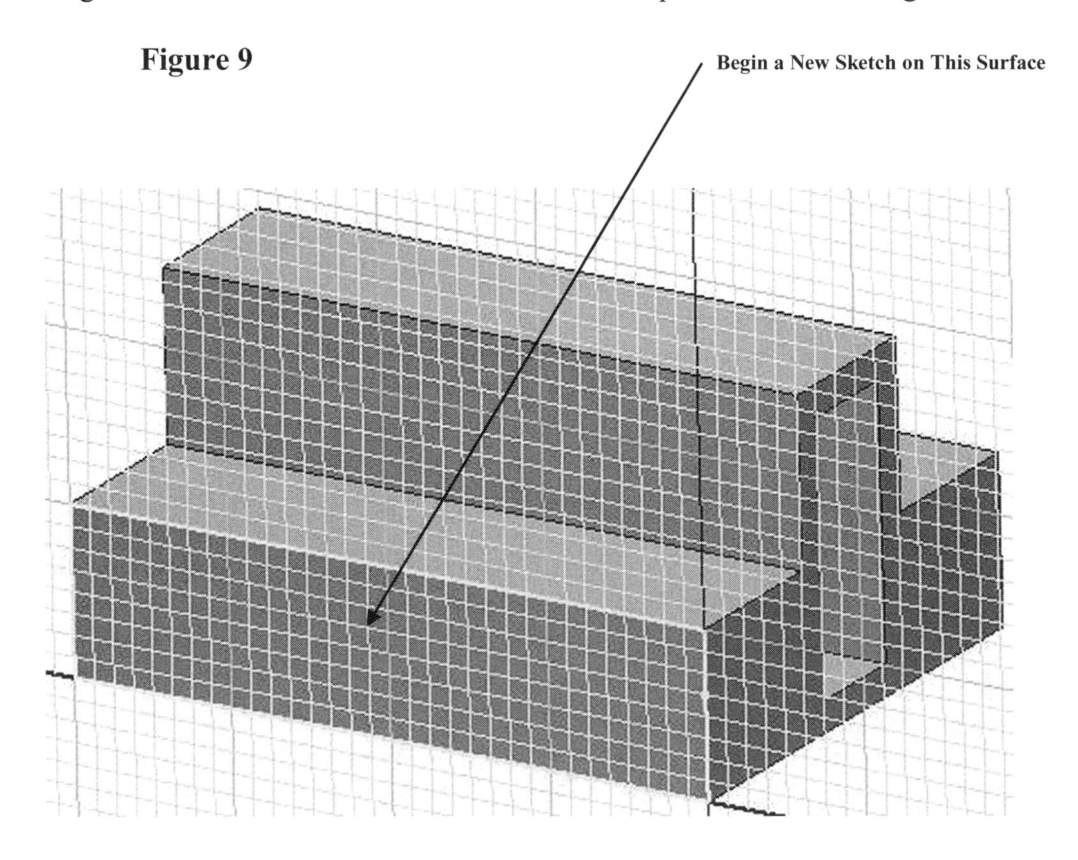

11. Complete the sketch (2 points) as shown in Figure 10.

Figure 10

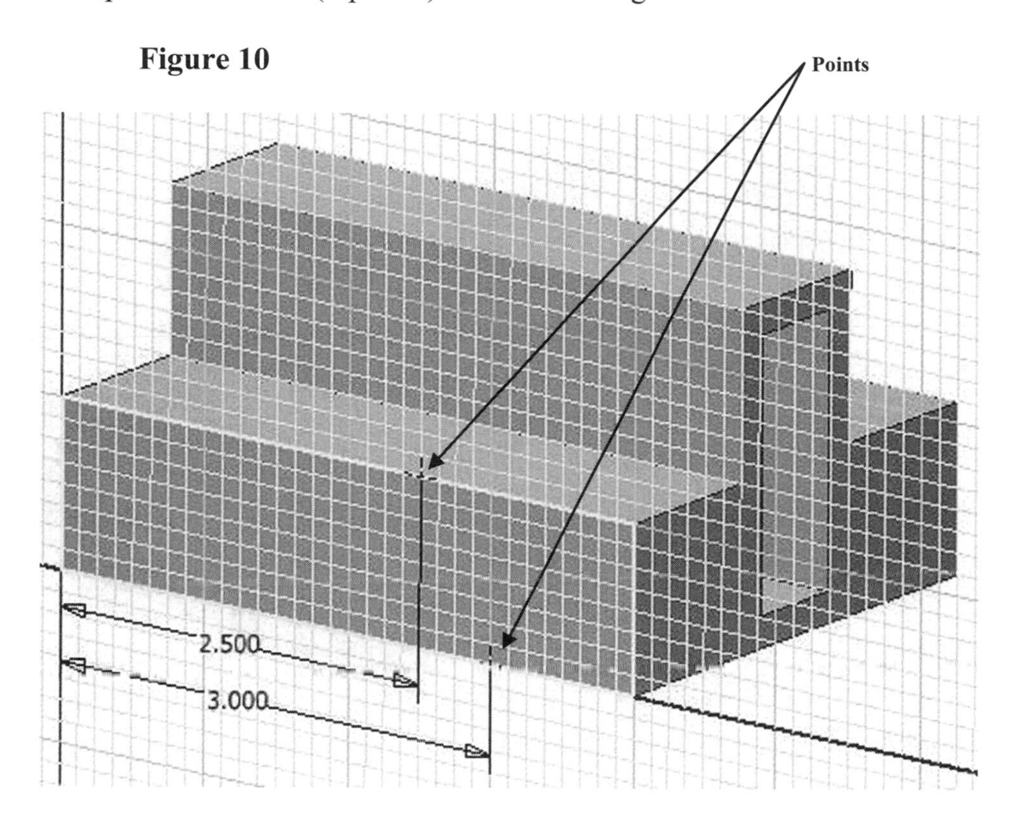

12. Exit out of the Sketch as shown in Figure 11.

Figure 11

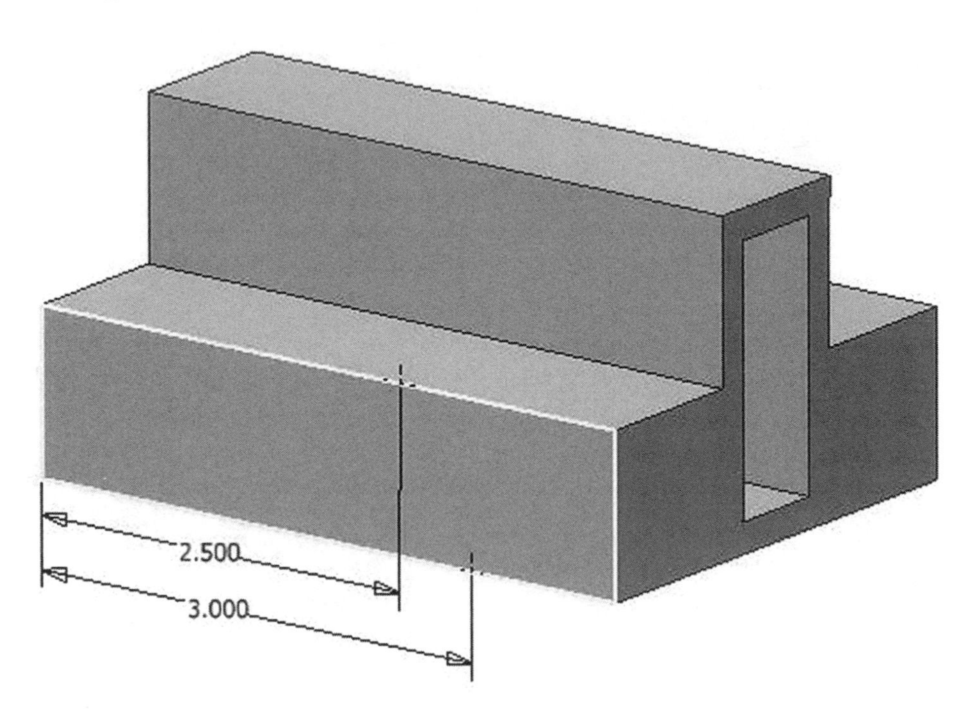

13. Rotate the part upward as shown in Figure 12.

Figure 12

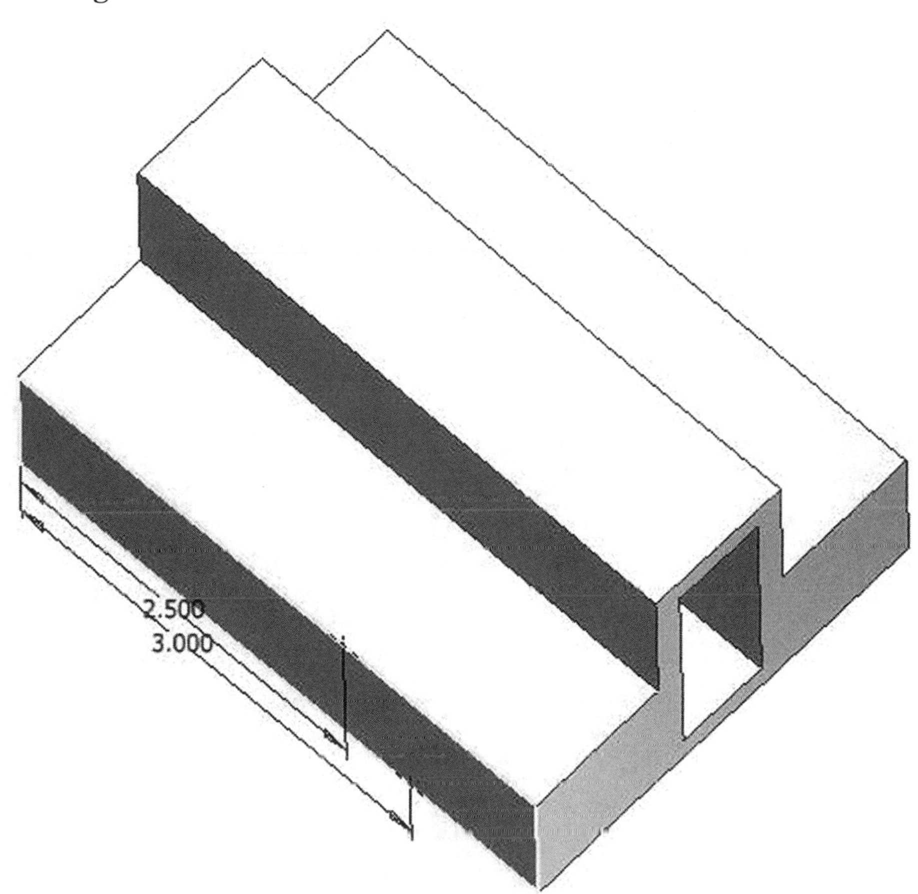

14. Begin a **New Sketch** on the surface shown in Figure 13.

Figure 13

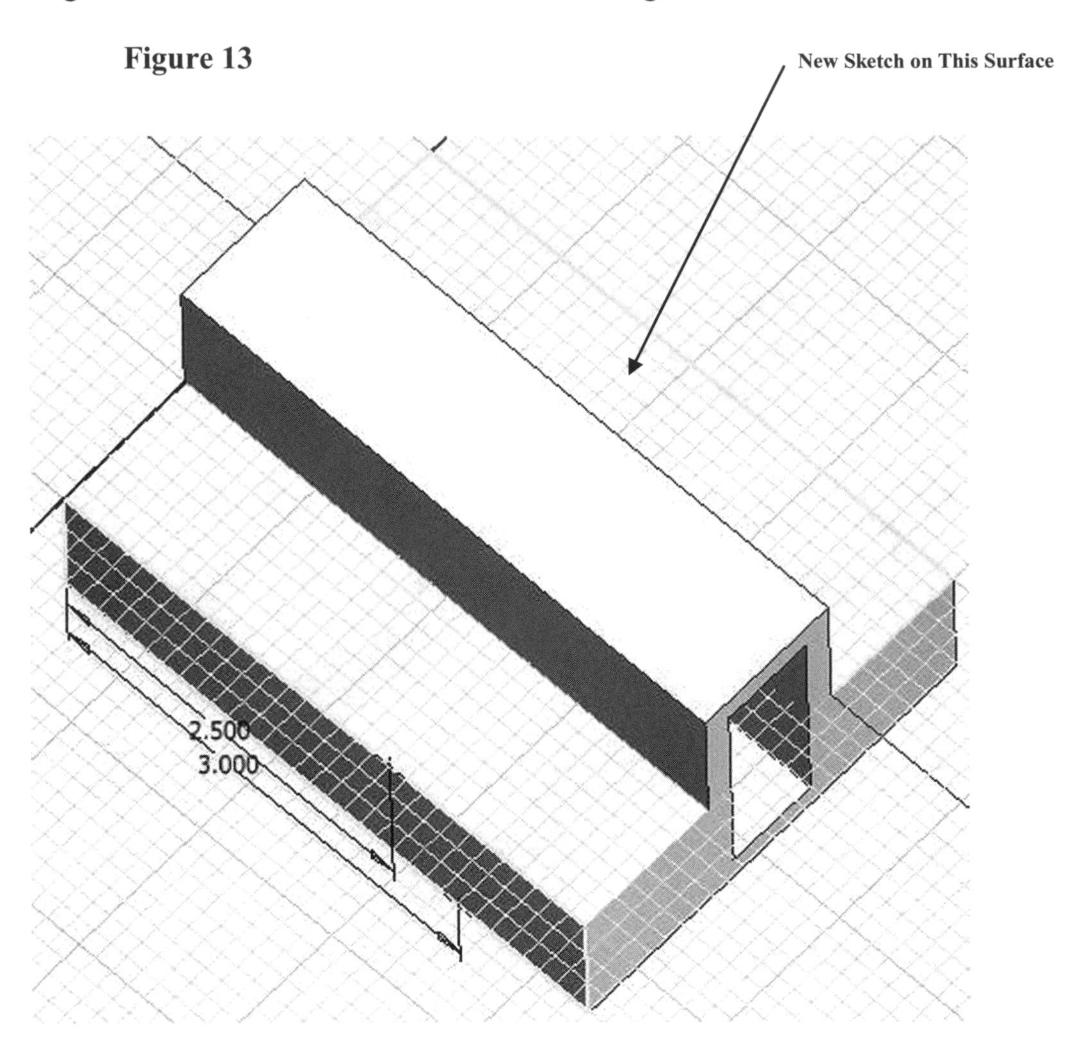

15. Create a **Point** on the surface as shown in Figure 14.

Figure 14

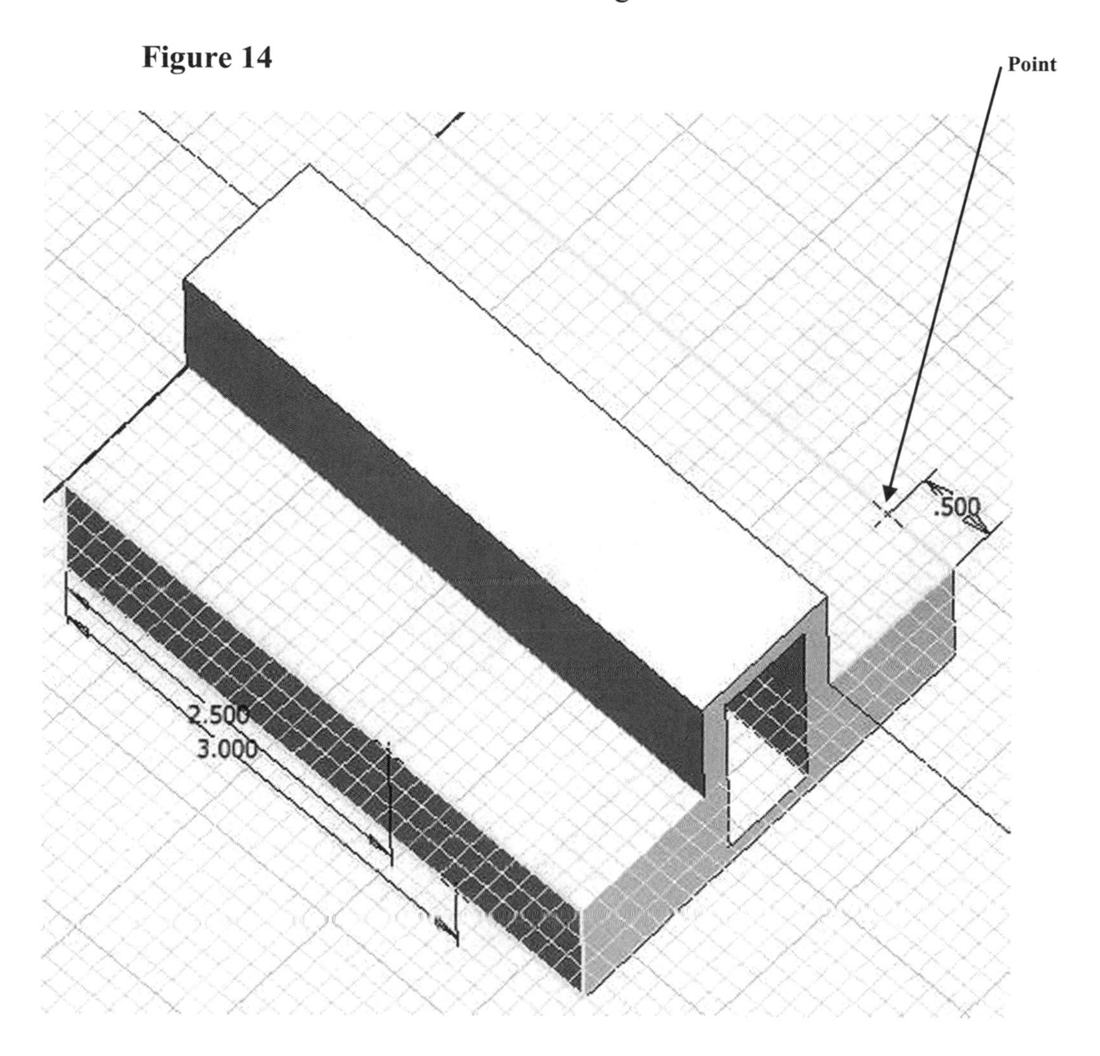

16. Exit the Sketch as shown in Figure 15.

Figure 15

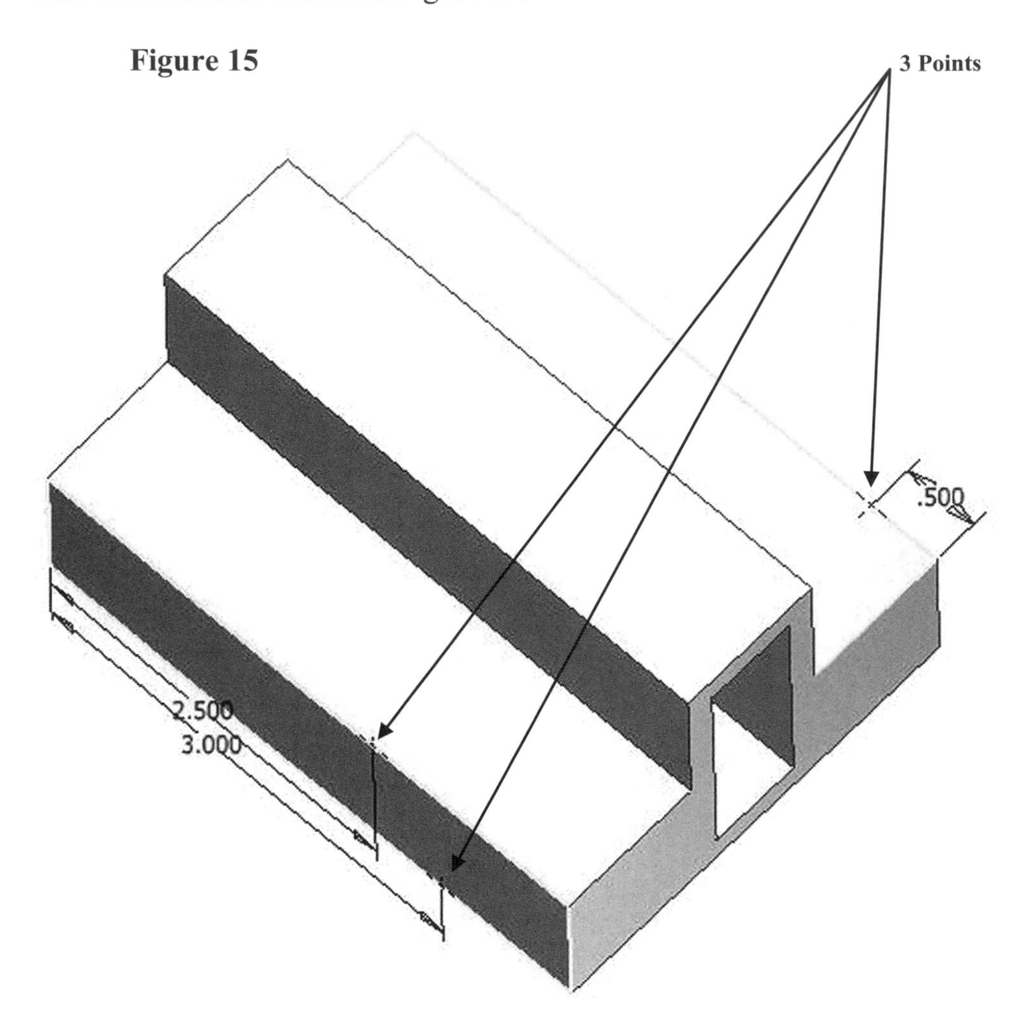

17. Move the cursor to the upper middle portion of the screen and left click on the arrow below Reference Geometry. A drop down menu will appear. Left click on **Plane** as shown in Figure 16.

Figure 16

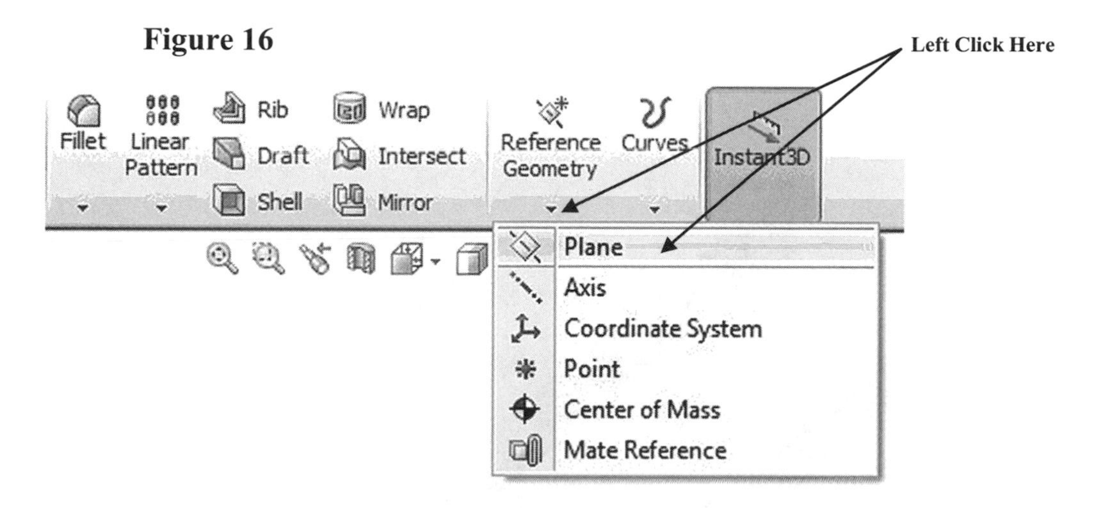

18. Left click on each of the 3 points. The dimensions are included in the figure below for instructional purposes. They may not be visible as shown in Figures 17- 23.

Figure 17

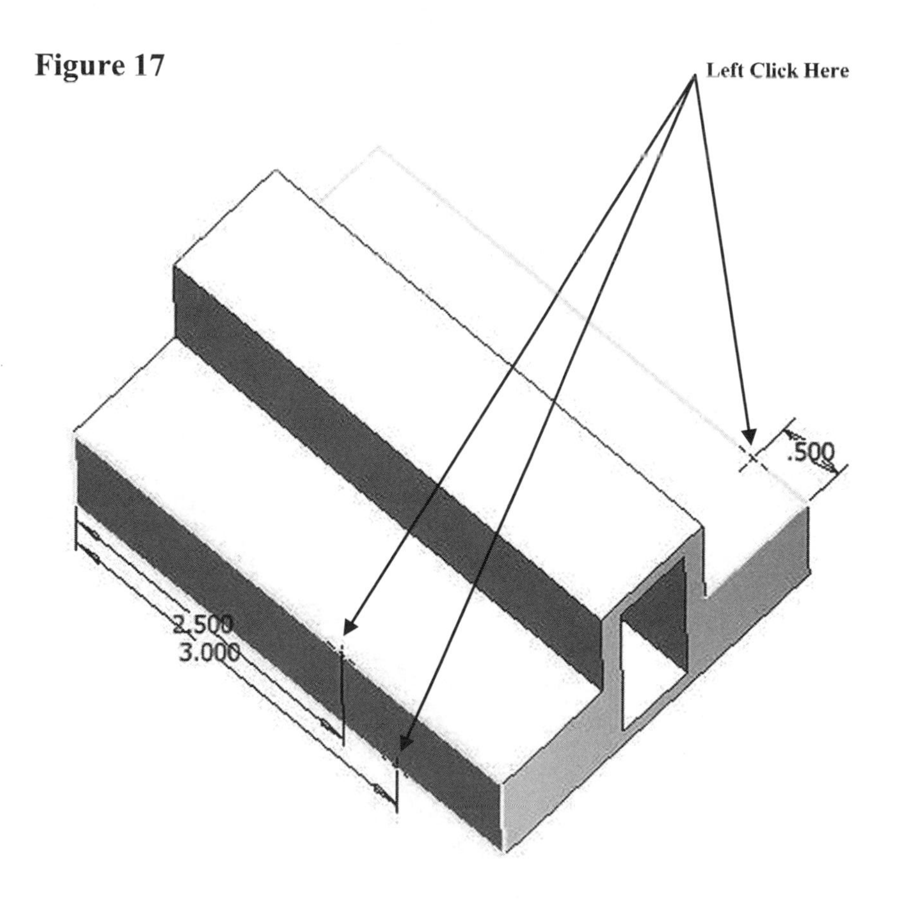

19. SolidWorks will create a Work Plane from the 3 points. Your screen should look similar to Figure 18.

Figure 18

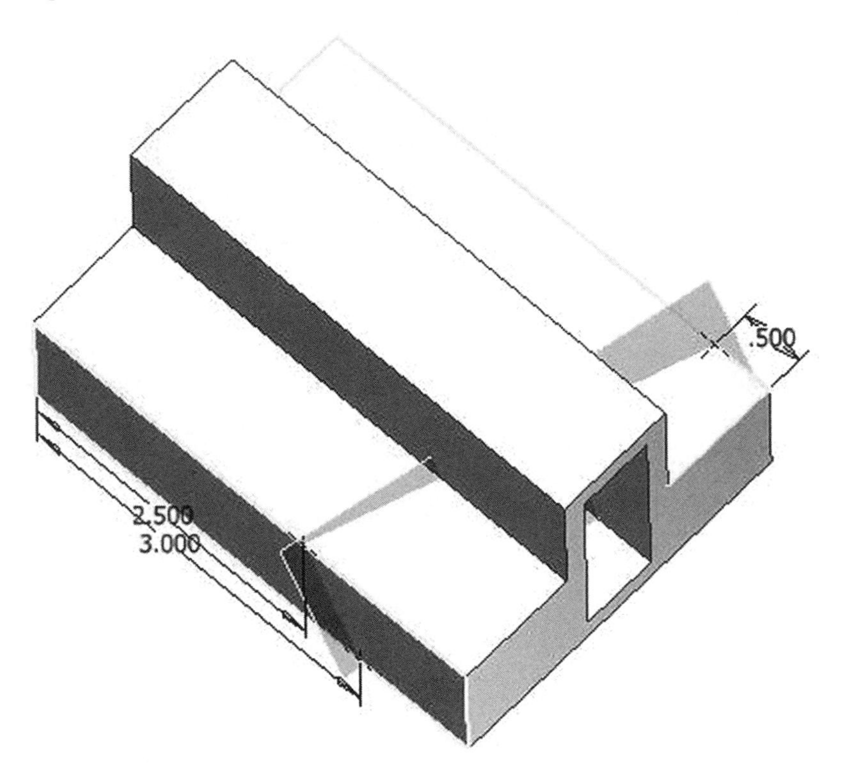

20. Rotate the newly created work plane upward perpendicular as shown in Figure 19.

Figure 19

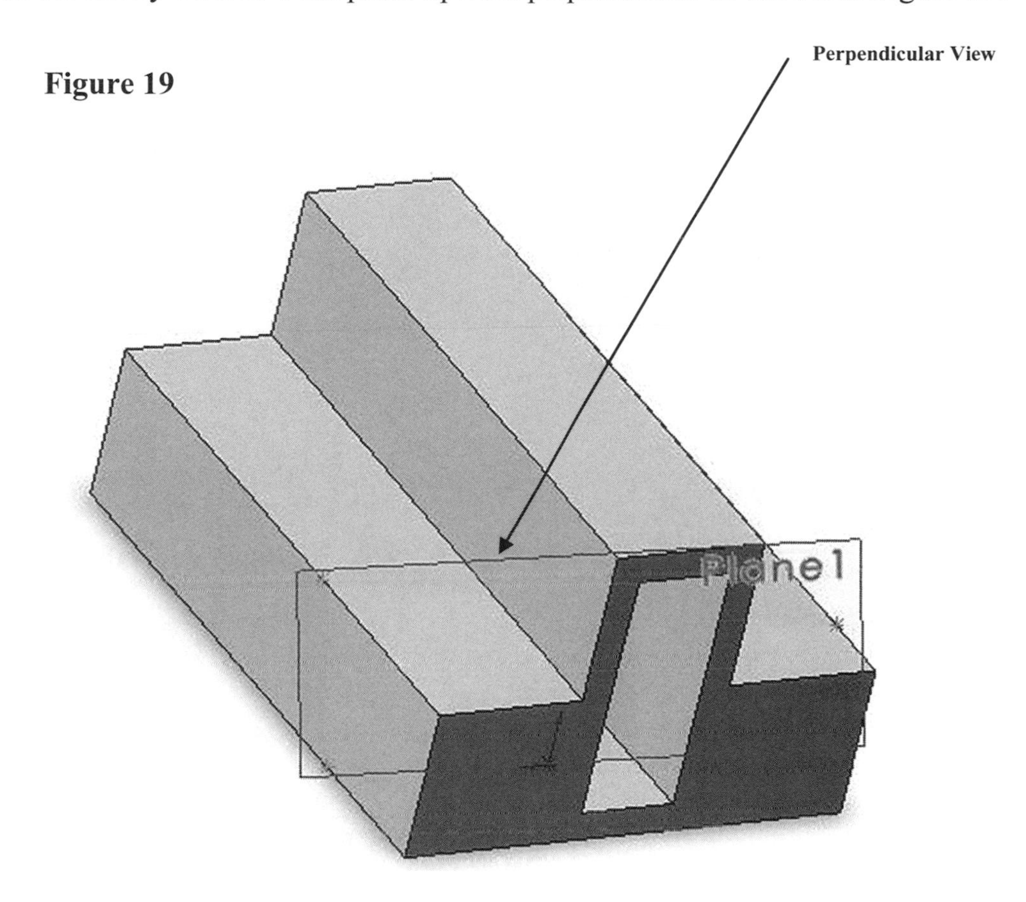

21. Move the cursor over the edge of the newly created plane causing the edges to turn red and create a New Sketch (right click, etc) as shown in Figure 20.

Figure 20

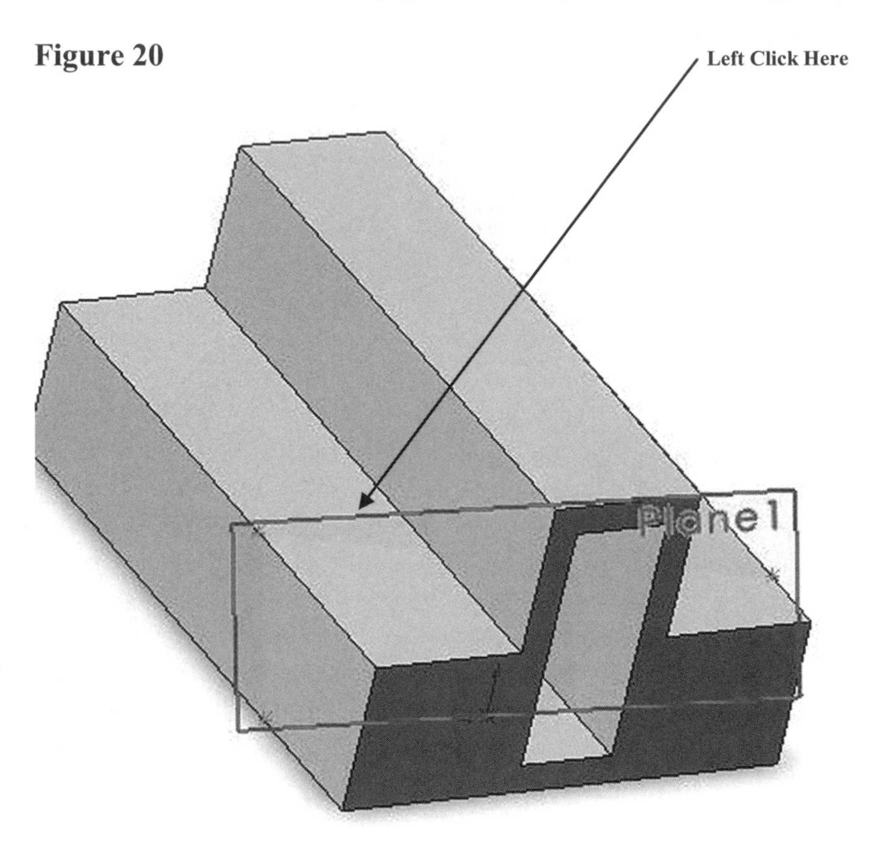

22. SolidWorks will create a New Sketch on the newly created work plane as shown in Figure 21.

Figure 21

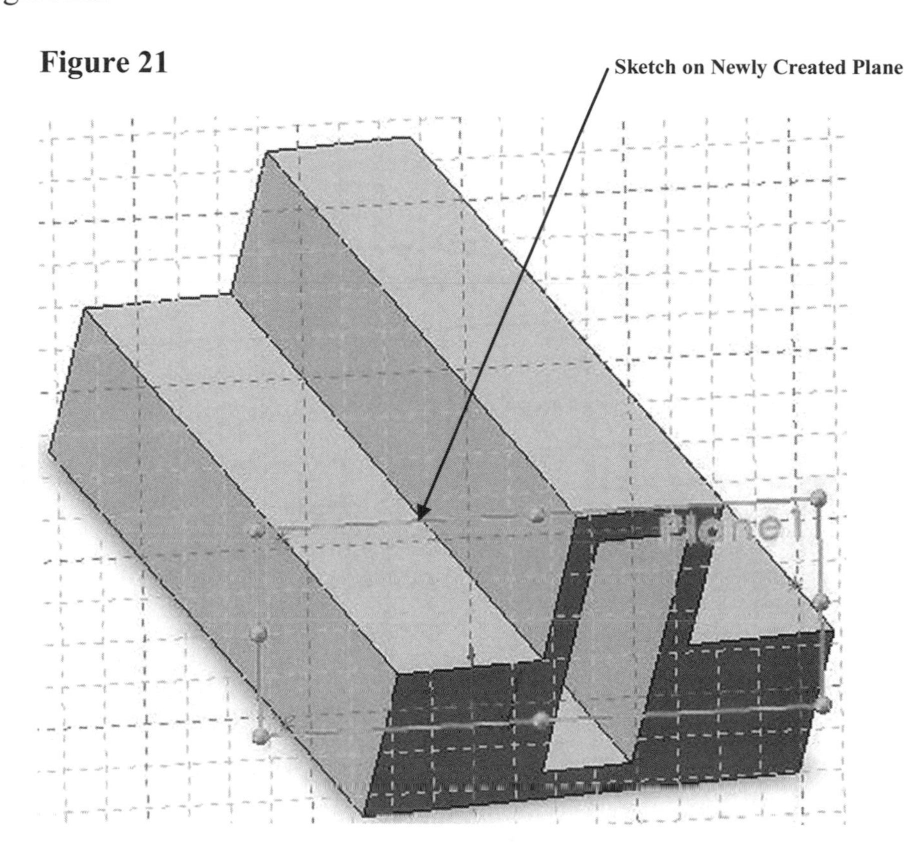

23. Draw a large rectangle that extends out past all for edges of the part as shown in Figure 22.

Figure 22

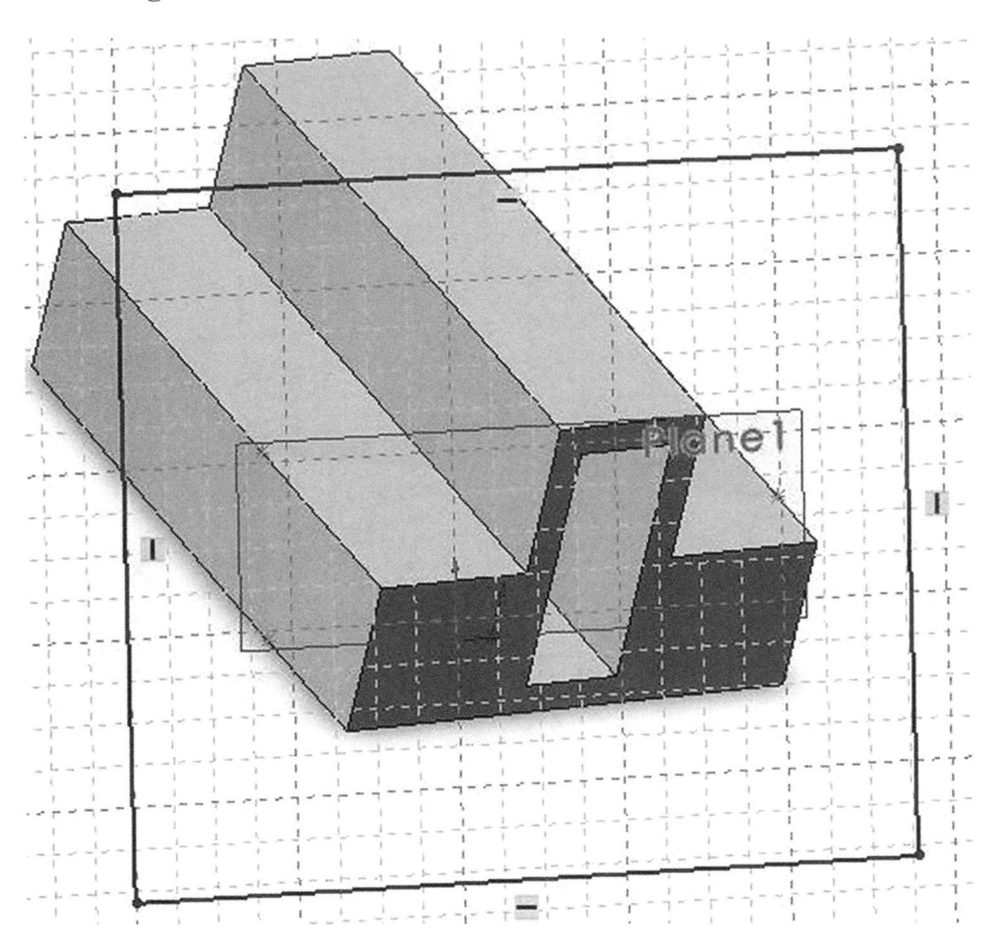

24. Exit the Sketch and use the **Extruded Cut** command to create a removal portion similar to Figure 23.

Figure 23

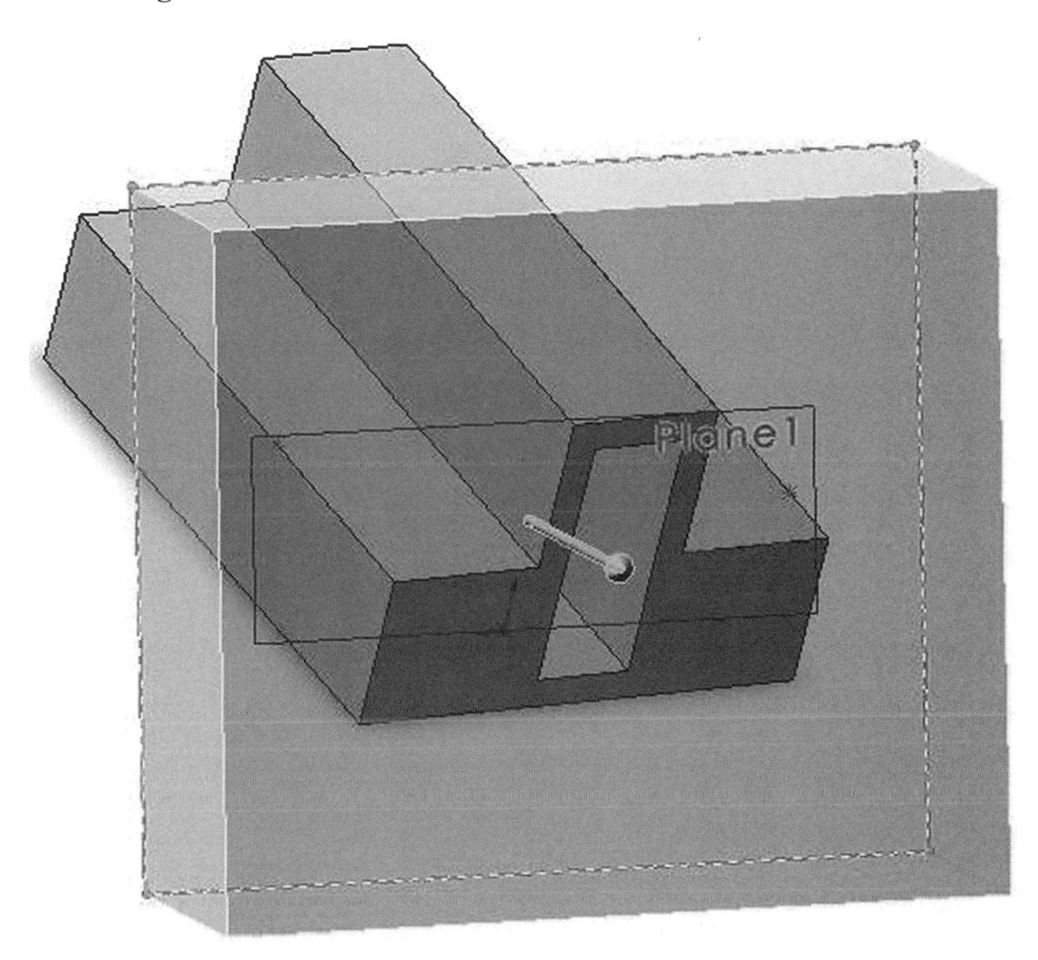

25. Your screen should look similar to Figure 24.

Figure 24

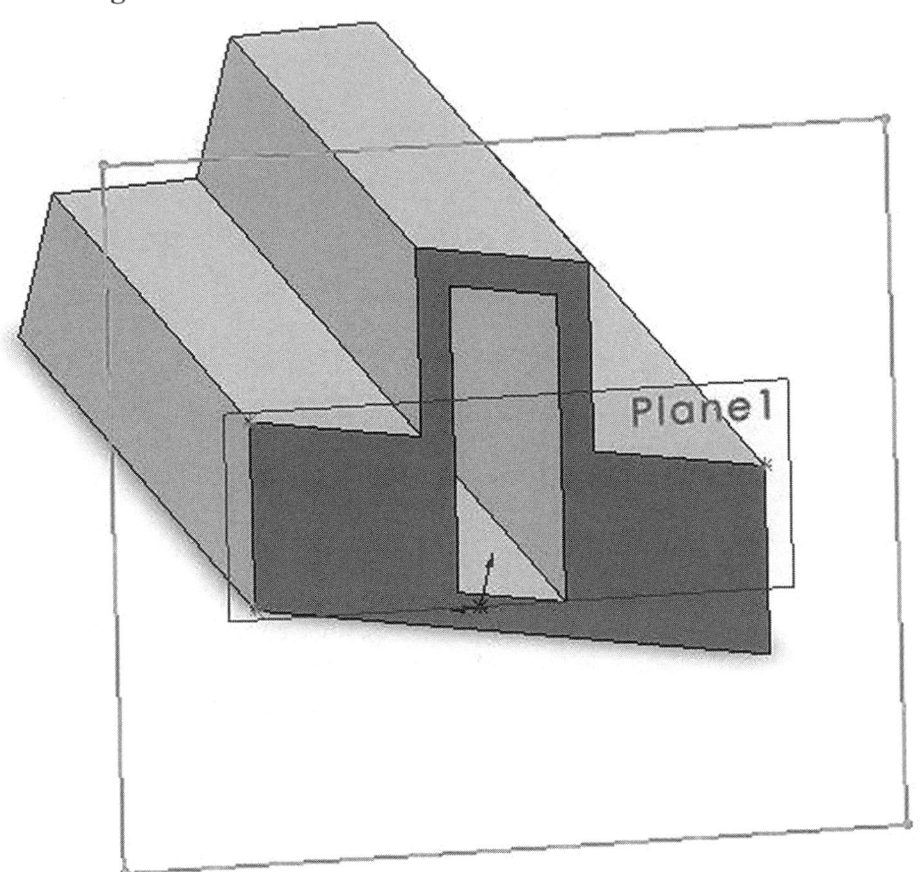

26. To clean up the appearance of the part, move the cursor over the left portion of the screen and left click, then right click on **Plane1**. A pop up menu will appear. Left click on the glasses (the hide command) as shown in Figure 25.

Figure 25

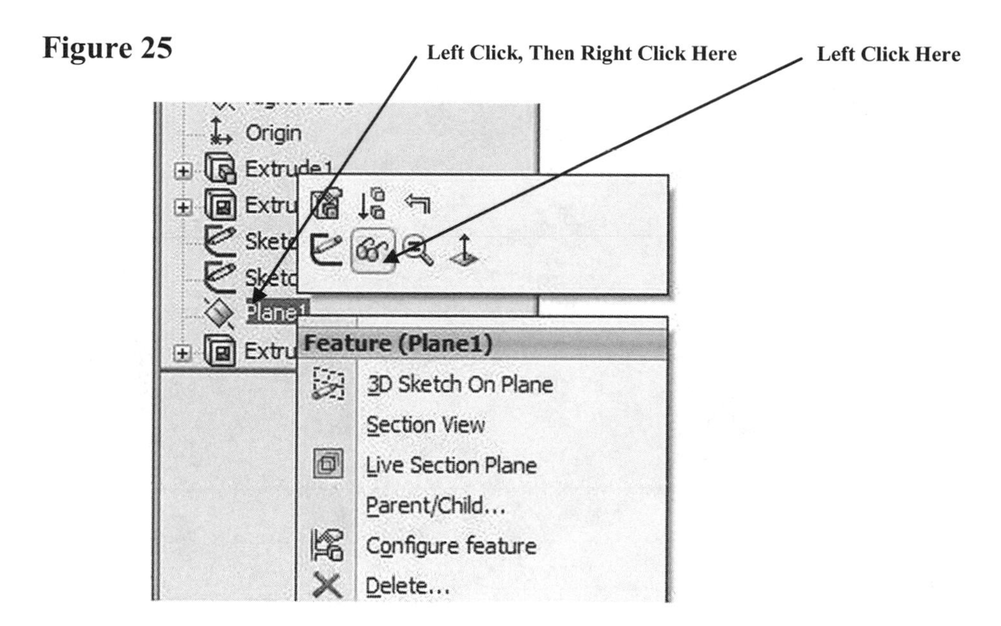

27. Your screen should look similar to Figure 26.

Figure 26

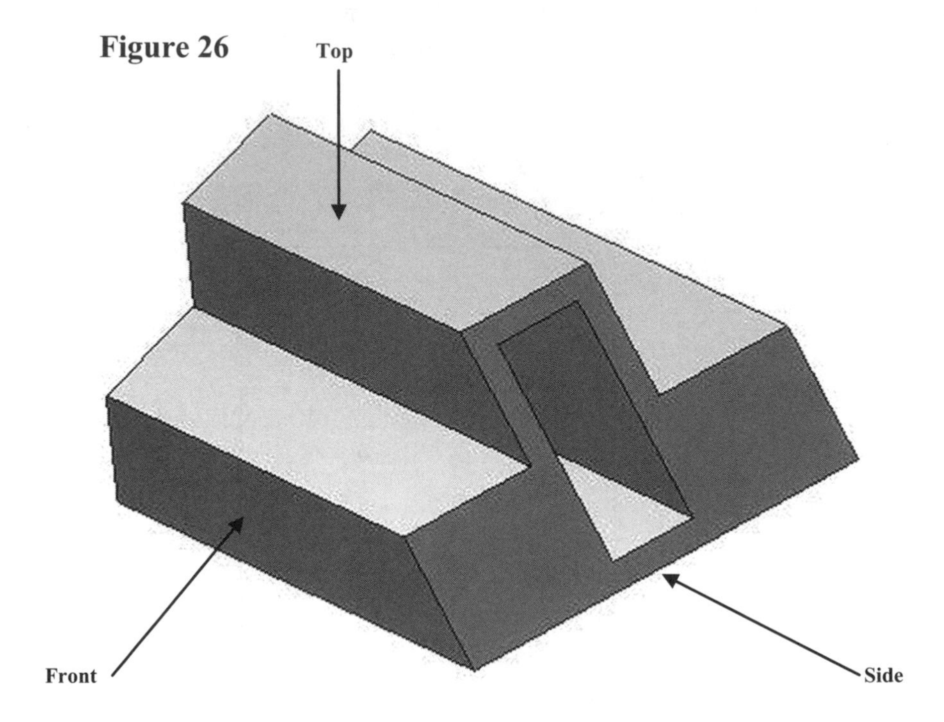

28. To test your Orthographic skills, create a 3 view Orthographic drawing of the part shown above (Top, Front and Right Hand side views) in either a 2 Dimensional CAD package or by hand on a Drafting Board. Show all hidden lines in your 2 Dimensional drawing. Once you have completed your 2 Dimensional drawing, refer to Chapter 3 on how to create a 3 View drawing in SolidWorks, (showing all hidden lines) and then compare your "hand drawing" to the SolidWorks 3 view drawing.

Chapter 12 Introduction to Creating a Helix

Objectives:

- Learn to create a Helix

Chapter 12 includes instruction on how to create a Helix

Chapter 12 includes instruction on how to design the part shown below.

.

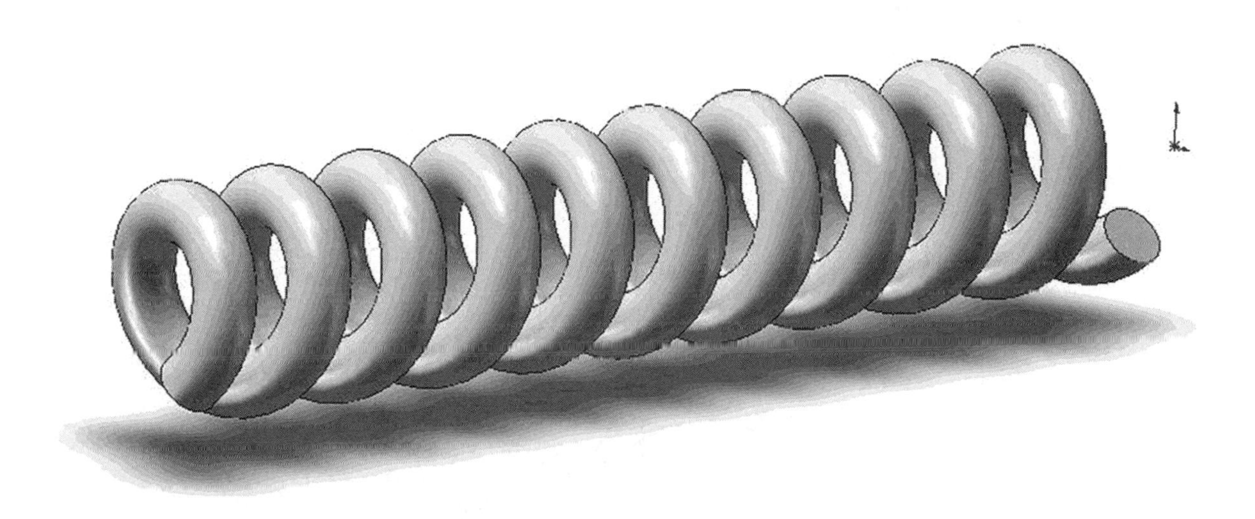

1. Start SolidWorks by referring to "Chapter 1 Getting Started". After SolidWorks is running, begin a New Sketch.

2. Complete the sketch shown in Figure 1.

Figure 1

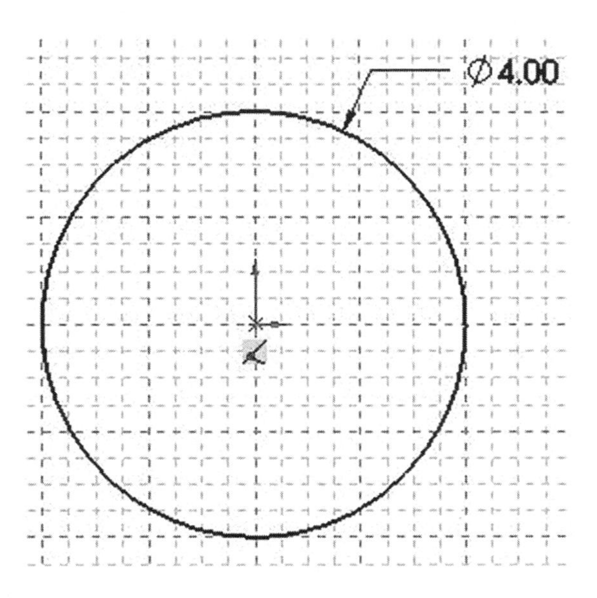

3. Rotate the drawing around as shown in Figure 2.

Figure 2

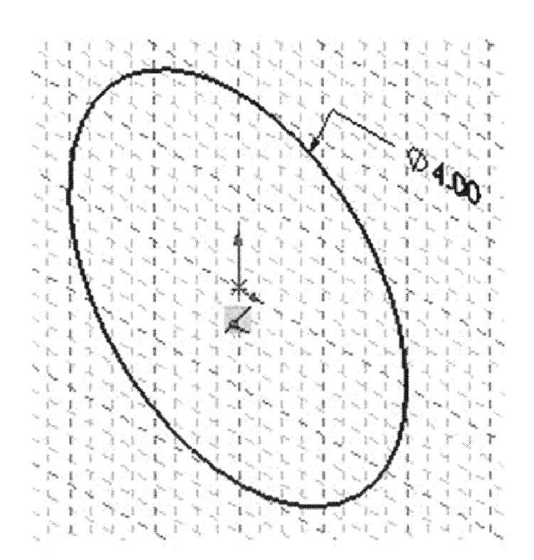

4. Exit the Sketch Panel. Your screen should look similar to Figure 3

Figure 3

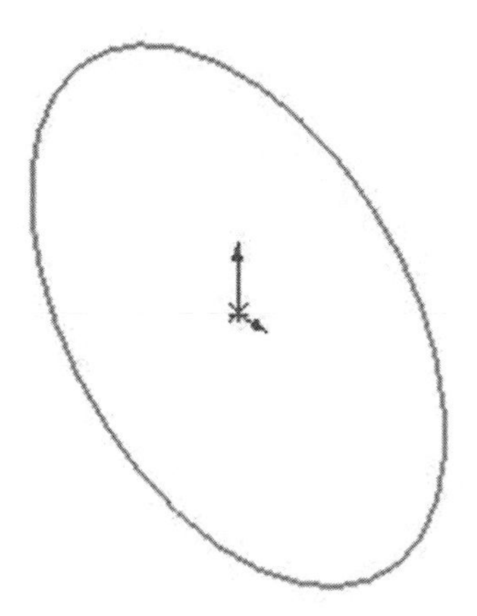

5. Move the cursor to the upper middle portion of the screen and left click on the arrow below Curves. A drop down menu will appear. Left click on **Helix and Spiral** as shown in Figure 4.

Figure 4

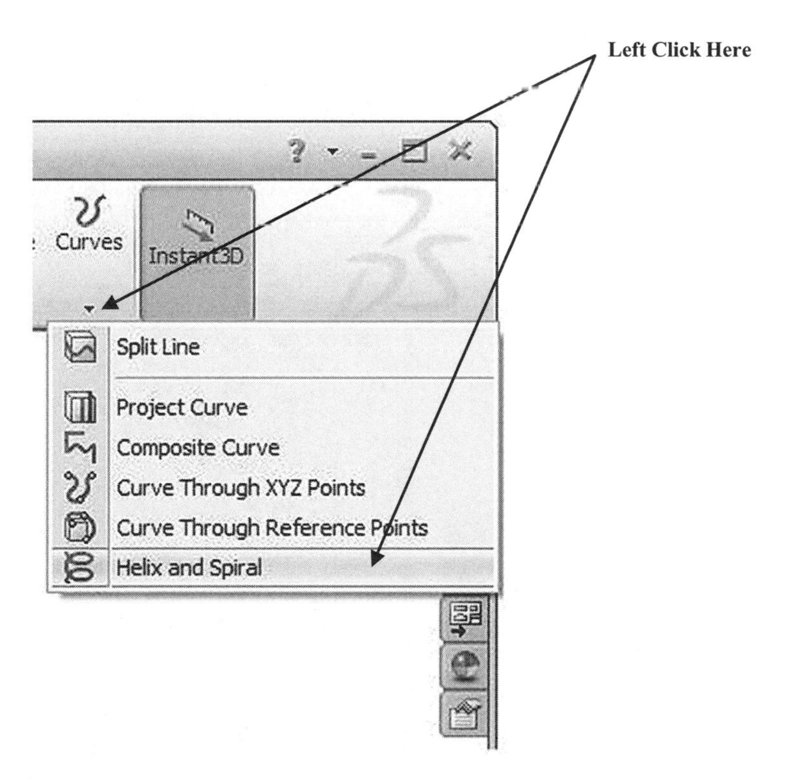

6. Left click on the circle as shown in Figure 5. SolidWorks will provide a preview of the Helix.

Figure 5

Left Click Here

7. Enter **2.00** for the Pitch and **10** for the Revolutions. Left click on the green check mark as shown in Figure 6.

Figure 6

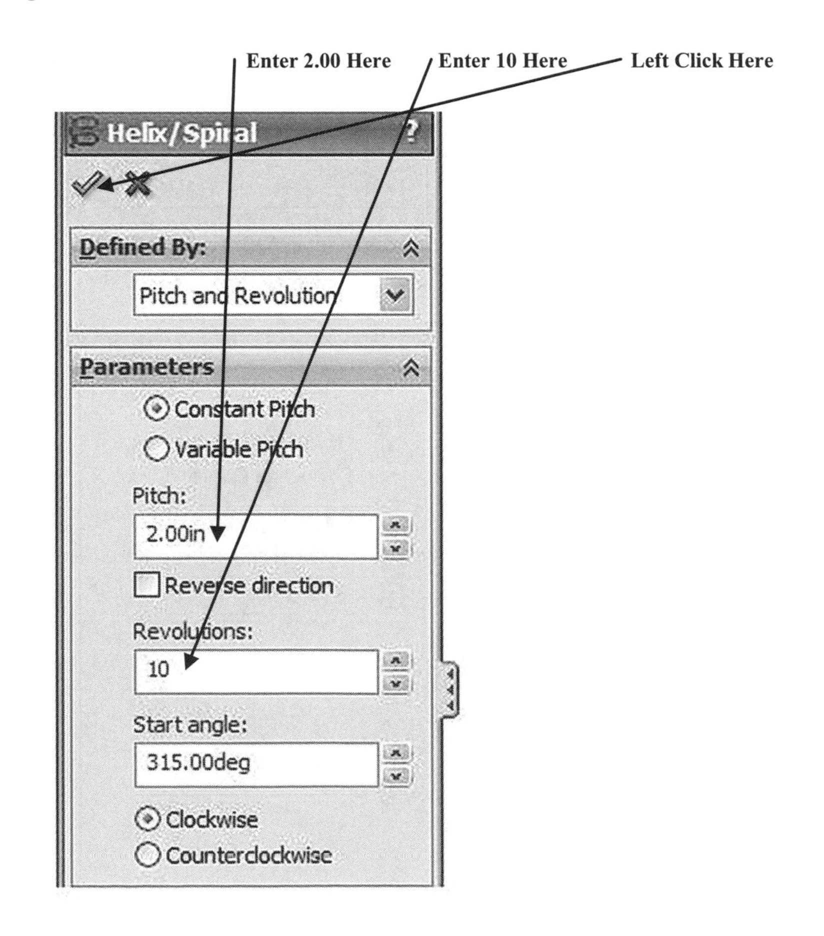

8. Your screen should look similar to Figure 7.

Figure 7

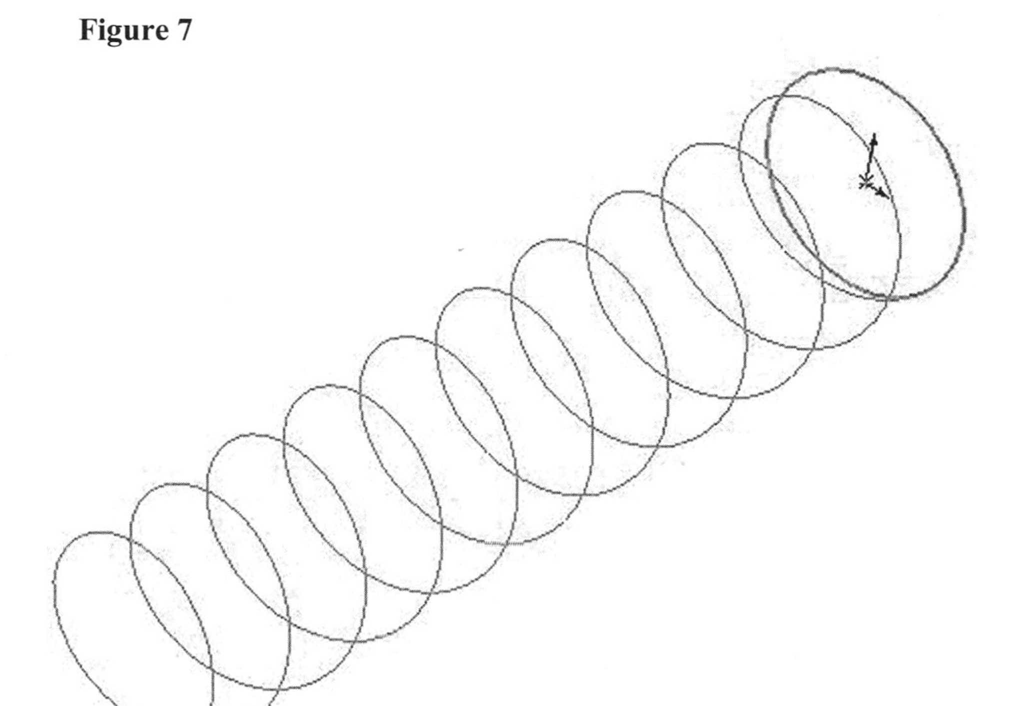

9. Move the cursor to the upper left portion of the screen and begin a **New Sketch** on the **Right Plane** as shown in Figure 8.

Figure 8

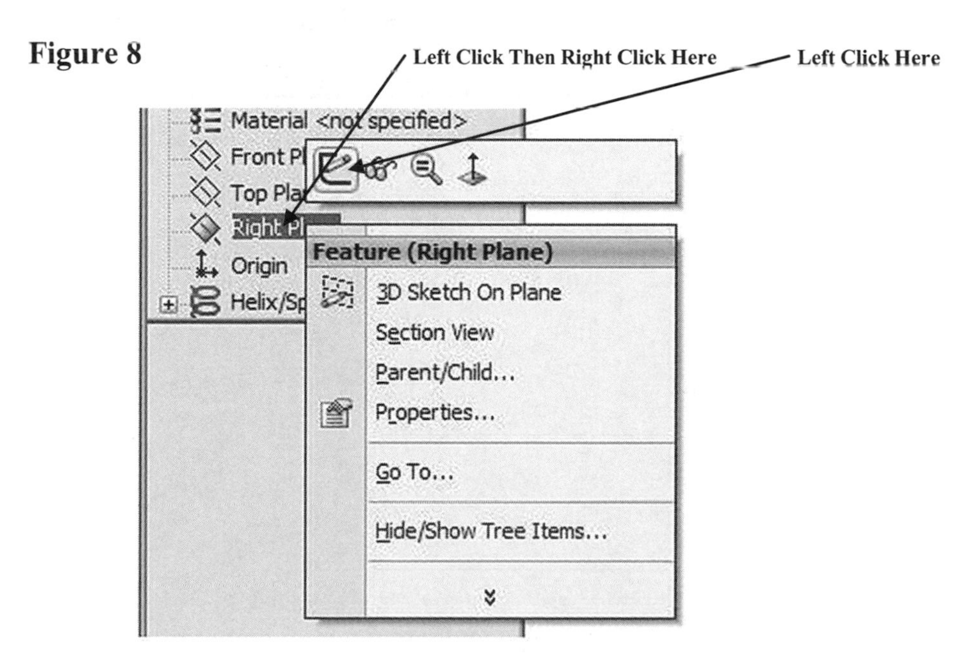

10. Rotate the view slightly Isometric. Left click on the Helix as shown in Figure 9.

Figure 9

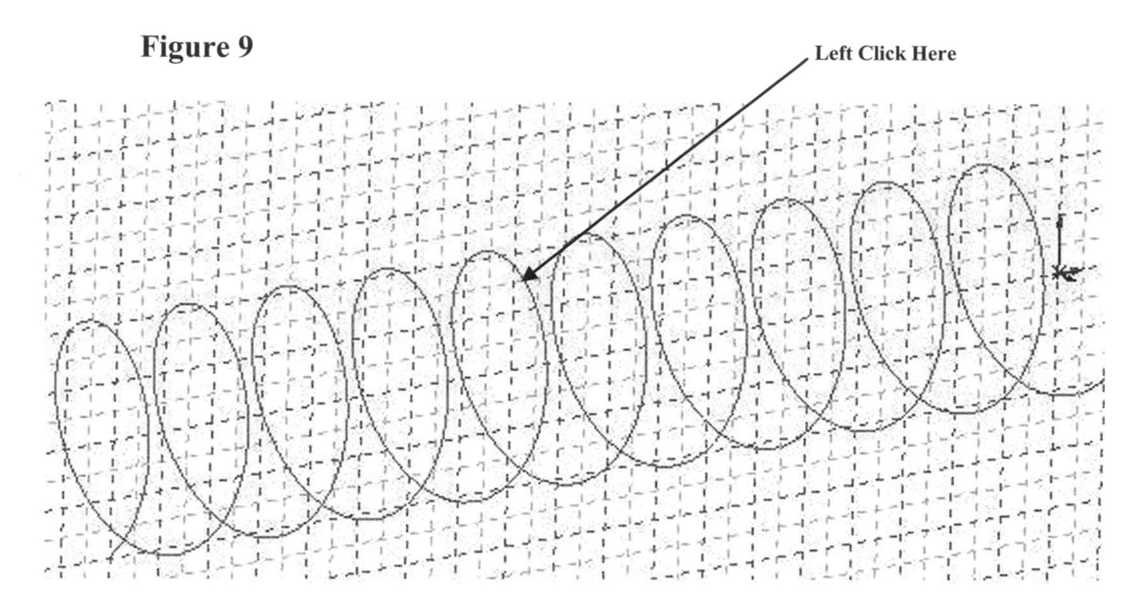

11. Move the cursor to the upper middle portion of the screen and left click on **Convert Entities** as shown in Figure 10.

Figure 10

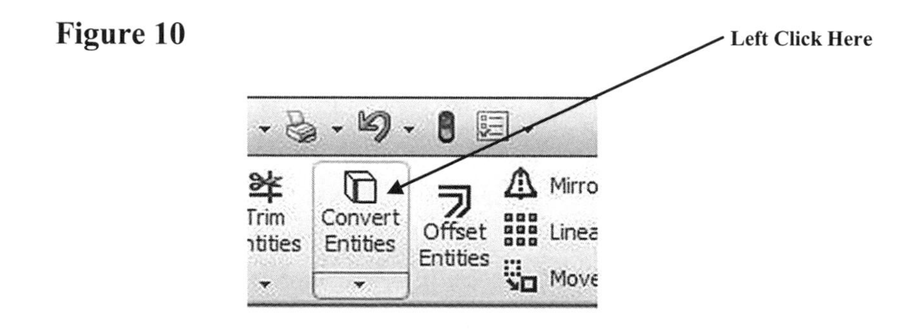

12. SolidWorks will convert the Helix into a 2 dimensional drawing as shown in Figure 11.

Figure 11

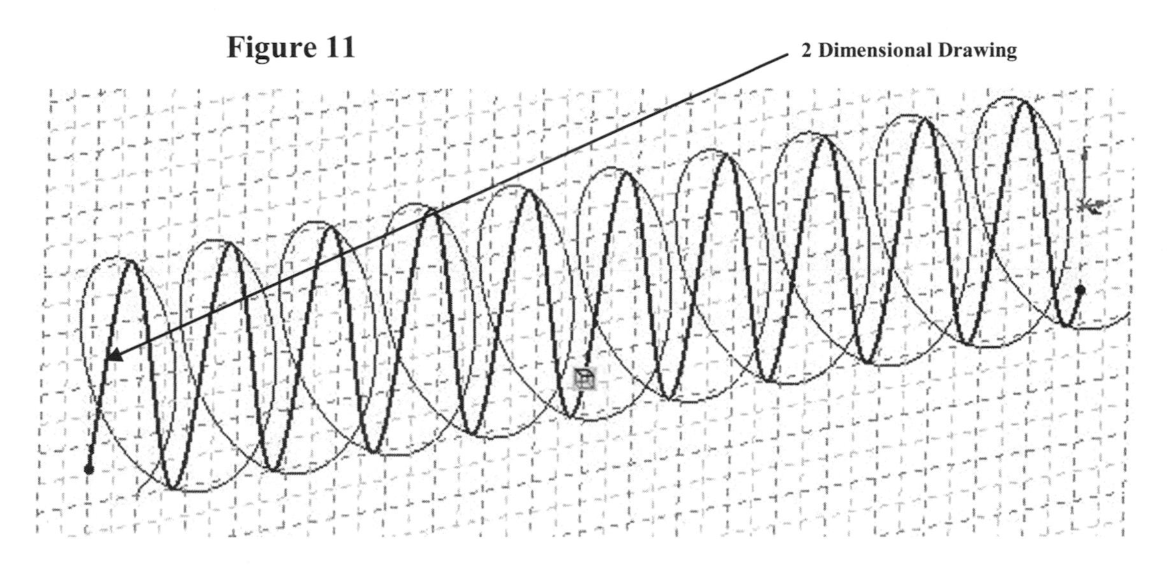

13. Draw a 1.00 inch circle at the 2 Dimensional end point of the Helix as shown in Figure 12.

Figure 12

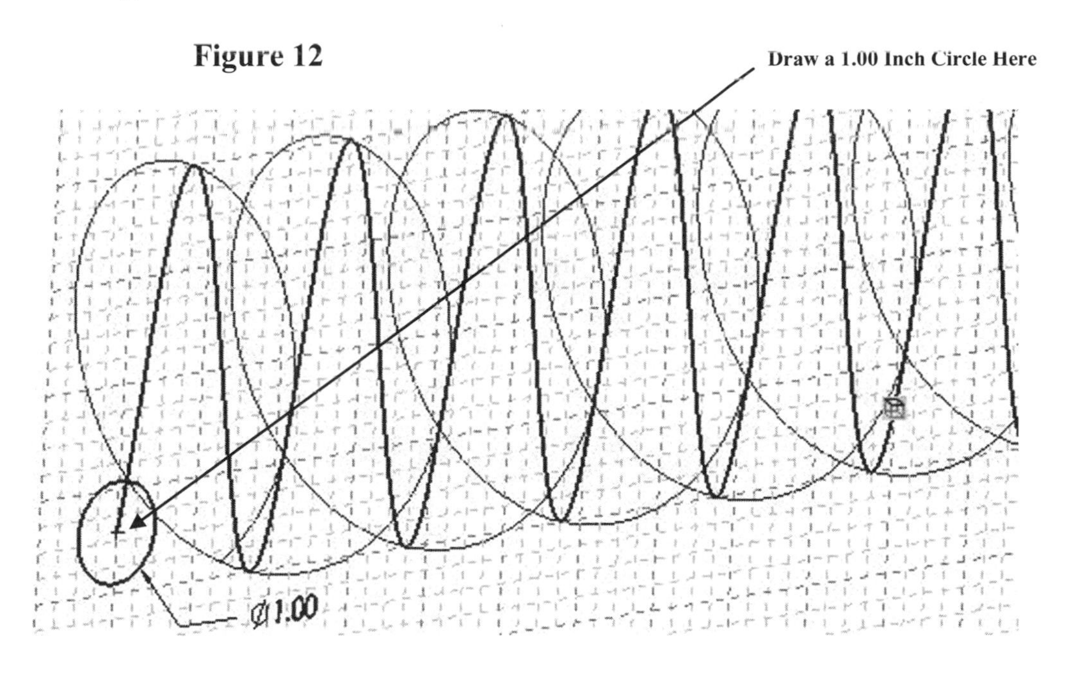

14. Delete the 2 Dimensional Helix drawing of the spring as shown in Figure 13.

Figure 13

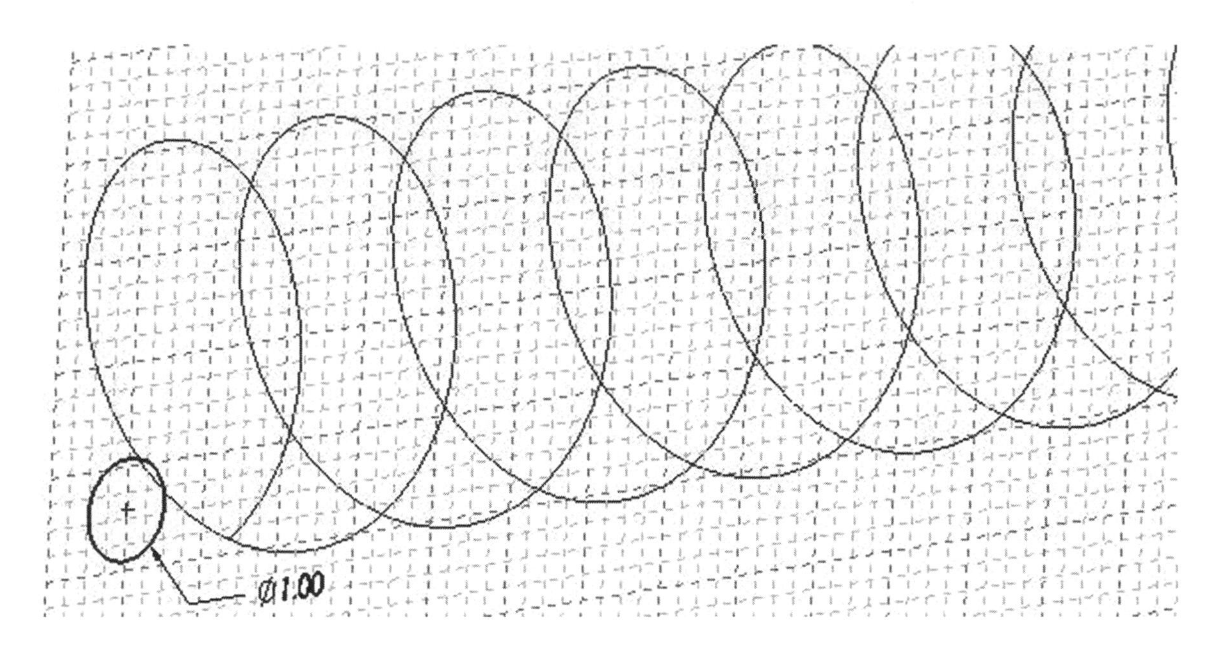

15. Move the cursor to the upper left portion of the screen and left click on **Swept Boss/Base** as shown in Figure 14.

Figure 14

Left Click Here

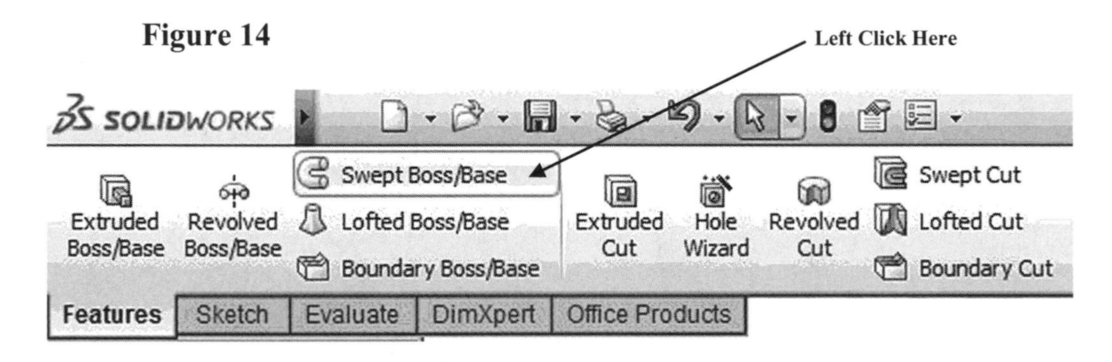

16. Left click on the small circle first. Then left click on the Helix as shown in Figure 15.

Figure 15

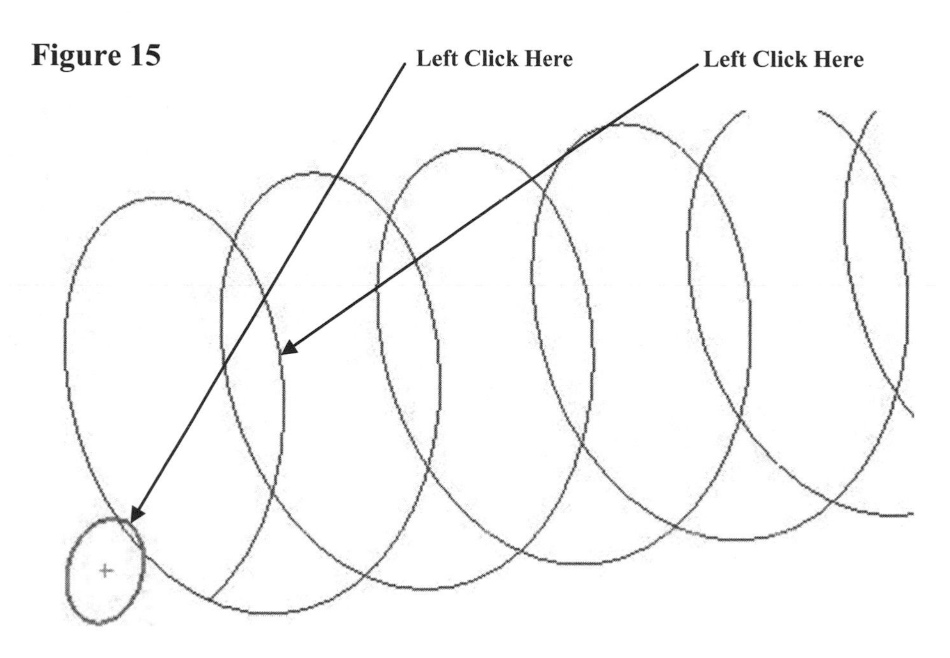

17. SolidWorks will provide a preview of the Helix as shown in Figure 16. Left click on the green check mark (not shown).

Figure 16

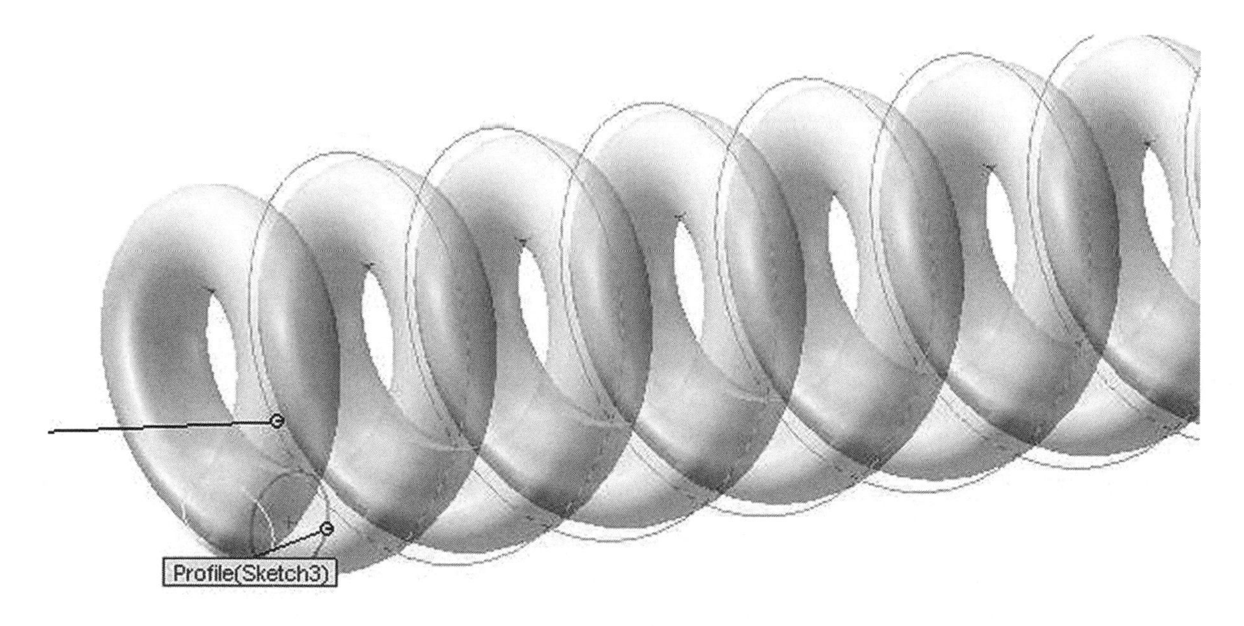

18. Your screen should look similar to Figure 17.

Figure 17

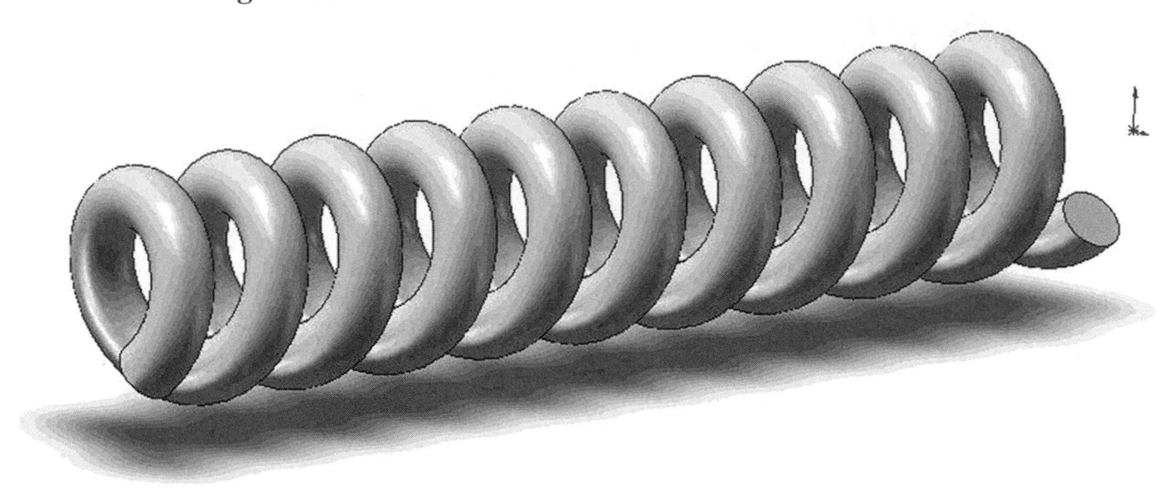

Chapter 13 Introduction to Importing DWG files

Objectives:

- Learn to create a simple .DWG file
- Learn to import a .DWG file into a 1 view drawing
- Learn to import a .DWG file into the solid model area and create a solid model

Chapter 13 includes instruction on how to import a .DWG file into either the solid model area or the orthographic drawing area of SolidWorks.

Chapter 13 includes instruction on how to design the part shown below.

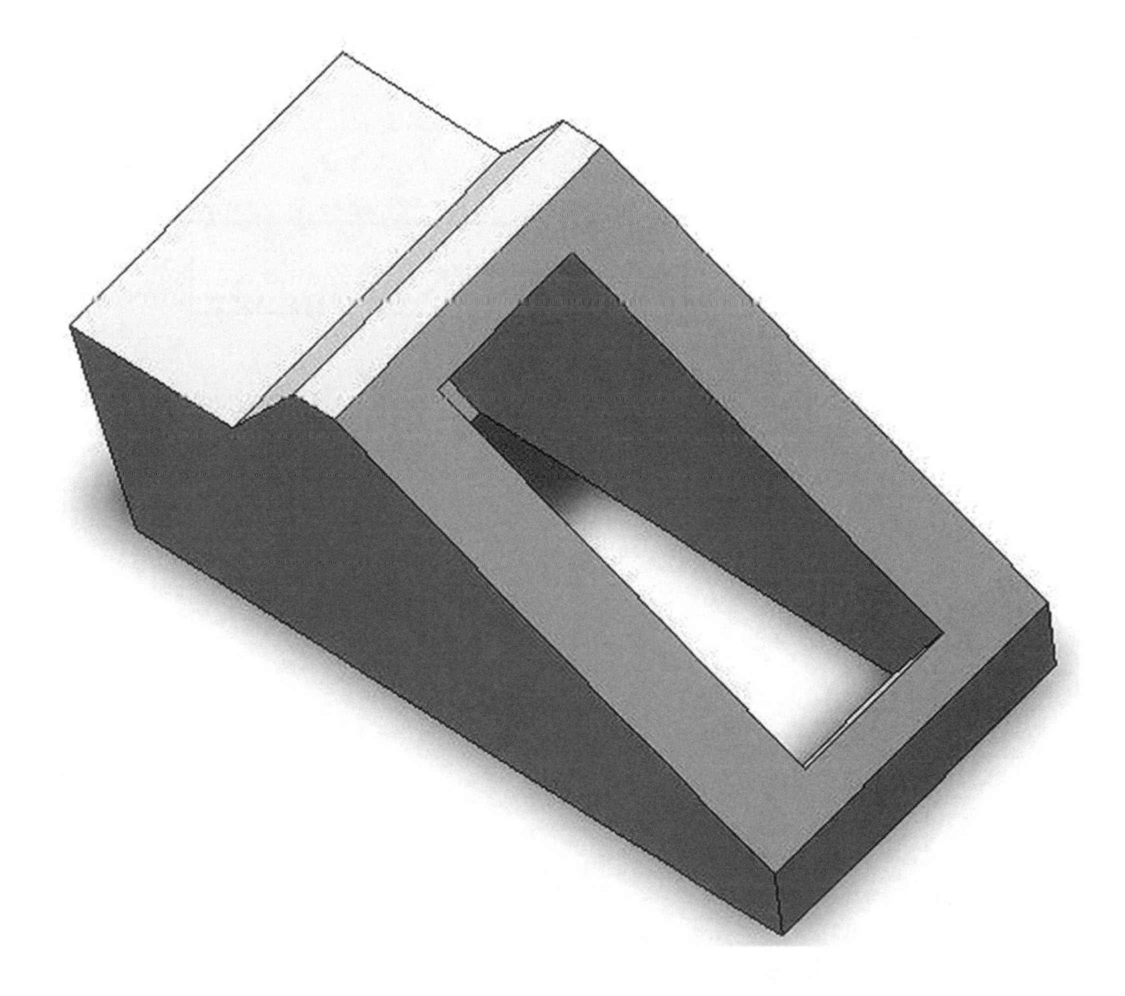

1. Create the following drawing in some type of .DWG editing software (software that is used to create and edit .DWG files such as AutoCAD, DWG Editor, etc) as shown in Figure 1. **If using AutoCAD the file must be saved as a 2010 or older .DWG file format. Screen captures for this chapter are from the SolidWorks 2011 software version.**

Figure 1

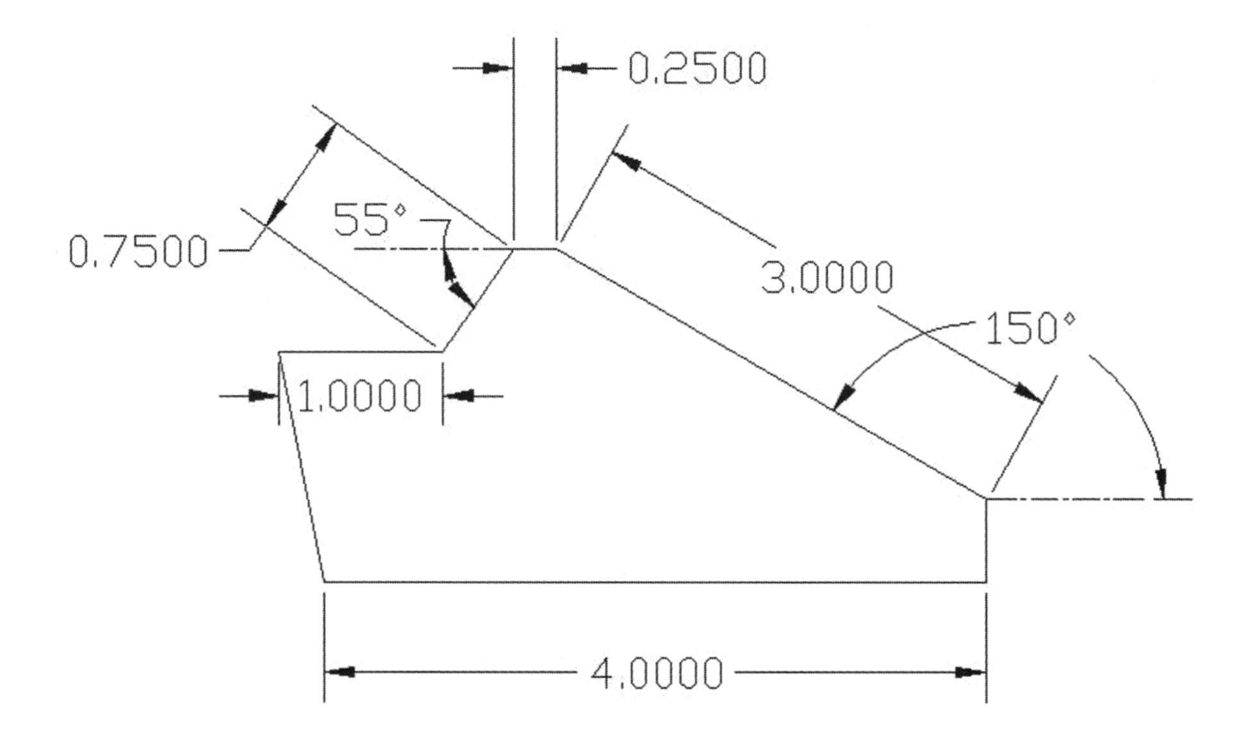

2. Save the drawing where it can be retrieved later. Start SolidWorks by referring to "Chapter 1 Getting Started" with the exception of selecting the "New" icon in the upper left corner of the screen (do not select the "New" icon as in previous chapters). Your screen should look similar to Figure 2.

Figure 2

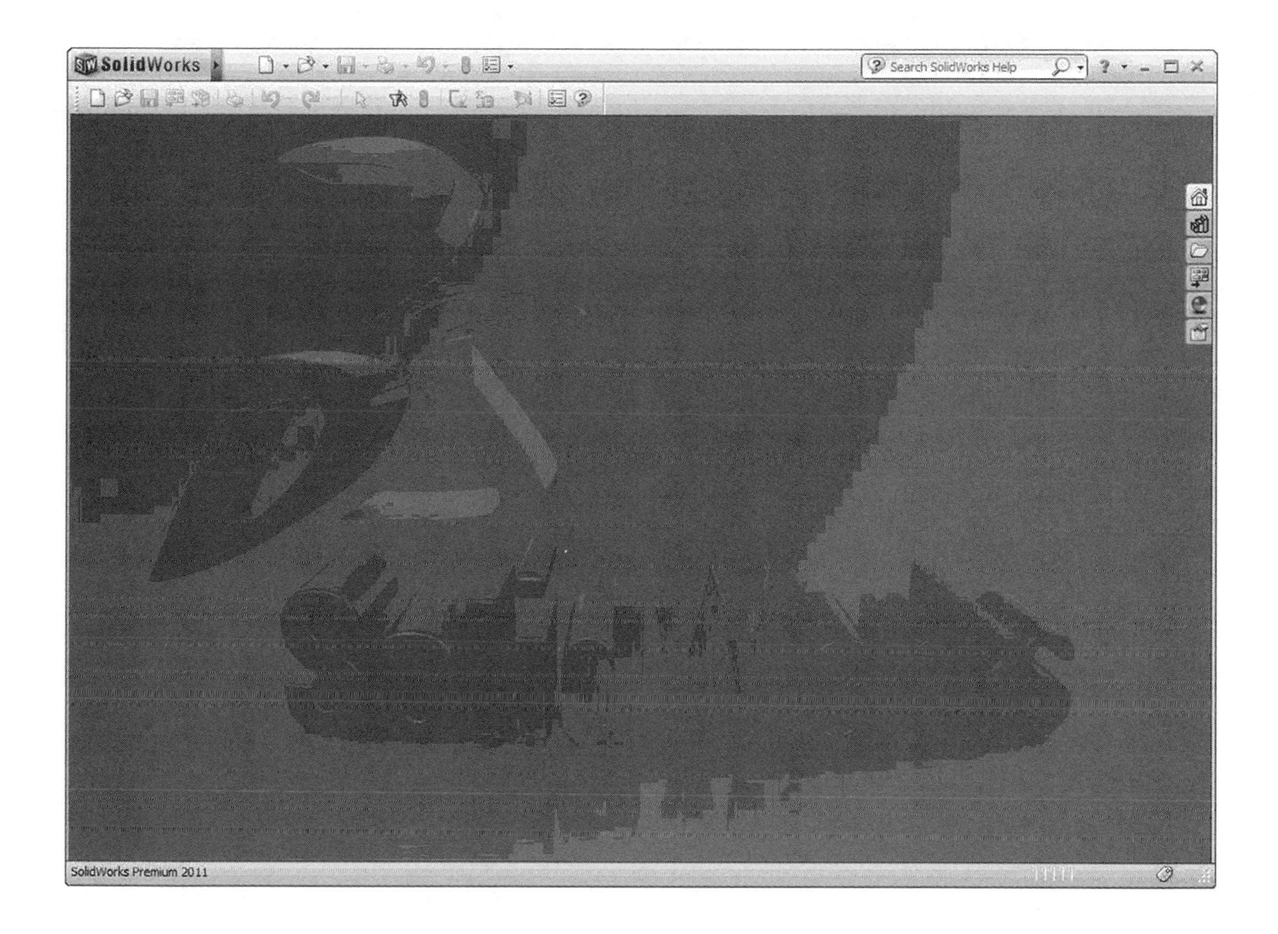

3. Left click on the "Open" icon as shown in Figure 3.

Figure 3

4. The Open dialog box will appear. Left click on the drop down arrow at the top of the dialog box to locate the drawing file. Left click on the drop down arrow to the right of "Files of Type" to show the different types of files that can be imported. Left click on **.DWG**. The file you just created and saved should appear in the dialog box. Left click on **Open** as shown in Figure 4.

Figure 4

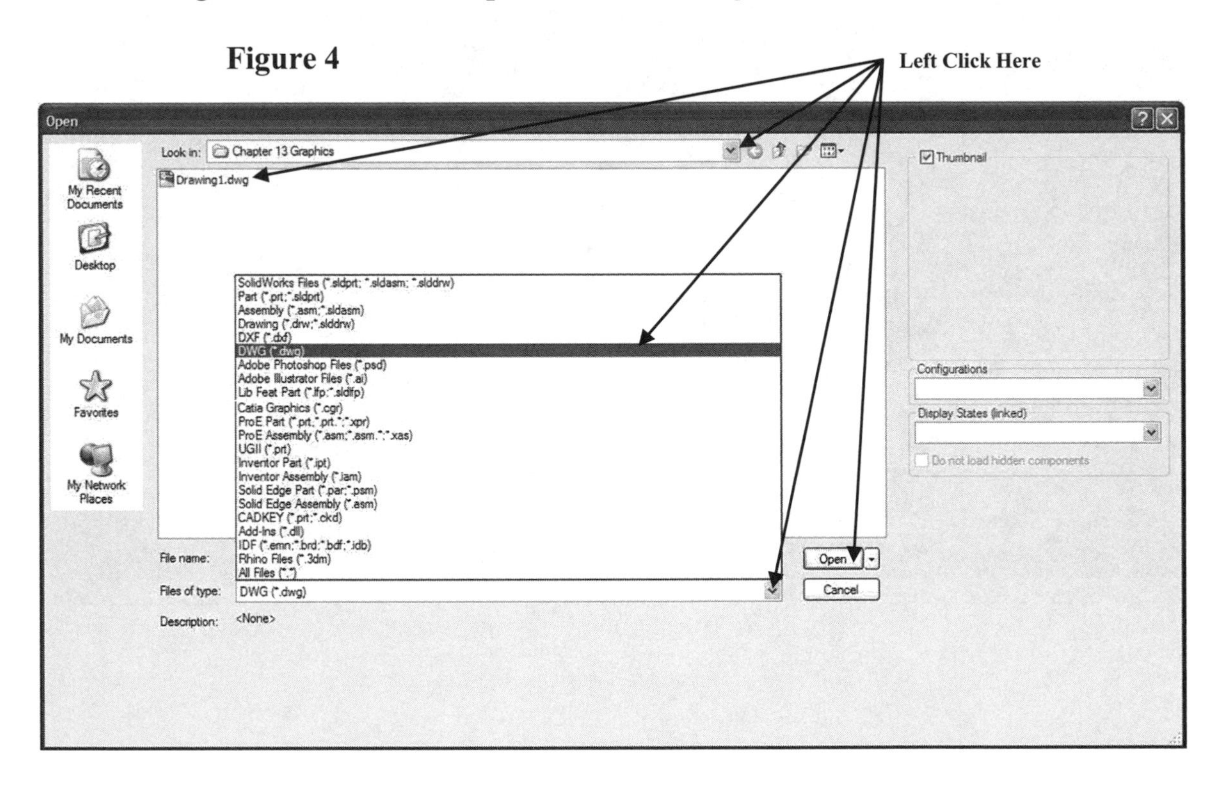

5. The DXF/DWG Import dialog box will appear. Left click next to Create new SolidWorks drawing and Convert to SolidWorks entities. Left click on **Next** as shown in Figure 5.

Figure 5

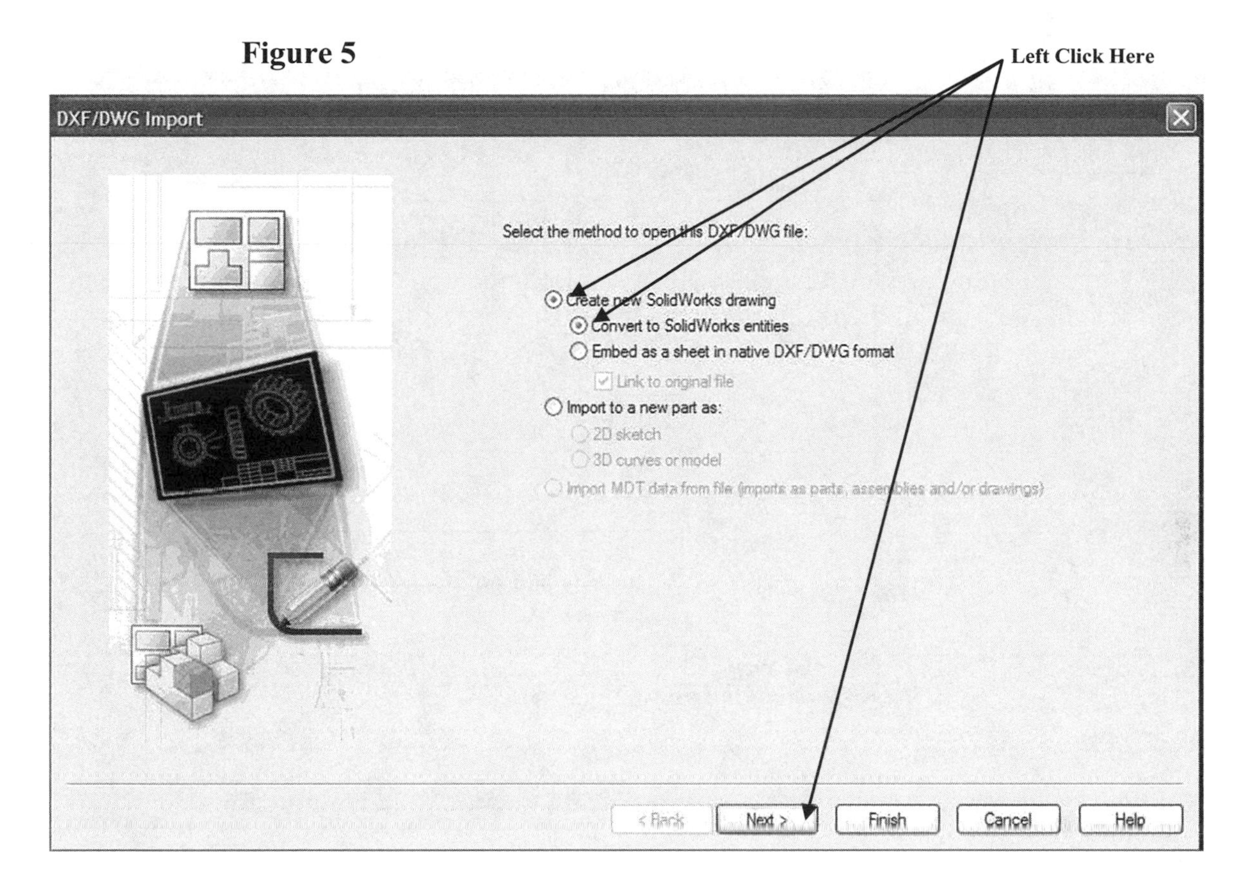

6. The DXF/DWG Import – Drawing Layer Mapping dialog box will appear. Left click on **Next** as shown in Figure 6.

Figure 6

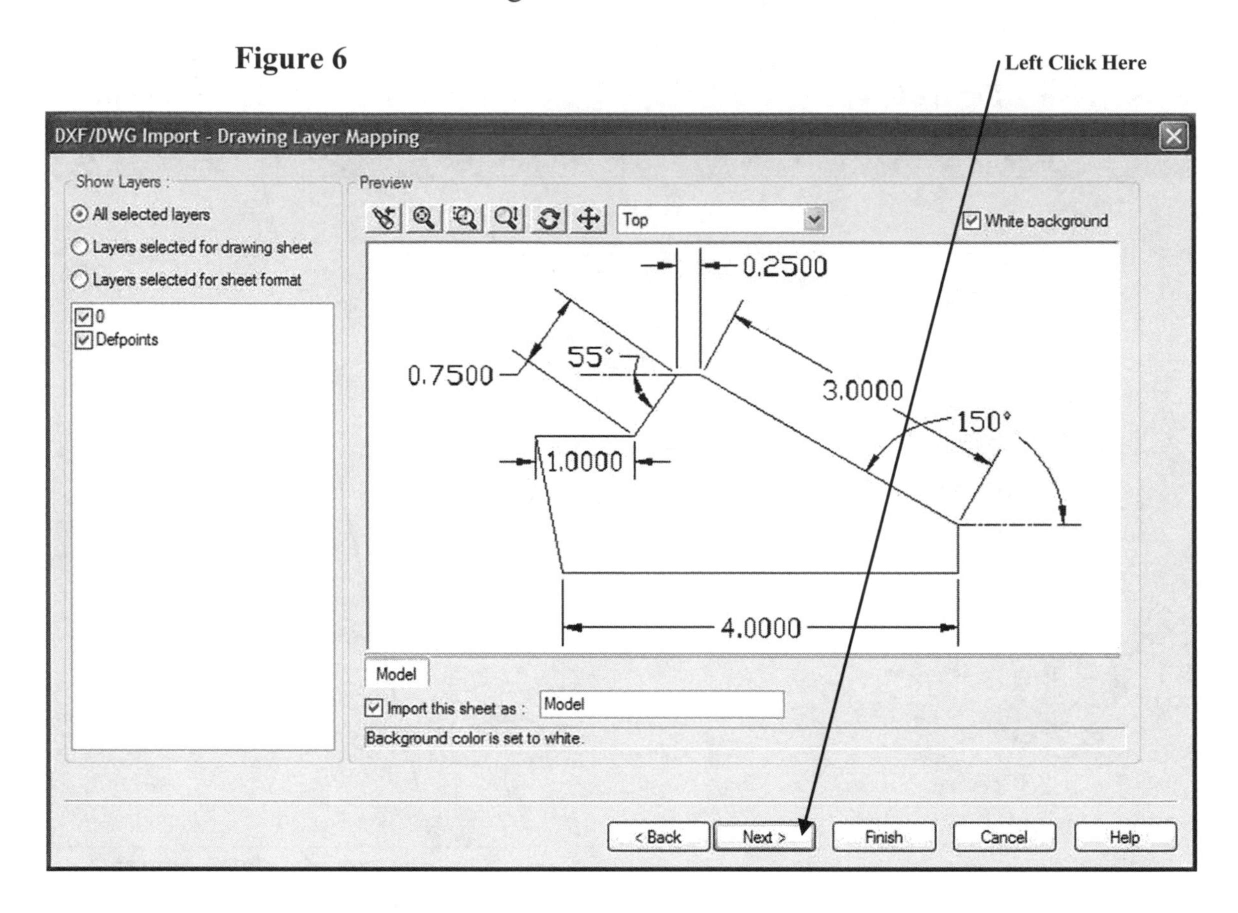

7. The DXF/DWG Import – Document Settings dialog box will appear. Left click on **Finish** as shown in Figure 7.

Figure 7

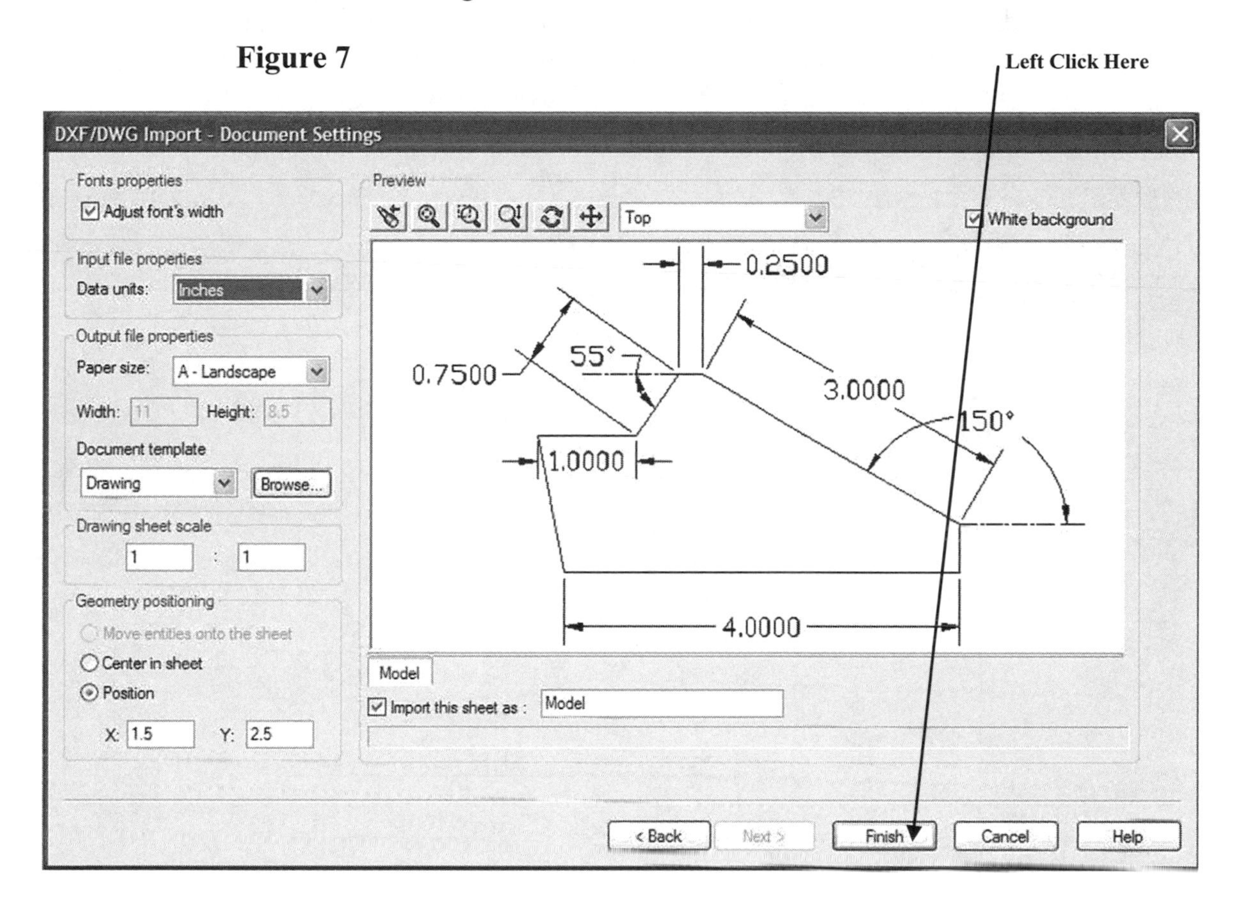

8. SolidWorks will convert the drawing to a 2 Dimensional one view drawing. The dimensions that were created in the DWG editor are not changeable but portions of the drawing that do not contain dimensions can have dimensions added to them. Create a 100 degree angle dimension as shown in Figure 8.

Figure 8

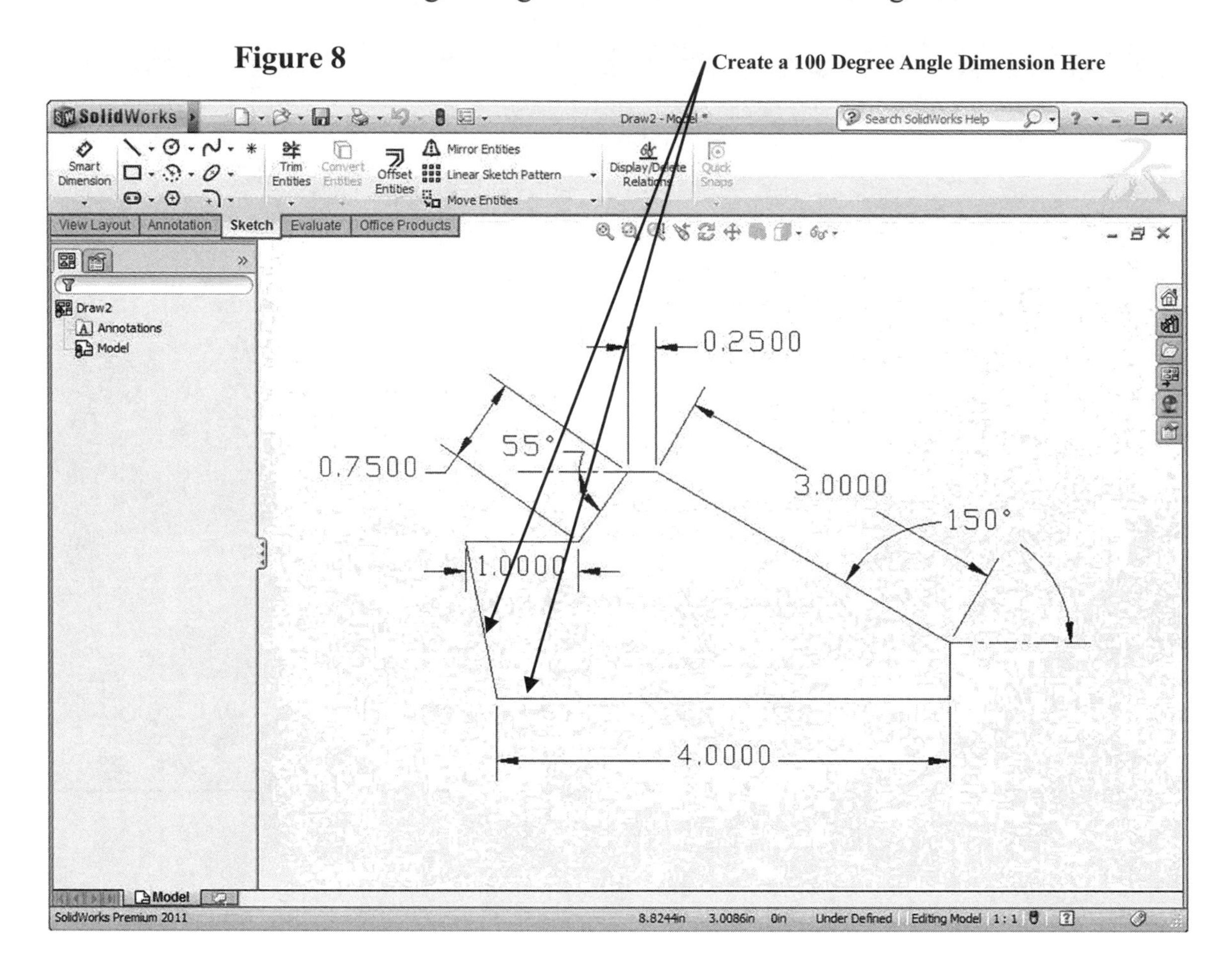

9. Enter **100** in the Modify dialog box as shown in Figure 9.

Figure 9

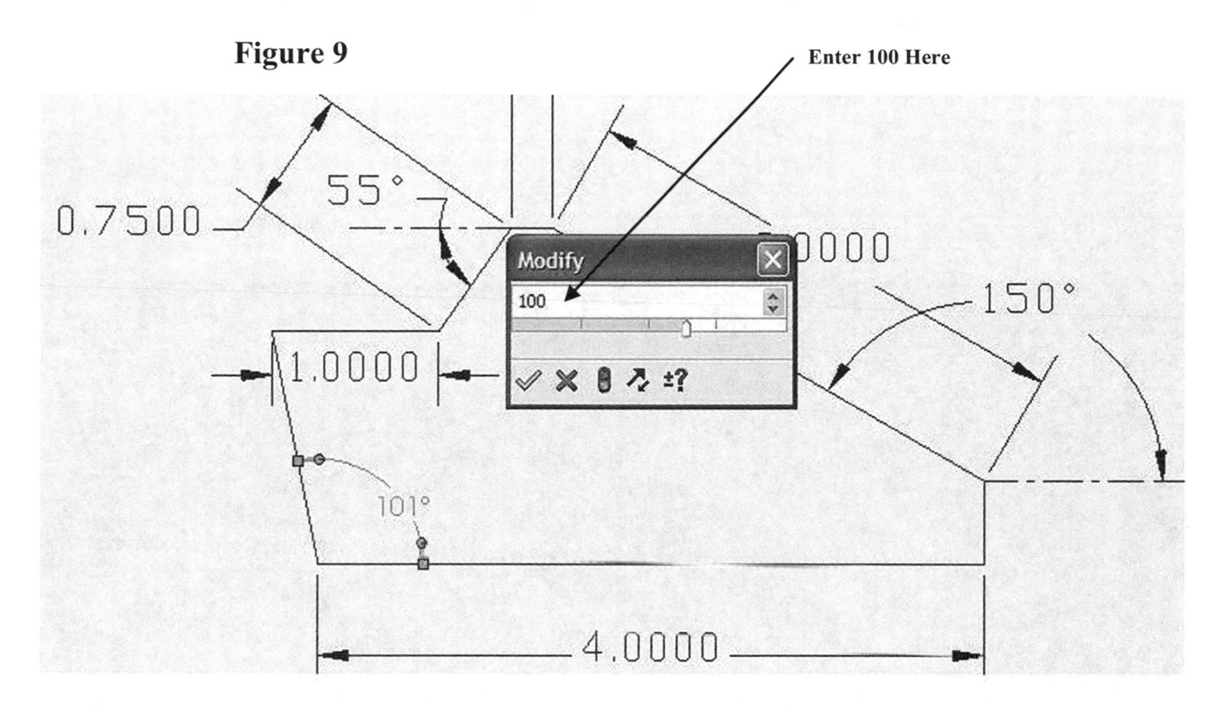

10. Your screen should look similar to Figure 10. Feel free to create other dimensions as desired. Disregard the fact that the addition of the dimension will cause the bottom line to not be parallel.

Figure 10

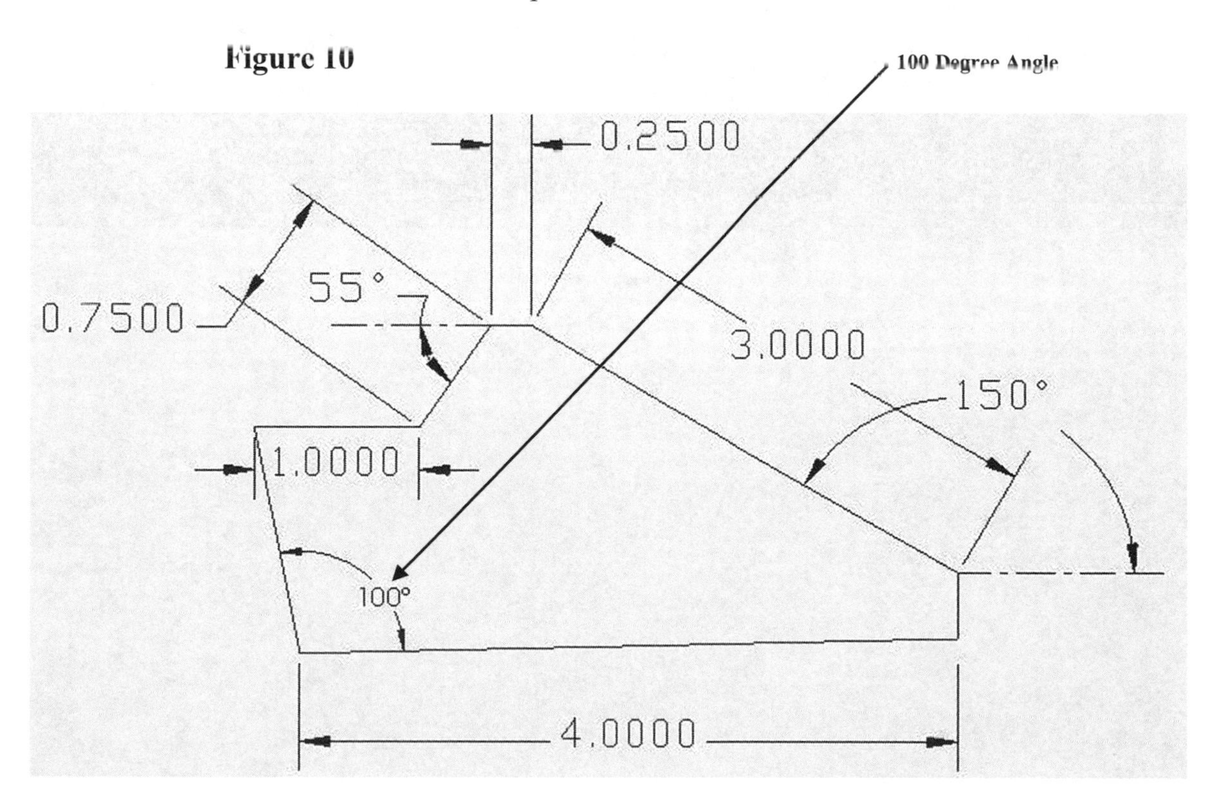

11. Save the drawing where it can be retrieved later. Start SolidWorks by referring to "Chapter 1 Getting Started" with the exception of selecting the "New" icon in the upper left corner of the screen (do not select the "New" icon as in previous chapters). Your screen should look similar to Figure 11.

Figure 11

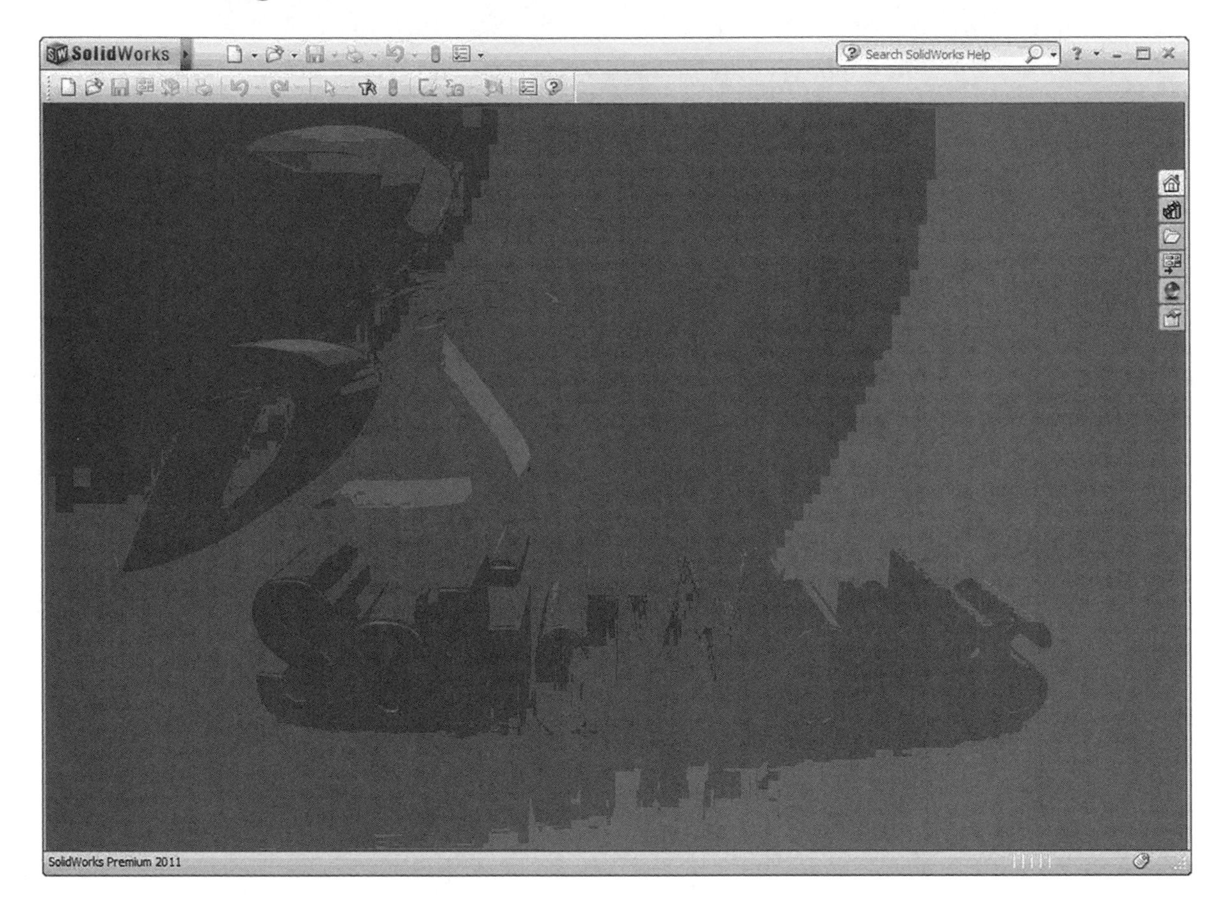

12. Left click on the **Open** icon as shown in Figure 12.

Figure 12

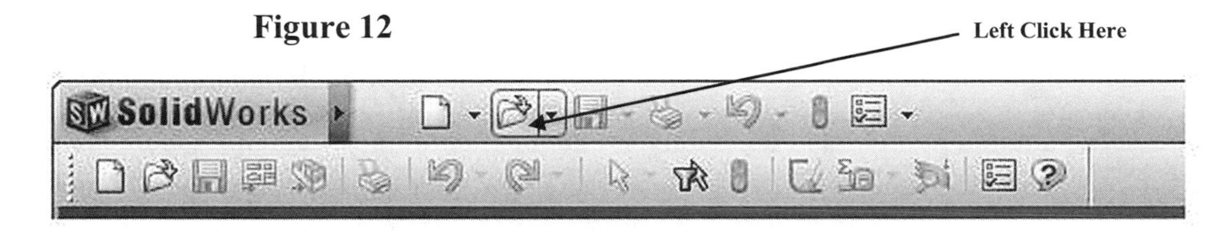

13. The Open dialog box will appear. Left click on the drop down arrow at the top of the dialog box to locate the file. Left click on the drop down arrow to the right of Files of Type to show the different types of files that can be imported. Left click on **.DWG**. The file that was originally created and saved should appear in the dialog box. Left click on **Open** as shown in Figure 14.

Figure 13

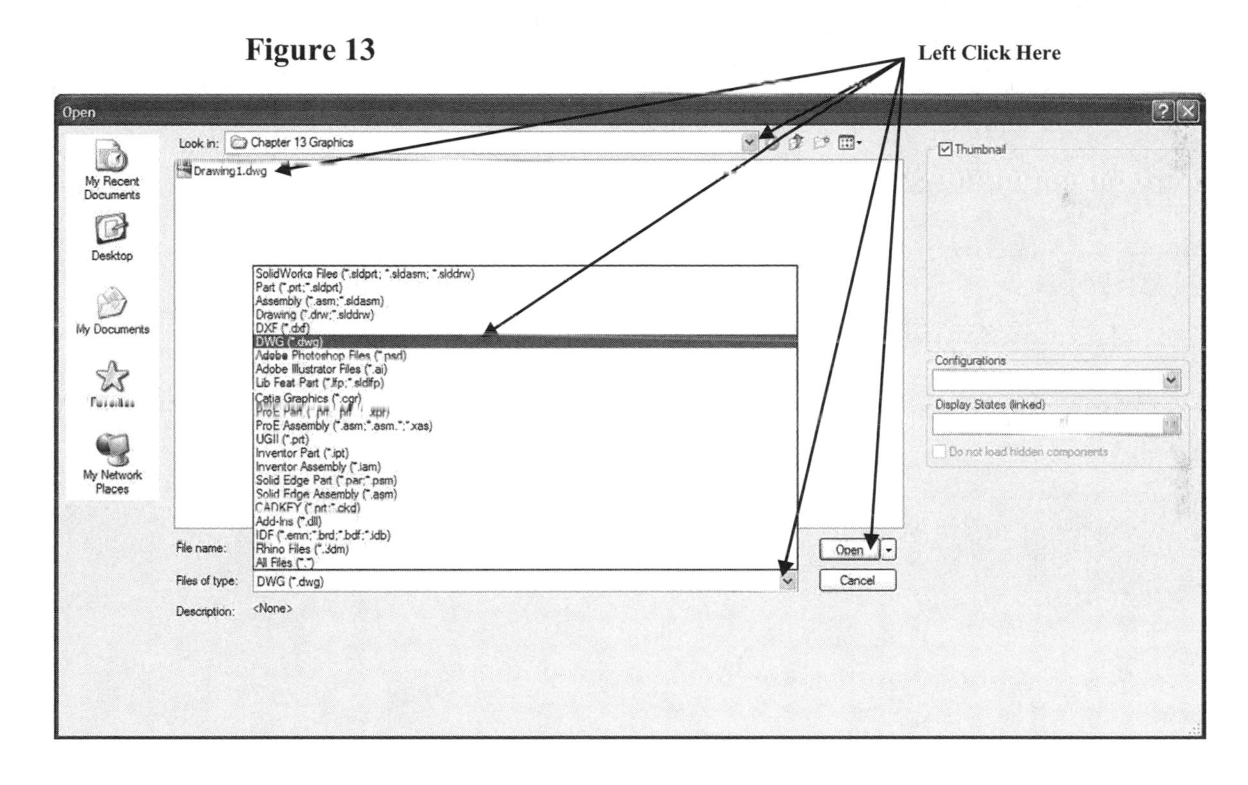

14. The DXF/DWG Import dialog box will appear. Left click next to Import to a new part. Left click on **Next** as shown in Figure 14.

Figure 14

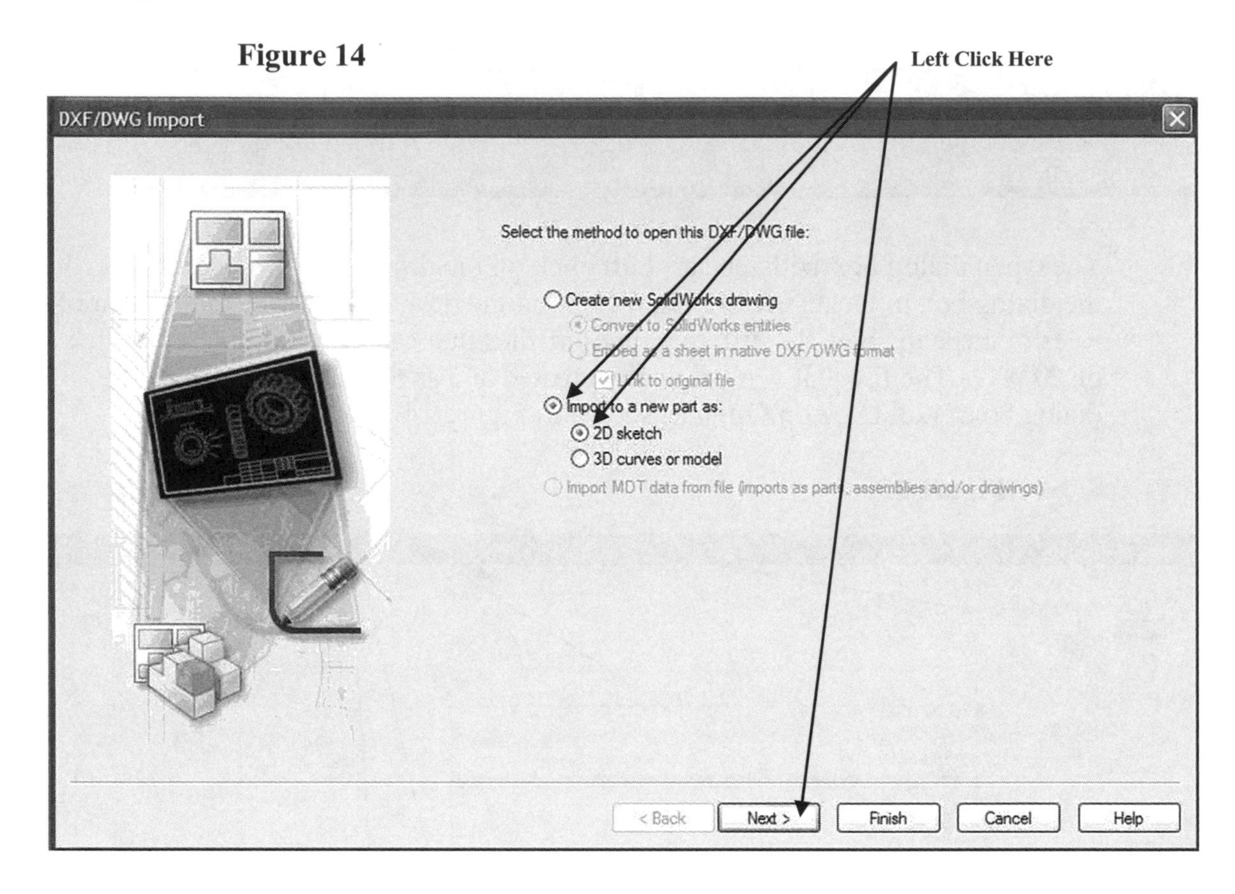

15. The DXF/DWG Import – Document Settings dialog box will appear. Left click on **Finish** as shown in Figure 6.

Figure 15

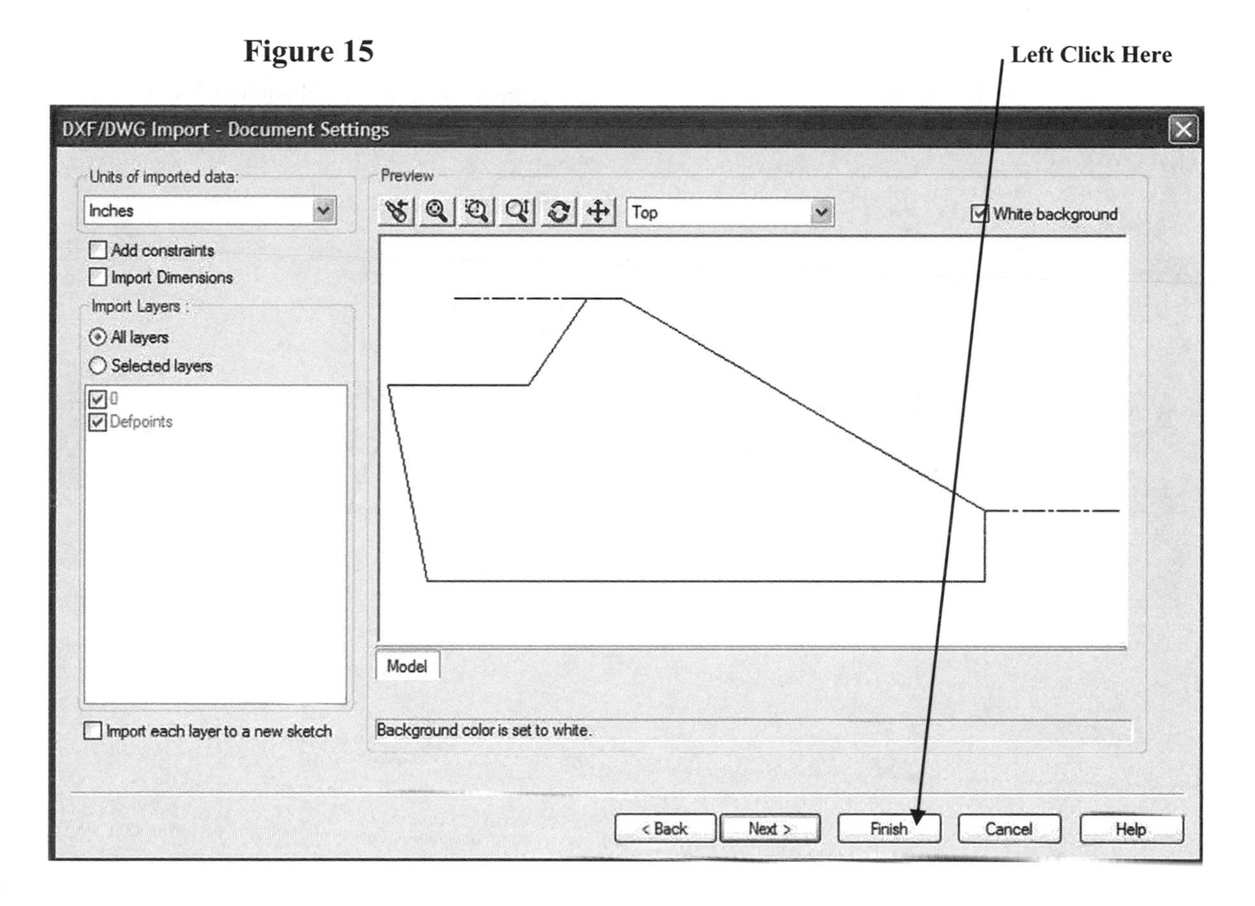

16. SolidWorks will convert the .DWG drawing file into a 2 dimensional sketch that can be dimensioned and extruded into a solid. Delete the 2 construction lines as shown in Figure 16.

Figure 16

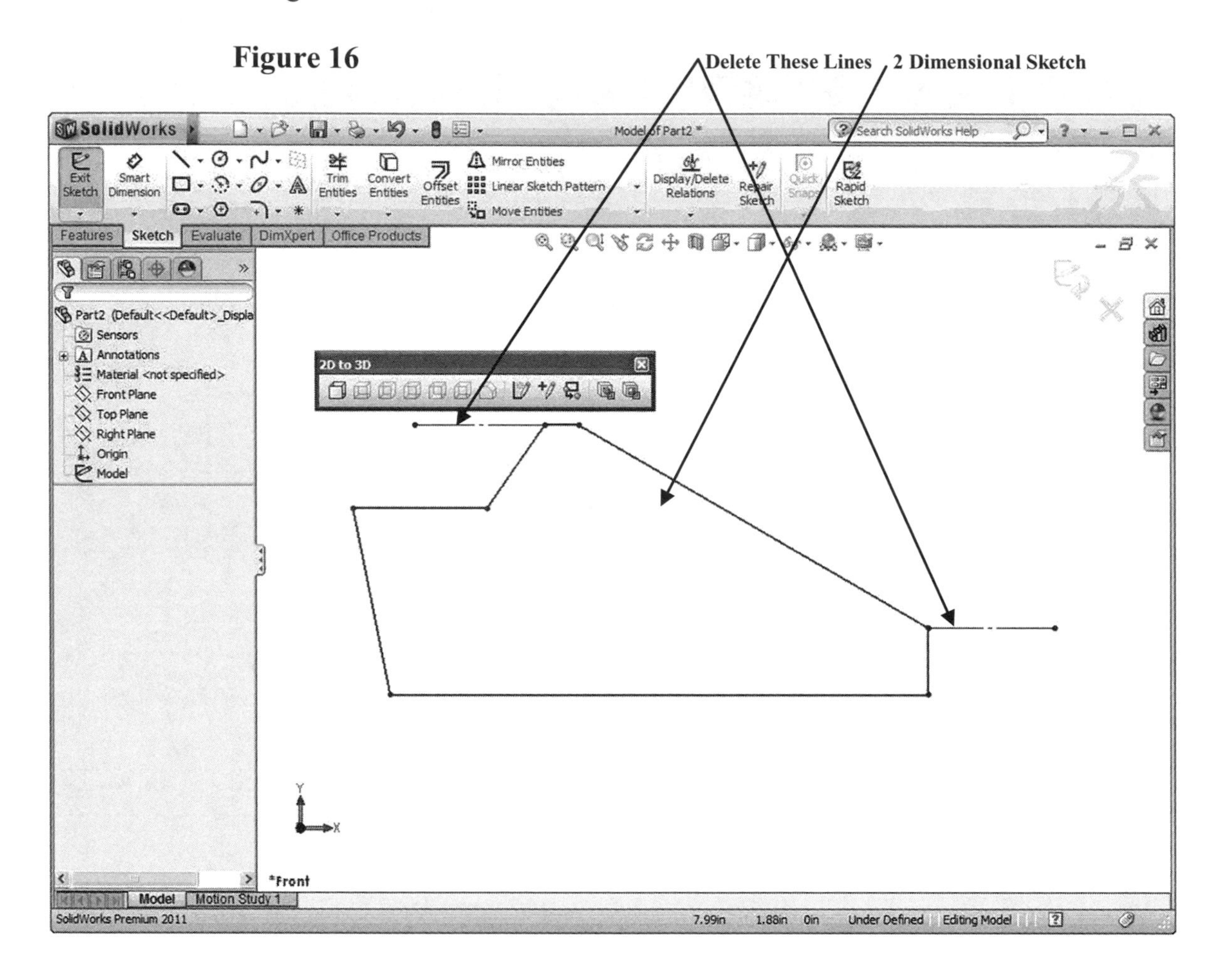

17. Dimension the sketch as shown in Figure 17.

Figure 17

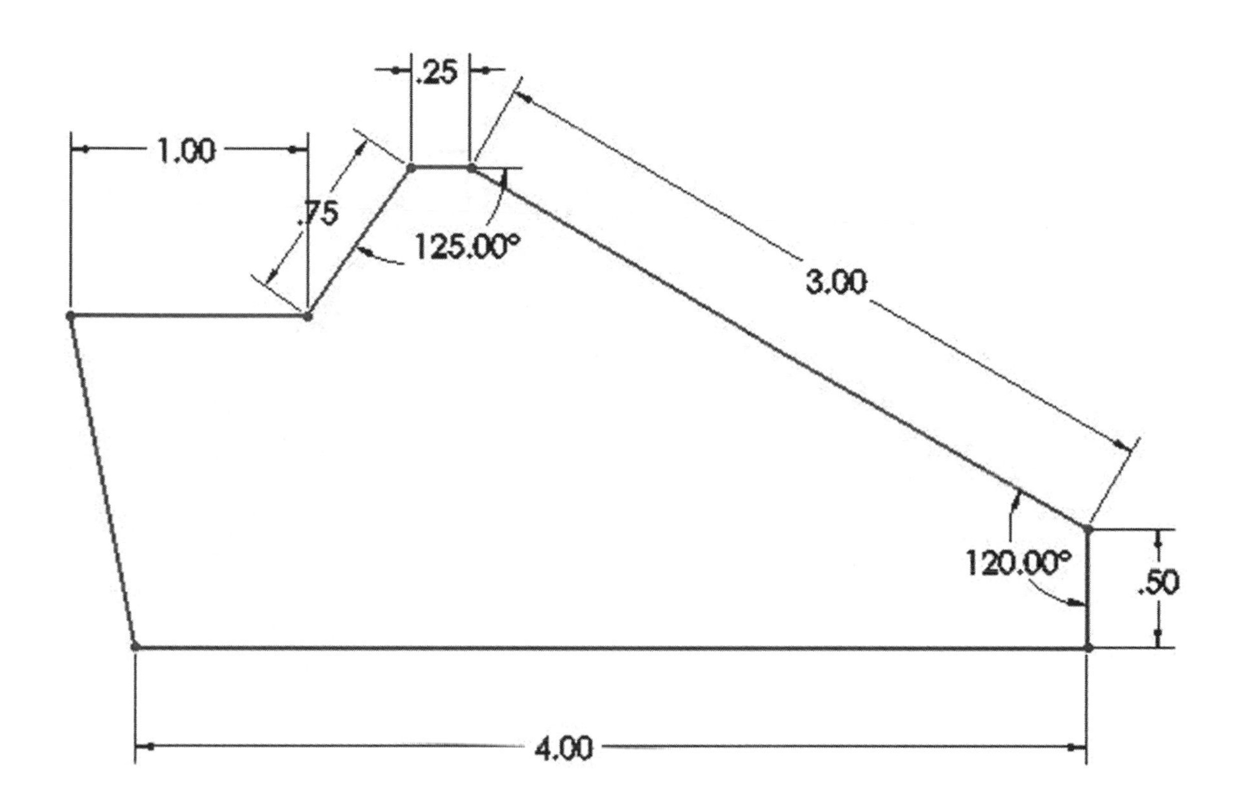

18. Extrude the sketch into a solid with a distance of 2 inches as shown in Figure 18.

Figure 18

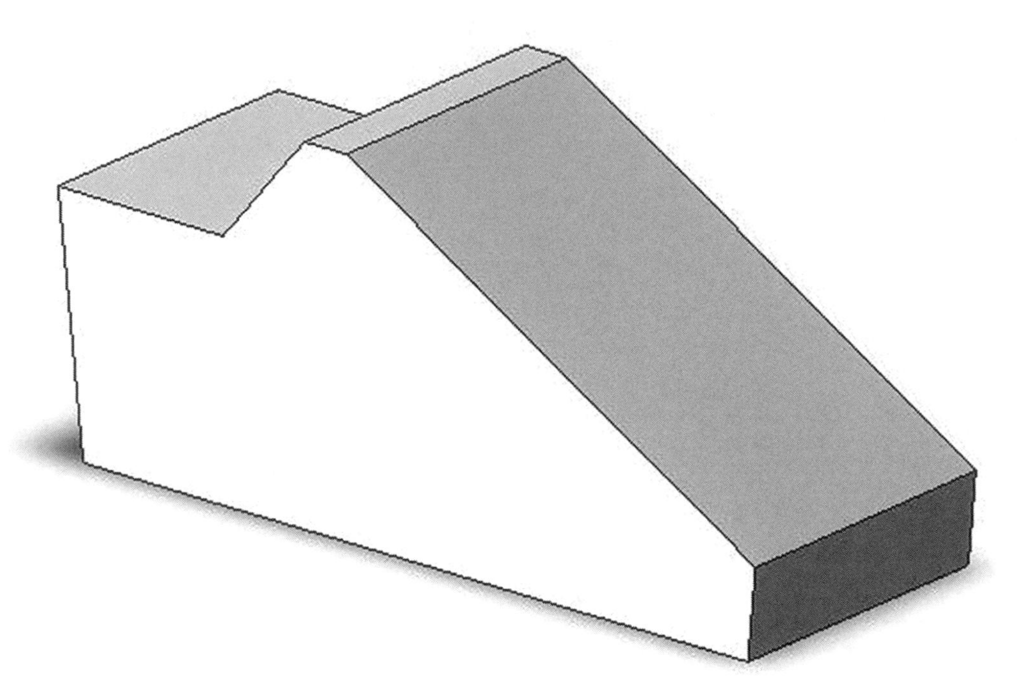

19. Complete the following sketch as shown in Figure 19.

Figure 19

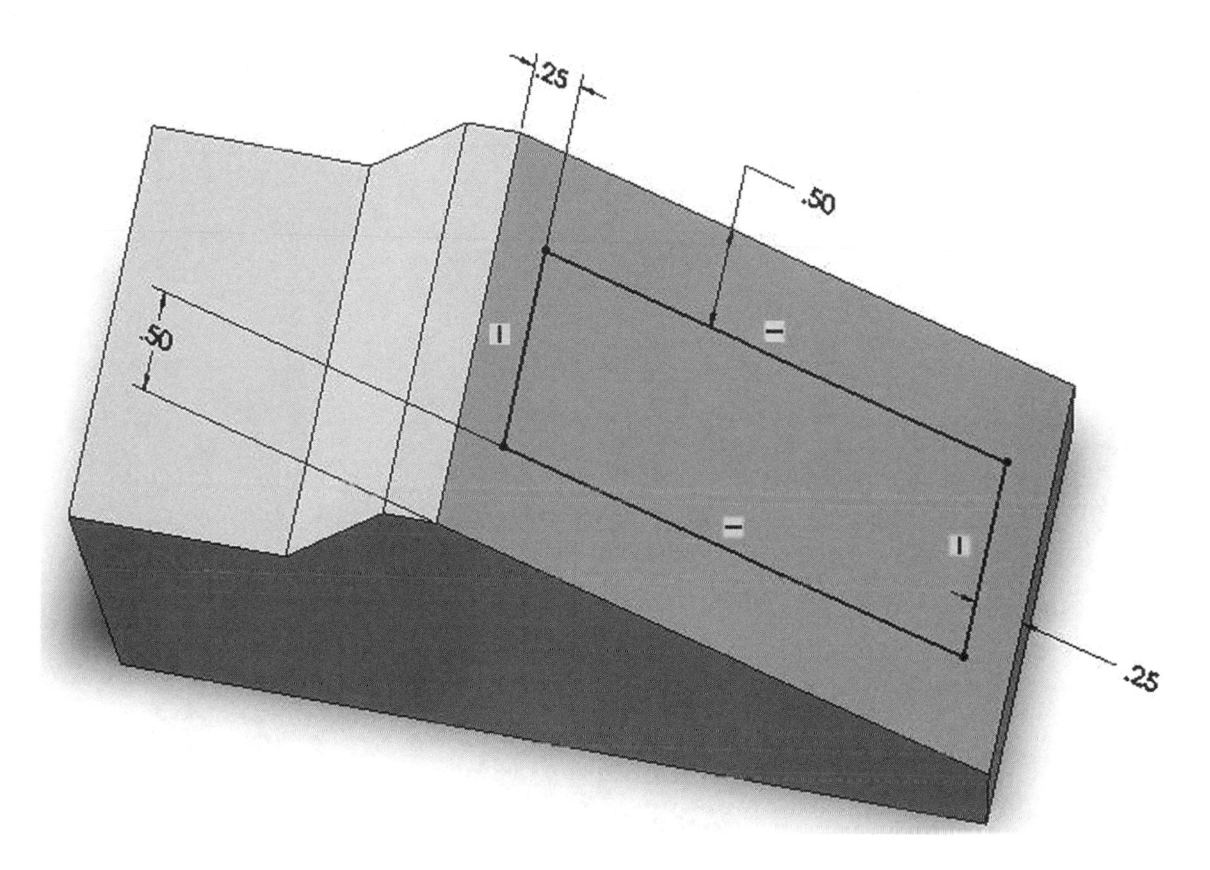

20. Extrude-Cut the rectangle to create a thru hole in the part as shown in Figure 20. You will see that the solid behaves as any solid that was initially created in SolidWorks.

Figure 20

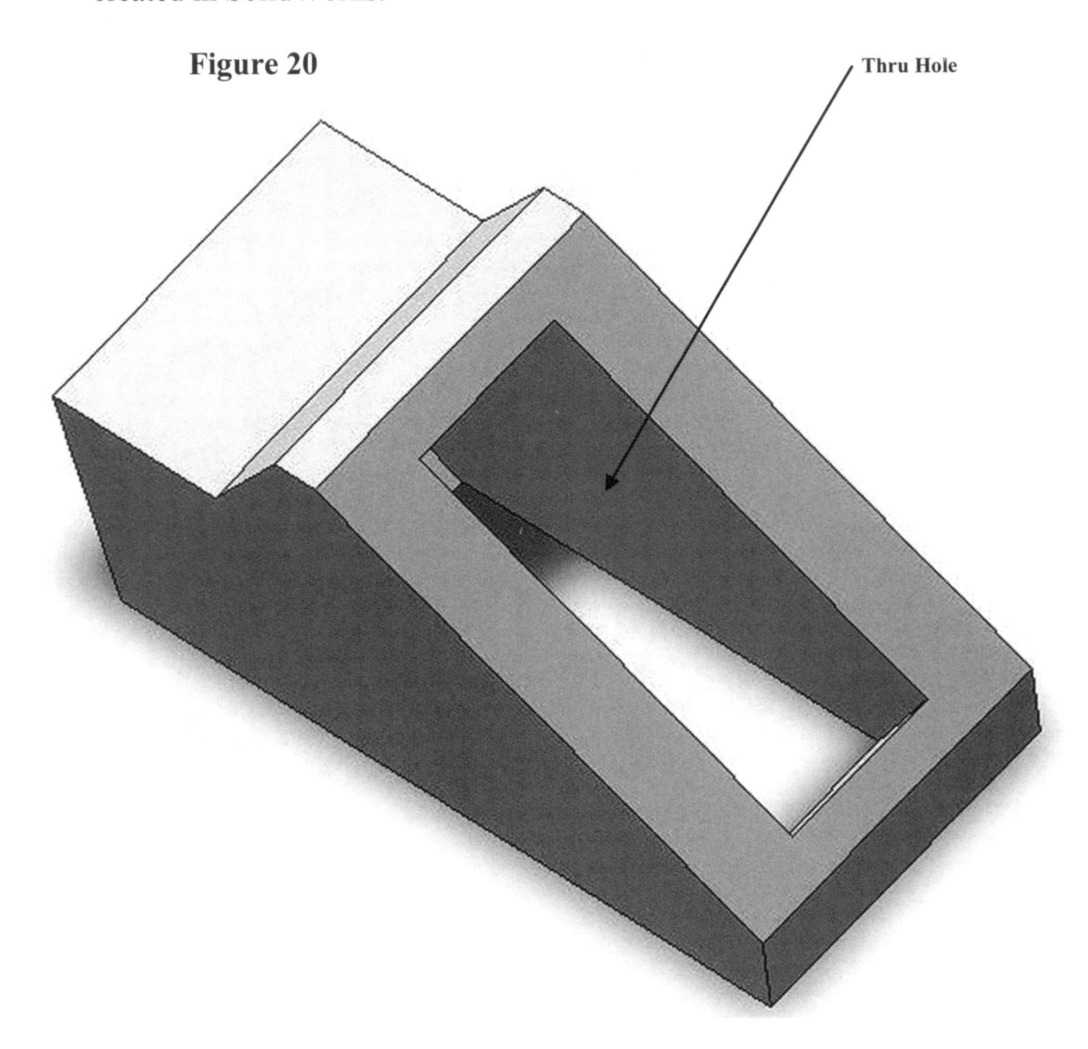

Chapter 14 Introduction to the Design Library

Objectives:

- Learn to import pre-designed parts from the Design Library Toolbox

Chapter 14 includes instruction on how to import pre-designed parts from the Design Library Toolbox and create a Part drawing file.

Chapter 14 includes instruction on how to use the Design Library Toolbox to design the part shown below.

1. SolidWorks has many pre-designed parts that can be imported to create a Part drawing file. These parts are located in the Design Library Toolbox. The purpose of the Design Library Toolbox is to reduce the amount of time required to design parts that are already available through a part supplier. Common items such as bearings, C-clip's bolts, nuts and washers are typically produced in mass quantity by part suppliers. For this reason designers typically do not design the above mentioned items. They simply import/create them in multiple numbers from the Design Library Toolbox.

2. Start SolidWorks by referring to "Chapter 1 Getting Started". Do NOT begin a New Drawing. You screen should look similar to Figure 1

Figure 1

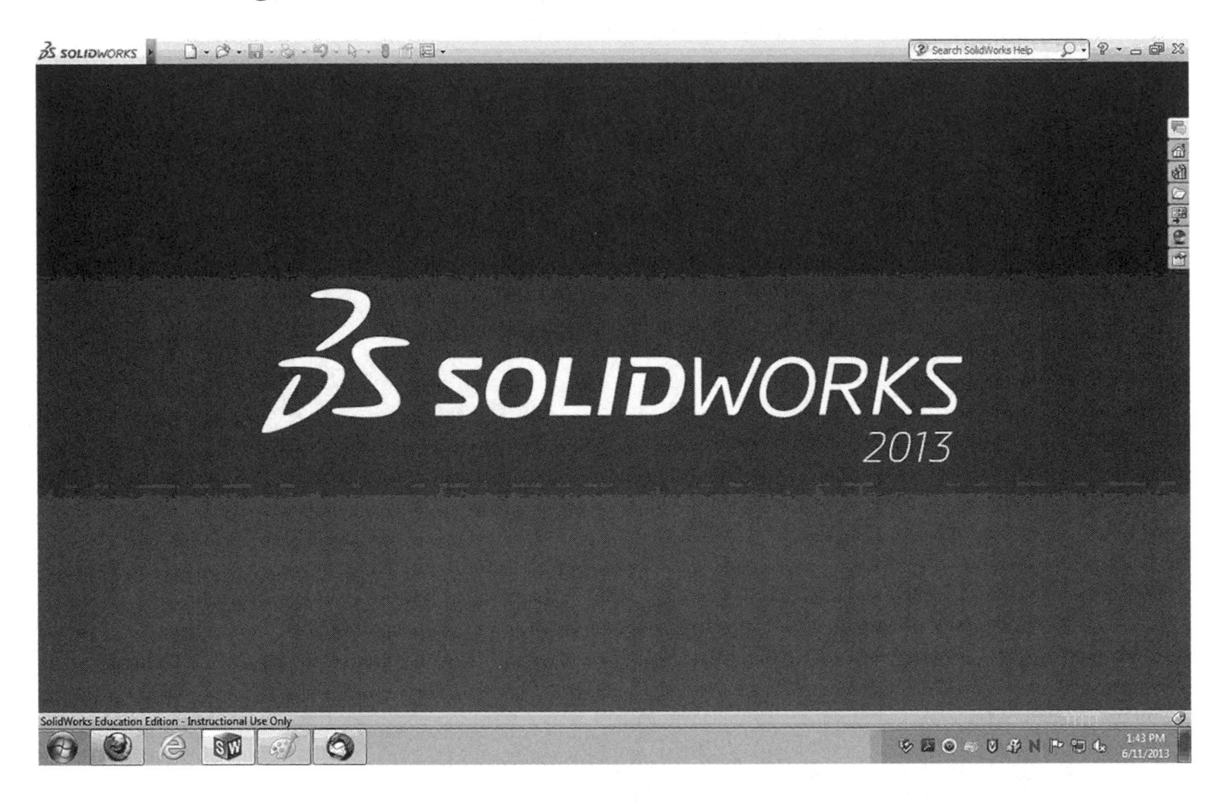

3. Move the cursor to the upper right portion of the screen and left click on the **Design Library** icon if the Design Library is not already open as shown in Figure 2.

Figure 2

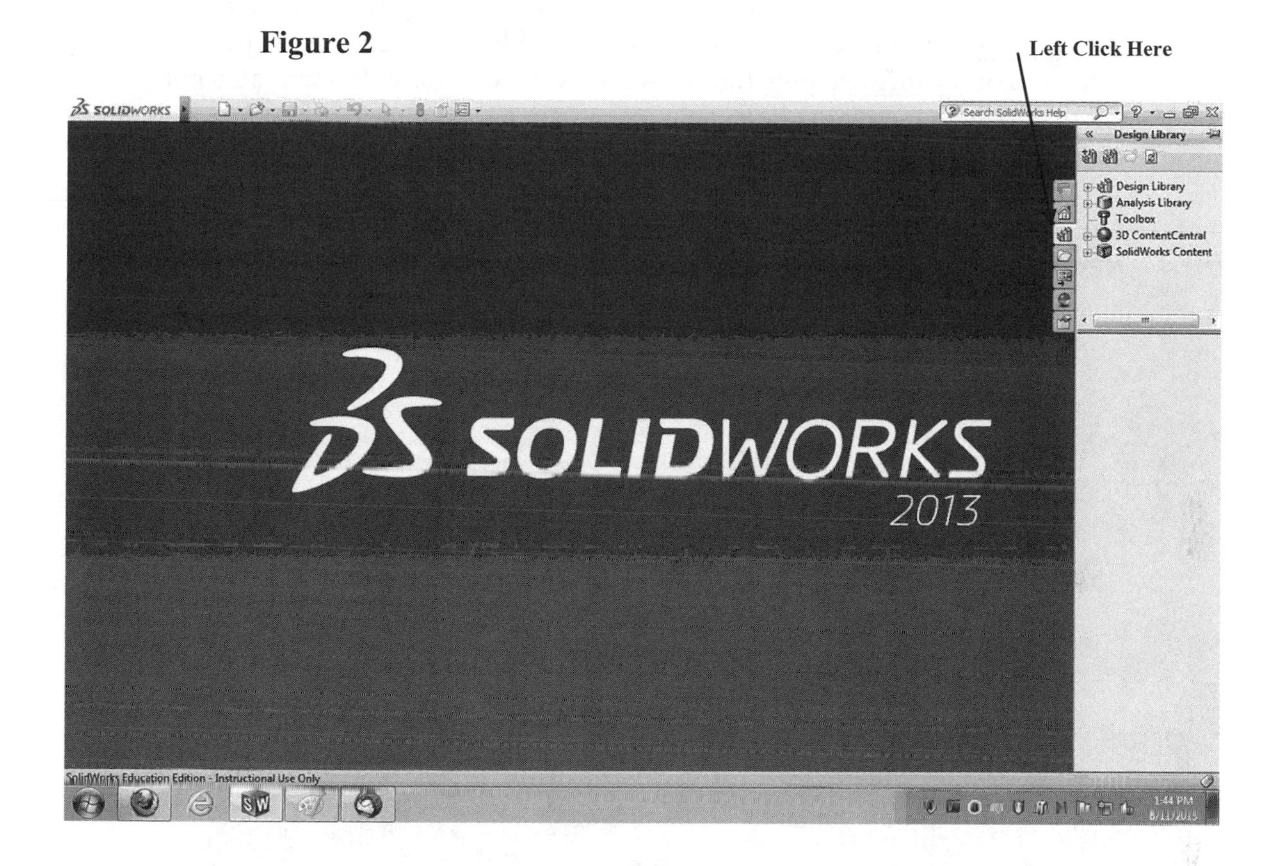

4. Move the cursor to the upper right portion the screen and left click on **Toolbox.** Move the cursor to the lower right portion of the screen and left click on ***Add in now***. You may have to raise the border to gain access to the "Add in now" text. This will load the Toolbox for use as shown in Figure 3. After selecting ***Add in now*** you may have to configure the tool box. To configure the tool box, left click on ***Configure now*** (not shown). A dialog box will open (also not shown). Close the dialog box.

Figure 3

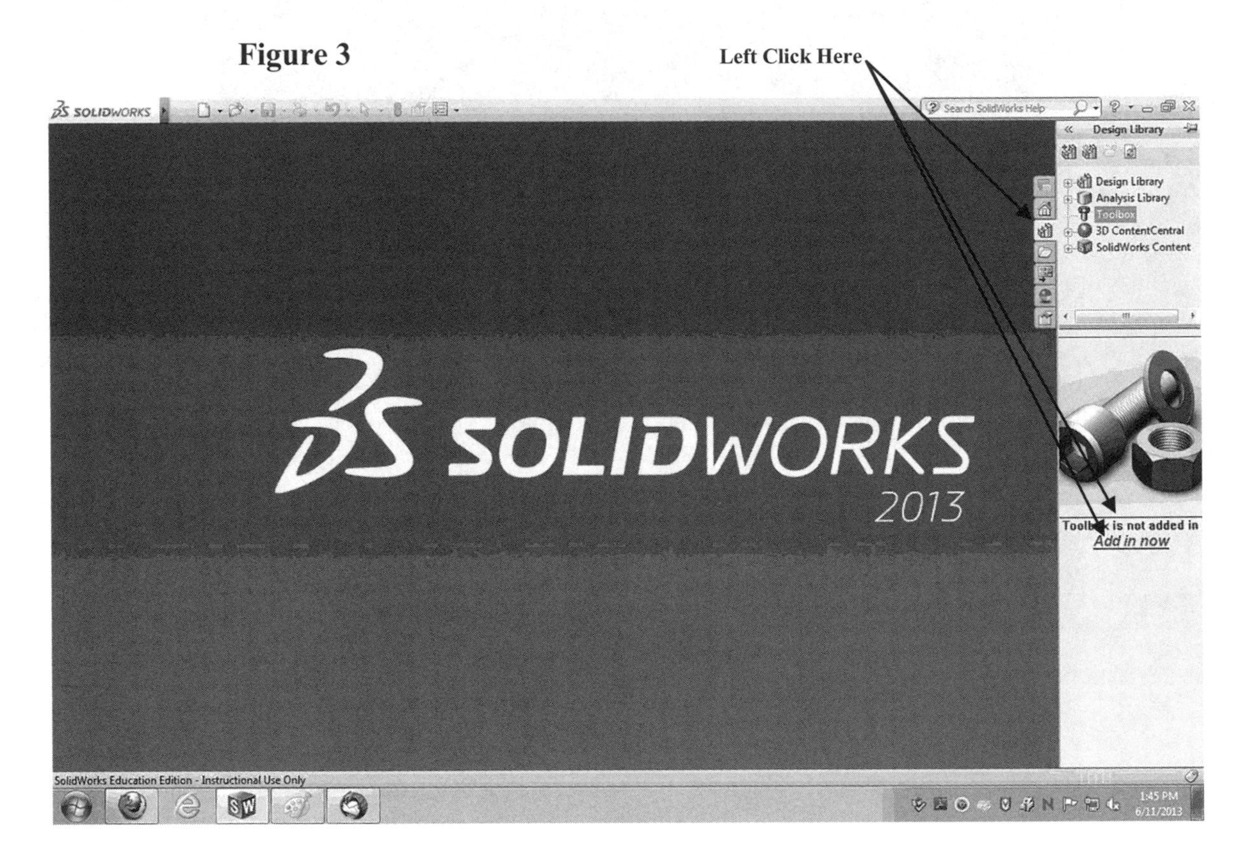

5. Move the cursor to the lower right portion of the screen and left click on the **Ansi Inch** folder as shown in Figure 4

Figure 4

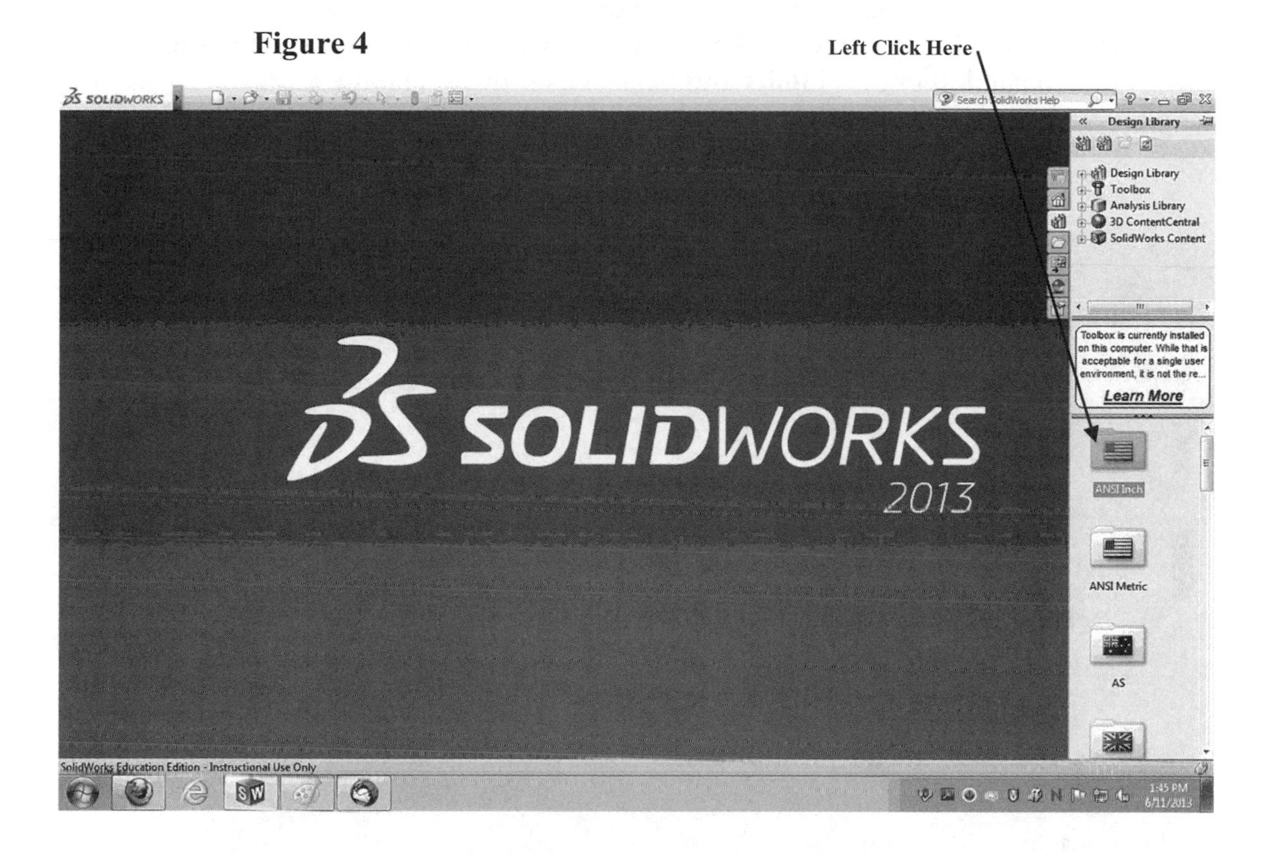

6. The Toolbox will open at the right of the screen as shown in Figure 5. The upper left portion of the screen will display all the different types of items available in the Toolbox. Some items in the Toolbox will need additional editing once they are imported into a drawing file. Move the cursor to the lower right portion of the screen and left click on **Bolts and Screws** as shown in Figure 5.

Figure 5

7. Move the cursor to the lower right portion of the screen and left click on **Hex Head** as shown in Figure 6.

Figure 6

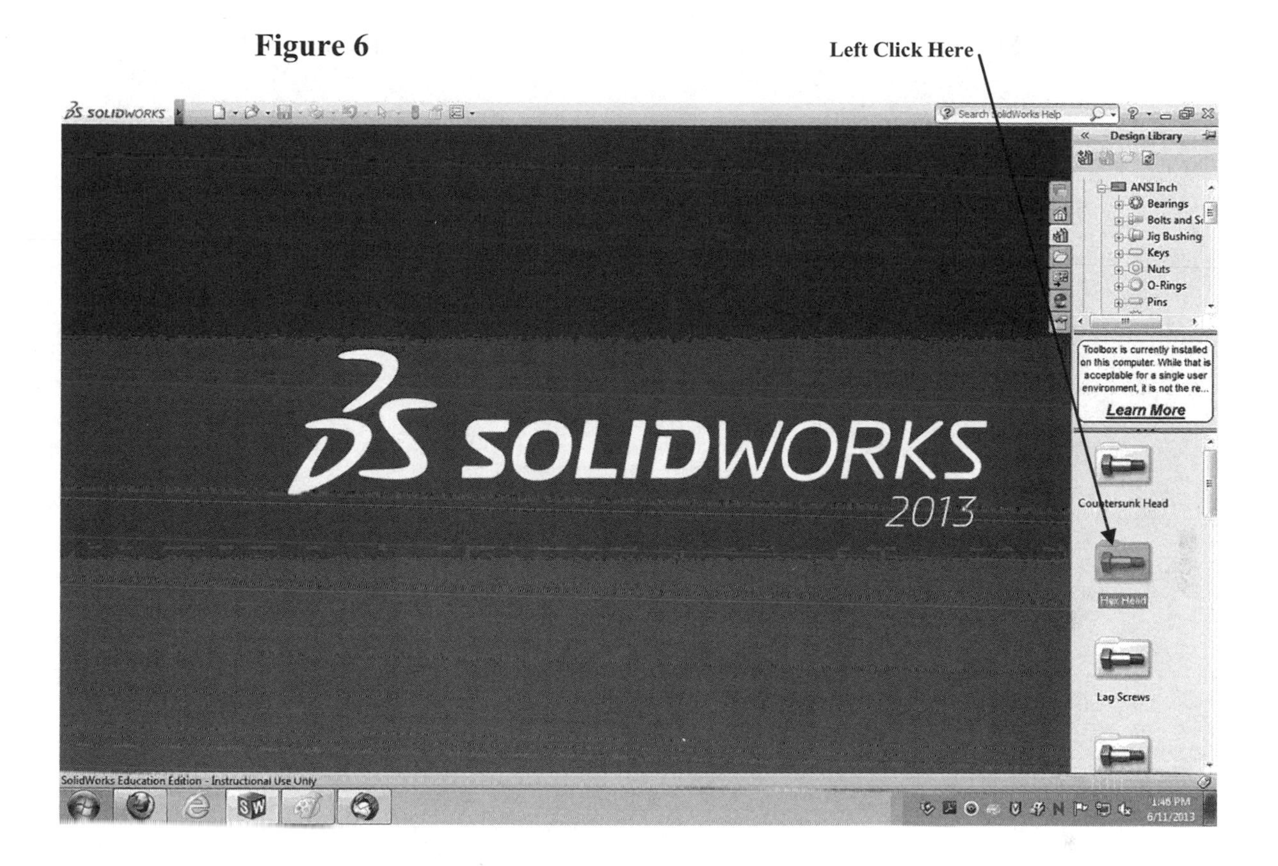

8. Move the cursor to the lower right portion of the screen and left click on **Heavy Hex Bolt** as shown in Figure 7.

Figure 7

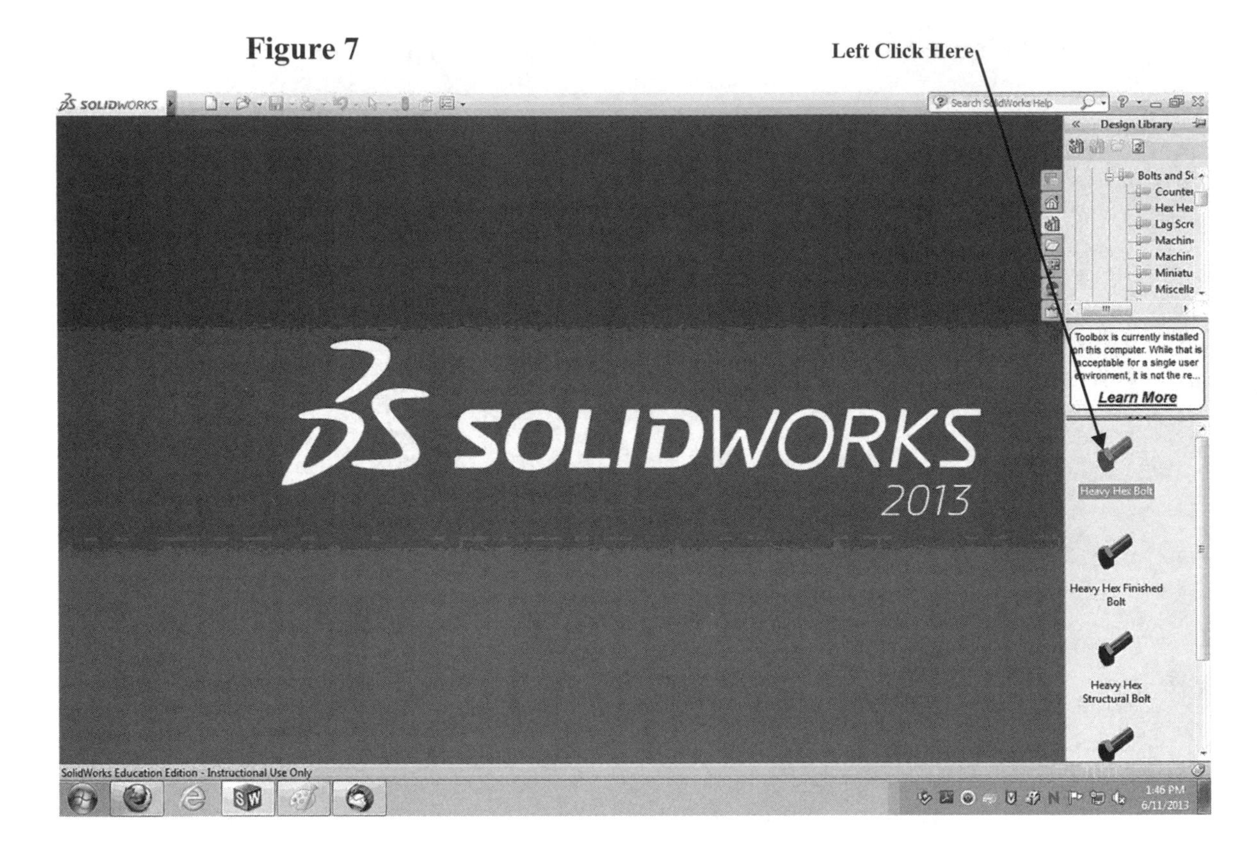

9. Right click on **Heavy Hex Bolt.** A pop up menu will appear. Left click on **Create Part** as shown in Figure **8**.

Figure 8

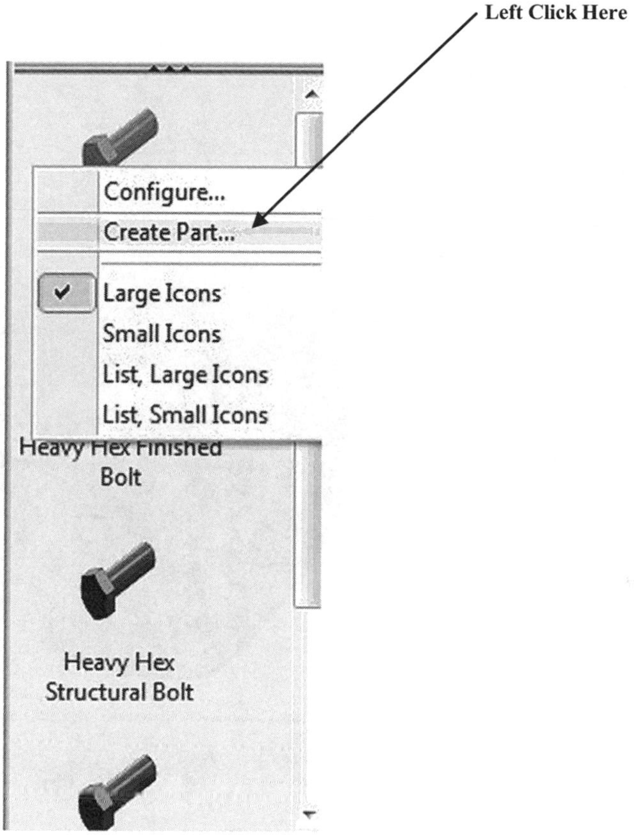

10. SolidWorks will create a Heavy Hex bolt. Your screen should look similar to Figure 9.

Figure 9

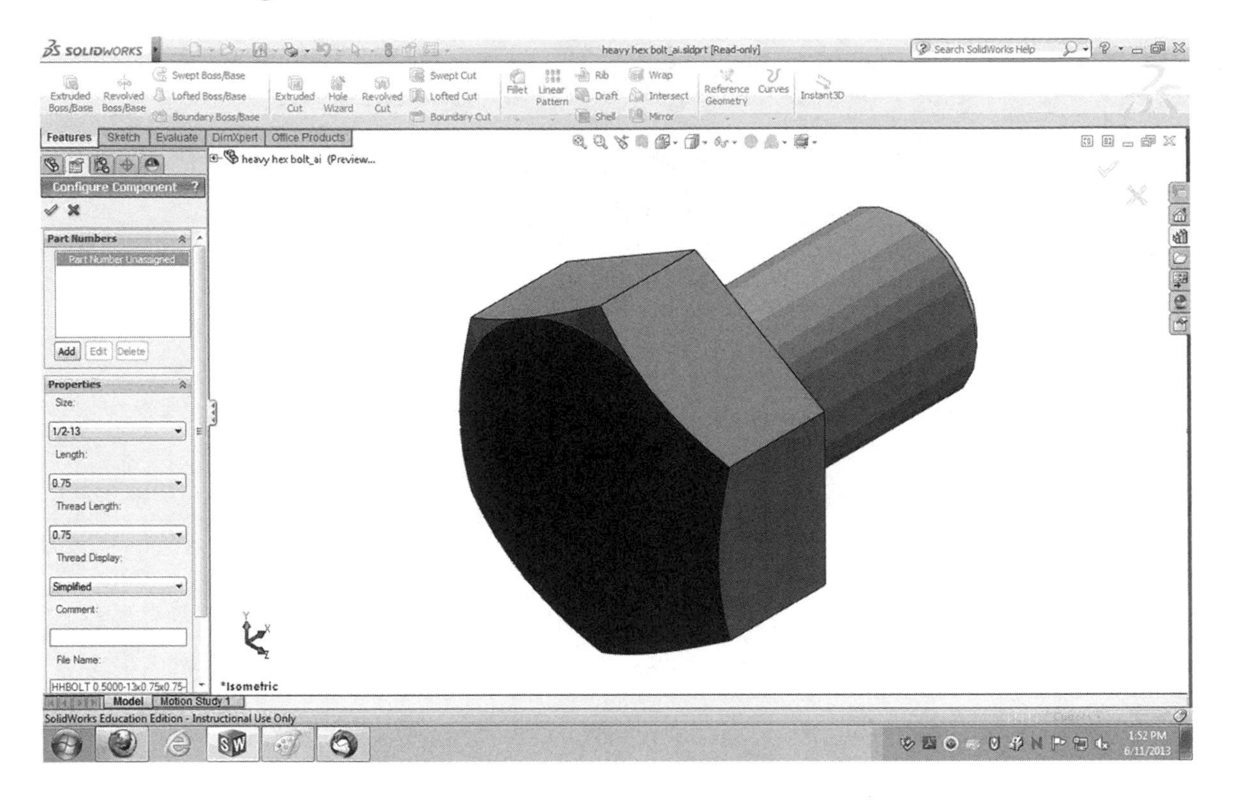

11. The Properties of the bolt can be edited. Move the cursor to the middle left portion of the screen and left click on the drop down arrow under "Size". Left click on **¾ -16**. This refers to the diameter and pitch of the bolt. This particular bolt is ¾ (.75) inches in diameter and has 16 threads per inch. Notice that the length of the bolt will change. Threads will NOT be displayed.

Figure 10

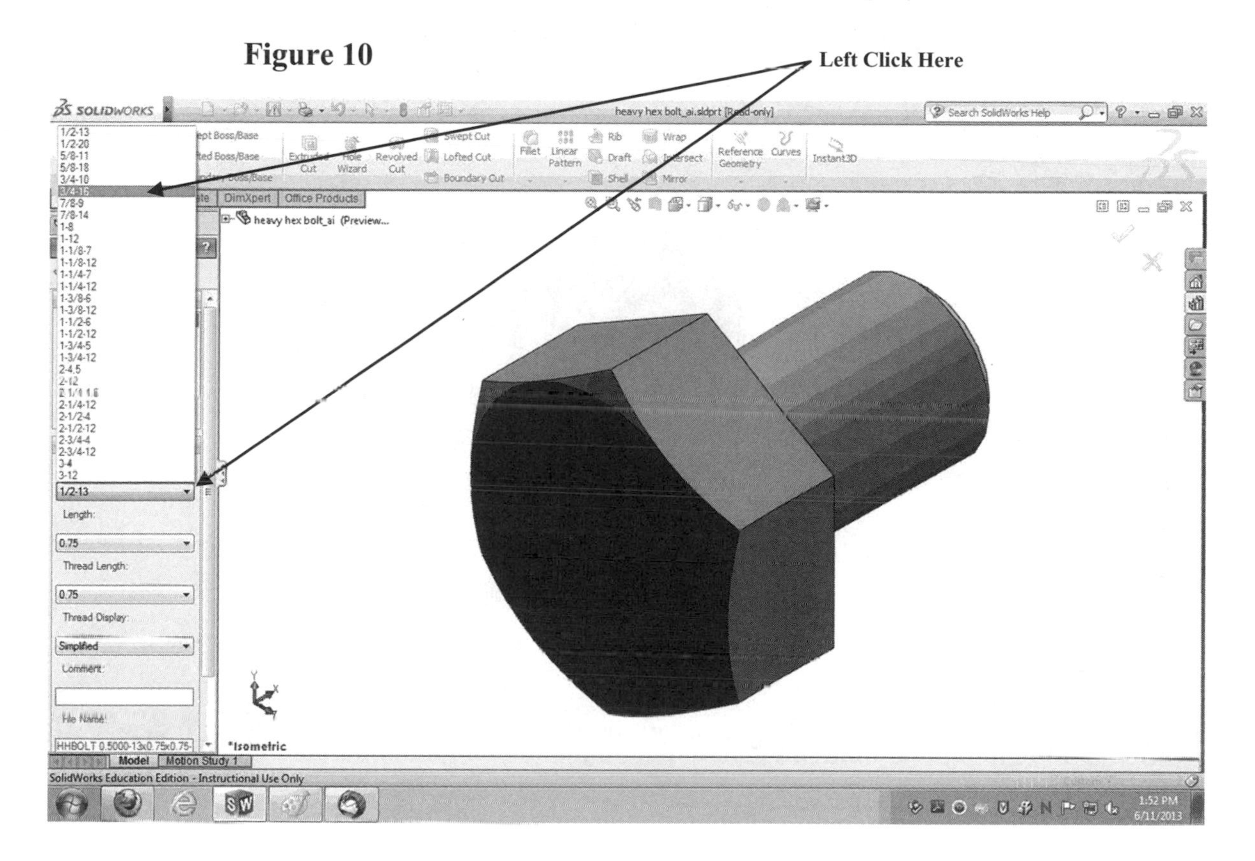

12. Move the cursor to the middle left portion of the screen and left click on the drop down arrow under "Length". Left click on **1.5**. This refers to the length of the bolt. This particular bolt is 1.5 inches in length as shown in Figure 11.

Figure 11

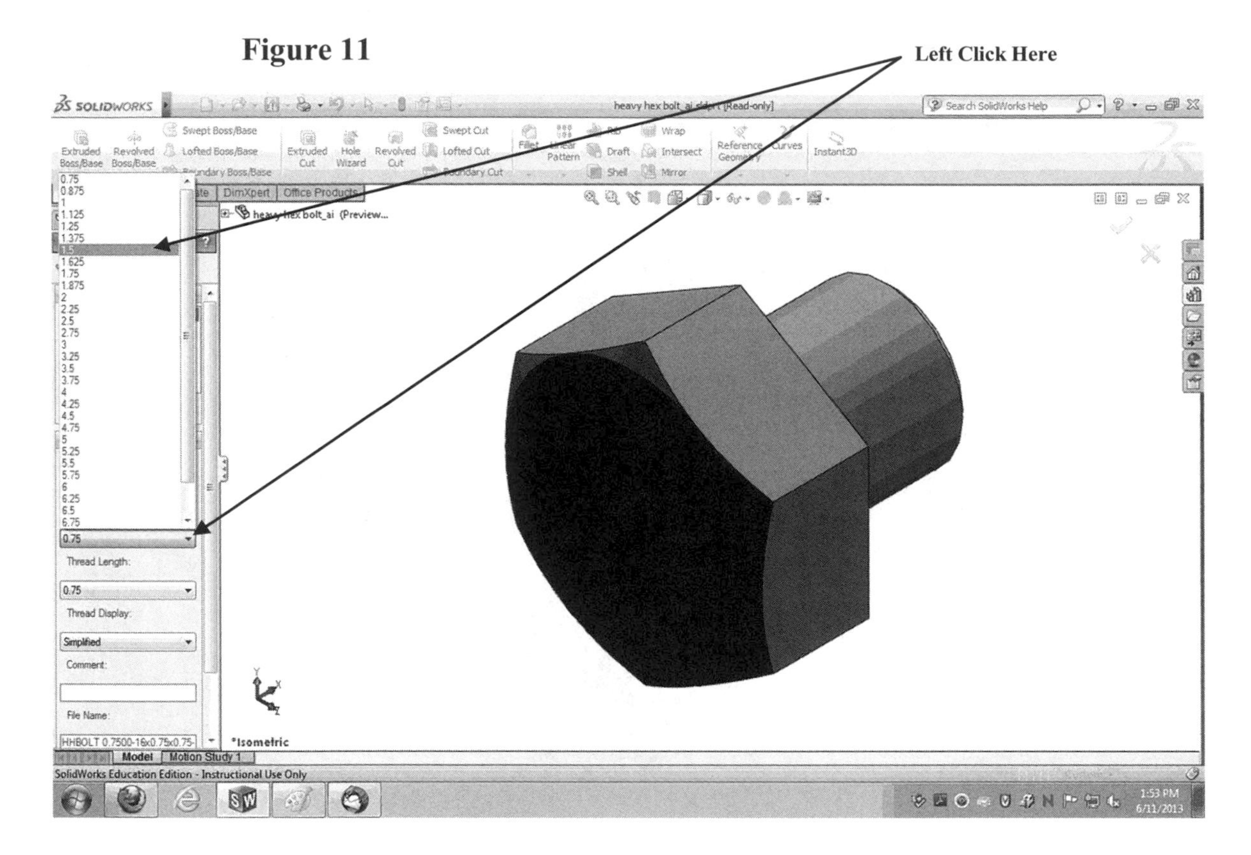

13. Left click on the green check mark as shown in Figure 12.

Figure 12

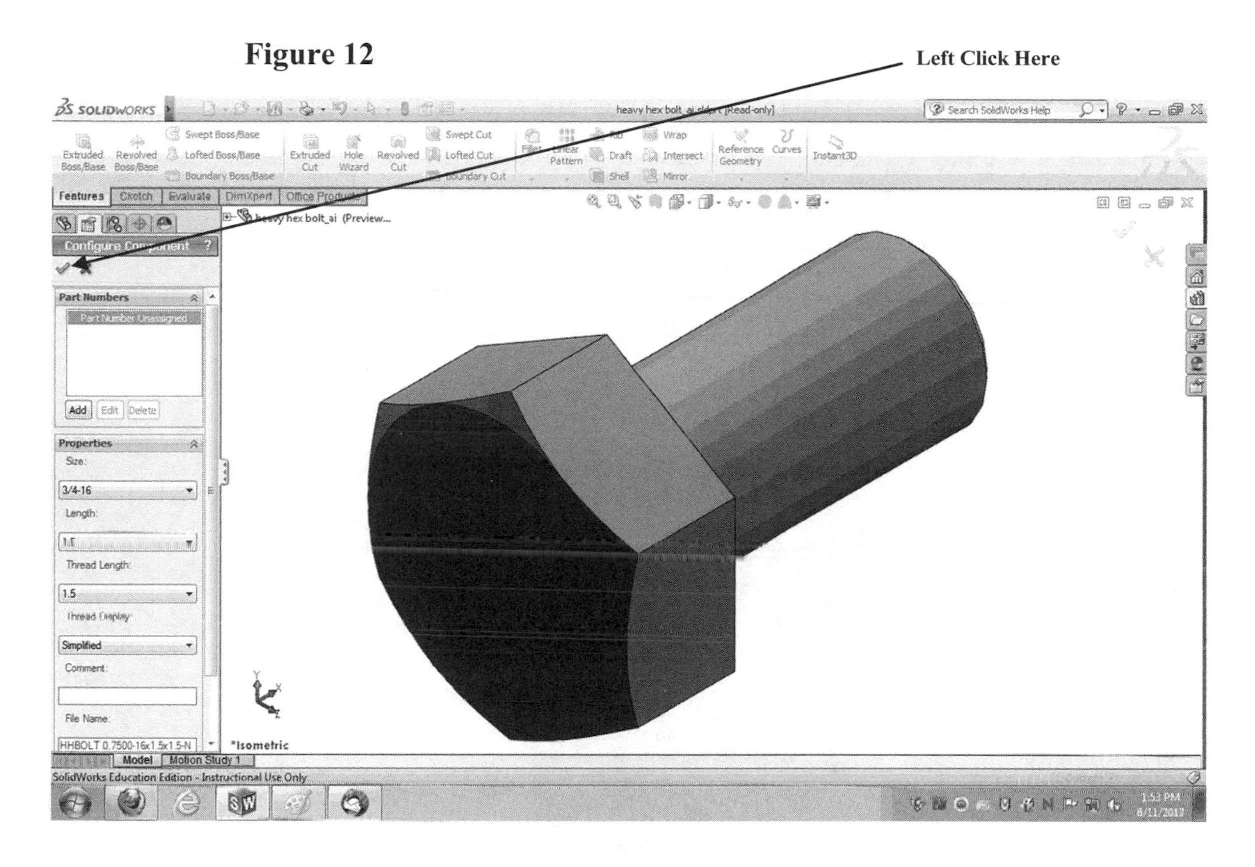

14. SolidWorks will calculate the properties that were entered, and generate a bolt from these properties. The bolt is ¾ inches in diameter with 16 threads per inch and 1 and ½ inches in length as shown in Figure 13. If there is a need, changes to the bolt can be accomplished in the same manner as editing all other existing solid models (Refer to Chapter 5 Editing Existing Solid Models).

Figure 13

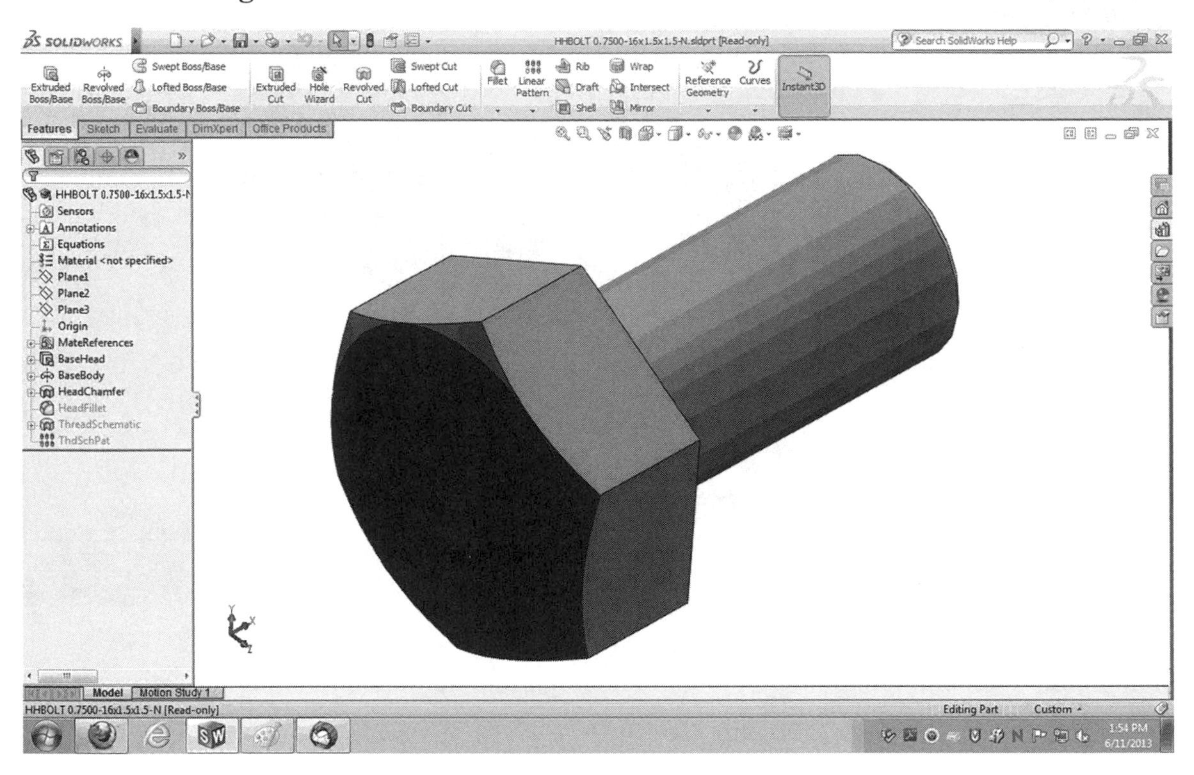

15. Save the bolt for use in a future Assembly drawing file.

16. Import other parts from the Toolbox for additional practice.

Chapter 15 Introduction to the DFMXpress Analysis Wizard

Objectives:

- Learn to create a simple part with the idea of designing for manufacture
- Learn to use the DFMXpress Analysis Wizard

Chapter 15 includes instruction on how to design and evaluate the part shown below.

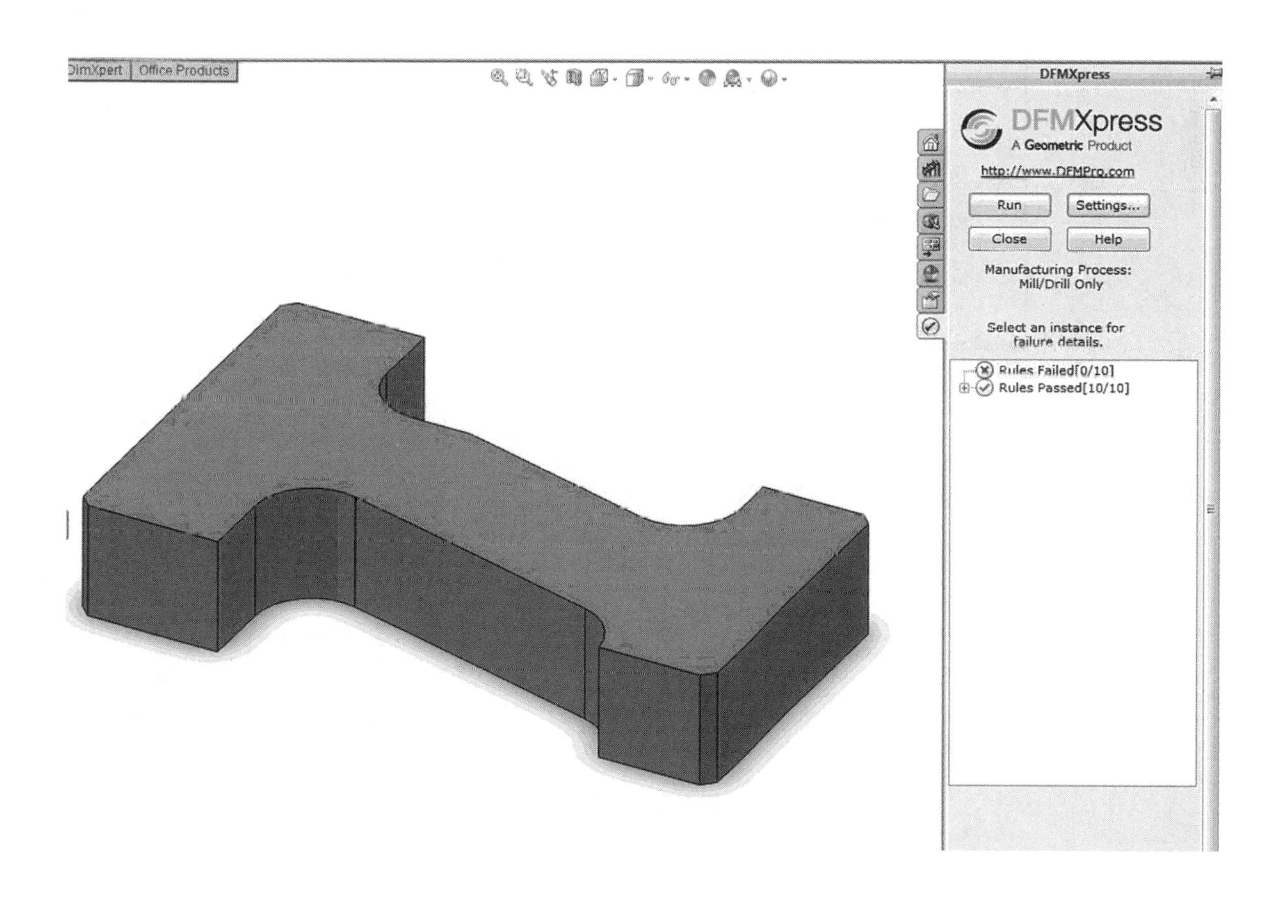

1. Complete the following sketch on the Top Plane and Extrude it to a distance of 1 inch as shown in Figure 1.

Figure 1

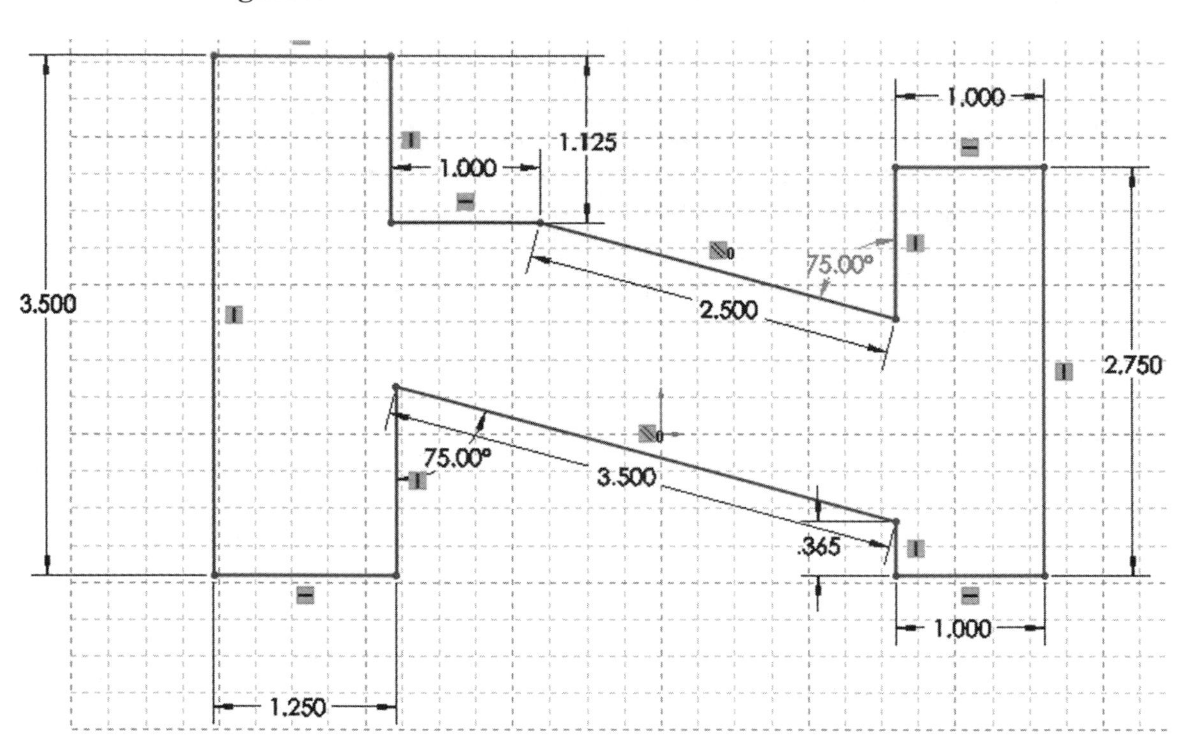

2. Use the **Fillet** command to create .5 inch radius fillets on the outside corners of the part as shown in Figure 2. Save the file where it can be retrieved later.

Figure 2

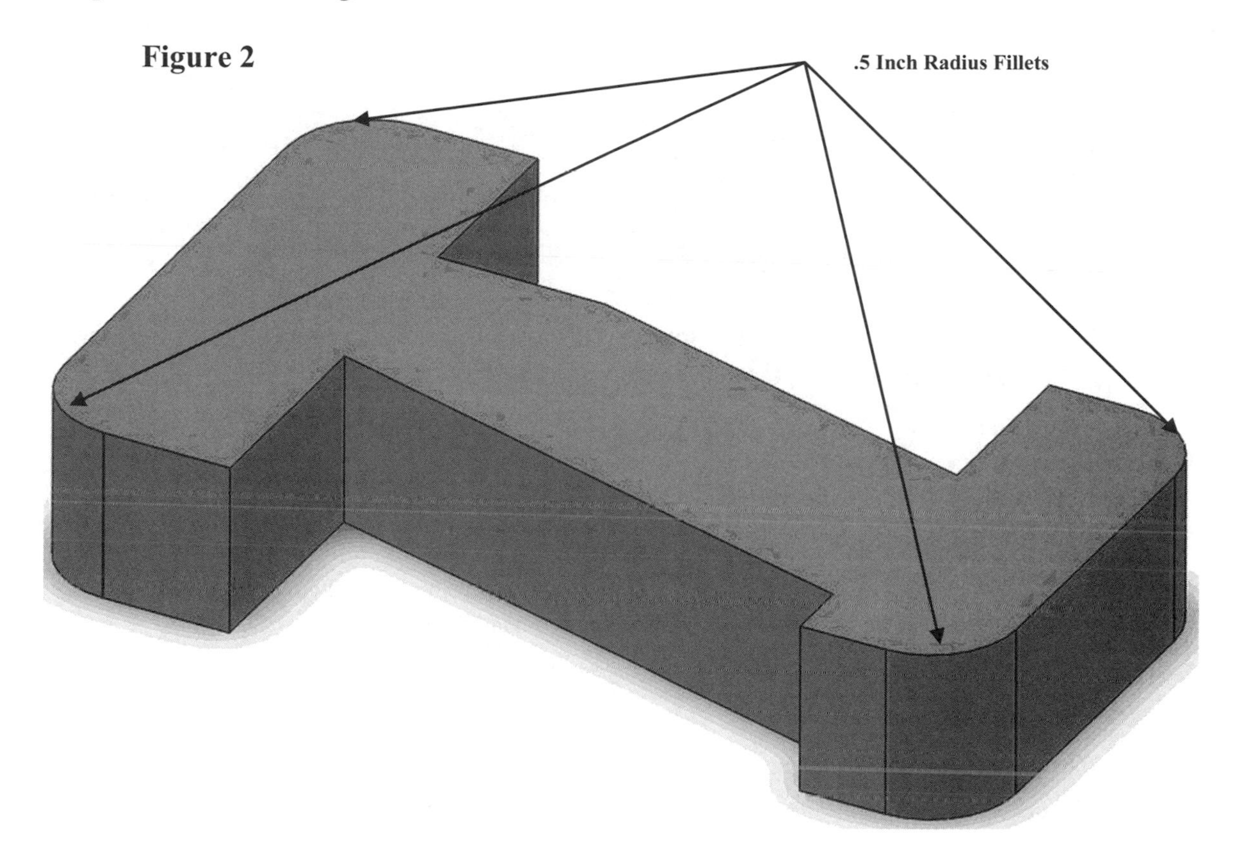

3. Move the cursor to the upper left portion of the screen and left click on the **Evaluate** tab as shown in Figure 3.

Figure 3

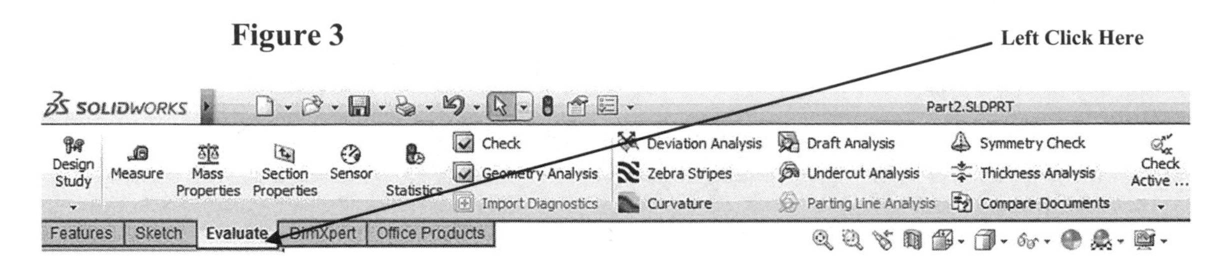

4. Move the cursor to the upper right portion of the screen and left click the on the **DFMXpress Analysis Wizard** icon as shown in Figure 4.

Figure 4

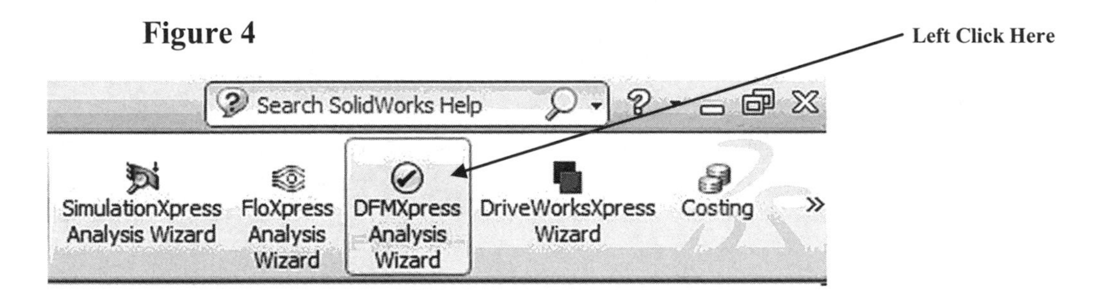

5. The DFMXpress Analysis Wizard will appear as shown in Figure 5.

Figure 5

6. Left click on **Run** as shown in Figure 6. The DFMXpress Analysis Wizard will begin evaluating the part for manufacture.

Figure 6

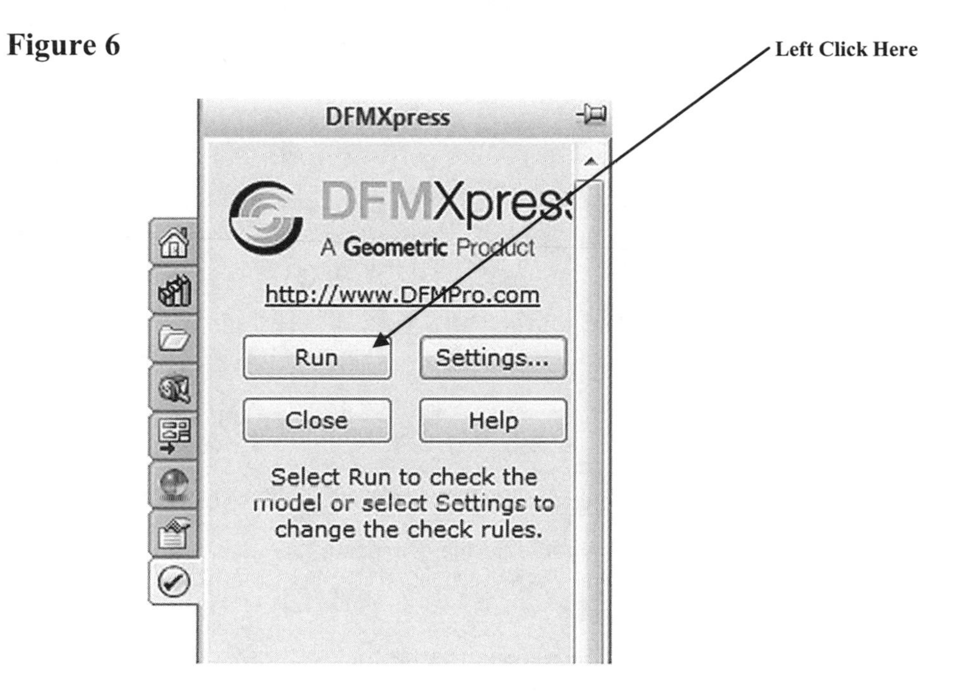

7. Once the evaluation is complete the DFMXrpress Analysis Wizard will reveal any problems that might occur with the manufacture of the part. Left click on the "plus" signs to show a complete evaluation. The evaluation was performed for milling and drilling processes only as shown in Figure 7.

Figure 7

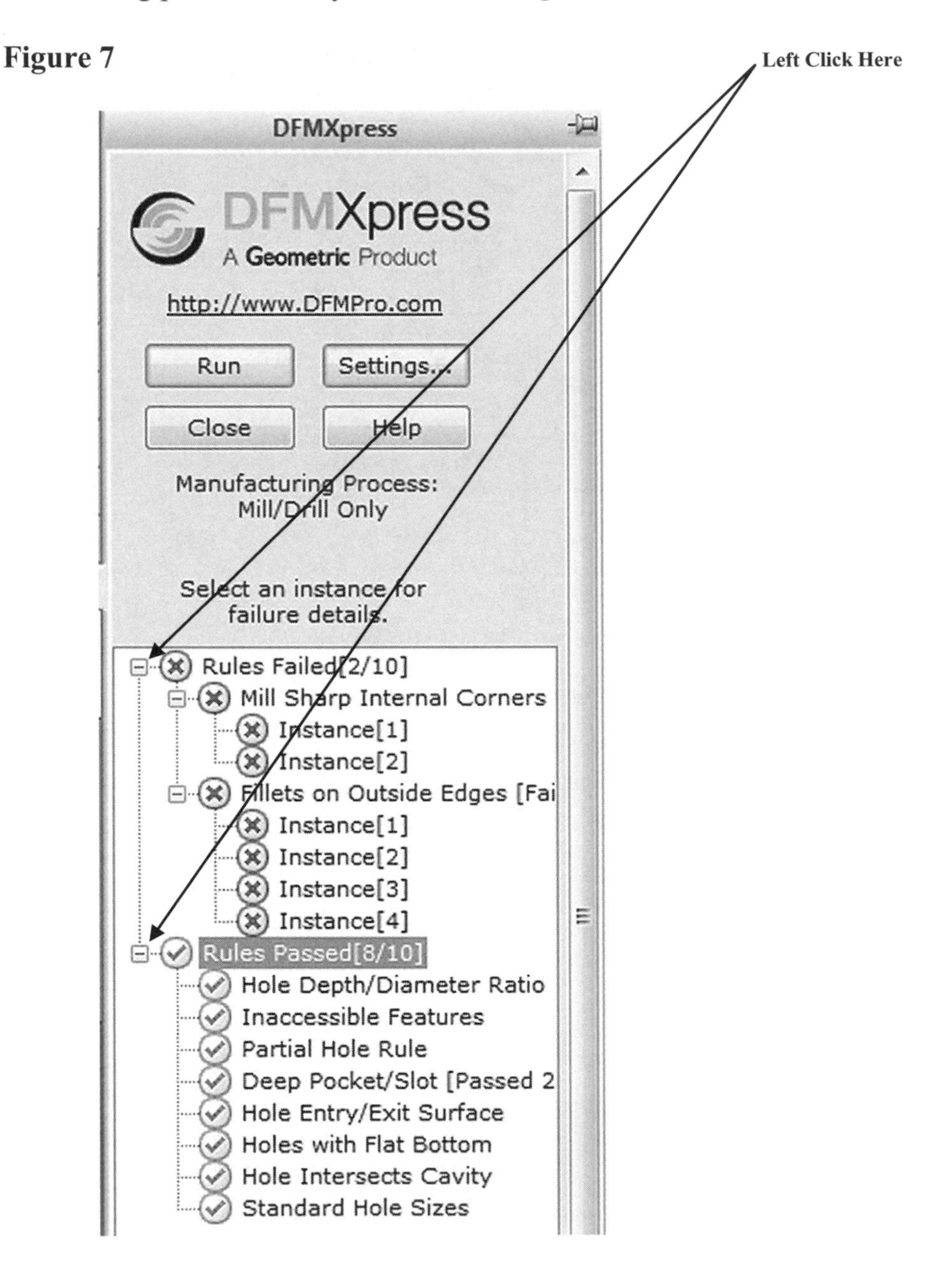

8. Out of 10 Rules the current design failed 2 of them. Move the cursor over one of the "Rules Failed" in the list. A detailed pop up message will appear explaining the problem and a possible solution as shown in Figure 8.

Figure 8

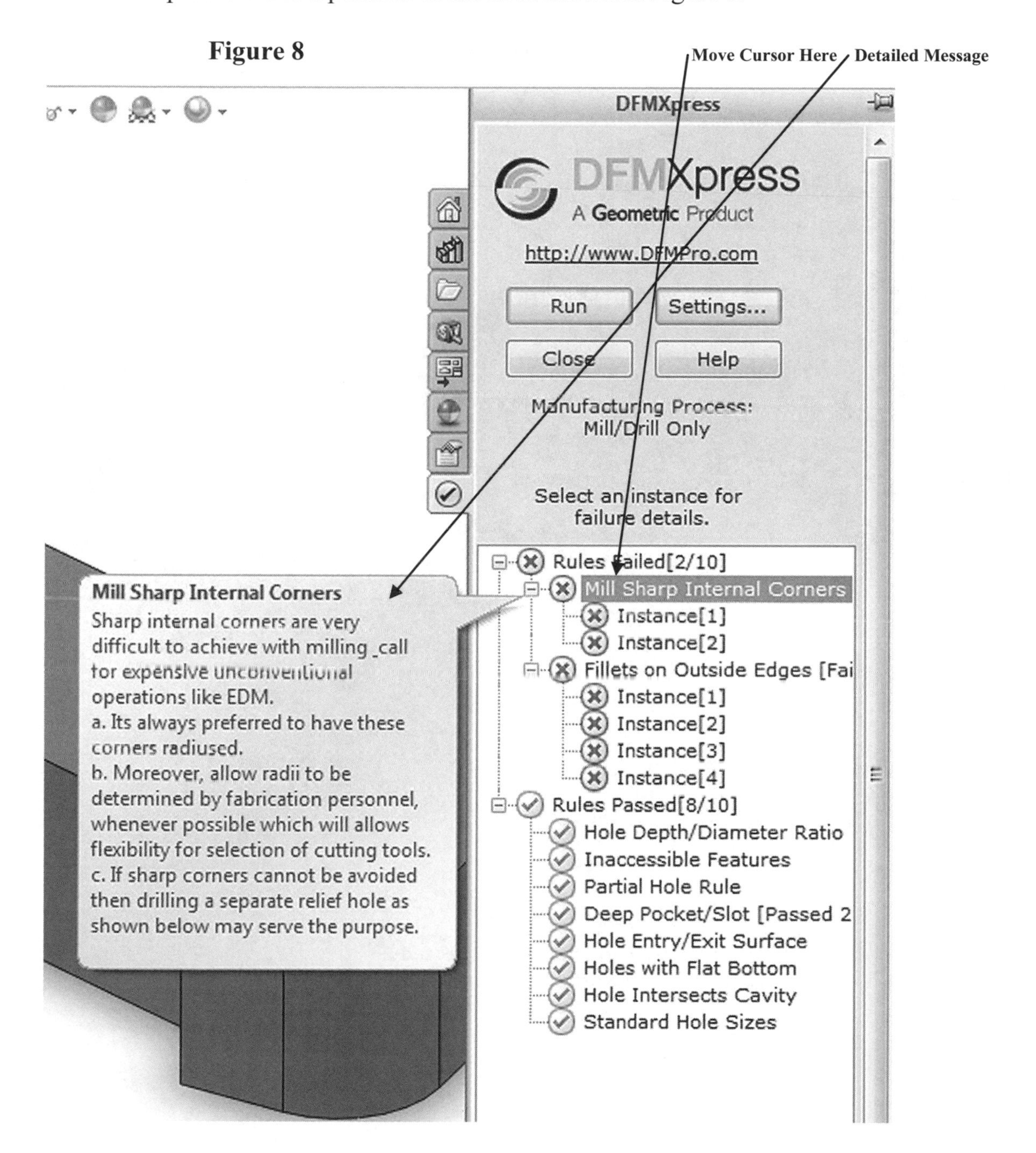

9. To address concerns the DFMXpress Analysis Wizard identified, **c**reate inside fillets with a .5 inch radius. Also delete the outside fillets and replace them with .5 equal distance chamfers (not shown) as shown in Figure 9.

Figure 9

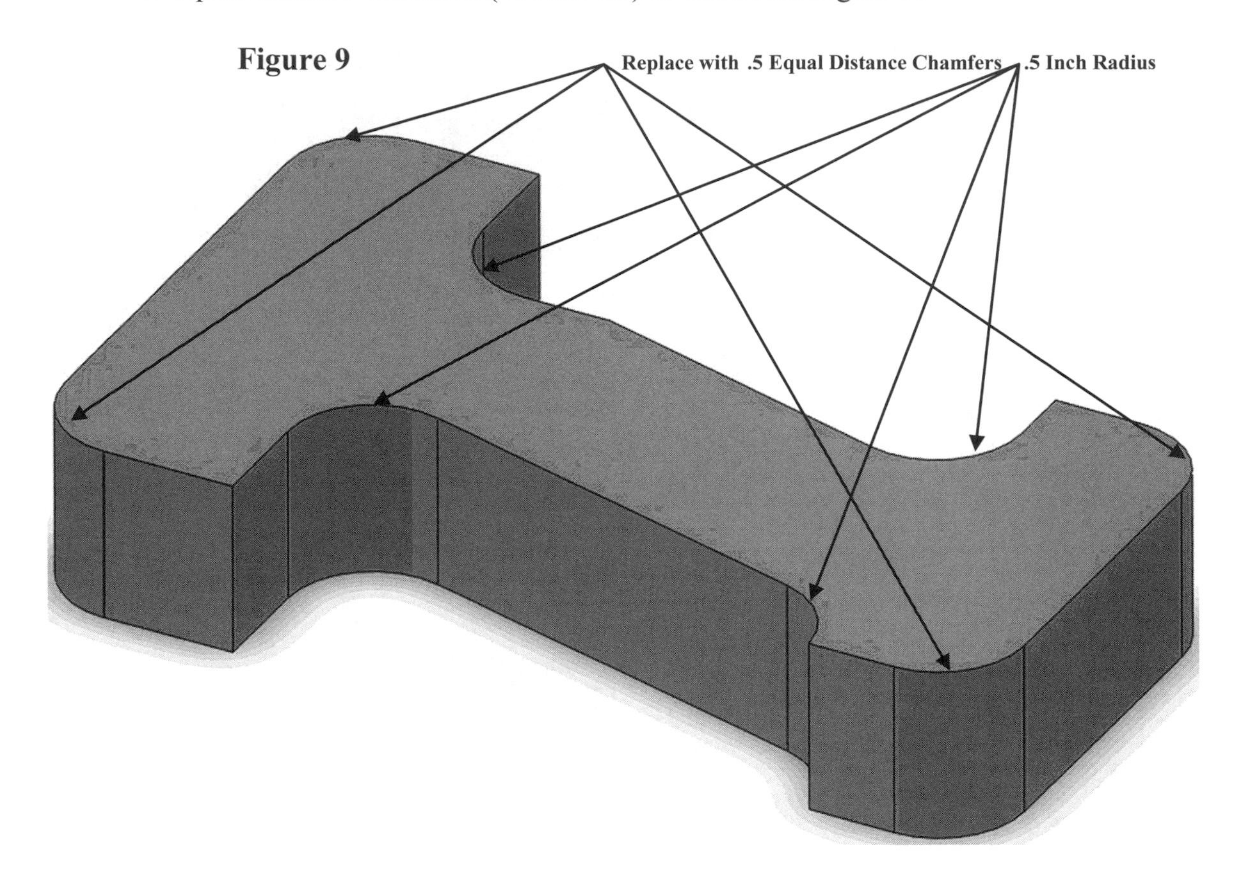

10. Left click on **Run**. All 10 Rules passed the evaluation. The DMFXpress Analysis Wizard found no concerns with the current design as shown in Figure 10.

Figure 10

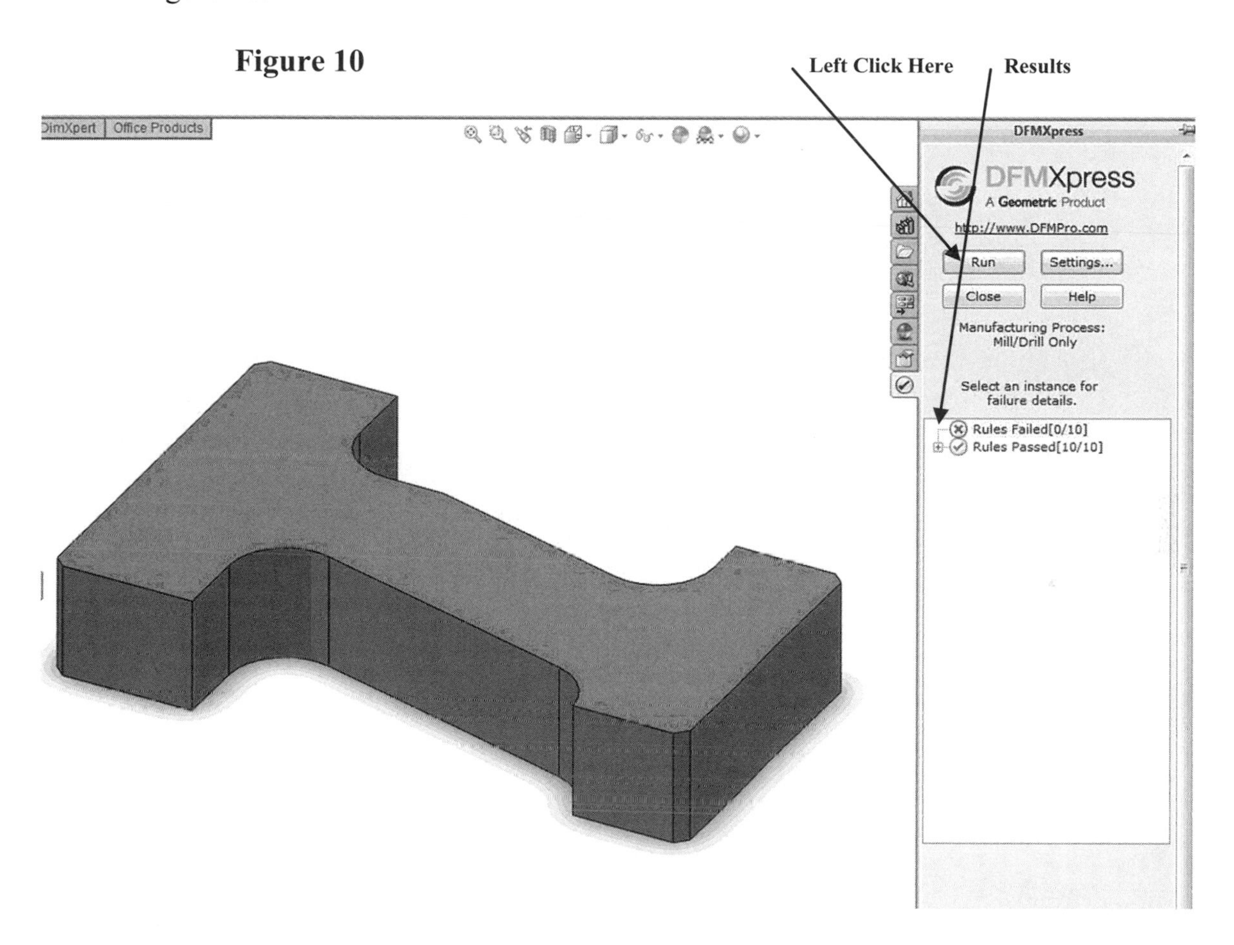

11. Feel free to create other parameters for evaluation. This is done by left clicking on **the Settings/Back button** and creating your own parameters as shown in Figure 11.

Figure 11

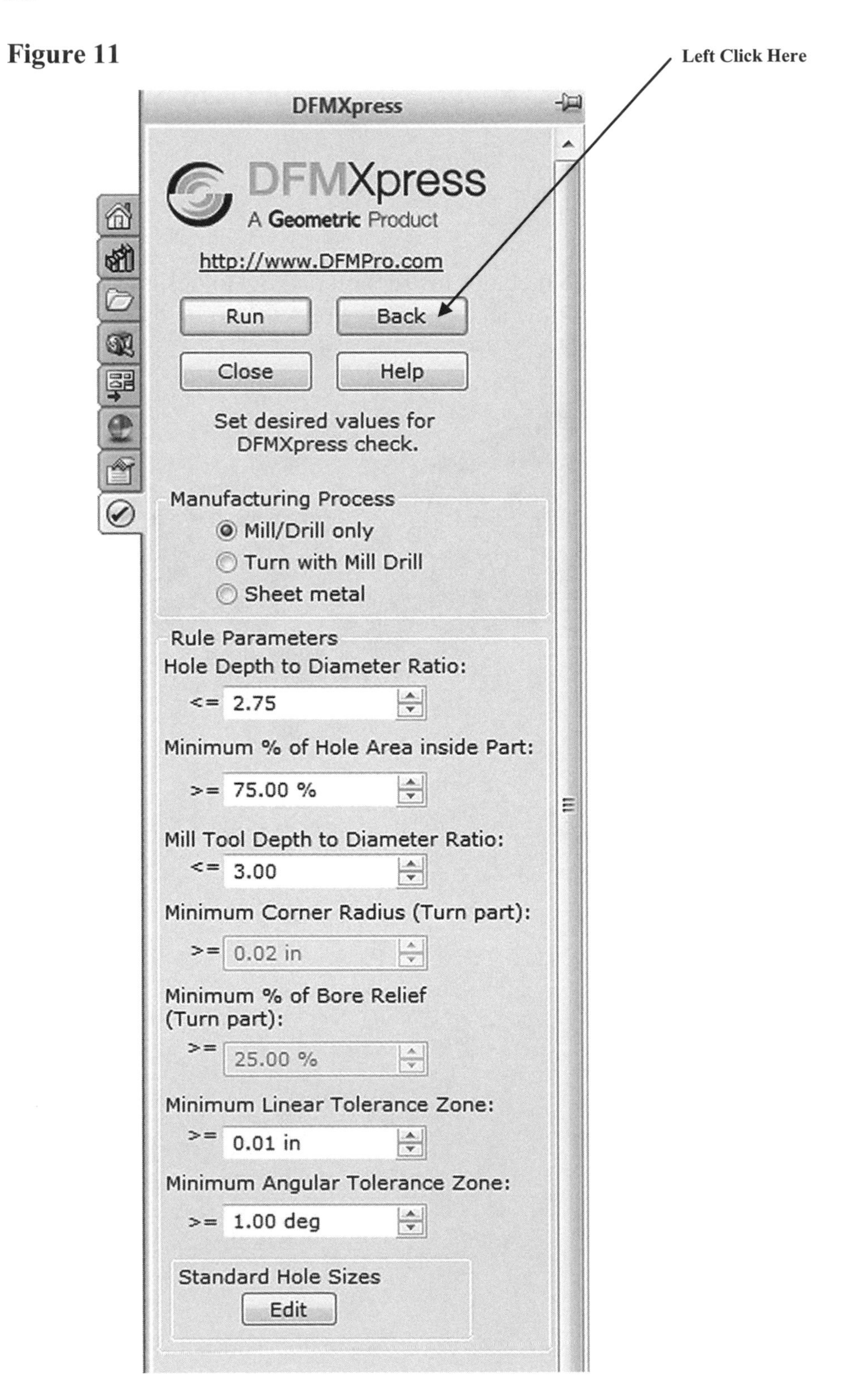

Chapter 16 Introduction to the FloXpress Analysis Wizard

Objectives:

- Learn to create a simple part with the idea of performing a flow analysis
- Learn to use the FloXpress Analysis Wizard

Chapter 16 includes instruction on how to design a simply part and then perform a flow analysis on the part shown below.

.

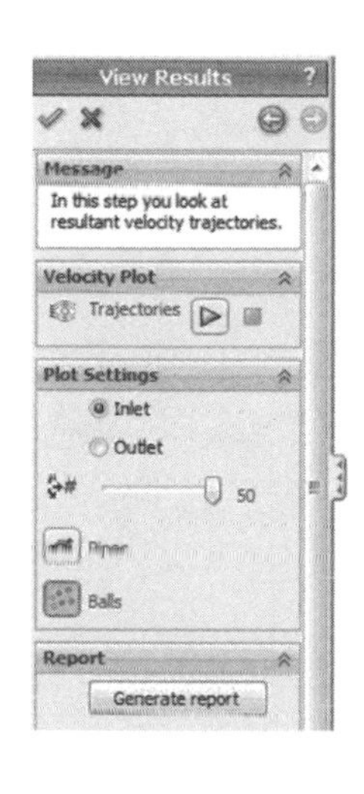

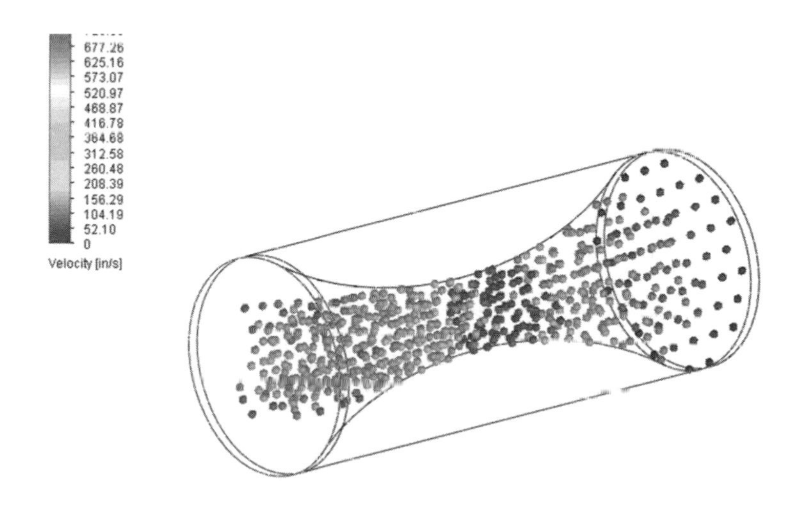

1. Using the 3 Point Arc command, complete the following sketch as shown in Figure 1.

Figure 1

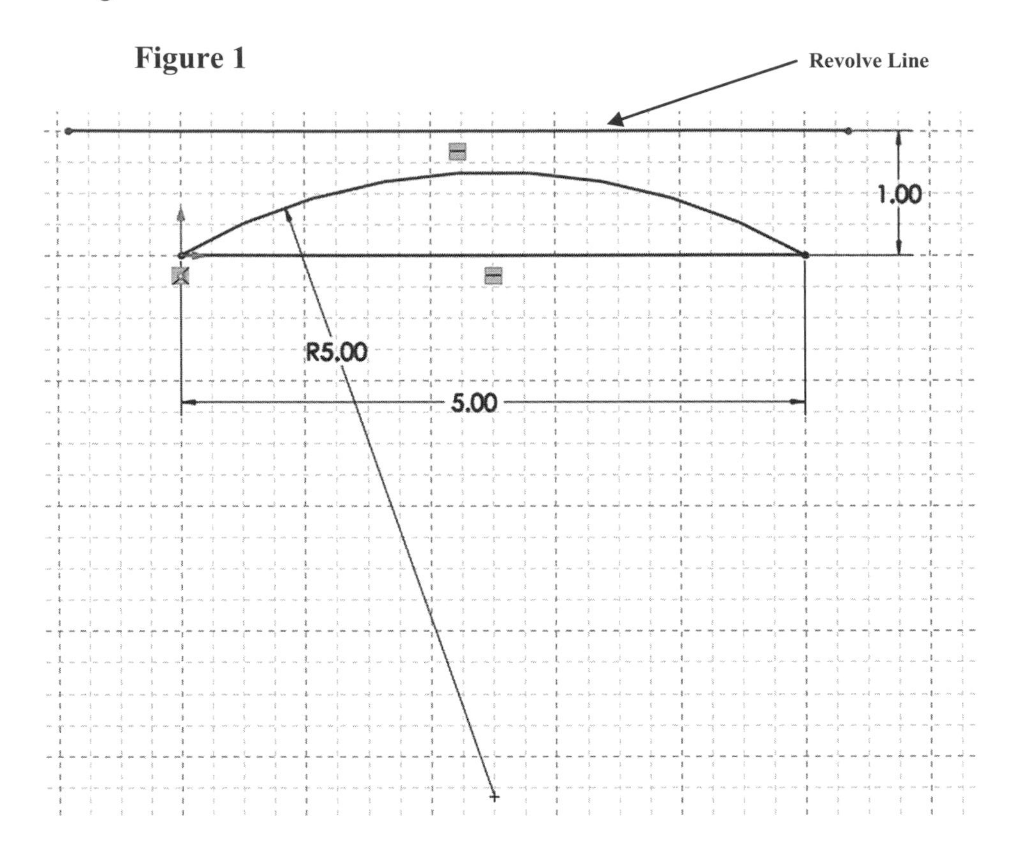

2. Use the **Revolve** command to create a cylinder as shown in Figure 2. Make sure to remove the checkmark from "Thin Feature" if needed.

Figure 2

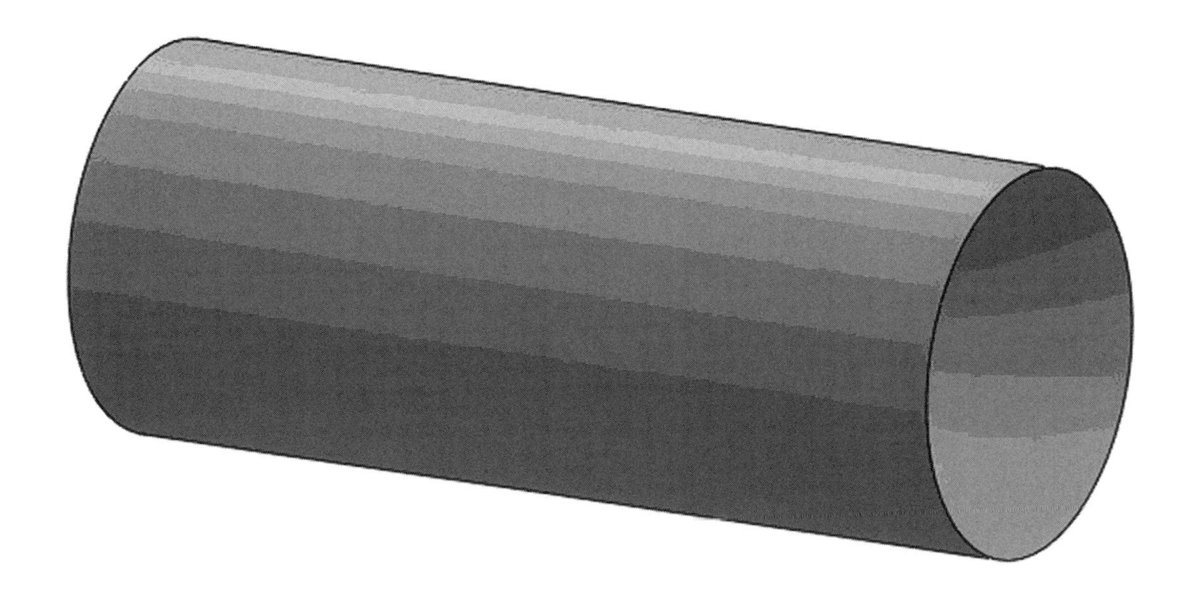

3. Use the Rotate command to rotate the cylinder around. The center of the cylinder should narrow. This is known as a Venturi as shown in Figure 3.

Figure 3

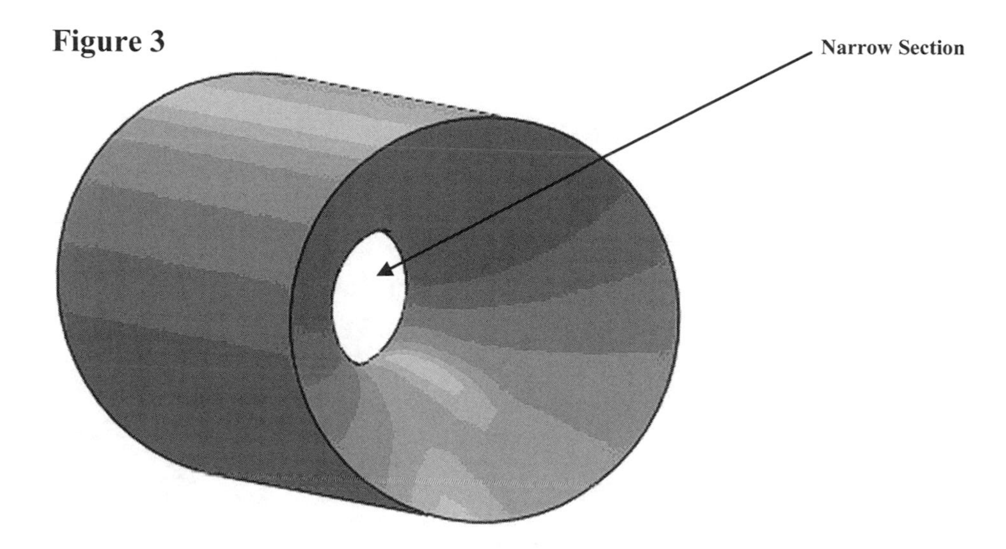

4. Begin a new drawing. Complete the following sketch and extrude to a distance of .010. Once extruded, this part will be known as a Lid as shown in Figure 4.

Figure 4

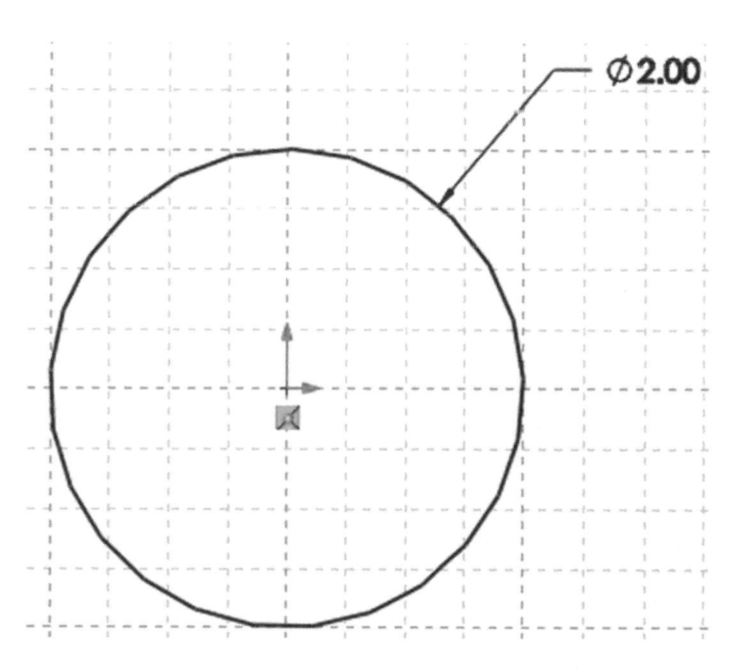

5. Begin a new Assembly drawing. Insert the cylinder and two Lids as shown in Figure 5.

Figure 5

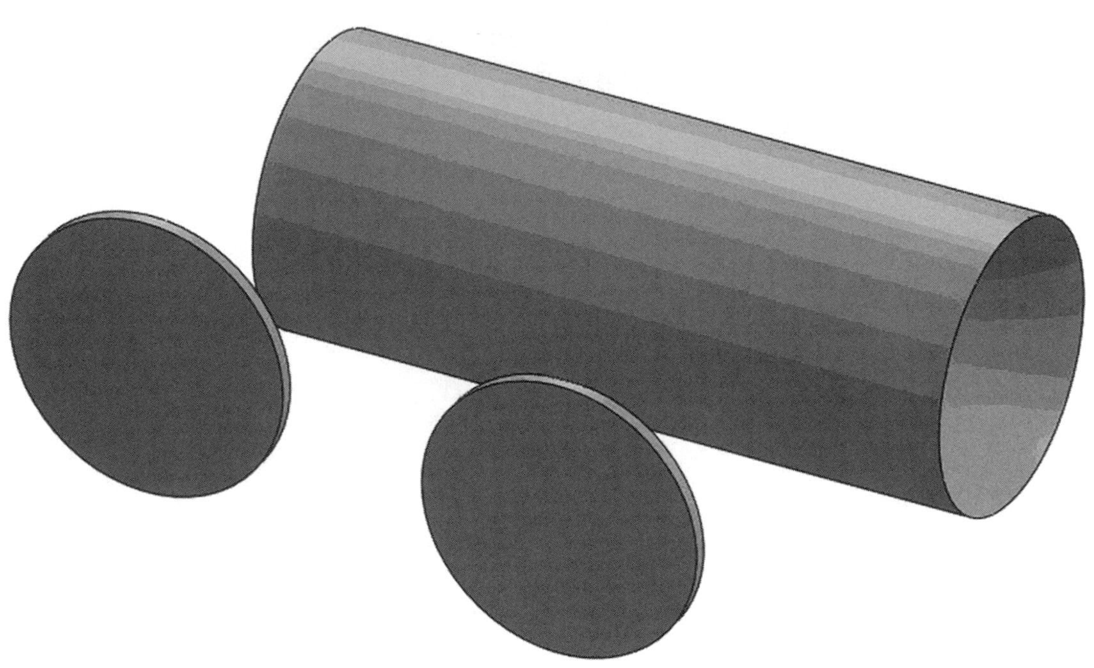

6. Mate the edges of the Lids to the edge of each end of the cylinder as shown in Figure 6. The diameter of the Lids must match the diameter of the cylinder. If the Lids will not Mate to the cylinder, the diameter of the Lids is incorrect.

Figure 6

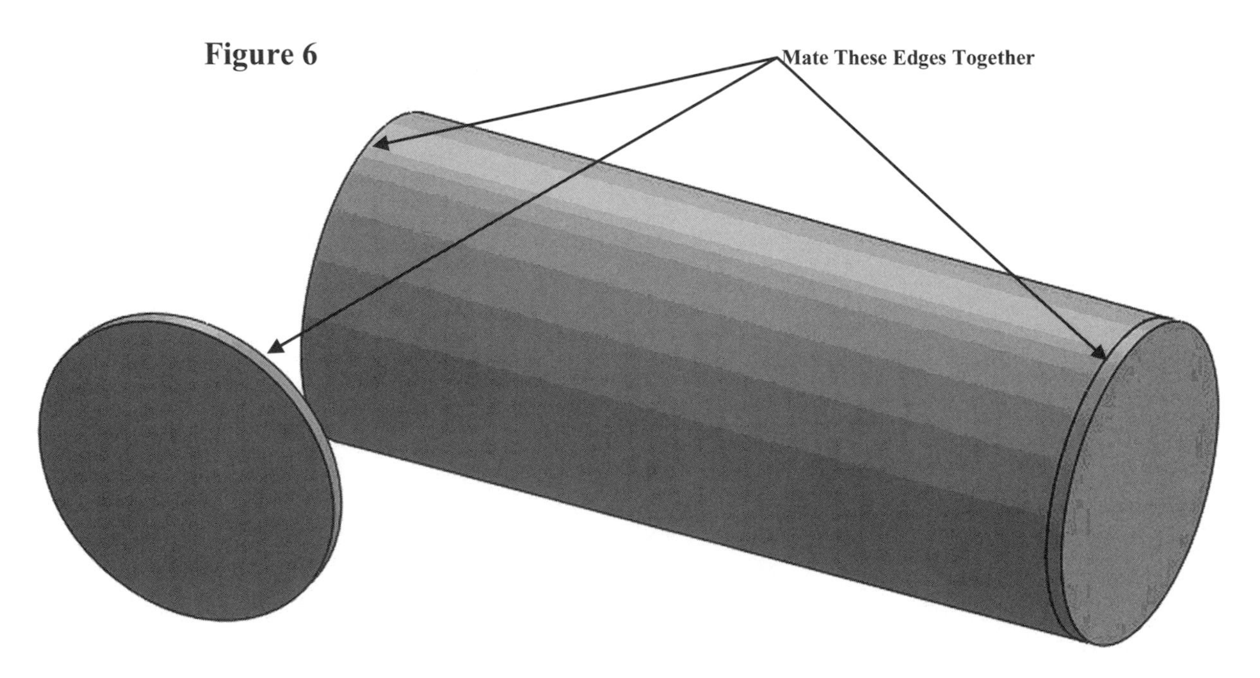

7. When fully Mated the Lids should cap off each end as shown in Figure 7.

Figure 7

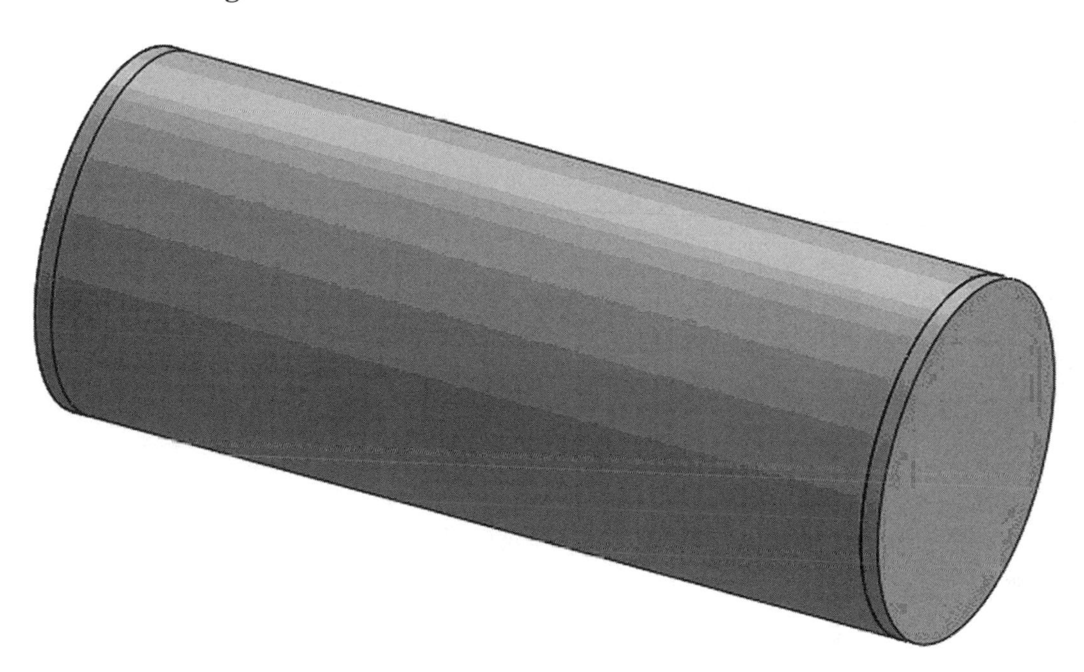

8. Change the view of the assembly to Wireframe as shown in Figure 8.

Figure 8

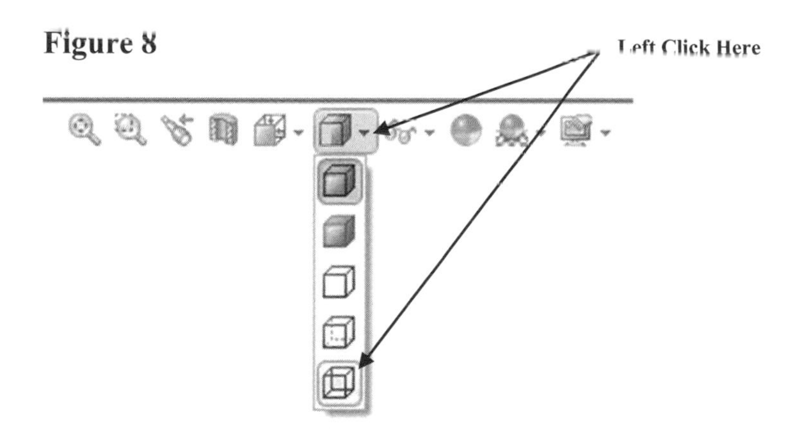

9. Left click on the **Evaluate** tab as shown in Figure 9.

Figure 9

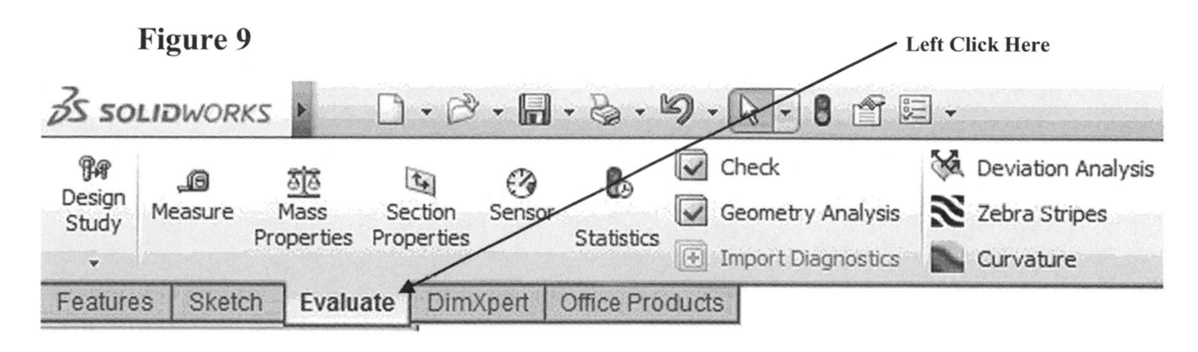

10. Left click on the **FloXpress Analysis Wizard** as shown in Figure 10.

Figure 10

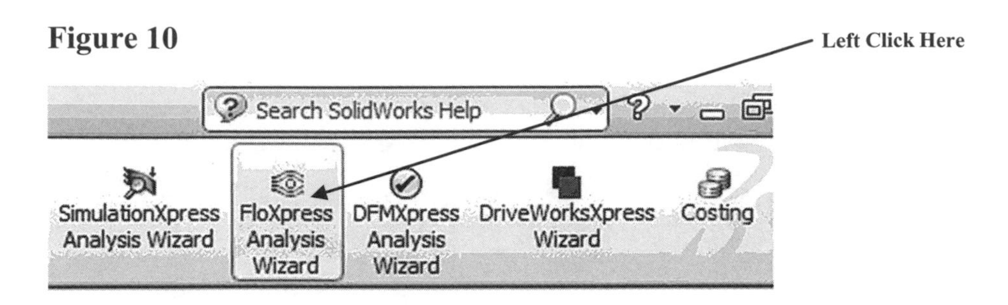

11. The FloXpress Analysis Wizard will begin running. The Welcome dialog box will appear as shown in Figure 11.

Figure11

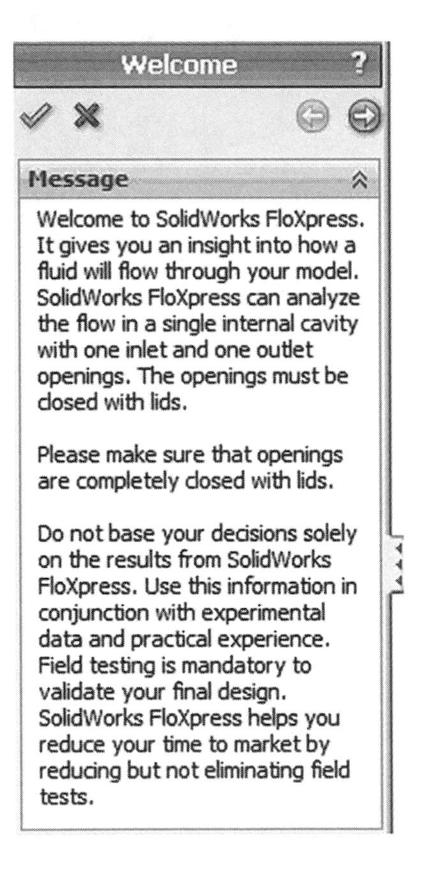

12. The Check Geometry dialog box will appear. Left click on the Next arrow at the right as shown in Figure 12.

Figure 12

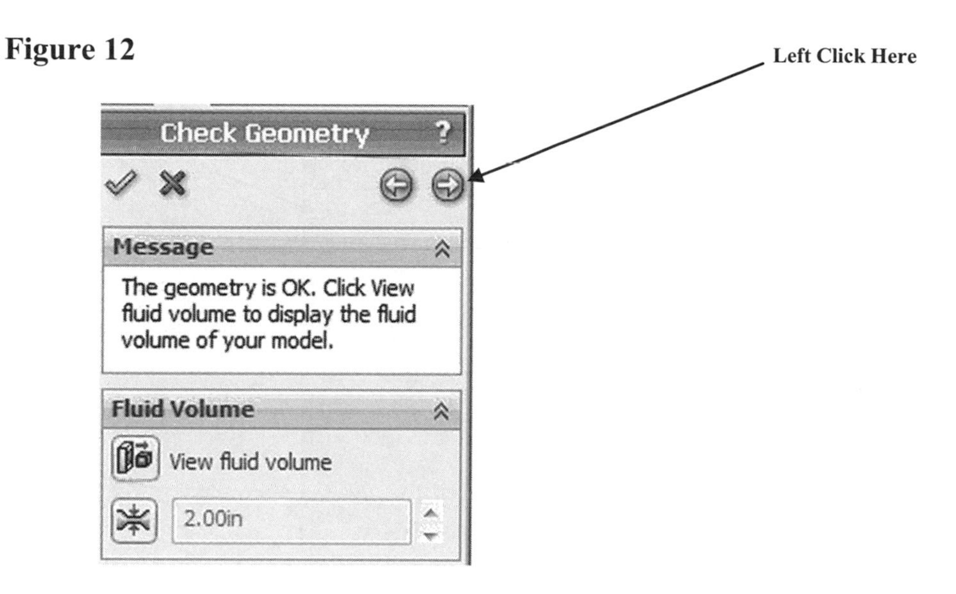

13. The Fluids dialog box will appear. Left click next to **Water** and then left click on the Next arrow at the right as shown in Figure 13.

Figure 13

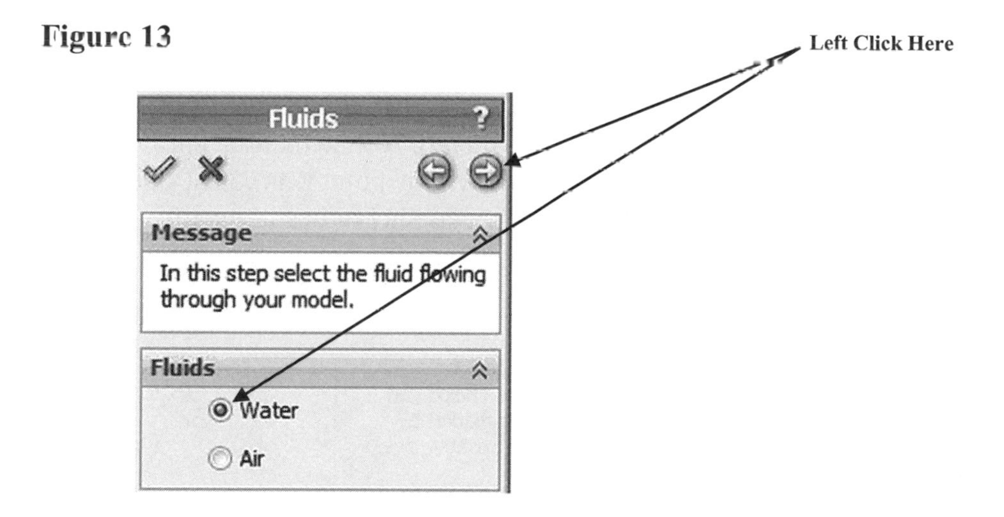

14. The Flow Inlet pressure dialog box will appear. Use the default of **Pressure** and the default inlet pressure of 14.6959473 lbf/in^2 or 1 Atmosphere. Left click on the Next arrow at the right as shown in Figure 14.

Figure 14

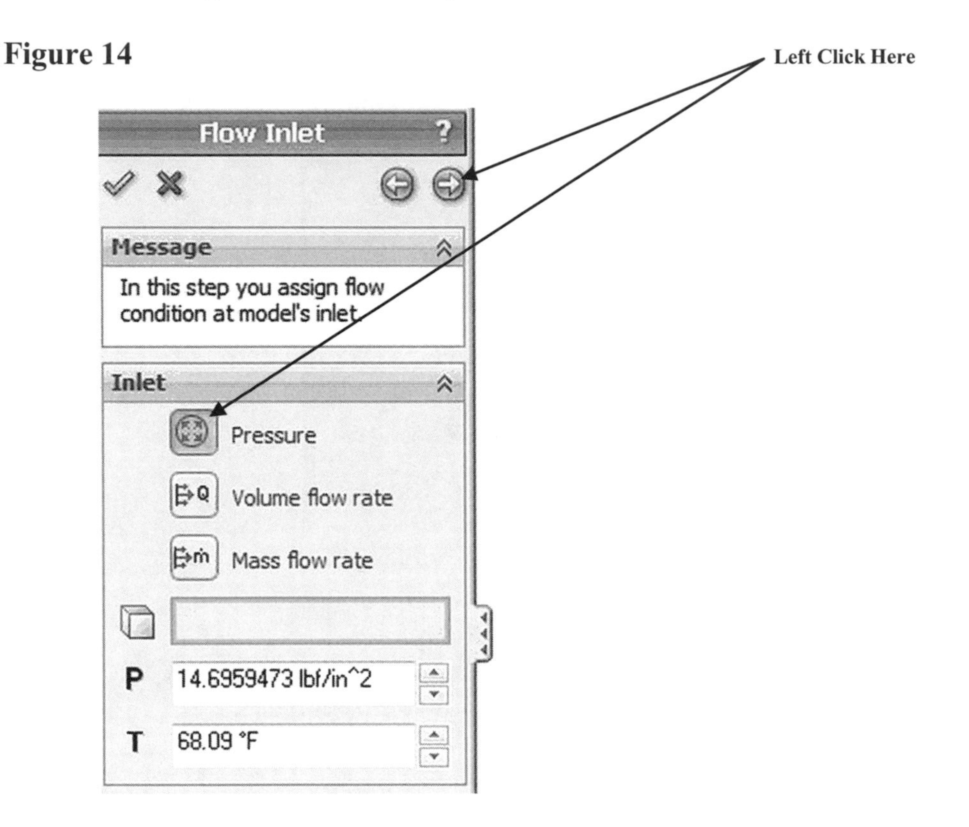

15. Left click on the inside of the right lid. To do this you will need to move your cursor over the inside center of the lid. Notice that the entire cylinder will become highlighted. Ignore this for now. At this point right click once. A popup menu will appear. Left click on **Select Other** as shown in Figure 15.

Figure 15

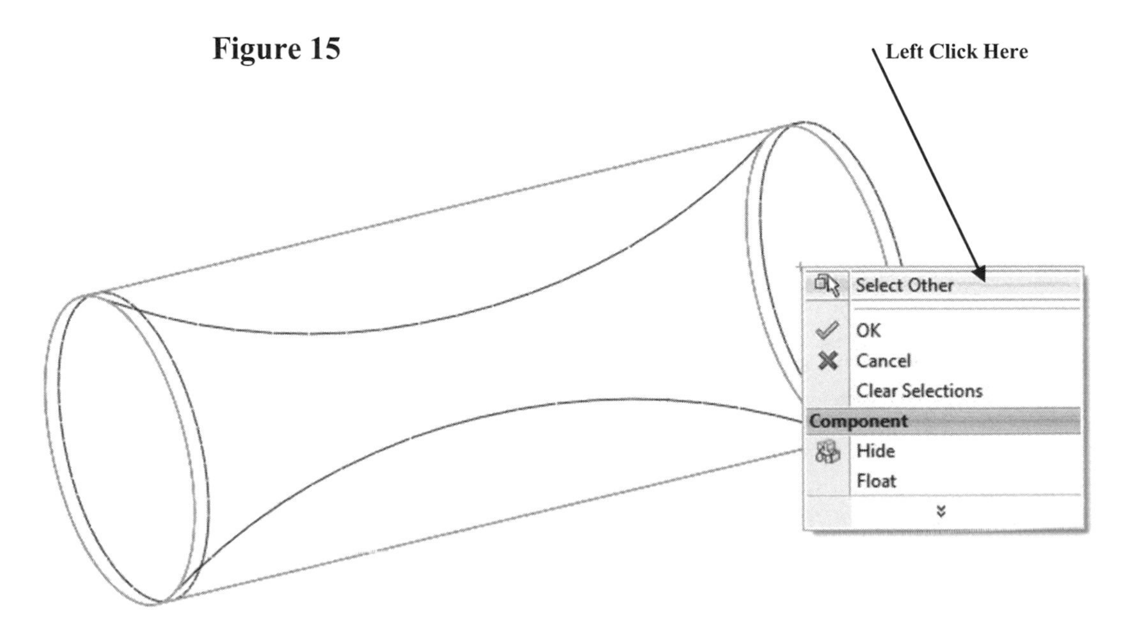

16. A popup menu will appear. Left click on **Face@Boss-Exturde1**, also known as the inside of the lid. The inside edges of lid must become highlighted as shown in Figure 16.

Figure 16

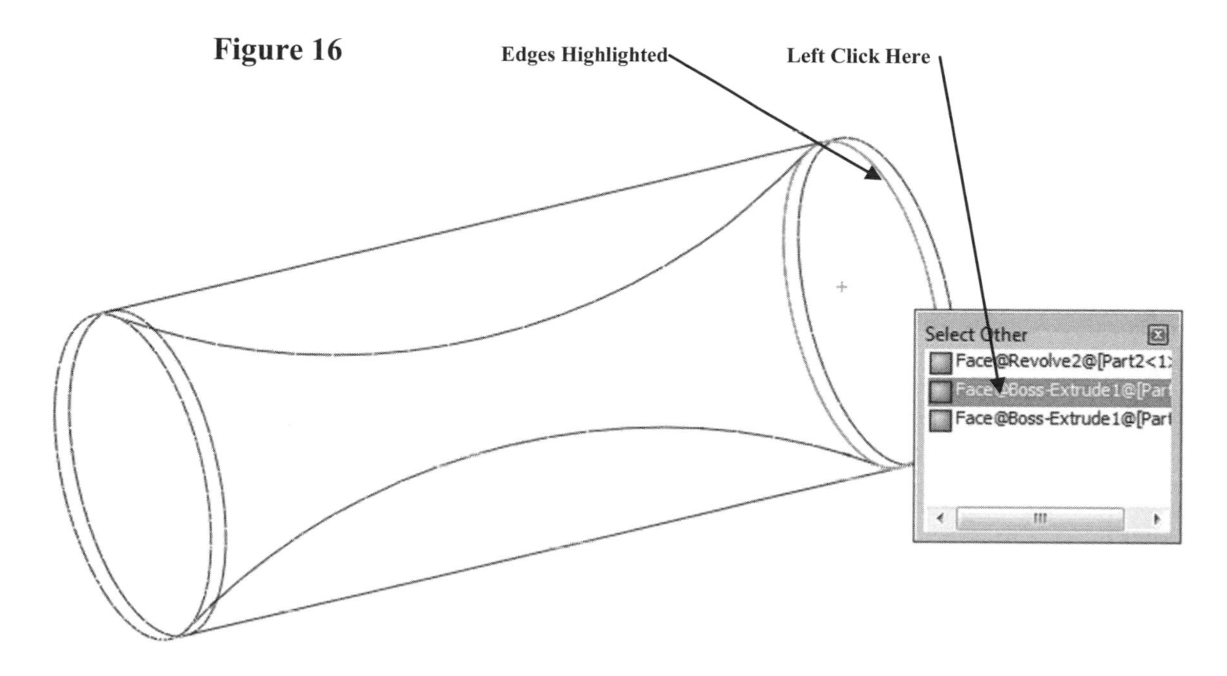

17. Once the inside of the lid has been selected, arrows will appear as shown in Figure 17. If arrows do not appear, you will need to repeat the process before the analysis can be performed.

Figure 17

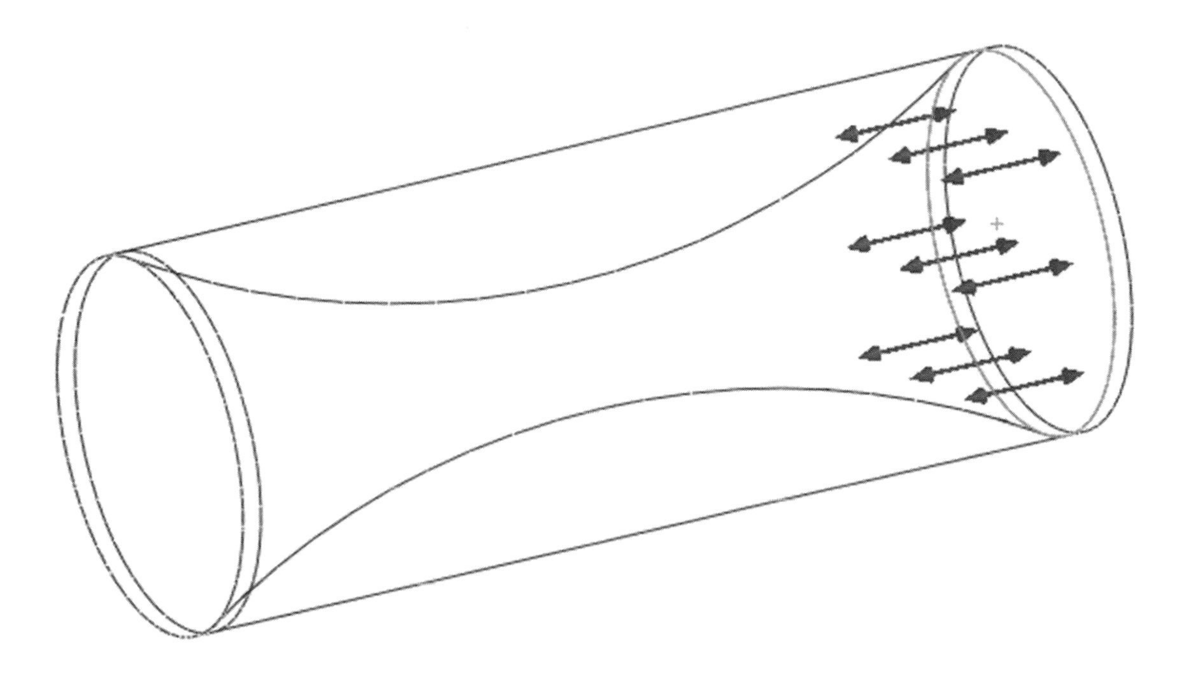

18. Next, the Flow Outlet dialog box will appear. Because there must be a difference between the Flow Inlet pressure and Flow Outlet pressure, simply delete the 1 from the Outlet Pressure value of 14.6959473 leaving 4.6959473 as shown in Figure 18. This will ensure adequate pressure difference between the Flow Inlet and Flow Outlet.

Figure 18

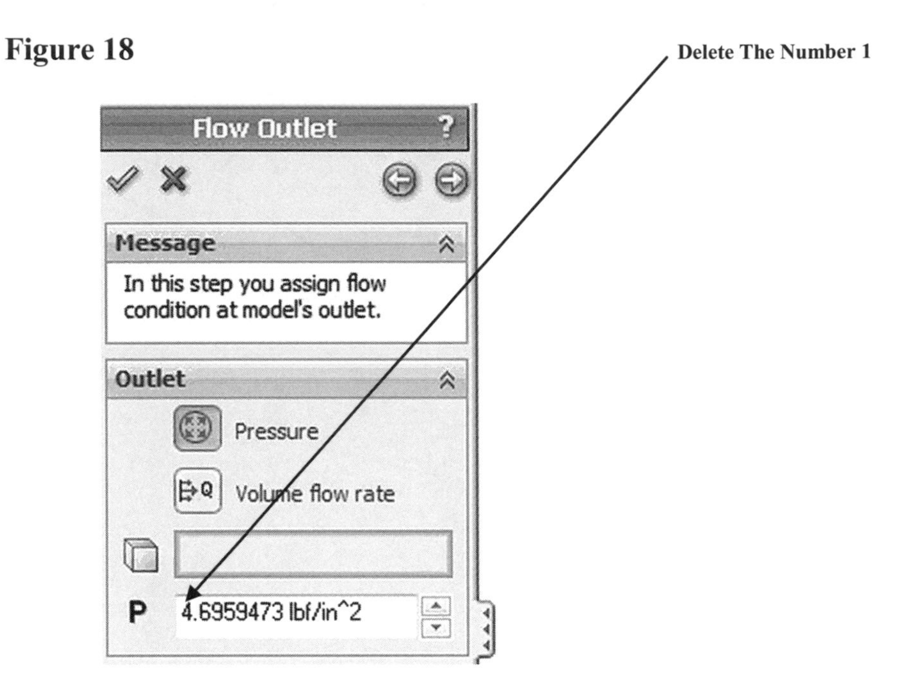

19. Rotate the cylinder around to gain access to the inside of the left lid. Left click on the inside of the left lid. To do this move your cursor over the inside center of the lid. Notice that the entire cylinder will become highlighted. At this point right click once. A popup menu will appear. Left click on **Select Other** as shown in Figure 19.

Figure 19

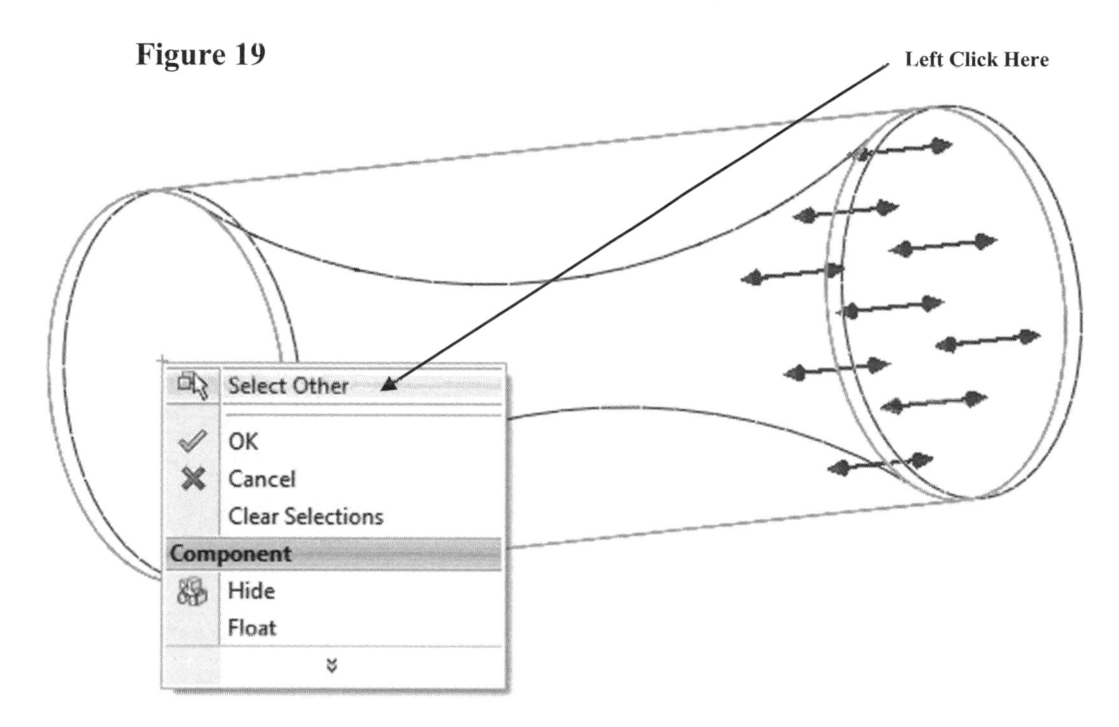

20. A popup menu will appear. Left click on **Face@Boss-Exturde1**, also known as the inside of the lid. The <u>inside</u> edges of lid must become highlighted as shown in Figure 20.

Figure 20

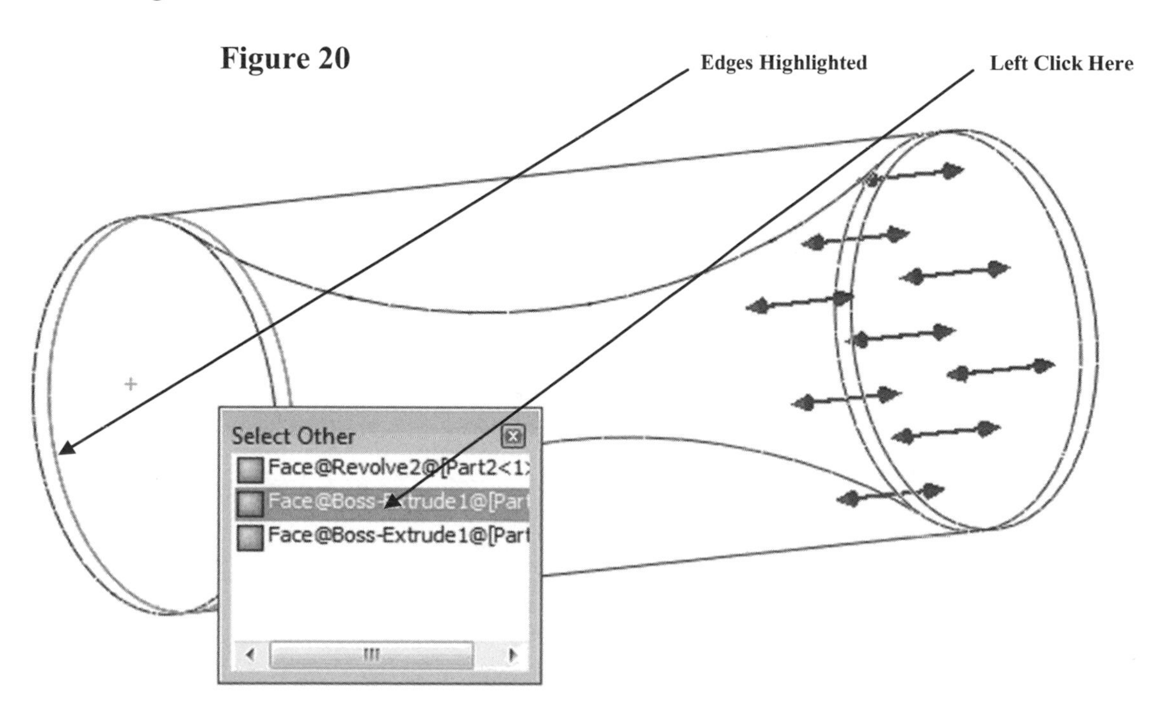

21. Once the inside of the lid has been selected, arrows will appear as shown in Figure 21. If arrows do not appear, you will need to repeat the process before the analysis can be performed.

Figure 21

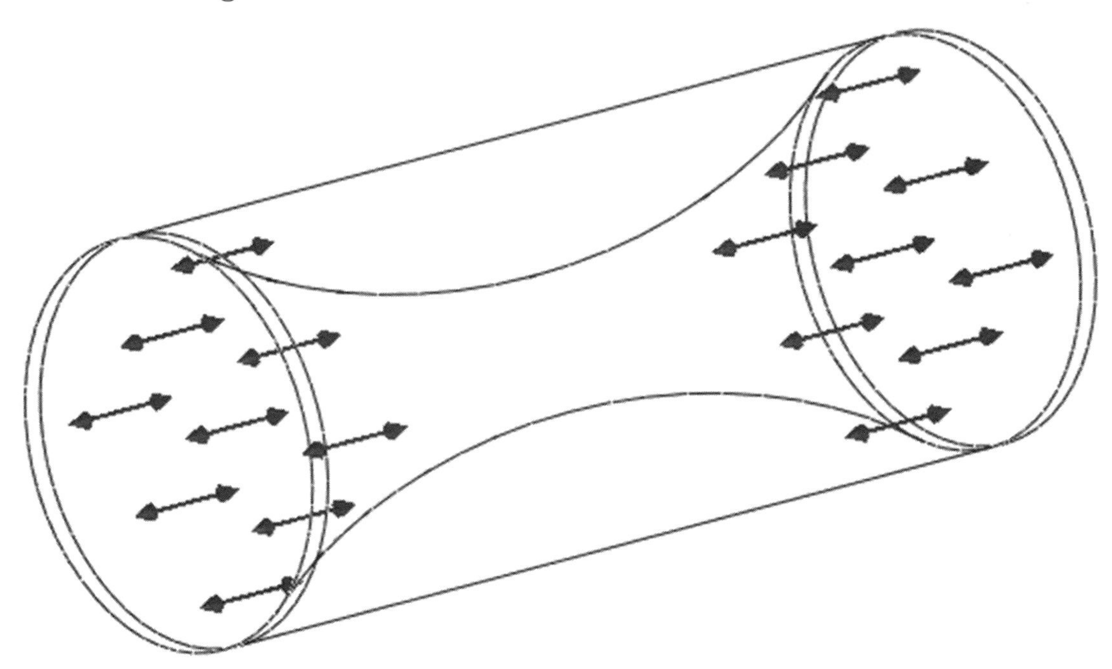

22. The Solve dialog box will appear. Left click on the Play icon to begin the analysis as shown in Figure 22.

Figure 22

Left Click Here

Solve

Message

In this step you solve the model.

Solve

23. SolidWorks will begin the flow analysis. It will take several minutes for the analysis to be completed as shown in Figure 23.

Figure 23

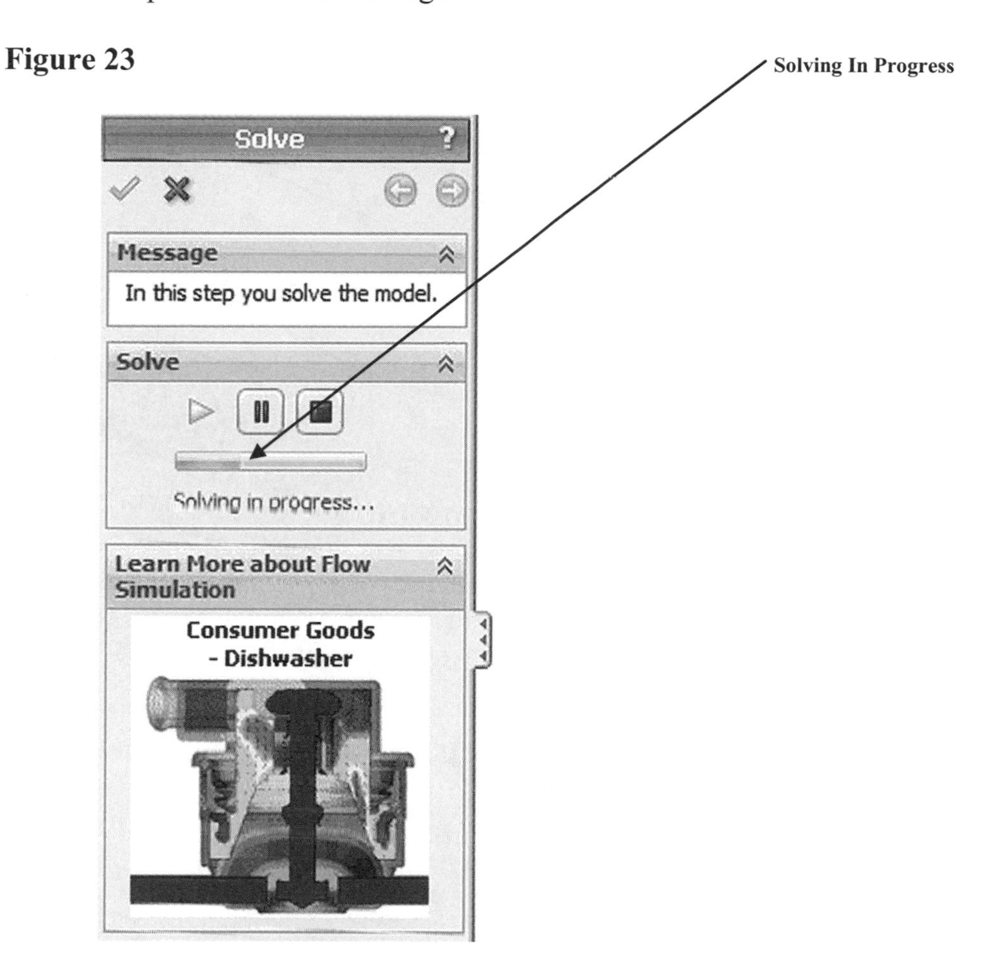

24. Once SolidWorks has completed the Flow analysis a plot will be displayed. Left click on the Play icon to the right of Trajectories. SolidWorks will illustrate the flow analysis as shown in Figure 24.

Figure 24

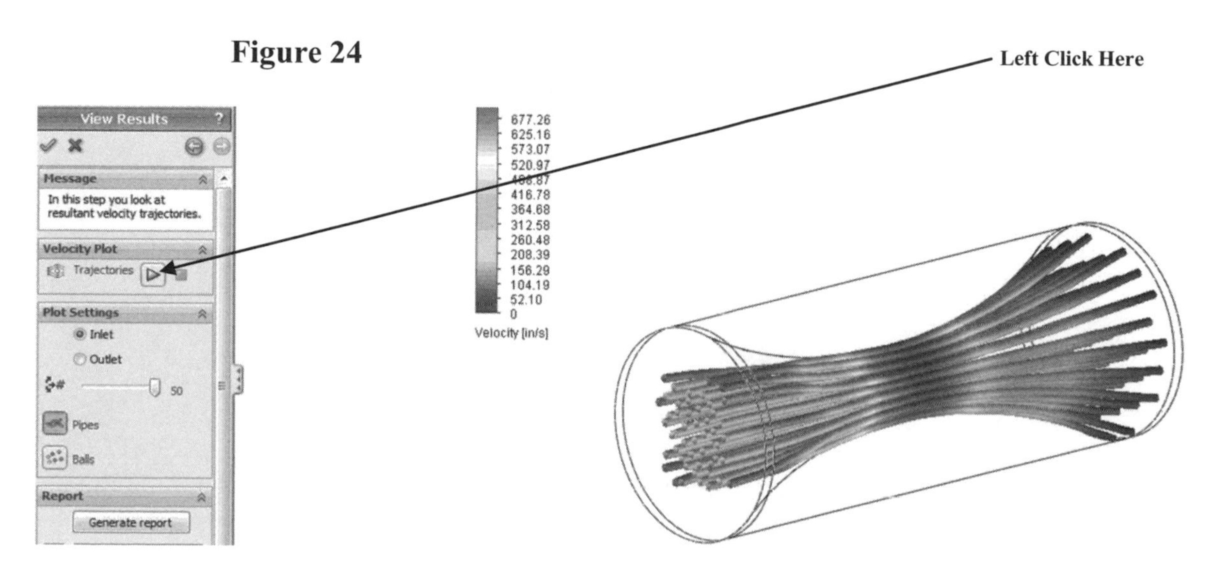

25. To change the Plot Settings, Left click on the Stop icon. Then left click on the Balls icon as shown in Figure 25.

Figure 25

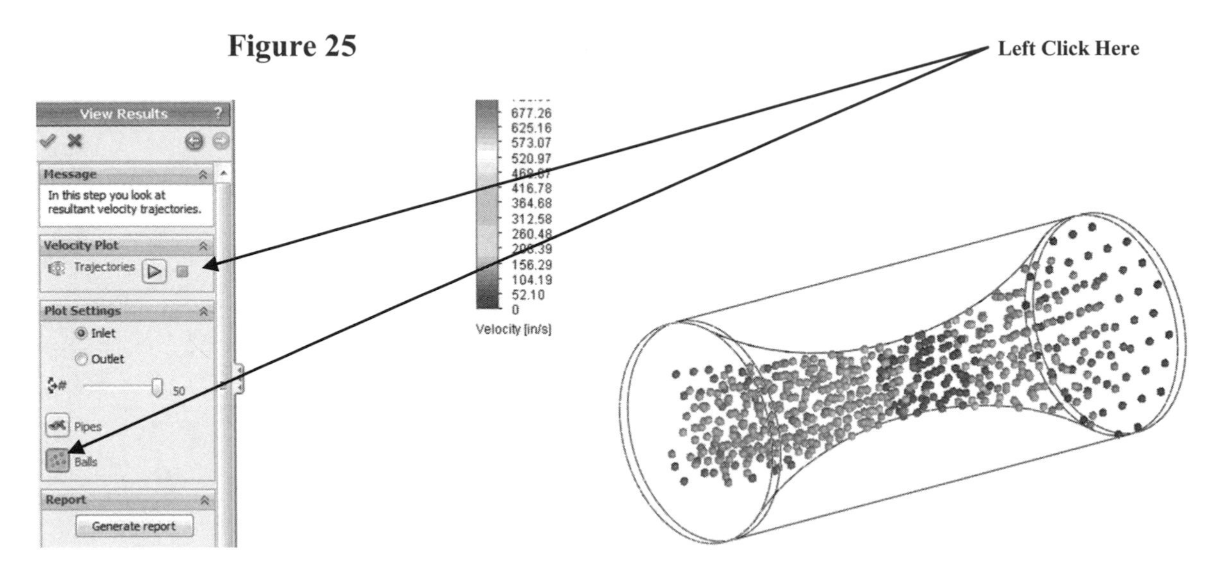

26. The plot illustrates velocity in inches per second. Changes to the inside of the cylinder will increase or decrease the velocity.

Chapter 17 Introduction to Equations

Objectives:

- Learn to create a simple part with the idea of creating simple equations

Chapter 17 includes instruction on how to design a simply part and then create simply equations for the part shown below.

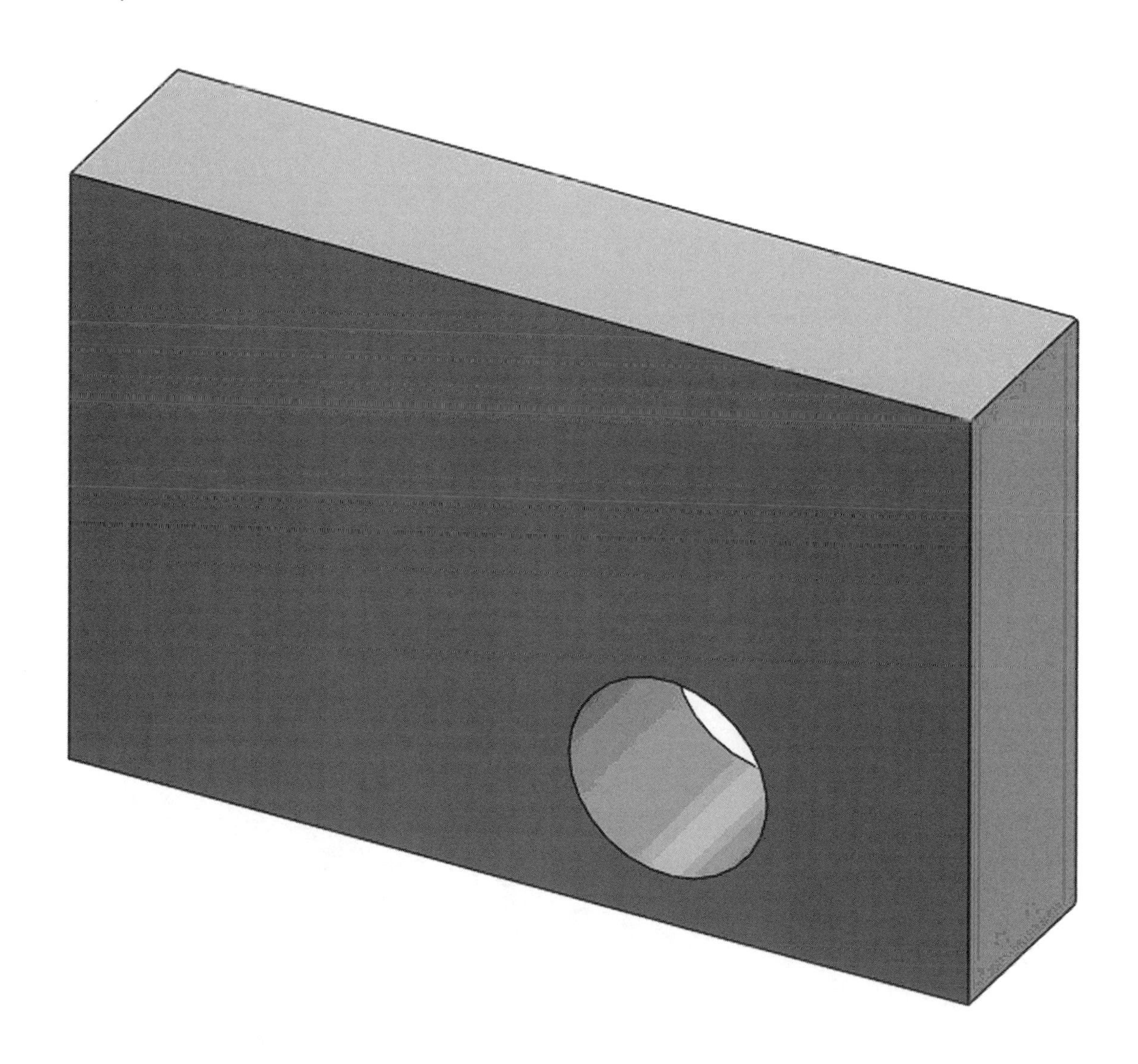

1. Complete the following sketch as shown in Figure 1.

Figure 1

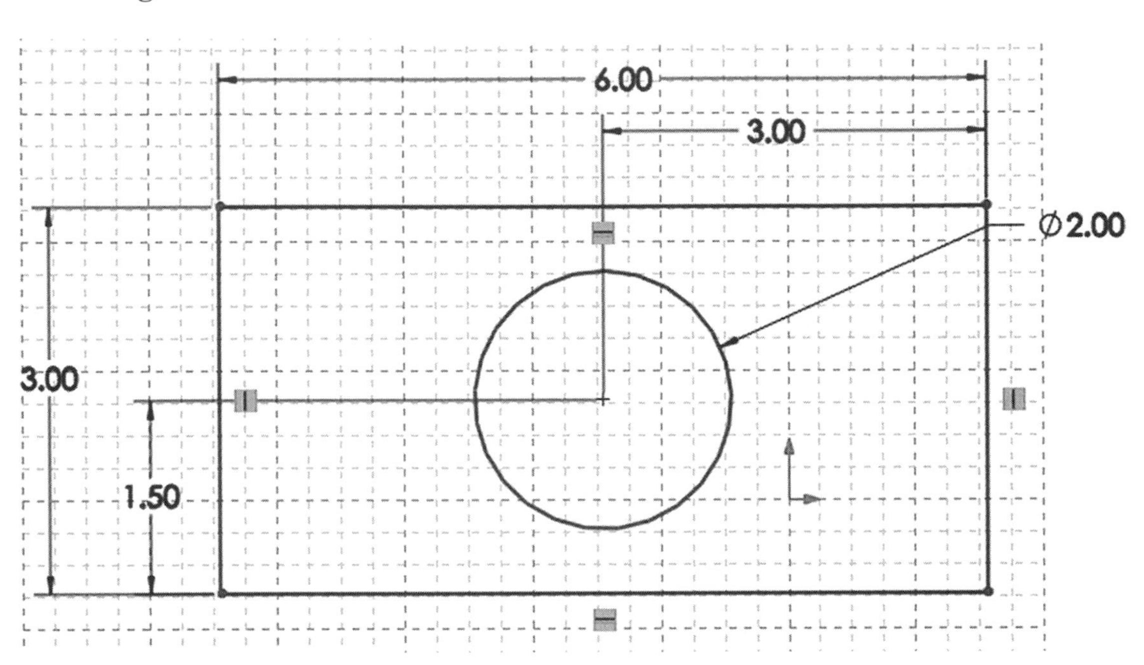

2. Extrude the sketch a distance of 2 inches as shown in Figure 2.

Figure 2

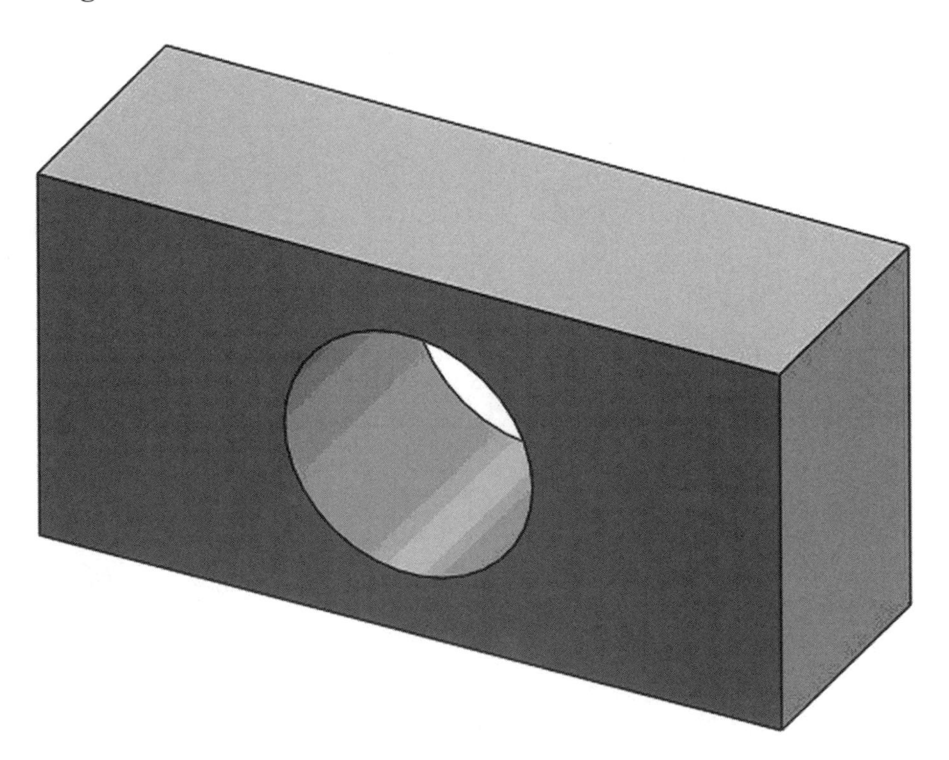

3. Move the cursor to the upper left portion of the screen and left click on the arrow to the right of the SolidWorks text. A fly out menu will appear. Left click on **Tools** (not show). Scroll down and left click on **Equations** as shown in Figure 3.

Figure 3

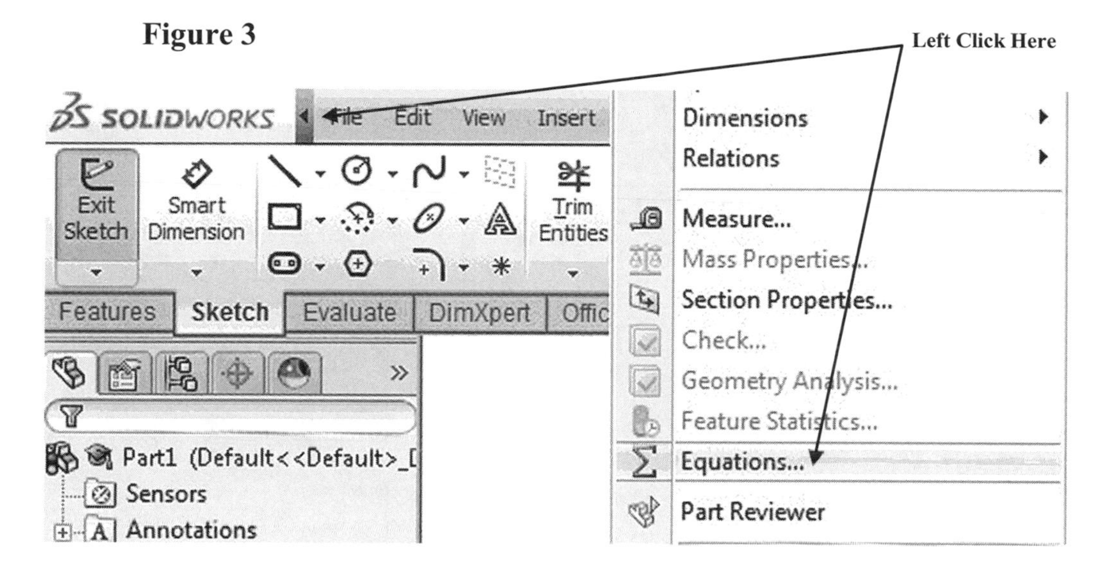

4. The Equations, Global Variables and Dimensions dialog box will appear. Left click on the Dimension View icon. Under the Dimensions text are all the dimensions used to create this part. The Value/Equation column illustrates the dimension values as shown in Figure 4.

Figure 4

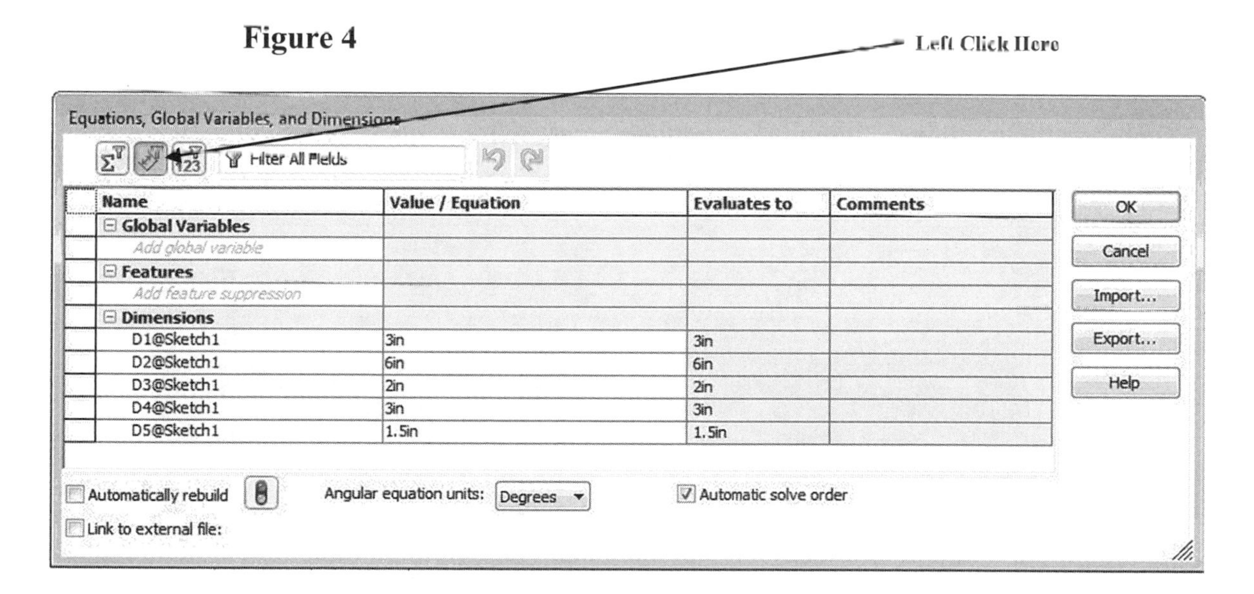

Name	Value / Equation	Evaluates to	Comments
Global Variables			
Add global variable			
Features			
Add feature suppression			
Dimensions			
D1@Sketch1	3in	3in	
D2@Sketch1	6in	6in	
D3@Sketch1	2in	2in	
D4@Sketch1	3in	3in	
D5@Sketch1	1.5in	1.5in	

5. Highlight the first dimension in the Value/Equation column. Enter *2 behind the current value (2*3) and press the **Enter** key as shown in Figure 5. SolidWorks will update the dimension in the model area.

Figure 5

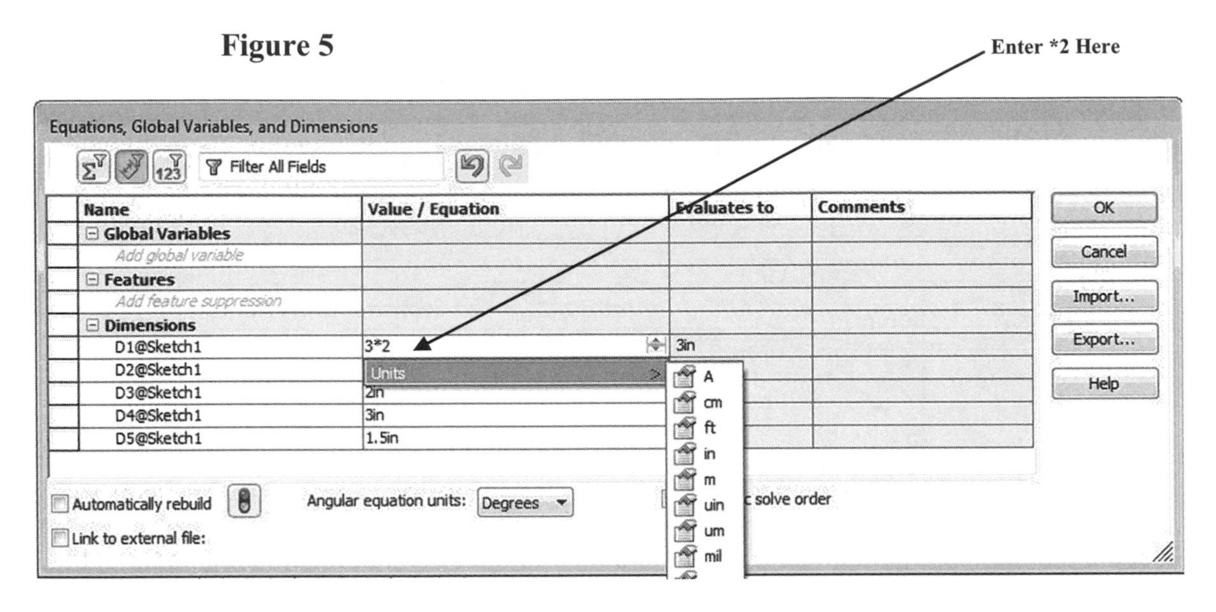

6. SolidWorks will update the value in the Value/Equation column as shown in Figure 6.

Figure 6

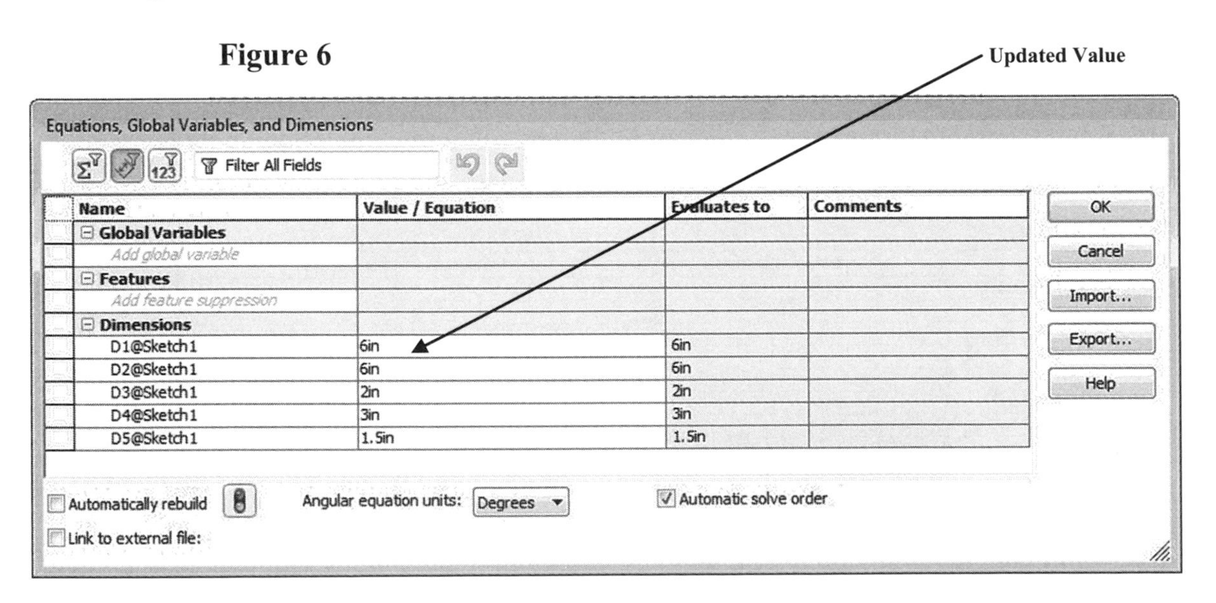

7. Highlight the second dimension in the Value/Equation column. Enter *1.5 behind the current value (6*1.5) and press the **Enter** key as shown in Figure 7.

Figure 7

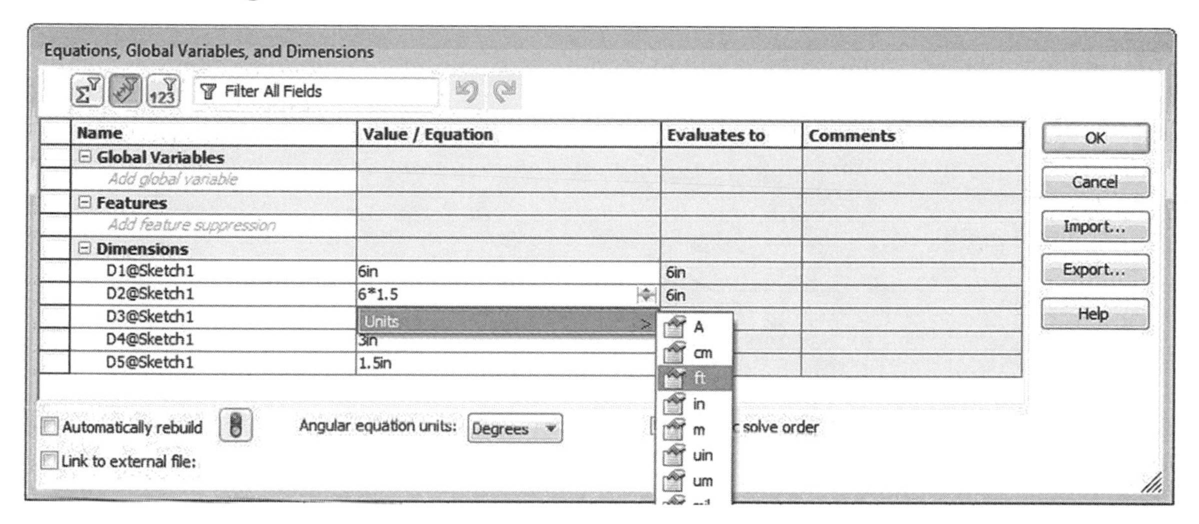

8. SolidWorks will update the dimension to 9 (6*1.5) in the model area as shown in Figure 8. Feel free to update other dimensions of the sketch using the same technique.

Figure 8

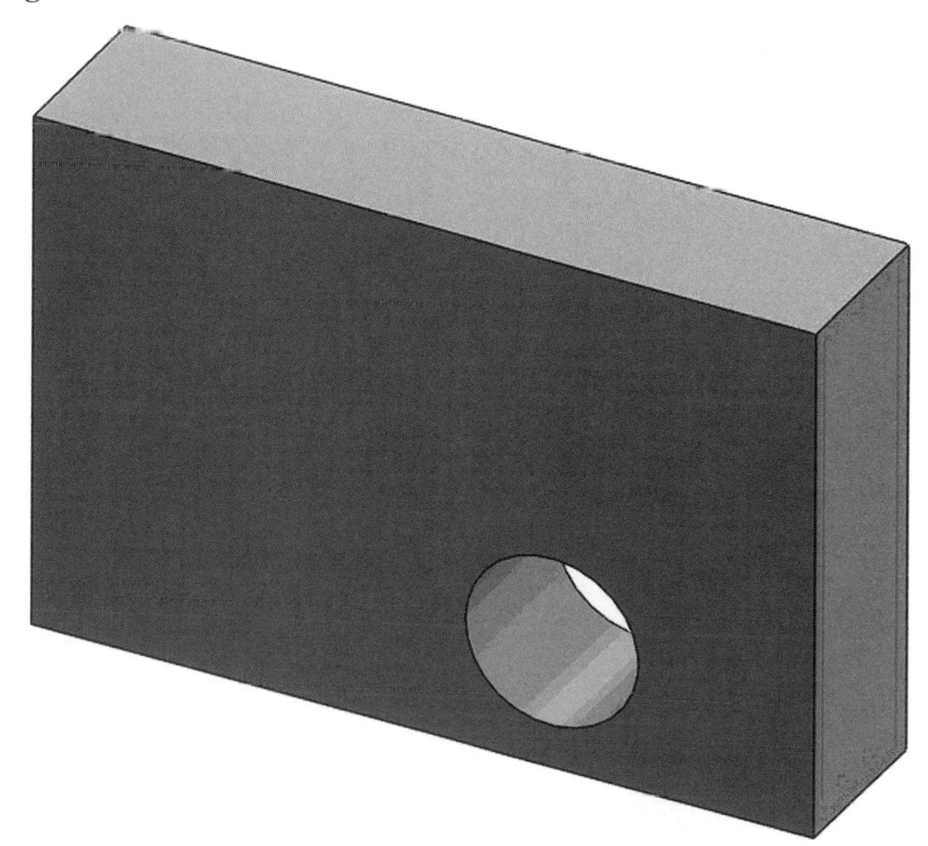

Additional Practice Problems

Problem 1-1

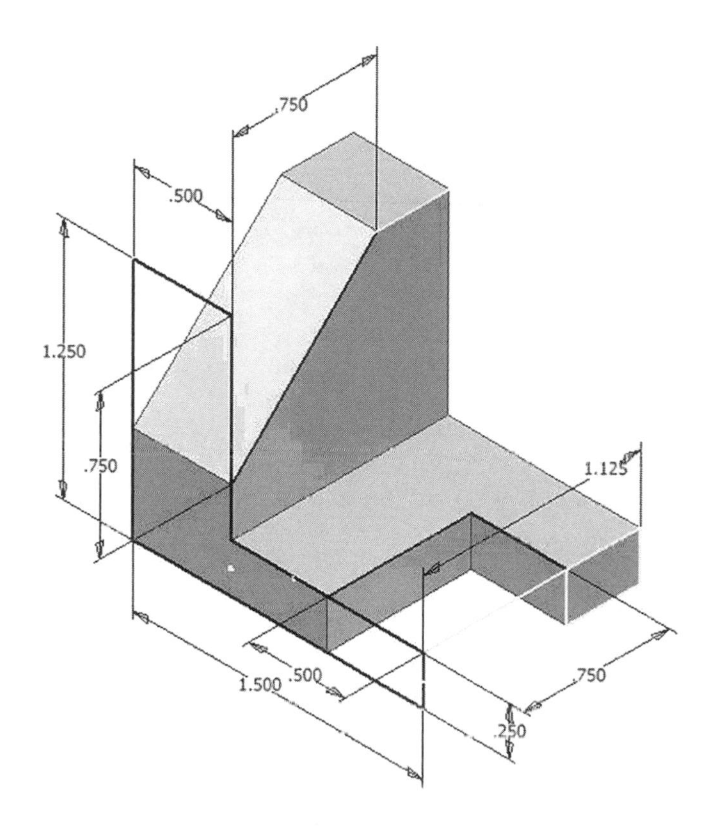

Problem 1-2

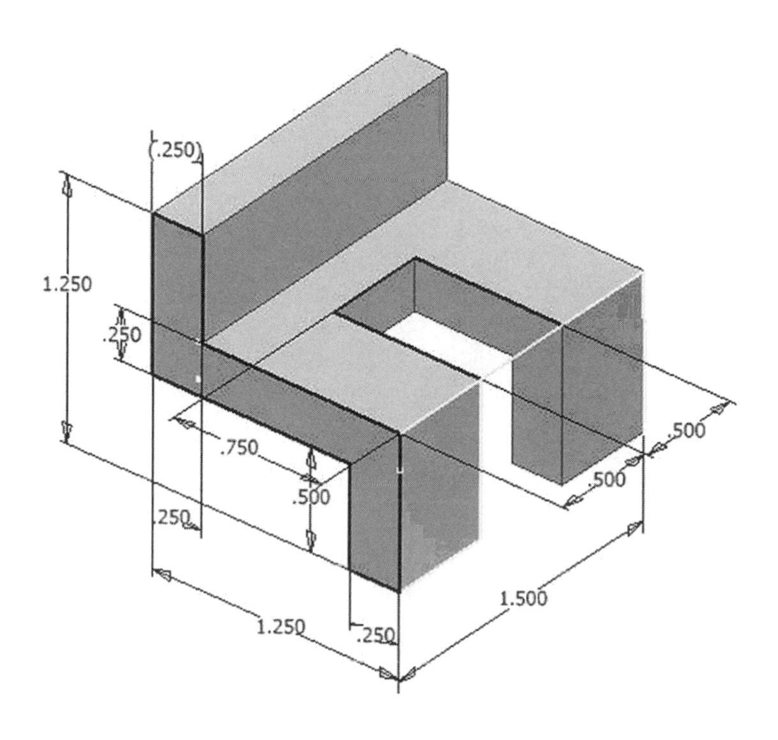

Problem 1-3

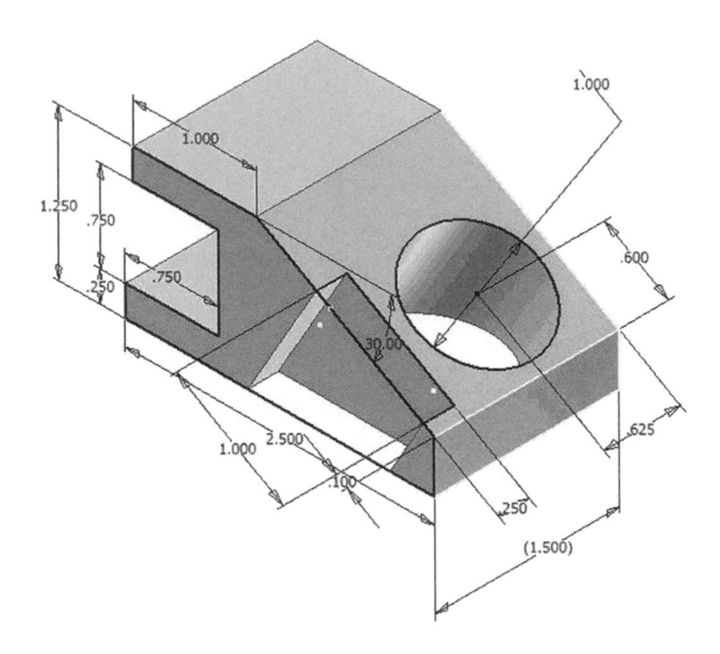

Problem 1-4

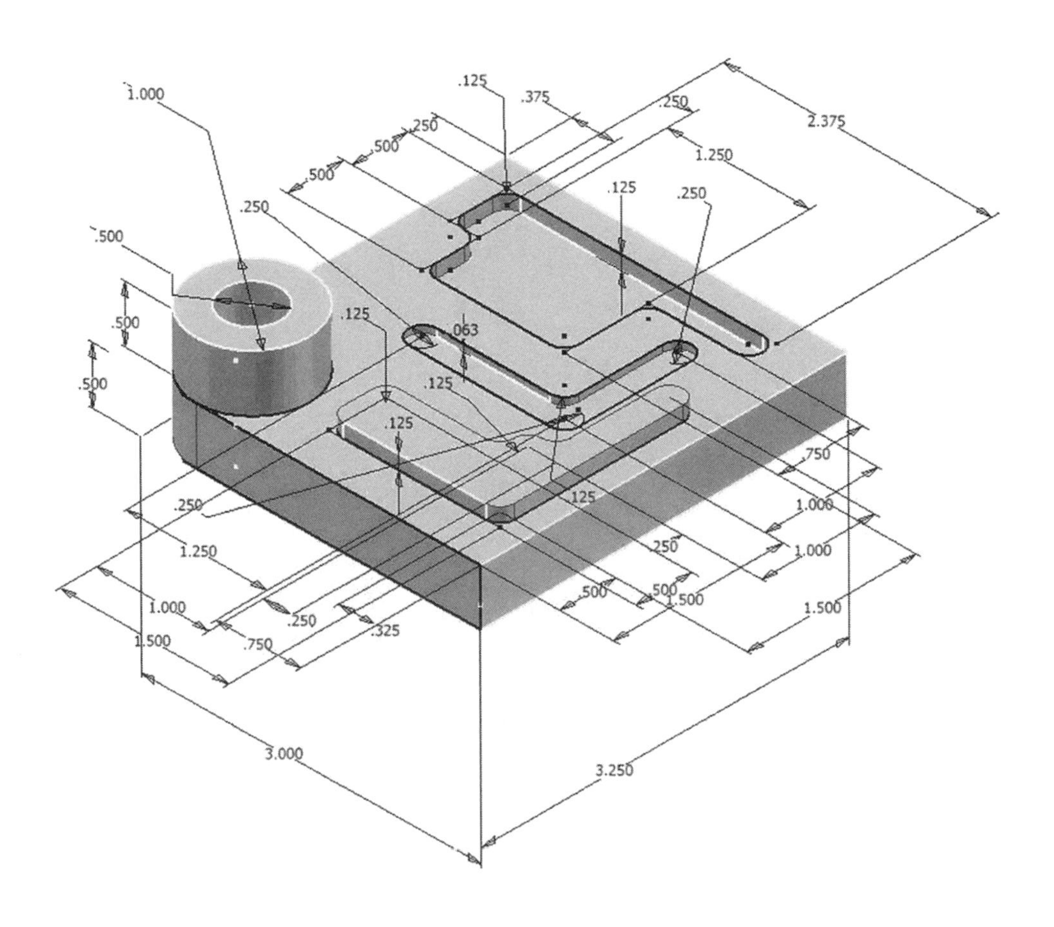

Problem 1-5

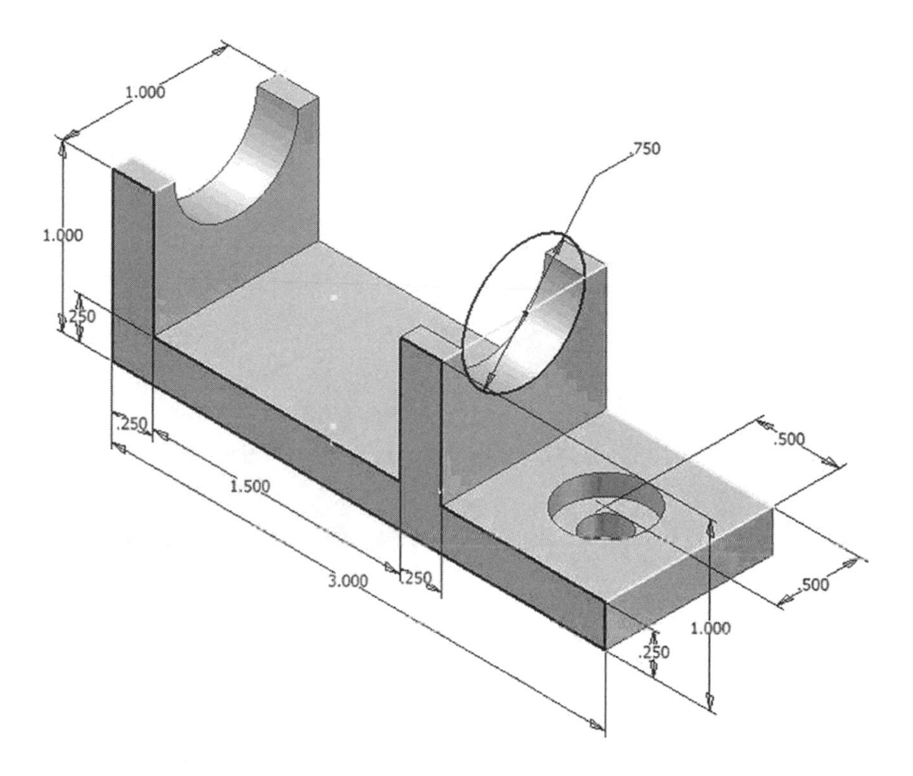

Problem 1-6

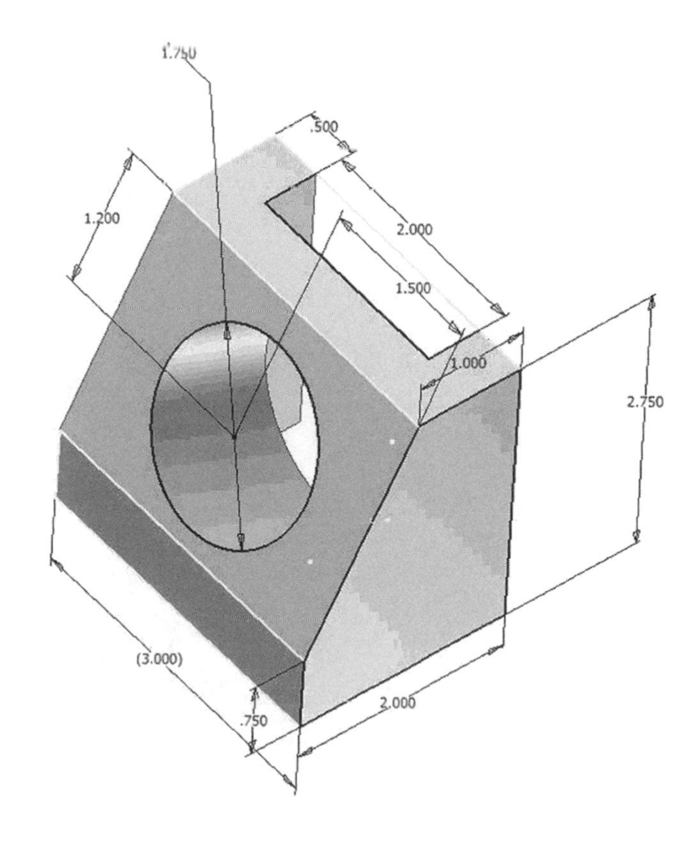

Problem 1-7

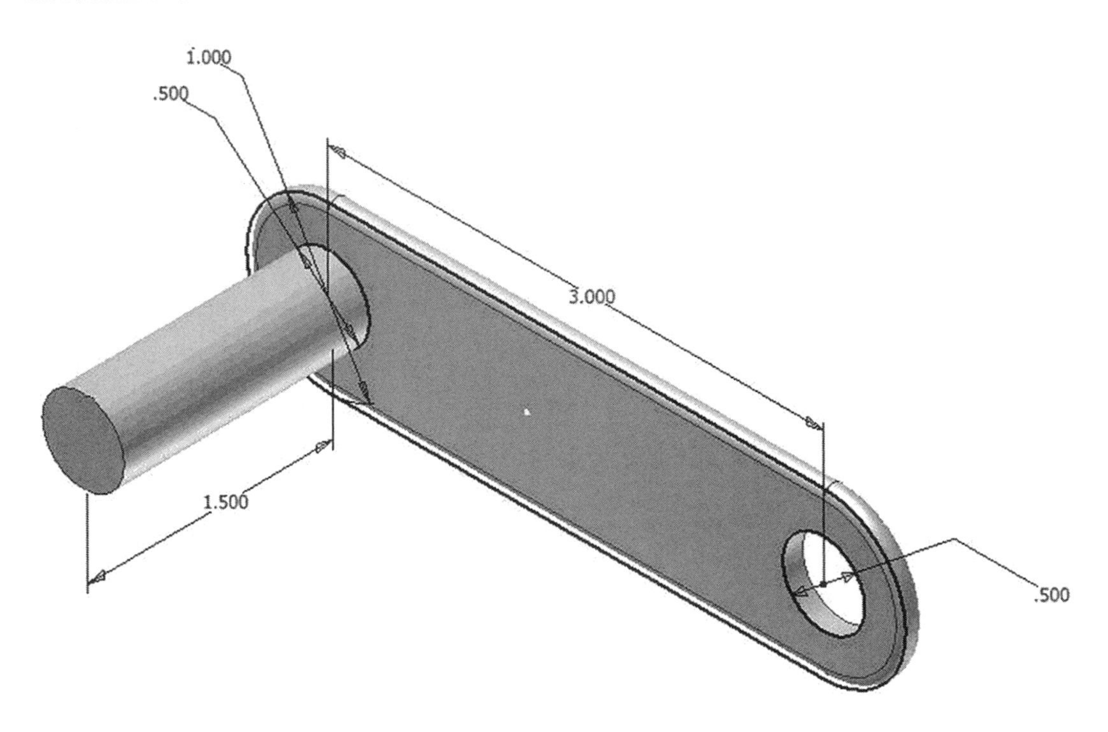

Problem 1-8

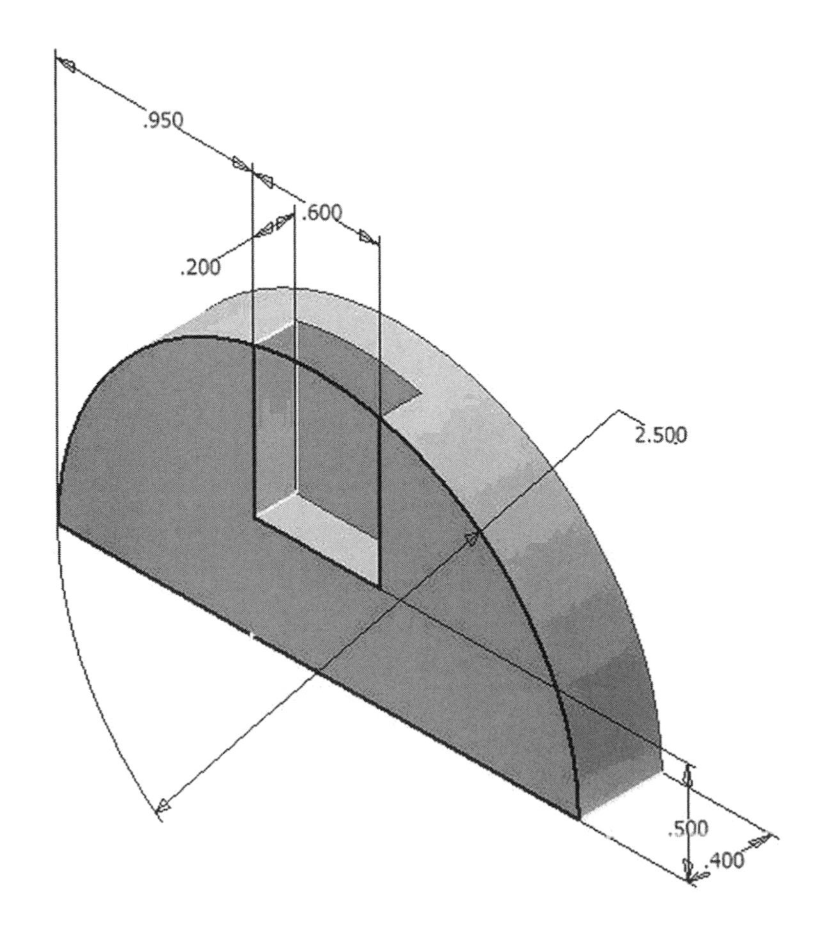

Problem 1-9

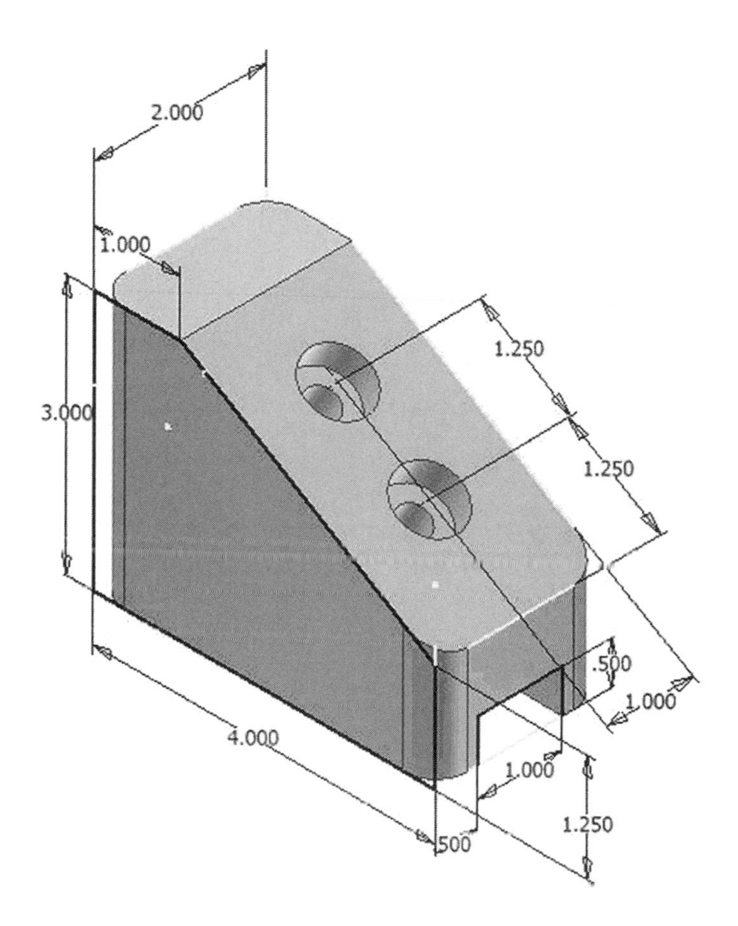

Problem 2-1

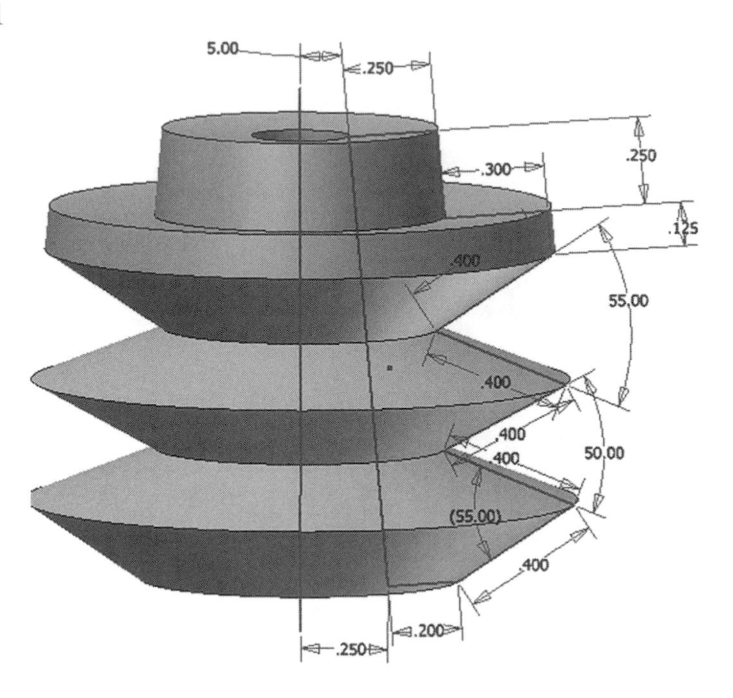

Problem 2-2

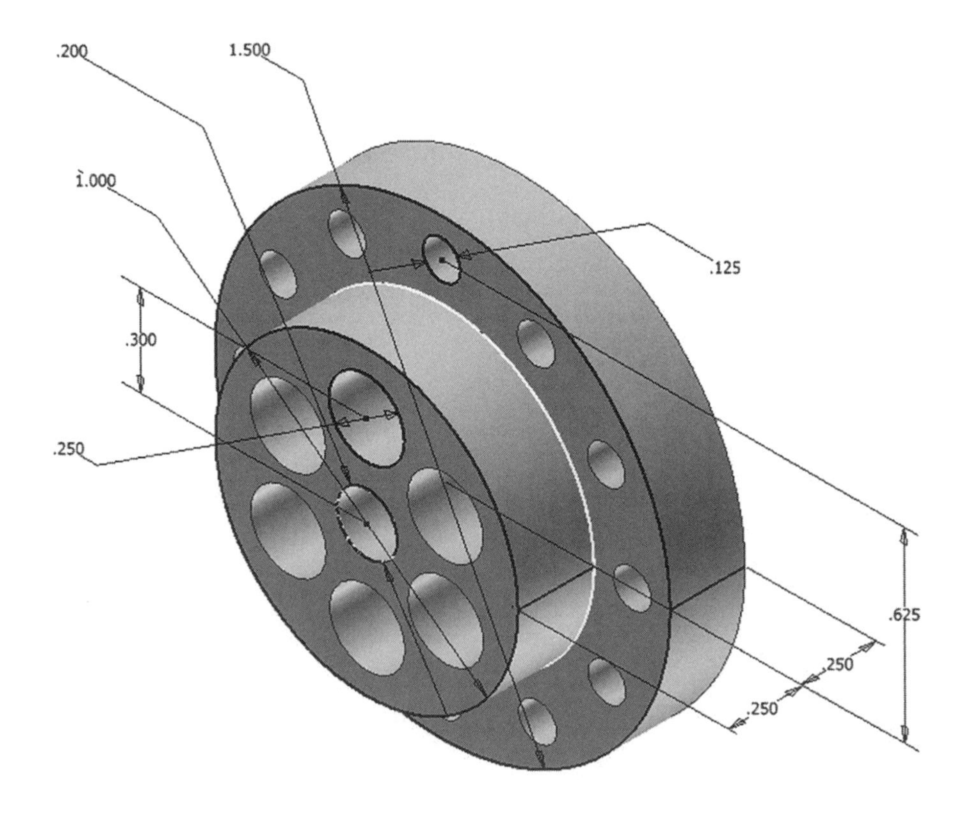

Problem 2-3

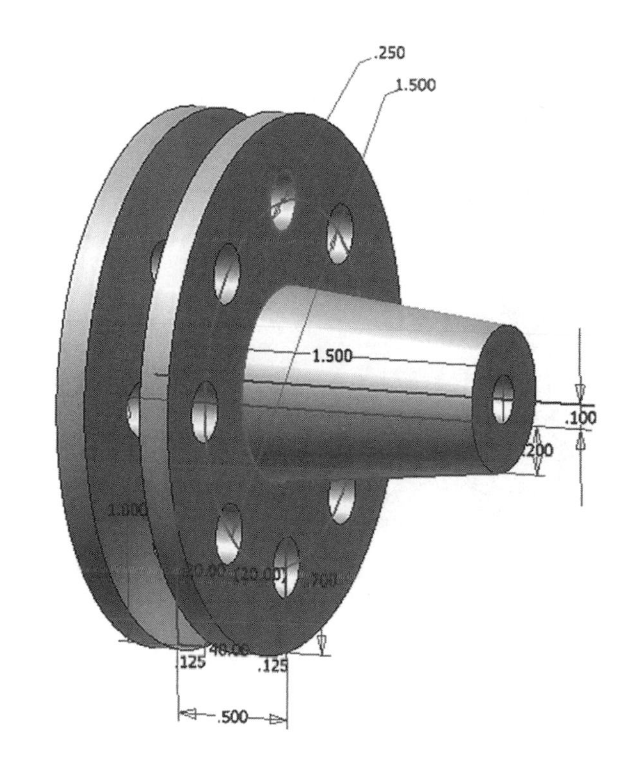

Problem 2-4

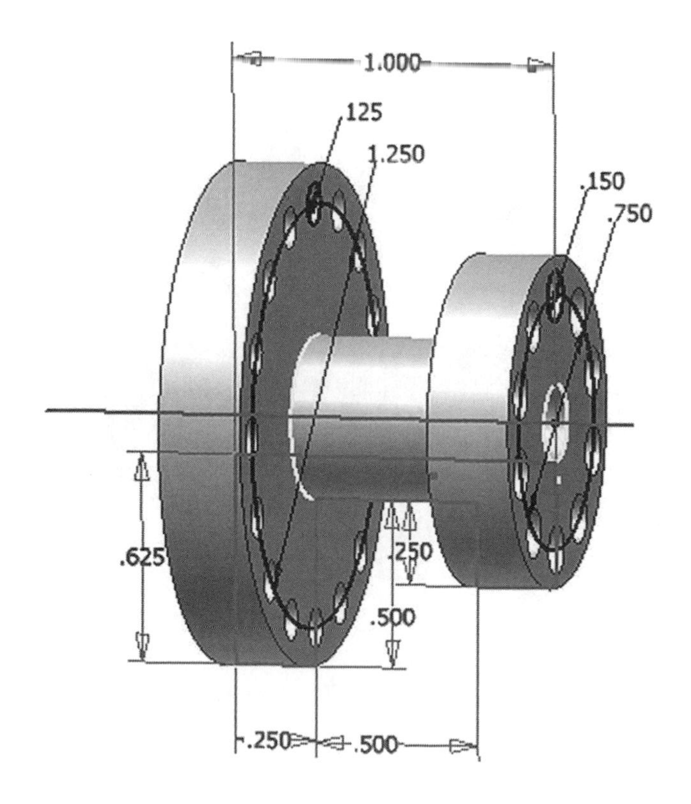

Problem 2-5

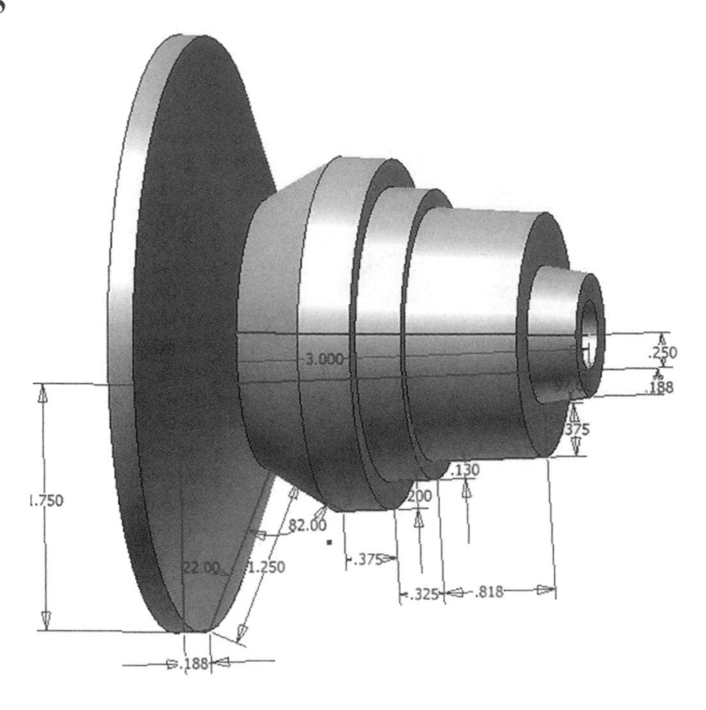

Problem 2-6

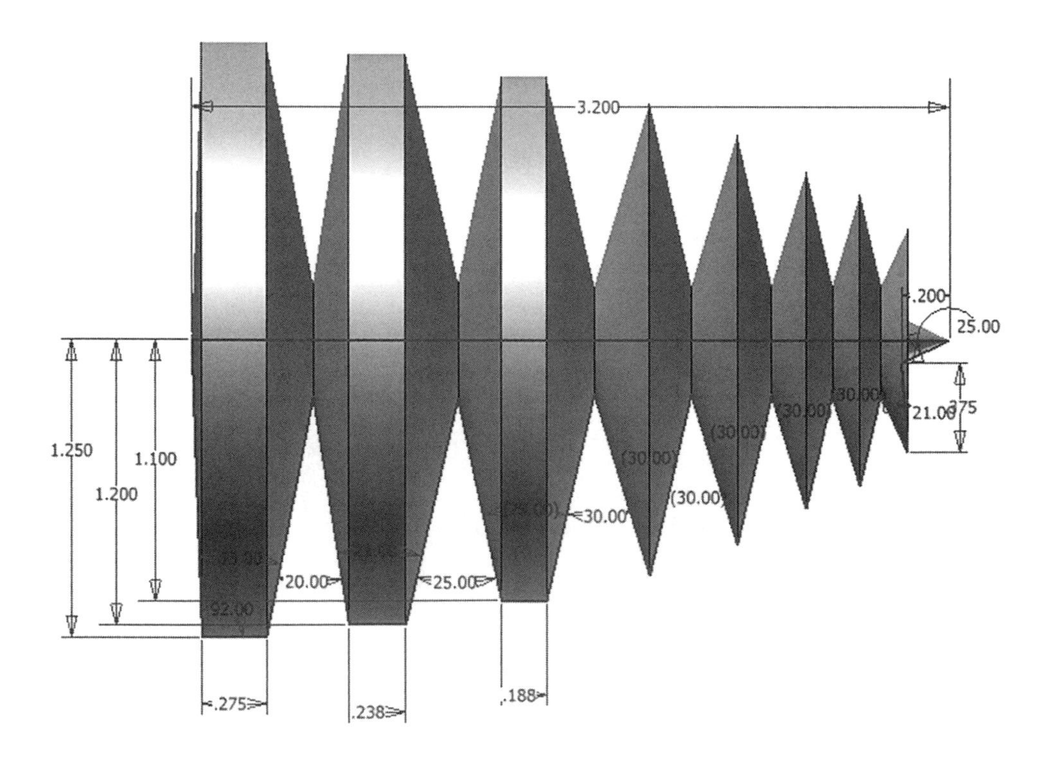

Problem 3-1

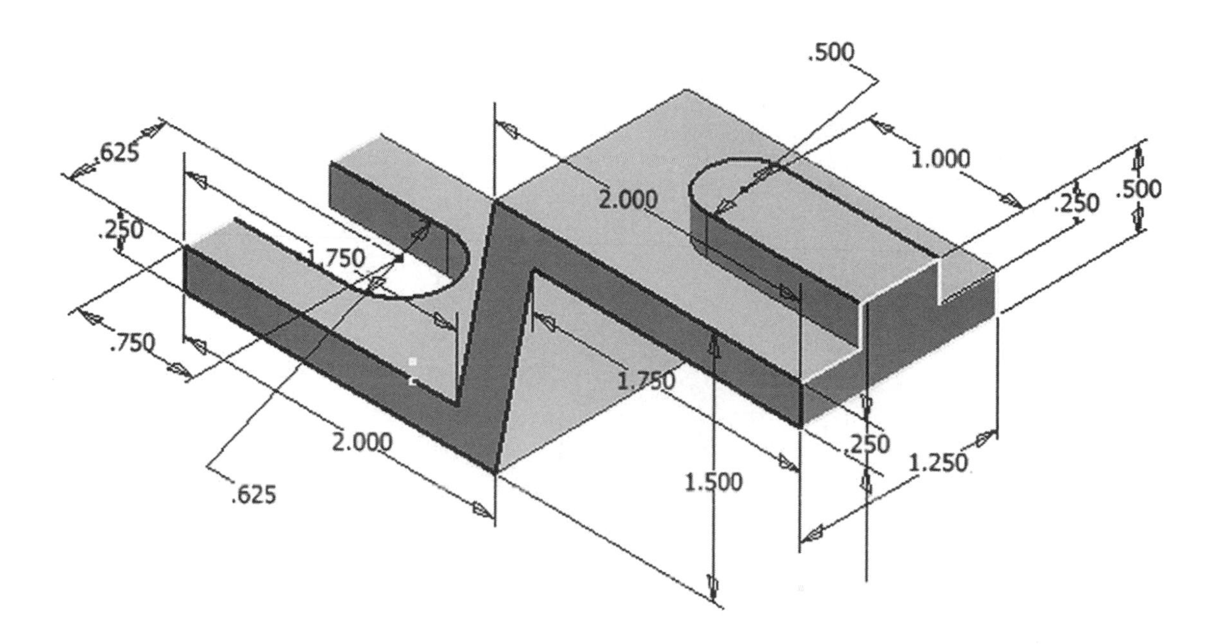

Problem 3-2

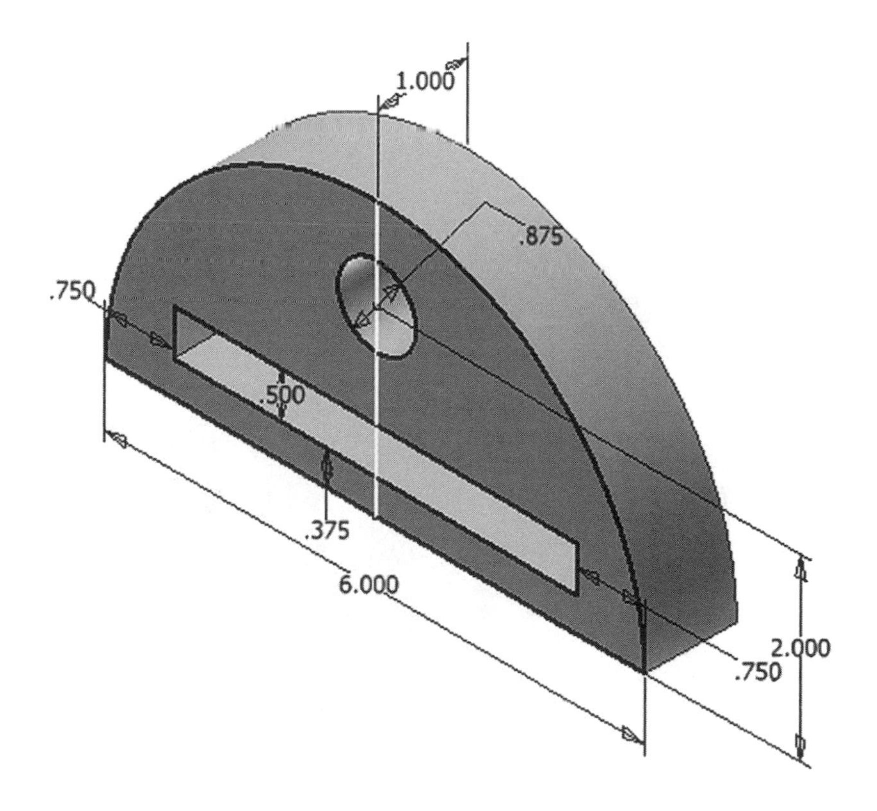

Problem 3-3

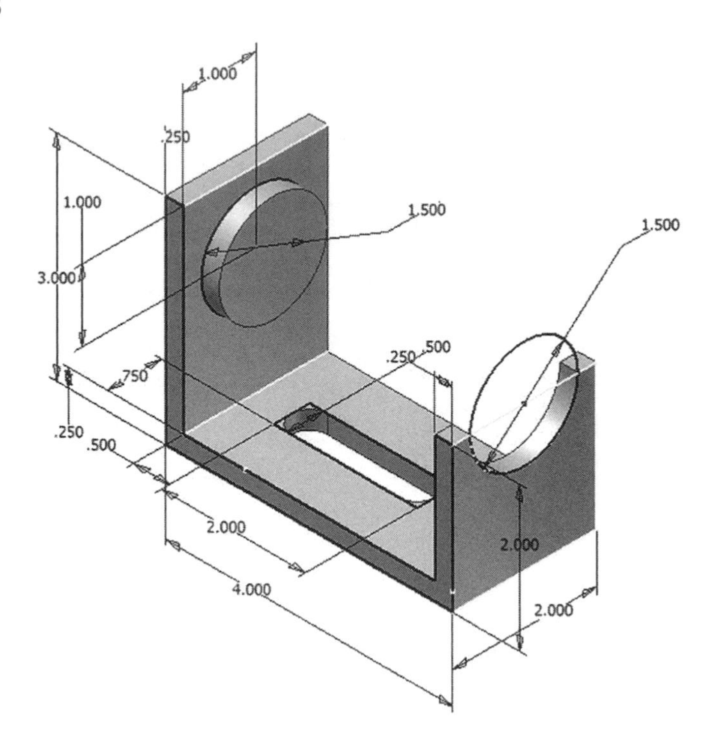

Problem 3-4

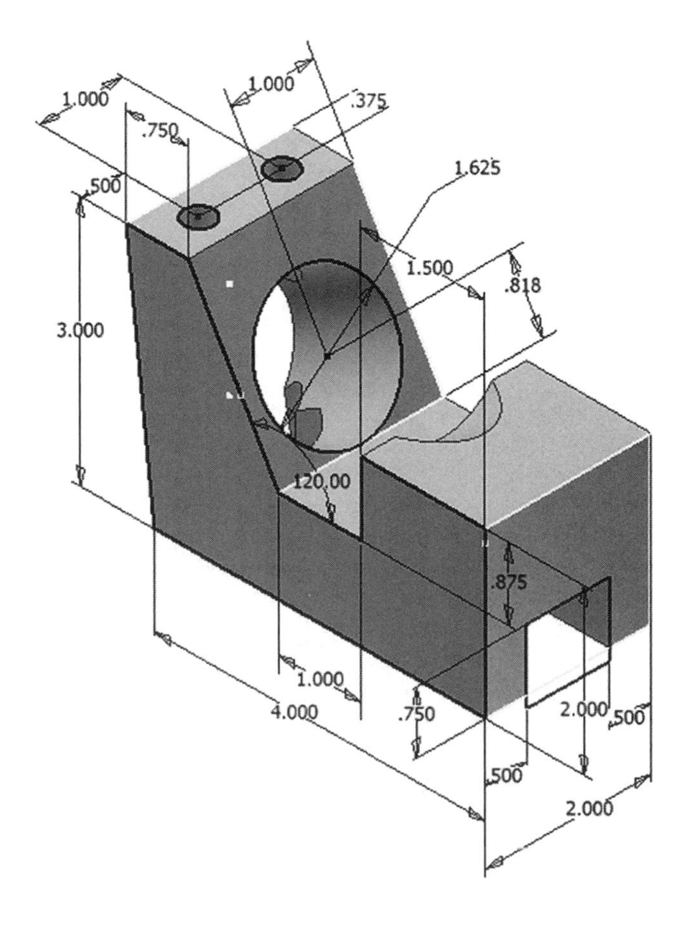

Problem 3-5

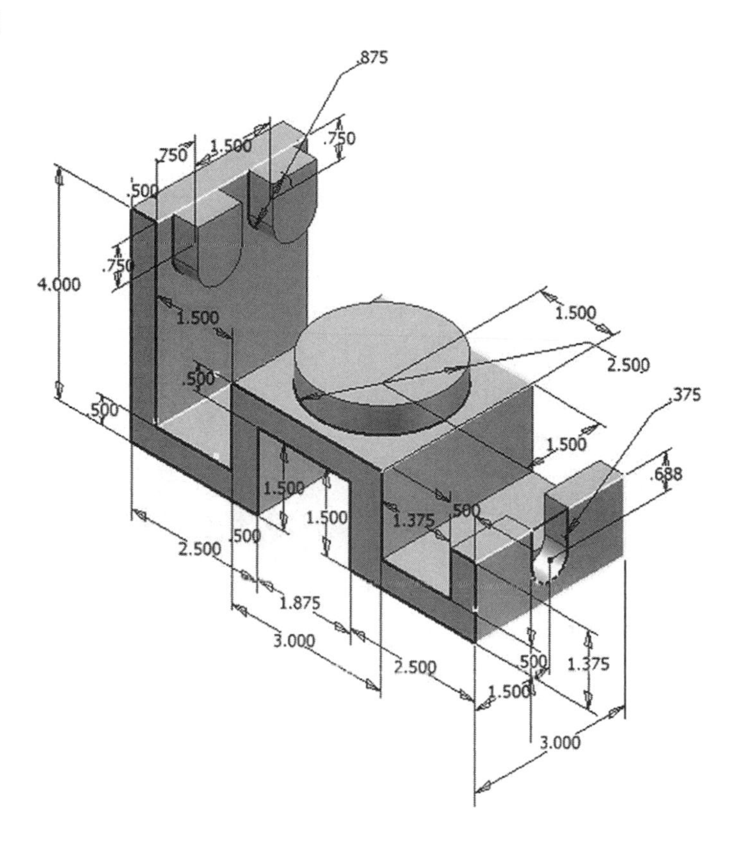

Problem 3-6

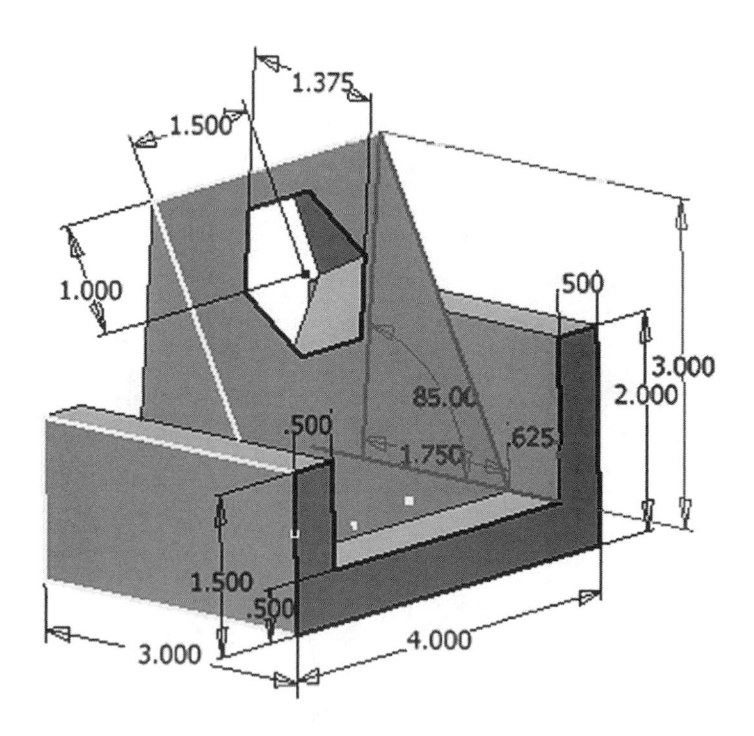

Problem 3-7

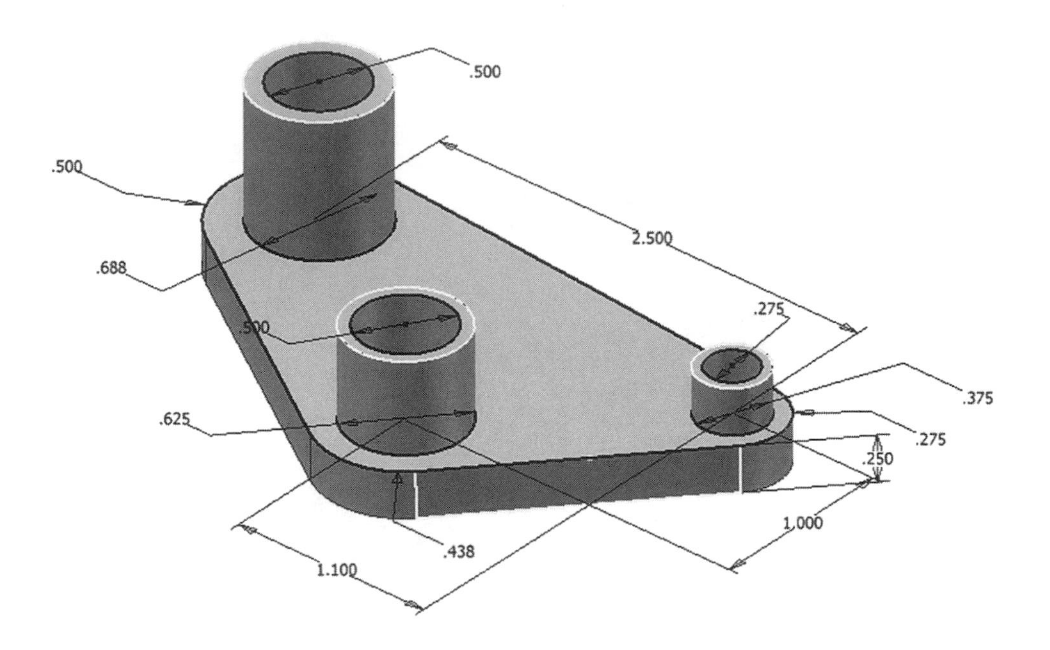

Problem 3-8

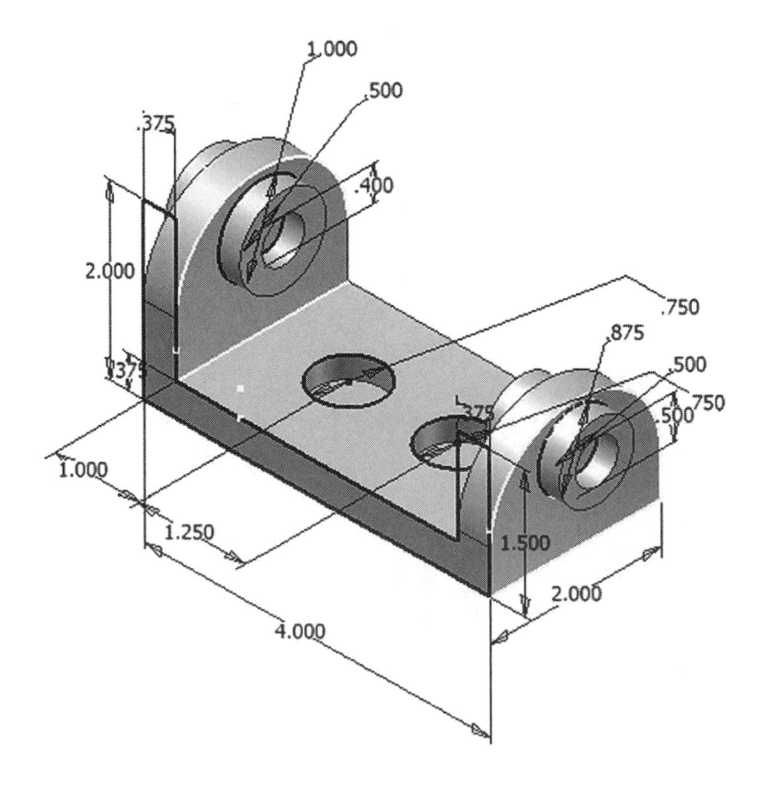

Problem 4-1

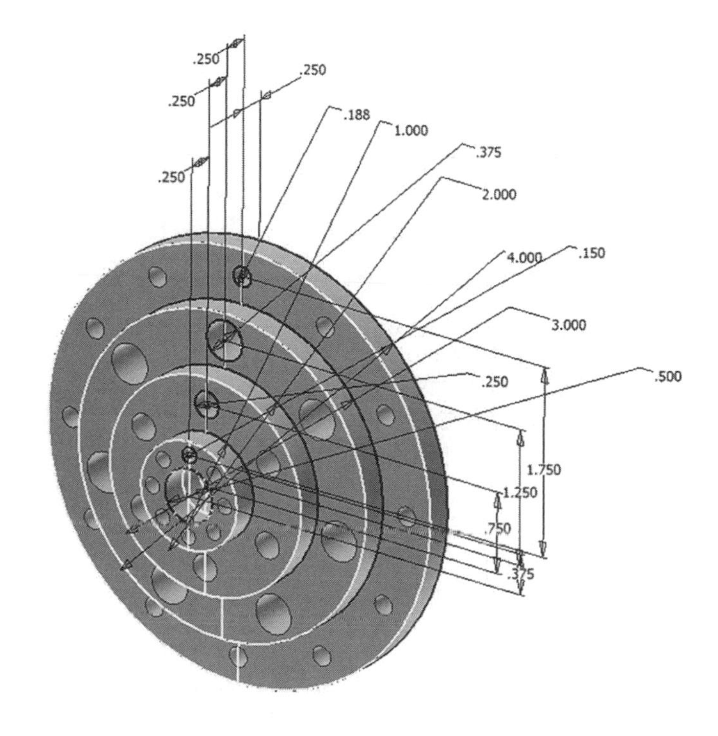

Problem 4-2

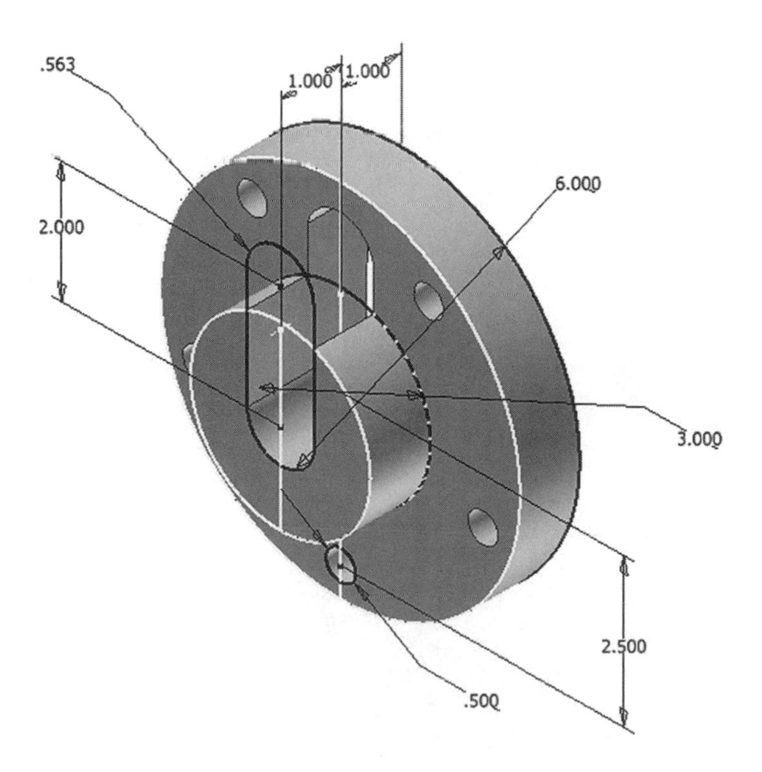

Problem 4-3a

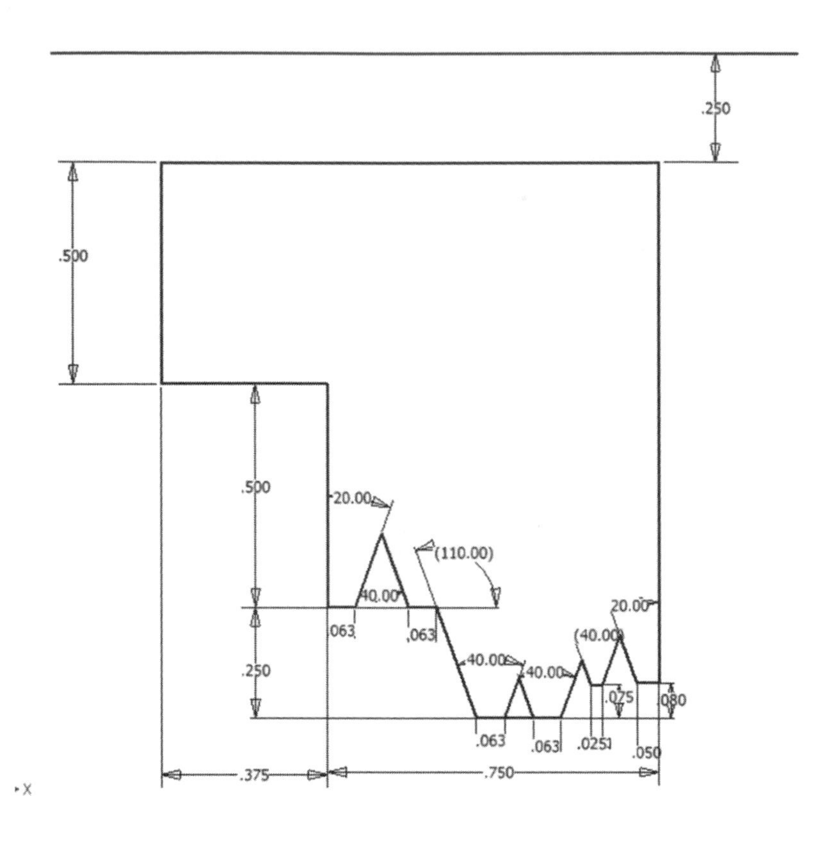

Problem 4-3b

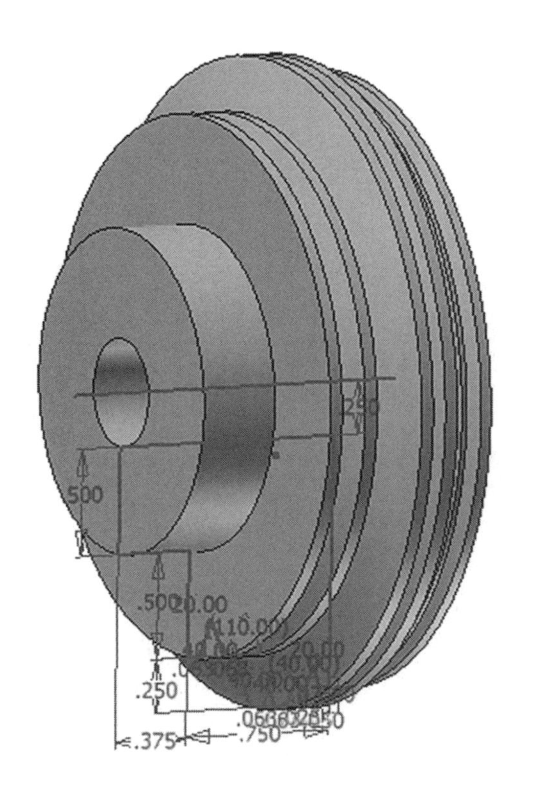

Problem 4-4a

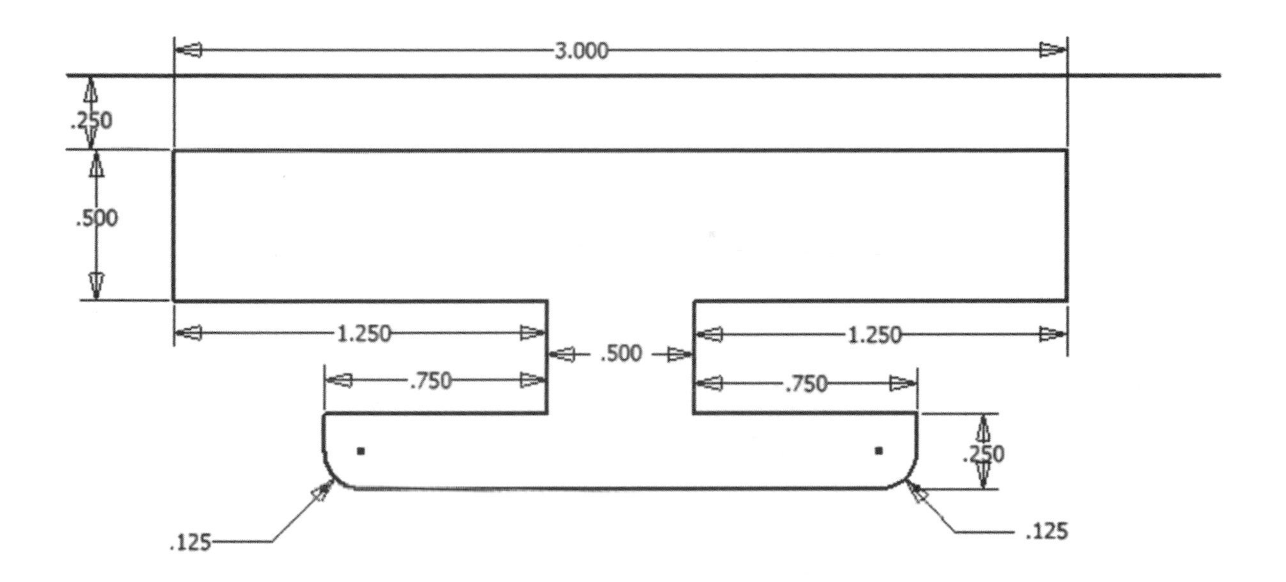

Problem 4-4b

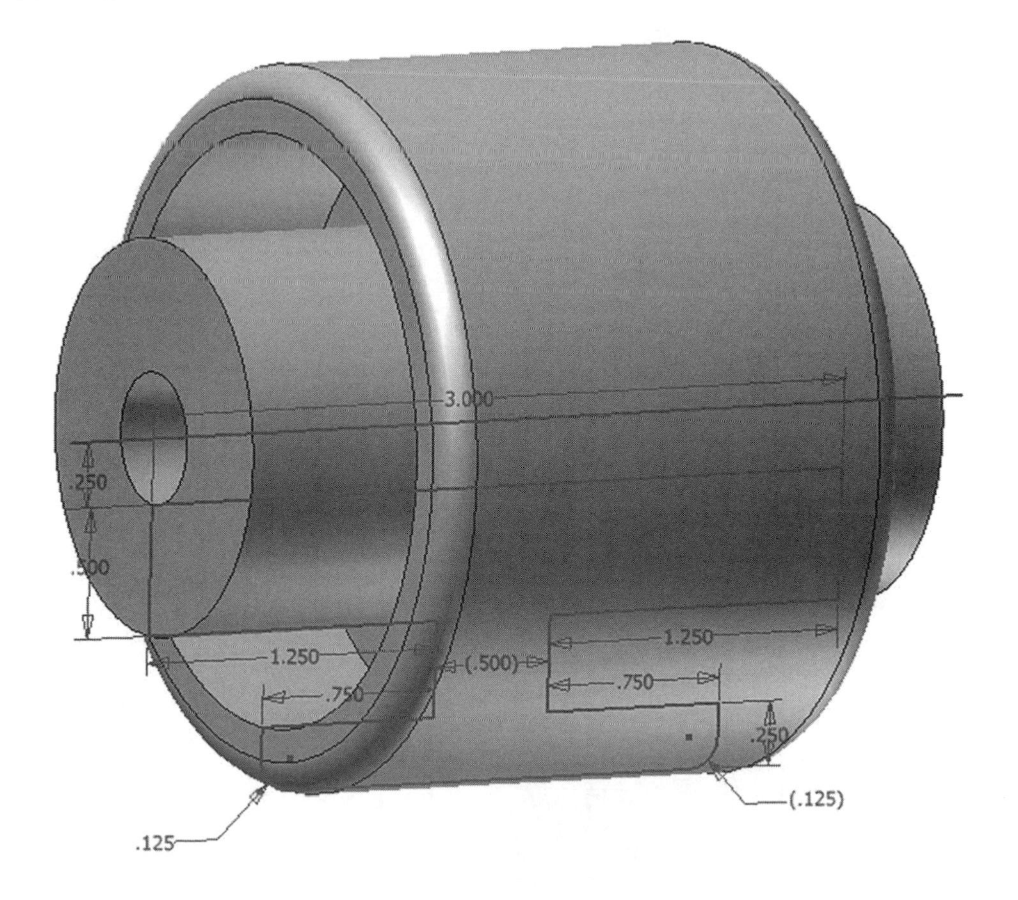

Problem 4-5

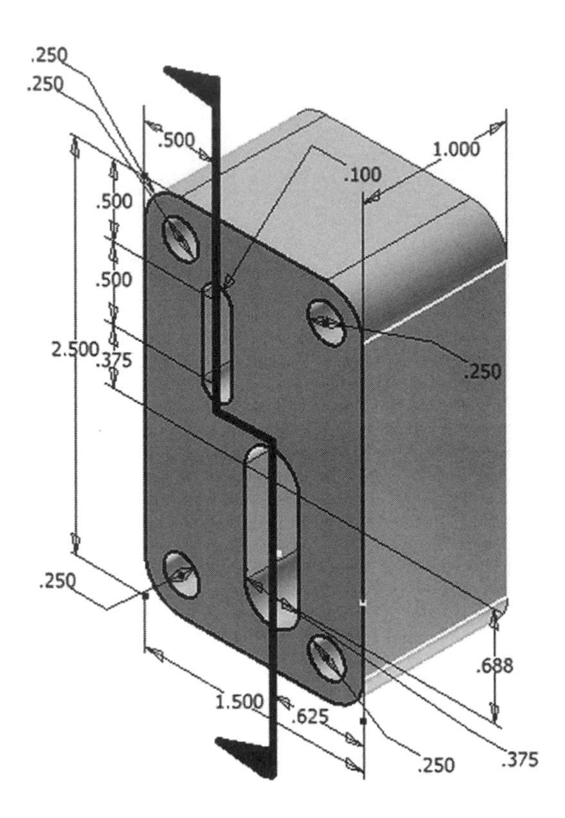

Problem 4-6

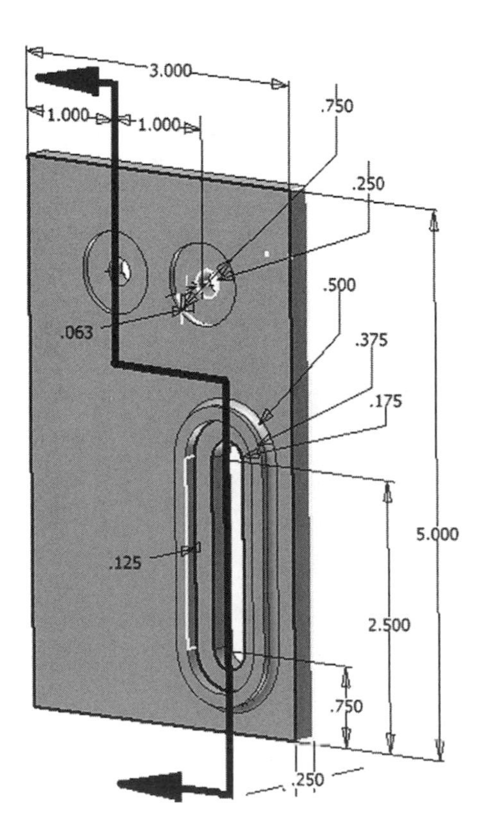

Problem 4-7

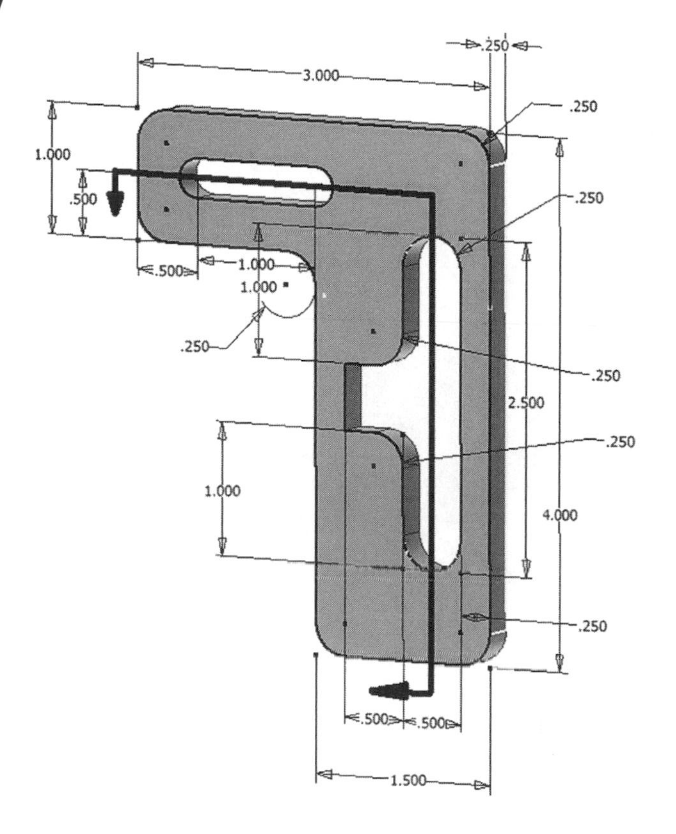

Problem 4-8

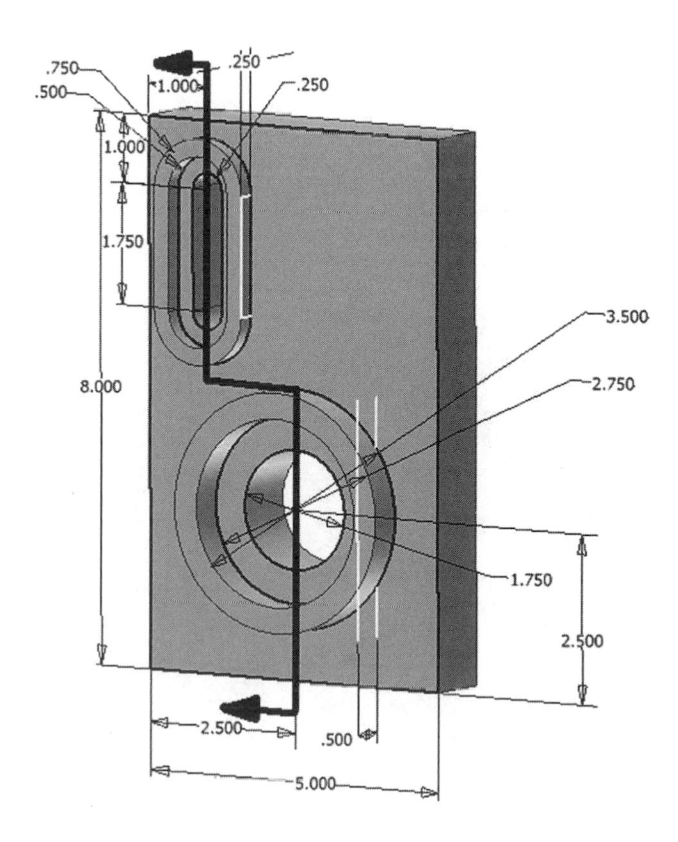

Problem 5-1

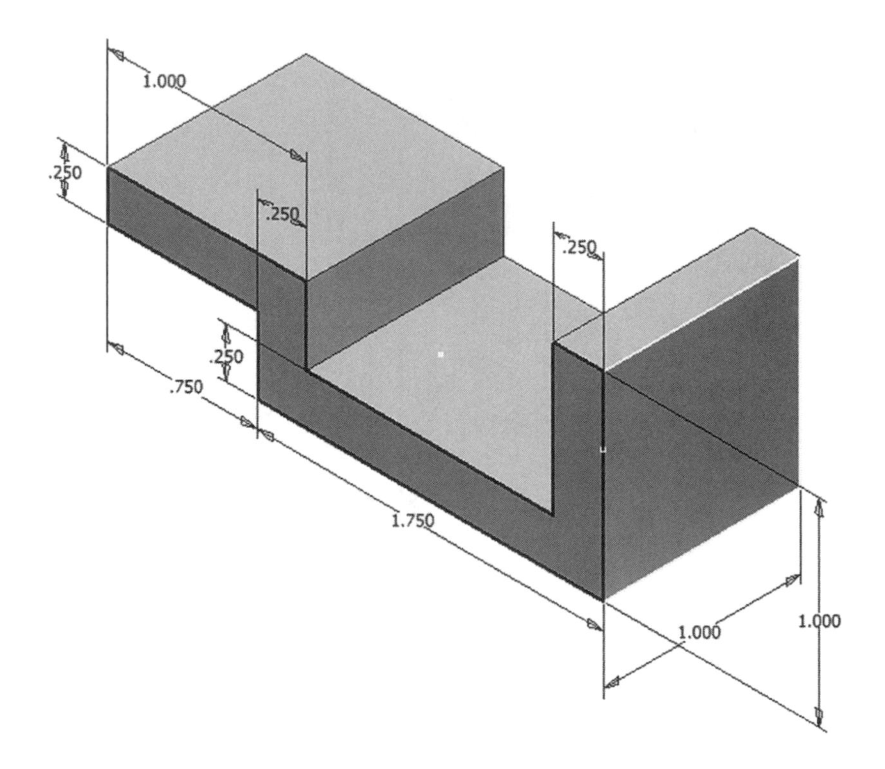

Problem 5-2

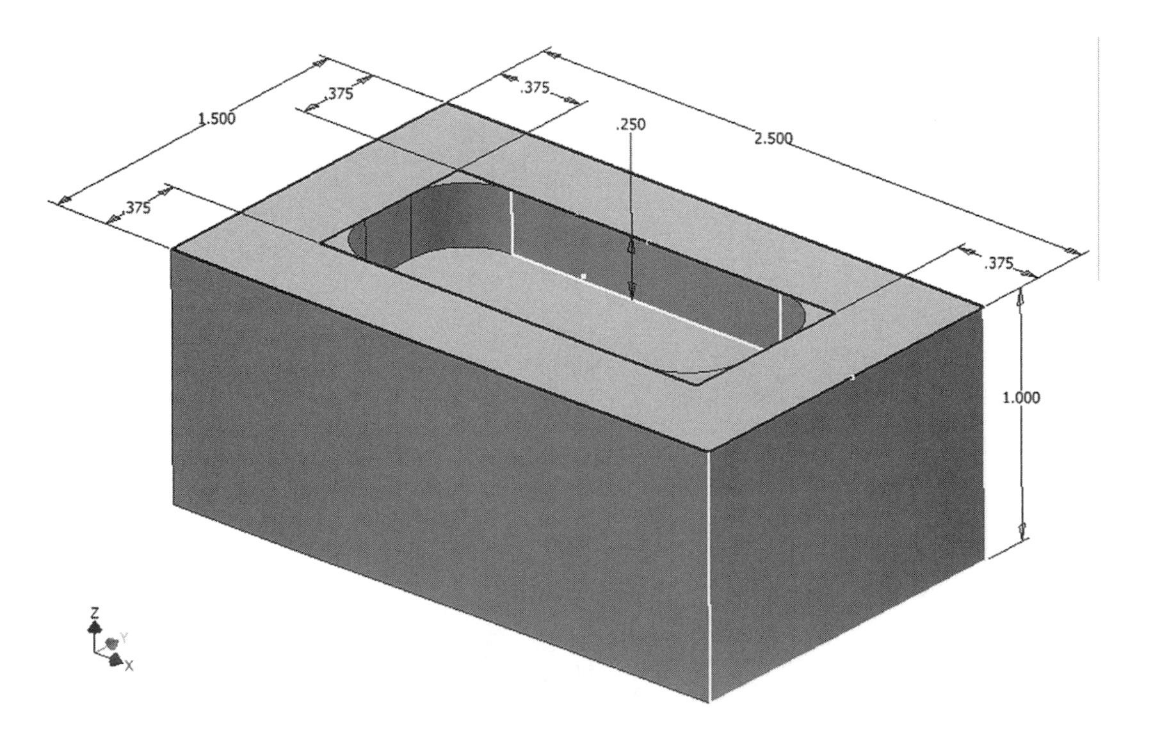

Problem 5-3

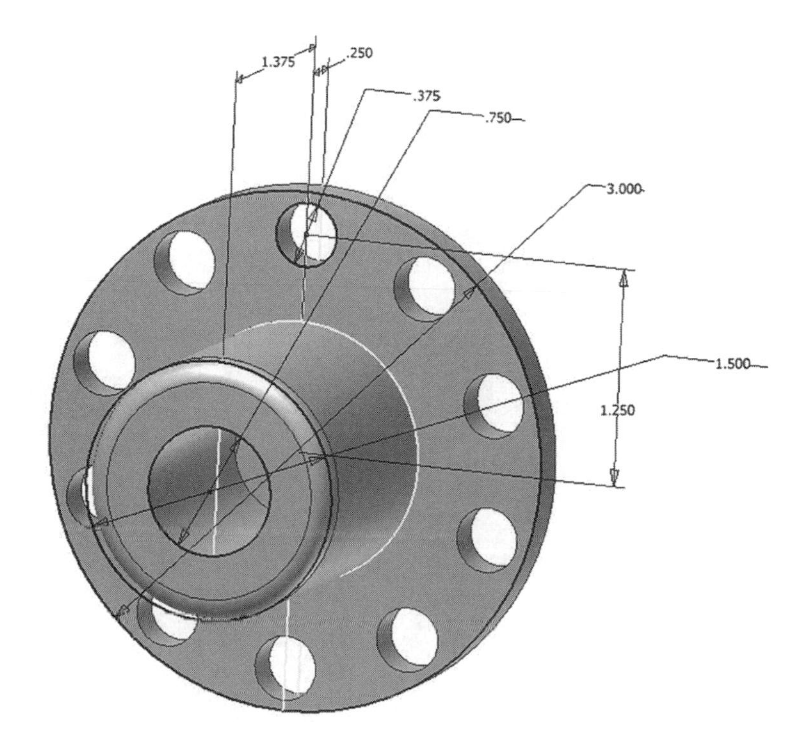

Problem 5-4

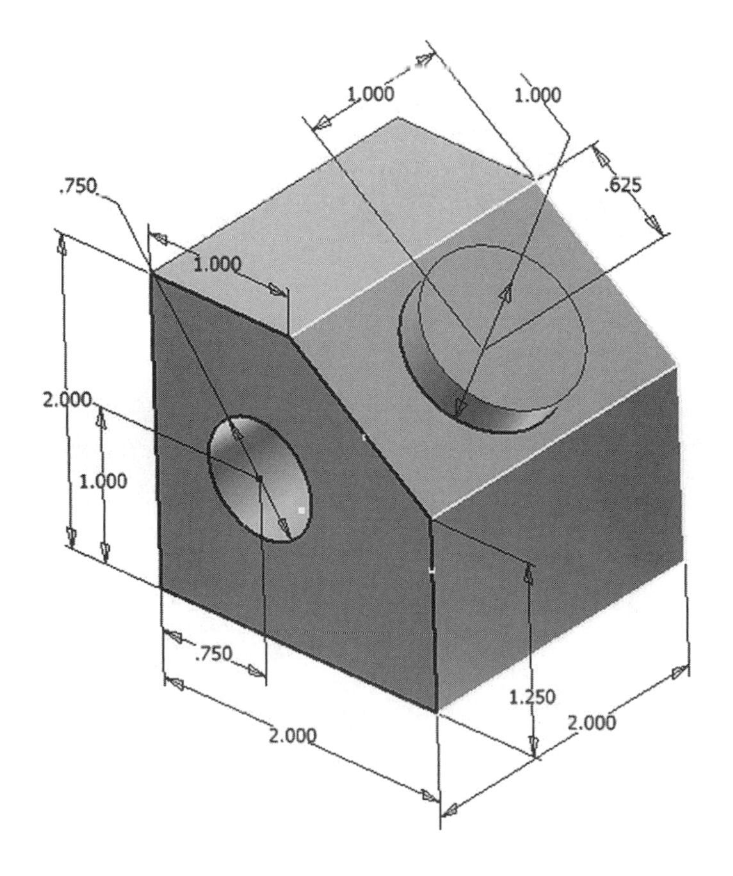

Problem 5-5

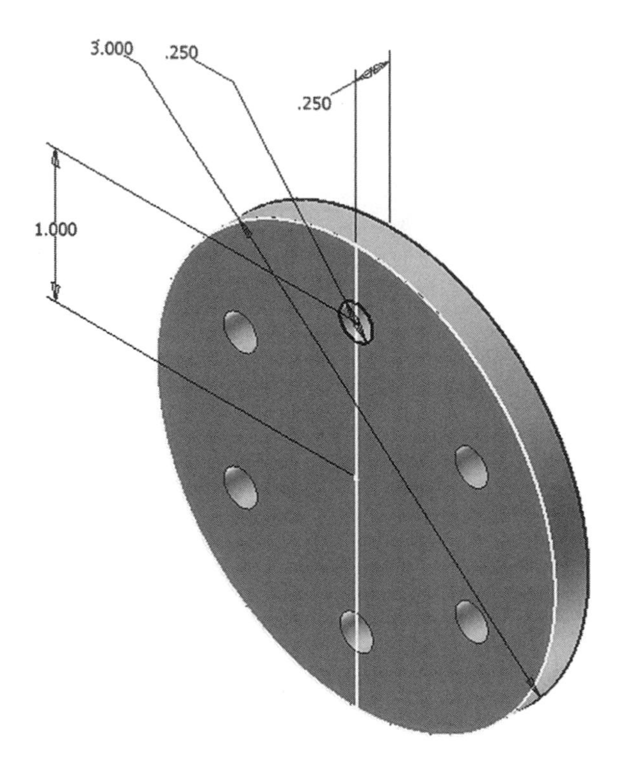

Problem 5-6

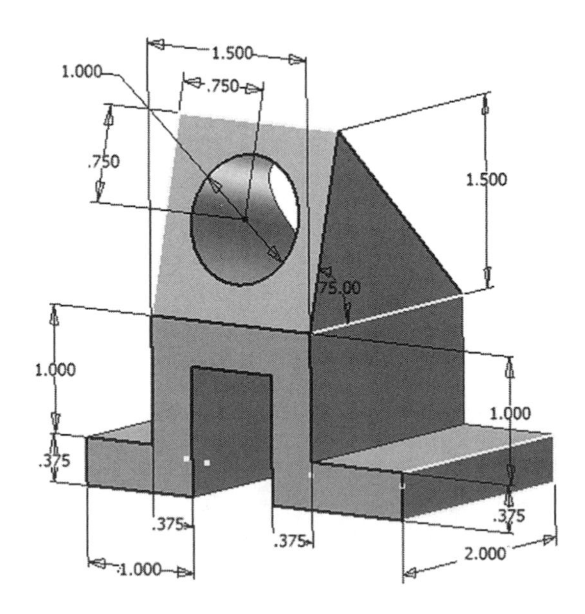

Problem 5-7

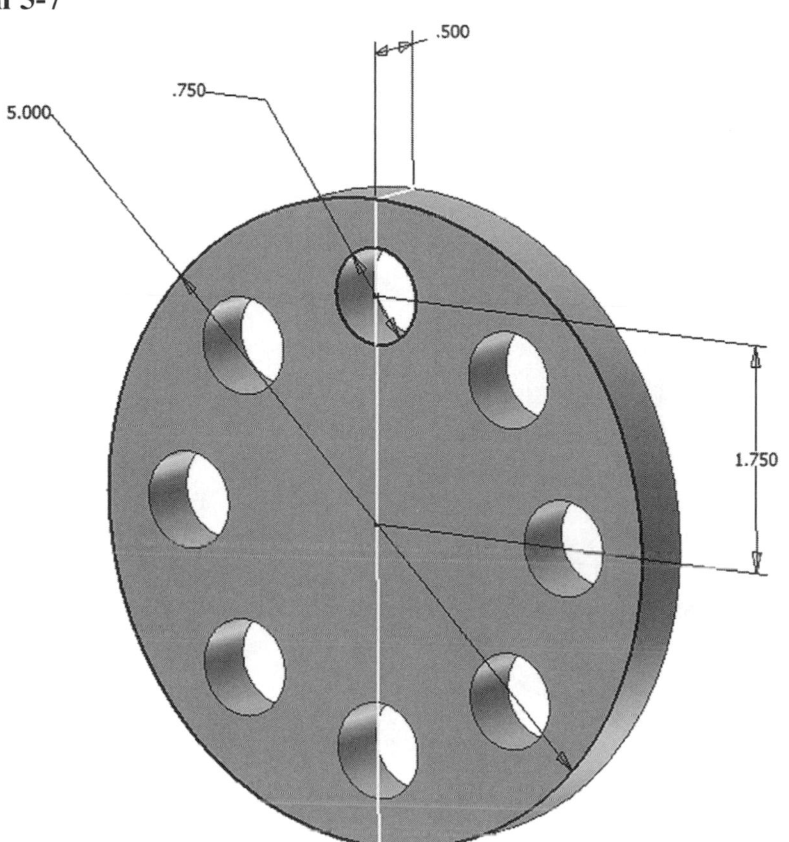

Problem 6-1

Using the existing parts from Chapter 6 to complete the following modification to the piston assembly.

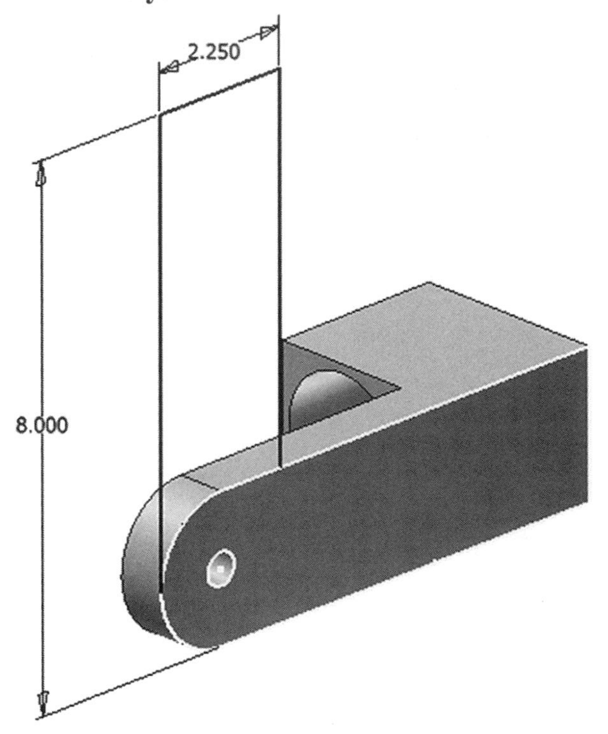

Problem 6-2

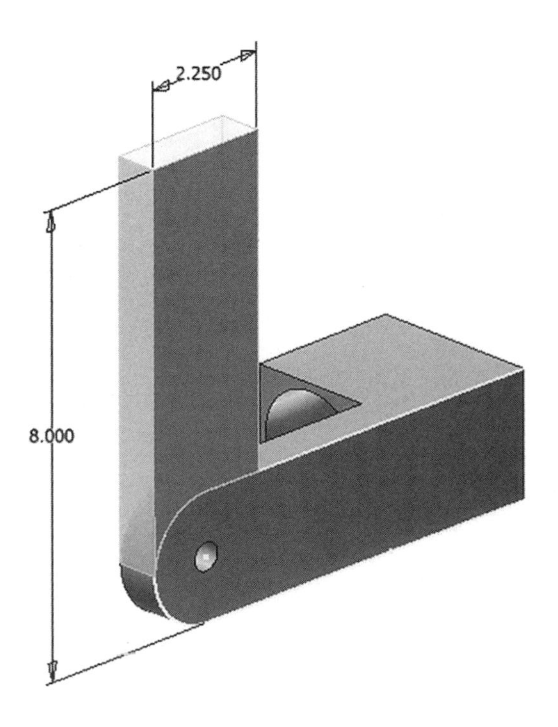

Problem 6-3

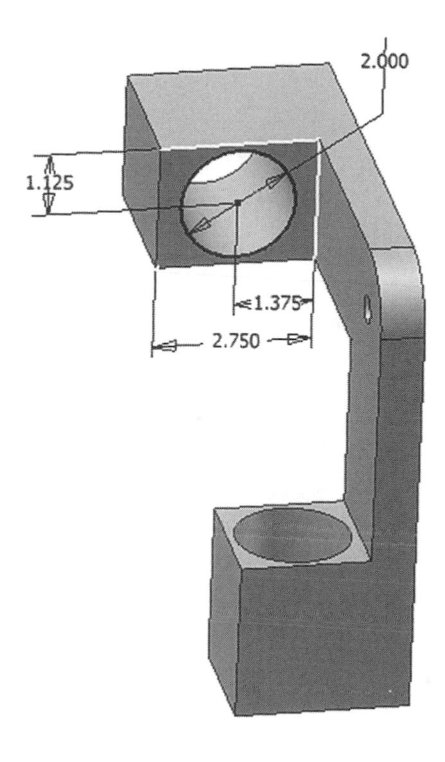

Problem 6-4

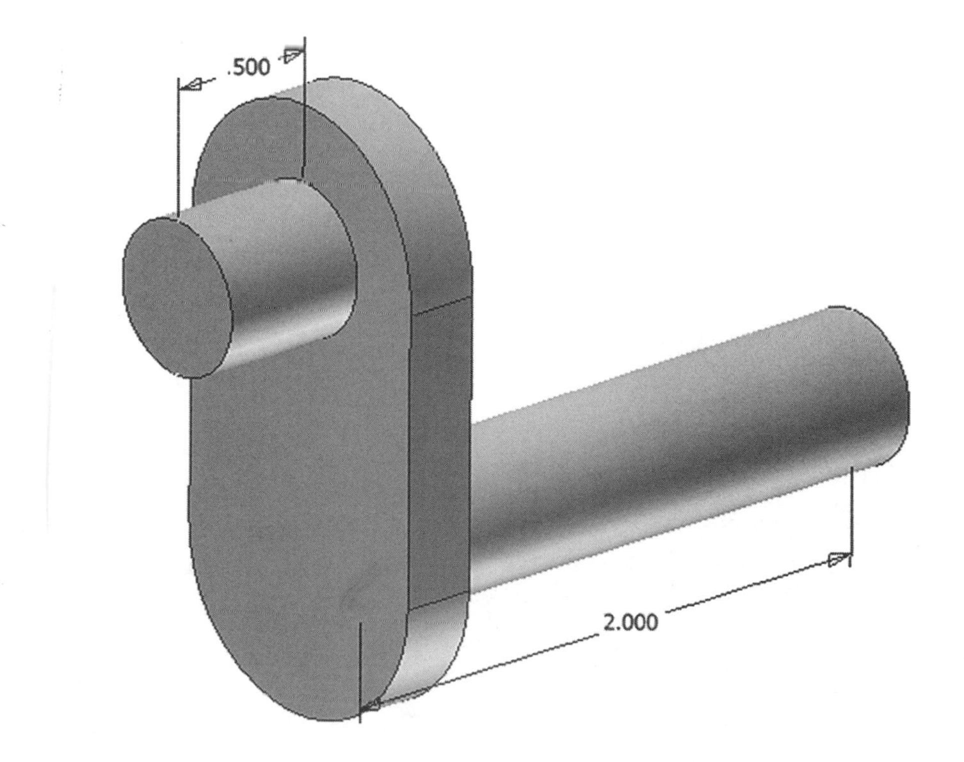

Problem 6-5

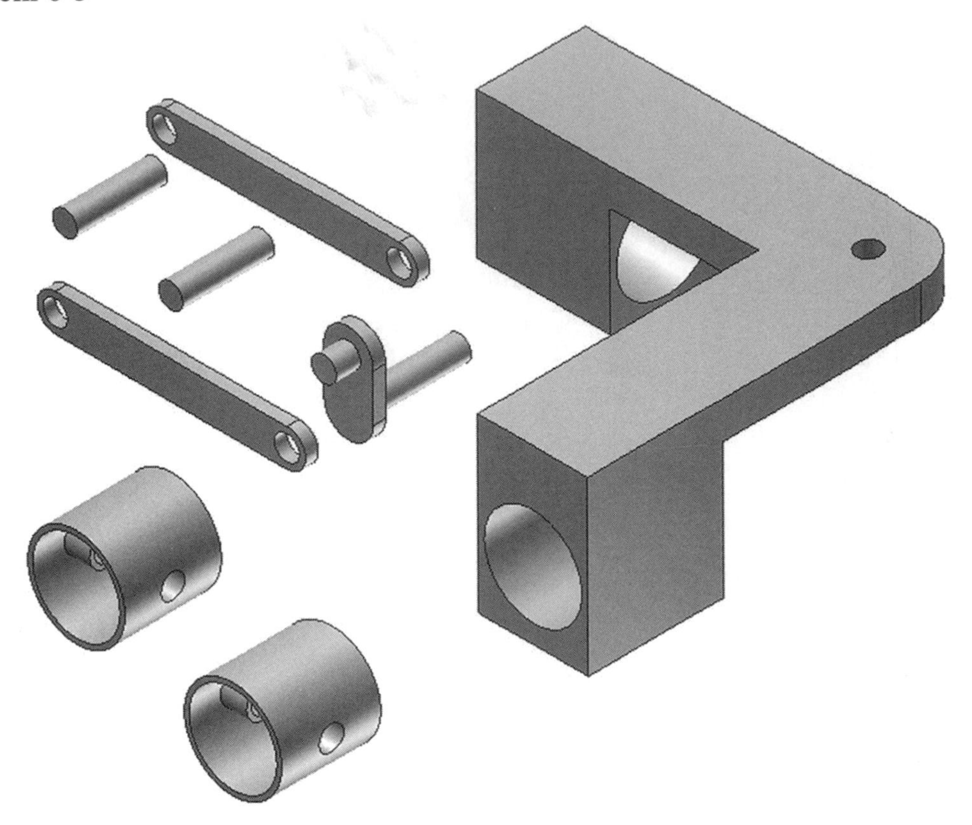

Problem 6-6

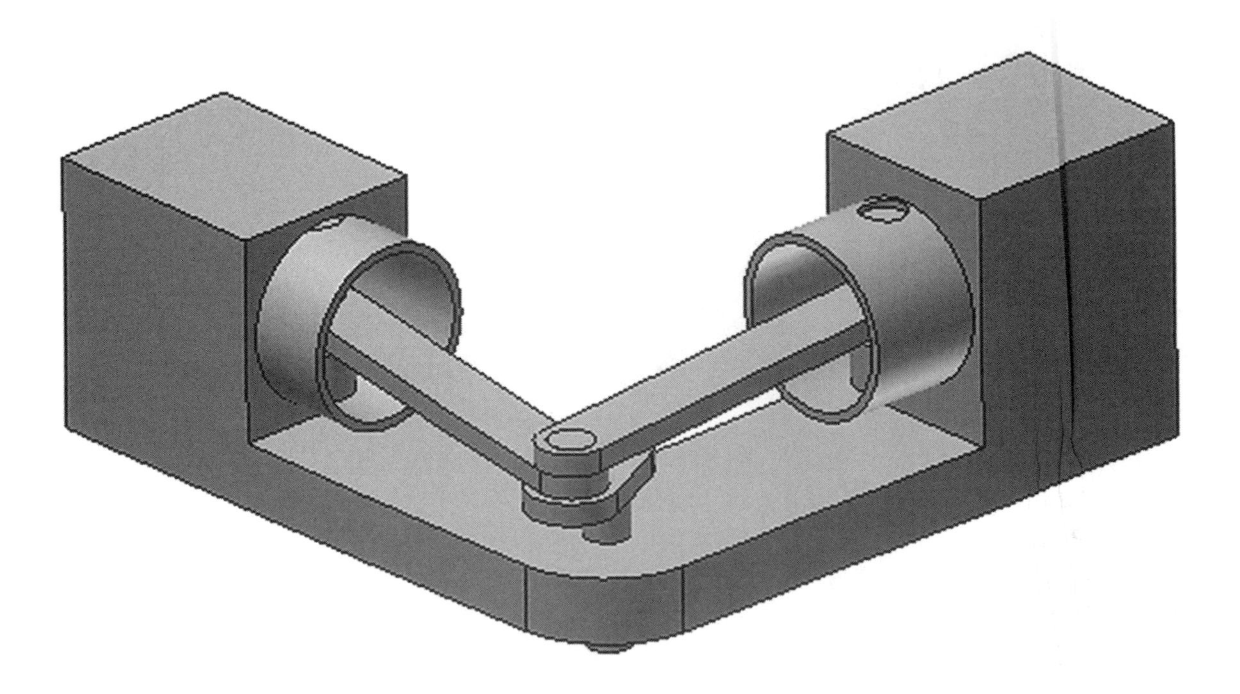

INDEX